Value-added Products from Algae

Abdelfatah Abomohra • Stephan Ende
Editors

Value-added Products from Algae

Phycochemical Production and Applications

Springer

Editors
Abdelfatah Abomohra
Aquatic Ecophysiology and Phycology,
Institute of Plant Science and
Microbiology
University of Hamburg
Hamburg, Germany

Stephan Ende
Alfred-Wegener-Institute AWI
Helmholtz Centre for Polar and Marine Research
Bremerhaven, Germany

ISBN 978-3-031-42025-2 ISBN 978-3-031-42026-9 (eBook)
https://doi.org/10.1007/978-3-031-42026-9

This Springer imprint is published by the registered company Springer Nature Switzerland AG
The registered company address is: Gewerbestrasse 11, 6330 Cham, Switzerland

Paper in this product is recyclable.

To the curious minds of tomorrow, may this book illuminate our path towards a sustainable future.

Foreword

In an era full of challenges regarding the growing global population, food security, pollution, and the rapidly changing climate, humanity finds itself in desperate need of innovative solutions and sustainable resources. Nowadays, the balance on the planet is being disrupted at an alarming rate, threatening life on Earth. However, humanity also holds the key to mitigate these challenges and nature provides keys to our salvation. To combat the global warming and revolutionizing our approach to sustainability, unleashing the power of algae (including cyanobacteria) for a sustainable future is one remarkable source that stands out above the rest.

Welcome to this enlightening journey through the world of algae to explore their untapped potential as a game-changer in providing abundant food and countless other resources with simultaneous mitigation of the global warming. Within the pages of this book, we delve into the mysteries and marvels of these seemingly humble organisms, unlocking the vast possibilities it holds for our sustainable future.

Through the process of photosynthesis, algae absorb vast amounts of carbon dioxide, a primary driver of global warming, while releasing oxygen back into the atmosphere, and providing the very basis of the aquatic food web. Harnessing this natural resource, algae present a promising solution to combat global warming, acting as nature's own carbon sink and helping to restore equilibrium to our planet's climate system. But the potential of algae extends far beyond their role as environmental guardians. As we progress further into this book, we uncover their incredible applications through production of majority of value-added products that are introduced in the book as "Phycochemicals" and how to enhance the production of these compounds in both a sustainable and economically feasible way through biorefinery and harnessing of wastes.

As you embark on this enlightening journey through the pages ahead, prepare to be captivated by the transformative potential of algae. Be ready to delve into the applications of algae reach into numerous industries, from pharmaceuticals and cosmetics to biofuels and wastewater treatment. Their versatility knows no bounds, providing renewable resources and solutions that can transform entire sectors, fostering a more sustainable and resilient future. This book in eighteen chapters aims to inspire change and prompt action, driving us towards a future where algae play a central role in mitigating global warming, ensuring food security, and shaping a world that thrives in harmony with nature.

May this exploration of algae's hidden treasures ignite the flames of curiosity and innovation within each of us, empowering us to embrace the possibilities that lie beneath the water surface. Together, let us unleash the power of algae and forge a path towards a sustainable future for all of us and for next generations.

University of Gothenburg
Gothenburg, Sweden

Angela Wulff

Swedish Algae Factory
Gothenburg, Sweden

Preface

Photosynthetic algae have gained attention as “cell factories” due to their fast growth rate and capacity to produce valuable compounds using sunlight and CO_2. Algal cells, including microalgae and seaweeds, are rich sources of essential nutrients such as minerals, carbohydrates, pigments, polyunsaturated fatty acids, and proteins. They generate various secondary metabolites with potential health benefits and therapeutic applications. Furthermore, algal biomass and/or residues can serve as potential feedstock for biofuel production through different conversion methods.

Numerous research papers have proposed innovative approaches to enhance the economic feasibility of value-added products derived from algal biomass, known for the first time in this book as “phycochemicals.” These approaches include high-throughput screening, bioprocess engineering, biorefinery, integrated sequential biofuel production, combined waste valorization, utilization of artificial intelligence, using algae as a superfood, and even the use of algae to reduce methane emission by rumens. Despite recent advancements in algal biotechnology, it is crucial to focus on integrated and cutting-edge research to present the latest novel technologies. This book aims to provide a comprehensive review of the fundamentals of bioprocess and biotechnology engineering for enhanced production of value-added products from algae and their impacts on the environment.

The book delves into the basics of algal cultivation, metabolism, harvest, and cellular pathways of phycochemicals biosynthesis, offering sufficient details for both experts and non-experts to grasp the recent progress in this field. It discusses new phycochemicals and advancements in technology development, from separation to scale-up commercialization. In summary, this book presents and evaluates recent research and development results that demonstrate notable competences in designing, performance, efficiency, and implementation of algal biotechnology for enhanced phycochemicals production. It also explores the latest technologies for the newly proposed integrated approaches.

The cutting-edge research topics covered in this book were accomplished through contributions from professionals and experts engaged in research, education, and industry related to the subject matter. This book serves as a comprehensive resource for readers with diverse scientific backgrounds who are interested in algal biotechnology, sustainability, biomass conversion, and new algal products from any perspective. It caters to a wide range of audiences, including undergraduate students, teachers, researchers, and consulting professionals in the aforementioned fields. By covering various aspects of these topics, this book acts as a primary reference for individuals seeking knowledge and insights in these areas.

Hamburg, Germany
Bremerhaven, Germany
Germany, 2023

Abdelfatah Abomohra
Stephan Ende

Contents

About the Editors

Abdelfatah Abomohra, PhD, is a researcher at Aquatic Ecophysiology and Phycology, Institute of Plant Science and Microbiology, Faculty of Mathematics, Informatics and Natural Sciences (MIN), University of Hamburg, Germany. He received his PhD in "Phycology" through Deutscher Akademischer Austauschdienst (DAAD) in 2012. He performed postdoctoral research at Hamburg University (2013-2014), Harbin Institute of Technology, China (2015-2017), and Alfred-Wegener-Institute, Germany (2023). Before joining University of Hamburg, he worked as professor and head of New Energy and Environmental Laboratory (NEEL) at Chengdu University, China. He was awarded Alexander-von-Humboldt fellowship for experienced researchers in 2022. His major is interdisciplinary of phycology and environmental engineering. He has stupendous research experience and finite vision in algal biotechnology and value-added products through processing of waste streams. According to Web of Science, he has published 165 articles, with 9 papers as "Highly Cited Paper" per September 2023. In 2020–2023, he is one of the World's Top 2% of Scientists List published by Stanford University. Moreover, he received a number of awards such as *Provincial award* of 2021 Sichuan Provincial Basic Education Teaching Achievement Award (China), *Young Scientist Medal* from International Association of Advanced Materials (Sweden), and Honor of "*Green Pepper Program*" for Outstanding Young Teachers at Chengdu University (China).

Stephan Ende, PhD, is a researcher at Alfred-Wegener-Institute (AWI)—Helmholtz Centre for Polar and Marine Research, Bremerhaven, Germany. He received his PhD from Wageningen University (Netherlands) in 2015. After having worked at Christian-Albrechts-University Kiel (Germany), he joined the aquaculture research unit (sustainable marine bioeconomy section) at Alfred-Wegener-Institute in 2016. His research areas focus on, e.g., multitrophic aquaculture systems, the utilization of aqua- and agricultural waste streams, and digitalization of land-based aquaculture systems. He has a special interest in unlocking the potential of marine microalgae for industrial applications. Dr. Ende works as a lecturer at several universities teaching basics of aquaculture and practical training in various fields such as aquaponics and microalgae cultivation. He has supervised multiple BSc and MSc students and has published 21 articles.

Opening the Algal Gateway

Stephan Ende and Abdelfatah Abomohra

Abstract Microalgae and seaweeds have been discussed as a promise of bolstering the blue economy and green technologies, potentially emerging as a primary future source for food, animal feed, and biofuel. Additionally, they are seen as a naturally sustainable reservoir of bioactive compounds, described for the first time in this book as "phycochemicals", possessing a wide array of applications for the betterment of humanity. Despite these hopeful prospects, the reality presents a different picture, where many of the established production facilities, particularly in Europe, remain modest in scale, and the sector as a whole is still in its developmental stages. The question arises, why this is still the case considering the vast amount of research conducted over decades? Extraction of valuable compounds from algae through integrated biorefinery approach could offer a solution to this quandary, creating a resource-efficient economy. Algae exhibit the capacity to yield high-value phycochemicals including pigments, polysaccharides, essential amino acids, polyunsaturated fatty acids, and can be used as feedstock for biofuels. This piece of work serves as a compendium of the most recent research, primarily centered around the realm of value-added compounds derived from algae. The overarching aim is to champion an integrated biorefinery system that could make algal production economically viable in the future. A worldwide multidisciplinary research is needed in order to cover the research on this topic from algae cultivation to different phycochemicals production and commercialization. This could allow decision makers, stakeholders, and producers to enhance the legislations and production system. The primary objective of this chapter is to wrap up and summarize the information that has been discussed throughout the 17 additional chapters of this book.

S. Ende
Alfred Wegener Institute – Helmholtz Centre for Polar and Marine Research, Bremerhaven, Germany

A. Abomohra (✉)
Institute of Plant Science and Microbiology, University of Hamburg, Hamburg, Germany
e-mail: abdelfatah.abomohra@uni-hamburg.de

A. Abomohra, S. Ende (eds.), *Value-added Products from Algae*,
https://doi.org/10.1007/978-3-031-42026-9_1

Keywords phycochemicals · value-added products · artificial intelligence · climate changes · microalgae · seaweeds

1 Introduction

Created by nuclear fusion, solar energy is generated which is necessary for life on the Earth to provide the foundation of the food chain and produce oxygen. The total incident solar power at the Earth's surface is estimated at 124,000 terrawatt-year, of which, only about 0.07% is utilized by all photosynthetic organisms (Junge 2019). Photosynthesis is a fundamental process that converts sunlight into energy stored in organic molecules. It occurs in chloroplasts in the presence of chlorophyll and other accessory pigments. Simply, light is absorbed by chlorophyll to initiate a complex series of chemical reactions that result in the production of ATP and NADPH as energy storage and electron carrier/donor molecule, which are used further to drive the synthesis of organic molecules through CO_2 fixation. Algae play a vital role in this process as they can be grown in various aquatic environments using only sunlight, CO_2, and water, converting solar energy into biomass with a CO_2 fixation rate at efficiencies that are often higher than land plants (Benedetti et al. 2018).

For decades, algae have been considered to potentially boost the blue economy, to be the future source of biofuel, food, and feed, as well as a natural sustainable source of bioactive compounds, with a wide range of applications for humankind. Therefore, algae are considered by the EU to play a significant role in the transition to sustainable economy balancing the growth of economic activities, the protection of natural resources, and the needs of a growing world population. Therefore, sustainable and circular bioeconomy is gaining importance around the world, where more than 60 countries have established bioeconomy-related strategies contributing efforts to meet the Paris Agreement targets (FAO 2022). For instance, the EU Bioeconomy Strategy, adopted in 2012 and updated in 2018 (European Commission 2018), aims at implementing a sustainable and circular bioeconomy throughout Europe. In addition, the United States Department of Agriculture (USDA) has launched a Bioeconomy Initiative to support the development of sustainable bioeconomy. Moreover, Japan has implemented policies to promote a sustainable and circular bioeconomy, such as the Basic Act on the Promotion of a Recycling-Oriented Society and the Biomass Nippon Strategy (Environment Agency 2000). Algae are also considered to play a significant role in ecosystem services, contributing significantly to global primary production, while also playing an important role in the uptake of dissolved nutrients from the surrounding environment, coastal defense from hazardous waves, and potentially in carbon sequestration. Algae also have received attention for their ability to produce high-value chemicals, including pigments, antioxidants, antimicrobial, and biofuels precursors (Abomohra et al. 2016; Beyer et al. 2023; Ebaid et al. 2017; El Zawawy et al. 2020; Shao et al. 2019). The potential for the production of valuable compounds from algae could create a resource-efficient economy. In addition, algae contain essential amino acids, fatty acids, and vitamins, making them valuable ingredients for aquaculture feed or biofertilizers (Alprol et al. 2021; Ashour et al. 2023).

Despite all aforementioned promises and expectations, however, many of the compiled production units especially in Europe are of small size and the sector is still immature. The question arises, why this is still the case considering the vast amount of research conducted over decades? Are there already solutions in place and is it simply a question of knowledge transfer from research into the industry, or, does even the latest research not provide any solution for the fast growth of this intriguing blue economy sector? In addition, the efforts from the governments to establish legislation in that regard could enhance or inhibit the development of the technology.

Still, algae-based industry holds great promise for a brighter and more sustainable future. This book summarizes the latest research mainly around the topic of value-added compounds produced from algae (firstly introduced as "phycochemicals" in this book) as a key aspect of the biorefinery approach aiming at making algal production economically viable. The emerging sector of the Blue Bioeconomy has identified a lack of technical innovations to upscale the algal production in order to reduce the production costs. The book brings together a worldwide multidisciplinary researchers to cover the topic from algae cultivation to different phycochemicals and commercialization, which allows stakeholders and producers to enhance their production system, e.g., from pure biomass or certain phycochemical to a biorefinery approach.

2 Algae Cultivation and Harvest

The development of an efficient microalgae cultivation system should overcome a number of challenges in order to comprehend photosynthetic processes, structural characteristics of algae, the appearance of the desired species in an axenic form, and biomass output per unit area. In addition, macroalgae (seaweeds) cultivation varies by nation and is dependent on seasonal variation in the marine habitat as well as the local legislation. Based on the wild species of their respective countries, only a small number of macroalgal species are cultivable worldwide. Similar to microalgae, seaweed farming is a promising industry because it does not compete for freshwater or arable territory, with many other advantages. However, cost-effective techniques should be created and used in order to make microalgal/macroalgal production commercially viable. The different cultivation systems for microalgae and seaweeds, as well as the challenges in scale-up, are discussed in Chapter "Algae Cultivation Systems".

Algae harvesting refers to the separation or detachment of algae from its growth medium. It also poses a huge challenge to the development and commercialization of scale-up technology for algal biomass production, especially for microalgae where the algal cell density is almost equal to that of the water. Comparatively, seaweeds have several advantages compared to microalgae in terms of harvesting, where harvesting and dewatering of seaweeds require far less energy. However, there are some ecological and environmental effects that are associated with seaweeds harvesting. In general, harvesting methods that require high energy input, expensive equipment, or a huge quantity of chemicals, like that in the case of microalgae, can increase the overall cost of algal biomass production, reducing the economic feasibility of the process. The harvesting method to be employed depends on the species, habitat, biomass density, scale of cultivation, and governmental regulations. Chapter "Algae Harvesting" provides a summary of the most recent developments in the harvesting techniques of microalgae and seaweeds, addressing advantages and challenges still related to this cost-driving process. Special attention is given to the problem of chemicals used in current harvesting techniques raising environmental concerns.

3 Phycochemicals

The diverse range of algal species results in a wide range of phycochemicals being produced, which can be applied in various industrial sectors. However, selecting species that will boost the production of the appropriate chemicals is essential when employing algae for different applications.

Many strategies have been designed to increase the algal growth rate and enhance value-added phycochemicals production. Controlling the substrate and light is the main strategy that may take place by the enhancement and modification of the algal cultivation systems such as cultivation modes and reactor design in order to provide a sufficient supply of light and nutrients based on the microbial physiology of the targeted algal cells (Abomohra et al. 2019; Tawfik et al. 2022). Chapter "Phycochemicals" provides an overview of recent knowledge gain in the field of physiology and metabolic pathways of different value-added chemicals in algal cells. In addition, recent research on nano-bionics, indirect/direct biosynthesis of lipids, proteins, vitamins, polysaccharides, and antioxidants from algae as well as biofuel production are reviewed.

Overcoming laborious and cost-intensive research is still a bottleneck for phycochemical production and other algal biotechnologies. Extracting relevant data and solving complex problems due to recent advances in computing power, e.g., high-performance computing, and improvements in technologies such as Deep Learning and Random Forest, has the potential to face these challenges. Chapter "Artificial Intelligence in Phycochemicals Recognition" describes the recent approaches on how artificial intelligence can be used to overcome laborious and cost-intensive research for exploring and enhancing the production of established or new phycochemicals. An overview of metabolomics research including artificial intelligence in microalgae research is discussed.

To strengthen the production processes, conventional engineering approaches have been established in which nature has provided a wide range of metabolites with the ability to be altered through changes of the growth conditions or the genetic code (Boodhoo et al. 2022). The efficiency of a certain system could be enhanced by intensifying the use of highly stable biological catalysts or developing continuous bioprocesses using novel approaches. Thus, it is possible to lower the unit cost by optimizing the growth conditions and downstream processing (which are discussed in Chapter "Overview of Bioprocess Engineering"), picking organisms that can overcome environmental restrictions, and selecting strains with high targeted compounds.

4 Biorefinery

It is preferable to obtain different products by integrating different processes for the same biomass in a single growing cycle, which is known as biorefinery. For example, lipids utilization for biodiesel coupled with extraction of value-added phycochemicals could ensure economical production. Chapter "Overview of Biorefinery Technology" gives an overview of the technologies/possible routes of algal biomass biorefinery for enhanced biomass utilization, most especially for the production of several industrially important phycochemicals and bioenergy. Besides, a general concept of biorefinery and a summary of the challenges, recent trends, and opportunities of algal biomass biorefinery are also provided in this chapter.

Due to the importance of biofuel production from algal biomass in biorefinery, Chapter "Biofuel-Integrated Routes" discusses in detail the possible routes integrated with biofuel production. The integration of various high-value phycochemicals with algal biofuel refineries was reviewed. The recent works in phycoremediation with microalgae that resulted in the production of biomass for assorted biofuels were discussed. Simultaneous bioelectricity generation with wastewater treatment, as well as biofuel production, have also been highlighted.

In context with integrated routes, the cultivation of algae on waste streams is a promising approach to reduce the overall cost. Chapter "The Use of Wastewater for Algal Growth" provides an overview for the application of microalgae in wastewater treatment systems and highlights its advantages over conventional methods. Various mechanisms of wastewater treatment by microalgae are discussed, including nutrient, phosphorus, and nitrogen recovery, as well as the removal of heavy metal compounds. The chapter also examines the factors that influence wastewater treatment by microalgae. The chapter addresses the challenges and future perspectives of microalgae-based wastewater treatment systems in a comprehensive manner.

5 Enhanced Production

Optimization of microalgae biomass production and enhancement of valuable compounds require extensive experimentation to achieve high yields and productivity. Traditional optimization approaches involve the screening of large numbers of samples in parallel, which can be costly and time consuming. Chapter "High Throughput Screening to Accelerate Microalgae-based Phycochemicals" reviews the latest development in high-throughput screening (HTS) methods as an alternative technology to improve the understanding and control of critical process parameters. It reviews the innovation pathways (miniaturization, automation), latest advances and bottlenecks in experimental design and data management, and provides an outlook for their potential to allow the production of new species and cell lines with molecules of added value. It will help to overcome such "manual" approaches to further accelerate processes and substantially reduce the costs.

Catalysts play a vital role in the overall economics and yield of a target product from algae. The product conversion efficiency, sustainability, and economics of the production processes depend on the type of reactions opted and the catalyst used for producing the targeted product. For biofuel production, chemical and biological catalysts were extensively researched. Specifically, green catalysts, mainly made from biomass, especially enzymes, are increasingly attractive and environmentally friendly. Recently, both nano-catalyst and biocatalysts are widely attractive due to plentiful advantages such as easy synthesis, simple disposal, and high reusability, along with enhanced yield of the desired product. Chapter "Catalyst in Action" discusses the potential of catalysts to significantly influence the yield and process economy in every algal product synthesis/recovery.

Many studies thrived to improve the efficiency of algal growth as well as to enhance phycochemicals, especially via genetic engineering. In that context, "Omics" approaches have greatly contributed to expanding the knowledge on reporting large data sets on microalgae as well as seaweed's genome, transcriptome, proteome and metabolome. Chapter "Omics Approaches for Algal Applications" provides an overview of the prospect of algae and their role in symbiotic association using different Omics approaches. In addition, algae-based Omics and multi-Omics approaches are critically analyzed in wastewater treatment, metal toxicity and remediation, biofuel production, and therapeutics.

6 Industrial Applications

In context with the previous section, cultivation of algae for the production of algae-based biopolymers that are biodegradable have gained increasing attention due to their potential to replace traditional polymers, which is discussed in Chapter "Algal-based Biopolymers". Diatoms represent a unique group of microalgae with exceptional ecological roles in many aquatic ecosystems. They have also a variety of phycochemicals that can be used for many applications. After extraction, diatoms biosilica could also represent a potential sustainable feedstock for various industrial and commercial applications. Chapter "Diatom Nanostructured Biosilica" overviews the wide applicability of nanostructured siliceous cell walls of diatoms, especially in the fossil form (i.e., diatomaceous earth), as a functional additive and filler in various industries. In addition, the chapter sheds light on the biorefinery approach of large-scale cultivated diatoms as a sustainable source of biosilica, besides extracting valuable metabolites.

Besides the extraction of certain phycochemical or biorefinery of algal biomass, the whole biomass can also be utilized for many applications such as food, feed, or industry. Chapter "Potential of Seaweeds to Mitigate Methane Emissions" discusses the perspective of seaweed-based feed additives to reduce livestock emissions from the agricultural sector, particularly arising from ruminant animals. The ongoing research and development efforts, coupled with the commitment of companies in this field, are discussed, highlighting the potential to make a meaningful contribution to reduce the emissions of livestock.

Chapter "Algae for Aquaculture: Recent Technological Applications" provides an overview of the role of algae as an aquaculture feed and explores the recent technological applications that have facilitated their integration into the industry. It discusses various types of algae, their cultivation methods, and the processing techniques used to produce algal-based feeds. Furthermore, the nutritional benefits of algal feeds and the environmental/sustainability considerations are discussed. The challenges and future prospects for the continued integration of algae into aquaculture are evaluated. In context with that, Spirulina as a superior food with nutritional value, that has proven health benefits and beneficial interaction in fermentation and preservation processes, is discussed in Chapter "Algae as a Functional Food: A Case Study on Spirulina". Some examples of cultivation parameters' affecting Spirulina's nutritional value are presented. In addition, the most significant producers with market share of powder, tablets, and capsules are summarized. Moreover, fresh frozen Spirulina is presented as a new, rapidly growing trend in Western countries due to its milder taste. Spirulina-containing products in the market, and many commercial products of snacks, drinks, pasta, and dairy products are presented.

7 Legislation and Biosecurity

With the mounting demand for a healthy lifestyle, industries have been innovating their products to cater the human needs and interests, given the ease of producing them on a large scale in a reduced physical and temporal space. Nonetheless, in parallel, legislative bodies are becoming increasingly concerned about the use of algae in industry, as well as ensuring food safety during their commercialization. This is due to the peril of contamination from the growth to the storage of these species, as well as the accumulation of heavy metals by many algal species due to high adsorption/absorption capacity. With the final chapter entitled "Legislation and Biosecurity", authors provide examples of approval procedures according to the novel food regulations including safety assessment of foods or food ingredients to protect public health. The legal issues related to the use of algae, as well as food safety, are outlined and discussed.

Acknowledgements The authors extend their heartfelt gratitude to the Alexander von Humboldt Foundation (Germany) for granting the fellowship which initiated such collaboration, a testament to their support for scholarly pursuits. Additionally, profound appreciation is reserved for all the esteemed contributors of this book, whose valuable insights, efforts, and commentary have greatly enriched the content of the book.

References

Abomohra A, Jin W, Tu R, Han S-F, Eid M, Eladel H (2016) Microalgal biomass production as a sustainable feedstock for biodiesel: current status and perspectives. Renew Sust Energ Rev 64: 596–606

Abomohra A, Shang H, El-Sheekh M, Eladel H, Ebaid R, Wang S, Wang Q (2019) Night illumination using monochromatic light-emitting diodes for enhanced microalgal growth and biodiesel production, vol 288. Bioresour, Technol, p 121514

Alprol AE, Heneash AMM, Ashour M, Abualnaja KM, Alhashmialameer D, Mansour AT, Sharawy ZZ, Abu-Saied MA, Abomohra AEF (2021) Potential applications of *Arthrospira platensis* lipid-free biomass in bioremediation of organic dye from industrial textile effluents and its influence on marine rotifer (*Brachionus plicatilis*). Mater 14(16):4446. https://doi.org/10.3390/MA14164446

Ashour M, Al-Souti AS, Hassan SM, Ammar GAG, Goda AMAS, El-Shenody R, Abomohra AEF, El-Haroun E, Elshobary ME (2023) Commercial seaweed liquid extract as strawberry biostimulants and bioethanol production. Life 13(1):85. https://doi.org/10.3390/life13010085

Benedetti M, Vecchi V, Barera S, Dall'Osto L (2018) Biomass from microalgae: the potential of domestication towards sustainable biofactories. Microb Cell Factories 17:173. https://doi.org/10.1186/S12934-018-1019-3

Beyer AS, Meier J, Jiménez-Muñoz M, Meixner R, Ende SSW, Abomohra A, Henjes J (2023) New microalgae media formulated with completely recycled phosphorus originating from agricultural sidestreams. J Appl Phycol 1:1–16. https://doi.org/10.1007/S10811-023-03005-Z/TABLES/6

Boodhoo KVK, Flickinger MC, Woodley JM, Emanuelsson EAC (2022) Bioprocess intensification: a route to efficient and sustainable biocatalytic transformations for the future. Chem Eng Process - Process Intensif 172:108793. https://doi.org/10.1016/J.CEP.2022.108793

Ebaid R, Elhussainy E, El-Shourbagy S, Ali S, Abomohra A (2017) Protective effect of *Arthrospira platensis* against liver injury induced by copper nanoparticles. Orient Pharm Exp Med 17:203–210

El Zawawy N, El Shafay S, Abomohra A (2020) Macroalgal activity against fungal urinary tract infections: in vitro screening and evaluation study. Rend Lincei 31:165–175. https://doi.org/10.1007/s12210-019-00856-y

Environment Agency (2000) The basic law for establishing the recycling-based society [WWW Document]. Law No.110 2000, Environ. Agency, Japan. https://www.env.go.jp/recycle/low-e.html. Accessed 27 June 2023

European Commission (2018) Commission Communication COM/2018/673. A Sustainable Bioeconomy for Europe: Strengthening the Connection between Economy, Society and the Environment. Available online at: https://eurlex.europa.eu/legal-content/en/ALL/?uri=CELEX%3A52018DC0673 (accessed April, 2023).

FAO (2022) Sustainable and circular bioeconomy in the climate. Opportunities To Transform Agrifood Systems, Agenda. https://doi.org/10.4060/cc2668en

Junge W (2019) Oxygenic photosynthesis: history, status and perspective. Q Rev Biophys 52:e1. https://doi.org/10.1017/S0033583518000112

Shao W, Ebaid R, El-Sheekh M, Abomohra A, Eladel H (2019) Pharmaceutical applications and consequent environmental impacts of spirulina (*Arthrospira*): an overview. Grasas Aceites 70: 292

Tawfik A, Ismail S, Elsayed M, Qyyum MA, Rehan M (2022) Sustainable microalgal biomass valorization to bioenergy: key challenges and future perspectives. Chemosphere 296:133812. https://doi.org/10.1016/J.CHEMOSPHERE.2022.133812

European Commission (2018) Commission Communication COM/2018/673. A Sustainable Bioeconomy for Europe: Strengthening the Connection between Economy, Society and the Environment. Available online at: https://eur-lex.europa.eu/legal-content/en/ALL/?uri=CELEX%3A52018DC0673 (accessed April, 2023).

Algae Cultivation Systems

Wafaa F. Zohir, Mofida E. M. Makhlof, Ahmed M. Abdallah, and Mostafa M. El-Sheekh

Abstract Algae are a diverse group of micro- and macroscopic, highly efficient photosynthetic organisms with rapid growth and flourish in a variety of environments where many land plants are uninhabitable. Algal biomass contains a wide range of biochemical components such as pigments, lipids, carbohydrates, and proteins (known as phycochemicals), making them a promising and renewable feedstock for different industrial/commercial applications such as biofuels, nutraceuticals, pharmaceuticals, and environmental sectors. The potential of algae as a renewable and eco-friendly bioresource has received much attention in recent years and made it the most promising and sustainable source of useful products. The general requirements for successful algae cultivation include many growth factors such as nutrients, temperature, light, pH, and different cultivation systems. There are two main cultivation systems used widely for algae cultivation, namely open raceway ponds (OWPs) and closed photobioreactors (PBRs). Each one of these systems has a series of advantages and disadvantages. This chapter will discuss the different cultivation systems used for seaweeds and microalgae for biomass production, including OWPs, PBRs, and mixed systems. Also, the pros and cons of each system will be evaluated. The natural growth and distribution of seaweeds, as well as recent advancements in seaweed farming technologies, will be discussed.

Keywords Microalgae · Seaweed farming · Algal growth · Open raceway ponds (OWPs) · Photobioreactors

---

W. F. Zohir (✉) · M. E. M. Makhlof
Department of Botany and Microbiology, Faculty of Science, Damanhour University, Damanhour, Egypt
e-mail: wafaa.zuhair@sci.dmu.edu.eg

A. M. Abdallah
Natural Resources and Agricultural Engineering Department, Faculty of Agriculture, Damanhour University, Damanhour, Egypt

M. M. El-Sheekh
Botany and Microbiology Department, Faculty of Science, Tanta University, Tanta, Egypt

A. Abomohra, S. Ende (eds.), *Value-added Products from Algae*,
https://doi.org/10.1007/978-3-031-42026-9_2

1 Introduction

Algae represent a widely varied group of photoautotrophic organisms that have adapted to various environmental conditions, and they constitute more than half of the total primary productivity, measured as carbon, on the earth (Raven and Giordano 2014; Fayyad et al. 2020). Macroalgae (or seaweeds) and microalgae are mostly eukaryotic, aquatic organisms that vary in shape, colour, and size from single cells to highly differentiated multicellular structures. Only a single group of algae known as blue-green algae, or Cyanophyta/Cyanobacteria, are the only prokaryotic algae which represent a large, heterogeneous photosynthetic microorganisms. Algae can fix carbon using CO_2 through photosynthesis and release O_2 into the atmosphere, which is important for the planet as they contribute more than 50% of the total photosynthetic activity, forming the main and principal foundation of the aquatic food chain (Singh and Singh 2015). Algae represent a highly promising feedstock for biofuels (El-Sheekh et al. 2020), cosmetics (Gokare 2022), pharmaceuticals (Zohir et al. 2022), nutrition and food additives (EL-Mohsnawy et al. 2020), and aquaculture (Perez-Legaspi et al. 2019). Moreover, it can be used in agriculture as bio-stimulants and bio-fertilizers (Ammar et al. 2022) and bioremediates (Chinnasamy et al. 2010). Microalgae and seaweeds are the two primary categories that make up the algal biomass, and both may be grown in a lab (small or large scale) or collected from the environment (Dębowski et al. 2013; Rosenberg et al. 2011).

Many studies have evaluated algae as a potential source of many biological products with their variable applications. Compared to terrestrial plants, algae have desirable features including shorter doubling times, the ability to grow in waste- or marine water, no need for arable land, do not require fertilizers, and have high photosynthetic efficiency (3–4 folds higher), the capability of surviving and thriving in harsh conditions, and the ability to accumulate unique and crucial bioactive compounds (Prasad Behera et al. 2022). The natural substances derived from algae are being researched and currently used in different fields. In addition, microalgae can be directly extracted or undergo combined sequential biorefinery to extract a wide range of promising biological compounds. Their molecular constituents, mainly carbohydrates and lipids, have been discussed as promising feedstocks for biofuel production.

Diverse algae offer numerous opportunities as a source of useful foods towards sustainable development. The development of an efficient mass cultivation system for microalgae must overcome a number of challenges in order to comprehend photosynthetic processes, structural characteristics of algae, the appearance of the desired species in an axenic form, and biomass output per unit area. Furthermore, seaweed cultivation varies by nation and is dependent on seasonal variation in the marine habitat. Based on the wild species of their respective countries, only a small number of macroalgal species are cultivable worldwide. Therefore, the original genome of different macroalgal species should be considered for essential preservation through tissue culture, which is still in its infancy stage (Kumar et al. 2021).

Similar to microalgae, seaweed farming is a promising industry because it does not compete for freshwater or arable territory. Importantly, their farming offers low eutrophication bioremediation in waterbodies presently damaged by over-accumulation of nutrients through adsorption and absorption mechanism (Jiang et al. 2020). Despite their numerous advantages, they could decrease water flows, seaweed farming cannot be permitted to flourish without restrictions because they compete with naturally occurring primary nutrients producers, absorb contaminants that could decrease its suitability for food or feed, shade out light and change the physiochemical environment, and they may bring parasites or alien species that may change the genetic makeup of the local ecosystem (Campbell et al. 2019). So, there are many aspects that should be taken into consideration during seaweed cultivation such as offshore system design, turbulence strength, and type/characteristics of the used seaweed species.

Regarding microalgae cultivation in open raceway ponds (OWPs), the low density of microalgal cells and low photosynthetic activity are the greatest issues in this cultivation system. It is attributed to a limited amount of available light and an increased risk of contamination. In closed photobioreactors (PBRs), such systems have been suggested to address the issues of low density and pollution, with a geometric design of the reactor to enhance illumination as its primary goal. Three approaches have been devised to enhance microalgal growth in PBRs for enhanced valuable chemical yield (Abomohra et al. 2016; Bumbak et al. 2011; Madkour et al. 2012). (1) The design of PBRs is very important because it is necessary to provide a source of energy and carbon depending on the type of microalgae and the desired by-products to be produced. (2) Minimizing unfavourable growth limiting factors for PBRs, such as light penetration and CO_2 and O_2 accumulation, for enhanced culture density. Thus, the PBR prototype should be studied for this purpose. (3) Once it functions well, the key factor criteria for the scale-up bioengineering process should be established.

Nutrients, CO_2 concentration, pH value, salinity, light intensity/quality, temperature, and mixing rate are the main factors that influence algal growth. Therefore, a thorough investigation of these growth parameters is required prior to designing a large-scale cultivation system (Chowdury et al. 2020). Also, the metabolic pathways of algae should be explored in order to prepare a suitable growth environment for the cultivated algal species. The availability and type of nutrients, notably nitrogen and phosphorus, in the growth medium have been discussed in relation to the metabolic pathways (Lowrey et al. 2015). Carbon is particularly significant because it is necessary for the creation and growth of every living thing, where algae can assimilate carbon uniquely from CO_2 through photosynthesis or another carbon source. Thus, algae can fix carbon through two different growth modes: (1) autotrophic growth, which involves fixing inorganic carbon (i.e. CO_2) through photosynthesis and (2) heterotrophic growth, which involves the assimilation of organic carbon in the absence of light. Additionally, some algae can typically exist in either metabolic state and are, therefore, referred to as mixotrophic or, alternatively, photoheterotrophic (Lee et al. 1996; Marquez et al. 1995; Ogawa and Aiba 1981).

The mode of cultivation as well as the cultivation system greatly influence the final production cost. Cost-effective techniques should be created and used in order to make microalgae/seaweed production commercially viable. It is necessary to be considered since algal cultivation and harvest have been estimated to represent the big share of the overall cost of a certain algal product. This chapter will discuss different cultivation modes and systems used for biomass production from seaweeds and microalgae, including OWPs, PBRs, mixed cultivation systems, ring systems, floating design, bottom culture technique (BCT), floating rafts for vertical rope culture (FRVRC), integrated MultiTrophic Aquaculture (IMTA), and long line square raft (LLSR). Also, the advantages and disadvantages of each cultivation system will be evaluated.

2 Seaweed Cultivation Techniques

There are three main scales used for seaweed study/cultivation based on the amount cultivated per unit area including lab-scale, pilot-scale, and large-scale (Fig. 1). Large outdoor cultivation systems include floating and submerged designs, while the latter includes different designs such as ring system, polyethylene net, LLSR, FRVRC, IMTA and BCT.

For large-scale outdoor cultivation, seaweeds can be cultivated on land, sea, desert, and in integrated aquaculture systems. To meet the growing demand for seaweed products, the field of seaweed farming has extended greatly in last years, and there is an interest to utilize the offshore deep-water for seaweed farming (Zhang et al. 2022) that relies on pillars, floating rafts, or long ropes (Fig. 2).

2.1 Lab-Scale (Macroalgal Photobioreactors)

Lab-scale cultivation is not used for mass production of seaweeds, but for experimental studies and to evaluate the optimum cultivation conditions required further in pilot- or large-scales. For example, *Ulva intestinalis* pilot-scale production showed growth rates of 7.72 day^{-1} in a cultivation cycle of 28 days. Based on the space demand of the production system, this resulted in fresh mass productivity of 3.0 kg m^{-2} and average annual production of 1.1 kg m^{-2} per year (Schmitz and Kraft 2022). Despite the importance of the lab-scale production in determining the appropriate conditions for growing seaweeds on a large scale, there are many drawbacks to this system. For example, the growth of kelp in deep water is affected by canopy shading's reduction of light, nutrient drawdown's reduction of nitrogen accessible for uptake, and slowed flow caused by kelp drag's increase in diffusive boundary layers (Frieder et al. 2022). However, PBR systems are still with the advantage of enabling the observation and adjustment of the growth environment, which lead to results that are predictable in terms of both quantity and quality.

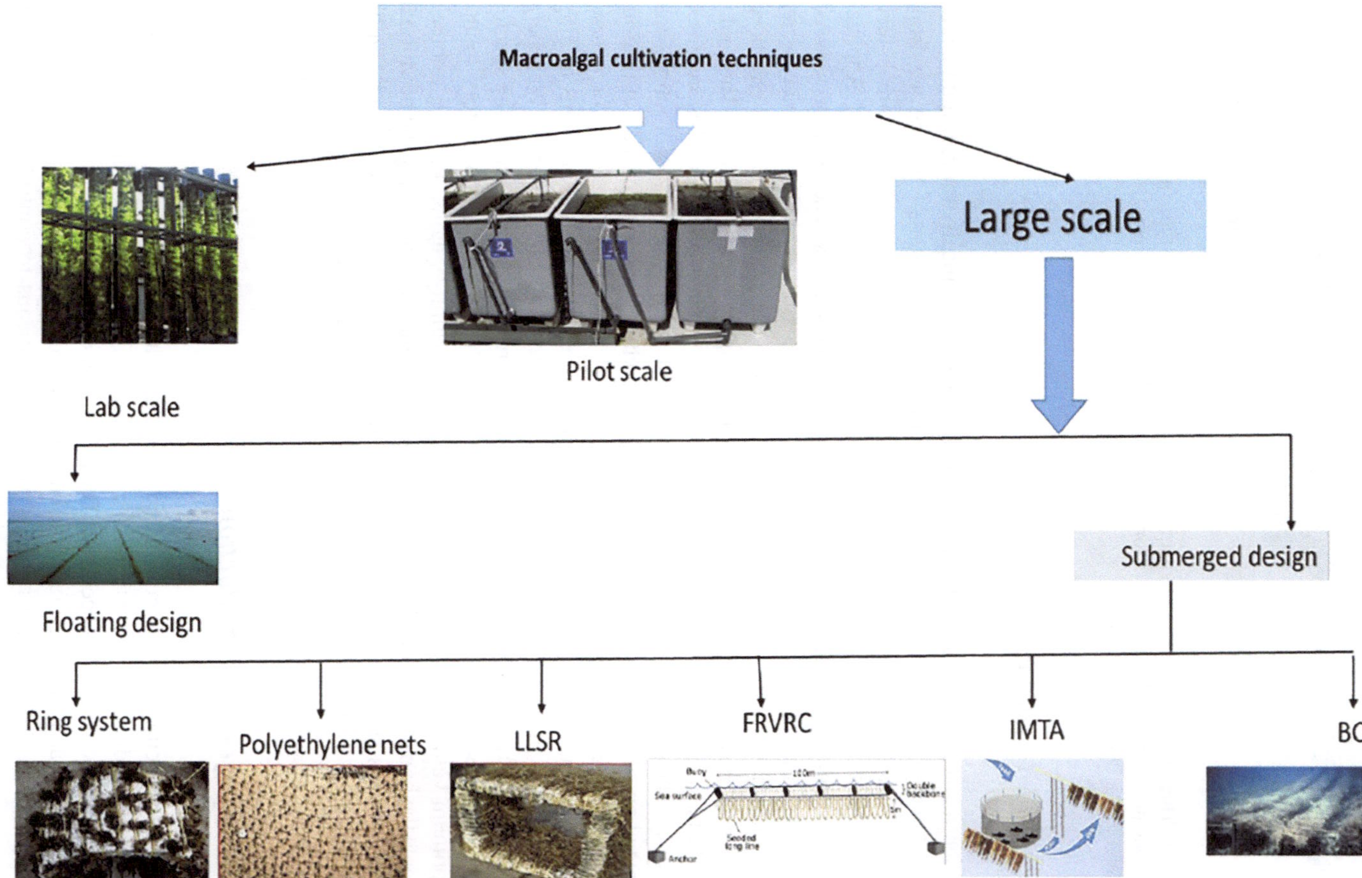

Fig. 1 Schematic diagram showing different types of seaweed cultivation techniques. LLSR (long line square raft), FRVRC (floating rafts for vertical rope culture), IMTA represents (integrated MultiTrophic aquaculture), and BCT (bottom culture technique)

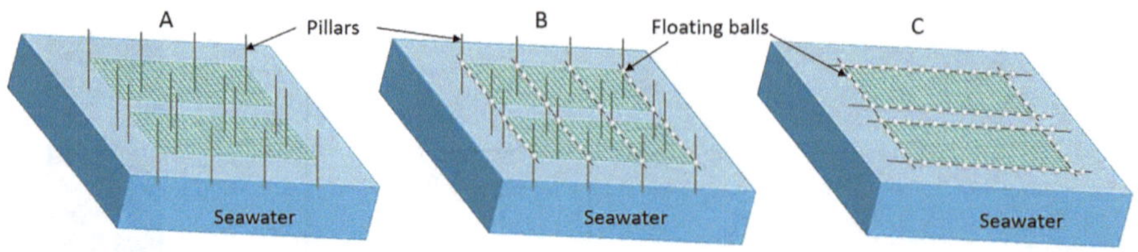

Fig. 2 Three offshore farming techniques of seaweeds, namely pillar type (**a**), semi-floating raft (**b**), and full floating raft composed of long rope and floating balls (**c**). Source: Modified from Zhang et al. (2022), Open access article

2.2 Pilot-Scale Production

When seaweeds are grown in tanks or small ponds, it is typically for premium goods like *Porphyra* sp. (Israel et al. 2006), or for bioremediation of wastewater from aquacultures, such as those growing shellfish, abalone, and as animal feed (as discussed in Chapter "Algae for Aquaculture Recent Technological Applications"). Tanks may be connected to water pipes and each tank is aerated for seaweed growth, which benefits water recirculation. According to aeration turbulence, water residence duration is controlled. Air diffusers rotate the algae culture for enhanced nutrient uptake and flashing light effect (Jun Pang and Lüning 2004). The tanks could be placed in greenhouses with recirculated water. Intense systems in tanks/ponds or land-based tank/pond cultivation exhibit a lot of epiphytes, fouling, and crucial nutrient needs, which are drawbacks and challenges in seaweed cultivation. For example, the algal species, most often *Ectocarpus, Pylaiella, Sagassum, Ceramium,* and *Polysiphonia*, constitute competitors for light and nutrients leading to a decrease in growth rates and final biomass yield of the target species (Wegeberg and Felby 2010). Site selection, tank design/construction, knowledge of the species biology, choosing the best strains, controlling the environmental factors to the optimum values (like temperature, pH, light, and salinity), agitating thalli to break down the barriers to nutrient uptake, seawater exchange, and adjusting the nutritional requirements are the main challenges for successful tank cultivation, which can be optimized in pilot-scale trials before applying in large scale.

2.3 Large-Scale Production

Large-scale cultivation is considered the main way for benefitting from seaweed massive production. For offshore seaweed farming in deeper water, suspended cultivation systems are necessary. These systems offer novel and complex hydrodynamics that can affect farm's nutrient availability and, consequently, system performance. As open sea aquaculture lacks the means to control the cultivation parameters, algal growth rates are influenced by oceanographic, meteorological, and water quality factors in the production region (Frieder et al. 2022). In addition, the impacts of climate change and ocean acidification, as well as spreading of pathogens, epiphytes, and herbivores, are increasingly limiting the biomass yield and quality (Vergara-Fernández et al. 2008). For example, the densities of two epiphyte-feeding stream insects (*Baetis* sp. and *Agapetus celata*) were experimentally manipulated on the seaweed *Cladophora glomerata*, and it was found that greater loss by 10% occurred in case of *Agapetus celata* (Dudley 1992). In an attempt to overcome such defects, many designs have been created to cultivate seaweeds, including floating and submerged designs. This section focuses on the vast majority of these types, mentioning examples, with the advantages and disadvantages of each system. The biomass yield varies widely based on the system used (Table 1). It is worth to mention that these types are not completely separated from each other, but it is possible to find a specific design for seaweed cultivation that includes a mixture of more than one type.

2.3.1 Floating Design

The idea of free-floating seaweed cultivation imitates their natural environment and habitats. As a result, it does not only prevent the permanent occupation of coastal space, but also saves money that would be spent on expensive buildings and the potential loss of large algae caused by strong current and wave forces during storms. Understanding possible pathways for free-floating seaweed cultivation in the coastal ocean is essential. A difficult job that has traditionally been taken into consideration to prevent marine debris, such as plastics from attaching to the seaweeds and tracking floating objects in the ocean (Van Sebille et al. 2020) and oil spills (Liu et al. 2011). Global drifting networks have been employed by researchers (Ebbesmeyer and Ingraham 1994), including satellite measurements (Mulet et al. 2012), models, and even accidental spills, to map out the seasonality of global currents and changes in seaweeds. The capacity to foresee potential algal biomass yield along the floating technology is crucial for achieving a large-scale and fiscally feasible seaweed production method. The seaweed growth model that includes environmental elements like light, temperature, and nutrition information for biomass growth prediction can be utilized when combined with the trajectory model for maximum biomass yield. The ideal time and location for discharge and collection of seaweeds may then be determined using this model (Brooks et al. 2018).

2.3.2 Submerged Systems

Anchor farms in coastal areas are the traditional method of cultivating seaweeds, where seaweeds are supported by long cultivation ropes and other structures. Such stationary farming methods require significant infrastructure investments and typically encompass a sizable coastal region. The plants in open water seaweed farms must be supported by a base structure. When used in open seas, many of these

Table 1 Examples for biomass yield in different large-scale seaweed production systems

Species name	Division	Purpose of cultivation	Best design for cultivation	Yield	References
Saccharina latissima	Phaeophyta	Food, raw material for various agricultural uses, alginate extraction, biogas production, and many other applications	IMTA combined with mussel rafts	20.67 wet kg m^{-2}	Freitas et al. (2016)
Fucus vesiculosus	Phaeophyta	Organic cosmetics and as food supplement, canopy forming species	Spray cultivation system with circulating brackish water	6.65–6.76 kg m^{-2}	Meichssner et al. (2021)
Furcellaria lumbricalis	Rhodophyta	Extraction of furcellaran for usage in food and cosmetic industries as stabilizing, thickening and gelling agents	Open water with all its designs	100,000–150,000 tonnes of wet weight every year	Weinberger et al. (2020)
Palmaria palmata	Rhodophyta	Food markets, and as feed, biofuel, or for bioactive compounds	Integrated multitrophic aquaculture (IMTA)	510 g m^{-2}	Grote (2019)
Ulva spp.	Chlorophyta	Human consumption, bioenergy or for bioremediation to reduce nutrients in highly eutrophicated waters	In herring production process waters	27.17–37.07 g m^{-2} fresh weight	Li et al. (2014)

structures have been destroyed. When the system is permitted to drift in a circular pattern, the forces associated with "orbiting" motions require a successful structure to be able to dissipate very high levels of energy in order to resist them. It follows that a seaweed farm system with an adequate structural design is required in order to maintain the algal mat and resist the immense forces imposed by the sea. Numerous seaweed farm designs have been suggested and, in few instances, tried to be used on continental shelves or the open ocean. The designs for submerged system aim to achieve the desired maximum yield, as these designs take in consideration the effect of kelp shading, kelp drag, kelp nutrient drawdown, and modified nutrient transport on kelp growth (Frieder et al. 2022). The different designs/techniques include the following.

Long-Line Technique

Because of its simplicity and low costs, this technique is considered as the most common one that regularly uses ropes, rings, or nets arranged in various ways as substrates for producing algae. The rope lines could be made of monofilament, nylon, rope, thin high-density polyethylene (HDPE) thread, in addition to many other kinds (Sahoo and Yarish 2005). Young seaweed gametophytes raised in onshore hatcheries are often used to seed the appropriate materials before they are released offshore, supported by networks of stakes or suspended below floats. Recent research has revealed that textile chemistry and structure have a significant impact on the adhesion and proliferation of seaweeds. Before being harvested, the seaweed may be allowed to develop for several months to a year, depending on the species. These fields may be found in open waters, enclosed bays, lagoons, and estuaries. The long line technique may be co-cultured with floating mussel lines by twisting the long lines of algal seeding put 0.5 m below the surface, surrounding the floating mussel lines that were already there. Also, there are different designs of structures on which offshore long lines are suspended, such as mesh bags, spider webs, square rafts, triangle rafts, PVC pipe rafts, cage systems, multiple rafts long line, moline tubular net, tubular square rafts, net pouches, net bags, and basket method (Redmond et al. 2014).

Bottom Culture Technique

Bottom planting (direct method), which entails inserting thalli into a sandy bottom using various tools, is a common cultivation technique, especially in warm temperate climates, which includes two technique. (1) Bottom stocking technique, which is the simplest strategy for transporting vegetative thalli that develop in a natural field environment, and rubber bands can also be used to fasten the species to rocks, stabilizing the thalli in soft sands. (2) The plastic tube technique, which is typically used in subtidal areas, entails fixing bundles of thalli to plastic tubes filled with sand to anchor the algae to the sea bottom. Using this technique, divers arrange the plastic tubes filled with sand in parallel rows, perpendicular to the ocean floor, to establish an underground thallus system that will eventually support the production (Brzeska-Roszczyk 2017).

Integrated Multi-Trophic Aquaculture (IMTA)

As bio-extractive organisms, seaweeds absorb additional nutrients produced by other marine organisms like fish or shrimp. The term "IMTA" refers to an integrated culture that combines extractive aquaculture (seaweed and shellfish) with fed aquaculture (fish and shrimp). This culture can occur in coastal waterways and can be greatly intensified. Cultivating seaweeds in the open ocean close to animal enclosures is one of the cultivation techniques used in integrated seaweed mariculture. Since seaweeds can remove up to 90% of the nutrients released from an intensive fish farm, combining the farming of fish and seaweeds may help to address the issue of animal farming effluents polluting the ecosystem. For use as potential biofilters of animal effluents in IMTA systems, more than 20 species, primarily brown algae, have been tried. The algae could be planted in a straight-line pattern close to the mariculture (Redmond et al. 2014).

Floating Rafts for Vertical Rope Culture

The floating lines making up the culture raft are connected to structural ropes that are fastened to the seabed by concrete blocks and hung horizontally by marker buoys. Cork floats attached to the rope at predetermined intervals make the floating lines float at the surface. By putting weight on one of the ropes connected to the structural ropes, these culture lines can be stretched horizontally and parallel to the main direction of the tidal current (Peteiro and Freire 2011).

Polyethylene Nets and Ropes Hanging on a Floating Rope

In this design, the polyethylene nets are cut into panels with specific dimensions for each measure. Melting is used to strengthen the nets' sides. Every cultivation unit uses pieces of rope that are cut into pieces of a set length. The foot and top of each unit have fixed metal bars that hold it to the main line. This technique could be also applied using fishing nets which could be used for polluted water treatment via seeding the nets through biofiltrations (using mussels, barnacles, and algae) before deployment in the polluted environment.

Ring System

In this technique, the alga sporophytes are suspended in ring construction with a specific diameter and consists of a polyethylene tube with a specific thick wall and diameter that is welded to rings. The rings consequently are floated at a certain depth and supported with carrier ropes (Buck and Buchholz 2004).

3 Microalgae Cultivation Modes and Techniques

3.1 Cultivation Modes and Growth Conditions

Microalgal growth characteristics and cell composition are significantly dependent on the cultivation conditions that provides different nutrient and energy sources. The growth of microalgae under different cultivation conditions causes a great variation in bio-products/compounds and biomass productivity. Microalgae are capable of metabolic shift depending on environmental conditions. Assuming that there are four main microalgae cultivation modes, photoautotrophic, heterotrophic, mixotrophic, and photoheterotrophic, these modes are further discussed in the following sections and illustrated in Fig. 3 (de Freitas Coêlho et al. 2019).

3.1.1 Photoautotrophic Cultivation

In this mode, microalgal cells utilize light (natural sunlight or artificial light) and inorganics (H_2O, salts, and CO_2) as the energy and nutrient sources for photosynthesis to produce organics, which is commonly the most approved and oldest cultivation mode/system for microalgae (Yun et al. 2021). Photoautotrophic cultivation contributes to global CO_2 reduction as the algae consume it from the surrounding atmosphere as the only C-source for cell growth. The cultivation site must be constructed close to businesses or power stations that produce high amounts of CO_2 for improved microalgal growth. Therefore, the cultivation site is encouraged to be constructed near to factories or power stations that produce high amounts of CO_2 for improved microalgal growth. The integration of autotrophic cultivation of microalgae with other heterotrophic microbes has also been performed as the CO_2 released from the cultivation of other microbes can be supplied for microalgae. This has shown to promote the sustainable process of microalgae cultivation through assimilation of CO_2 derived from yeast (Sun et al. 2018). This mixed culture of microalgae and yeast cultivation has also successfully enhanced the CO_2 bio-fixation rate and formation of lipids, where the biodiesel produced will have better oxidative stability as compared to those from microalgae monocultures. Moreover, contamination risk is less severe in photoautotrophic cultivation than compared with the other three cultivation modes (Shu et al. 2013). Hence, an outdoor scale-up cultivation system for microalgae (PBRs and OWPs) is always recommended to be done using photoautotrophic cultivation to minimize the contamination problems. However, biomass harvesting cost of microalgae cultivated under phototrophic cultivation is increased as the biomass concentration of microalgae is relatively lower under this cultivation mode (Chew et al. 2018; Yen et al. 2016).

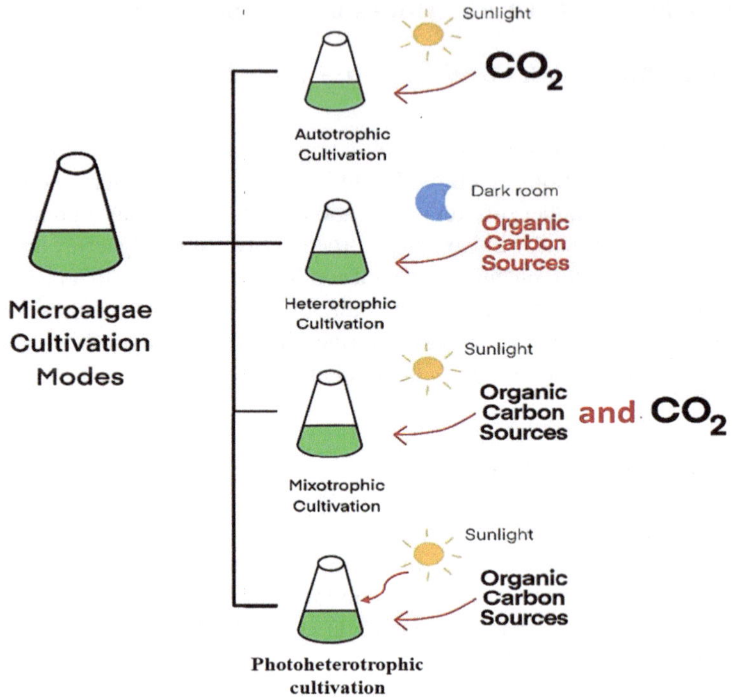

Fig. 3 Different microalgae cultivation modes showing the main principles of autotrophic, heterotrophic, mixotrophic, and photoheterotrophic cultivation conditions

3.1.2 Heterotrophic Cultivation

Heterotrophic mode is a cultivation technique whereby microalgae use organic compounds as both carbon and energy sources (Fig. 3). Some microalgae species can grow and adapt in both photoautotrophic conditions (in the presence of a light source) and in heterotrophic conditions (where light is absent) (Chen et al. 2011). This cultivation mode can be used to overcome the limitations with restricted light source that prevent high cell density in large-scale PBRs during photoautotrophic growth (Ruiz et al. 2022). According to numerous studies, heterotrophic growing conditions can produce microalgae with better biomass and lipid productivity (Abomohra et al. 2013, 2018; Almutairi et al. 2021). For instance, *Chlorella protothecoides* showed 40% increase in lipid content after shifting the cultivation conditions from photoautotrophic to heterotrophic (Liang et al. 2009). Several organic carbon sources can be used and consumed by microalgae for growth such as glucose, sucrose, lactose, galactose, glycerol, and fructose. This culture has the limitation that the sugar-based organic carbon supply usually has contamination problems (Chen et al. 2011). As a result, closed fermenters or traditional microbial bioreactors provide a better cultivation medium for this mode of operation than outdoor scale-up processes. Also, the cost of the substrate needed for microalgal growth is increased when organic compounds are used, raising the overall cost of a heterotrophic cultivation system (Ruiz et al. 2022).

3.1.3 Mixotrophic Cultivation

Mixotrophic cultivation occurs when microalgae are capable of using both inorganic (CO_2) and organic carbon sources for growth (Fig. 3.). In addition, a variety of microalgae species that are suitable for mixotrophic cultivation can exist either in a photoautotrophic or heterotrophic state, or even both, depending on the proportion of organic compounds and available light intensity (Castillo et al. 2021). When sufficient light is present during phototrophic cultivation, the CO_2 that the microalgae release during respiration is retained and used again. Similar to phototrophic cultivation, mixotrophic cultivation can reduce the global CO_2 levels since microalgae use it as a carbon source during their growth cycle. Because using organic materials may result in contamination issues, a closed photobioreactor is employed for this culture mode, which may supply a light source while minimizing the risk of contamination. Additionally, as compared to heterotrophic growth conditions, mixotrophic microalgae have relatively high lipid and biomass productivities but are typically still less employed to grow microalgae for oil production (Castillo et al. 2021; Shu and Tsai 2016).

3.1.4 Photoheterotrophic Cultivation

Photoheterotrophic cultivation is a cultivation condition that is also known as photo-assimilation or photo-metabolism, where microalgae require light and use organic compounds as the carbon source (Fig. 3). The major difference between photoheterotrophic and mixotrophic cultivation is that the first needs both sugars and a light source at the same time for energy production, whereas the later can use of organic molecules or CO_2 (Abreu et al. 2022; Chen et al. 2011). Similar to mixotrophic and heterotrophic cultivation conditions, contamination problems occur in photoheterotrophic cultivation due to the presence of sugar-based organic compounds. Moreover, this mode needs a special design of PBRs during scale-up, which increases the capital and operating cost of cultivation.

3.2 Microalgae Cultivation Systems

Many techniques have been applied for microalgae cultivation that can be classified into five categories such as open, closed, hybrid, dark, and offshore systems (Fig. 4). The used system is defined based on the type of strain and it needs source of nutrients

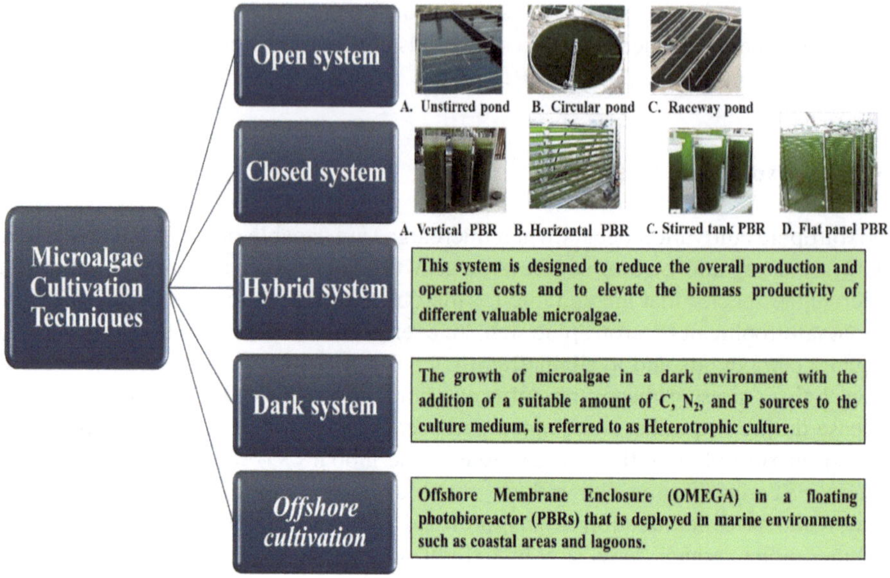

Fig. 4 Different microalgae cultivation systems including open ponds (A. unstirred pond; B. circular pond; C. raceway pond), closed photobioreactors (A. vertical PBR; B. horizontal PBR; C. stirred tank PBR; D. flat panel PBR), and hybrid systems, in addition to dark and offshore systems

and investment cost. The different cultivation systems are discussed in the following sections.

3.2.1 Open Systems

Open ponds represent one of the cheapest, most widely, and least complicated cultivation systems/methods in the large-scale microalgae culture techniques (Suali and Sarbatly 2012). The benefits of using this system are supported by cheaper construction, maintenance, low operation cost and energy demand, and applicable to scale-up. In open ponds, the culture medium is directly in contact with different external environmental conditions (climate variation, temperature, light duration, and intensity), and the possibility of contamination by protozoa, bacteria, and insects which lead to toxicity and ineffectiveness of the final products (Singh and Sharma 2012). For artificial water bodies, microalgae can be cultured in a container or tank (made of plastic, fibreglass, or cement materials) with smooth internal surfaces to reduce friction damage and clean easily. The depth of the tanks or containers varies between 10 and 50 cm and it is mixed with large rotational arms (extend up to 45 m in diameter and 30 cm deep) or aerated by air bubbles, which help in the dissemination of CO_2 and penetration of the sunlight to promote the biomass yield in the culture medium (Amaral et al. 2020). In some cases of circular ponds/tanks, the dry biomass yield can reach 1.5–16.5 g m^{-2} d^{-1} (Suali and Sarbatly 2012), while in raceway ponds, the productivity can vary from 0.19 to 23.5 g m^{-2} d^{-1} of dry biomass (Al Hattab et al. 2014). There are different types of open ponds, which include unstirred lakes/ponds (natural water bodies) and artificial water bodies such as circular and raceway ponds (Fig. 5).

Unstirred Ponds

Unstirred open ponds are simple, shallow, and widely found in natural water systems (lakes or lagoon ponds) for microalgae cultivation (Fig. 5). The depth of the pond is about 0.5 m that helps and accelerates light diffusion, while the lack of a stirring or mixing unit makes microalgae cultivation affordable and more practical. For instance, *Dunaliella salina* is one of the suitable, distinctive microalgal species to be cultivated under these circumstances for β-carotene commercial production (Chew et al. 2018; Converti et al. 2009). The disadvantages of this system include the poor atmospheric CO_2 mass transfer/diffusion to the water column/culture in the ponds, contaminations caused by the presence of other microorganisms in the culture that cause microalgae to grow in a competitive manner, low nutrient circulation, and flotation of dead and living algae. These all factors will slow down and curb the growth rate of the microalgae and effect dramatically the biomass, lipids, and pigment final productivity. With the addition of mechanical stirring equipment or agitation, this system can be improved and broadly employed in different commercial industries (Casanova et al. 2022; Tan et al. 2020).

Circular Ponds

These are the first artificial ponds designed to be used in large-scale microalgae cultivation, consisting of a circular-shaped tank/container, with a depth ranging from 30 to 50 cm, a width of 45 m along with a rotating arm/agitator fixed in the centre of the pond to mix efficiently, and prevent the sedimentation of the algal biomass. However, the design of this cultivation system is restricted by the pond's size because large pond may propose a stronger water resistance, and therefore causes strain on the mechanical parts of agitator (Iluz and Abu-Ghosh 2016). Currently, Japan and Taiwan used this cultivation system to culture *Chlorella* for different commercial applications (Tan et al. 2020). However, this design has many disadvantages such as high energy cost required for the agitation process, the lack of temperature control, contaminations, and relatively high construction cost (Kumar et al. 2021).

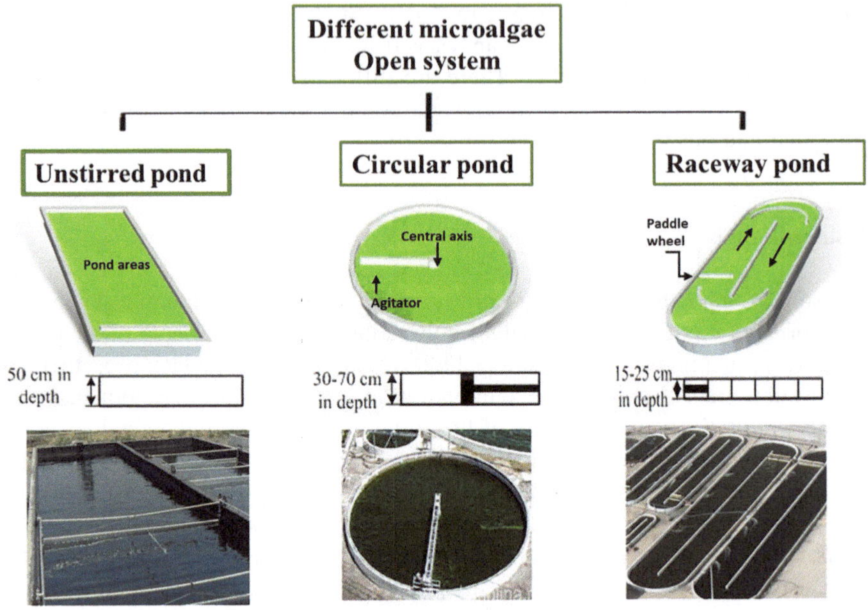

Fig. 5 Different open ponds used for cultivation of microalgae to enhance and optimize the biomass and valuable phycochemicals productivity. Source: Modified from Shen et al. (2009) and Zerrouki and Henni (2019)

Racetrack/Way Pond

This system (Fig. 5) is one of the most used open pond systems for microalgae cultivation in commercial scale. It consists of a single channel or groups of the closed-loop channel with a depth of 15–40 cm. The paddlewheel is used as a mixing unit to recirculate the microalgal culture in the pond and reduce the residence time of the algae in dark regions. The single paddlewheel can be adequate to serve and cover a large area of cultivation pond, prevent biomass sedimentation, promote sunlight penetration, and allow atmospheric CO_2 exchange with culture in the pond (Chew et al. 2018; Iluz and Abu-Ghosh 2016; Kumar et al. 2021). Raceway pond is recognized as one of the most excellent designs for open pond microalgae cultivation due to its energy efficiency since one paddlewheel can cover about 5-hectare racetrack pond (Tan et al. 2020). The most promising and appropriate microalgae reported to this system are *Spirulina* sp., *Dunaliella* sp., *Chlorella* sp., and *Haematococcus* sp. Sapphire Energy's Algal Biomass Farm, in the United States, is one of the successful commercial raceway pond cultivation systems that over the course of two years has fruitfully produced 520 metric tonnes of dried biomass without encountering any technical difficulties (Tan et al. 2020; White and Ryan 2015).

3.2.2 Closed Systems

Closed PBRs are closed systems to cultivate microalgae, where the culture has no direct contact with the external environment. This system has the ability to overcome several limitations and challenges facing the open cultivation systems. The main advantages that this system provides is a high degree of control on the microalgae culture that leads to high biomass productivity with high quality. They contain several designs/systems (e.g. vertical, horizontal, flat plate, and stirred tank) under higher nutrients, metabolic efficiency, and countless combinations of light sources and sunlight that can be applied (Fig. 6). However, the main bottlenecks in closed PBRs application are the various design flaws (the restricted scalability), high capital, and high operating costs.

Vertical Photobioreactors

There are two designs for vertical PBR designs, namely bubble and airlift columns. The bubble column PBR has a cylindrical form (Fig. 6), if the reactor is divided into two parts, one with aeration and one without, it is called airlift PBR (Benner et al. 2022). The bubble column PBR is used widely in laboratory and commercial scales due to its changeability and flexibility of operation, and simplicity in handling and building structure. It consists of a cylindrical long shape tube/vessel with no internal structure, made of transparent materials such as glass or polymers such as acrylics or polyvinyl chlorides (Huang et al. 2017). Air bubbles are used for mixing, while fluorescent tubes or LED lamps are employed to light the bubble column PBR with a diameter ranging from 7 to 24 cm. In order to provide homogeneous illumination, light sources are dispersed all across the cylindrical bioreactor. From the surface to the centre of the cylindrical bubble column containing the microalgae suspension, the incident light intensity diminishes, which may cause light-limited microalgal growth in the centre and photoinhibition effects near the surface (Benner et al. 2022). The main advantages of the bubble column PBR include reasonable mass transfer and a great surface to volume ratio (Ting et al. 2017).

Generally, vertical PBRs consist of two components, the solar receiver and the airlift system, which are connected by water pumps. The majority of the photosynthetic reactions take place in the solar receiver, a long tube installed in a variety of geometries. The tube diameter used in this part is typically optimized for the best sunlight capture by the microalgae suspension pushed through the tube (while reducing the area needed for the tube's installation). The airlift mechanism transports extra oxygen produced by the cells' metabolism outside of the reactor. In order to extend CO_2 stay in the solar receiver section, CO_2 is typically also injected at the end of the airlift system and the beginning of the solar receiver. To have better control over the culture temperature, this component may also include a heat exchanger. Simpler variations of this architecture, such as those with simply a solar receiver portion and a simpler pumping framework, can also be used to create laboratory-

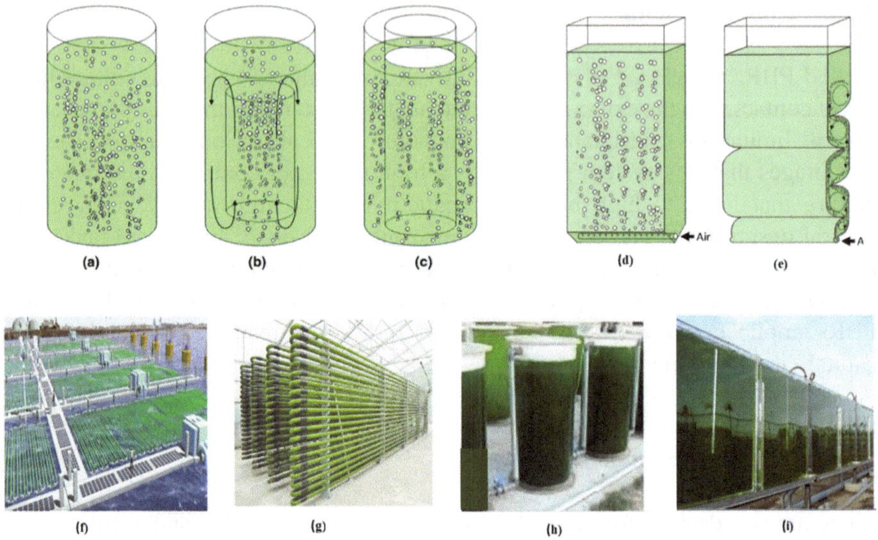

Fig. 6 Illustration showing different microalgae closed cultivation systems. (**a**) Normal column reactor, (**b**) Column reactor with concentric airlift (**c**) Annular reactor, (**d**) Flat plate aeration tube (**e**) FPA photobioreactor (**f**) Offshore cultivation (**g**) Tubular reactor (**h**) stirred tank (**i**) Flat reactor. Source: Modified from Pal and Satpati (2018) and Yusoff et al. (2019)

scale vertical PBRs. This lowers the expenses and accurately replicates a tubular PBR. Because of their modular construction approach, the design is also easily scalable to industrial capacities (Mendes and Badino 2016; Ting et al. 2017).

Additionally, the cost of building and maintaining a tubular PBR is high, especially when it comes to mixing and cooling. The maintenance of high CO_2 concentrations after the injection has taken place is one of the key difficulties for culture in tubular PBRs. The productivity and growth rate of the microalgal cells rise in direct proportion to the amount of CO_2 present in the solution. Additionally, the lipid content within the microalgae rises in direct proportion to the CO_2 concentration, which may be significant to produce biofuels from microalgae. Different approaches are being thought about to increase the availability of CO_2 and nutrients for microalgae cells along the length of the tubes (De Vree et al. 2015; Huang et al. 2017). The tube-in-tube design has been investigated with the inclusion of a porous tube (dialysis tube) in between the culture tube, in which these substrates flow in opposite directions. Utilizing nanofibres or other substances in the solution that absorb more CO_2 than water is another option for increasing CO_2 concentration in the tubular PBR, such that the CO_2 transfer into the liquid phase along the bioreactor's length resembles the injection site (Comitre et al. 2021). To maintain the best CO_2 circulation throughout the entire reactor, the nanofibres must be replaced after a few days since biomass adsorption can prevent CO_2 release (Moraes et al. 2020). On a lab-scale, new tubular photobioreactor designs have been tested, i.e. Fibonacci-type bioreactor system, which reaches a higher light utilization efficacy. These vertical PBRs, which contain tubes with a vertical orientation, intercepted solar radiation by 1.4 times more, while maintaining the necessary pH, temperature, and oxygen concentration levels for microalgae (Benner et al. 2022; Díaz et al. 2019).

Horizontal Photobioreactors (HPBRs)

Horizontal PBRs are closed tubular reactors, consisting of a parallel set of connected loops of tubes and placed horizontally or inclined at an angle to the horizontal (Singh and Sharma 2012). Because tubular types produce a large amount of biomass in a short period of cultivation, they have been reported to be the most popular HPBRs. Transparent materials, made of plastic or glass, are used in the parallel construction of the tubes to maximize the effectiveness of solar absorption. The tubes have a diameter of 0.2 meters or smaller (Al-Dailami et al. 2022; Benner et al. 2022). They have higher solar absorption capacity, where the average energy requirement is about 4000% more than that of the flat panel PBRs (FpPBRs) and bubble columns (Singh and Sharma 2012). They are connected to an airlift or pump system that pumps microalgae through the tubes. In outdoor culture, the reactor's design is essential for converting light energy based on its orientation towards the sun. A special gas exchange system is used to introduce the CO_2 gas mixture into the tube connection. However, robust deoxygenation processes are required to improve the dissolved oxygen content because photobleaching occurs when oxygen is accumulated in the tubes. It is important to keep and sustain the highly turbulent flow regime in the HPBRs to avoid flocculation of microalgae (Klinthong et al. 2015). Furthermore, the laying of the tubes closer to each other within long smaller diameter tubes can provide a higher areal and volumetric productivity, but needs to be maintained to inhibit oxygen accumulation and keep sufficient light access to each tube. Replacing natural sunlight with artificial light has shown an increase in the yield and produced more valuable bioproducts (Al Hattab et al. 2014; Wang et al. 2012; Yen and Chiang 2012), which could be a solution for countries with limited natural light such as in Europe.

Stirred Photobioreactors (SPBRs)

The SPBRs are derived from the fermentation tank design and improved to be suitable for microalgae cultivation. For this reactor, an external light source and agitator for mixing are added to control and enhance the optimal temperature and biomass transfer. The use of mixing and aeration for indoor cultivation are features of this reactor that results in excellent biomass productivity through improved gas exchange and solubility (Kong et al. 2021). Currently, these PBRs are applied on a small laboratory-scale only, and not suitable nor preferable for larger scale that may not encompass sufficient illumination. Eventually, this type of PBRs needs more development of a continuously stirred tank reactor to increase the biomass concentration and enhance the surface-to-volume ratio that could improve the photosynthetic efficiency of the microalgae (Tan et al. 2016).

Flat Panel Photobioreactors (FpPBRs)

The most popular type of closed PBRs is the FpPBRs, which are made of two sheets of transparent or semi-transparent glass that are joined in a cascade towards the light source (Fig. 6). FpPBRs can be placed vertically or at specific angles to absorb most of the light intensity from sunlight or different light sources (Ting et al. 2017). It has a very short light path which helps light penetration to microalgae cells, making it appropriate for both indoor and outdoor cultivations. The PBR's air sparger creates air bubbles that are sent to the pump to help mix and circulate the culture media. By speeding up the mixing process, adding sufficient CO_2, and eliminating the accumulated O_2, this technique has the ability to increase the microalgal biomass production and would magnify the effect of the flashing light (Dębowski et al. 2013; Klinthong et al. 2015; Ting et al. 2017). This reactor will increase the biomass production and be beneficial for the extraction of microalgal phycochemicals linked to light-harvesting complexes (Lin et al. 2015). This design, however, has several known drawbacks, including a potential high hydrodynamic stress damage from the aeration, biofouling formation at the surface of the culture, and the possibility of biomass attachment and adherence to the walls. Another issue with the FpPBRs is temperature control, but this issue can be resolved by installing a sprinkler or heat exchanger systems. Overall, the biomass productivity will be affected by the type of cultivation system used and by the type of cultivated microalgae. Table 2 lists some examples for different closed PBRs systems as well as open ponds that are employed for microalgae cultivation along with details about capacity and biomass yield using different microalgae.

3.2.3 Hybrid Cultivation System

The hybrid cultivation system comprises two-stage culture, aiming to reduce and address the drawbacks of both open and closed cultivation systems (Narala et al. 2016; Tan et al. 2016). In a photobioreactor, the first stage of cultivation is carried out, where the growth parameters are maintained to promote cell division and to reduce the contamination from other microorganisms (Brennan and Owende 2010; Zakariah et al. 2015). High inoculum is required in the second phase to make sure that the microalgal population rather prevails in the cultivation medium and no other microbes. In the second phase, the growing microalgae are shifted to open ponds and confronted with nutrient depletion and certain additional environmental stresses. Huntley and Redalje (2007) used the hybrid system for the production of astaxanthin and oil from *Haematococcus pluvialis* to be used in aquaculture for salmon feed. Depending on the local climatic circumstances, this system produced an annual average of 10 to 24 tonnes ha^{-1} of lipids from *Haematococcus pluvialis*. The study

Table 2 Microalgae cultivation and corresponding biomass yield in different cultivation systems including photobioreactors (PBRs) and open ponds

Cultivation system	Strain	Capacity	Biomass yield
Open ponds	*Haematococcus pluvialis*	20 m^3	0.107 g L^{-1} d^{-1}
	Haematococcus pluvialis	100 m^3	0.122 g L^{-1} d^{-1}
	Dictyosphaerium sp.	8 m^3	5.8 g m^{-2} d^{-1}
	Pediastrun boryanum	8 m^3	9.2 g m^{-2} d^{-1}
	Botryococcus braunii	0.080 m^3	0.1 g L^{-1} d^{-1}
	Scenedesmus sp.	20 m^3	17 g m^{-2} d^{-1}
Tubular PBRs	*Chlorella sorokiniana*	–	1.47 g L^{-1} d^{-1}
	Spirulina sp.	5.5–100 L	0.62–0.01 g L^{-1}
	Phaeodactylum tricornutum	75–200 L	1.19–1.38 g L^{-1}
Flat panel	*Dunaliella tertiolecta*	3.4–30 L	1.5 g/L–3.42 g L^{-1}
	Phaeodactylum sp.	5 L	1.38 g L^{-1}
Stirred tank	*Chlorella vulgaris*	1.5 L	0.027–0.045 g L^{-1} h^{-1}
Bubble PBRs	*Cyanobium* sp.	1.8 L	0.071 g L^{-1}
	Aphanothece microscopica	3 L	0.77 g L^{-1} d^{-1}
	Monodus sp.	64 L	0.03–0.2 g L^{-1}
	Spirulina sp.	3.5 L	4.13 g L^{-1}
Airlift PBRs	*Chaetoceros* sp.	170 L	0.8 g L^{-1}
	Haematococcus pluvialis	3 L	4.06 g L^{-1}
	Botryococcus branuii	3 L	2.31 g m^{-3} d^{-1}
	Porphyridium cruentum	–	1.5 g L^{-1} d^{-1}

Modified from Chew et al. (2018) and Kumar et al. (2021)

revealed that when microalgal species with higher lipid content and photosynthetic efficiency such as *Chlorella vulgaris* are utilized, it is possible to reach up to 76 tonnes ha^{-1} per year under similar conditions. An initial conceptual investigation concerning this two-stage cultivation procedure was also carried out by Rodolfi et al. (2009), which showed that 90 kg ha^{-1} of lipid might be produced every day (10 and 8 kg ha^{-1} in the first and second phases, respectively). In climates like the Mediterranean, this system produced 20 tonnes of lipids ha^{-1}; while in bright tropical regions, it might yield up to 30 tonnes ha^{-1}. A 0.2 m^3 hybrid-type PBR was created by Acién Fernández et al. (2001) by combining an airlift system and an external tubular loop. In order to reduce dead and dark zones and remove the oxygen created during photosynthesis, a degasser was utilized in the airlift section. When *Phaeodactylum tricornutum* was cultivated using this method, a biomass productivity of 1.20 g L^{-1} d^{-1} was attained at a dilution rate of 0.050 h^{-1}. It is important to note that the proper selection microalgae culture system is essential for ensuring the output of high biomass productivity.

3.2.4 Dark System

The vast majority of microalgal species demand photoautotrophic conditions for growth, but some can be cultivated in a dark system, additionally referred to as heterotrophic culture, in which the culture medium is supplemented with a suitable amount of C, N, and P sources (Chen et al. 2014). The dark system involves standard components including no light, a limited range of 200–480 rpm agitation, a pH range of 6.1–6.5, growth medium with different organic carbon sources containing glucose or acetate, and a sufficient amount of N and P. Commercially, long-chain unsaturated fatty acids (LC-USFAs) are produced from *Crypthecodinium cohnii* in a heterotrophic growth environment. When microalgae are grown under heterotrophic circumstances, light pigmentation on the algae fades away, which gradually raises volumetric biomass yield as well as protein and lipid productivities. Because of the increased energy concentration in the given organic carbon source than in CO_2, far more cell mass is observed when utilizing this heterotrophic culture strategy compared to the photoautotrophic approach (Rastogi et al. 2018). In this strategy of cultivation, microalgae are grown using high cellular growth methods using a fed-batch culture approach that makes use of a large amount of carbon supply while removing the self-shading effect of concentrated biomass, which will lead to a significant cost saving. There are some drawbacks to dark system growth, including the fact that only a few species of algae can be cultivated heterotrophically, that adding organic materials raises the expense, and the algae cannot create light-induced metabolites due to the absence of light (Barros et al. 2015; Rastogi et al. 2018).

3.2.5 Offshore Cultivation

In a floating photobioreactor that is placed in marine environments such as coastal areas and lagoons, microalgae are primarily cultivated offshore utilizing Offshore Membrane Enclosure (OMEGA), according to recent reports (Rashid et al. 2014). The OMEGA system's architecture cultivates microalgae in different types of wastewater from the land that are transported and anchored offshore in different estuaries/bays, where the source for the algal growth is easily accessible inside the membrane. Due to the offshore location, agricultural usage no longer competes with other uses of land and water. Culture media circulate in the PBRs under a turbulent system in place, helped by mixing through swirl vanes that provide a whorled flow pattern. The flow enhances the turbulence, which raises exchange rates and maximizes the amount of light exposure in the culture media. This technology has some drawbacks, including the possibility of biofouling on the PBR's wall. To prevent this, higher pumping rates are needed, which increases the need for energy because of the higher drag produced (Rashid et al. 2014).

4 Advantages and Drawbacks of Different Microalgae Cultivation Systems

Algae producers have expressed interest in synthesizing algal biomass using more cost-effective and efficient methods to achieve optimal productivity. However, production intensities of each system showed different results and were primarily affected by the microalgal species, growth medium, surrounded environmental conditions, and types of the mixing system. The comparison of open and closed systems is done in order to find the cultivation technique that will best achieve a given aim. Due to their significant simplicity and economic benefits, open systems are frequently employed in large-scale applications. Basically, with the help of a mixing mechanism and sunlight absorption, this technology enables algae to grow naturally. Nevertheless, growing algae in open environment has a number of difficulties, including contamination of the culture and evaporation loss that reduces productivity.

Most of the challenges associated with open cultivation system, like contamination and evaporation rate, are eliminated using the closed cultivation systems. Additionally, PBRs are utilized for producing specialized bioproducts and phycochemicals with high-added value. In general, PBRs are able to provide larger yields than open systems, and the harvesting methods are also not too expensive. Algae need CO_2 and sunlight to grow, the process by which photosynthesis occurs. Because a PBR is enclosed, there is a higher potential for O_2 accumulation, which will impede the algal growth. In addition, a closed system needs a constant flow of CO_2 induced through the culture medium. However, while CO_2 bubbles in open systems have a short contact time to be absorbed before being released into the atmosphere, closed systems have a far higher CO_2 absorption rate. Thus, closed systems recycle CO_2 until equilibrium is reached, making them the ideal cultivation mode for this parameter. The growth rate of microalgae can be adjusted by experimentation due to the tightly regulated surroundings of closed systems. The system is fitted with the ideal conditions, causing the algae to produce more biomass per unit volume. It is attributed to the water's capacity to hold a higher concentration of algal biomass and the presence of more healthy algae. Open systems are more difficult to regulate, particularly when it comes to temperature as a crucial factor for algal development. Open systems are still in use even though closed systems have much more benefits to offer. The primary issue with closed systems is the investment and maintenance costs associated with ongoing closed cultivation. Energy is needed to keep characteristics like illumination and temperature constant, and the technology is more expensive. However, open ponds are simply strategically placed holes in the ground, while PBRs and closed systems need a lot of resources in addition to the building effort needed to develop the systems. Thus, OWPs are more preferred than the expensive reactors because of their lower cost (Abomohra et al. 2016). However, the low biomass yield, high evaporation rate, and sever contamination remain the main challenges towards outdoor applications.

5 Challenges and Future Perspectives

Despite the fact that algae are promising organisms with industrial biotechnology and environmental applications, the cost of producing algae-derived products is still higher than the cost of synthetic compounds, e.g. biofuel compared to fossil fuel. In other words, the basic processes used for the manufacture and marketing of algal products continue to be a significant barrier, including the use of chemicals, high technology, electricity, and labour. Wastewater and flue gas have been studied to minimize the cost required for nutrients used for algal cultivation. Microalgal culture carried out in closed and open cultivation systems that have optimal pH, temperature, and light aid in fostering rapid doubling of microalgae that produce high biomass. To raise the compound of interest in microalgae, the optimization of growth conditions, nutritional stress, and physical modification should be developed into a more effective technique. In addition, the low productivity of bio-compounds and potentials metabolites in algae place restrictions on the production of value bio-products including pigments, lipids, enzymes, carbohydrates, proteins, fatty acids, and other long/short polysaccharides. So, it is necessary to come up with different methods for increasing the value of algal products.

Using algal extracts or whole algal cells for the green production of many phycochemicals has been practised recently, which will be discussed in different chapters of this book. In comparison to physical and chemical approaches, algae-based synthesis is eco-friendly, even though the cost is relatively high. Thus, there must be enough research to solve the related issues including low production yield, elevated costs, and marketing of microalgae-based resources and products. To lower the price of high-value chemicals, it is possible to develop novel extraction techniques, increase microalgae application in aquaculture, and utilize the whole microalgal biomass, and apply biorefinery routes (Abomohra and Hanelt 2022; Wang et al. 2022; Zaky and Abomohra 2023). To make genetic engineering successful in microalgal growth and production, more information need to be explored to the full genome data of the species, that could also enhance the algal CO_2 sequestration efficiency (Barati et al. 2021). The two main strategies for getting a desired product in a molecular approach are strain selection and transformation. For genetic engineering to be successful in the microalgal industry, more thoroughness must be added to the full genomic data of the organism. Microalgal nuclear, mitochondrial, and chloroplast genomes can all be edited using contemporary gene editing technologies including CRISPR-Cas9, TALEN, and ZFN 17. In addition, Omics methods will thereby revolutionize the development of cutting-edge technologies. However, safety issues and possible impact of outdoor cultivation of genetically modified organisms should be considered.

6 Conclusions

Recently, algae have attracted great attention due to the need for sustainable resources and the rise in the global warming awareness, since it can be employed for a variety of environmentally friendly applications with coupled CO_2 reduction. The success of algae cultivation depends critically on selecting the cultivation system under the proper and optimal cultural conditions. Algal strain plays also a major role in determining the ideal growth parameters, and there are many difficulties involved in outdoor cultivation of different algal species. Despite having several drawbacks, OWPs are still advised for use in treating wastewater, producing biofuel, and reducing the amount of CO_2 that is produced from power plants and industrial facilities. It is suited to produce low-value algae-based products due to the lower cost compared to PBRs. But because it is not affected by the environment, cultivation of microalgae in closed systems produces good quality and high biomass yield, making it unquestionably suitable for high-value products such as application in cosmetics, medicinal products, nutritional compounds, and pharmaceutical applications. However, PBRs merit additional work to overcome the difficulties and reduce the overall cost.

References

Abomohra A, Hanelt D (2022) Recent advances in micro-/Nanoplastic (MNPs) removal by microalgae and possible integrated routes of energy recovery. Microorg 10:2400. https://doi.org/10.3390/MICROORGANISMS10122400

Abomohra A, Wagner M, El-Sheekh M, Hanelt D (2013) Lipid and total fatty acid productivity in photoautotrophic fresh water microalgae: screening studies towards biodiesel production. J Appl Phycol 25:931–936. https://doi.org/10.1007/s10811-012-9917-y

Abomohra A, Jin W, Tu R, Han S-F, Eid M, Eladel H (2016) Microalgal biomass production as a sustainable feedstock for biodiesel: current status and perspectives. Renew Sust Energ Rev 64: 596–606

Abomohra A, Eladel H, El-Esawi M, Wang S, Wang Q, He Z, Feng Y, Shang H, Hanelt D (2018) Effect of lipid-free microalgal biomass and waste glycerol on growth and lipid production of *Scenedesmus obliquus*: innovative waste recycling for extraordinary lipid production. Bioresour Technol 249:992–999

Abreu AP, Morais RC, Teixeira JA, Nunes J (2022) A comparison between microalgal autotrophic growth and metabolite accumulation with heterotrophic, mixotrophic and photoheterotrophic cultivation modes. Renew Sust Energ Rev 159:112247. https://doi.org/10.1016/J.RSER.2022.112247

Acién Fernández FG, Fernández Sevilla JM, Sánchez Pérez JA, Molina Grima E, Chisti Y (2001) Airlift-driven external-loop tubular photobioreactors for outdoor production of microalgae: assessment of design and performance. Chem Eng Sci 56:2721–2732. https://doi.org/10.1016/S0009-2509(00)00521-2

Al Hattab M, Ghaly A, De AP, Ríos L (2014) Production of biodiesel from marine and freshwater microalgae: a review. Adv Res 3:107–155. https://doi.org/10.9734/AIR/2015/7752

Al-Dailami A, Koji I, Ahmad I, Goto M (2022) Potential of Photobioreactors (PBRs) in cultivation of microalgae. J Adv Res Appl Sci Eng Technol 27:32–44. https://doi.org/10.37934/ARASET.27.1.3244

Almutairi AW, Al-Hasawi ZM, Abomohra A (2021) Valorization of lipidic food waste for enhanced biodiesel recovery through two-step conversion: a novel microalgae-integrated approach. Bioresour Technol 342:125966. https://doi.org/10.1016/J.BIORTECH.2021.125966

Amaral MS, Loures CCA, Naves FL, Samanamud GL, Silva MB, Prata AMR, Amaral MS, Loures CCA, Naves FL, Samanamud GL, Silva MB, Prata AMR (2020) Microalgae cultivation in Photobioreactors aiming at biodiesel production. Biotechnol Appl Biomass. https://doi.org/10.5772/INTECHOPEN.93547

Ammar EE, Aioub AAA, Elesawy AE, Karkour AM, Mouhamed MS, Amer AA, El-Shershaby NA (2022) Algae as bio-fertilizers: between current situation and future prospective. Saudi J Biol Sci 29:3083–3096. https://doi.org/10.1016/j.sjbs.2022.03.020

Barati B, Zeng K, Baeyens J, Wang S, Addy M, Gan SY, Abomohra A (2021) Recent progress in genetically modified microalgae for enhanced carbon dioxide sequestration. Biomass Bioenergy 145:105927. https://doi.org/10.1016/J.BIOMBIOE.2020.105927

Barros AI, Gonçalves AL, Simões M, Pires JCM (2015) Harvesting techniques applied to microalgae: a review. Renew Sust Energ Rev 41:1489. https://doi.org/10.1016/j.rser.2014.09.037

Benner P, Meier L, Pfeffer A, Krüger K, Oropeza Vargas JE, Weuster-Botz D (2022) Lab-scale photobioreactor systems: principles, applications, and scalability. Bioprocess Biosyst Eng 45(5):791–813. https://doi.org/10.1007/S00449-022-02711-1

Brennan L, Owende P (2010) Biofuels from microalgae—a review of technologies for production, processing, and extractions of biofuels and co-products. Renew Sust Energ Rev 14:557–577

Brooks MT, Coles VJ, Hood RR, Gower JFR (2018) Factors controlling the seasonal distribution of pelagic Sargassum. Mar Ecol Prog Ser 599:1–18. https://doi.org/10.3354/MEPS12646

Brzeska-Roszczyk P (2017) A review of the selected methods of macroalgae cultivation in marine waters. Biul Inst Morskiego 32:108–115

Buck BH, Buchholz CM (2004) The offshore-ring: a new system design for the open ocean aquaculture of macroalgae. J Appl Phycol 16:355–368. https://doi.org/10.1023/B:JAPH.0000047947.96231.EA/METRICS

Bumbak F, Cook S, Zachleder V, Hauser S, Kovar K (2011) Best practices in heterotrophic high-cell-density microalgal processes: achievements, potential and possible limitations. Appl Microbiol Biotechnol 91(1):31–46. https://doi.org/10.1007/S00253-011-3311-6

Campbell I, Macleod A, Sahlmann C, Neves L, Funderud J, Øverland M, Hughes AD, Stanley M (2019) The environmental risks associated with the development of seaweed farming in Europe - prioritizing key knowledge gaps. Front Mar Sci 6:107. https://doi.org/10.3389/fmars.2019.00107

Casanova LM, Mendes LBB, Corrêa T d S, da Silva RB, Joao RR, Macrae A, Vermelho AB (2022) Development of microalgae biodiesel: current status and perspectives. Microorg 11(1):34. https://doi.org/10.3390/MICROORGANISMS11010034

Castillo T, Ramos D, García-Beltrán T, Brito-Bazan M, Galindo E (2021) Mixotrophic cultivation of microalgae: an alternative to produce high-value metabolites. Biochem Eng J 176:108183. https://doi.org/10.1016/J.BEJ.2021.108183

Chen C-YY, Yeh K-LL, Aisyah R, Lee D-JJ, Chang J-SS (2011) Cultivation, photobioreactor design and harvesting of microalgae for biodiesel production: a critical review. Bioresour Technol 102:71–81. https://doi.org/10.1016/j.biortech.2010.06.159

Chen CY, Yeh KL, Chang HY, Chang JS (2014) Strategies to improve oil/lipid production of microalgae in outdoor cultivation using vertical tubular-type Photobioreactors. Energy Procedia 61:2755–2758. https://doi.org/10.1016/J.EGYPRO.2014.12.298

Chew KW, Chia SR, Show PL, Yap YJ, Ling TC, Chang JS (2018) Effects of water culture medium, cultivation systems and growth modes for microalgae cultivation: a review. J Taiwan Inst Chem Eng 91:332–344. https://doi.org/10.1016/J.JTICE.2018.05.039

Chinnasamy S, Bhatnagar A, Hunt RW, Das KC (2010) Microalgae cultivation in a wastewater dominated by carpet mill effluents for biofuel applications. Bioresour Technol 101:3097–3105. https://doi.org/10.1016/j.biortech.2009.12.026

Chowdury KH, Nahar N, Deb UK (2020) The growth factors involved in microalgae cultivation for biofuel production: a review. Comput Water Energy Environ Eng 9:185–215

Comitre AA, Vaz B d S, Costa JAV, de Morais MG (2021) Renewal of nanofibers in *Chlorella fusca* microalgae cultivation to increase CO2 fixation. Bioresour Technol 321:124452. https://doi.org/10.1016/J.BIORTECH.2020.124452

Converti A, Casazza AA, Ortiz EY, Perego P, Del Borghi M (2009) Effect of temperature and nitrogen concentration on the growth and lipid content of *Nannochloropsis oculata* and *Chlorella vulgaris* for biodiesel production. Chem Eng Process Process Intensif 48:1146–1151. https://doi.org/10.1016/j.cep.2009.03.006

de Freitas Coêlho D, Tundisi LL, Cerqueira KS, da Silva Rodrigues JR, Mazzola PG, Tambourgi EB, de Souza RR (2019) Microalgae: cultivation aspects and bioactive compounds. Brazilian Arch Biol Technol 62:e19180343. https://doi.org/10.1590/1678-4324-2019180343

De Vree JH, Bosma R, Janssen M, Barbosa MJ, Wijffels RH (2015) Comparison of four outdoor pilot-scale photobioreactors. Biotechnol Biofuels 8:1–12. https://doi.org/10.1186/S13068-015-0400-2/FIGURES/5

Dębowski M, Zieliński M, Grala A, Dudek M (2013) Algae biomass as an alternative substrate in biogas production technologies—review. Renew Sust Energ Rev 27:596–604. https://doi.org/10.1016/J.RSER.2013.07.029

Díaz JP, Inostroza C, Acién Fernández FG (2019) Fibonacci-type tubular photobioreactor for the production of microalgae. Process Biochem 86:1–8. https://doi.org/10.1016/J.PROCBIO.2019.08.008

Dudley TL (1992) Beneficial effects of herbivores on stream macroalgae via epiphyte removal. Oikos 65:121. https://doi.org/10.2307/3544894

Ebbesmeyer CC, Ingraham WJ (1994) Pacific toy spill fuels ocean current pathways research. Eos Trans Am Geophys Union 75:425–430. https://doi.org/10.1029/94EO01056

EL-Mohsnawy E, El-sheekh M, Mabrouk M, Zohir W (2020) Enhancing accumulation of omega 3 and 9 fatty acids in *Chlorella vulgaris* under mixotrophic nutrition. J Anim Plant Sci 30:485–492. https://doi.org/10.36899/JAPS.2020.2.0047

El-Sheekh MM, El-Mohsnawy E, Mabrouk ME, Zoheir WF (2020) Enhancement of biodiesel production from the green microalga Micractinium reisseri via optimization of cultivation regimes. J Taibah Univ Sci 14(1):437–444

Fayyad RJ, Muslim SN, Ali ANM (2020) Application strategies for using fungi and algae as bioremediators: a review. Plant Arch 20:788–792

Freitas JRC, Salinas Morrondo JM, Cremades Ugarte J (2016) *Saccharina latissima* (Laminariales, Ochrophyta) farming in an industrial IMTA system in Galicia (Spain). J Appl Phycol 28:377–385. https://doi.org/10.1007/S10811-015-0526-4/TABLES/6

Frieder CA, Yan C, Chamecki M, Dauhajre D, McWilliams JC, Infante J, McPherson ML, Kudela RM, Kessouri F, Sutula M, Arzeno-Soltero IB, Davis KA (2022) A macroalgal cultivation modeling system (MACMODS): evaluating the role of physical-biological coupling on nutrients and farm yield. Front Mar Sci 9:752951. https://doi.org/10.3389/FMARS.2022.752951/BIBTEX

Gokare ARA (2022) Global perspectives on Astaxanthin: from industrial production to food, health, and pharmaceutical applications eBook : Ravishankar, Gokare A., Ambati, Ranga Rao: Amazon.in: Kindle Store. https://doi.org/10.1016/C2019-0-01800-6

Grote B (2019) Recent developments in aquaculture of *Palmaria palmata* (Linnaeus) (Weber & Mohr 1805): cultivation and uses. Rev Aquac 11:25–41. https://doi.org/10.1111/RAQ.12224

Huang Q, Jiang F, Wang L, Yang C (2017) Design of Photobioreactors for mass cultivation of photosynthetic organisms. Engineering 3:318. https://doi.org/10.1016/J.ENG.2017.03.020

Huntley ME, Redalje DG (2007) CO2 mitigation and renewable oil from photosynthetic microbes: a new appraisal. Mitig Adapt Strateg Glob Chang 12:573–608. https://doi.org/10.1007/S11027-006-7304-1/METRICS

Iluz D, Abu-Ghosh S (2016) A novel photobioreactor creating fluctuating light from solar energy for a higher light-to-biomass conversion efficiency. Energy Convers Manag 126:767–773. https://doi.org/10.1016/J.ENCONMAN.2016.08.045

Israel A, Levy I, Friedlander M (2006) Experimental tank cultivation of *Porphyra* in Israel. In: Eighteenth International Seaweed Symposium. Springer, Dordrecht, pp 9–14. https://doi.org/10.1007/978-1-4020-5670-3_2

Jiang Z, Liu J, Li S, Chen Y, Du P, Zhu Y, Liao Y, Chen Q, Shou L, Yan X, Zeng J, Chen J (2020) Kelp cultivation effectively improves water quality and regulates phytoplankton community in a turbid, highly eutrophic bay. Sci Total Environ 707:135561. https://doi.org/10.1016/j.scitotenv.2019.135561

Jun Pang S, Lüning K (2004) Breaking seasonal limitation: year-round sporogenesis in the brown alga *Laminaria saccharina* by blocking the transport of putative sporulation inhibitors. Aquaculture 240:531–541. https://doi.org/10.1016/J.AQUACULTURE.2004.06.034

Klinthong W, Yang YH, Huang CH, Tan CS (2015) A review: microalgae and their applications in CO2 capture and renewable energy. Aerosol Air Qual Res 15:712–742. https://doi.org/10.4209/AAQR.2014.11.0299

Kong W, Shen B, Lyu H, Kong J, Ma J, Wang Z, Feng S (2021) Review on carbon dioxide fixation coupled with nutrients removal from wastewater by microalgae. J Clean Prod 292:125975. https://doi.org/10.1016/j.jclepro.2021.125975

Kumar BR, Mathimani T, Sudhakar MP, Rajendran K, Nizami AS, Brindhadevi K, Pugazhendhi A (2021) A state of the art review on the cultivation of algae for energy and other valuable products: application, challenges, and opportunities. Renew Sust Energ Rev 138:110649. https://doi.org/10.1016/J.RSER.2020.110649

Lee YK, Ding SY, Hoe CH, Low CS (1996) Mixotrophic growth of *Chlorella sorokiniana* in outdoor enclosed photobioreactor. J Appl Phycol 8:163–169. https://doi.org/10.1007/BF02186320/METRICS

Li J, Kangas P, Terlizzi DE (2014) A simple cultivation method for Chesapeake Bay *Ulva intestinalis* for algal seed stock. N Am J Aquac 76:127–129. https://doi.org/10.1080/15222055.2013.869287

Liang Y, Sarkany N, Cui Y (2009) Biomass and lipid productivities of *Chlorella vulgaris* under autotrophic, heterotrophic and mixotrophic growth conditions. Biotechnol Lett 31:1043–1049

Lin JH, Lee DJ, Chang JS (2015) Lutein in specific marigold flowers and microalgae. J Taiwan Inst Chem Eng 49:90–94. https://doi.org/10.1016/J.JTICE.2014.11.031

Liu Y, Weisberg RH, Hu C, Zheng L (2011) Tracking the Deepwater horizon oil spill: a modeling perspective. Eos Trans Am Geophys Union 92:45–46. https://doi.org/10.1029/2011EO060001

Lowrey J, Brooks MS, McGinn PJ (2015) Heterotrophic and mixotrophic cultivation of microalgae for biodiesel production in agricultural wastewaters and associated challenges—a critical review. J Appl Phycol 27:1485–1498. https://doi.org/10.1007/S10811-014-0459-3/TABLES/3

Madkour FF, Kamil AEW, Nasr HS (2012) Production and nutritive value of *Spirulina platensis* in reduced cost media. Egypt J Aquat Res 38:51–57. https://doi.org/10.1016/j.ejar.2012.09.003

Marquez FJ, Nishio N, Nagai S, Sasaki K (1995) Enhancement of biomass and pigment production during growth of *Spirulina platensis* in mixotrophic culture. J Chem Technol Biotechnol 62:159–164. https://doi.org/10.1002/JCTB.280620208

Meichssner R, Krost P, Schulz R (2021) Experimental testing of density- and season-dependent growth in vegetative Fucus aquaculture and modelling of growth over one year for different cultivation scenarios. J Appl Phycol 33:3939–3950. https://doi.org/10.1007/S10811-021-02597-8/TABLES/1

Mendes CE, Badino AC (2016) Hydrodynamics of Newtonian and non-Newtonian liquids in internal-loop airlift reactors. Biochem Eng J 109:137–152. https://doi.org/10.1016/J.BEJ.2016.01.007

Moraes L, Rosa GM, Cara IM, Santos LO, Morais MG, Grima EM, Costa JAV, Fernández FGA (2020) Bioprocess strategies for enhancing the outdoor production of *Nannochloropsis gaditana*: an evaluation of the effects of pH on culture performance in tubular photobioreactors. Bioprocess Biosyst Eng 43:1823–1832. https://doi.org/10.1007/S00449-020-02373-X/FIGURES/4

Mulet S, Rio MH, Mignot A, Guinehut S, Morrow R (2012) A new estimate of the global 3D geostrophic ocean circulation based on satellite data and in-situ measurements. Deep Sea Res Part II Top Stud Oceanogr 77–80:70–81. https://doi.org/10.1016/J.DSR2.2012.04.012

Narala RR, Garg S, Sharma KK, Thomas-Hall SR, Deme M, Li Y, Schenk PM (2016) Comparison of microalgae cultivation in photobioreactor, open raceway pond, and a two-stage hybrid system. Front Energy Res 4:29. https://doi.org/10.3389/FENRG.2016.00029/BIBTEX

Ogawa T, Aiba S (1981) Bioenergetic analysis of mixotrophic growth in *Chlorella vulgaris* and *Scenedesmus acutus*. Biotechnol Bioeng 23:1121–1132. https://doi.org/10.1002/BIT.260230519

Pal R, Satpati GG (2018) Microalgae-biomass to biodiesel: a review. J Algal Biomass Utln 2018: 11–37

Perez-Legaspi IA, Valadez-Rocha V, Ortega-Clemente LA, Jimenez-Garcıa MI (2019) Microalgal pigment induction and transfer in aquaculture. Rev Aquac 12:1–21. https://doi.org/10.1111/raq.12384

Peteiro C, Freire Ó (2011) Effect of water motion on the cultivation of the commercial seaweed *Undaria pinnatifida* in a coastal bay of Galicia, Northwest Spain. Aquaculture 314:269–276. https://doi.org/10.1016/J.AQUACULTURE.2011.02.009

Prasad Behera D, Vadodariya V, Veeragurunathan V, Sigamani S, Moovendhan M, Srinivasan R, Kolandhasamy P, Ingle KN (2022) Seaweeds cultivation methods and their role in climate mitigation and environmental cleanup. Total Environ Res Themes 3–4:100016. https://doi.org/10.1016/J.TOTERT.2022.100016

Rashid N, Ur Rehman MS, Sadiq M, Mahmood T, Han JI (2014) Current status, issues and developments in microalgae derived biodiesel production. Renew Sust Energ Rev 40:760–778. https://doi.org/10.1016/J.RSER.2014.07.104

Rastogi RP, Pandey A, Larroche C, Madamwar D (2018) Algal green energy–R&D and technological perspectives for biodiesel production. Renew Sust Energ Rev 82:2946–2969

Raven AJ, Giordano M (2014) Algae. CurrBio 24(13):590–595. ISSN09609822. https://doi.org/10.1016/j.cub.2014.05.039. https://www.sciencedirect.com/science/article/pii/S0960982214006046

Redmond S, Green L, Yarish C, Kim J, Neefus C (2014) New England seaweed culture handbook: nursery systems

Rodolfi L, Zittelli GC, Bassi NN, Padovani G, Biondi N, Bonini G, Tredici MR, Chini Zittelli G, Bassi NN, Padovani G, Biondi N, Bonini G, Tredici MR (2009) Microalgae for oil: strain selection, induction of lipid synthesis and outdoor mass cultivation in a low-cost photobioreactor. Biotechnol Bioeng 102:100–112. https://doi.org/10.1002/bit.22033

Rosenberg JN, Mathias A, Korth K, Betenbaugh MJ, Oyler GA (2011) Microalgal biomass production and carbon dioxide sequestration from an integrated ethanol biorefinery in Iowa: a technical appraisal and economic feasibility evaluation. Biomass Bioenergy 35:3865–3876. https://doi.org/10.1016/j.biombioe.2011.05.014

Ruiz J, Wijffels RH, Dominguez M, Barbosa MJ (2022) Heterotrophic vs autotrophic production of microalgae: bringing some light into the everlasting cost controversy. Algal Res 64:102698. https://doi.org/10.1016/J.ALGAL.2022.102698

Sahoo D, Yarish C (2005) Mariculture of seaweeds. In: Phycological methods: algal culturing techniques. Academic, New York, pp 219–237

Schmitz TA, Kraft E (2022) Pilot scale photobioreactor system for land-based macroalgae cultivation. J Appl Phycol 34:507–516. https://doi.org/10.1007/S10811-021-02617-7/TABLES/3

Shen Y, Yuan W, Pei ZJ, Wu Q, Mao E (2009) Microalgae mass production methods. Trans ASABE 52:1275–1287. https://doi.org/10.13031/2013.27771

Shu CH, Tsai CC (2016) Enhancing oil accumulation of a mixed culture of *Chlorella* sp. and Saccharomyces cerevisiae using fish waste hydrolysate. J Taiwan Inst Chem Eng 67:377–384. https://doi.org/10.1016/J.JTICE.2016.08.022

Shu CH, Tsai CC, Chen KY, Liao WH, Huang HC (2013) Enhancing high quality oil accumulation and carbon dioxide fixation by a mixed culture of *Chlorella* sp. and Saccharomyces cerevisiae. J Taiwan Inst Chem Eng 44:936–942. https://doi.org/10.1016/J.JTICE.2013.04.001

Singh RN, Sharma S (2012) Development of suitable photobioreactor for algae production–a review. Renew Sust Energ Rev 16:2347–2353

Singh SP, Singh P (2015) Effect of temperature and light on the growth of algae species: a review. Renew Sust Energ Rev 50:431–444. https://doi.org/10.1016/j.rser.2015.05.024

Suali E, Sarbatly R (2012) Conversion of microalgae to biofuel. Renew Sust Energ Rev 16:4316–4342

Sun Y, Huang Y, Martin GJO, Chen R, Ding Y (2018) Photoautotrophic microalgal cultivation and conversion. Green Energy Technol 0:81–115. https://doi.org/10.1007/978-981-10-7677-0_3/TABLES/2

Tan CH, Chen CY, Show PL, Ling TC, Lam HL, Lee DJ, Chang JS (2016) Strategies for enhancing lipid production from indigenous microalgae isolates. J Taiwan Inst Chem Eng 63:189–194. https://doi.org/10.1016/J.JTICE.2016.02.034

Tan JS, Lee SY, Chew KW, Lam MK, Lim JW, Ho SH, Show PL (2020) A review on microalgae cultivation and harvesting, and their biomass extraction processing using ionic liquids. Bioengineered 11:116. https://doi.org/10.1080/21655979.2020.1711626

Ting H, Haifeng L, Shanshan M, Zhang Y, Zhidan L, Na D (2017) Progress in microalgae cultivation photobioreactors and applications in wastewater treatment: a review. Int J Agric Biol Eng 10:1–29. https://doi.org/10.25165/IJABE.V10I1.2705

Van Sebille E, Aliani S, Law KL, Maximenko N, Alsina JM, Bagaev A, Bergmann M, Chapron B, Chubarenko I, Cózar A, Delandmeter P, Egger M, Fox-Kemper B, Garaba SP, Goddijn-Murphy L, Hardesty BD, Hoffman MJ, Isobe A, Jongedijk CE, Kaandorp MLA, Khatmullina L, Koelmans AA, Kukulka T, Laufkötter C, Lebreton L, Lobelle D, Maes C, Martinez-Vicente V, Morales Maqueda MA, Poulain-Zarcos M, Rodríguez E, Ryan PG, Shanks AL, Shim WJ, Suaria G, Thiel M, Van Den Bremer TS, Wichmann D (2020) The physical oceanography of the transport of floating marine debris. Environ Res Lett 15:023003. https://doi.org/10.1088/1748-9326/AB6D7D

Vergara-Fernández A, Vargas G, Alarcón N, Velasco A (2008) Evaluation of marine algae as a source of biogas in a two-stage anaerobic reactor system. Biomass Bioenergy 32:338–344

Wang B, Lan CQ, Horsman M (2012) Closed photobioreactors for production of microalgal biomasses. Biotechnol Adv 30:904–912. https://doi.org/10.1016/j.biotechadv.2012.01.019

Wang S, Mukhambet Y, Esakkimuthu S (2022) Integrated microalgal biorefinery–routes, energy, economic and environmental perspectives. J Clean Prod 348:131245

Wegeberg S, Felby C (2010) Algae biomass for bioenergy in Denmark. Biological/technical challenges and opportunities. University of Copenhagen, Copenhagen

Weinberger F, Paalme T, Wikström SA (2020) Seaweed resources of the Baltic Sea, Kattegat and German and Danish North Sea coasts. Bot Mar 63:61–72. https://doi.org/10.1515/BOT-2019-0019/ASSET/GRAPHIC/J_BOT-2019-0019_CV_003.JPG

White RL, Ryan RA (2015) Long-term cultivation of algae in open-raceway ponds: lessons from the field. Ind Biotechnol 11:213–220. https://home.liebertpub.com/ind. https://doi.org/10.1089/IND.2015.0006

Yen HW, Chiang WC (2012) Effects of mutual shading, pressurization and oxygen partial pressure on the autotrophical cultivation of *Scenedesmus obliquus*. J Taiwan Inst Chem Eng 43:820–824. https://doi.org/10.1016/J.JTICE.2012.06.002

Yen HW, Hsu CY, Chen PW (2016) An integrated system of autotrophic *Chlorella vulgaris* cultivation using CO2 from the aerobic cultivation process of *Rhodotorula glutinis*. J Taiwan Inst Chem Eng 62:158–161. https://doi.org/10.1016/J.JTICE.2016.01.025

Yun HS, Kim YS, Yoon HS (2021) Effect of different cultivation modes (photoautotrophic, Mixotrophic, and heterotrophic) on the growth of *Chlorella* sp. and biocompositions. Front Bioeng Biotechnol 9:774143. https://doi.org/10.3389/FBIOE.2021.774143/BIBTEX

Yusoff FM, Nagao N, Imaizumi Y, Toda T, Yusoff FM, Nagao N, Imaizumi Y, Toda T (2019) Bioreactor for microalgal cultivation systems: strategy and development. In: Rastegari A, Yadav A, Gupta A (eds) Prospects of renewable bioprocessing in future energy systems. Biofuel and biorefinery technologies, vol 10. Springer, Cham, pp 117–159. https://doi.org/10.1007/978-3-030-14463-0_4

Zakariah N, Rahman N, Hamzah F, Teknologi TJ-J (2015) Cultivation system of green microalgae, *Botryococcus braunii*: a review. J Teknol 76:1–5. https://doi.org/10.11113/jt.v76.5517

Zaky A, Abomohra A (2023) Marine-based biorefinery: a path forward to a sustainable future. Ferment 9(6):554. https://doi.org/10.3390/FERMENTATION9060554

Zerrouki D, Henni A (2019) Outdoor microalgae cultivation for wastewater treatment. In: Application of Microalgae in Wastewater Treatment. Springer, Cham, pp 81–99. https://doi.org/10.1007/978-3-030-13913-1_5

Zhang L, Liao W, Huang Y, Wen Y, Chu Y, Zhao C (2022) Global seaweed farming and processing in the past 20 years. Food Prod Process Nutr 41(4):1–29. https://doi.org/10.1186/S43014-022-00103-2

Zohir WF, Kapase VU, Kumar S (2022) Identification and characterization of a new Microalga Dysmorphococcus globosus-HI from the Himalayan Region as a potential source of natural Astaxanthin. Biology 11:884. https://doi.org/10.3390/biology11060884

Algae Harvesting

Benjamin Bernard Uzoejinwa and Felix Uzochukwu Asoiro

Abstract Algae harvesting poses a huge challenge to the development and commercialization of the technology of algal biomass production, especially for microalgae where the algal cell density is almost equal to that of water. Macroalgae, which are popularly known as seaweeds, have several advantages compared to microalgae in terms of harvesting. For example, harvesting and dewatering require far less energy compared to that of microalgae. It is therefore pertinent to note that harvesting methods that require high energy input, expensive equipment, or huge quantity of chemicals can increase the overall cost of algal biomass production, reducing the economic feasibility of the process. Thus, cost-effective, environmentally friendly, and best-of-all new hybrid algal harvesting techniques are urgently needed. This chapter discusses the various methods of harvesting algae (both seaweeds and microalgae), their economic feasibility, environmental impacts, as well as the possible constraints to algal harvesting for commercial utilization.

Keywords Economic feasibility · Environmental impacts · Harvesting techniques · Microalgae · Seaweeds

1 Introduction

Generally, the detachment or separation of algal cells from their cultivation broth medium is known as harvesting. It also involves the removal of moisture from diluted algal suspension into a concentrated and thick algal paste, with the intention to obtain a slurry of not less than 2–7% dry matter basis (dm) of algal suspension (Barros et al. 2015; Show et al. 2015; Singh and Patidar 2018). As a result of the relatively low cell density of algae in the culture media, with a low concentration of algal slurry, the commercial harvesting and development technology of algae are fraught with a wide range of challenges. In terms of harvesting, seaweeds have

B. B. Uzoejinwa (✉) · F. U. Asoiro
Department of Agricultural and Bio-resources Engineering, University of Nigeria, Enugu, Nigeria
e-mail: benjamin.uzoejinwa@unn.edu.ng

A. Abomohra, S. Ende (eds.), *Value-added Products from Algae*,
https://doi.org/10.1007/978-3-031-42026-9_3

several advantages compared to microalgae. For example, harvesting and dewatering of seaweeds require far less energy, less expensive equipment, and chemicals compared to that of microalgae, thus reducing the overall cost of algal biomass production, and increasing the economic feasibility of the process.

Several methods, such as physical/mechanical (filtration, sedimentation, centrifugation, ultrasound, and flotation), electrical, chemical (flocculation/coagulation), and biological methods, as well as a possible combination of any of these techniques, have been applied for algal biomass harvesting and concentrating (Ananthi et al. 2021), with the mechanical method being the most reliable and commonly used method (Ananthi et al. 2021). In order to improve the microalgae harvesting efficiency and reduce cost, flocculation and coagulation are often carried out before filtration and centrifugation (Das et al. 2016; Gorin et al. 2015). Seaweed harvesting demands simple but laborious work, whereas microalgae harvesting involves the use of elaborate chemicals and costly mechanical procedures. Results from several studies in the literature revealed that harvesting is the most challenging activity in the downstream processing of algal biomass for biofuel production, owing to the associated high cost or high power/energy input (Caetano et al. 2020; Milledge and Heaven 2013). Amer et al. (2011) elucidated that the cost of dewatering and harvesting equipment may account for over 90% of the total cost of producing microalgal biomass from raceway/open pond cultivation systems. Thus, the efficient harvesting of algal biomass for commercial utilization depends largely on the type and effectiveness of harvesting technique employed, the type of algae to be processed, cell density and size, the allowable degree of moisture, salt concentrations, and cell damage (Ananthi et al. 2021; Barros et al. 2015). The need for additional processing of the algal biomass must also be considered while choosing an effective harvesting method. Therefore, appropriate harvesting methods must not be harmful to the algal biomass or health.

The chosen harvesting technique should also permit the reuse of the culture media. For the majority of applications, the two-step concentration approach of thickening and dewatering is used for microalgal harvesting. To increase the solid concentration of the suspended microalgae and decrease the quantity or volume that needs to be handled, thickening is done. It consists of gravity sedimentation, flotation, coagulation/flocculation (both chemically and biologically based), or electrical techniques, such as electroporation, ultrasounds, pulsed electric fields (Guldhe et al. 2016), and low or moderate electric fields (Lucakova et al. 2021). Dewatering increases the concentration of microalgal slurry to 15–25% of total suspended solids, which consists of various types of centrifugations and filtrations (Show et al. 2015; Singh and Patidar 2018). Likewise, harvesting is the major obstacle to the economic production of seaweeds on a large scale for biofuels and/or phycochemicals.

Currently, seaweed harvesting can be achieved manually or mechanically, depending on the type of species, location, biomass density, and scale of cultivation (Alam et al. 2021; Fernand et al. 2017; Stagnol et al. 2013). Harvesting of seaweeds makes up 20–30% of the total costs of biomass production and further downstream processing (Fernand et al. 2017; Mac Monagail et al. 2017). Experience has shown that there is not yet a universal seaweed harvesting technique. This is still an active area of research which may lead to the development of an appropriate, efficient, and cost-effective harvesting method for diverse species of seaweeds. Besides, there are some ecological and environmental effects that are associated with seaweed harvesting (Campbell et al. 2019). For example, the marine environment is a symbiotic ecosystem such that improper harvesting methods can lead to severe adverse effects on some other marine species beyond those directly targeted for harvest. Thus, the harvesting methods, frequency of harvest, seasonality, and magnitude of harvest strongly impact the ecological effects of commercial scale harvesting, resulting in instability of the fauna and flora community, and low survival rate of the affected population. These factors should be carefully selected and must be in combination with the biological characteristics of the target species to allow for sufficient recovery, hence maintaining sustainable biomass production (Kumar et al. 2021; Mac Monagail et al. 2017). This chapter identified and discusses the various methods of harvesting algae (both seaweeds and microalgae), their economic feasibility, environmental impacts, as well as the possible constraints to algal harvesting for commercial utilization.

2 Microalgae Harvesting

2.1 Methods of Microalgae Harvesting

An overview of various methods used for harvesting microalgae, including chemical flocculation/ coagulation, gravity sedimentation, floatation, centrifugation, filtration, and electrical techniques, is given in Fig. 1. Their respective merits and demerits as well as the mode of operation are summarized in Table 1. The total cost for the most promising microalgae harvesting method (floatation) was estimated as \$1.93–\$2.16-USD kg^{-1} dry matter or concentrated sludge (Sharma et al. 2013). Besides, the total costs (AU\$) of harvesting 10,000 L of *Chlorella* sp. with different harvesting techniques (including centrifugation, sedimentation, primary floatation, secondary floatation, and secondary centrifugation) were estimated as 12.10, 3.70–4.10, 1.86–2.08, 0.833, and 1.21, respectively (Sharma et al. 2013).

2.1.1 Flocculation

Flocculation is a technique that is often carried out using flocculants (natural or synthetic origin) or flocculating agents, which cause the coagulation of algal biomass cells into minute clumps, regarded as flocs, which will aid sedimentation and easy extraction from the culture medium (Show et al. 2015; Singh and Patidar 2018). Flocculation is a preparatory step for thickening prior to dewatering algal biomass using different methods such as filtration, flotation, or centrifugation (Vandamme et al. 2013). The ability to harvest microalgae on a large scale and with a variety of microalgal species has led to the proposal that flocculation is a superior method for

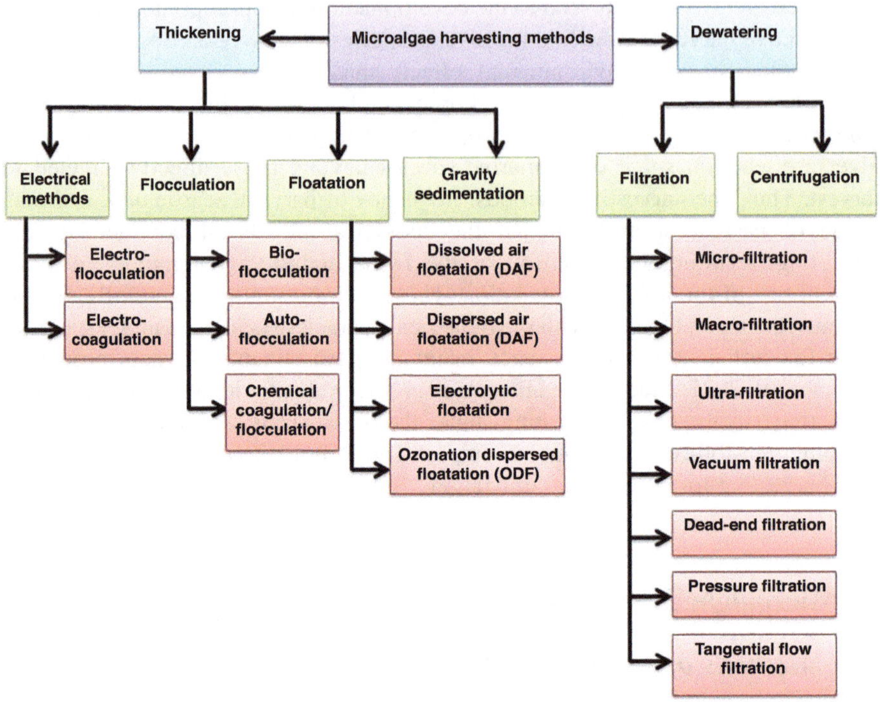

Fig. 1 Schematic diagram showing the various methods used for harvesting microalgae

microalgae harvesting. There are three (3) types of flocculation including chemical coagulation/flocculation, auto-flocculation, and bio-flocculation (Matter et al. 2019).

Chemical Coagulation/Flocculation

Chemical coagulation/flocculation involves the use of chemicals as flocculants or coagulants for the negative charge neutralization and permission of microalgae agglomeration. Suitable chemical coagulants are known for their renewability and sustainability characteristics, without contaminating the biomass but permit the recycling of culture medium. They should also be economical in commercial applications, extracted from renewable bioresources and efficient in low quantities (Barros et al. 2015; Tran et al. 2013). A wide range of salts have been utilized as coagulants in the harvesting of microalgae by numerous researchers (Abomohra et al. 2018; Kurniawan et al. 2022; Wang et al. 2019). Though this is an economic technique, chemicals employed during flocculation could contaminate the algal biomass and thus compromise product quality (Branyikova et al. 2018; Zhao et al. 2018). The three main types of flocculants generally employed in chemical flocculation include, organic polymers (e.g., certain synthetic fibers, surfactants, chitosan,

Table 1 Advantages and disadvantages of various algae harvesting techniques

Harvesting technique	Advantages	Disadvantages
Coagulation/ Flocculation	• The method is easy and fast. • Adaptable to large-scale operation. • Limited damage of seaweed cells. • Can be used in varieties of species. • The energy requirement is low. • Bio-flocculation and auto-flocculation are normally inexpensive.	• Possible use of expensive chemicals. • Depends on relatively high pH. • Challenges in recovering coagulants from harvested biomass. • Type of coagulants normally affect the efficiency. • No possibility in recycling media used for culture. • Microbial or chemical contamination is possible.
Flotation	• Usually adaptable in areas with limited space and low cost. • Has short operating duration.	• Require surfactants. • Other type, e.g., ozonation dispersed floatation may be expensive. • Require high energy input.
Electrical-based processes	• Can be used for all microalgal species. • Require no chemical.	• Needs metal electrodes with may be scarce or expensive. • Equipment and energy required may be costly. • Can cause contamination.
Filtration	• High biomass harvesting efficiency. • Require little cost. • Require no chemical. • Energy consumption is low. • Low shear stress. • Water can be recycled.	• Slow operation. • Needs vacuum or pressure. • Not used for algae with small cell diameters. • High operational and maintenance costs due to membrane replacement occasioned by fouling/clogging. • Use of vacuum filter cause high energy consumption.
Centrifugation	• Method is effective and fast. • Has high biomass harvesting efficiency (>91%). • Most adaptable to laboratory studies. • Can be used for all species of microalgae.	• Very expensive with high power demand. • Has high maintenance cost. • Most useful in recovering high value-added products. • Very expensive and time consuming at industrial level. • Can cause damage to cell walls.

cellulose), inorganic polymers (e.g., polyaluminum chloride, polyelectrolyte), and inorganic flocculants (e.g., metallic salts such as $Al_2(SO_4)_3$, $FeCl_3$, $Fe_2(SO_4)_3$, $AlCl_3$) (Chu et al. 2022).

Among different flocculants, $FeCl_3$ and $Al_2(SO_4)_3$ with ionic charges have found wide applications in the algal biomass harvesting process. The inorganic flocculants which gained huge research attention are $MgSO_4$, $Al_2(SO_4)_3$, $ZnCl_2$, $ZnSO_4$, $Fe_2(SO_4)_3$, $AlCl_3$, NH_4Cl, $(NH_4)_2SO_4$, $MgCl_2$, $FeCl_3$, $CaCl_2$, and $CaSO_4$ (Mathimani and Mallick 2018). Solubility and electronegativity are the parameters that determine the effectiveness of these multivalent salts. Generally, salts with electronegative ions and lower solubility tend to be more effective and faster coagulating agents (Papazi et al. 2010). Studies on 12 inorganic salts to harvest *Chlorella minutissima* confirmed that $FeCl_3$, $Fe_2(SO_4)_3$, $AlCl_3$, $Al_2(SO_4)_3$, $ZnCl_2$, and $ZnSO_4$ could flocculate *C. minutissima* to an optimal concentration of 0.75 g L^{-1} (sulfate salts) and 0.5 g L^{-1} (chloride salts) within 24 h (Papazi et al. 2010). However, six other salts showed no significant changes in the efficiency of flocculation. Aluminum salts, such as $Al_2(SO_4)_3$ and $AlCl_3$, resulted in the destruction of the specific methanogenic activity of the acetogenic bacteria cultured in wastewater sludge (Papazi et al. 2010). Ferric ($FeCl_3$, $Fe_2(SO_4)_3$) and zinc ($ZnCl_2$, $ZnSO_4$) salt coagulants resulted in no cell lysis. Besides, change in algal cells' color could be noticed by using ferric salts such as $Fe_2(SO_4)_3$ and $FeCl_3$ at a concentration greater than 1 g L^{-1} (Mathimani and Mallick 2018).

Organic flocculants include chitosan, modified cationic chitosan-polyacrylamide, Greenfloc 120®, and combinations of starch and chitosan. They can be anionic, cationic, or non-ionic. While anionic or non-ionic polymers are unable to produce microalgae flocs as a result of electro-repulsion, cationic polymers do so because they physically join the microalgae cells together (Singh and Patidar 2018). The charge and functional groups on the surface of microalgae, the pH of the growth medium, and the density of the algal culture are all factors that affect the polyelectrolyte's ability to cause flocculation (Branyikova et al. 2018; Matter et al. 2019). Anionic polyelectrolytes do not flocculate, whereas cationic polyelectrolytes having high charge density are more effective flocculants to harvest microalgae, and the effective dose reduces as the coagulant molecular weight increases (Granados et al. 2012). Letelier-Gordo et al. (2014) reported a flocculation efficiency of 70–90% and a harvesting duration of 60 minutes for *Chlorella protothecoides* using chitosan as an inorganic flocculant. Due to their pH sensitivity, many organic flocculants, such as cationic starch and chitosan, have limited application for microalgae harvesting (Yin et al. 2020). Chemical flocculation is frequently too expensive for large-scale operations due to the excessive flocculants needed (Singh and Patidar 2018). Other drawbacks include high sensitivity to pH, contamination of harvested biomass, and flocculants recycling, which restricts its suitability in downstream food or feed processing and likely renders it uneconomical for large-scale commercial applications (Matter et al. 2019). Due to its high harvesting efficiency and biodegradability, inorganic flocculant is a well-established and secure substance. It is important to note that adding organic polymers costs more than adding inorganic flocculants.

Auto-Flocculation

The term auto-flocculation is the natural flocculation of certain microalgae species without the addition of supplementary chemicals. This process occurs as a result of some environmental stresses, e.g., increase in magnesium or calcium ions, changes in pH levels, variations in nutrient composition, and the dissolved oxygen contents in the culture media (Kushwaha et al. 2020). As an inexpensive, low-energy, non-toxic to microalgae alternative that does not require the use of flocculants and allows for straightforward medium recycling, auto-flocculation (flocculation solely by pH increase) is appealing (Horiuchi et al. 2003). Calcium and phosphate ions will become extremely saturated as the pH of the solution rises. Microalgal cells can create a stable suspension because the negatively charged surface allows for pH-adjustable flocculation. The electrostatic connections between anionic algae can be destroyed by changing the medium's Mg^{2+} concentration and H^+/OH^- ratio (Branyikova et al. 2018).

Different microalgal species, including *Chlorella vulgaris* JSC-7, *Scenedesmus obliquus* AS-6-1, *Ankistrodesmus falcatus* SAG202–9, and *Ettlia texensis* SAG79.80, have also been found to exhibit the auto-flocculation mechanism (García-Pérez et al. 2014). Horiuchi et al. (2003) investigated recovery of *Dunaliella tertiolecta* by the addition of a solution of NaOH, and obtained over 90% biomass recovery efficiency within few minutes of settling time and 8.6– 10.5 range of pH. Salim et al. (2014) observed that the auto-flocculation of the green microalga *Ettlia texensis* SAG79.80 is as a result of glycoprotein. García-Pérez et al. (2014) employed a solution of $Mg(OH)_2$ at pH 10.5 for harvesting of *Chlorella vulgaris and obtained over* 95% recovery efficiency within 30 min. Wu et al. (2012) increased the pH up to 10.6 in their studies and also obtained over 90% biomass recovery efficiency for both marine and freshwater microalgae. Several studies have reported some pH values where outstanding biomass recovery via auto-flocculation was recorded, by applying some quantity of ammonia to vary the pH of the culture. However, results of many research works have revealed that a rise in pH produces precipitation of calcium phosphate, calcium hydroxide, magnesium hydroxide, and other insoluble particles covering the surface of the cell (Lei et al. 2018; Vandamme et al. 2013). These techniques are not chosen for large-scale microalgae pre-concentration in the industry in spite of their merits because they are not dependable for controlled flocculation and can result in variations in the composition of the biomass (Branyikova et al. 2018).

Bio-Flocculation

Bio-flocculation refers to the process whereby bio-flocculants used for the flocculation of the algal cells in the suspension are produced by other microorganisms. The bio-flocculation process involves the accumulation of microalgae triggered by secreted biopolymers, such as the extracellular polymeric substances (EPS) (Christenson and Sims 2011). Fungal flocculation, bacterial flocculation, plant-based flocculation, and actinomycetes flocculation are the four main categories of bio-flocculation methods for algal cells harvesting. Several factors such as the microalgae's ability to be attached onto their surface for the formation of flocs and the production of high concentration of EPS by the bacteria affect the microalgae bio-flocculation (Lee et al. 2010). For instance, bacterial strain (*Solibacillus silvestris*) was used to harvest *Nannochloropsis oceanica* with an efficiency of 88% (Wan et al. 2013). Oh et al. (2001) harvested a number of green algae (*Botryococcus braunii, Chlorella vulgaris, Scenedesmus quadricauda,* and

Selenastrum capricornutum) using facultative anaerobic bacterium (*Paenibacillus* sp.) within 91–95% range of efficiency. Wang et al. (2015) reported a flocculation efficiency of 45–50% when *Chlorella vulgaris* was co-cultured with bacteria-producing bioflocculant at a bacteria:microalgae cultivation ratio of 0.2:0.25. In a related study, 10 mg L^{-1} chitosan alone produced flocculation efficiency of 89% for *Chlorella vulgaris* (Xu et al. 2021). Using only walnut protein extract (WPE), modest flocculation efficiency was attained (40%). However, the combination of WPE with chitosan (6 mg L^{-1}) doubled the flocculation efficiency for *Chlorella vulgaris* (Xu et al. 2021). Xie et al. (2013) and Zhou et al. (2013) employed filamentous fungus (*C. echinulata*) and pallet forming fungus (*Aspergillus oryzae*) in the flocculation of *Chlorella vulgaris,* respectively, and obtained less than 97% removal efficiency. Salim et al. (2012) used the green microalga (*Ettlia texensis*) to harvest 55% of *Chlorella vulgaris,* however only 34% of the biomass was harvested when they employed *Scenedesmus obliquus.* Nguyen et al. (2019) carried out bio-flocculation and cultivation with untreated seafood wastewater using *Chlorella vulgaris* and obtained a nutrient removal efficiency of 88.0 ± 2.2% and flocculation activity of 92.0 ± 6.0%. This was as a result of the bacteria attachment on the microalgae cells for formation of bioflocs. The results also revealed that the bio-flocculation process formation was promoted by the presence of bacteria. Currently, most research studies employed bacteria (as the microorganism) in bio-flocculation of microalgae while fungi and actinomycetes applications are still limited.

2.1.2 Gravity Sedimentation

Gravity sedimentation is a solid–liquid separation technique which employs the action of gravity in the precipitation of concentrated slurry, leaving a clear liquid supernatant (Nie et al. 2018). This technique can be employed when the economic value of end products is very low. The application of sedimentation tanks during process makes it very simple and cheap, however, a very low concentration is obtained without preceding coagulation/flocculation (Nie et al. 2018). Sedimentation has proven to be highly effective and energy efficient for different kinds of microalgae in spite of the process' rudimental characteristics (Rawat et al. 2011). Gravity sedimentation with the best microalgal harvesting results was obtained using the sedimentation tanks and lamella-type separators of recovery efficiencies of 3% and 1.6% TSS, respectively, owing to microalgal auto-flocculation (Show and Lee 2014). A coagulation/flocculation method is commonly applied prior to gravity sedimentation in order to facilitate the settling process of the microalgae. In general, gravity sedimentation is a very cheap and energy efficient method, but with low harvesting efficiency.

2.1.3 Flotation

Flotation refers to as "inverted" sedimentation employs gas or air bubbles in moving the suspended matter to the top of a liquid surface for collection by the skimming process. Several factors (e.g., the likelihood of collision, size of suspended particle, and adhesion) affect the suspended particles' attachment to the air or gas bubbles. Flotation is more effective in the removal of microalgae than sedimentation owing to its self-float characteristics and microalgal low density (Hanotu et al. 2012). Flotation can be categorized into four majorly types (namely ozonation dispersed flotation (ODF), dispersed air flotation (DiAF), dissolved air flotation (DAF), and electrolytic flotation) based on the bubble size (Barros et al. 2015). DAF is an innovative water treatment method which clarifies waste water by removing suspended biomass, e.g. solids and oil (Wang et al. 2021). This method involves dissolving air in the water or wastewater under pressure and then releasing the air at atmospheric pressure in a flotation tank basin. DiAF is a specifically designed system used for physical/chemical pre-treatment of wastewater. In this process, extremely fine bubbles (20–50 μm) are produced and moved into the water via special ejectors for the pretreatment of wastewater prior to discharging (Barros et al. 2015). This method helps to greatly reduce the load for further microalgae downstream processing facilities (Barros et al. 2015). In electrolytic flotation, the method of electrolysis is employed to generate the tiny bubbles needed for air flotation, which will adhere to the suspended biomass and colloidal particles in the microalgal broth, making them float to the liquid surface where they can be scraped off. ODF utilizes charged bubbles resulting from the disintegration of ozone (O_3) to produce hydroxyl radical (OH^-) for the neutralization of the negative charge of the algal surface. This technique is very effective in disinfection and wastewater treatment (Barros et al. 2015). In comparison, although DiAF requires expensive instruments, but it is more energy efficient than the electrolytic flotation process (Vandamme et al. 2011). Electrolytic flotation is the most effective and largely used method, however, in this process, the air in the water is needed to be dissolved at a very high pressure in order to make the solution supersaturated and result in the nucleation of bubbles as the pressure in the nozzle is lowered (Vandamme et al. 2011), while in DiAF, bubbles are produced by constantly passing air via a porous material. It requires less energy but involves costlier instruments and greater demand of pressure drop for the production of bubbles (Vandamme et al. 2011). Electrolytic flotation relies on fine hydrogen bubbles formation by electrolysis (Show and Lee 2014). ODF is usually achieved when ozone gas is passed to the suspension for the production of charged bubbles as the microalgal cells are negatively charged, enhancing the flotation of cells on the surface. In ODF, the mechanism involves the interaction of the charged bubbles with cells (Rawat et al. 2011).

2.1.4 Electrical Methods

The electrical method is a process that involves the concentration of the microalgae by the movement of electric field owing to the negative charge on the microalgal cells' surface (Branyikova et al. 2018). This process is applicable to several microalgal species and does not need the addition of chemicals. In this process, precipitates are formed on the electrodes by the microalgal cells (via electrophoresis) and also on the bottom of the vessel by electro-flocculation. Sacrificial and non-sacrificial electrodes are the two kinds of electrodes employed in this technique (Show and Lee 2014). Sacrificial electrodes lead to the release of metal ions to the broth based on the quantity of electricity that is passed through the electrolytic solution (Show and Lee 2014; Zenouzi et al. 2013). Non-sacrificial electrode application solely depends on the movement of the negatively charged microalgae toward the anode, where cells lose their charge and form aggregates (Guldhe et al. 2016). The electrical method is largely affected by the kind of materials (mainly aluminum and iron) employed in production of the electrode (Kobya et al. 2011). Ferric electrodes consume more energy than the aluminum electrodes and lead to the production of a brown slurry owing to the formation of ferric oxide (Barros et al. 2015). Generally, the harvesting efficiency and energy consumption of the process are affected by the current density which also affects the reaction kinetics. Current density (A m^{-2}) is the amount of current traveling per unit cross-section area of a conductor. It is a vector quantity, having both direction and magnitude. Also, the current density and electrolysis time normally increase with the aluminum ions formed at the anode by the oxidation process, and the increase in the current density lowers the harvesting period (Vandamme et al. 2011). Likewise, electro-flocculation accompanied by settling and mixing reduces the energy required for the harvesting of the microalgae. Electro-flocculation will also require higher concentrations with more energy but reduced time of processing when accompanied by flotation (Zenouzi et al. 2013). Temperature is a vital factor that increases the particle collision and transport rate (Naje et al. 2016). Also, the time of electrolysis can be reduced when the process temperature is increased, thus enhancing the aluminum dissolution rate (Naje et al. 2016). Likewise, the medium initial pH affects the recovery of the microalgal cells because it determines the kind of aluminum ions that is formed (Vandamme et al. 2011). Results of studies reported in the literature unveiled that the electrodes' power consumptions for *Chlorella vulgaris* and *Phaeodactylum tricornutum* were calculated to be 2 and 0.3 kWh kg^{-1}, respectively. This shows that the electrical harvesting method has huge potential as an economic energy efficient process, in comparison with centrifugation (Lee et al. 2013; Vandamme et al. 2011).

2.1.5 Centrifugation

Centrifugation is a method of separating two kinds of immiscible matters. The cell size, retention time of the cell slurry in the centrifuge, and negligible density difference of the microalgal cells to their culture medium are some of the important parameters that affect the microalgae centrifugal separation (Singh and Patidar 2018). Centrifugation harvesting involves high efficiency of separation (of over 90%) under reduced rate of flow and high consumption of energy (Singh and Patidar 2018). Centrifugation is the fastest harvesting method employed in the production of high-value products. It can destroy the cellular structure as it is associated with high force of gravitation and shear stresses. It is also the costliest method owing to its high rate of consumption of energy. Thus, longer time of retention in the bowl is required to aid their sedimentation as a result of the small size of the cells in order to obtain high efficiency of harvesting. The high energy consumption of the process can be reduced by applying coagulation/flocculation prior to centrifugation since this will aid in the reduction of the volume to be processed (Vandamme et al. 2011). The imperforated basket centrifuges, perforated basket centrifuges, decanters, disk stack centrifuges, and hydrocyclones are kinds of centrifuges that have been tested for the separation of microalgae (Fig. 2).

Disc stack centrifuges are industrial centrifuges which concentrate microalgae with sizes between 3 and 30 mm to 0.02 and 0.05% and the force applied is equal to 4000–14,000 times the gravitational force (Najjar and Abu-Shamleh 2020). The energy requirement of this type of centrifuge is very high (Rawat et al. 2011). Among the various types of the disc stack centrifuges, the nozzle type is the most promising, considering its microalgae removal efficiency and harvested biomass quantity (Mathimani and Mallick 2018). However, the solid ejecting disc type has a wide commercial application.

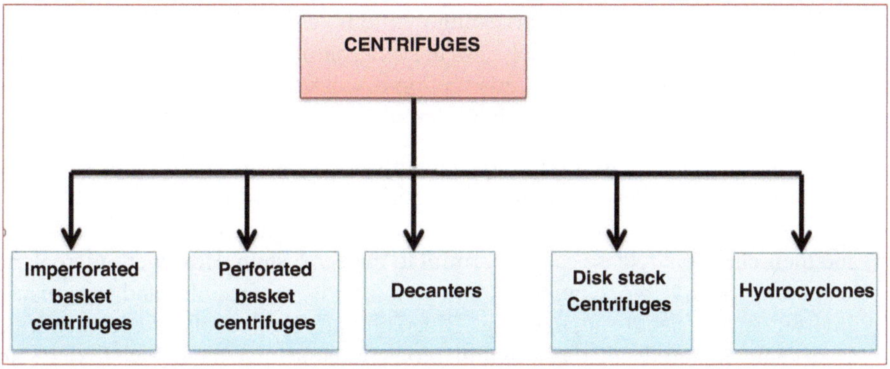

Fig. 2 Different types of centrifuges used for microalgae harvest

2.1.6 Filtration

Filtration refers to a process whereby solid is separated from liquid or gases by the creation of a medium that permits the passage of only fluid. The process is applied in dewatering often accompanied by coagulation/flocculation for the improvement of the efficiency of harvesting. In this method, the algae suspension is subjected to filtration by means of a porous membrane or a filter, retaining the algae slurry as water escapes (Barros et al. 2015). The following types of filtration processes have been successfully applied for microalgal harvesting (Fig. 1): ultra-filtration, dead-end filtration, pressure filtration, vacuum filtration, macro-filtration, and tangential flow filtration (TFF) (Show and Lee 2014). Microalgae species of smaller dimensions with the pore size of 0.1–10 mm (e.g., *Scenedesmus, Dunaliella, Cyclotella and Chlorella*) are not adaptable to this process unlike the microalgae of larger sizes (Yin et al. 2020). The harvesting of fragile cells and small-scale microalga production can be achieved by either ultra-filtration and the membrane micro-filtration method owing to their low cost, the absence of chemicals' application, low rate of consumption of energy (Rubio et al. 2002), and due to the slow process of harvesting involved with this technique. However, the harvesting of microalgae of larger cell sizes is best achieved by macro-filtration, which is basically employed for filtration of the flocs (i.e., biomass obtained via flocculation) or filamentous algae (Yin et al. 2020). Ultra-filtration has a huge potential for fragile algal cells recovery, however, it is not commonly employed in microalgae dewatering. Owing to certain factors such as intensive power consumption, frequent membrane replacement, and higher operation and maintenance costs (Hanotu et al. 2012), tangential flow filtration (TFF) is employed for smaller suspended algae owing to negligible fouling challenges. TFF can recover freshwater algal cells of 70–89% (Show and Lee 2014).

2.2 *Environmental and Economic Impacts*

The environmental impact of the microalgae harvesting process is often a constraint that must be considered. Some harvesting methods of microalgae can result in the release of chemicals or heavy metals into the environment, contributing to pollution and ecosystem degradation. Thus, this section discusses the environmental impacts of various methods used for harvesting microalgae.

1. *Flocculation*: The main environmental impact of chemical flocculation is the potential release of large quantities of residual sludge formed at the end of the process into the environment. Some chemical flocculants can impact aquatic life, and their chemical release can be harmful to the ecosystem. However, microalgae harvesting using biocoagulants/bioflocculants is a feasible and promising environmental-friendly technology. Biocoagulants/bioflocculants are identified as organic materials that are highly biodegradable and have low toxicity to aquatic organisms (Choudhary et al. 2019; Lee et al. 2013). However, on the

other hand, it might have the issue of pathogens release in case of using the whole microbe as a biocoagulant.

2. *Gravity sedimentation*: The environmental impact of this method is minimal since it does not require any energy or chemicals. However, it can be slow and may not be suitable for large-scale operations. Also, this method requires a lot of space and may impair product quality if the sedimented bacteria grow in the settled fraction.
3. *Filtration*: The environmental impact of this harvesting technique depends on the type of filter used. Some filters can be energy-intensive, while others may require chemicals for cleaning. Additionally, the disposal of the filter waste can be an issue. The filters also require regular maintenance to prevent clogging.
4. *Centrifugation*: This method can be energy-intensive and can generate a significant amount of heat, which may require cooling. Additionally, the disposal of centrifuge waste can be an issue. The high energy consumption results in high carbon emissions, contributing to climate change.

The cost of harvesting can be a significant constraint for microalgae production. Harvesting methods that require expensive equipment, chemicals, or energy can increase the overall cost of production, reducing the economic feasibility of the process. The economic feasibility of each of the various methods of harvesting microalgae is discussed in this section:

1. *Centrifugation*: It is a highly efficient method for microalgae harvesting, resulting in high yields of biomass. However, centrifugation is also an expensive method due to the high cost of equipment, maintenance, and energy consumption. Therefore, this process is typically only economically feasible for large-scale operations to produce high-value products.
2. *Flocculation*: It is a cost-effective method for microalgae harvesting, as it does not require expensive equipment and can be scaled up easily. However, the cost of flocculants can vary depending on the type and quantity used, and the use of some chemicals may require additional processing steps to remove them from the biomass. Therefore, the economic feasibility of flocculation depends on the cost of the flocculants and the target product's value. For example, 4–38 mg flocculent g^{-1} chitosan is required for 85–98% of maximum microalgae biomass recovery, using membrane filter or pressure filter with an operating cost of 0.43 (€ kg^{-1}) and 0.23 (€ kg^{-1}) for open system and tubular system, respectively, whereas using vacuum filter will require 0.24 (€ kg^{-1}) and 0.18 (€ kg^{-1}) operating cost for the tubular system and flat panel system, respectively (Fasaei et al. 2018).
3. *Sedimentation*: It is a low-cost method for microalgae harvesting, as it does not require expensive equipment or chemicals. However, sedimentation can be slow, and the biomass yield is often lower than other methods. Therefore, sedimentation is typically only economically feasible for small-scale operations producing low-value products.
4. *Filtration*: It is a highly efficient method for microalgae harvesting, resulting in high yields of biomass. However, filtration requires expensive equipment and may require additional processing steps to remove the biomass from the filter.

Therefore, filtration is typically only economically feasible for large-scale operations producing high-value products.

5. *Electro-coagulation*: It is a cost-effective method for microalgae harvesting, as it does not require expensive equipment or chemicals. However, the effectiveness of this method can be affected by the composition of the microalgae culture, and the process may require additional steps to remove the biomass from the electrode. Therefore, the economic feasibility of electro-coagulation depends on the cost of the electrodes and the target product's value. Another disadvantage of this process includes the need of electrode replacement and maintenance.
6. *Floatation*: This method has a short operation time, low space requirement, large scale harvesting, and high flexibility with low initial cost. The total cost for microalgae harvesting using the floatation method is estimated as $1.93–$2.16 (Sharma et al. 2013). This method is already a common practice but it is limited due to the need for small quantity of chemical coagulant to enhance the efficiency of floatation. Similarly, it has been reported that DiAF suffers from floc shearing as a result of weak hydrophilic bond associated with metal hydroxide (Sharma et al. 2013).

2.3 Constraints to Microalgae Harvesting

Some of the constraints to microalgae harvesting are identified and discussed as follows:

1. *Low biomass concentration*: Microalgae cells are typically small and have a low density, which can make them challenging to harvest. The low biomass concentration in the culture can result in low yields and make the harvesting process more energy-intensive. Recovery efficiency of 99.9% had been reported for *Chlorella* sp. using a cell concentration of 1.43 g L^{-1} with 1432 mg L^{-1} and 360 mg L^{-1} for cetylpyridinium chloride and poly (acrylic acid), respectively (Wu et al. 2015). At pH of 4.5, the flocculation efficiencies of *Scenedesmus* sp., *Chlorococcum nivale*, and *Chlorococcum ellipsoideum* were reported to be between 92 and 98% for a 4.17–6.94 g L^{-1} range of cell concentration (Liu et al. 2013).
2. *High water content*: Microalgae cells contain high water content, which can make it challenging to separate the cells from the culture medium. High water content can also result in a low concentration of the harvested biomass, reducing the overall yield.
3. *Cell fragility*: Some microalgae have fragile cell walls and can be easily damaged during the harvesting process, resulting in a lower yield and reduced quality of the harvested biomass.
4. *Presence of impurities*: The presence of impurities, such as other microorganisms, debris, and dissolved organic matter, can make it challenging to separate the microalgae cells from the culture medium. Impurities can also reduce the quality of the harvested biomass, making it less suitable for certain applications. Gravity

sedimentation and centrifugal force are physical treatments normally employed in separating insoluble impurities like debris from microalgae biomass.

3 Seaweed Harvesting

3.1 Methods of Seaweed Harvesting

Commercial harvest of seaweeds, which is particularly grown for the purposes of biofuels production, industrial uses, fine chemicals synthesis, culinary, or for removing invasive planktons, has currently increased across countries (Nakai 2018). Harvesting partially or totally removing seaweed dominant population results in changing the distribution and the abundance of associated species. Seaweed harvesting is mainly done by manual and mechanical methods (Alam et al. 2021), as discussed further in this section. This section concentrated more on the harvest of naturally grown seaweeds (not those in seaweed farms), owing to the fact that a large proportion of seaweed harvest are naturally grown

3.1.1 Manual Harvesting

The manual method of harvesting seaweeds includes selective hand cutting and gathering. Hand cutting is employed to harvest seaweeds in close inshore areas. It is currently limited in scale and method, with generally low production output. Hand gathering is predominantly used in some gathering of cast seaweeds from shorelines. Hand cutting and gathering include baling of the seaweeds to be towed by boat, and the use of boat to rake the seaweeds. Hand cutting involves the removal of part or all of the living seaweed from its position of growth on the foreshore or seabed by hand. Manual harvesting, which involves manual cutting or pulling the seaweeds from its natural environment, is typically used for small-scale operations and for species that grow close to the shore, such as dulse, nori, and wakame. Manual harvesting has been a normal age-long technique for seaweed harvesting worldwide. Some simple gadgets like wires, nets, and manual cutters are predominantly used in the manual method of harvesting which are often used in small water bodies such as coastal areas, narrow streams, canals, and small ponds (Madsen 2000). Mechanical and manual methods are usually employed in the removal of seaweeds and aquatic weeds. However, mechanical methods are normally preferred for seaweed harvesting since the harvest is mostly carried out in the ocean or cultivated ponds and it involves large quantity of the biomass (Datta 2009). Emergent and floating weeds are harvested differently. Emergent seaweeds should be harvested with caution because they are needed for re-growth, whereas floating seaweeds which easily form dense mats on the marine surface can be manually harvested by cutting and separating into smaller parts. Generally, seaweed harvesting is conducted by using cutting tools like cutlass, blades, or manually operated motorized cutters. The harvest is later transported to shore using manually operated boats. Considerable safety precautions and constant monitoring are needed as parasitic and venomous organisms are often found in marine ecosystems (Alam et al. 2021).

3.1.2 Mechanical Harvesting

Seaweeds are largely harvested in huge quantity using mechanized harvesters, which showed higher efficiency compared to the manual method (Roesijadi et al. 2008). Mechanical harvesting of seaweeds uses equipment such as boats, mechanical cutters, saw, choppers, and mowing bar installed in a land-based long arm vehicle. Others include vacuum suction equipment to suck and collect small plants especially seaweeds, rake for collection, conveyer belts, trailers, loading cranes, etc. (Fig. 3).

Mechanical harvesting technology is commonly utilized when handling emergent and rooted submerged seaweeds in larger water bodies and manual removal can be difficult and inconvenient (Alam et al. 2021). It entails the use of plant harvesters/cutters designed and constructed specifically for harvesting aquatic flora more efficiently in deep and large oceanic water bodies and coastal waters where manual harvesting can be tedious and nearly impossible (Alam et al. 2021; Ananthi et al. 2021). The choice of the harvesting method depends on the type/species of seaweeds in cultivation and the surroundings. Dredging, suction, and cutting may also be used

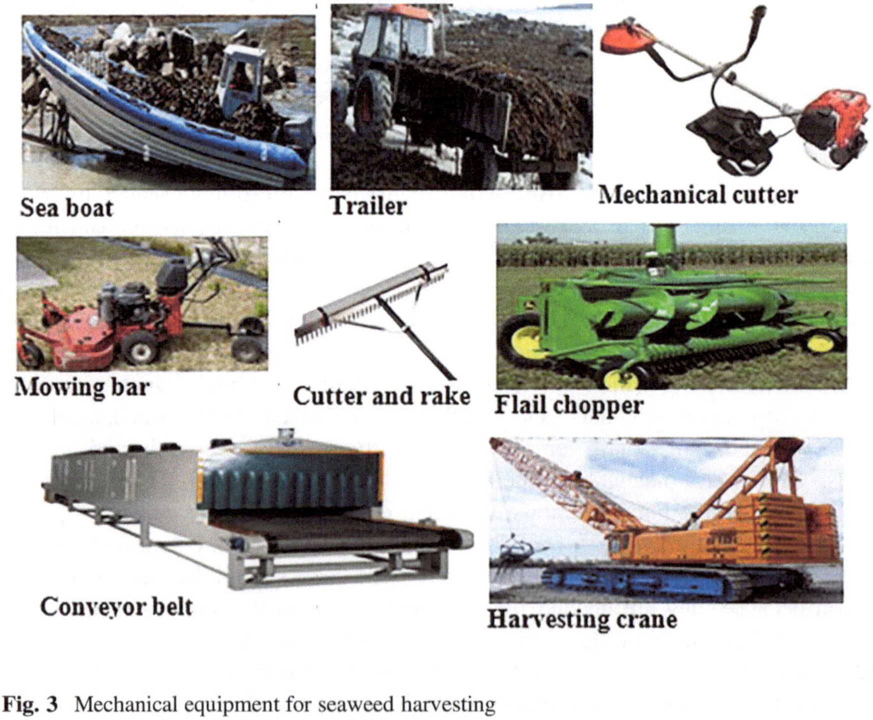

Fig. 3 Mechanical equipment for seaweed harvesting

for harvesting free-floating seaweed species while grown seaweeds attached to nets, ropes, or lines are best removed by using rotating blades (Alam et al. 2021; Das et al. 2016). Mechanical equipment, accessories, and harvested seaweed biomass are conveyed onshore and offshore by collection boats, nets, and dredgers (Peteiro and Freire 2012; Roesijadi et al. 2010). For increased harvesting efficiency, chemicals such as flocculants and herbicides are often applied before harvesting small-sized seaweed biomass (Gupta et al. 2018; Sahoo et al. 2017). The mechanized harvesters are usually mounted in boats which may be operated from water body or shores and once harvested, seaweeds are pumped through pipes directly into dredges or nets from where they can be transported to the required locations (Potts et al. 2012; Roesijadi et al. 2010).

Other approaches/methods of harvesting seaweeds include floating harvest technique, harvesting by collecting from shorelines, dredging method, rack and line harvesting, harvesting with floating mats, and farming with harvesting infrastructure (Araújo et al. 2021; Roesijadi et al. 2008; Wilding et al. 2021). Floating harvest involves cutting the top portion of the seaweed and allowing it to float to the surface, where it can be collected using nets or other equipment. This method is typically used for species that have gas-filled bladders, such as kelp (Troell et al. 2022). Collecting seaweeds that have naturally been washed up on shorelines is another method of harvesting. This method is typically used for small-scale operations and for species that have a tendency to wash up on beaches, such as kelp and seaweeds. Dredging is a method of harvesting seaweeds that involves dragging a rake or comb through the water to collect the seaweeds. This method is typically used for species that grow on rocky substrates, such as Irish moss (Kraan 2020b).

Rack and line harvesting involves suspending ropes or lines in the water, to which the seaweeds attach and grow. The biomass can then be harvested by hand or using machinery. This method is suitable for certain seaweed species and can be environmentally friendly, as it does not disrupt the seabed (Kraan 2020a). Harvesting floating mats are floating mats that are made of biodegradable material, which attracts *Ulva*, a fast-growing and high-nutrient seaweed known also as sea lettuce. The mats are left in the water for *Ulva* to grow, and once they are harvested, the mats can be disposed-off. This method is suitable for low-volume production or for small-scale cultivation (Smetacek and Zingone 2013). Farming with harvesting infrastructure which popularly refers to as seaweed farming involves growing the seaweeds on ropes, nets, or other structures in the ocean. This method is typically used for species that have high growth rates and are in high demand for food, fertilizer, and other applications. The seaweeds can be harvested by cutting the ropes or nets and collecting the biomass. Seaweed farming is an increasingly popular method of harvesting seaweeds as it is a sustainable and scalable approach (Heery et al. 2020; Tullberg et al. 2022).

3.2 Environmental and Economic Impacts of Seaweed Harvesting

It is important to consider the environmental impacts of each of the various methods of harvesting seaweeds. Generally, improper harvesting methods can lead to severe adverse effects on the marine environment and on some other marine species beyond those directly targeted for harvest as the marine environment is a symbiotic ecosystem (Lotze et al. 2019). Many seaweeds create conducive habitats for other species of conservative and commercial importance (Hinojosa et al. 2015). Thus, the harvesting of habitat-creating seaweed species could lead to a breakdown in the ecosystem structure and functioning of harvested area and can result in mortality of other species which are directly dependent on those seaweeds for their survival. The environmental impacts of each of harvesting methods are discussed as follows;

1. *Manual harvesting*: This method has typically low impact as it involves minimal equipment and disturbance to the natural environment. However, overharvesting of seaweeds can deplete natural populations and negatively impact the marine ecosystem. Besides, the removal of seaweeds by the manual harvesting method has the potential to spread diseases and pests present in the wild population.
2. *Mechanical harvesting*: It has more significant environmental impact than the manual harvesting, as it can cause physical damage to the seafloor and disturb the natural habitat of other marine organisms. The use of heavy machinery can also result in noise pollution and carbon emissions. Also, wide-scale mechanical harvesting of seaweeds in an unsustainable manner has the potential for the displacement of the ecosystem balance, as most seaweed species are important food resources for animals of higher trophic levels. For instance, the large-scale mechanical harvest of kelp (*Laminaria hyperboreal*) was reported to lower the foraging efficiency of seabirds in Norway (Lorentsen et al. 2010). Environmental conditions such as currents, sun light, and temperature, as well as the ecological communities, could also be affected by the removal of canopy-forming seaweeds via mechanical harvesting. It has been reported that kelp harvesting using rake-type dredge, pulled by boat in the coast of Central Norway significantly affected fish population and reduced coastal seabird foraging efficiency up to one year following kelp removal. The number of cods (*Gadus microcephalus)* fingerlings (less than 15 cm) in harvested areas is 92% lower than in unharvested areas. Seaweed harvesting on a large scale using mechanical means could also affect the provision of the following services: (1) biodiversity support through habitat provision, (2) bioremediation of nutrients and contaminants, (3) climate regulation via carbon sequestration, and (4) coastal protection (Lotze et al. 2019; Smale et al. 2013).
3. *Floating harvest*: This method can have a relatively low environmental impact, as it typically involves cutting the top portion of the seaweed and allowing it to float to the surface, where it can be collected. However, if not properly managed, the use of nets or other equipment can result in entanglement and bycatch of other

marine organisms. For example, 40.4% of global aquatic harvest, amounting to 38.5 million tons are bycatch (Ashour et al. 2023; Davis et al. 2021) These bycatch include fishes (sword fish, tuna, halibut, starfish, sea urchins, pufferfish, and betta fish), birds (albatross), sea turtles, eels, crabs, lobsters, sea lions, penguins, and other higher sea animals (dolphins, whales, etc.) that get entangled or hooked in the harvesting gadgets (Tahiluddin et al. 2022). Estimated 40% of fish catch worldwide is unintentional, and are usually thrown back into the sea either dying or dead; 53% of crustaceans in bycatch are crabs while shrimps and stomatopods account for 18% and 23%, respectively (Moore 2012). Every year, 100,000 small whales and dolphins, and 200,000 seabirds are killed as bycatch which represents 23 birds hour^{-1} (Christensen and Trites 2011).

4. *Collecting from shorelines*: Collecting seaweed that has naturally been washed up on shorelines can have a minimal environmental impact. However, excessive collection can disrupt the natural balance of the shoreline ecosystem. For example, the accumulation of seaweeds on shorelines helps to reduce wave energy and protect shorelines from erosion catastrophic effects (Lorentsen et al. 2010). Depending on the soil type, wave current and sensitivity to erosion, excessive seaweed removal has been reported to cause severe erosion of the shorelines (Grue et al. 2012).
5. *Harvesting with farming infrastructure*: Seaweed farming with harvesting infrastructure is a sustainable and low-impact method of seaweed harvesting. However, it can have negative impacts on the surrounding environment if not properly managed. For example, the use of large-scale farming infrastructure can cause physical damage to the seafloor (Lorentsen et al. 2010; Quiñones et al. 2019), and can result in eutrophication and harmful algal blooms.

Results of several research studies in the literature unveiled that the productivity and economics of seaweed harvesting can adversely be affected by contaminants as well as inefficient methods of harvesting which often require significant investments in new stock, remediation, and improved protection of farm environments, leading to an added cost (Barkia et al. 2019; Cai et al. 2021). For example, Valderrama et al. (2015)) reported the cost of protective farm gate for *Kappaphycus* seaweed farm is 0.85, 1.09, 0.27, 0.35, and $1 USD kg^{-1} biomass for Indonesia, Philippines, Tanzania, India, and Mexico respectively, while St-Gelais et al. (2022) reported a prices range from $1.10 to $2.20 USD kg^{-1} wet weight for farm gate. Generally, the poorer smallholders are severely constrained in both knowledge and resources when faced with such threats. Thus, this section presents the economic feasibility of each of the various methods used for harvesting seaweeds.

1. *Manual harvesting*: It is generally not economically feasible for large-scale operations as it is labor-intensive and time-consuming. However, for small-scale operations and for species that grow close to the shore, it can be a cost-effective method.
2. *Mechanical harvesting*: It is an efficient method of harvesting seaweeds and is typically used for large-scale operations. The cost-effectiveness of this method depends on the type of machinery used, the volume of seaweeds harvested, and

the distance of the harvesting location from the processing plant. Depending on the type of vegetation, size of farm, accessibility, and distance to site of disposal, costs of mechanical harvesting of tussock (*Poa labillardierei)* commonly range \$3000–\$12,000 USD acre^{-1} of biomass. The final cost involved in harvesting seaweeds from a 160 acre of lake in 2021 was approximately \$2400 USD acre^{-1} of biomass (Show et al. 2017).

3. *Floating harvest*: It is an efficient method for harvesting seaweeds with gas-filled bladders, such as kelp and *Sargassum.* The cost-effectiveness of this method depends on the type of equipment used and the volume of seaweeds harvested (Ma'ruf Kasim et al. 2020; Smetacek and Zingone 2013).
4. *Collecting from shorelines*: Collecting seaweeds that has naturally been washed up on shorelines can be a cost-effective method for small-scale operations. However, it is not a reliable method of harvesting as it is dependent on natural conditions such as currents and tides.
5. *Harvesting with farming infrastructure*: Seaweed farming (as discussed in Chapter "Algae Cultivation Systems") is an increasingly popular method of harvesting seaweeds as it is a sustainable and scalable an approach. The cost-effectiveness of this method depends on the cost of the farming infrastructure, such as ropes, nets, or other structures, as well as the cost of labor and maintenance.

3.3 Constraints to Seaweed Harvesting

1. *Socio-economic and technological constraints*: The cost of seaweed harvesting can be a constraint, especially if the method of harvesting is labor-intensive or requires expensive equipment (Mac Monagail et al. 2017; Moffitt and Cajas-Cano 2014). Some of the socio-economic and technological constraints to the seaweed manual method of harvesting include (1) low harvest efficiency (Cai and Chen 2000), (2) exposure to dangerous venomous reptiles, (3) possibility of strike by laborers during harvest, (4) poor status of naturally occurring seaweeds (Fakoya et al. 2011), (5) high capital cost of equipment, especially in developing countries (Kite-Powell et al. 2022), and (6) paucity of basic information on the biomass quantitative assessment, taxonomy, population biology, and seaweed cultivation and harvesting (Buschmann et al. 2017). Though mechanical harvesting operation for seaweeds leads to an increase in the efficiency, however, (1) the high capital initial investment required (Kumar et al. 2021); (2) lack of economically viable species derived from genetically improved and novel strains with improved yield and capacity to produce new bio-based products, needed to offset the huge capital cost of investment (Charrier et al. 2015); (3) high maintenance cost (Kumar et al. 2021); and (4) the possibility of accumulating so much unwanted trash and debris during operation are some of the bottlenecks militating against the mechanical method of seaweed harvesting.

2. *Legal regulations*: Seaweed harvesting is regulated in many countries, and obtaining the necessary permits and licenses can be a time-consuming process. Harvesting without proper authorization can result in fines and legal consequences. Also, information on seaweed harvest regulations and management vary from country to country. Findings showed that existing harvest management and regulation plans for most countries are often sketchy without sufficient details. Most countries utilize some form of single-species harvest management practice. Such management focus on the regeneration of the seaweed resource itself, with limited or no consideration of other species that are associated with the target species and are often affected by bycatch or habitat loss and alterations. However, few countries outlaw seaweed harvesting in areas important to other species, such as in seabird protection areas in Norway.
3. *Environmental constraints*: Seaweed harvesting can have environmental impacts, and in some cases, it may not be permitted in sensitive marine ecosystems or protected areas. In addition, adverse weather conditions, such as storms or strong currents, can prevent or limit seaweed harvesting.
4. *Seasonality*: The growth and harvesting of some seaweed species are seasonal, which can limit the availability of seaweeds for harvesting at certain times of the year.
5. *Infrastructure*: Some seaweed harvesting methods, such as farming or mechanical harvesting, require specialized infrastructure such as moorings, buoys, or boats. The construction and maintenance of such infrastructure can be costly.
6. *Availability*: The availability of seaweeds can be a constraint, especially for species that are rare or only grow in specific locations. In some cases, the availability of seaweeds may not be sufficient to support large-scale harvesting.

4 Conclusions

Various methods of harvesting algae, their economic feasibility, environmental impacts, and possible constraints have been identified and discussed in this chapter. The lack of efficient and sustainable harvesting techniques has also been identified as a major obstacle toward the full utilization of algal biomass on an industrial scale for the production of biofuels and other value-added products. The total cost for the most promising microalgae harvesting method (floatation) was estimated as \$1.93–\$2.16-USD kg^{-1} dry matter or concentrated sludge. Besides, the total costs (AU\$) of harvesting 10,000 L of *Chlorella* sp. with different harvesting techniques (including centrifugation, sedimentation, primary floatation, secondary floatation, and secondary centrifugation) were also estimated as 12.10, 3.70–4.10, 1.86–2.08, 0.833, and 1.21, respectively. Despite extensive research, to date, energy costs associated with harvesting of micro- and macroalgae still remain a constraint for the industry.

References

Abomohra AE-F, Jin W, Sagar V, Ismail GA (2018) Optimization of chemical flocculation of *Scenedesmus obliquus* grown on municipal wastewater for improved biodiesel recovery. Renew Energy 115:880–886

Alam SN, Khalid Z, Guldhe A, Singh B, Korstad J (2021) Harvesting and pretreatment techniques of aquatic macrophytes and macroalgae for production of biofuels. Environ Sustain 4(2): 299–316

Amer L, Adhikari B, Pellegrino J (2011) Technoeconomic analysis of five microalgae-to-biofuels processes of varying complexity. Bioresour Technol 102(20):9350–9359

Ananthi V, Balaji P, Sindhu R, Kim S-H, Pugazhendhi A, Arun A (2021) A critical review on different harvesting techniques for algal based biodiesel production. Sci Total Environ 780: 146467

Araújo R, Vázquez Calderón F, Sánchez López J, Azevedo IC, Bruhn A, Fluch S, Garcia Tasende M, Ghaderiardakani F, Ilmjärv T, Laurans M (2021) Current status of the algae production industry in Europe: an emerging sector of the blue bioeconomy. Front Mar Sci 7: 626389

Ashour M, Al-Souti AS, Hassan SM, Ammar GA, Goda A, El-Shenody R, Abomohra AE-F, El-Haroun E, Elshobary ME (2023) Commercial seaweed liquid extract as strawberry biostimulants and bioethanol production. Life 13(1):85

Barkia I, Saari N, Manning SR (2019) Microalgae for high-value products towards human health and nutrition. Mar Drugs 17(5):304

Barros AI, Gonçalves AL, Simões M, Pires JC (2015) Harvesting techniques applied to microalgae: a review. Renew Sust Energ Rev 41:1489–1500

Branyikova I, Prochazkova G, Potocar T, Jezkova Z, Branyik T (2018) Harvesting of microalgae by flocculation. *Fermentation* 4(4):93

Buschmann AH, Camus C, Infante J, Neori A, Israel Á, Hernández-González MC, Pereda SV, Gomez-Pinchetti JL, Golberg A, Tadmor-Shalev N (2017) Seaweed production: overview of the global state of exploitation, farming and emerging research activity. Eur J Phycol 52(4): 391–406

Caetano N, Martins A, Gorgich M, Gutiérrez D, Ribeiro T, Mata T (2020) Flocculation of *Arthrospira maxima* for improved harvesting. Energy Rep 6:423–428

Cai J, Chen CW (2000) Robust joint source-channel coding for image transmission over wireless channels. IEEE Trans Circuits Syst Video Technol 10(6):962–966

Cai J, Lovatelli A, Aguilar-Manjarrez J, Cornish L, Dabbadie L, Desrochers A, Diffey S, Garrido Gamarro E, Geehan J, Hurtado A (2021) Seaweeds and microalgae: an overview for unlocking their potential in global aquaculture development. FAO Fisheries and Aquaculture Circular (1229). FAO, Rome

Campbell I, Macleod A, Sahlmann C, Neves L, Funderud J, Øverland M, Hughes AD, Stanley M (2019) The environmental risks associated with the development of seaweed farming in Europe-prioritizing key knowledge gaps. Front Mar Sci 6:107

Charrier B, Rolland E, Gupta V, Reddy C (2015) Production of genetically and developmentally modified seaweeds: exploiting the potential of artificial selection techniques. Front Plant Sci 6: 127

Choudhary M, Ray MB, Neogi S (2019) Evaluation of the potential application of cactus (Opuntia ficus-indica) as a bio-coagulant for pre-treatment of oil sands process-affected water. Sep Purif Technol 209:714–724

Christensen V, Trites AW (2011) Predation on Fraser River sockeye salmon. Cohen Commission Tech Rep 8:129

Christenson L, Sims R (2011) Production and harvesting of microalgae for wastewater treatment, biofuels, and bioproducts. Biotechnol Adv 29(6):686–702

Chu R, Hu D, Zhu L, Li S, Yin Z, Yu Y (2022) Recycling spent water from microalgae harvesting by fungal pellets to re-cultivate *Chlorella vulgaris* under different nutrient loads for biodiesel production. Bioresour Technol 344:126227

Das P, Thaher MI, Hakim MAQMA, Al-Jabri HMS, Alghasal GSH (2016) Microalgae harvesting by pH adjusted coagulation-flocculation, recycling of the coagulant and the growth media. Bioresour Technol 216:824–829

Datta S (2009) Aquatic weeds and their management for fisheries, 1–22

Davis D, Simister R, Campbell S, Marston M, Bose S, McQueen-Mason SJ, Gomez LD, Gallimore WA, Tonon T (2021) Biomass composition of the golden tide pelagic seaweeds *Sargassum fluitans* and *S. natans* (morphotypes I and VIII) to inform valorisation pathways. Sci Total Environ 762:143134

Fakoya K, Owodeinde F, Akintola S, Adewolu M, Abass M, Ndimele P (2011) An exposition on potential seaweed resources for exploitation, culture and utilization in West Africa: a case study of Nigeria. J Fish Aquat Sci 6(1):37

Fasaei F, Bitter J, Slegers P, Van Boxtel A (2018) Techno-economic evaluation of microalgae harvesting and dewatering systems. Algal Res 31:347–362

Fernand F, Israel A, Skjermo J, Wichard T, Timmermans KR, Golberg A (2017) Offshore macroalgae biomass for bioenergy production: environmental aspects, technological achievements and challenges. Renew Sust Energ Rev 75:35–45

García-Pérez JS, Beuckels A, Vandamme D, Depraetere O, Foubert I, Parra R, Muylaert K (2014) Influence of magnesium concentration, biomass concentration and pH on flocculation of *Chlorella vulgaris*. Algal Res 3:24–29

Gorin KV, Sergeeva YE, Butylin VV, Komova AV, Pojidaev VM, Badranova GU, Shapovalova AA, Konova IA, Gotovtsev PM (2015) Methods coagulation/flocculation and flocculation with ballast agent for effective harvesting of microalgae. Bioresour Technol 193:178–184

Granados M, Acién F, Gómez C, Fernández-Sevilla J, Grima EM (2012) Evaluation of flocculants for the recovery of freshwater microalgae. Bioresour Technol 118:102–110

Grue J, Gjevik B, Weber JE (2012) Waves and nonlinear processes in hydrodynamics. Springer Science & Business Media, Berlin

Guldhe A, Misra R, Singh P, Rawat I, Bux F (2016) An innovative electrochemical process to alleviate the challenges for harvesting of small size microalgae by using non-sacrificial carbon electrodes. Algal Res 19:292–298

Gupta SK, Kumar NM, Guldhe A, Ansari FA, Rawat I, Nasr M, Bux F (2018) Wastewater to biofuels: comprehensive evaluation of various flocculants on biochemical composition and yield of microalgae. Ecol Eng 117:62–68

Hanotu J, Bandulasena HH, Zimmerman WB (2012) Microflotation performance for algal separation. Biotechnol Bioeng 109(7):1663–1673

Heery EC, Lian KY, Loke LH, Tan HT, Todd PA (2020) Evaluating seaweed farming as an eco-engineering strategy for 'blue'shoreline infrastructure. Ecol Eng 152:105857

Hinojosa IA, Green BS, Gardner C, Jeffs A (2015) Settlement and early survival of southern rock lobster, *Jasus edwardsii*, under climate-driven decline of kelp habitats. ICES J Mar Sci 72 (suppl_1):i59–i68

Horiuchi J-I, Ohba I, Tada K, Kobayashi M, Kanno T, Kishimoto M (2003) Effective cell harvesting of the halotolerant microalga *Dunaliella tertiolecta* with pH control. J Biosci Bioeng 95(4):412–415

Kite-Powell HL, Ask E, Augyte S, Bailey D, Decker J, Goudey CA, Grebe G, Li Y, Lindell S, Manganelli D (2022) Estimating production cost for large-scale seaweed farms. Appl Phycol 3(1):435–445

Kobya M, Gebologlu U, Ulu F, Oncel S, Demirbas E (2011) Removal of arsenic from drinking water by the electrocoagulation using Fe and Al electrodes. Electrochim Acta 56(14): 5060–5070

Kraan S (2020a) Concise review of the genus Alaria Greville, 1830. J Appl Phycol 32(6): 3543–3560

Kraan S (2020b) Seaweed resources, collection, and cultivation with respect to sustainability. In: *Sustainable seaweed technologies*. Elsevier, Amsterdam, pp 89–102

Kumar D, Pugazhendi A, Bajhaiya AK, Gugulothu P (2021) Biofuel production from macroalgae: present scenario and future scope. Bioengineered 12(2):9216

Kurniawan SB, Ahmad A, Imron MF, Abdullah SRS, Othman AR, Hasan HA (2022) Potential of microalgae cultivation using nutrient-rich wastewater and harvesting performance by biocoagulants/bioflocculants: mechanism, multi-conversion of biomass into valuable products, and future challenges. J Clean Prod 132806:132806

Kushwaha P, Kumari S, Singh K, Kumar KS (2020) Bioflocculation: A potential means of harvesting microalgae. J Emerg Technol Innov Res 7(10):2593–2609

Lee AK, Lewis DM, Ashman PJ (2010) Energy requirements and economic analysis of a full-scale microbial flocculation system for microalgal harvesting. Chem Eng Res Des 88(8):988–996

Lee AK, Lewis DM, Ashman PJ (2013) Harvesting of marine microalgae by electroflocculation: the energetics, plant design, and economics. Appl Energy 108:45–53

Lei X, Zheng W, Ding H, Zhu X, Chen Q, Xu H, Zheng T, Tian Y (2018) Effective harvesting of the marine microalga *Thalassiosira pseudonana* by *Marinobacter* sp. FL06. Bioresour Technol 269:127–133

Letelier-Gordo CO, Holdt SL, De Francisci D, Karakashev DB, Angelidaki I (2014) Effective harvesting of the microalgae *Chlorella protothecoides* via bioflocculation with cationic starch. Bioresour Technol 167:214–218

Liu J, Zhu Y, Tao Y, Zhang Y, Li A, Li T, Sang M, Zhang C (2013) Freshwater microalgae harvested via flocculation induced by pH decrease. Biotechnol Biofuels 6:1–11

Lorentsen S-H, Sjøtun K, Grémillet D (2010) Multi-trophic consequences of kelp harvest. Biol Conserv 143(9):2054–2062

Lotze HK, Milewski I, Fast J, Kay L, Worm B (2019) Ecosystem-based management of seaweed harvesting. Bot Mar 62(5):395–409

Lucakova S, Branyikova I, Kovacikova S, Pivokonsky M, Filipenska M, Branyik T, Ruzicka MC (2021) Electrocoagulation reduces harvesting costs for microalgae. Bioresour Technol 323: 124606

Ma'ruf Kasim AMB, Mustafa A, Nurdin R, Patadjai RS, Jalil W (2020) Floating cage: a new innovation of seaweed culture. Emerging technologies, environment and research for sustainable aquaculture, 103. https://doi.org/10.5772/intechopen.90035

Mac Monagail M, Cornish L, Morrison L, Araújo R, Critchley AT (2017) Sustainable harvesting of wild seaweed resources. Eur J Phycol 52(4):371–390

Madsen JD (2000) Advantages and disadvantages of aquatic plant management techniques. Lakeline 20(1):22–34

Mathimani T, Mallick N (2018) A comprehensive review on harvesting of microalgae for biodiesel–key challenges and future directions. Renew Sust Energ Rev 91:1103–1120

Matter IA, Bui VKH, Jung M, Seo JY, Kim Y-E, Lee Y-C, Oh Y-K (2019) Flocculation harvesting techniques for microalgae: a review. Appl Sci 9(15):3069

Milledge JJ, Heaven S (2013) A review of the harvesting of micro-algae for biofuel production. Rev Environ Sci Biotechnol 12:165–178

Moffitt CM, Cajas-Cano L (2014) Blue growth: the 2014 FAO state of world fisheries and aquaculture. Fisheries (Bethesda) 39(11):552–553

Moore AB (2012) Elasmobranchs of the Persian (Arabian) gulf: ecology, human aspects and research priorities for their improved management. Rev Fish Biol Fish 22(1):35–61

Naje AS, Chelliapan S, Zakaria Z, Abbas SA (2016) Electrocoagulation using a rotated anode: a novel reactor design for textile wastewater treatment. J Environ Manag 176:34–44

Najjar YS, Abu-Shamleh A (2020) Harvesting of microalgae by centrifugation for biodiesel production: a review. Algal Res 51:102046

Nakai J (2018) Food and agriculture Organization of the United Nations and the sustainable development goals. Sustain Dev 22:1–450

Nguyen TDP, Le TVA, Show PL, Nguyen TT, Tran MH, Tran TNT, Lee SY (2019) Bioflocculation formation of microalgae-bacteria in enhancing microalgae harvesting and nutrient removal from wastewater effluent. Bioresour Technol 272:34–39

Nie C, Pei H, Jiang L, Cheng J, Han F (2018) Growth of large-cell and easily-sedimentation microalgae Golenkinia SDEC-16 for biofuel production and campus sewage treatment. Renew Energy 122:517–525

Oh H-M, Lee SJ, Park M-H, Kim H-S, Kim H-C, Yoon J-H, Kwon G-S, Yoon B-D (2001) Harvesting of *Chlorella vulgaris* using a bioflocculant from *Paenibacillus* sp. AM49. Biotechnol Lett 23:1229–1234

Papazi A, Makridis P, Divanach P (2010) Harvesting *Chlorella minutissima* using cell coagulants. J Appl Phycol 22(3):349–355

Peteiro C, Freire Ó (2012) Outplanting time and methodologies related to mariculture of the edible kelp *Undaria pinnatifida* in the Atlantic coast of Spain. J Appl Phycol 24:1361–1372

Potts T, Du J, Paul M, May P, Beitle R, Hestekin J (2012) The production of butanol from Jamaica bay macro algae. Environ Prog Sustain Energy 31(1):29–36

Quiñones RA, Fuentes M, Montes RM, Soto D, León-Muñoz J (2019) Environmental issues in Chilean salmon farming: a review. Rev Aquac 11(2):375–402

Rawat I, Kumar RR, Mutanda T, Bux F (2011) Dual role of microalgae: phycoremediation of domestic wastewater and biomass production for sustainable biofuels production. Appl Energy 88(10):3411–3424

Roesijadi G, Copping A, Huesemann M, Forster J, Benemann J (2008) *Techno-economic feasibility analysis of offshore seaweed farming for bioenergy and biobased products*. Battelle Pacific Northwest Division Report Number PNWD-3931, 1e115

Roesijadi G, Jones SB, Snowden-Swan LJ, Zhu Y (2010) *Macroalgae as a biomass feedstock: a preliminary analysis*. Pacific northwest National lab. (PNNL), Richland, WA (United States)

Rubio J, Souza M, Smith R (2002) Overview of flotation as a wastewater treatment technique. Miner Eng 15(3):139–155

Sahoo NK, Gupta SK, Rawat I, Ansari FA, Singh P, Naik SN, Bux F (2017) Sustainable dewatering and drying of self-flocculating microalgae and study of cake properties. J Clean Prod 159:248–256

Salim S, Vermuë M, Wijffels R (2012) Ratio between autoflocculating and target microalgae affects the energy-efficient harvesting by bio-flocculation. Bioresour Technol 118:49–55

Salim S, Kosterink N, Wacka NT, Vermuë M, Wijffels R (2014) Mechanism behind autoflocculation of unicellular green microalgae *Ettlia texensis*. J Biotechnol 174:34–38

Sharma KK, Garg S, Li Y, Malekizadeh A, Schenk PM (2013) Critical analysis of current microalgae dewatering techniques. Biofuels 4(4):397–407

Show K-Y, Lee D-J (2014) Algal biomass harvesting. In: *Biofuels from algae*. Elsevier, North Holland, pp 85–110

Show K-Y, Lee D-J, Mujumdar AS (2015) Advances and challenges on algae harvesting and drying. Dry Technol 33(4):386–394

Show PL, Tang MS, Nagarajan D, Ling TC, Ooi C-W, Chang J-S (2017) A holistic approach to managing microalgae for biofuel applications. Int J Mol Sci 18(1):215

Singh G, Patidar S (2018) Microalgae harvesting techniques: a review. J Environ Manag 217:499–508

Smale DA, Burrows MT, Moore P, O'Connor N, Hawkins SJ (2013) Threats and knowledge gaps for ecosystem services provided by kelp forests: a northeast Atlantic perspective. Ecol Evol 3(11):4016–4038

Smetacek V, Zingone A (2013) Green and golden seaweed tides on the rise. Nature 504(7478): 84–88

Stagnol D, Renaud M, Davoult D (2013) Effects of commercial harvesting of intertidal macroalgae on ecosystem biodiversity and functioning. Estuar Coast Shelf Sci 130:99–110

St-Gelais AT, Fredriksson DW, Dewhurst T, Miller-Hope ZS, Costa-Pierce BA, Johndrow K (2022) Engineering a low-cost kelp aquaculture system for community-scale seaweed farming at nearshore exposed sites via user-focused design process. Front Sustain Food Syst 6:875509

Tahiluddin AB, Nuñal SN, Santander–de Leon, S.M.S. (2022) Inorganic nutrient enrichment of seaweed Kappaphycus: farmers' practices and effects on growth and ice-ice disease occurrence. Reg Stud Mar Sci 55:102593

Tran D-T, Le B-H, Lee D-J, Chen C-L, Wang H-Y, Chang J-S (2013) Microalgae harvesting and subsequent biodiesel conversion. Bioresour Technol 140:179–186

Troell M, Henriksson PJ, Buschmann A, Chopin T, Quahe S (2022) Farming the ocean–seaweeds as a quick fix for the climate? Taylor & Francis 31:1–11

Tullberg RM, Nguyen HP, Wang CM (2022) Review of the status and developments in seaweed farming infrastructure. J Mar Sci Eng 10(10):1447

Valderrama D, Cai J, Hishamunda N, Ridler N, Neish IC, Hurtado AQ, Msuya FE, Krishnan M, Narayanakumar R, Kronen M (2015) The economics of Kappaphycus seaweed cultivation in developing countries: a comparative analysis of farming systems. Aquac Econ Manag 19(2): 251–277

Vandamme D, Pontes SCV, Goiris K, Foubert I, Pinoy LJJ, Muylaert K (2011) Evaluation of electro-coagulation–flocculation for harvesting marine and freshwater microalgae. Biotechnol Bioeng 108(10):2320–2329

Vandamme D, Foubert I, Muylaert K (2013) Flocculation as a low-cost method for harvesting microalgae for bulk biomass production. Trends Biotechnol 31(4):233–239

Wan C, Zhao X-Q, Guo S-L, Alam MA, Bai F-W (2013) Bioflocculant production from *Solibacillus silvestris* W01 and its application in cost-effective harvest of marine microalga *Nannochloropsis oceanica* by flocculation. Bioresour Technol 135:207–212

Wang S-K, Stiles AR, Guo C, Liu C-Z (2015) Harvesting microalgae by magnetic separation: a review. Algal Res 9:178–185

Wang S, Yerkebulan M, Abomohra AE-F, El-Khodary S, Wang Q (2019) Microalgae harvest influences the energy recovery: a case study on chemical flocculation of *Scenedesmus obliquus* for biodiesel and crude bio-oil production. Bioresour Technol 286:121371

Wang Q, Oshita K, Takaoka M (2021) Flocculation properties of eight microalgae induced by aluminum chloride, chitosan, amphoteric polyacrylamide, and alkaline: life-cycle assessment for screening species and harvesting methods. Algal Res 54:102226

Wilding C, Tillin H, Corrigan S, Stuart E, Ashton I, Felstead P, Lubelski A, Burrows M, Smale D (2021) Seaweed aquaculture and mechanical harvesting: an evidence review to support sustainable management. Natural England, York

Wu Z, Zhu Y, Huang W, Zhang C, Li T, Zhang Y, Li A (2012) Evaluation of flocculation induced by pH increase for harvesting microalgae and reuse of flocculated medium. Bioresour Technol 110:496–502

Wu J, Liu J, Lin L, Zhang C, Li A, Zhu Y, Zhang Y (2015) Evaluation of several flocculants for flocculating microalgae. Bioresour Technol 197:495–501

Xie S, Sun S, Dai SY, Yuan JS (2013) Efficient coagulation of microalgae in cultures with filamentous fungi. Algal Res 2(1):28–33

Xu K, Zou X, Chang W, Qu Y, Li Y (2021) Microalgae harvesting technique using ballasted flotation: a review. Sep Purif Technol 276:119439

Yin Z, Zhu L, Li S, Hu T, Chu R, Mo F, Hu D, Liu C, Li B (2020) A comprehensive review on cultivation and harvesting of microalgae for biodiesel production: environmental pollution control and future directions. Bioresour Technol 301:122804

Zenouzi A, Ghobadian B, Hejazi M, Rahnemoon P (2013) Harvesting of microalgae Dunaliella salina using electroflocculation. J Agric Sci Technol 15(5):879–887

Zhao F, Zhang Y, Chu H, Jiang S, Yu Z, Wang M, Zhou X, Zhao J (2018) A uniform shearing vibration membrane system reducing membrane fouling in algae harvesting. J Clean Prod 196: 1026–1033

Zhou W, Min M, Hu B, Ma X, Liu Y, Wang Q, Shi J, Chen P, Ruan R (2013) Filamentous fungi assisted bio-flocculation: a novel alternative technique for harvesting heterotrophic and autotrophic microalgal cells. Sep Purif Technol 107:158–165

Phycochemicals

Mahdy Elsayed, Mohamed Eraky, Shah Faisal, and Jing Wang

Abstract Circular platforms are becoming more popular around the world. It is an eco-friendly approach that promotes economic growth through resource utilization while ensuring environmental sustainability. The multiple industrial applications of algal cells (microalgae and seaweeds) have received great attention due to their high content of essential nutrients and elements. This chapter gives a deep overview of the physiology and pathways of different value-added chemicals in algal cells. Recent development has been conducted on indirect/direct biosynthesis of lipids, proteins, vitamins, polysaccharides, and antioxidants from algal cells for the production of biofuel and high-value products. The potential of algae as a sustainable source of bioactive compounds for application in many bio-based industries such as food, feed, cosmetics, nutraceuticals, and pharmaceuticals is discussed in this chapter. The metabolic system and pathways employed as well as the potential technical challenges for the production and extraction of commercial phycochemicals from algal cells cultivated on waste streams are compared. The nano-bionics strategy used in microalgal cells, their uptake and interaction with chloroplasts during photosynthesis, and their impact on phycochemicals are introduced. Finally, a brief introduction of a possible integrated route from algae nanotechnology integration for value-added industrial products with bioenergy production is highlighted.

Keywords Algal cells · Value-added products · Lipids · Polysaccharides: Circular platforms

---

M. Elsayed (✉)
Department of Environmental Engineering, School of Architecture and Civil Engineering, Chengdu University, Chengdu, China

Department of Agricultural Engineering, Cairo University, El-Gamma Street, Giza, Egypt
e-mail: mahdy.mahdy@agr.cu.edu.eg

M. Eraky
College of Engineering, Huazhong Agricultural University, Wuhan, China

S. Faisal · J. Wang
Department of Environmental Engineering, School of Architecture and Civil Engineering, Chengdu University, Chengdu, China

A. Abomohra, S. Ende (eds.), *Value-added Products from Algae*,
https://doi.org/10.1007/978-3-031-42026-9_4

1 Introduction

Phytochemicals are the natural bioactive compounds found in plants that are not essential nutrients but play a crucial role in promoting human health. These compounds are responsible for the vibrant colors, flavors, and aromas of fruits, vegetables, herbs, and spices. Phytochemicals exhibit various beneficial properties, including antioxidant, anti-inflammatory, antimicrobial, and anticancer activities (Kumar et al. 2023). They can help protect against chronic diseases, such as cardiovascular diseases, cancer, diabetes, and neurodegenerative disorders. Different classes of phytochemicals include polyphenols, carotenoids, flavonoids, terpenoids, and glucosinolates, among others. Each class has unique characteristics and potential health benefits. In that context, seaweeds and microalgae are desirable organisms due to their high growth rate and ease of cultivation on different waste streams, as well as their wide applications ranging from food production to bioenergy generation. Natural bioactive compounds or products that are produced from algae are described, for the first time, as "Phycochemicals." Algae have remarkably few development requirements and can survive in a wide range of habitats, including those with a variety of pH levels, both freshwater or saltwater environments, and conditions with reduced C/N ratios (Pandit et al. 2017). Algae have a higher nutritional value since they may create a variety of bio-components, including pigments, proteins, lipids, and carbohydrates. In addition, a revolution in bioenergy production technology has been launched by the utilization of algal biomass to produce biofuels, which helps to offset the depletion of fossil fuels (Vassilev and Vassileva 2016). Acetone, ethanol, and furan are just a few examples of value-added compounds that have been created using algal carbohydrates and cellulose (González Fernández et al. 2023; Tawfik et al. 2022a).

The diverse range of algal species results in a wide range of phycochemicals being produced, which can be applied in various sectors including medicine, pharmaceuticals, agriculture, and manufacturing (Li et al. 2019). Several types of algae have been realized to produce high levels of value-added products, like carotenoids, vitamins, and sterols, that are in demand in cosmetic, nutraceutical, and pharmaceutical markets (Spolaore et al. 2006). However, selecting species and strains that will boost the production of the appropriate chemicals is essential when employing algae for different applications. It is also preferable to obtain different products by integrating different processes for the same biomass in a single growing cycle, such as lipids utilization for biodiesel coupled with the extraction of value-added phycochemicals, to ensure economical production. For sectors that depend on algae, such as biofuels and food, it is possible to lower the unit cost by optimizing the growth conditions and downstream processing (as discussed in Chapter "Overview of Bioprocess Engineering"), picking organisms that can overcome environmental restrictions, and selecting strains with high targeted compounds (Abomohra et al. 2013; El-Sheekh et al. 2018; Na et al. 2021).

Many strategies have been designed to increase the algal growth rate and enhance value-added phycochemicals production. The optimization of growth medium and conditions is the main strategy that may take place for growth enhancement, together with the modification of the algal cultivation systems such as cultivation modes and reactor design in order to provide a sufficient supply of light and nutrients based on the microbial physiology of the targeted algal cells (Abomohra et al. 2016, 2019; Ak et al. 2022; Tawfik et al. 2022b). The heterotrophic cultivation strategy was established to increase the algal product yields (Abomohra et al. 2014; Lakshmikandan et al. 2020). Despite having a sufficient supply of resources, algal cells have adapted systems and metabolic pathways that result in relatively lower conversion efficiency, with only about 2–6% of photosynthetic efficiency being used to fix CO_2 and 5–10% of carbon being used to produce energy and metabolites (Subramanian et al. 2013). The algal physiology and biochemical pathways influence the efficiency and productivity of the phycochemicals and biofuels produced. Carbon metabolism and energy conversion occur during the light reaction, glycolysis, the Calvin–Benson–Bassham (CBB) cycle, and the tricarboxylic acid (TCA) cycle in algae. In the past, efforts in metabolic engineering for enhanced carbon fixation in microalgae have focused on improving the efficiency of the CBB cycle and light reactions in the chloroplast. However, recent advances have shown that engineering of the TCA cycle can also significantly increase carbon fixation (Bailleul et al. 2015; Sun et al. 2018a, 2018b). Consistent with the previous chapters of algae cultivation and harvest, this chapter provides an in-depth discussion of the physiological processes and pathways involved in the biosynthesis of numerous compounds with high added value within algal cells. The most recent research on the ways to enhance the biosynthesis of lipids, proteins, vitamins, polysaccharides, and antioxidants from algal cells to produce valuable phycochemicals and biofuels is discussed. The potential of algae as an environmentally benign source of bioactive substances suited for use in a variety of bio-based economy, including food, feed, cosmetics, nutraceuticals, and medicines, is highlighted. The nanobionic strategy used in algae cultivation/processing, their uptake, and interaction with chloroplasts during photosynthesis, and their impact on the high value-added compounds in algal cells are introduced. Finally, the suggested integrated route from algae–nanotechnology integration for value-added industrial products with bioenergy production (that are discussed further in Chapter "Biofuel-Integrated Routes") is introduced.

2 The Strength of Microalgal Metabolism

Algal metabolism is complicated pathways resulting in the wide spreading and high biodiversity of algae. Algae have the ability and flexibility to adapt their metabolism according to the cultivation conditions (Treves et al. 2017). Reserving the nutrients and energy for biomass production is the first option of the algal cells factories before dividing into new cells. Theoretically, microalgae can produce 100–200 g of dry-weight biomass per square meter per day, while in practice this output is closer to 15–30 g, with protein and carbohydrates serving as the main metabolites (Sun et al. 2018a, 2018b). Through photosynthesis, algae can utilize sunlight to create both energy and organic substances including proteins, lipids, and carbohydrates. Algae can use this mechanism to transform carbon dioxide into organic material and create oxygen as a byproduct, which is necessary for the life on Earth.

Algae have a strong preference for nutrients like nitrogen, phosphorus, and sulfur that are necessary for their growth and development. These elements can be ingested by algae, and then used to carry out metabolic functions and create new cellular components. Algae grow more quickly than other photosynthetic creatures like plants. As a result, they can quickly colonize new areas and engage in resource competition (Beardall and Raven 2021). They are capable of adapting to shifting environmental factors like temperature, light, and nutrient supply. Certain algae are able to survive in hostile habitats because they can withstand extreme conditions, such as high salt or acidity (Abiusi et al. 2022; Abomohra et al. 2017, 2020; Desjardins et al. 2021). The strength of algal cells can be emphasized by the ability to synthesize a number of biological substances, including pigments, polyunsaturated fatty acids, and polysaccharides as shown in Fig. 1, thanks to their vast range of metabolic processes. The ability to manufacture a variety of value-added products with a wide range of uses is provided by the metabolic diversity of algae (De Bhowmick et al. 2015; Li et al. 2019).

3 Types of Phycochemicals

The concept of "high-value molecules" encompasses a diverse collection of bioactive compounds, leading to the subdivision of these compounds into multiple subcategories. These subcategories consist of polyunsaturated fatty acids (PUFAs), pigments, carbohydrates, peptides, vitamins, polyphenols, phytosterols, and hormones. As mentioned previously, the term "phycochemicals" is newly suggested in the present book to describe such molecules produced from algal biomass. The majority of the current and future high-value phycochemicals have alternative natural sources, as shown in Table 1. In that context, the production of many products that are currently produced by chemical synthesis competes with several of the products produced from algae. These alternative sources pose a challenge to the manufacturers of phycochemicals, who must either compete on price or set itself apart from the external option in the market to command a higher price.

3.1 Polyunsaturated Fatty Acids (PUFAs)

The primary group of lipids found in microalgae can be divided into two categories based on the target of production; storage and structural lipids. PUFAs are primary metabolites which make up storage lipids or triglycerides that have a particular importance. They are macromolecules with a long unsaturated hydrocarbon chain that has more than one double bond. PUFAs offer appealing qualities for the food

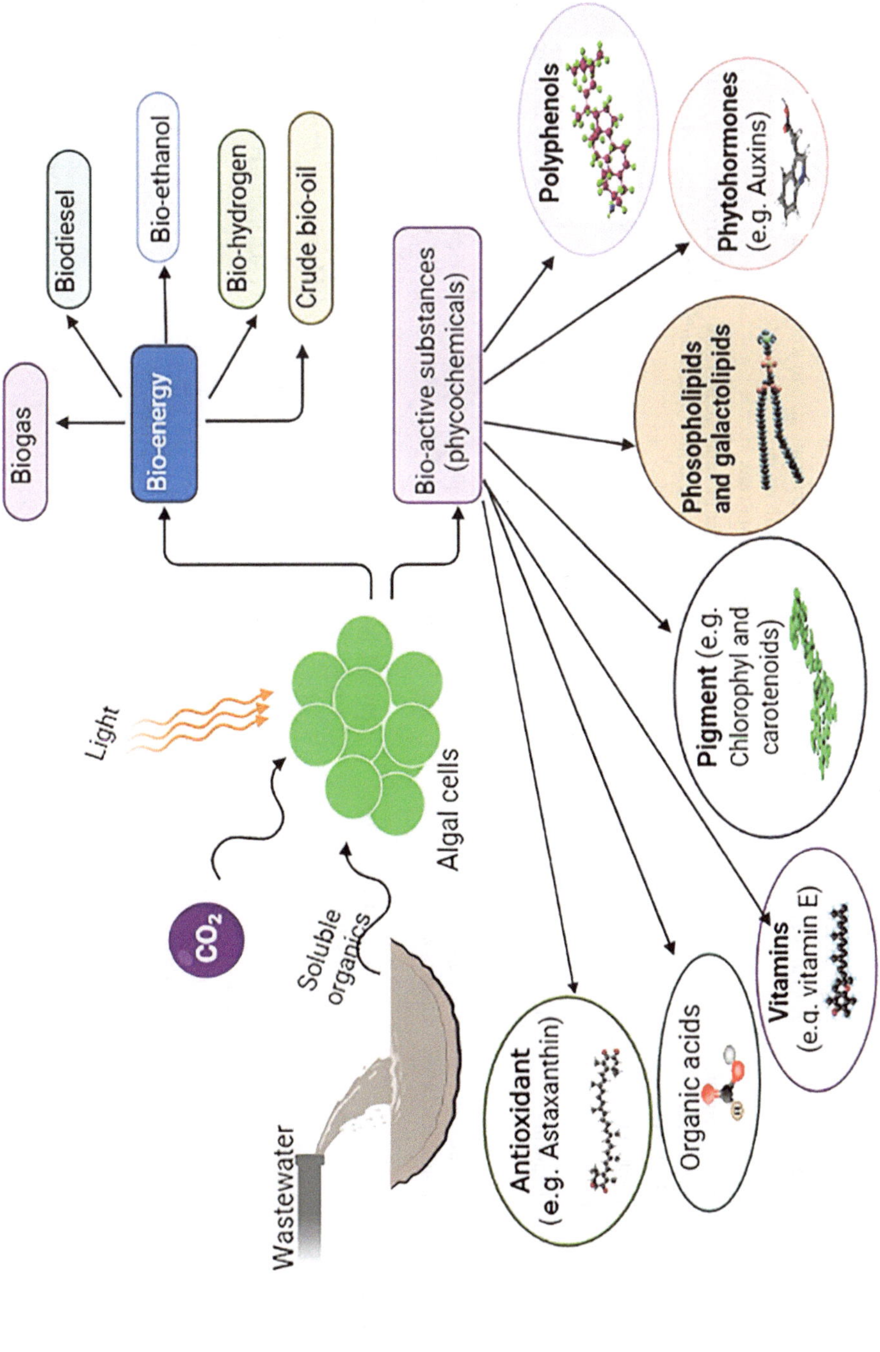

Fig. 1 Various natural phycochemicals and biofuels produced from algal biomass that created in BioRender

Table 1 A summary of the use of value-added products from algae, showing the current alternative resources

Product	Type of the product	Algal Species	Alternative Sources	Applications	References
Astaxanthin	Pigment	*Haematococcus pluvialis*	Synthetic or from other microorganisms	Pharmaceutical, nutraceutical, cosmetic, and food industries	Oslan et al. (2021)
Omega-3 fatty acids	Polyunsaturated fatty acids (PUFAs)	*Nannochloropsis, Schizochytrium, Ulkenia, Isochrysis*	Fish oil, flax-seeds, chia seeds, canola oil	Pharmaceutical, nutraceutical, and functional food industries	Balakrishnan et al. (2019), Katiyar and Arora (2020), Schambach et al. (2020)
β-Carotene	Pigment	*Dunaliella salina*	Synthetic or from other microorganisms	Pharmaceutical, nutraceutical, and food industries	Xu and Harvey (2019)
Phycocyanin	Pigment	*Arthrospira platensis, Spirulina maxima*	N/A	Food colorant, pharmaceutical, and cosmetic industries	Abd El-Baky and El-Baroty (2012), Khandual et al. (2021)
Carrageenan	Sulfated polysaccharides	*Kappaphycus, Eucheuma, Gigartina, Chondrus*	Seaweed, agar, pectin, starch	Food, pharmaceutical, and cosmetic industries	Lipinska et al. (2020), Ninghidayati et al. (2017), Pereira (2013), Rupert et al. (2022)
Alginate	Anionic polysaccharide	*Laminaria, Ascophyllum, Macrocystis, Fucus*	Seaweed, bacteria, fungi	Food, pharmaceutical, and biotechnology industries	Fertah et al. (2017), Malvis Romero et al. (2023), McKee et al. (1992), Moen et al. (1999)
Fucoidan	Sulfated polysaccharides	*Fucus vesiculosus, Undaria pinnatifida, Cladosiphon okamuranus*	Brown seaweed, animals, fungi	Pharmaceutical and nutraceutical industries	Oliveira et al. (2018), Tomori et al. (2021). Zhao et al. (2018)
Indole-3-acetic acid (IAA	Phytohormone	*Cladophora glomerata, Spirulina* sp.	Oat, coconut, soybean	Plant growth promotion	Górka and Wieczorek (2017)
Vitamin E	Vitamin	*Schizochytrium* sp.	Animal and plant sources	Dietary supplements, fortification	Valença et al. (2022)
Vitamin 12	Vitamin	*Chlorella* sp., *Spirulina* sp.	Animal and plant sources	Dietary supplements, fortification	van den Oever and Mayer (2022)

and pharmaceutical industries, and research has proven that they have positive impact on health, including promoting the growth of the human neurological system and lowering the incidence of chronic diseases (Hamed 2016; Haznedaroglu et al. 2016).

The anti-inflammatory, anti-proliferative, and anticachectic properties of PUFAs have also been confirmed. Omega-3 and omega-6 PUFAs must be received by the human body from an external source because they cannot be synthesized in the body. Microalgae represent a prospective source of omega-3 PUFAs, including eicosapentaenoic acid (EPA) and docosahexaenoic acid (DHA). Given the constraints of wild-caught fisheries, the ability of microalgae to produce these necessary compounds for commercial purposes is crucial (Levasseur et al. 2020). Although microalgae have a fairly stable fatty acid profile, the fatty acid content and profile change from species to species and per growth conditions (Abomohra et al. 2014; Almutairi et al. 2021; Battah et al. 2015; Bellou et al. 2014). Although microalgae naturally contain little EPA and DHA, techniques to increase their productivity have been developed that either directly alter microalgal metabolism or modify ambient factors to affect internal metabolism. One of the most popular methods for increasing lipid output is to deprive microalgae of nitrogen, which was shown lower biomass yield resulting in relatively low lipid productivity. Other technologies have been suggested to increase both biomass yield and lipid content, such as treatment with cold plasma (Almarashi et al. 2020), using phytohormones or plant growth regulators (Esakkimuthu et al. 2020), and cultivation on organic waste products (Abomohra et al. 2018). In addition, studies showed that an optimal quantity of intracellular reactive oxygen species (ROS) may enhance lipid formation (Raghukumar 2008; Sun et al. 2018a, 2018b).

3.2 Pigments

Microalgae and blue-green algae are able to produce a number of pigments with great commercial value, such as carotenoids, phycobiliproteins, and chlorophylls. Carotenoids act as auxiliary light-harvesting pigments and for defense against photooxidative damage, whereas chlorophylls and phycobiliproteins are primarily involved in the light reaction of photosynthesis (Jeon et al. 2017). Chlorophylls, carotenoids, and phycobiliproteins are three groups of pigments that provide green, yellow/orange, and red/blue colors, respectively. These substances are used as natural colorants, food supplements, and as a source of bioactive molecules in food, pharmaceutical, and cosmetic applications as they have beneficial properties for human health, such as antioxidant effects, vitamin precursors, immune system activators, and anti-inflammatory agents (López and Soto 2020).

3.2.1 Chlorophylls

Chlorophyll pigments, present in all photosynthetic organisms including higher plants and algae, are essential for capturing and transferring light energy to reaction centers during photosynthesis. Water-soluble semi-synthetic chlorophyllin showed potential as an anticancer drug due to its antimutagenic qualities, and natural chlorophylls from algae are also utilized as food colorings (Pucci et al. 2021). Chlorophylls from the miscella fraction of photosynthetic algal biomass could be isolated and used as high-value phycochemicals in biorefineries before further oil purification (Chiu et al. 2003). Due to rising consumer demand, chlorophylls are also becoming more important as natural coloring agents in the food industry, medicines, and cosmetics (Begum et al. 2015). Despite algal biomass containing a considerable quantity of chlorophylls and providing an alternative source for sustainable chlorophyll extraction, chlorophylls are typically derived from low-cost sources like grass or alfalfa. In algae, the chlorophyll content can range from 0.5 to 4% of dry weight, depending on the strain and the environment (Huang et al. 2004). This demonstrates the possibility of co-valorizing chlorophylls during the manufacture of biofuel and other valuable phycochemicals (Huang et al. 2021; Sun et al. 2018a, 2018b).

Carotenoids, which range in color from yellow to red, are highly colored molecules and abundant in microalgae. These compounds are useful in a number of industries, including food, feed, cosmetics, and medicines because they possess both antioxidant and coloring capabilities. Carotenoids can be categorized into two families called carotenes and xanthophylls, and are the most diverse and common class of pigments with over 600 varieties (Igielska-Kalwat et al. 2015). Chlorophyceae are the primary producers of carotenoids, while some varieties can also be produced by other algal phyla. Carotenoids typically make up 0.1–0.2% of the dry matter in algal biomass, however under unfavorable conditions, this percentage can reach up to 12% (Gimpel et al. 2015; Spolaore et al. 2006).

Microalgae utilize the advantages of the 2-C-methyl-d-erythritol-4-phosphate/1-deoxy-d-xylulose-5-phosphate (MEP/DOXP) route for the production of carotene, which takes place in the plastids as part of the production of terpenoid backbones (Treves et al. 2017). In this pathway, carotene as well as other carotenoids including lycopene, zeaxanthin, and astaxanthin are synthesized. Isopentenyl pyrophosphate (IPP) and dimethylallyl pyrophosphate (DMAPP) are the main isoprenoid terpenoids that plants use as building blocks for the production of carotenoids. IPP and DMAPP are generated in higher plants via two pathways; the cytosolic mevalonic acid (MVA) pathway and the chloroplastic MEP/DOXP pathway. However, as there has been no discovery of MVA route-directed isoprenoid synthesis in the cytosol, it is thought that IPP and DMAPP are exclusively generated in microalgae through the MEP/DOXP pathway in the plastids. The fact that microalgae lack essential MVA pathway enzymes lends credence to this notion. The enzyme 1-deoxy-d-xylulose-5-phosphate synthase *dxs* produces MEP from 1-deoxy-d-xylulose 5-phosphate in the MEP/DOXP pathway, which also produces it from pyruvate and

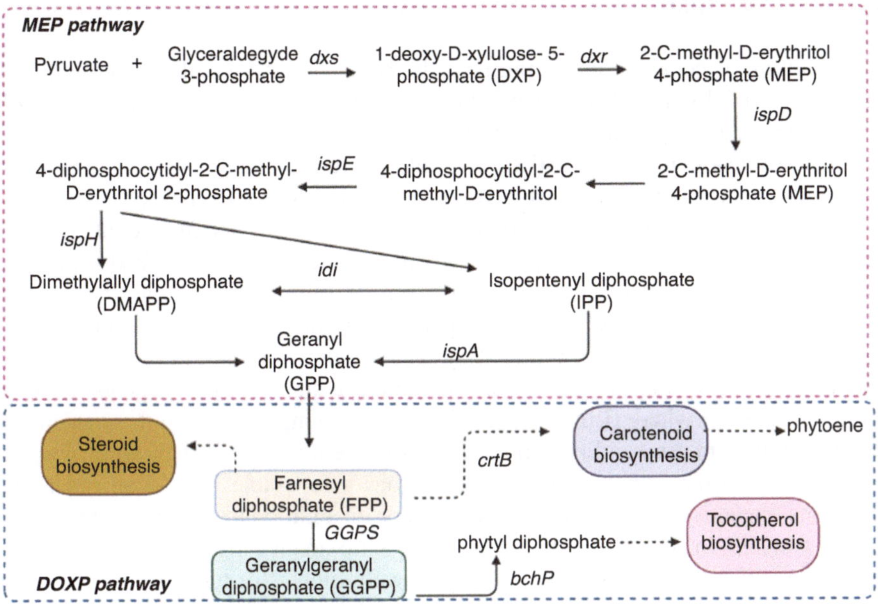

Fig. 2 The 2-C-methyl-d-erythritol-4-phosphate/1-deoxy-d-xylulose 5-phosphate (MEP/DOXP) biosynthesis pathway modified from Haznedaroglu et al. (2016). Enzymes depicted within boxes include the following: dxs (1-deoxy-d-xylulose-5-phosphate synthase), dxr (1-deoxy-d-xylulose-5-phosphate reductoisomerase), ispD (2-C-methyl-d-erythritol 4-phosphate cytidylyltransferase), ispE (4-diphosphocytidyl-2-C-methyl-d-erythritol kinase), idi (isopentenyl-diphosphate Δ-isomerase), ispH (4-hydroxy-3-methylbut-2-enyl diphosphate reductase), FDPS (farnesyl diphosphate synthase), ispA (geranyltranstransferase), GGPS (geranylgeranyl diphosphate synthase, type II), and bchP (geranylgeranyl reductase and crtB (phytoene synthase), that created in BioRender

d-glyceraldehyde-3P. IPP is created from MEP by further processes involving reductases, dehydratases, a kinase, and cofactors (Haznedaroglu et al. 2016).

In Fig. 2, The MEP/DOXP pathway initiates with the enzymatic synthesis of 1-deoxy-d-xylulose 5-phosphate using pyruvate and d-glyceraldehyde-3P. This process is facilitated by the enzyme 1-deoxy-d-xylulose-5-phosphate synthase (dxs). Subsequently, 1-deoxy-d-xylulose-5-phosphate reductoisomerase (DXR) transforms it into MEP. The conversion of MEP into isopentenyl pyrophosphate (IPP) occurs through a series of reactions involving reductases, dehydratases, a kinase, and cofactors. Additional conversions take place, leading to the production of significant precursor molecules like 2-C-methyl-d-erythritol 2,4-cyclodiphosphate and 1-hydroxy-2-methyl-2-butenyl 4-diphosphate. IPP can be synthesized directly or indirectly via dimethylallyl pyrophosphate (DMAPP). IPP and/or DMAPP contribute to the formation of geranyl diphosphate (GPP) and geranylgeranyl diphosphate (GGPP). In the synthesis of carotenoids, GGPP is converted into phytoene, followed by desaturation and cyclization reactions, ultimately resulting in the synthesis of various carotenoid pigments (Haznedaroglu et al. 2016; Lichtenthaler 1999).

3.2.2 Astaxanthin

Astaxanthin, a red xanthophyll pigment, is the second most utilized carotenoid in industry (Griffiths et al. 2016). Nevertheless, the microalga *Haematococcus pluvalis* is the most efficient astaxanthin producer, accounting for up to 81% of its total carotenoids (Siqueira et al. 2018). It is also naturally produced by a variety of microorganisms, including yeasts. Due to its strong ability to dye red, astaxanthin is frequently used in aquaculture to color fish and shellfish (Griffiths et al. 2016). A large portion of astaxanthin's applications in commerce are in the feed sector. For producing the desired coloration in the meat of salmon, trout, and shrimp, it is thought to be the most precious and expensive pigment used in aquaculture. Since astaxanthin is not naturally produced by these aquatic animals, supplementation is essential. The addition of astaxanthin to feed has a significant impact on consumer preferences everywhere (Lim et al. 2018). Large ornamental fish and aquarium fish both benefit from astaxanthin in their diets because it improves their color. Furthermore, research has shown that astaxanthin has a favorable impact on the broiler chickens' meat tissues, skin, and egg yolk color. In addition to its abilities to produce color, astaxanthin is well known for its strong antioxidant capacities and other health advantages (Griffiths et al. 2016). It is also well known for having ten times more effective antioxidants than other carotenoids, which may have a beneficial impact on human health. Notwithstanding these advantages, the cost of manufacturing natural astaxanthin is higher than that of synthetic one, which may limit its economic utilization. Astaxanthin and β-carotene are the two carotenoids with the largest demand on the world market. The commercialization of carotenoids as value-added goods is a result of their vital biological roles in microalgae and the nutraceutical benefits they provide to humans and other animals (Griffiths et al. 2016).

3.3 Organic Acids

The production of organic acids like succinic and malic acids, which may be made from carbohydrates supplied from biorefineries, is regarded as an important platform chemical. These opportunities rank among the top 15 according to the US Department of Energy (Pleissner et al. 2019). Maleic acid or anhydride from the petrochemical industry is traditionally catalyzed into succinic acid through hydrogenation (Delhomme et al. 2009). However, in the near future, it is anticipated that microbial activities will produce more biosuccinic acid. Through the use of metabolically engineered bacteria, algal biomass with fermentable sugars can be used to produce succinic acid. However, in the food and beverage sectors, malic acid is mostly employed as an acidulant and a flavor enhancer (Haznedaroglu et al. 2016). Several metabolic processes can create it in bacterial, yeast, and fungal species as well. By the oxidation phases of the TCA cycle or glyoxylate shunt, followed by the conversion of oxaloacetate to malate by NADP+ malate dehydrogenase, algae can also be employed to create malic acid (Haznedaroglu et al. 2016; Kuo et al. 2013).

3.4 Vitamins

Although it is commonly accepted that some algae need specific forms and dosages of vitamins B12 (cobalamin), B1 (thiamine), and B7 (biotin) to thrive, they can also synthesize a number of other vitamins, including A, C, and E. As was already established, carotenoids can act as building blocks for the production of vitamin A. The commercial production of vitamin E and C from microalgae is well-established and ongoing, although research is still being done on the production of other vitamins from algae (Haznedaroglu et al. 2016). Vitamins are essential anti-oxidants in the removal of reactive oxygen species (ROS). The majority of studies on vitamins generation by algae have therefore concentrated on strains employed in aquaculture. Examples of these algal strains *Pavlova pinguis, Stichococcus* sp., and *Tetraselmis* sp. (Levasseur et al. 2020).

3.4.1 Vitamin E

Vitamin E is an important lipid-soluble metabolite with antioxidant and radical scavenging activities that is synthesized only by photosynthetic organisms. It is made up of several different tocopherols and tocotrienols, with -tocopherol being the most potent. Researchers are keen to develop alternate sources of vitamin E due to its critical roles in both human and animal health, growing demand in numerous industries, and low extraction efficiency from vegetable oil (Durmaz 2007; Ogbonna 2009). Vitamin E may be found in microalgae, some of which produce higher concentrations and more varied forms of the vitamin than conventional dietary sources (Durmaz 2007). Research has been done to improve the growth circumstances and culture media to improve the composition and accumulation of vitamin E in microalgae. Despite the success of genetic engineering in raising vitamin E levels in many plants, microalgae have not been the subject of genetic engineering due to a lack of understanding of the molecular genetics and biochemistry of vitamin E biosynthesis in these eukaryotic organisms (DellaPenna and Pogson 2006).

3.4.2 Vitamin C

L-ascorbic acid, also known as vitamin C, is a type of water-soluble metabolite that has strong antioxidant properties and acts as a cofactor for many biological enzymes. While plants and certain microorganisms can synthesize vitamin C, humans and animals must obtain it through their diet (Martí et al. 2009). While the majority of commercially available vitamin C is produced using the synthetic Reichstein

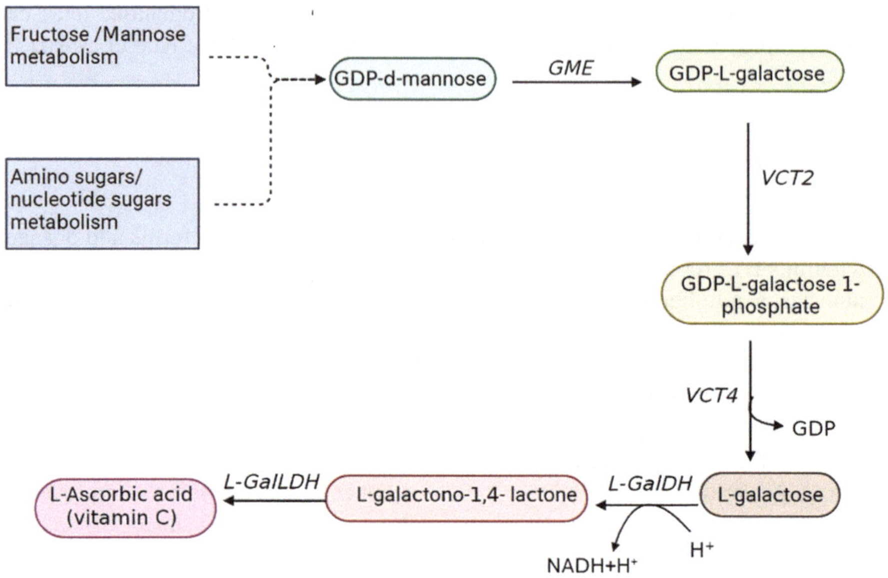

Fig. 3 Pathway of L-ascorbic acid (vitamin C) biosynthesis (Haznedaroglu et al. 2016). *GDP* (guanosine diphosphate), *GME* (GDP-d-mannose 3′, 5′-epimerase), VCT2 (GDP-l-galactose phosphorylase), VCT4 *(l-galactose 1-phosphate phosphatase), L-GalDH* (l-galactose dehydrogenase), and *L-GalLDH* (l-GalLDH,l-galactono-1,4-lactone dehydorogenase), that created in BioRender

method, there is growing interest in exploring alternative, microbiological methods due to economic, energy efficiency, and environmental sustainability concerns (Bremus et al. 2006; Hancock and Viola 2005). Only a limited number of microalgal species, such as *D. tertiolecta*, have been shown to produce L-ascorbic acid under nitrogen deprivation and osmotic stress, where there is ongoing research in this area. The vitamin C biosynthesis pathway, substrate is converted into a product in a reaction catalysed by an enzymes. The initial step is the formation of GDP-L-galactose by the enzymatic transformation of GDP-d-mannose. Successive enzymatic reactions occur till vitamin C biosynthesis as shown in Fig. 3.

3.5 Hormones

Phytohormones are chemical messengers that higher plants create in little amounts to coordinate different cellular activities. Auxin, abscisic acid, cytokinin, ethylene, gibberellins, polyamines, jasmonides, salicylates, signal peptides, and brassinosteroids are only few of phytohormones that are produced by the majority of microalgae and seaweeds (Wang et al. 2022). Despite the fact that we still do not fully understand what they do, current research has helped to clarify their activities. Various strains of microalgae react to phytohormones in diverse ways, including development, growth, light response, stress tolerance, secondary metabolite synthesis, and senescence. These responses help microalgae control homeostasis to adapt to changing environmental conditions. Hence, choosing microalgal species for particular commercial uses, such as increased biofuel production, can be done using an approach based on an understanding of phytohormone pathways (Stirk and Van Staden 1997). Several industrial and economic characteristics of microalgae, such as higher biomass productivity and secondary metabolite production, as well as better resistance to unfavorable environmental conditions, can be improved by modulating phytohormone metabolism. Hormones derived from algae may also be helpful in cosmetics to combat skin aging symptoms.

3.5.1 Auxins

Auxins function similarly in algae and higher plants, with indole-3-acetic acid being the most significant auxin in plant cells. Forty-six microalgae species of Cyanophyta and Chlorophyta have been identified to produce indole-3-acetic acid (IAA), phenyleacetic acid (PAA), naphtyleacetic acid (NAA), and indolbutyric acid (IBA) (Romanenko et al. 2015; Stirk and Van Staden 1996). Auxins have diverse roles in microalgal growth and metabolism, with low concentrations promoting growth and increasing biomass and biosynthesis of valuable biomolecules (Wang et al. 2022). However, higher concentrations can hinder the cellular growth.

3.5.2 Cytokinin

In microalgae and seaweeds, the major type of cytokinin, which is generated from purine, is zeatin (Stirk et al. 2002). The cytokinin molecule possesses both *cis* and *trans* arrangements. In microalgae, cytokinin has been demonstrated to drive cell division, increase the accumulation of photosynthetic pigments, and improve photosynthetic efficiency, all of which boost biomass output (Mousavi et al. 2016; Stirk et al. 2013). Cytokinin levels are low during night, but they were reported to rise during the day (Schmülling et al. 2003). Under unfavorable climatic conditions, cytokinins exert additional protective effects on the physiological functions of algal cells, particularly photosynthesis.

3.6 Polyphenols

In microalgae, phytosterols, which are steroid alcohols, are crucial parts of biomembranes and have a significant impact on the signal transduction and cell growth processes (Volkman 2003). Due to these substances' potential to decrease cholesterol and their usage as therapeutic agents to treat hypercholesterolemia, they have garnered commercial interest. Since their presence impacts how desirable these organisms are as food, microalgae's phytosterols are particularly valuable

(Haznedaroglu et al. 2016). Some microalgal species have been discovered to produce particular phytosterols, including ergosterol, stigmasterol, and β-sitosterol, which are used to treat hypercholesterolemia and to prepare the body for the synthesis of vitamin D2, D3, and calcium (Voshall et al. 2021).

Polyphenols have an aromatic ring with one or more hydroxyl groups attached, making them polar molecules. Antioxidant, anti-inflammatory, anti-cancer, anti-allergic, anti-diabetic, anti-aging, and antibacterial capabilities are few of biological effects of polyphenols. Seaweed polyphenols may ameliorate cardiovascular-related illnesses, according to a comprehensive assessment of preclinical and clinical trials, but further studies are required to substantiate these claims (Galasso et al. 2019). A recent study investigated antifungal and antimycotoxigenic effects of polyphenols derived from two different microalgal species, *Nannochloropsis* sp. and *Spirulina* sp., on *Trichothecenes mycotoxins* from in vitro cultures of *Fusarium graminearum* (Scaglioni et al. 2019). Results showed that 40 g mL^{-1} of phenolic extracts from *Nannochloropsis* sp. totally inhibited nivalenol and deoxynivalenol after 168 hours of growth and markedly decreased the formation of acetylates by 98%. Moreover, *Spirulina* sp. polyphenol extracts suppressed the synthesis of nivalenol and deoxynivalenol by 62% and 78%, respectively, and lowered the acetylate levels.

3.7 Miscellaneous Phycochemicals

There are thought to be approximately 300,000 species of algae, including cyanobacteria. The algal biorefinery has the chance to take advantage of the variety of species in this group and their distinctive chemicals. However, for widespread market acceptability, such items might need further development and study. Future algae biofuels may be economically stable, thanks to the potential of these products. This section will discuss some of common phycochemicals including botryococcene, sporopollenin, and polyhydroxyalkanoates.

3.7.1 Polyhydroxyalkanoates

Polyhydroxyalkanoates (PHAs) are distinctive biopolymers produced by cyanobacteria that can be used to create biodegradable plastics as explained in Chapter “Algal-based Biopolymers”. Particularly interesting bioplastics for biomedical, culinary, packaging, textile, and domestic materials are medium chain length PHAs with 6–14 carbon atoms. Different development strategies can change the material characteristics of bioplastics, such as their processing simplicity, strength, and flexibility (Haznedaroglu et al. 2016).

3.7.2 Sporopollenin

Sporopollenin, found in the cell walls of certain algal species, is a resilient compound that resists decomposition. Its composition can differ, typically encompassing lignin, long chain fatty acids, and carotenoids. In the nutraceutical industry, sporopollenin has been utilized as an immune stimulator and is believed to provide protection against UV damage and help prevent the formation of wrinkles on the skin (Thompson 1996).

3.7.3 Botryococcene

Botryococcene is a hydrocarbon produced by *Botryococcus braunii* Race B strains that is unique from other algal lipids used in transesterification to create biodiesel. It is a crucial feedstock that can be utilized in hydrocracking to create a variety of drop-in liquid fuels. However, because of slow natural growth rate of *B. braunii* species, optimization studies are necessary to increase productivity for commercial purposes (Haznedaroglu et al. 2016).

4 Microalgae Nanobionics for Enhanced Phycochemical Production

Microalgae nanobionics is a cutting-edge research direction that combines plant biology with nanotechnology to produce new biotechnological materials (Parveen et al. 2023). The study of electrical interactions at the nanoscale in biological systems is known as nanobionics, where the words "nano" and "bionics" are combined to refers to the study of science in the nanoscale, within size range of 1–100 nm (Siddiqui et al. 2015). Microalgae species can be used as a valuable source of active chemicals by enhancing photosynthesis through nanoparticle processing and creating and modifying the biochemical functions of organelles within the algae ecosystem (Agarwal et al. 2022). These nanomaterials can increase the microbial activity; so by adding the active chemicals to microalgae cultivation media, they can significantly affect pigments and targeted metabolites accumulation (Fig. 4). As discussed earlier, microalgae are a source of important phycochemicals that can be enhanced by bioprocess engineering and downstream processing (Chapter "Overview of Bioprocess Engineering"), where integrated algae nanobionics blends algal nanotechnology as well. Metals, metalloids, and metallic nanoparticles (e.g., silver, zinc, and copper nanoparticles) can have an impact on the development and metabolism of microalgae, and their dispersion or transportation within the cell can have both positive or negative effects. Optimizing the synthesis and production of various chemicals from microalgae cells is the focus of current research in the area (Arif et al. 2018; Gonzalez et al. 2014).

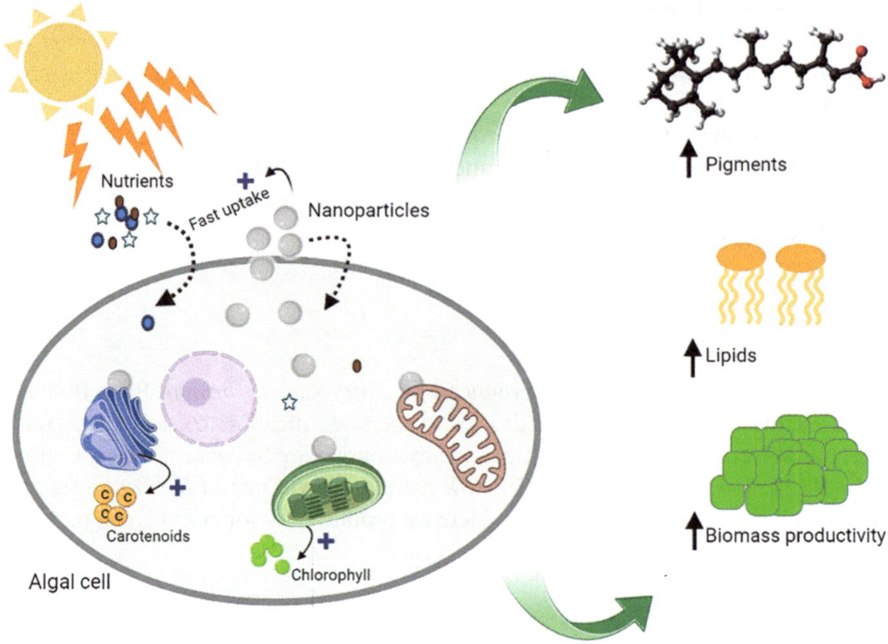

Fig. 4 The interaction of nanomaterial with the algal system "Algal nanobiohybrid" that created in BioRender

Due to its antimicrobial qualities, silver nanoparticles are frequently used in a variety of industries and applications. Recent research has demonstrated that these nanoparticles can reduce the photosynthetic activity and growth of freshwater green microalgae (Parveen et al. 2023). The freshwater green alga *Pithophora oedogonium* experienced morphological deformities and a decrease in chlorophyll content as a result of silver nanoparticles exposure (Parveen et al. 2023). In addition, the microalgae *Chlorella vulgaris* and *Dunaliella tertiolecta* displayed increased ROS generation, lipid peroxidation, and decreased chlorophyll content after being treated by silver nanoparticles for 24 hours (He et al. 2012; Oukarroum et al. 2012). On the other hand, the marine flagellate *Chattonella marina* showed higher antioxidant activity when exposed to silver nanoparticles (Moreno-Garrido et al. 2015).

Copper (Cu) is another crucial nutrient for plant growth and development because it activates a number of enzymes, which improves the efficiency of the photosynthesis. Copper ions can be introduced into aquatic settings and are beneficial in antifouling paints due to their strong reactivity and antibacterial capabilities. It has been demonstrated that copper ions (Cu^{2+}), present in *Chlamydomonas acidophilla* at concentrations ranging from 0.05 to 0.2 g L^{-1}, increase β-carotene production (Jamers et al. 2013). Moreover, there was a recorded increase in biomass density and cell lipid content when *C. minutissima* was grown in the presence of Cd (0.2–0.4 mM) or Cu (0.2–1 mM). Lipid productivity increased by 2.17-fold with the supplementation of 0.4 mM Cd and by 34% with 0.4 mM Cu (de la Rosa et al. 2013; Miazek et al. 2015).

In addition, studies have been done on the beneficial effects of zinc oxide nanoparticles (ZnO NPs) on plant growth since zinc is another crucial micronutrient. At lower concentrations of approximately 10–50 mg L^{-1}, ZnO NPs have been proven to be quite effective (Lin et al. 2023). They boost chlorophyll levels by 40–49%, and they also boost photosynthetic effectiveness by 35%. In addition, stomatal conductance and enhanced CO_2 absorption were reported (Ainsworth and Rogers 2007). Reactive oxygen species (ROS) are regulated by antioxidant enzymes such as superoxide catalase and dismutase, where zinc is necessary for their composition (Meitha et al. 2020). Zinc is also needed for other enzymes such as polymerase, kinase, dehydrogenase, and phosphatase, which are all involved in photosynthesis (Parveen et al. 2023; Rajput et al. 2021a, 2021b). As a summary, nanoparticles significantly contribute to increasing the algal growth. They can function as effective carriers of vital nutrients, ensuring a steady release that promotes healthy growth. Nanoparticles also function as light-harvesting substances, enhancing the absorption of sunlight for improved photosynthesis (Sarkar et al. 2021). By preventing the formation of hazardous germs, their antimicrobial qualities aid in maintaining a suitable environment. Additionally, nanoparticles facilitate the effective use of carbon dioxide and reduce a number of stressors that might obstruct algal growth, including oxidative stress and the toxicity of heavy metals. These methods allow nanoparticles to increase algal production and show promise for biotechnological and sustainable algae farming (He et al. 2012). Nanoparticles have positive effects on the algal biomass and algal value-added products. The role of nanoparticles is considered in increasing the nutrient uptake and absorption by the algal cell. In addition, nanoparticles can activate metabolic enzymes and hence enhance the biomass production. Furthermore, the nanoparticle exposure can trigger the algal cell to produce chemicals such as phyto-hormones and pigments as defense mechanisms (Sarkar et al. 2021).

4.1 Algae–Nanotechnology Integration

It is critical to shift the focus of research on nanoparticles and microalgae from toxicity to potential applications to promote microalgal growth and yield as well as to enhance the downstream processes and biomass utilization. Algal nanobionics can be regarded as a strategy to produce biostimulant products to encourage sustainable food choices, economic growth, and environmental responsibility. Bioactive compounds created with this method can also be utilized to prevent and possibly treat a wide range of human ailments and health issues. Although there is a lot of promise for creating biomass, biofuels, and high-value goods using nanoparticles in microalgae production, more research is required to fully realize this potential. Studies should also look into how nanoparticles affect harvesting and how to use microalgal growth in conjunction with harvesting to increase the energy efficiency

that are discussed further in this book. In addition, microalgae-based biorefineries could play a significant role in providing a wide range of bioproducts, including biopolymer, pharmaceuticals, pigments, cosmetics, biofuels, biofertilizers, animal feed, and food additives, which is discussed further in Chapter "Overview of Biorefinery Technology".

5 Conclusions

There are a considerable number of high-value algae bioproducts that are well established in the industry sector, and there are obviously great opportunities for more new products. However, there are crucial factors to consider when developing such phycochemicals such as size of the potential market, potential rivalry from non-algae sources, the time and expense required to obtain regulatory approval for the new products, and the likelihood that consumers will accept them. However, new bioproducts are being developed, thanks to the massively increased efforts being made right now to commercialize algae cultivation. In addition, nanobionics and cascading approach of value-added industrial phycochemicals together with bioenergy production from algal biomass will reduce both energy inputs and carbon footprint of the feedstock, while improving the profit yield.

References

Abd El-Baky HH, El-Baroty GS (2012) Characterization and bioactivity of phycocyanin isolated from *Spirulina maxima* grown under salt stress. Food Funct 3:381–388. https://doi.org/10.1039/C2FO10194G

Abiusi F, Trompetter E, Pollio A, Wijffels RH, Janssen M (2022) Acid tolerant and acidophilic microalgae: an underexplored world of biotechnological opportunities. Front Microbiol 13:51. https://doi.org/10.3389/FMICB.2022.820907/BIBTEX

Abomohra A, Wagner M, El-Sheekh M, Hanelt D (2013) Lipid and total fatty acid productivity in photoautotrophic fresh water microalgae: screening studies towards biodiesel production. J Appl Phycol 25:931–936. https://doi.org/10.1007/s10811-012-9917-y

Abomohra A, El-Sheekh M, Hanelt D (2014) Extracellular secretion of free fatty acids by the chrysophyte *Ochromonas danica* under photoautotrophic and mixotrophic growth. World J Microbiol Biotechnol 30:3111. https://doi.org/10.1007/s11274-014-1738-5

Abomohra A, Jin W, Tu R, Han S-F, Eid M, Eladel H (2016) Microalgal biomass production as a sustainable feedstock for biodiesel: current status and perspectives. Renew Sust Energ Rev 64: 596–606

Abomohra A, El-Sheekh M, Hanelt D (2017) Screening of marine microalgae isolated from the hypersaline Bardawil lagoon for biodiesel feedstock. Renew Energy 101:1266–1272

Abomohra A, Eladel H, El-Esawi M, Wang S, Wang Q, He Z, Feng Y, Shang H, Hanelt D (2018) Effect of lipid-free microalgal biomass and waste glycerol on growth and lipid production of *Scenedesmus obliquus*: innovative waste recycling for extraordinary lipid production. Bioresour Technol 249:992–999

Abomohra A, Shang H, El-Sheekh M, Eladel H, Ebaid R, Wang S, Wang Q (2019) Night illumination using monochromatic light-emitting diodes for enhanced microalgal growth and biodiesel production, vol 288. Bioresour, Technol, p 121514

Abomohra A, El-Naggar AH, Alaswad SO, Elsayed M, Li M, Li W (2020) Enhancement of biodiesel yield from a halophilic green microalga isolated under extreme hypersaline conditions through stepwise salinity adaptation strategy. Bioresour Technol 310:123462

Agarwal A, Jeevanandham S, Sangam S, Chakraborty A, Mukherjee M (2022) Exploring the role of carbon-based nanomaterials in microalgae for the sustainable production of bioactive compounds and beyond. ACS Omega 7:22061–22072. https://doi.org/10.1021/ACSOMEGA.2C01009/ASSET/IMAGES/LARGE/AO2C01009_0005.JPEG

Ainsworth EA, Rogers A (2007) The response of photosynthesis and stomatal conductance to rising [CO2]: mechanisms and environmental interactions. Plant Cell Environ 30:258–270. https://doi.org/10.1111/J.1365-3040.2007.01641.X

Ak İ, Çankırılıgil EC, Türker G, Sever O, Abomohra A (2022) Enhancement of antioxidant properties of *Gongolaria barbata* (Phaeophyceae) by optimization of combined light intensity and salinity stress. Phycologia 1–11:584. https://doi.org/10.1080/00318884.2022.2099136

Almarashi JQM, El-Zohary SE, Ellabban MA, Abomohra A (2020) Enhancement of lipid production and energy recovery from the green microalga *Chlorella vulgaris* by inoculum pretreatment with low-dose cold atmospheric pressure plasma (CAPP). Energy Convers Manag 204:112314. https://doi.org/10.1016/j.enconman.2019.112314

Almutairi AW, Al-Hasawi ZM, Abomohra A (2021) Valorization of lipidic food waste for enhanced biodiesel recovery through two-step conversion: a novel microalgae-integrated approach. Bioresour Technol 342:125966. https://doi.org/10.1016/J.BIORTECH.2021.125966

Arif N, Yadav V, Singh S, Tripathi DK, Dubey NK, Chauhan DK, Giorgetti L (2018) Interaction of copper oxide nanoparticles with plants: uptake, accumulation, and toxicity. Nanomater Plants Algae Microorg 1:297–310. https://doi.org/10.1016/B978-0-12-811487-2.00013-X

Bailleul B, Berne N, Murik O, Petroutsos D, Prihoda J, Tanaka A, Villanova V, Bligny R, Flori S, Falconet D, Krieger-Liszkay A, Santabarbara S, Rappaport F, Joliot P, Tirichine L, Falkowski PG, Cardol P, Bowler C, Finazzi G (2015) Energetic coupling between plastids and mitochondria drives CO2 assimilation in diatoms. Nature 524:366–369. https://doi.org/10.1038/NATURE14599

Balakrishnan J, Dhavamani S, Sadasivam SG, Arumugam M, Vellaikumar S, Ramalingam J, Shanmugam K (2019) Omega-3-rich *Isochrysis* sp. biomass enhances brain docosahexaenoic acid levels and improves serum lipid profile and antioxidant status in Wistar rats. J Sci Food Agric 99:6066–6075. https://doi.org/10.1002/JSFA.9884

Battah M, El-Ayoty Y, Abomohra A, El-Ghany SA, Esmael A (2015) Effect of Mn2+, Co2+ and H2 O2 on biomass and lipids of the green microalga *Chlorella vulgaris* as a potential candidate for biodiesel production. Ann Microbiol 65:155–162. https://doi.org/10.1007/s13213-014-0846-7

Beardall J, Raven JA (2021) Algal metabolism. eLS. https://doi.org/10.1002/9780470015902.a0029261

Begum H, Yusoff FMD, Banerjee S, Khatoon H, Shariff M (2015) Availability and utilization of pigments from microalgae. Crit Rev Food Sci Nutr 56:2209–2222. https://doi.org/10.1080/10408398.2013.764841

Bellou S, Baeshen MN, Elazzazy AM, Aggeli D, Sayegh F, Aggelis G (2014) Microalgal lipids biochemistry and biotechnological perspectives. Biotechnol Adv 32:1476–1493. https://doi.org/10.1016/j.biotechadv.2014.10.003

Bremus C, Herrmann U, Bringer-Meyer S, Sahm H (2006) The use of microorganisms in L-ascorbic acid production. J Biotechnol 124:196–205. https://doi.org/10.1016/J.JBIOTEC.2006.01.010

Chiu L, Kong C, Ooi V (2003) Antiproliferative effect of chlorophyllin derived from a traditional Chinese medicine Bombyx mori excreta on human breast cancer MCF-7 cells. Int J Oncol. https://doi.org/10.3892/ijo.23.3.729

De Bhowmick G, Koduru L, Sen R (2015) Metabolic pathway engineering towards enhancing microalgal lipid biosynthesis for biofuel application—a review. Renew Sust Energ Rev 50: 1239–1253

de la Rosa G, López-Moreno ML, de Haro D, Botez CE, Peralta-Videa JR, Gardea-Torresdey JL (2013) Effects of ZnO nanoparticles in alfalfa, tomato, and cucumber at the germination stage: root development and X-ray absorption spectroscopy studies. Pure Appl Chem 85:2161–2174. https://doi.org/10.1351/PAC-CON-12-09-05/MACHINEREADABLECITATION/RIS

Delhomme C, Weuster-Botz D, Kühn FE (2009) Succinic acid from renewable resources as a C 4 building-block chemical—a review of the catalytic possibilities in aqueous media. Green Chem 11:13–26. https://doi.org/10.1039/B810684C

DellaPenna D, Pogson BJ (2006) Vitamin synthesis in plants: tocopherols and carotenoids. Annu Rev Plant Biol 57:711. https://doi.org/10.1146/annurev.arplant.56.032604.144301

Desjardins SM, Laamanen CA, Basiliko N, Scott JA (2021) Selection and re-acclimation of bioprospected acid-tolerant green microalgae suitable for growth at low pH. Extremophiles 25:129. https://doi.org/10.1007/s00792-021-01216-1

Durmaz Y (2007) Vitamin E (α-tocopherol) production by the marine microalgae *Nannochloropsis oculata* (Eustigmatophyceae) in nitrogen limitation. Aquaculture 272:717. https://doi.org/10.1016/j.aquaculture.2007.07.213

El-Sheekh M, Abomohra A, Eladel H, Battah M, Mohammed S (2018) Screening of different species of *Scenedesmus* isolated from Egyptian freshwater habitats for biodiesel production. Renew Energy 129:114–120. https://doi.org/10.1016/j.renene.2018.05.099

Esakkimuthu S, Krishnamurthy V, Wang S, Hu X, Swaminathan K, Abomohra A (2020) Application of p-coumaric acid for extraordinary lipid production in *Tetradesmus obliquus*: a sustainable approach towards enhanced biodiesel production. Renew Energy 157:368–376

Fertah M, Belfkira A, Dahmane EM, Taourirte M, Brouillette F (2017) Extraction and characterization of sodium alginate from Moroccan *Laminaria digitata* brown seaweed. Arab J Chem 10: S3707–S3714. https://doi.org/10.1016/J.ARABJC.2014.05.003

Galasso C, Gentile A, Orefice I, Ianora A, Bruno A, Noonan DM, Sansone C, Albini A, Brunet C (2019) Microalgal derivatives as potential nutraceutical and food supplements for human health: a focus on cancer prevention and interception. Nutrients 11(6):226. https://doi.org/10.3390/nu11061226

Gimpel JA, Henríquez V, Mayfield SP (2015) In metabolic engineering of eukaryotic microalgae: potential and challenges come with great diversity. Front Microbiol 6:1376. https://doi.org/10.3389/FMICB.2015.01376/BIBTEX

González Fernández LA, Castillo Ramos V, Sánchez Polo M, Medellín Castillo NA (2023) Fundamentals in applications of algae biomass: a review. J Environ Manag 338:117830. https://doi.org/10.1016/J.JENVMAN.2023.117830

Gonzalez V, Vignati DAL, Leyval C, Giamberini L (2014) Environmental fate and ecotoxicity of lanthanides: are they a uniform group beyond chemistry? Environ Int 71:148–157. https://doi.org/10.1016/J.ENVINT.2014.06.019

Górka B, Wieczorek PP (2017) Simultaneous determination of nine phytohormones in seaweed and algae extracts by HPLC-PDA. J Chromatogr B 1057:32–39. https://doi.org/10.1016/J.JCHROMB.2017.04.048

Griffiths M, Harrison STL, Smit M, Maharajh D (2016) Major commercial products from micro- and macroalgae. In: Green energy and technology. Springer, Cham. https://doi.org/10.1007/978-3-319-12334-9_14

Hamed I (2016) The evolution and versatility of microalgal biotechnology: a review. Compr Rev Food Sci Food Saf 15:1104–1123. https://doi.org/10.1111/1541-4337.12227

Hancock RD, Viola R (2005) Improving the nutritional value of crops through enhancement of L-ascorbic acid (vitamin C) content: rationale and biotechnological opportunities. J Agric Food Chem 53:5248–5257. https://doi.org/10.1021/JF0503863

Haznedaroglu BZ, Rismani-Yazdi H, Peccia J, Allnutt FCT, Reeves D (2016) Algal biorefinery for high-value platform chemicals. In: Platform Chemical biorefinery: future green chemistry. https://doi.org/10.1016/B978-0-12-802980-0.00018-3

He D, Dorantes-Aranda JJ, Waite TD (2012) Silver nanoparticle-algae interactions: oxidative dissolution, reactive oxygen species generation and synergistic toxic effects. Environ Sci Technol 46:8731–8738. https://doi.org/10.1021/ES300588A/SUPPL_FILE/ES300588A_SI_001.PDF

Huang XD, El-Alawi Y, Penrose DM, Glick BR, Greenberg BM (2004) Responses of three grass species to creosote during phytoremediation. Environ Pollut 130:453–463. https://doi.org/10.1016/J.ENVPOL.2003.12.018

Huang Y, Lou C, Luo L, Wang XC (2021) Insight into nitrogen and phosphorus coupling effects on mixotrophic *Chlorella vulgaris* growth under stably controlled nutrient conditions. Sci Total Environ 752:141747. https://doi.org/10.1016/j.scitotenv.2020.141747

Igielska-Kalwat J, Gościańska J, Nowak I (2015) Carotenoids as natural antioxidants. Postepy Hig Med Dosw (Online) 69:418–428. https://doi.org/10.5604/17322693.1148335

Jamers A, Blust R, De Coen W, Griffin JL, Jones OAH (2013) Copper toxicity in the microalga *Chlamydomonas reinhardtii*: an integrated approach. Biometals 26:731–740. https://doi.org/10.1007/S10534-013-9648-9

Jeon S, Jeong B, Chang YK (2017) Chemicals and fuels from microalgae, in: consequences of microbial interactions with hydrocarbons, oils, and lipids: production of fuels and chemicals. In: Handbook of hydrocarbon and lipid microbiology. Springer, Cham. https://doi.org/10.1007/978-3-319-50436-0_384

Katiyar R, Arora A (2020) Health promoting functional lipids from microalgae pool: a review. Algal Res 46:101800. https://doi.org/10.1016/J.ALGAL.2020.101800

Khandual S, Sanchez EOL, Andrews HE, de la Rosa JDP (2021) Phycocyanin content and nutritional profile of *Arthrospira platensis* from Mexico: efficient extraction process and stability evaluation of phycocyanin. BMC Chem 15:1–13. https://doi.org/10.1186/S13065-021-00746-1/FIGURES/6

Kumar A, Kumar M, Jose A, Tomer V, Oz E, Proestos C, Zeng M, Elobeid T, Oz F (2023) Major phytochemicals: recent advances in health benefits and extraction method. Molecules 28:887

Kuo RC, Zhang H, Zhuang Y, Hannick L, Lin S (2013) Transcriptomic study reveals widespread spliced leader trans-splicing, short 5′-UTRs and potential complex carbon fixation mechanisms in the Euglenoid alga *Eutreptiella* sp. PLoS One 8:e60826. https://doi.org/10.1371/journal.pone.0060826

Lakshmikandan M, Murugesan AG, Wang S, Abomohra A (2020) Optimization of acid hydrolysis on the green seaweed *Valoniopsis pachynema* and approach towards mixotrophic microalgal biomass and lipid production. Renew Energy 164:1052–1061

Levasseur W, Perré P, Pozzobon V (2020) A review of high value-added molecules production by microalgae in light of the classification. Biotechnol Adv 41:107545. https://doi.org/10.1016/j.biotechadv.2020.107545

Li S, Ji L, Shi Q, Wu H, Fan J (2019) Advances in the production of bioactive substances from marine unicellular microalgae *Porphyridium* spp. Bioresour Technol 292:122048. https://doi.org/10.1016/J.BIORTECH.2019.122048

Lichtenthaler HK (1999) The 1-deoxy-D-xylulose-5-phosphate pathway of isoprenoid biosynthesis in plants. Annu Rev Plant Biol 50:47–65. https://doi.org/10.1146/ANNUREV.ARPLANT.50.1.47

Lim KC, Yusoff FM, Shariff M, Kamarudin MS (2018) Astaxanthin as feed supplement in aquatic animals. Rev Aquac 10:738–773. https://doi.org/10.1111/RAQ.12200

Lin Y, Wang J, Dai H, Mao F, Chen Q, Yan H, Chen M (2023) Salinity moderated the toxicity of zinc oxide nanoparticles (ZnO NPs) towards the early development of *Takifugu obscurus*. Int J Environ Res Public Health 20(4):3209. https://doi.org/10.3390/IJERPH20043209

Lipinska AP, Collén J, Krueger-Hadfield SA, Mora T, Ficko-Blean E (2020) To gel or not to gel: differential expression of carrageenan-related genes between the gametophyte and tetasporophyte life cycle stages of the red alga *Chondrus crispus*. Sci Rep 10:11498. https://doi.org/10.1038/s41598-020-67728-6

López Y, Soto SM (2020) The preventing usefulness biofilm microalgae infections compounds for preventing biofilm infections. Antibiotics 9:9. https://doi.org/10.3390/antibiotics9010009

Malvis Romero A, Brozio F, Kammler S, Burkhardt C, Baruth L, Kaltschmitt M, Antranikian G, Liese A (2023) Enzyme-assisted extraction of alginate from beach wrack *Fucus vesiculosus*. Chemie Ing Tech 95:549–556. https://doi.org/10.1002/CITE.202200173

Martí N, Mena P, Cánovas JA, Micol V, Saura D (2009) Vitamin C and the role of citrus juices as functional food. Nat Prod Commun 4:1934578X0900400. https://doi.org/10.1177/1934578x0900400506

McKee JWA, Kavalieris L, Brasch DJ, Brown MT, Melton LD (1992) Alginate content and composition of *Macrocystis pyrifera* from New Zealand. J Appl Phycol 4:357–369. https://doi.org/10.1007/BF02185794

Meitha K, Pramesti Y, Suhandono S (2020) Reactive oxygen species and antioxidants in postharvest vegetables and fruits. Int J Food Sci 2020:1. https://doi.org/10.1155/2020/8817778

Miazek K, Iwanek W, Remacle C, Richel A, Goffin D (2015) Effect of metals, metalloids and metallic nanoparticles on microalgae growth and industrial product biosynthesis: a review. Int J Mol Sci 16(10):23929–23969. https://doi.org/10.3390/IJMS161023929

Moen E, Larsen B, Østgaard K, Jensen A (1999) Alginate stability during high salt preservation of *Ascophyllum nodosum*. In: Kain JM, Brown MT, Lahaye M (eds) Sixteenth International Seaweed Symposium. Developments in Hydrobiology, vol 137. Springer, Dordrecht, pp 535–539. https://doi.org/10.1007/978-94-011-4449-0_66

Moreno-Garrido I, Pérez S, Blasco J (2015) Toxicity of silver and gold nanoparticles on marine microalgae. Mar Environ Res 111:60–73. https://doi.org/10.1016/J.MARENVRES.2015.05.008

Mousavi P, Morowvat MH, Montazeri-Najafabady N, Abolhassanzadeh Z, Mohagheghzadeh A, Hamidi M, Niazi A, Ghasemi Y (2016) Investigating the effects of phytohormones on growth and β-carotene production in a naturally isolates stain of *Dunaliella salina*. J Appl Pharm Sci 6:164. https://doi.org/10.7324/JAPS.2016.60826

Na H, Jo SW, Do JM, Kim IS, Yoon HS (2021) Production of algal biomass and high-value compounds mediated by interaction of microalgal oocystis sp. knua044 and bacterium sphingomonas knu100. J Microbiol Biotechnol 31:387. https://doi.org/10.4014/JMB.2009.09055

Ninghidayati S, Arina Nurlaili R, Gunardi I, Roesyadi A (2017) Production of carrageenan from seaweed (*Eucheuma cottoni*) with KOH treatment. AIP Conf Proc 1840:060008. https://doi.org/10.1063/1.4982288

Ogbonna JC (2009) Microbiological production of tocopherols: current state and prospects. Appl Microbiol Biotechnol 84:217–225. https://doi.org/10.1007/S00253-009-2104-7

Oliveira RM, Câmara RBG, Monte JFS, Viana RLS, Melo KRT, Queiroz MF, Filgueira LGA, Oyama LM, Rocha HAO (2018) Commercial Fucoidans from Fucus vesiculosus can be grouped into Antiadipogenic and Adipogenic agents. Mar Drugs 16:193. https://doi.org/10.3390/MD16060193

Oslan SNH, Tan JS, Oslan SN, Matanjun P, Mokhtar RAM, Shapawi R, Huda N (2021) *Haematococcus pluvialis* as a potential source of Astaxanthin with diverse applications in industrial sectors: current research and future directions. Molecules 26(21):6470. https://doi.org/10.3390/MOLECULES26216470

Oukarroum A, Bras S, Perreault F, Popovic R (2012) Inhibitory effects of silver nanoparticles in two green algae, *Chlorella vulgaris* and *Dunaliella tertiolecta*. Ecotoxicol Environ Saf 78:80–85. https://doi.org/10.1016/J.ECOENV.2011.11.012

Pandit PR, Fulekar MH, Karuna MSL (2017) Effect of salinity stress on growth, lipid productivity, fatty acid composition, and biodiesel properties in *Acutodesmus obliquus* and *Chlorella vulgaris*. Environ Sci Pollut Res 24:13437–13451

Parveen A, Bhatnagar P, Bisht B, Kumar S, Joshi S, Gautam P, Kumar V, Irina AA, Anna IK (2023) Algal nanobionics to enhance value added products – a review. Curr Res Biotechnol 5:100124. https://doi.org/10.1016/J.CRBIOT.2023.100124

Pereira L (2013) Population studies and carrageenan properties in eight Gigartinales (Rhodophyta) from Western coast of Portugal. Sci World J 2013:1. https://doi.org/10.1155/2013/939830

Pleissner D, Dietz D, van Duuren JBJH, Wittmann C, Yang X, Lin CSK, Venus J (2019) Biotechnological production of organic acids from renewable resources. Adv Biochem Eng Biotechnol 166:373–410. https://doi.org/10.1007/10_2016_73/COVER

Pucci C, Martinelli C, Degli'nnocenti A, Desii A, De Pasquale D, Ciofani G, Pucci C, Degli'nnocenti A, Desii A, De Pasquale D, Ciofani G, Martinelli C (2021) Light-activated biomedical applications of chlorophyll derivatives. Macromol Biosci 21, 2100181. https://doi.org/10.1002/MABI.202100181

Raghukumar S (2008) Thraustochytrid marine protists: production of PUFAs and other emerging technologies. Mar Biotechnol 10:631. https://doi.org/10.1007/s10126-008-9135-4

Rajput VD, Minkina T, Fedorenko A, Chernikova N, Hassan T, Mandzhieva S, Sushkova S, Lysenko V, Soldatov MA, Burachevskaya M (2021a) Effects of zinc oxide nanoparticles on physiological and anatomical indices in spring barley tissues. Nano 11:1722. https://doi.org/10.3390/NANO11071722/S1

Rajput VD, Minkina T, Kumari A, Harish Singh VK, Verma KK, Mandzhieva S, Sushkova S, Srivastava S, Keswani C (2021b) Coping with the challenges of abiotic stress in plants: new dimensions in the field application of nanoparticles. Plants 10:1221. https://doi.org/10.3390/PLANTS10061221

Romanenko EA, Kosakovskaya IV, Romanenko PA (2015) Phytohormones of microalgae: biological role and involvement in the regulation of physiological processes Pt I. auxins, abscisic acid. Ethylene Int J Algae 17:275–289. https://doi.org/10.1615/INTERJALGAE.V17.I3.80

Rupert R, Rodrigues KF, Thien VY, Yong WTL (2022) Carrageenan from *Kappaphycus alvarezii* (Rhodophyta, Solieriaceae): metabolism, structure, production, and application. Front Plant Sci 13:1437. https://doi.org/10.3389/FPLS.2022.859635/BIBTEX

Sarkar RD, Singh HB, Kalita MC (2021) Enhanced lipid accumulation in microalgae through nanoparticle-mediated approach, for biodiesel production: a mini-review. Heliyon 7:e08057. https://doi.org/10.1016/J.HELIYON.2021.E08057

Scaglioni PT, Pagnussatt FA, Lemos AC, Nicolli CP, Del Ponte EM, Badiale-Furlong E (2019) Nannochloropsis sp. and spirulina sp. as a source of antifungal compounds to mitigate contamination by fusarium graminearum species complex. Curr Microbiol 76:930–938. https://doi.org/10.1007/S00284-019-01663-2

Schambach JY, Finck AM, Kitin P, Hunt CG, Hanschen ER, Vogler B, Starkenburg SR, Barry AN (2020) Growth, total lipid, and omega-3 fatty acid production by *Nannochloropsis* spp. cultivated with raw plant substrate. Algal Res 51:102041. https://doi.org/10.1016/J.ALGAL.2020.102041

Schmülling T, Werner T, Riefler M, Krupková E, Manns IBY (2003) Structure and function of cytokinin oxidase/dehydrogenase genes of maize, rice, Arabidopsis and other species. J Plant Res 116:241–252. https://doi.org/10.1007/S10265-003-0096-4

Siddiqui MH, Al-Whaibi MH, Firoz M, Al-Khaishany MY (2015) Role of nanoparticles in plants. In: Nanotechnology and plant sciences: nanoparticles and their impact on plants. Springer International Publishing, Berlin, pp 19–35. https://doi.org/10.1007/978-3-319-14502-0_2

Siqueira SF, Queiroz MI, Zepka LQ, Jacob-Lopes E (2018) Introductory chapter: microalgae biotechnology - a brief introduction, in: microalgal biotechnology. IntechOpen. https://doi.org/10.5772/intechopen.73250

Spolaore P, Joannis-Cassan C, Duran E, Isambert A (2006) Commercial applications of microalgae. J Biosci Bioeng 101:87–96

Stirk WA, Van Staden J (1996) Comparison of cytokinin- and auxin-like activity in some commercially used seaweed extracts. J Appl Phycol 8:503. https://doi.org/10.1007/BF02186328

Stirk WA, Van Staden J (1997) Isolation and identification of cytokinins in a new commercial seaweed product made from *Fucus serratus* L. J Appl Phycol 9:327–330. https://doi.org/10.1023/A:1007910110045/METRICS

Stirk WA, Ördög V, Van Staden J, Jäger K (2002) Cytokinin- and auxin-like activity in Cyanophyta and microalgae. J Appl Phycol 14:215–221. https://doi.org/10.1023/A:1019928425569/METRICS

Stirk WA, Ördög V, Novák O, Rolčík J, Strnad M, Bálint P, van Staden J (2013) Auxin and cytokinin relationships in 24 microalgal strains1. J Phycol 49:459. https://doi.org/10.1111/jpy.12061

Subramanian S, Barry AN, Pieris S, Sayre RT (2013) Comparative energetics and kinetics of autotrophic lipid and starch metabolism in chlorophytic microalgae: implications for biomass and biofuel production. Biotechnol Biofuels 6:150. https://doi.org/10.1186/1754-6834-6-150

Sun H, Zhao W, Mao X, Li Y, Wu T, Chen F (2018a) High-value biomass from microalgae production platforms: strategies and progress based on carbon metabolism and energy conversion. Biotechnol Biofuels 11:227. https://doi.org/10.1186/s13068-018-1225-6

Sun XM, Geng LJ, Ren LJ, Ji XJ, Hao N, Chen KQ, Huang H (2018b) Influence of oxygen on the biosynthesis of polyunsaturated fatty acids in microalgae. Bioresour Technol 250:868. https://doi.org/10.1016/j.biortech.2017.11.005

Tawfik A, Eraky M, Alhajeri NS, Osman AI, Rooney DW (2022a) Cultivation of microalgae on liquid anaerobic digestate for depollution, biofuels and cosmetics: a review. Environ Chem Lett 20:3631. https://doi.org/10.1007/s10311-022-01481-2

Tawfik A, Ismail S, Elsayed M, Qyyum MA, Rehan M (2022b) Sustainable microalgal biomass valorization to bioenergy: key challenges and future perspectives. Chemosphere 296:133812. https://doi.org/10.1016/J.CHEMOSPHERE.2022.133812

Thompson GA (1996) Lipids and membrane function in green algae. Biochim Biophys Acta - Lipids Lipid Metab 1302(1):17–45. https://doi.org/10.1016/0005-2760(96)00045-8

Tomori, M., Nagamine, T., Miyamoto, T., Iha, M., 2021. Effects of ingesting fucoidan derived from cladosiphon *Okamuranus tokida* on human NK cells: a randomized, double-blind, parallel-group, placebo-controlled pilot study. Mar Drugs 19(6), \. https://doi.org/10.3390/MD19060340/S1

Treves H, Murik O, Kedem I, Eisenstadt D, Meir S, Rogachev I, Szymanski J, Keren N, Orf I, Tiburcio AF, Alcázar R, Aharoni A, Kopka J, Kaplan A (2017) Metabolic flexibility underpins growth capabilities of the fastest growing alga. Curr Biol 27:2559–2567.e3. https://doi.org/10.1016/J.CUB.2017.07.014

Valença R d L, Silva Sobrinho AGD, Silva LG, Borghi TH, De Andrade N, Soares MS, Cirne LGA, Castro Filho ES, Ezequiel JMB (2022) Marine algae meal (*Schizochytrium* sp.) and vitamin E in lamb nutrition: intake, digestibility, nitrogen balance, ingestive behavior, ruminal parameters, performance, and carcass characteristics. Livest Sci 258:104877. https://doi.org/10.1016/J.LIVSCI.2022.104877

van den Oever SP, Mayer HK (2022) Biologically active or just "pseudo"-vitamin B12 as predominant form in algae-based nutritional supplements? J Food Compos Anal 109:104464. https://doi.org/10.1016/J.JFCA.2022.104464

Vassilev SV, Vassileva CG (2016) Composition, properties and challenges of algae biomass for biofuel application: an overview. Fuel 181:1–33. https://doi.org/10.1016/J.FUEL.2016.04.106

Volkman JK (2003) Sterols in microorganisms. Appl Microbiol Biotechnol 60:495–506. https://doi.org/10.1007/S00253-002-1172-8

Voshall A, Christie NTM, Rose SL, Khasin M, Van Etten JL, Markham JE, Riekhof WR, Nickerson KW (2021) Sterol biosynthesis in four green algae: a Bioinformatic analysis of the Ergosterol versus Phytosterol decision point. J Phycol 57:1199. https://doi.org/10.1111/jpy.13164

Wang C, Qi M, Guo J, Zhou C, Yan X, Ruan R, Cheng P (2022) The active Phytohormone in microalgae: the characteristics, efficient detection, and their adversity resistance applications. Molecules 27(1):46. https://doi.org/10.3390/MOLECULES27010046

Xu Y, Harvey PJ (2019) Carotenoid production by *Dunaliella salina* under red light. Antioxidants 8(5):123. https://doi.org/10.3390/ANTIOX8050123

Zhao Y, Zheng Y, Wang J, Ma S, Yu Y, White WL, Yang S, Yang F, Lu J (2018) Fucoidan extracted from *Undaria pinnatifida*: source for nutraceuticals/functional foods. Mar Drugs 16(9):321. https://doi.org/10.3390/MD16090321

Artificial Intelligence in Phycochemicals Recognition

Anja Dullius, Giseli Buffon, Milton Faria Junior, and Silvana Giuliatti

Abstract Given the exponential growth of data in biotechnological processes, artificial intelligence (AI) and machine learning applications are getting much attention in both academic and industrial research, where the rapid product development in a constantly changing environment is challenging, as quality, speed, and efficiency are crucial. Since the commercial value of microalgae and seaweeds has been recognized due to their unique chemical spectrum of phytochemicals, such as pigments, fatty acids, phenolic compounds, polysaccharides, and proteins, the systematic search for new innovative and sustainable inputs is increasingly ongoing. However, overcoming laborious and cost-intensive research is still a bottleneck. Extracting relevant data and solving complex problems due to recent advances in computing power, e.g., high-performance computing and improvements in technologies such as Deep Learning and Random Forest, have the potential to face these challenges. In chemodiversity research, AI has been widely applied in drug development, omics data analysis, as well as system design and optimization, so far. In this context, this chapter introduces the basic concept of computational methods, with a focus on the recent achievements and research trends proposed for the discovery of bioactive algal compounds and their pathways, including the elucidation of their structural information.

Keywords Microalgae · Seaweeds · Chemodiversity · Bioactive molecules · Machine learning

---

A. Dullius · G. Buffon
Syntalgae Ltda, Lajeado, Brazil

M. F. Junior
Department of Exact Sciences, University of Ribeirão Preto - UNAERP, Ribeirão Preto, SP, Brazil

S. Giuliatti (✉)
Department of Genetics, Faculty of Medicine of Ribeirão Preto, University of São Paulo - USP, São Paulo, SP, Brazil
e-mail: silvana@fmrp.usp.br

A. Abomohra, S. Ende (eds.), *Value-added Products from Algae*,
https://doi.org/10.1007/978-3-031-42026-9_5

1 Introduction

Artificial Intelligence (AI) is an interdisciplinary field of computer science that originated in the history of human thought and the evolution of machines and computers. From the conception of mechanical machines by Charles Babbage and Ada Lovelace in the nineteenth century, through the creation of the first mathematical model of an artificial neuron by Warren McCulloch and Walter Pitts in 1943 (McCulloch and Pitts 1943), to the advances promoted by Alan Turing, who established the foundations of AI and computer science with his Turing Machine and Turing Test (Turing 1950). In 1956, the Dartmouth Summer Research Project on Artificial Intelligence brought together influential researchers, who were thinking about "*how to make machines use language, form abstractions and concepts, solve kinds of problems now reserved for humans, and improve themselves,*" consolidating AI as an independent research field and introducing the term "artificial intelligence" coined by McCarthy (McCarthy et al. 2006). Over time, AI has experienced remarkable growth and evolved through different paradigms, resulting in the development of modern algorithms and architectures that shape the field today.

One subfield of AI is machine learning (ML), which encompasses a variety of data-driven algorithms, and has a long history dating back to the pioneering work of Rosenblatt (1958). Rosenblatt's major achievement was to show that artificial neurons are able to learn from training data by themselves. It must be emphasized that he was already using a supervised algorithm at that time and thus gave the practical initiation of ML. According to the training method, ML is categorized in supervised (classification and regression methods where the predictive model is developed based on input and output data), unsupervised (clustering and feature-finding methods based on only input data), and reinforcement learning (decision making based on trial and error) (Mak and Pichika 2019). In 1986, Rumelhart published together with Hinton and Williams a milestone paper in *Nature*, introducing the backpropagation algorithm (Rumelhart et al. 1986). Today seen as the probably most important building block in a neural network, that guarantees effective learning of the model, backpropagation performs a backward pass after each forward pass through a network, while adjusting the model's parameters.

Deep Learning (DNNs) is a Machine Learning technique that has been developed by several researchers over the last few decades. However, the work of Professor Geoffrey Hinton from the University of Toronto and his research team, especially from 2006 onward, was fundamental for developing and popularizing the technique (Hinton et al. 2006). Hinton and his team were responsible for significantly advancing the training of deep neural networks using the backpropagation technique, as well as introducing new architectures of deep neural networks such as Convolutional Neural Networks (CNNs) (Krizhevsky et al. 2017) and Recurrent Neural Networks (RNNs) (Yu et al. 2019). Their work was instrumental in driving research in Deep Learning and making the technique more effective and widely applicable in various areas such as computer vision, natural language processing, speech recognition, and more.

In the 1980s, the general enthusiasm for AI was pretty much on the decline as progress fell short of expectations, but this was not true for medical biotechnology, being the most developed field in biotechnology where AI is used. Perhaps the best-known example is the drug development process, including drug screening and drug target identification as reviewed in detail by Mak and Pichika (2019). Other examples of applied AI techniques include image screening to analyze CT or MRI images to identify abnormalities and predictive modeling to predict the effectiveness of a person's treatment based on data analysis from wearable devices (Haghi et al. 2017). The latest trend is AI-based in vitro diagnostics (IVDs), which analyze data from smart diagnostics-integrated biosensors and allow early disease detection in the coming years, for example for cancer or cardiovascular diseases (McRae et al. 2022).

Not far behind medical biotechnology, bioprocess control and optimization have made great progress in microbiology in the past years. Usually, a strain's performance is limited by its metabolic characteristics and environmental conditions, resulting in reduced economic benefits when a bioprocess has a non-optimized performance (Wang et al. 2020). Orthogonal experimental design (OED) or response surface methodology (RSM) was the classical method applied to bioprocess modeling for a few decades (Bernaerts and Van Impe 2004; Nwabueze 2010). However, due to the complexity of the bioprocesses, such models led to accuracy deficiencies for which ML techniques were found to solve these problems (Schubert et al. 1994). While microbes such as bacteria and yeasts were initially used for the production of medical metabolites, today the production pressure is increasingly coming from the bioeconomy sector, due to the transition from a fossil-based to a bio-based economy (Kawaguchi et al. 2017; Ioannidou et al. 2020). For this purpose, algae were recently included in the biorefinery concept and researchers have applied machine learning techniques with notable success in identifying species and monitoring growth processes, as demonstrated by the work of Carleo et al. (2019) and intensively reviewed by Lim et al. (2022). As discussed in previous chapters, microalgae are single-celled and photosynthetic organisms that showed great potential for applications in areas such as agriculture, biofuel production, animal and human nutrition, health care, and wastewater treatment (Fernández et al. 2021). Farming of seaweeds, which are multicellular large-size organisms, additionally contributes to ecological aspects, as nutrient and CO_2 uptake prevents coastal eutrophication and ocean acidification (van Hal et al. 2014; Hasselström et al. 2020).

Machine Learning is an area of Artificial Intelligence that includes several subdivisions; such as Supervised Learning, where the model is trained with labeled data; Unsupervised Learning, which seeks patterns and structures in unlabeled data; and Reinforcement Learning, where the model learns through interaction with an environment; receiving rewards or penalties for its actions, with the aim of making the best decision in each situation. Figure 1 illustrates a diagram of the most applied techniques in recent years for problems related to microalgae, which are within the field of Machine Learning. These techniques include Deep Learning with Recurrent Neural Networks (RNNs) and CNNs, as well as Machine Learning techniques such as Multilayer Perceptron (MLP) and Random Forest (RF). The use of these AI models can facilitate automation of microalgae detection and utilization, and can

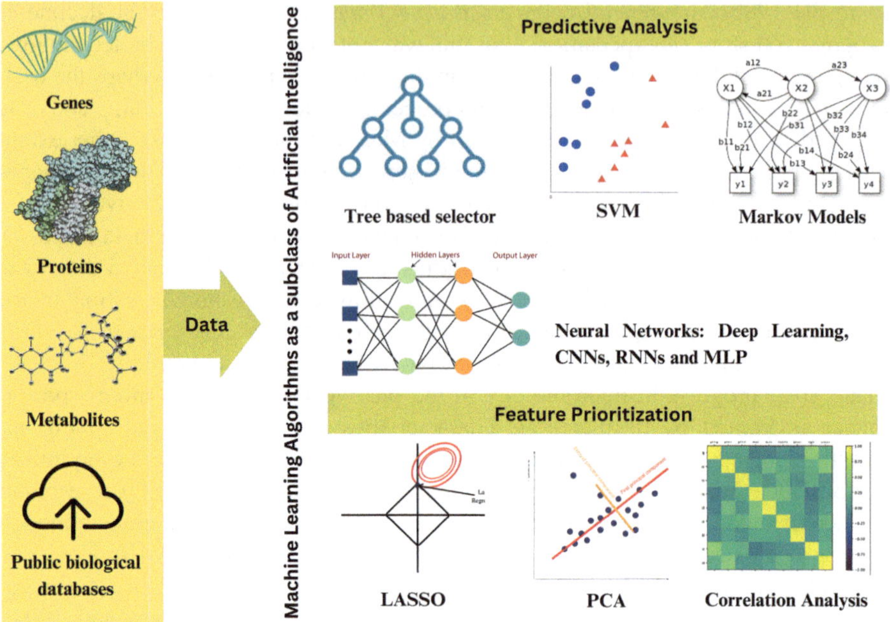

Fig. 1 Diagram of the most applied AI techniques for predictive analysis and feature prioritization related to omics data obtained from microalgae and seaweed, which are within the field of Machine Learning

also optimize the conditions of microalgae cultivation, such as pH, temperature, and nutrient concentration, to enhance biomass or compound production.

This chapter presents an overview of recent advances in the application of machine learning algorithms to microalgae and seaweed research. Topics include the use of various AI models for processing derived from metabolomics, including pathway and structure elucidation. Moreover, the inclusion of mathematical modeling is discussed by way of examples published in the past 5 years.

2 Artificial Intelligence Algorithms

Artificial intelligence algorithms can be divided into two types; supervised and unsupervised. In the case of unsupervised algorithms, the analysis is carried out without the use of labeled data, meaning that there is no prior indication of categories or classes to be identified. Within this context, three unsupervised techniques commonly used in metabolomic analysis stand out; namely principal component analysis (PCA), clustering, and self-organizing maps (SOMs) (Corsaro et al. 2022; Kikuchi et al. 2018).

2.1 Principal Component Analysis (PCA)

It is a technique used to reduce the dimensionality of a high-dimensional dataset. It involves projecting the data into a lower-dimensional space while preserving as much information as possible. Through this technique, it is possible to identify patterns and groupings in data with multiple variables. PCA works by transforming the original variables in a dataset into a new set of uncorrelated variables called principal components (PCs). These PCs are linear combinations of the original variables and are ordered by the amount of variance they explain in the data. The first principal component (PC1) captures the maximum variance in the data, followed by the second principal component (PC2), which captures the maximum remaining variance orthogonal to PC1, and so on. This process continues until all original variables are represented by PCs.

In practice, only few PCs (usually two or three) are needed to explain most of the variance in the dataset, which allows for a significant reduction in dimensionality. By plotting the data points using the first few PCs as axes, it becomes possible to visualize high-dimensional data in a lower-dimensional space, facilitating the identification of patterns, trends, and potential outliers in the data. PCA is an unsupervised learning method, meaning that it does not require any prior knowledge or labeled data to perform the analysis. This makes PCA a useful exploratory tool for understanding the structure and relationships within complex datasets before applying more advanced machine learning techniques or hypothesis-driven statistical tests. PCA is a practical algorithm for reducing the dimensionality of complex data and facilitating visualization and interpretation, providing researchers with valuable insights into the underlying structure of their datasets.

2.2 Clustering

This technique is used to group data into homogeneous sets based on their common characteristics. This technique is particularly useful in metabolomic analysis, as it allows for the identification of groups of metabolites with similar characteristics, such as chemical structures and biological functions. The goal of clustering is to identify patterns and trends in the data by dividing them into groups (clusters) so that the elements within each group are more similar to each other than to elements from other groups. This can help researchers uncover relationships, natural groupings, and potential outliers in metabolomic data. Some of the most common clustering algorithms, such as *k*-means and hierarchical clustering analysis (HCA), can be applied to metabolomic analysis. There are several clustering algorithms that can be applied to the study of metabolism. Some of the most common algorithms include:

2.2.1 Hierarchical Clustering Analysis (HCA)

HCA is a method that constructs a tree-like structure (dendrogram) to represent the hierarchy and similarity among the data. There are two main approaches in HCA: agglomerative (bottom-up) and divisive (top-down). The agglomerative approach starts with each element as an individual cluster and iteratively combines them based on similarity. The divisive approach, on the other hand, starts with a single cluster containing all elements and iteratively splits them into subgroups.

2.2.2 *K*-Means Clustering

It is a widely used partition-based clustering algorithm that seeks to divide a dataset into a predetermined number of clusters (*k*) based on the similarity of the data points. The primary goal of *K*-means clustering is to minimize the sum of squared distances between the data points and their corresponding cluster centroids, which represent the centers of the clusters. The *K*-means clustering algorithm operates through the following steps:

- Initialization: Randomly select '*k*' initial centroids from the dataset. These centroids serve as the starting centers of the clusters.
- Assignment: Allocate each data point to the nearest centroid. The distance between the data points and the centroids is typically measured using Euclidean distance, although other distance metrics may also be employed.
- Update: Compute the new centroids by calculating the mean of all data points assigned to each cluster. This step relocates the centroids to the center of their respective clusters.
- Iterate: Repeat the assignment and update steps until the centroids' positions converge or a maximum number of iterations is reached.

The *K*-means algorithm's sensitivity to the initial placement of the centroids can influence the final clustering outcomes. To mitigate this issue, the algorithm is frequently executed multiple times with different initial centroid positions, and the solution with the lowest sum of squared distances is chosen as the final result.

A prevalent challenge in using *K*-means clustering is determining the optimal number of clusters (*k*). Various methods can estimate the best value for k, such as the elbow method, silhouette analysis, and gap statistic. These methods assess the quality of clustering solutions for different *k* values and aid in selecting the one that best represents the underlying structure of the data. *K*-means clustering is a versatile and efficient algorithm that performs well on large datasets and is extensively employed in diverse fields, including metabolomics studies. By grouping similar data points, *K*-means clustering can unveil patterns and trends in the data, which are crucial for comprehending complex biological systems and facilitating meaningful discoveries.

2.2.3 Self-Organizing Maps (SOM)

SOM is a neural network-based machine learning algorithm, developed by Teuvo Kohonen in the 1980s. It maps high-dimensional data onto a two-dimensional or three-dimensional space, preserving the topology and relationships among the data points. SOM is useful for visualizing the distribution and patterns in complex high-dimensional data, such as those found in metabolomics studies. SOMs are an unsupervised learning technique that allows for the visualization of large datasets in a two-dimensional space. Through this technique, it is possible to identify groups of metabolites with common characteristics, as well as the relationships between them. This technique is particularly useful in the analysis of large datasets, such as those generated by nuclear magnetic resonance (NMR) spectroscopy. The SOM algorithm consists of the following steps:

- Initialization: Create a two-dimensional grid of nodes, each with a weight vector of the same dimensionality as the input data. The weight vectors are often initialized with small random values.
- Competitive Phase: For each input data point, calculate the distance between the data point and all weight vectors in the grid. The node with the weight vector closest to the input data point is considered the Best Matching Unit (BMU). The distance is typically calculated using the Euclidean distance, but other distance metrics can also be employed.
- Cooperative Phase: Define a neighborhood around the BMU, which includes all nodes within a certain radius. The size of the neighborhood typically decreases over time, allowing the algorithm to refine the map as it iterates.
- Adaptive Phase: Update the weight vectors of the nodes within the BMU's neighborhood by moving them closer to the input data point. The learning rate, which determines how much the weight vectors are adjusted, also decreases over time, allowing for more precise refinements.
- Iterate: Repeat the competitive, cooperative, and adaptive phases for a predetermined number of iterations or until convergence is reached.

After the training process, the SOM forms a grid where similar data points are mapped to nearby nodes, preserving the topological structure of the input data. The resulting two-dimensional representation can be visualized using various techniques, such as U-Matrix or component planes, to reveal patterns and clusters within the data. One of the advantages of SOM over other clustering algorithms, such as *K*-means, is its ability to handle nonlinear relationships between data points. Moreover, SOM does not require the user to specify the number of clusters beforehand, as the topological structure of the data is inherently represented in the resulting grid. However, it is essential to note that SOM is sensitive to the choice of parameters, such as the learning rate, neighborhood size, and grid dimensions. Proper parameter tuning is crucial for obtaining meaningful results and accurate representations of the data (Helmy et al. 2023).

By applying clustering algorithms to metabolomic data, researchers can identify patterns and relationships in the data that can lead to important discoveries, such as biomarkers, metabolic pathways, and differences between biological conditions. This information can be valuable in better understanding metabolic processes, developing personalized diagnostics and treatments, and improving the overall understanding of biological systems.

2.3 Supervised Learning

It is used in classification or regression problems, where the desired output is known. Some examples of supervised algorithms applied to metabolomic analysis include *k*-nearest neighbors (KNN), random forest (RF), principal component regression (PCR), support vector machines (SVM), multi-layer perceptron (MLP), and CNN.

2.3.1 *k*-Nearest Neighbors

KNN is a classification algorithm that is commonly used in machine learning and can be applied to metabolomics studies. The KNN algorithm works by determining the *k*-nearest data points to a new, unlabeled data point and classifying it based on the majority class of those *k* neighbors. In metabolomics, KNN can be used to classify samples based on their metabolic profiles. For example, KNN has been applied to classify different types of tea based on their NMR spectra and to classify different types of wine based on their metabolic profiles. The KNN algorithm consists of the following steps:

- Choose the number of neighbors (*k*) to consider.
- Calculate the distance between the new, unlabeled data point and all other data points in the dataset.
- Select the *k* nearest data points to the new data point based on the calculated distances.
- Classify the new data point based on the majority class of the *k* neighbors.

The distance between data points can be measured using various metrics, such as Euclidean distance or Manhattan distance. One challenge with KNN is determining the optimal value for *k*. A small value of *k* may lead to overfitting, while a large value of *k* may lead to underfitting. Cross-validation techniques can be used to select the best value for *k*.

Overall, KNN is a simple and effective algorithm for classification tasks in metabolomics studies. By using the metabolic profiles of samples to classify them into different groups, KNN can help researchers gain insights into the underlying biology of complex systems.

2.3.2 Random Forest

RF algorithm is a supervised machine learning approach commonly used in the field of metabolomics for classification and feature selection. It operates by creating a collection of decision trees, each independently predicting a class, and combining these predictions to obtain the final output. The algorithm begins by creating a selection tree and using observations to obtain different sets using different splitting criteria that operate on the considered vectors. The dataset is then divided into two subsets using each criterion.

Next, the algorithm creates multiple decision trees, each using a randomly selected subset of the features and training them using a subset of the training data. The trees are used to predict the class of new data points, and the final output is obtained by taking the average of the predictions of all the trees. RF can handle both categorical and continuous features, making it suitable for metabolomics data, which often contain both types of features. It is also capable of dealing with high-dimensional spaces and many training examples, making it a powerful tool for analyzing large datasets. Figure 3 shows the structure of RF algorithms used to handle large datasets.

The Random Forest algorithm has been extensively used in metabolomics studies, including the analysis of microalgae metabolomics data. In this study, RF was used to identify the most significant metabolites in relation to biomass production. RF has also been used for feature selection and variable importance analysis in metabolomics studies, where it has been shown to outperform other feature selection methods. Additionally, RF was used to evaluate the classification performance of NMR spectroscopic data in a metabolomics study, where it was compared to five other binary classification algorithms. RF is a widely used and powerful algorithm in the field of metabolomics due to its ability to handle large datasets, feature selection, and classification tasks (Corsaro et al. 2022).

2.3.3 Support Vector Machine

SVM is a supervised machine learning algorithm that aims to classify data into two or more classes. In SVM, each data point is represented as a single point in a high-dimensional space. The algorithm tries to find the optimal hyperplane that can separate the two classes with maximum margin, which is defined as the maximum distance between the hyperplane and the closest data points on either side. SVM can handle linearly separable as well as nonlinearly separable data by using a technique called the kernel trick. The kernel function maps the data points into a higher-dimensional space where they can be linearly separated. Some common kernel functions used in SVM include linear, polynomial, radial basis function (RBF), and sigmoid. The SVM algorithm works in the following steps:

- Select the kernel function and its parameters.
- Find the hyperplane that maximizes the margin between the two classes.
- Classify new data points based on their position relative to the hyperplane.

One of the advantages of SVM is that it can handle datasets with many features and still maintain high accuracy. SVM has been widely used in the field of metabolomics for classification and feature selection. For example, SVM has been used to classify metabolomic data in studies on liver diseases, where it achieved high accuracy in distinguishing between healthy and diseased subjects. However, one of the main drawbacks of SVM is its sensitivity to the choice of kernel function and its parameters, which can lead to overfitting or underfitting of the data. To mitigate this issue, various techniques such as cross-validation and grid search have been employed to optimize the hyperparameters.

SVM is a powerful algorithm for classification tasks in metabolomics, and its ability to handle nonlinearly separable data makes it a popular choice for various applications. By utilizing the kernel trick, SVM can accurately classify and select features in high-dimensional metabolomic datasets (Corsaro et al. 2022).

SVM is another ML algorithm that is widely used in metabolomics. It uses the kernel method to perform linear and nonlinear classification and regression tasks. SVM is known for its superior classification accuracy and feature selection compared to other approaches. SVM has been used to identify important variables in biochemical samples, to evaluate the cardiotoxicity of traditional Chinese medicine compatibility, and to normalize and integrate large-scale metabolomics data. SVM is particularly useful in studies involving small sample sizes, and it can handle overlearning issues that arise in other ML approaches (Kikuchi et al. 2018).

2.3.4 Multi-Layer Perceptron

MLP is a type of feedforward neural network used in supervised learning, which involves an input layer, an output layer, and several hidden layers that serve as the computational engine. The MLP is an artificial neural network with multiple layers of neurons that allow for learning complex patterns. In terms of applications in metabolomics, supervised learning algorithms such as MLP have been used to classify and predict metabolites based on different features extracted from metabolomics data. For instance, an MLP-based approach was used to identify potential biomarkers of liver cancer based on the analysis of NMR-based metabolomics data (Kikuchi et al. 2018).

2.3.5 Convolutional Neural Networks

CNNs are a type of neural network that is highly suited for complex data processing, especially when it comes to image recognition. The architecture of a CNN takes into consideration the neuron organization found in the visual cortex of an animal brain.

It typically consists of four stages: feature map deduction from the input using a convolution function, revealing an image after given changes using max-pooling, flattening the data for analysis, and compiling the loss function by a hidden layer using the full connection. Figure 1 shows the structure of a Convolutional Neural Network (CNN) for metabolic synthesis identification.

Convolutional neural network has been applied in NMR spectra to achieve accurate peak identification in complex mixtures. The results reveal that the neural network was successfully trained on metabolite identification from these 2D NMR spectra and achieved very good performance compared with other NMR-based metabolomic tools. The model effectively identified metabolites in a high-throughput manner, allowing for quick and efficient analysis of metabolomics data (Helmy et al. 2023).

Hybrid learning approaches also exist, which combine supervised and unsupervised techniques, such as deep transfer learning (DTL) and deep reinforcement learning (DRL). DTL is a technique that utilizes a pretrained neural network to improve accuracy in a related machine learning task. DRL is a technique that uses reward-based learning to improve performance in a decision-making task.

3 Seventy Years to Advance the Development of Microalgae Chemodiversity

Since the early 1970s, pigment studies from microalgae for industrial applications had begun (Tett et al. 1975; Scheibe 1972; Norgård et al. 1974). However, the developed techniques required large biomass quantities for molecule purification, thus they were not suitable for high-throughput screening purposes (Serive et al. 2012). A decade later, in the 1980s, research was ready to extract some temperature-sensitive molecules, but even these techniques were still far from being used as high-throughput screening (Rowan 1989). The first commercialized high-value product was the pigment β-carotene from the microalga *Dunaliella salina* in the 1980s followed by astaxanthin from *Haematococcus pluvialis* in the 1990s. Using microalgae as an industrial source of carotenoids was first proposed in 1964 and less than 20 years later, it was done. This depended on several circumstances and was made possible by the fact that the cultivation of *Dunaliella* is relatively easy due to its resistance to high salinity which avoids contamination. Since then, the scientific community has accumulated a detailed understanding of metabolism and chemodiversity, although this knowledge is limited to a few microalgal species.

Still in the 2010s, research in seaweeds and microalgal chemodiversity was overshadowed by research that focused on determination of commercial value, which could be noted in an explosive number of publications about applications of seaweeds and microalgae in agriculture, food, cosmetic, pharmaceutical, and bioenergy markets (Stengel et al. 2011). Research in this field was extended to increasingly look for genes that encode key enzymes for bioactive molecules and advances in analytical techniques, such as monoclonal antibody labeling, Fourier transform infrared (FTIR) spectroscopy, and Near-infrared (NIR) spectroscopy, allowed the first fully characterizations of algal molecules (Lu et al. 2009; Mangoni et al. 2011; Remias et al. 2012; Challagulla et al. 2017). It has been recognized that genetic variability is coherent with chemodiversity, and that a decade ago, scientists were scratching the tip of an iceberg. Metabolomics analysis, that are increasingly based on more environmentally friendly extraction techniques with higher yield of phytochemicals, and coupled with Nuclear magnetic resonance (NMR) spectroscopy, liquid chromatography with mass spectrometry coupling (LC-MS), or gas chromatography with mass spectrometry coupling (GC-MS) are one of the techniques recommended today to kick-start chemodiversity research (Andrich et al. 2005; Herrero et al. 2006; Lee et al. 2010; Pasquet et al. 2011; Liang et al. 2012; Adam et al. 2012; Réveillon et al. 2019; Bisht et al. 2021; Chen et al. 2023).

After almost 70 years of algal chemodiversity science (Fig. 2), research in physiology and metabolites has been limited to about 50 species, which is reflected in the current market (Stirk and van Staden 2022). Rumin and Nicolau (2020) reported that worldwide research and market restricted their activities to about 15 microalgal genera, which can differ according to geographic locality and legislation. These are *Chlorella* sp., *Scenedesmus* sp., *Chlamydomonas* sp., *Phaeodactylum* sp., *Nannochloropsis* sp., *Dunaliella* sp., *Isochrysis* sp., *Tetraselmis* sp., *Arthrospira* sp., *Selenastrum* sp., *Botryococcus* sp., *Haematococcus* sp., *Acutodesmus* sp., *Synechocystis* sp., and *Schizochytrium* sp. Furthermore, due to competing resources on the market established at lower costs, innovative developments tend to concentrate on human health with the application of microalgal products as supplements and in the dermocosmetic sector or development of innovative biomaterials, such as bioplastic or polymeric foams (Spolaore et al. 2006; Morocho-Jácome et al. 2022). In China, Japan, India, Europe, and North America, the main production of microalgae is of *Chlorella* and *Spirulina*, while Brazil is just discovering this possibility, due to favorable climate and substrate conditions, and possesses one of the world's largest biodiversities. First and foremost, it must be mentioned that there are pilot projects involving energy companies, the use of microalgae in aquaculture, and the emergence of start-ups coming up in the microalgae business (de Oliveira and Bragotto 2022).

The circumstances described above make it clear that bioprospecting for new bioactive compounds in microalgae and seaweed and the development of innovative and sustainable products on the market are still in their infancy. Thus, the number of new metabolites in a single study is surprising, for example, the work of Serive et al. (2017) showed that from 124 pigment species that were extracted out of 37 isolated microalgal strains, 98 pigments or derivatives were yet unidentified. Similarly, Brillatz et al. (2018) found 11 previously unknown secondary metabolites in a metabolite profiling study in the diatom *Skeletonema marinoi*, of which inosine was observed as a potent anticonvulsant compound in zebrafish being effective at a concentration of 10 $\mu g\ mL^{-1}$.

Regarding extrapolysaccharides (EPS) in microalgae, since their discovery in 1964 by Tischer and Moore (Tischer and Moore 1964), research has been moderate,

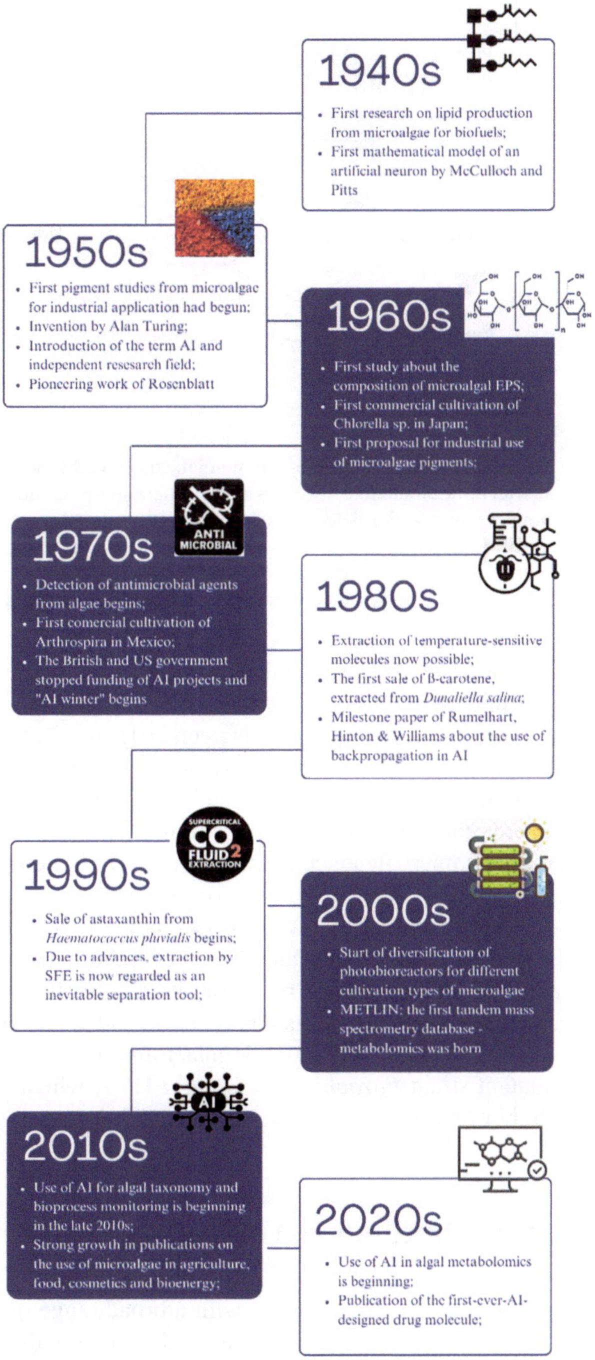

Fig. 2 Timeline of chemodiversity research in algae and important key development points for the concomitant evolution of AI in the past 70 years. Today's position announces that AI is a trend for now being used in chemodiversity research in microalgae

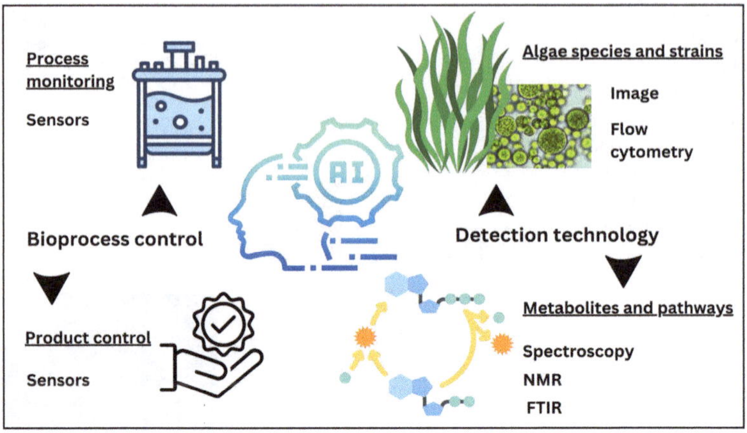

Fig. 3 The use of AI in algae-to-commercialization workflow. ML/DL has been successfully used in the screening of suitable species/strains, where the predominant techniques are image analysis and flow cytometry. For detection of metabolites and pathways, ML/DL is used for evaluation of data derived from spectroscopy, NMR, and FTIR. During manufacturing of algal biomass, ML/DL algorithms support sensor-driven bioprocess monitoring, thus evaluating data of biomass density and metabolites or substrate concentration. Sensor technology is also used for final quality control

with fewer than 200 structural features of their EPS currently described out of the 35,000 microalgal species we know of, to the point that only 12 microalgal EPS structures have been described between 2015 and 2020 (Halaj et al. 2022). Detection of new phytochemicals not only depends on instrumental techniques applied, but also on metabolic plasticity, indicating that careful screening of species is essential. In general, the biochemical composition of seaweed underlies seasonal fluctuations, whereas for microalgae, cultivation conditions are crucial (Li et al. 2015; Sudhakar et al. 2019; Rashidi et al. 2019). For instance, *Chlamydomonas reinhardtii* showed an increased concentration of polysaccharides and sulfation degree of polysaccharides, when exposed to elevated levels of sodium nitrate (Vishwakarma et al. 2019). In addition, mutations that result in altered metabolite expression must be considered. With this in mind, Sushytskyi et al. (2020) revealed branched non-sulphated α-L-rhamnopyranan and β-D-xylogalactofuranan of unusual structure produced by the mutant strain *Parachlorella kessleri* HY1, which is deficient by chlorophylls a and b, but rich in lutein, accumulated in chloroplasts.

4 Algae-Integrating Artificial Intelligence

Seaweeds and microalgae show with a broad range of unique metabolites that they are cellular biofactories of chemical compounds that can be processed in biorefineries into high-value products (Martínez-Ruiz et al. 2022). Such chemical bioactives of natural origin are characterized by a structural complexity with densely packed functional groups, which as a consequence allows for maximum selectivity and thus functionality with the target. To their downside, extraction of deeply inaccessible metabolites compromises success of high-throughput screenings, for which literature often reports on the biological activities of algae extracts, but neglects to take a closer look at which substances are responsible for the respective activities (Serive et al. 2012).

The application of artificial intelligence in phycology is a relatively new research trend that aims to remedy this and look forward to understanding metabolic synthesis pathways, as well as new discoveries of chemical substances. In other areas, AI is being used for some time already: image analysis of satellite pictures is used to detect algal carpets and to obtain information about water quality or potential toxin hazards by identifying their species based on their pigment content (Sita et al. 2021). Besides the use of algae as environmental indicators (Qian et al. 2020; Zhuo et al. 2022), commercial interest led above all to the screening of biotechnologically interesting species and the development of production processes becoming of new interest, where AI is used to remove seemingly insurmountable hurdles from the path. Teng et al. (2020) reviewed integration of AI in microalgae process system design and optimization under the aspect of genetic development and suggested the use of AI in the workflow microalgae-to-commercialization, including the following sections: (1) genetic engineering, (2) cultivation and conversion, (3) system design and control, and (4) optimal process design and integration. These steps necessarily integrate chemodiversity research and the use of AI for (1) discovery of phytochemicals and their respective chemical structure, (2) discovery of metabolic pathways, and (3) monitoring of bioprocesses being at a tenuous beginning from today's research standpoint, as will be discussed below and as summarized in Table 1 and Fig. 3.

4.1 ML/DL for Discovery of Phytochemicals and their Respective Chemical Structure

The scarce number of metabolites and databases to register algal chemodiversity not only reflects that metabolomics in microalgae is a late-comer, but additionally shows that despite developed Omics methodology (Chapter "Omics Approaches for Algal Applications"), the flood of data has to be managed by computer technology. Metabolomics analytical techniques encompass nuclear magnetic resonance (NMR), Fourier transform infrared spectroscopy (FTIR), and different separation methods (such as gas chromatography, liquid chromatography, capillary electrophoresis, and supercritical fluid chromatography) that are coupled to Mass Spectrometry. Interpretation of mass spectra is not only challenging, but also the most time-consuming task, due to the stochastic nature of the fragmentation process of metabolites (Nguyen et al. 2019). With this in mind, Nguyen et al. (2019) in their review article proposed the use of machine learning, and Lauritano et al. (2019)

Table 1 Overview of metabolomics including AI in microalgae research

Methodology	Key contributions	References
Microalgae metabolite identification using GC/MS and metabolomics statistical analysis programs as well as the NIST14 database.	Diagnose microalgal extracellular signaling molecules of cultures under stress conditions to save future bioprocesses from crashing.	Fisher et al. (2020)
Ramanome database constructed by machine learning algorithms and validation on a set of unknown species.	Approach for identification and metabolic profiling of hitherto uncultured microalgae.	Heidari Baladehi et al. (2021)
Bioinformatics approaches such as gene mining, subcellular localization, structural analysis, and physicochemical characterization.	Exploration of metabolic pathways for carotenoid biosynthesis across algal lineage.	Narang et al. (2022)
RNA-seq meta-analysis and supervised ML algorithms for data cleaning and classification.	Investigation of the salt stress response (metabolic pathways) in *Dunaliella salina* and *Dunaliella tertiolecta*.	Panahi et al. (2019)
Application of different ML algorithms to analyze a genome-scale metabolic model.	Investigation of the metabolic mechanisms of *Synechococcus* sp. PCC 7002 dealing with light intensity and salinity.	Vijayakumar and Angione (2020)
Multistep in silico analysis for selection of key enzymes and multiple alignments for generation of phylogenetic trees.	Exploitation of the biosynthetic pathways of polyphenolic compounds in microalgae.	Del Mondo et al. (2022)

focused in particular on deep learning, both of which are able to process unstructured large datasets to find the correlation between them. Raw data processing encompasses noise subtraction, chromatographic peak alignment, data normalization, and scaling. Then for analysis of metabolomics data, multivariate statistics or machine learning analysis with supervised or unsupervised algorithms are applied. Unsupervised methods can be principal component analysis (PCA), hierarchical clustering, or self-organizing maps. Supervised methods include ANOVA, discriminant function analysis (DFA), partial least squares (PLS), and support vector machines (SVM) (Shulaev and Isaac 2018).

NMR approaches provide data with high reproducibility and compatibility, so that further analysis using ML is often worthwhile. The most common regression and classification tools are PLS, as this approach enables analysis of large and highly complex datasets (Azizan et al. 2020). Parameter optimization is an important step in ML computation, as it improves its performance to avoid overfitting, thus resulting in very poor prediction for an unknown dataset. It can be done in a cross-validation (using a test- and a training set) or evaluation of model performance (using modeling data and evaluation data) (Kikuchi et al. 2018). Another widely used ML algorithm is SVM, which uses the kernel method. It performs linear and nonlinear regression/ classification, and besides NMR studies is applicable in FTIR and Raman spectroscopy. Regarding the big ML problem overfitting, it could be shown for SVM that it is applicable to metabolomics data, even when the sample size was small

(Heinemann et al. 2014). RF can deal with high-dimensional spaces and a large number of training examples, whereas the *k*-nearest neighbor algorithm considers that similar outcomes lie near each other (Corsaro et al. 2022).

Classification and identification of biotechnologically valuable algal species is another strategy for species screening where ML/DL can be successfully employed. The search for strains with better traits and capabilities to meet the specific requirements of biorefineries is proving to be extremely lengthy and in some cases unachievable without the application of AI. Heidari Baladehi et al. (2021) applied an approach for identification and metabolic profiling of hitherto uncultured microalgae, that was thought to significantly speed up the mining of microalgae and extraction of their metabolites. For the construction of a microalgae Ramanome database, that comprised >9000 cells from 27 phylogenetically diverse species, data were collected at exponential and stationary phase. The study showed that Raman spectra of pigments as well as the whole spectrum signals, when combined, are enough to classify species and growth states with 97% accuracy. Altogether, 12 machine learning algorithms were tested for ability to phylogenetic classification, from which Random Forest (RF), linear discriminant analysis (LDA), decision trees (C5.0), linear extreme gradient boosting (XGB), linear support vector machine (SVM), multilayer perceptron (MLP), and generalized linear model (GLMnet) had a better performance >80% than *k*-nearest neighbors (KNN), extreme learning machine (ELM), SVM Radial Basis Function (SVM RBF), or deep neural network (DNN). In a second step, the model was validated on a series of ML model "unknown cells" that were pure cultures but whose single-cell Raman spectra were not recorded in the database. As a third step, the model was extended to not-yet-cultured cells from environmental samples, where at the resolution of a single cell, a metabolic profile has been drawn up, followed by phylogenetic identification. In this way, two cells were identified as *Malassezia restricta*, one cell as uncultured choanoflagellate, where the fourth cell was recognized as a novel organism whose 18S rRNA did not match any known sequence.

Chong et al. (2023) studied image processing techniques coupled to ML/DL enabling classification of microalgae by morphological features as size, contours, ferret diameter, centroid, center of gravity coordinates, area, and the dissimilarity measurement. The challenges in classifying microalgae are due to the great diversity and the similarities in morphology (Hannon et al. 2010). As the classification of algae has so far been carried out with insufficient similarity to each other and too low a species number, Yadav et al. (2020) could achieve classification accuracy of 99.7% for 16 algal families, including for example Nostocaceae, Chlamydomonaceae, Euglenaceae, Scenedesmusceae, Dunaliellaceae, and Volvocaceae, that possess morphological similarities. The algorithm was a modified Convolution Neural Network (CNN), where the model has been altered by reducing kernel size and filter size among other parameters. Park et al. (2022) achieved a classification accuracy of 89.8% for distinguishing between 30 algal genera using CNN model, with the objective to reduce the interference time of the object detection process. With data generation by polarized light scattering, Zhuo et al. (2022) defined more than 10 categories out of 35 categories of microalgae with accuracy greater than 80%

using algorithms such as Linear Discriminant Analysis (LDA) and two types of SVM. A method for the automatic identification of *Scenedesmus coenobia* using microscopic image processing was developed by Giraldo-Zuluaga et al. (2018), achieving accuracy levels of 98.63% and 97.32% for SVM and ANN, respectively.

Image flow cytometry coupled with AI represents an alternative to distinguish between microalgal species; however, it was criticized that image resolution strongly depends on the volumetric throughput (Otálora et al. 2021). Nevertheless, a CNN model was successfully used to develop a deep learning-based digital holographic phase recovery method, with which image sampling of microalgae is possible without sacrificing the high-throughput operation (Göröcs et al. 2018).

4.2 ML/DL for Discovery of Metabolic Pathways

One would think that since carotenoids from microalgae are already being marketed from the beginning of the 1980s, their intracellular metabolic pathway would be known, especially as research has spent a good 20 years trying to elucidate it. However, this is not the case, as shown by Narang et al. (2022). Thus, from 403 enzymes involved in the synthesis of carotene, lutein, zeaxanthin, violaxanthin, canthaxanthin, and astaxanthin, a quantity of 85 enzymes were hypothetical proteins, whose biological functions have not been yet confirmed experimentally. Moreover, microalgae exhibit high metabolic plasticity in response to environmental influences, which means that profiles of secondary metabolites, carbohydrates, and lipids change depending on stress factors to which microalgae are exposed, including salt, light, temperature, pH, and nutrient starvation (Shi et al. 2020). Panahi et al. (2019) investigated salt stress responses in *Dunaliella salina* and *Dunaliella tertiolecta* at the level of the underlying metabolic pathways. The use of supervised ML algorithms with RNA-seq meta-analysis was performed for data cleaning, employing the algorithms, specifically the Support Vector Machine (SVM) algorithm, along with attribute selection using the Relief algorithm. The identified meta-genes were classified as species-dependent and species-independent, which verified that lipid and nitrogen metabolism, structural proteins of the photosynthesis apparatus, chaperone-mediated autophagy, and ROS-related genes are the keys of the *Dunaliella* salt stress response system.

ML is helpful, where transcriptomics alone is unable to elucidate metabolic mechanisms. Vijayakumar and Angione (2020) used different ML approaches to analyze a genome-scale metabolic model of the cyanobacteria *Synechococcus* sp. PCC 7002. Thus, Principal Component Analysis (PCA) was used to reduce data dimensionality; *k*-means clusters for growth conditions; LASSO regression eliminates redundant features, and Correlation Analysis associates OMICs predictors with growth. As a result, specific mechanisms of the cyanobacteria to cope with fluctuations in light intensity and salinity could be revealed.

The approach for solving research questions about detection of new microalgal metabolites in silico are diverse. Del Mondo et al. (2022) studied the polyphenolic compounds' biosynthetic pathways in microalgae, which were previously only known in higher plants. This knowledge was used to perform a multistep in silico analysis with a selection of key enzymes from the phenylpropanoid/flavonoid pathway described in land plants. Multiple alignments for generation of phylogenetic trees for core enzymes was applied, followed by maximum likelihood phylogenetic analysis. Other computational tools to investigate metabolite pathways were described by Passi et al. (2022). Pathway Tools is a software system that enables creating and management of pathway/genome databases (PGDBs). It offers the functions' reconstruction and viewing of PGDBs, for creating new PGDBs, for editing existing PGDBs, and running metabolic models (MetaFlux). The MetaFlux gap filler automatically identifies missing reactions, nutrients, and secretions (Karp et al. 2002). As an alternative for pathway reconstruction, Raven 2.0 MATLAB toolbox performs de novo pathway reconstruction at genome level or complete already existing pathways using template models or the database Kyoto Encyclopedia of Genes and Genomes (KEGG) for protein homology alignments (Kanehisa and Goto 2000; Wang et al. 2018).

PlantSEED is a toolbox including genome information of 39 plant and algae species for automated metabolic reconstruction from transcriptome data (Seaver et al. 2018). The annotation algorithm of PlantSEED propagates annotations using short amino acid sequences that are characteristic in particular proteins with an accuracy of about 97%. Together with the reconstruction algorithm, reconstructions without gaps and with more accurate compartmentalization are produced than in already known toolboxes. Specifically for microalgae, Romero-Losada et al. (2022) reported from the platform ALGAEFUN with MARACAS (microALGAE FUNctional enrichment tool with MicroAlgae RnA-seq and Chip-seq AnalysiS) as a tool for analysis of transcriptomic and cistromic high-throughput raw data. First, MARACAS streams as a pipeline the RNA-seq or CHIP-seq raw data and generates sets of differentially expressed genes or lists of genomic loci. Second, ALGAEFUN analyzes these sets or lists for pathway construction, using Gene Ontology or KEGG Orthology terms. If lacking these annotation systems, biological sequence analysis using profile Hidden Markov Models (HMMER) are applied. MetaCyc is a database of metabolic pathways and enzymes, currently containing 2749 pathways from 60.000 publications, thus making it the largest collection of metabolic pathways (Caspi et al. 2020). It contains data about chemical compounds, reactions, enzymes, and metabolic pathways, covering small and macromolecular metabolism, e.g., protein modification of all domains of life, thus also for microalgae. Jones et al. (2021) using CyanoMetDB reported a database for secondary metabolites from cyanobacteria that currently harbors 2010 entries and is dedicated to (1) facilitating the detection of cyanobacterial toxins and secondary metabolites, (2) identification of novel natural products, (3) research on pathways of cyanobacterial metabolites, and (4) investigation of abundance, toxicity, and persistence in natural environments.

4.3 ML/DL for Monitoring of Bioprocesses

As mentioned earlier, microalgae hold great promise as a source of biofuels, pharmaceuticals, and other valuable compounds. However, optimizing the growth parameters of microalgae is a complex and time-consuming process due to the high diversity of microalgae species and their varying responses to environmental conditions. Machine learning (ML) has emerged as a powerful tool for optimizing microalgal growth conditions and improving the efficiency of microalgae production.

Numerous studies have been conducted to optimize microalgal growth parameters using machine learning (ML). For example, Long et al. (2022) utilized artificial intelligence techniques to inform algal cultivation design. The study yielded quantitative insights into how light intensities and cell density affect light distribution patterns (LDP), and how LDPs, in turn, impact cyanobacterial growth rates. The findings demonstrated increased biomass yields in both indoor and outdoor systems with biomass productivity of 0.1 g L^{-1} $hour^{-1}$. Conventionally, growth is measured by the chlorophyll content. However, overlapping of the spectrum of other pigments, such as carotenoids can falsify the results. Tang et al. (2023) utilized linear regression (LR) and a multilayer perceptron algorithm for chlorophyll prediction. The results indicated that ANN model provides superior performance in predicting chlorophyll concentration, while the use of the red-green-blue (RGB) model can improve the accuracy of image-based analysis over the cyan-magenta-yellow-black model. The findings suggested that predictive models can offer a rapid, efficient, and cost-effective alternative to traditional methods and have implications for various fields, including biotechnology, environmental monitoring, and agriculture.

Understanding the interactions between different algal species or communities and the environment is also very important for microalgae cultivation in terms of survival traits under different biotic and abiotic conditions. Peters et al. (2018) described this so-called Eco-Metabolomics as "the application of metabolomics techniques in ecological studies to characterize biochemical mechanisms underlying interactions of organisms with the environment and with other organisms across different spatial and temporal scales.". This way, interactions can be understood as chemical cross talking or cross feeding through released metabolites in the environment and ranges from cooperation to competition, which can be species-specific or not. Related to microalgae, Fisher et al. (2020) investigated the occurrence of volatile organic compounds (VOCs) and breakdown products from carotenoids generated by the marine microalga *Microchloropsis salina* in the presence of the grazer *Brachionus plicatilis*. The authors explicitly pointed out that these results were found being important as they will aid machine learning algorithms to save algae cultures in industrial production from imminent crashes.

Moreover, computational tools are required to understand the complex datasets of metabolomic analysis from communities, from which the use of optimized-based algorithms was highlighted in the review of Daly et al. (2022). Traditionally applied ordinary differential equations (ODEs) are limited by processing the metabolism of each species involved due to the lack of kinetic information from the involved enzymes and because of the complexity of the genome-scale kinetic models. Seen as a new approach in this field, constrained-based metabolic models (CBMM) can replicate metabolomic experiments in silico and extend their applicability to metabolic pathways. The community can be investigated under different factors that may influence the metabolic network, such as temperature, pH, nutrient availability, or light (Daly et al. 2022).

Commercial cultivation of seaweeds requires rapid control of the biomass to maintain product quality, due to a short shelf life of the biomass of 3–14 days (Nayyar and Skonberg 2019). Lytou et al. (2022) used partial least square regression for microbiological quality assessment and shelf life of the brown marine alga *Alaria esculenta*. The potential of FTIR spectroscopy, e-nose, and multispectral imaging was analyzed to estimate microbial counts, where models developed using FTIR data showed best performance. Kwak et al. (2021) developed an object detection algorithm that performs superior to conventionally used CNN, due to the need for real-time inspections at a speed of one seaweed per second. With an accuracy of 95%, the developed algorithm was successfully used to detect foreign objects as screws and stones during the manufacturing process, to prevent injuries of the consumers. Cheng et al. (2022) provided a theoretical guidance for developing AI-guided bioprocess optimization and control technologies. In the examples given in the review, ML has demonstrated considerable potential in optimizing microalgal growth parameters, as well as detection of metabolites, thus resulting in enhanced production efficiency.

5 Conclusions

In summary, research for discovery of new metabolites of algae tends to focus on microalgae. In the last 5 years, there is a trend that AI subfields such as machine learning and deep learning will have significant access in metabolomics, discovery of new metabolic pathways, and structural exploration of molecules. Neural networks, RF, and SVM beside others are powerful and versatile machine learning algorithms that promise great potential in environmental metabolomics. They are useful for data classification, regression, and feature selection, and they can be applied to a wide range of spectroscopic techniques. The use of supervised and unsupervised machine learning algorithms is becoming increasingly necessary in metabolomic analysis, enabling sample classification and prediction of metabolite concentrations. Furthermore, multidisciplinary and integrated approaches seem to be a promising way forward, as coupling of AI with mathematical modeling showed satisfactory results in identifying key pathways. One cannot overestimate the importance of artificial intelligence in microalgae research, as it can enable breakthroughs through analysis and prediction that cannot be achieved in any other way.

Acknowledgments We would like to thank Conselho Nacional de Desenvolvimento Científico e Tecnológico (CNPq) for scholarships, grant #2022/03038-8, São Paulo Research Foundation (FAPESP).

References

Adam F, Abert-Vian M, Peltier G, Chemat F (2012, Jun) "Solvent-free" ultrasound-assisted extraction of lipids from fresh microalgae cells: a green, clean and scalable process. Bioresour Technol 114:457–465. https://doi.org/10.1016/j.biortech.2012.02.096

Andrich G et al (2005) Supercritical fluid extraction of bioactive lipids from the microalga Nannochloropsis sp. Eur J Lipid Sci Technol 107(6):381–386. https://doi.org/10.1002/ejlt.200501130

Azizan A, Maulidiani M, Shaari K, Ismail IS, Nagao N, Abas F (2020) Mass spectrometry-based metabolomics combined with quantitative analysis of the microalgal diatom (Chaetoceros calcitrans). Mar Drugs 18(8):403

Bernaerts K, Van Impe JF (2004) Data-driven approaches to the modelling of bioprocesses. Trans Inst Meas Control 26(5):349–372

Bisht B et al (2021) The potential of nuclear magnetic resonance (NMR) in metabolomics and lipidomics of microalgae-a review. Arch Biochem Biophys 710:108987

Brillatz T, Lauritano C, Jacmin M, Khamma S, Marcourt L, Righi D et al (2018) Zebrafish-based identification of the antiseizure nucleoside inosine from the marine diatom Skeletonema marinoi. PLoS One 13(4):e0196195

Carleo G, Cirac I, Cranmer K, Daudet L, Schuld M, Tishby N et al (2019) Machine learning and the physical sciences. Rev Modern Phys 91:045002. https://doi.org/10.1103/RevModPhys.91.045002

Caspi R, Billington R, Keseler IM, Kothari A, Krummenacker M, Midford PE, Ong WK, Paley S, Subhraveti P, Karp PD (2020, Jan 8) The MetaCyc database of metabolic pathways and enzymes-a 2019 update. Nucleic Acids Res 48(D1): D445–D453. https://doi.org/10.1093/nar/gkz862. PMID: 31586394; PMCID: PMC6943030.

Challagulla V, Nayar S, Walsh K, Fabbro L (2017, Aug) Advances in techniques for assessment of microalgal lipids. Crit Rev Biotechnol 37(5): 566–578. https://doi.org/10.1080/07388551.2016.1206058. Epub 2016 Jul 15

Chen Q et al (2023) Metabolomic analysis reveals astaxanthin biosynthesis in heterotrophic microalga Chromochloris zofingiensis. Bioresour Technol 374:128811

Cheng Y, Bi X, Xu Y, Liu Y, Li J, Du G, Lv X, Liu L (2022, Feb) Artificial intelligence technologies in bioprocess: opportunities and challenges. Bioresour Technol 369:128451. https://doi.org/10.1016/j.biotech.2022.128451. Epub 2022 Dec 9.

Chong JWR, Khoo KS, Chew KW, Vo DN, Balakrishnan D, Banat F, Munawaroh HSH, Iwamoto K, Show PL (2023, Feb) Microalgae identification: future of image processing and digital algorithm. Bioresour Technol 369:128418. https://doi.org/10.1016/j.biortech.2022.128418. Epub 2022 Dec 5

Corsaro C, Vasi S, Neri F, Mezzasalma AM, Neri G, Fazio E (2022) NMR in metabolomics: from conventional statistics to machine learning and neural network approaches. Appl Sci 12:2824. https://doi.org/10.3390/app12062824

Daly G, Ghini V, Adessi A, Fondi M, Buchan A, Viti C (2022) Towards a mechanistic understanding of microalgae–bacteria interactions: integration of metabolomic analysis and computational models. FEMS Microbiol Rev 46(5):fuac020

de Oliveira APF, Bragotto APA (2022) Microalgae-based products: food and public health. Future Foods 6:100157., ISSN 2666-8335. https://doi.org/10.1016/j.fufo.2022.100157

Del Mondo A, Sansone C, Brunet C (2022) Insights into the biosynthesis pathway of phenolic compounds in microalgae. Comput Struct Biotechnol J 20:1901–1913

Fernández FGA et al (2021) The role of microalgae in the bioeconomy. New Biotechnol 61:99–107

Fisher CL, Lane PD, Russell M, Maddalena R, Lane TW (2020) Low molecular weight volatile organic compounds indicate grazing by the marine rotifer brachionus plicatilis on the microalgae microchloropsis Salina. Meta 10(9):361

Giraldo-Zuluaga JH, Salazar A, Diez G et al (2018) Automatic identification of Scenedesmus polymorphic microalgae from microscopic images. Pattern Anal Applic 21:601–612. https://doi.org/10.1007/s10044-017-0662-3

Göröcs Z, Tamamitsu M, Bianco V et al (2018) A deep learning-enabled portable imaging flow cytometer for cost-effective, high-throughput, and label-free analysis of natural water samples. Light Sci Appl 7:66. https://doi.org/10.1038/s41377-018-0067-0

Haghi M, Thurow K, Stoll R (2017) Wearable devices in medical internet of things: scientific research and commercially available devices. Healthc Inform Res 23(1):4–15

Halaj M, Matulová M, Capek P (2022) Structural features of biologically active extracellular polysaccharide produced by green microalgae Dictyosphaerium chlorelloides. Int J Biol Macromol 214:152–161

Hannon M, Gimpel J, Tran M, Rasala B, Mayfield S (2010, Sep) Biofuels from algae: challenges and potential Biofuels 1(5):763–784. https://doi.org/10.4155/bfs.10.44. PMID: 21833344; PMCID: PMC3152439.

Hasselström L et al (2020) Socioeconomic prospects of a seaweed bioeconomy in Sweden. Sci Rep 10(1):1610

Heidari Baladehi M, Hekmatara M, He Y, Bhaskar Y, Wang Z, Liu L et al (2021) Culture-free identification and metabolic profiling of microalgal single cells via ensemble learning of ramanomes. Anal Chem 93(25):8872–8880

Heinemann J, Mazurie A, Tokmina-Lukaszewska M, Beilman GJ, Bothner B (2014) Application of support vector machines to metabolomics experiments with limited replicates. Metabolomics 10:1121–1128

Helmy M, Elhalis H, Liu Y, Chow Y, Selvarajoo K (2023) Perspective: multiomics and machine learning help unleash the alternative food potential of microalgae. Advances in Nutrition 14(1): 1–11, ISSN 2161-8313,. https://doi.org/10.1016/j.advnut.2022.11.002

Herrero M et al (2006) Optimization of the extraction of antioxidants from Dunaliella salina microalga by pressurized liquids. J Agric Food Chem 54(15):5597–5603

Hinton GE, Osindero S, Teh YW (2006) A fast learning algorithm for deep belief nets. Neural Comput 18(7):1527–1554

Ioannidou SM et al (2020) Sustainable production of bio-based chemicals and polymers via integrated biomass refining and bioprocessing in a circular bioeconomy context. Bioresour Technol 307:123093

Jones MR, Pinto E, Torres MA, Dörr F, Mazur-Marzec H, Szubert K et al (2021) CyanoMetDB, a comprehensive public database of secondary metabolites from cyanobacteria. Water Res 196: 117017

Kanehisa M, Goto S (2000) KEGG: Kyoto encyclopedia of genes and genomes. Nucleic Acids Res 28(1):27–30

Karp PD, Paley S, Romero P (2002) The pathway tools software. Bioinformatics 18(Suppl_1): S225–S232

Kawaguchi H, Ogino C, Kondo A (2017) Microbial conversion of biomass into bio-based polymers. Bioresour Technol 245:1664–1673

Kikuchi J, Ito K, Date Y (2018) Environmental metabolomics with data science for investigating ecosystem homeostasis. Prog Nucl Magn Reson Spectrosc 104:56–88., issn 0079-6565,. https://doi.org/10.1016/j.pnmrs.2017.11.003

Krizhevsky A, Sutskever I, Hinton GE (2017) Imagenet classification with deep convolutional neural networks. Commun ACM 60(6):84–90

Kwak D-H, Son G-J, Park M-K, Kim Y-D (2021) Rapid foreign object detection system on seaweed using VNIR hyperspectral imaging. Sensors 21(16):5279. https://doi.org/10.3390/s21165279

Lauritano C, Ferrante MI, Rogato A (2019) Marine natural products from microalgae: an -omics overview. Mar Drugs 17(5):269. https://doi.org/10.3390/md17050269

Lee J-Y et al (2010) Comparison of several methods for effective lipid extraction from microalgae. Bioresour Technol 101(1):S75–S77

Li L et al (2015) Screening and phylogenetic analysis of lipid-rich microalgae. Algal Res 11:381–386

Liang K, Zhang Q, Cong W (2012) Enzyme-assisted aqueous extraction of lipid from microalgae. J Agric Food Chem 60(47):11771–11776

Lim HR et al (2022) Smart microalgae farming with internet-of-things for sustainable agriculture. Biotechnol Adv 57:107931

Long B, Fischer B, Zeng Y, Amerigian Z, Li Q, Bryant H, Li M, Dai SY, Yuan JS (2022, Jan 27) Machine learning-informed and synthetic biology-enabled semi-continuous algal cultivation to unleash renewable fuel productivity. Nat Commun 13(1):541. https://doi.org/10.1038/s41467-021-27665-y. PMID: 35087023; PMCID: PMC8795378

Lu Y et al (2009) Molecular cloning and stress-dependent expression of a gene encoding Δ 12-fatty acid desaturase in the Antarctic microalga Chlorella vulgaris NJ-7. Extremophiles 13:875–884

Lytou AE, Tsakanikas P, Lymperi D, Nychas G-JE (2022) Rapid assessment of microbial quality in edible seaweeds using sensor techniques based on spectroscopy, imaging analysis and sensors mimicking human senses. Sensors 22(18):7018. https://doi.org/10.3390/s221870181

Mangoni O, Imperattore C, Tomas CR, Costantino V, Saggiomo V, Mangoni A (2011) The new carotenoid pigment moraxanthin is associated with toxic microalgae. Mar Drugs 9(2):242–255

Mak K-K, Pichika MR (2019) Artificial intelligence in drug development: present status and future prospects. Drug Discov Today 24(3):773–780

Martínez-Ruiz M, Martínez-González CA, Kim DH, Santiesteban-Romero B, Reyes-Pardo H, Villaseñor-Zepeda KR et al (2022) Microalgae bioactive compounds to topical applications products—a review. Molecules 27(11):3512

McCarthy J, Minsky ML, Rochester N, Shannon CE (2006, December) A proposal for the Dartmouth summer research project on artificial intelligence, August 31, 1955. AI Mag 27(4): 12. https://doi.org/10.1609/aimag.v27i4.1904

McCulloch WS, Pitts W (1943) A logical calculus of the ideas immanent in nervous activity. Bull Math Biophys 5:115–133

McRae MP et al (2022) Smart diagnostics: combining artificial intelligence and in vitro diagnostics. Sensors 22(17):6355

Morocho-Jácome AL, Santos BBD, Carvalho JCMD, Almeida TSD, Rijo P, Velasco MVR et al (2022) Microalgae as a sustainable, natural-oriented and vegan dermocosmetic bioactive ingredient: the case of neochloris oleoabundans. Cosmetics 9(1):9

Narang PK, Dey J, Mahapatra SR, Roy R, Kushwaha GS, Misra N et al (2022) Genome-based identification and comparative analysis of enzymes for carotenoid biosynthesis in microalgae. World J Microbiol Biotechnol 38:1–22

Nayyar D, Skonberg DI (2019) Contrasting effects of two storage temperatures on the microbial, physicochemical, and sensory properties of two fresh red seaweeds, Palmaria palmata and Gracilaria tikvahiae. J Appl Phycol 31:731–739. https://doi.org/10.1007/s10811-018-1545-8

Nguyen DH, Nguyen CH, Mamitsuka H (2019, Nov 27) Recent advances and prospects of computational methods for metabolite identification: a review with emphasis on machine learning approaches. Brief Bioinform 20(6):2028–2043. https://doi.org/10.1093/bib/bby066. PMID: 30099485; PMCID: PMC6954430

Norgård S et al (1974) Algal carotenoids and chemotaxonomy. Biochem Syst Ecol 2(1):7–9

Nwabueze TU (2010) Basic steps in adapting response surface methodology as mathematical modelling for bioprocess optimisation in the food systems. Int J Food Sci Technol 45(9): 1768–1776

Otálora P, Guzmán JL, Acién FG, Berenguel M, Reul A (2021) Microalgae classification based on machine learning techniques. Algal Res 55:102256. https://doi.org/10.1016/j.algal.2021.102256

Panahi B, Frahadian M, Dums JT, Hejazi MA (2019) Integration of cross species RNA-Seq meta-analysis and machine-learning models identifies the most important salt stress–responsive pathways in microalga Dunaliella. Front Genet 10:752

Park J, Baek J, Kim J, You K, Kim K (2022) Deep learning-based algal detection model development considering field application. Water 14(8):1275. https://doi.org/10.3390/w14081275

Pasquet V et al (2011) Study on the microalgal pigments extraction process: performance of microwave assisted extraction. Process Biochem 46(1):59–67

Passi A, Tibocha-Bonilla JD, Kumar M, Tec-Campos D, Zengler K, Zuniga C (2022) Genome-scale metabolic modeling enables in-depth understanding of big data. Meta 12(1):14

Peters K, Worrich A, Weinhold A, Alka O, Balcke G, Birkemeyer C et al (2018) Current challenges in plant eco-metabolomics. Int J Mol Sci 19(5):1385

Qian P, Zhao Z, Liu H, Wang Y, Peng Y, Hu S, Zhang J, Deng Y, Zeng Z. (2020) Multi-target deep learning for algal detection and classification. In Proceedings of the 2020 42nd Annual International Conference of the IEEE Engineering in Medicine & Biology Society (EMBC); Montreal, QC, Canada. 20–24 July 2020, pp 1954–1957. https://doi.org/10.1109/EMBC44109.2020.9176204.

Rashidi B, Dechesne A, Rydahl MG, Jorgensen B, Trindade LM (2019) Neochloris oleoabundans cell walls have an altered composition when cultivated under different growing conditions. Algal Res 40:101–482. https://doi.org/10.1016/j.algal.2019.101482

Remias D et al (2012) Characterization of an UV-and VIS-absorbing, purpurogallin-derived secondary pigment new to algae and highly abundant in M esotaenium berggrenii (Z ygnematophyceae, Chlorophyta), an extremophyte living on glaciers. FEMS Microbiol Ecol 79(3):638–648

Réveillon D et al (2019) Exploring the chemodiversity of tropical microalgae for the discovery of natural antifouling compounds. J Appl Phycol 31:319–333

Romero-Losada AB, Arvanitidou C, de Los Reyes P, García-González M, Romero-Campero FJ (2022, 2022 Mar 31) ALGAEFUN with MARACAS, “microALGAE FUNctional enrichment tool for MicroAlgae RnA-seq and Chip-seq AnalysiS.” BMC Bioinformatics 23(1):113. https://doi.org/10.1186/s12859-022-04639-5. PMID: 35361110; PMCID: PMC8973887

Rosenblatt F (1958) The perceptron: a probabilistic model for information storage and organization in the brain. Psychol Rev 65(6):386

Rowan KS (1989) Photosynthetic pigments of algae. CUP Archive

Rumelhart DE, Hinton GE, Williams RJ (1986) Learning representations by back-propagating errors. Nature 323(6088):533–536

Rumin J, Nicolau E, de Oliveira Junior RG, Fuentes-Grünewald C, Picot L (2020) Analysis of scientific research driving microalgae market opportunities in Europe. Mar Drugs 18(5):264. https://doi.org/10.3390/md18050264

Scheibe J (1972) Photoreversible pigment: occurrence in a blue-green alga. Science 176(4038):1037–1039

Schubert J et al (1994) Bioprocess optimization and control: application of hybrid modelling. J Biotechnol 35(1):51–68

Seaver SM, Lerma-Ortiz C, Conrad N, Mikaili A, Sreedasyam A, Hanson AD, Henry CS (2018) PlantSEED enables automated annotation and reconstruction of plant primary metabolism with improved compartmentalization and comparative consistency. Plant J 95(6):1102–1113

Serive B et al (2012) Selection and optimisation of a method for efficient metabolites extraction from microalgae. Bioresour Technol 124:311–320

Serive B et al (2017) Community analysis of pigment patterns from 37 microalgae strains reveals new carotenoids and porphyrins characteristic of distinct strains and taxonomic groups. PLoS One 12(2):e0171872

Shi TQ, Wang LR, Zhang ZX, Sun XM, Huang H (2020) Stresses as first-line tools for enhancing lipid and carotenoid production in microalgae. Front Bioeng Biotechnol 8:610

Shulaev V, Isaac G (2018) Supercritical fluid chromatography coupled to mass spectrometry–a metabolomics perspective. J Chromatogr B 1092:499–505

Sita K, Ricardo B, Robert W, Mac MM, Eve D, Mark H, Jenny H, Alastair M, Per-Erik M, Owen F, Liam M (2021) Mapping spatial distribution and biomass of intertidal ulva blooms using machine learning and earth observation. Front Mar Sci 8. https://www.frontiersin.org/articles/10.3389/fmars.2021.633128, ISSN=2296-7745 https://doi.org/10.3389/fmars.2021.633128

Spolaore P, Joannis-Cassan C, Duran E, Isambert A (2006) Commercial applications of microalgae. J Biosci Bioeng 101(2):87–96

Stengel DB, Connan S, Popper ZA (2011) Algal chemodiversity and bioactivity: sources of natural variability and implications for commercial application. Biotechnol Adv 29(5):483–501

Stirk WA, van Staden J (2022) Bioprospecting for bioactive compounds in microalgae: antimicrobial compounds. Biotechnol Adv 59:107977

Sudhakar MP et al (2019) A review on bioenergy and bioactive compounds from microalgae and macroalgae-sustainable energy perspective. J Clean Prod 228:1320–1333

Sushytskyi L, Lukáč P, Synytsya A, Bleha R, Rajsiglová L, Capek P et al (2020) Immunoactive polysaccharides produced by heterotrophic mutant of green microalga Parachlorella kessleri HY1 (Chlorellaceae). Carbohydr Polym 246:116588

Tang DYY, Wayne Chew K, Ting HY, Sia YH, Gentili FG, Park YK, Banat F, Culaba AB, Ma Z, Loke Show P (2023, Feb) Application of regression and artificial neural network analysis of Red-Green-Blue image components in prediction of chlorophyll content in microalgae. Bioresour Technol 370:128503. https://doi.org/10.1016/j.biortech.2022.128503. Epub 2022 Dec 16

Teng SY et al (2020) Microalgae with artificial intelligence: a digitalized perspective on genetics, systems and products. Biotechnol Adv 44:107631

Tett P, Kelly MG, Hornberger GM (1975) A method for the spectrophotometric measurement of chlorophyll a and pheophytin a in benthic microalgae. Limnol Oceanogr 20(5):887–896. https://doi.org/10.4319/lo.1975.20.5.0887

Tischer RG, Moore BG (1964) An extracellular polysaccharide produced by Palmella mucosa Kütz. Arch Mikrobiol 49:158–166

Turing AM (1950) Computing Machinery and Intellligence. Mind 59:433–460

Yu Y, Si X, Hu C, Zhang J (2019) A review of recurrent neural networks: LSTM cells and network architectures neural computation. 31(7):1235–1270. https://doi.org/10.1162/neco_a_01199

van Hal JW, Huijgen WJJ, López-Contreras AM (2014) Opportunities and challenges for seaweed in the biobased economy. Trends Biotechnol 32(5):231–233

Vijayakumar S, Angione C (2020) Protocol for hybrid flux balance, statistical, and machine learning analysis of multi-omic data from the cyanobacterium Synechococcus sp. PCC 7002. STAR Protoc 2(4):100837

Vishwakarma J, Parmar V, Vavilala SL (2019) Nitrate stress-induced bioactive sulfated polysaccharides from Chlamydomonas reinhardtii. Biomed Res J 6(1):7

Wang H, Marcišauskas S, Sánchez BJ, Domenzain I, Hermansson D, Agren R, Kerkhoven EJ (2018) RAVEN 2.0: "A versatile toolbox for metabolic network reconstruction and a case study on Streptomyces coelicolor.". PLoS Comput Biol 14(10):e1006541

Wang G et al (2020) Developing a computational framework to advance bioprocess scale-up. Trends Biotechnol 38(8):846–856

Yadav DP, . Jalal AS, Deviram Garlapati, Kaizar Hossain, Ayush Goyal, Gaurav Pant, (2020), "Deep learning-based ResNeXt model in phycological studies for future.", Algal Res, Volume 50, 102018, https://doi.org/10.1016/j.algal.2020.102018. ISSN 2211-9264

Zhuo Z, Wang H, Liao R, Ma H (2022) Machine learning powered microalgae classification by use of polarized light scattering data. Appl Sci 12:3422. https://doi.org/10.3390/app12073422

Overview of Bioprocess Engineering

Richard D. Ashby, Joseph Msanne, Mamoona Munir, Abrar Inayat, Carlo Pastore, and Ahmad Mustafa

Abstract Algal systems are gaining popularity in biosynthetic operations owing to their ability to capture and assimilate carbon into biomass through the photosynthetic processes, generate bioproducts and materials, and remove/recycle minerals from wastewater. However, their synthetic efficiency and economic outlook are dictated by growth rate, processing conditions, and the genetic capabilities associated with specific algal species. Because of their simple cultivation requirements, abundance, and ability to endure broad variations in temperature, salinity, pH, nutrients and oxygen availability, carbon source, and light intensity, algae are being adapted for use in a variety of industrial fields including both biofuel and bioproducts generation. This chapter explores algal biomass as a means for producing key value-added industrial products and describes effective processing parameters that are necessary for optimizing bioreactor efficiency leading to a more economic outlook. In addition, the chapter focuses on large-scale bioprocessing from bioreactor engineering to downstream protocols to enhance the desired product yield. Finally, the major obstacles and trends for future insights are discussed.

R. D. Ashby · J. Msanne
United States Department of Agriculture, Agricultural Research Service, Eastern Regional Research Center, Wyndmoor, PA, USA

M. Munir
Department of Botany, Rawalpindi Women University, Rawalpindi, Pakistan

A. Inayat
Department of Sustainable and Renewable Energy Engineering, University of Sharjah, Sharjah, United Arab Emirates

C. Pastore
Water Research Institute (IRSA), National Research Council (CNR), Bari, Italy

A. Mustafa (✉)
Faculty of Engineering, October University for Modern Sciences and Arts (MSA), Giza, Egypt

Center of Excellence, October University for Modern Sciences and Arts (MSA), Giza, Egypt
e-mail: ammhamed@msa.edu.eg

A. Abomohra, S. Ende (eds.), *Value-added Products from Algae*,
https://doi.org/10.1007/978-3-031-42026-9_6

Keywords Algal biomass · Microbial fermentation · Phytochemicals · Saccharification · Sustainability

1 Introduction

Process engineering is geared toward the development of specific procedures designed to synthesize, isolate, and purify desirable biological and/or chemical products from raw materials at an industrial scale in a cost-effective manner. Concerns regarding decreased fossil reserves, high carbon dioxide emissions, and the environmental persistence of many products have attracted interest toward exploring a more sustainable, eco-friendly production platform for the development of both improved protocols and generation of analogous replacement materials. An algae-based third-generation biorefinery concept seeks to create value-added products by utilizing a microbial cell factory concept incorporating whole cell biocatalysis. These efforts are supported by the abundance, carbohydrate-rich, and lignin-lacking properties of algae (Zhang et al. 2021). For an industrially relevant biobased process, it is pivotal to establish an efficient procedure whose economics are at least comparable with the petroleum industry to maximize application potential. Biomass energy is naturally stored within plants and algae as chemical energy (Munir et al. 2021b) and is a renewable, sustainable, and environment-friendly alternative to petroleum-based energy.

Biomass energy can be converted into solid, liquid, or gaseous fuels through a series of enzymatic and/or chemical conversions, and can be utilized to produce various chemical products in an effort to solve disposal problems associated with ecological persistence (Munir et al. 2023). Among the sustainable biomass, food crops have been regarded as first-generation feedstocks for the production of bioproducts and bioenergy and were initially used to produce bioethanol, biodiesel, and other biofuels (Chaudhry et al. 2022). However, producing biofuels and bioproducts directly from edible crops is costly and results in competition for food, feed, and arable land. Instead, crop straw and other agricultural waste products that contain lignocellulose were used in bioproduction strategies and were identified as second-generation feedstocks. However, the complex structure of lignocellulose led to relatively high costs and low yields during pretreatment, which limited large-scale and commercial applications (Voloshin et al. 2016). Therefore, the yield and purity of the hydrolysate have become the main issues for further development of second-generation feedstocks. In recent years, algal biomass has again been cautiously attracting attention because of the potential advantages that it provides including species abundance, rapid growth, high lipid and carbohydrate contents, lack of lignin, and limited competition for land/water (Wang et al. 2021a; Wang et al. 2021b). Therefore, algal biomass used for biofuel and bioproduct synthesis has been referred to as third-generation feedstocks with superior advantages over first and second generations.

Generally, algae can be divided into two broad groups, namely macroalgae and microalgae existing in multicellular and unicellular forms (Mostafa 2012). Macroalgae (or seaweed) are classified into three main groups including red, green, and brown algae, mainly dependent upon their natural color resulting from the presence of intracellular chlorophyll and/or other pigments. The polysaccharides present in different macroalgal species are highly variable (Beaumont et al. 2021) allowing different species to be considered as potential contributors of fermentable substrates for generation of value-added products (Lafarga et al. 2020), and for use as colloids or hydrogels (Beaumont et al. 2021). As for microalgae, only a few species (e.g., various species of *Chlorella, Scenedesmus, Botryococcus* among others) have been cultivated to extract sufficient lipids with amenable fatty acids for biodiesel production (Zhang et al. 2022). In some instances, extracellular polysaccharides are secreted by microalgae (Costa et al. 2021) under stressful physiological conditions, which can also be beneficial on their own for use as thickeners, dispersants, gelling agents, and/or bioactive molecules (Kumar et al. 2018), or can be used for biofuel conversion (Goo et al. 2013). Algal species including halophilic, halotolerant, and halo-adapted strains may be grown in marine water growth media and some of these strains have recently been researched as viable feedstocks for both biofuel (Abd El-Malek et al. 2022) and bioproduct synthesis (Cheng et al. 2022). With the renewable and sustainable properties of algal biomass, opportunities are expanding regarding their use in the production of various biochemicals, including advanced biofuels, organic acids, and biomaterials (Behera et al. 2019). However, more research is required since few studies have investigated algal biomass as microbial feedstocks to meet the growing need for value-added products at reduced costs (Siddiki et al. 2022).

Even with technological challenges and barriers, the third-generation biorefinery has its advantages and can contribute to the development of a renewable and sustainable bioeconomy (Gu et al. 2020). Further understanding of the challenges and possible solutions regarding biomass conversion and product synthesis should provide insights into balancing sustainability and economic profits. Future progress will require careful consideration toward species/feedstock selection, process optimization, as well as metabolic flux regulations to create cost-effective bioprocessing methods. Efforts to accomplish the goal of a net zero target to mitigate carbon emissions and climate change impacts, necessitate the development of sustainable processing protocols. Bioprocessing can potentially reduce energy consumption/carbon footprint and may offer advantageous attributes including the use of inexpensive renewable resources, natural production systems, and highly selective biocatalysts operating optimally under mild conditions (Ingle et al. 2022). However, with biomass productivity being far from optimal to meet the large-scale need for food, drugs, biofuels, and biobased chemicals, there has been tremendous interest in developing cost-effective bioprocesses (Behera et al. 2019).

Progress has been made in biocatalyst engineering and productivity enhancement as discussed in Chapter "Catalyst in Action". In conjunction with innovative bioreactor technologies and *in situ* product separations, bioprocesses are becoming more competitive. The prospects of bioprocess intensification are promising but there remain challenges to overcome to fully exploit this technology. Some examples of process intensification taken from the chemical industries are to increase the reaction rates and final product yields, and to significantly minimize power, and water use in biological and bio-solar processes which require enzymes (Boodhoo and Harvey 2013; Nguyen et al. 2023). Many value-added products, produced by industrial biotechnological processes can be economically valuable even with low efficiency batch manufacturing. The potential for increased efficiency using continuous bio-manufacturing is slowly developing for some bioproducts as their market penetration increases, but this is often hindered by the lack of continuous product recovery (Bux 2013).

Owing to continuous population growth, challenges to the food supply, access to clean water, public health threats, inadequate sanitation, and deteriorating air and water quality have been exacerbated. Therefore, there is a great need for engineering highly stable biocatalysts (Arevalo-Gallegos et al. 2017). Future biocatalysts must be as efficient as chemical catalysts in order to process large volumes of air, water, pollutants, and capture/recycle greenhouse gases (Sreekala et al. 2022). None of these global challenges are driven by subsist market economics, so addressing these global needs using the selectivity of biological catalysts will require a very different engineering approach. These significant global requirements will have to be addressed by intensifying the use of highly stable biological catalysts (which will be discussed in Chapter "Catalyst in Action") and developing continuous bioprocesses using novel approaches. To strengthen the production processes, conventional engineering approaches have been established in which nature has provided a wide range of metabolites with the ability to be altered through changes to the genetic code (Boodhoo et al. 2022). Using the whole range of recombinant DNA technologies (see Chapter "Omics Approaches for Algal Applications") allows not only new-to-nature products to be made, but also at new-to-nature rates and stabilities under the required industrial conditions (Sheldon and Brady 2019). Due to the big challenges of scaling-up microalgae for industrial phytochemical production, this chapter focuses on various biosynthesis, and downstream processes using microalgal systems as they pertain to cost-effective generation of algal biomaterials including biofuels and bioproducts.

2 Basics of Algal Production Bioprocesses

Many steps are required to effectively produce biofuels and bioproducts from algal biomass. The harvesting of biomass from liquid medium and extraction of the desired products can be complex and expensive processes, but these procedures are the key factors that dictate the economic viability and quality of algal products. One of the initial steps in downstream processing is harvesting to achieve solid content. This process typically can cost as much as 20–30% of overall processing expense (Karemore and Sen 2016). Constant harvesting utilizes dilute suspensions, which makes the process more expensive than harvesting terrestrial flora. Harvesting and dewatering together constitute about 90% of the total cost to produce algal biomass. Figure 1 shows a process scheme of microalgae downstream processing for possible combined biodiesel, biogas and bioethanol production.

One of the hurdles of commercializing algal biofuels revolves around their production costs. Within the past few years, a number of techno-economic and life-cycle analyses have been performed which have favorably indicated that algal biofuels are becoming more economically feasible at a relatively large scale (Kruger et al. 2022; Nodooshan et al. 2018; Quinn and Davis 2015; Wiatrowski et al. 2022), but the financial stability of biofuel production is dependent upon the synchronous processing of algal biomass to simultaneously produce additional value-added products (Sathya et al. 2023). As such, algal biorefinery offers potential for synthesizing a number of dissimilar products (as discussed in Chapter "Overview of Biorefinery Technology").

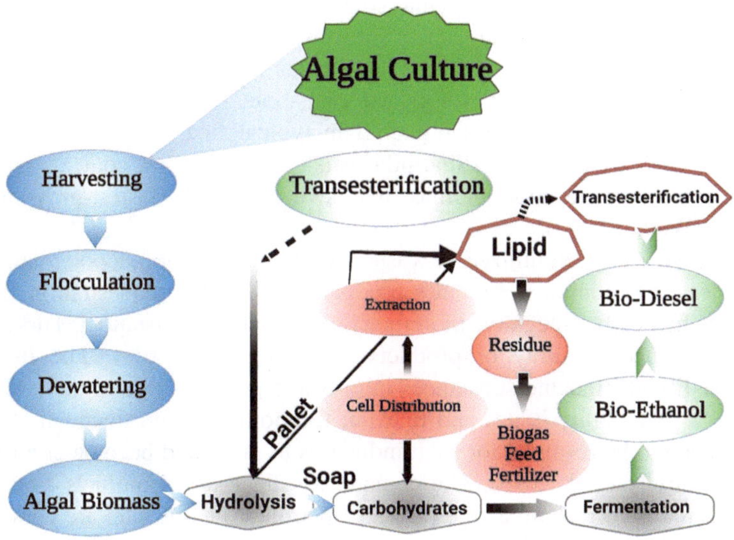

Fig. 1 Process scheme showing downstream processing of microalgal culture for combined biogas, biodiesel, and bioethanol production using microalgal culture

Modern industrialized zones across the globe face the challenges of excessive CO_2, which escalate climate change and contaminant levels, resulting in more individuals losing access to food, clean air, and arable land as the wealth gap broadens (Kumar et al. 2021). Although there is an increasing demand for energy on a broader scale, fossil fuels have limitations and a serious environmental concern (Munir et al. 2021b). The decree for reducing fossil fuel consumption has grown dramatically worldwide, since these types of fuels have potentially harmful effects on the environment (Munir et al. 2021a). In 2020, global CO_2 emissions decreased by about 5.8% (a reduction of about 2 billion tons) as COVID-19 lockdowns restricted global travel and economic activity (Le Quéré et al. 2020). Despite this reduction, the annual average CO_2 concentration in the atmosphere reached 412.5 parts per million (Sharmila et al. 2022). In 2021, demands for coal, oil, and gas increased by 4.8%, leading to a further rise in the global CO_2 emissions (Guan et al. 2023).

Process biotechnology using algal biomass can potentially play an important role in solving the global energy demand and protecting the environment (Wang et al. 2021a). Biorefineries permit the sustainable refining of biomass to generate a wide range of products that can be commercialized (Culaba et al. 2022). The diversity of products produced by algal systems makes these organisms an ideal choice for biomass and biofuel production (Wang et al. 2023). Whole algae cells may be converted into various fuel types such as biogas, liquid and gaseous transportation fuels, aviation fuel, and biohydrogen, using conversion systems including anaerobic digestion, catalytic conversion, and enzymatic or chemical transesterification. Moreover, recently developed methods of biomass extraction and product conversion may eliminate the use of organic solvents, and strong acids or bases allowing for more eco-friendly processes (Teo and Wahab 2020). Algae have the potential to reduce the dependency on fossil fuels, may act as substitutes for biofuel production, and have the capacity to reduce CO_2 emissions, as well as requirements for land and clean water. Other benefits include faster growth rate about 20–30 times the rate in plants and oil content about 30 times higher than typical first- and second-generation feedstocks on a weight basis (Sharmila et al. 2022). In that context, algae may generate about 50 times more biomass than switchgrass, the fastest-growing terrestrial plant (Walker 2009).

Algae have also been utilized in wastewater bioremediation, a potentially profitable process that combines biomass production and CO_2 sequestration (Li et al. 2022). The first carbon-negative biorefinery was launched in Istanbul, Turkey, using algae to produce an array of products for various industries. The biorefinery transformed biodegradable substances into usable products (Sharmila et al. 2022). However, sustainability issues and relevant technological and financial challenges will first need to be resolved for such industries to grow and become commercially feasible (Murphy et al. 2013). The demand for algae-based biofuels as a primary energy source is increasing as the prices of petroleum-based fuel are continuously rising (Sharmila et al. 2022).

The exponential growth in global population has fueled the demand for energy, food, water, medicine, and other essential products (Abd El-Malek et al. 2022).

Biomass and bioproducts that may be produced using a wide variety of substrates, including wastes generated from the agricultural, food, chemical, and pharmaceutical industries have recently gained more attention (Zhang et al. 2021). Environmentally friendly microbial compounds are highly attractive due to their safety and renewable nature (Jin et al. 2022). Over the past decade, remarkable progress has been made in the extraction of high-value bioenergy compounds from algae biomass (Abd El-Malek et al. 2022). However, fewer studies have investigated microbial conversion into value-added products or metabolic and pathway engineering that may be necessary for technology advancements (Zhang et al. 2021). Metabolic engineering strategies may provide a path to increase the yield and productivity of target compounds (Adegboye et al. 2021). Due to increased demands for pigments, biosurfactants, and natural plastic, several algal species have also been investigated for the production of these value-added products (Arora et al. 2021) and while increasing environmental awareness and demand for green products are becoming more compelling (Syrpas and Venskutonis 2020), the commercial feasibility of biobased products remains unfulfilled.

3 Commercial-Scale Production of Phytochemicals

Accumulation of high-energy molecules including lipid and starch is essential to identify algal species that are suitable for large-scale production processes. These molecules may be efficiently converted into biodiesel and bioethanol, respectively, which can then be readily employed into the existing transportation and energy sectors (Chen et al. 2015b; Chisti 2007; Karemore and Sen 2016; Khoo et al. 2020; Kumar et al. 2017). Although much research in recent years has been conducted on microalgae biofuel production, several strains of industrial and economic importance have also been used to manufacture bioactive compounds, supplements, nutraceuticals, and cosmetics including fatty acids, pigments, vitamins, and recombinant proteins (Callegari et al. 2020; Converti et al. 2009; Dixon and Wilken 2018). Ideal production strains should exhibit certain desirable properties and characteristics. They must be widely available, can sustain growth under variable environmental conditions (e.g., temperature and pH), and effectively utilize flue gas for high efficiency CO_2 fixation (de Alva et al. 2018; Ji et al. 2015). Currently, microalgae with diverse metabolisms are commercially used for high biomass productivity and recovery of added-value products. These organisms include members of the genera *Scenedesmus*, *Chlorella*, *Nannochloropsis*, *Dunaliella*, *Chlorococcum*, *Haematococcus*, *Botryococcus*, *Chlamydomonas*, and *Spirulina* (Chisti 2007; Dixon and Wilken 2018; Harun and Danquah 2011b; Karemore and Sen 2016).

The genus *Scenedesmus* contains some species with high lipid content, reportedly 20% to 50% of the dry biomass weight (Dasgupta et al. 2018; Yang et al. 2018), and desirable fatty acid profiles (Dasgupta et al. 2018). *Scenedesmus* is also able to accumulate large amounts of carbohydrate, which is highly beneficial for the bioethanol industry (de Alva et al. 2013; Rattanapoltee and Kaewkannetra 2014).

In addition, some strains can sustain growth under relatively low or high temperatures (14–30°C), which is an essential trait for outdoor cultivation (Xu et al. 2012). *Chlorella* strains show a tendency toward high carbohydrate and protein contents and can produce several bioactive and health products, which are important for nutraceutical and pharmaceutical industries (Ramaraj et al. 2016). The high lipid and carbohydrate content of *Chlorococcum* also makes it attractive for biofuel production (Harun et al. 2010; Karemore et al. 2013). In other instances, some species have been cultivated to produce essential fatty acids including eicosapentaenoic acid (EPA) from *Nannochloropsis*, and carotenoids including astaxanthin and β-carotene from *Haematococcus* and *Dunaliella*, respectively (Mourelle et al. 2017; Wang et al. 2015). These pigments have major health benefits and in fact, up to 30% of commercial β-carotene is produced using the halotolerant microalga *Dunaliella salina* (Borowitzka 2013; Ericksen 2016).

Green microalgae including *Chlamydomonas*, *Chlorella*, and *Dunaliella* have also been genetically engineered to produce antibodies, vaccines, and enzymes with wide industrial and pharmaceutical applications (Rasala and Mayfield 2015; Scranton et al. 2015; Yusibov et al. 2016). Additionally, cyanobacteria or blue-green algae (e.g., *Spirulina*) have been employed for the production of commercially important nitrogen storage compounds including phycocyanin, a member of the light-harvesting phycobiliprotein family of molecules, and the nitrogen-rich reserve polymer cyanophycin, with antioxidant activity and numerous other applications in health care products (Boussiba and Richmond 1980; Grossmann et al. 2020; Trautmann et al. 2016). Several microalgae strains have also been applied as bioremediation agents for wastewater and in biomass generation (Chisti 2007; Karemore and Sen 2016; Seo et al. 2015). Therefore, integrating wastewater treatment and CO_2 sequestration using microalgae allows the recycling of CO_2 into biomass *via* photosynthesis that, in turn, can be used to produce biofuels and biomaterials. In this section, the upstream microalgae cultivation and growth conditions, and the downstream processing including harvesting, disruption, and consequently the production of biofuel and value-added products are discussed.

3.1 Microalgae Cultivation Systems

An essential criterion in developing algal systems is to choose a species that can easily adapt to large-scale cultivation for industrial biomass production. Although all algal production strains are photoautotrophs, several microalgae can metabolize organic compounds such as monosaccharides (e.g., glucose) or other carbon sources (e.g., acetate), and can shift between photoautotrophic, heterotrophic, and mixotrophic growth (Rosenberg et al. 2014). Some microalgae can also directly utilize complex organic polymers derived from plant material (Patel et al. 2018). Photoautotrophic growth is usually slower than heterotrophic and mixotrophic growth since autotrophically grown microalgae cultures utilize sunlight and CO_2 to undergo photosynthesis and carbon assimilation into the organic matter (Su et al.

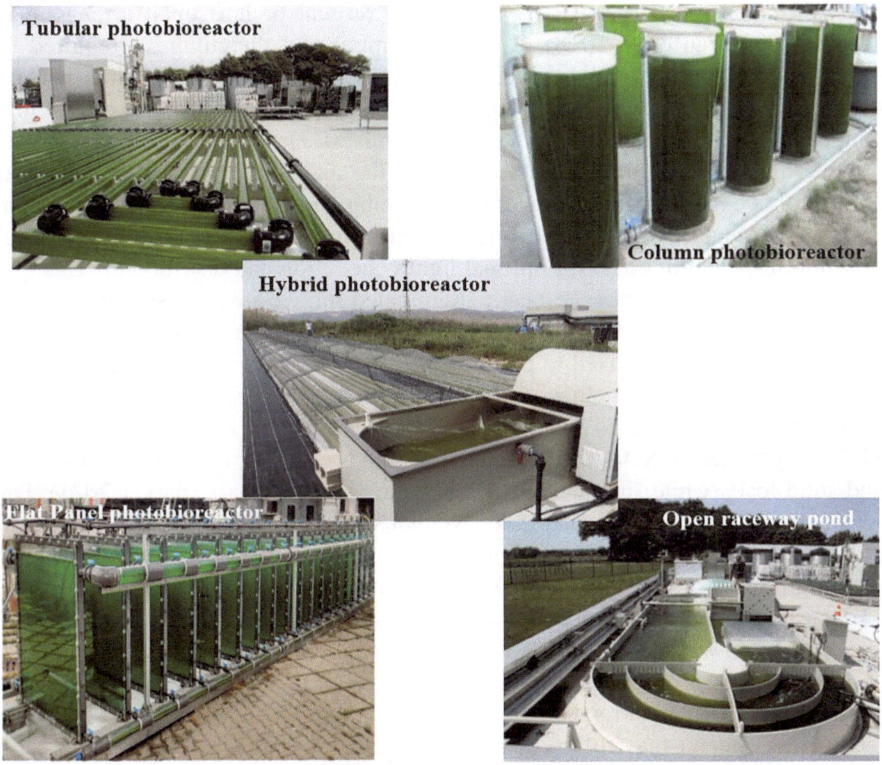

Fig. 2 Commonly used photobioreactor systems for the growth of microalgae. *Sources*: the open access references (Ahmad et al. 2021; De Vree et al. 2015); Lindblad et al. (2019) with License Number 5570200199862; and Belohlav et al. (2021) with License Number 5570190519426

2020; Vargas et al. 2018). Heterotrophic cultivation can increase cell density and growth rate, as well as biomass productivity and lipid content (Borowitzka 1999; Chen and Johns 1991; Hongjin and Guangce 2009). In this case, algae can metabolize exogenous organic compounds to produce energy and grow independent of light or to augment carbon assimilation for biomass production. These organic compounds may derive from inexpensive sources (An et al. 2003; Hongjin and Guangce 2009; Rosenberg et al. 2014; Vasudevan et al. 2012). Depending on the specific need and application, microalgae cultivation approaches utilize various types of bioreactors including closed systems (e.g., photobioreactors and fermenters), and open raceway ponds. Growth and operational parameters of various microalgae cultivation systems were comprehensively reviewed in previous reports (Abomohra et al. 2016; Ahmad et al. 2022).

As discussed in Chapter "Algae Cultivation Systems", different photobioreactor (PBR) designs are currently available for closed-system cultivation techniques. The common and variable PBRs shapes include tubular, column, and flat PBRs (Fig. 2), made from plastic or glass material. Tubular PBRs allow for effective light exposure, since they consist of transparent tubes with vertical or horizontal arrangements (Posten 2009; Tan et al. 2018). Appropriate mixing of the culture is required for optimal light distribution and reduction of cell adhesion (Huang et al. 2017). These types of PBRs are preferred in temperate climates where overheating due to high temperatures may not be a problem (Ahmad et al. 2022). Vertical column PBRs also provide microalgae with efficient culture mixing and light exposure. This type of bioreactor is easy to install and operate presenting a major advantage (Ahmad et al. 2022). Flat panel PBRs are usually rectangular in shape and made with a transparent or semitransparent material for optimal light penetration and distribution. This type of bioreactor offers microalgae a large illuminated surface area, is a simple structure, and is easily maintained (Suparmaniam et al. 2019; Wang et al. 2021b). Closed systems are typically utilized to produce pharmaceuticals, nutraceuticals, and other value-added compounds since contamination can be effectively controlled and axenic cultures are easier to maintain (Perez-Garcia et al. 2011). Furthermore, certain PBRs may be deployed for the treatment of some wastewater types with low organic and solid loads while simultaneously producing algal biomass (Sarker 2021). Different genera including *Scenedesmus*, *Nannochloropsis*, *Chlorella*, *Chlorococcum*, and *Spirulina*, characterized by their tolerance to stress conditions and/or to sparging with high levels of CO_2 have been cultivated and grown in these types of systems (Adamczyk et al. 2016; Zabed et al. 2020). Growth characteristics and biomass productivity of some representative microalgae cultivated in different systems are summarized in Table 1. Although closed PBRs require small areas to operate with minimal nutritional requirements for photosynthetic microalgae, biomass production using these systems may still be costly and time-consuming (Narala et al. 2016; Tan et al. 2018). However, algal heterotrophic growth potential using microbial fermentation can lower some of the production costs. Heterotrophic growth has been previously tested under controlled conditions using microalgae including *Scenedesmus*, *Chlorella*, *Nitzschia*, and *Tetraselmis* grown on exogenous organic compounds (Apt et al. 2011). A strain of *Scenedesmus obliquus* was experimentally used to study heterotrophic growth for biodiesel production. This green microalga can grow and increase biomass by utilizing glucose as carbon substrate (Abeliovich and Weisman 1978). *Chlorella saccharophila* and *Chlorella protothecoides* cultures can also grow heterotrophically when supplied with glucose (Miao and Wu 2006; Tan and Johns 1991b). Growth and lipid content of *Chlorella saccharophila* increased over 40% when supplemented with glucose at 2.5 g L^{-1}, compared to photoautotrophic growth (Tan and Johns 1991a).

Open raceway ponds (Fig. 2) are commonly available and widely used for large-scale commercial production of microalgae. These artificial ponds with variable depths ranging from 0.25 to 5 meters may be constructed in ground with concrete or made from plastic and equipped with paddlewheels for culture mixing (Tan et al. 2018). They require adequate amounts of water and have a widespread application in tropical environments. Microalgae and cyanobacteria with higher tolerance to environmental stress conditions including *Scenedesmus*, *Chlorella*, *Dunaliella*, *Haematococcus*, and *Spirulina* have been successfully grown in these ponds (Shekh et al. 2022). They are easy to operate and more cost-effective compared to

Table 1 Growth parameters of representative microalgae cultivated in photobioreactors (PBRs) and open raceway ponds

Reactor Type	Microalgae	Growth Media	Temp (°C)	CO_2 (%)	Volume (L)	Growth Duration (Days)	Productivity (g L^{-1} day^{-1})	Peak Biomass (g L^{-1})	References
Tubular Photobioreactor	*Haematococcus pluvialis*	BG-11	30	–	6000	10	–	1.057	Teng et al. (2023)
	Acutodesmus obliquus	Synthetic Medium	25	3	90	14	0.15	2.1	Sandmann et al. (2021)
	Chlorella pyrenoidosa	BG-11	18–40	3	60	8	0.107 ± 0.072	1.17 ± 0.58	Sukacová et al. (2021)
Vertical Column Photobioreactor	*Heterosigma akashiwo*	Seawater-F/2	21	–	10	51	0.132	1	Macías-de la Rosa et al. (2023)
	Scenedesmus acutus	Urea-based Medium	12–38	10–13	1200	120	0.1–0.2	1	Mohler et al. (2019)
	Chlorella sp.	Synthetic Medium	29	5	4	5	–	3.6	Nguyen and Hoang (2016)
Flat Panel Photobioreactor	*Chlorella sorokiniana*	BG-11	25.5–33	5	90	7	0.28–0.47	2.53	Do et al. (2022)
	Monoraphidium sp.	Synthetic Dairy wastewater	30	–	50	8	0.05	-	Kuravi and Mohan (2022)
	Chlorella pyrenoidosa	BG-11	25	3	25	8	0.199 ± 0.007	1.84 ± 0.12	Sukacová et al. (2021)
Open Raceway Ponds	*Chlorella vulgaris*	BG-11	–	29–33	4000	21	–	2.65 ± 0.3	Parashar et al. (2023)
	Chlorella pyrenoidosa	Secondary Wastewater Effluent	18–31	–	165	19	–	1.71 ± 0.04	Dahmani et al. (2016)
	Botryococcus braunii	Chu 13	23–32	–	80	18	–	2	Rao et al. (2012)

closed systems. However, open ponds have many disadvantages that influence microalgae biomass productivity, which is lower in most cases when compared to closed systems (Tan et al. 2018). For instance, monocultures are nearly impossible to achieve, since ponds are highly prone to contamination with other algae, bacteria, protozoans, and grazers (Assunçao and Malcata 2020). Environmental conditions including light and CO_2, nutrient availability, temperature, and evaporation also affect productivity (Tan et al. 2018). Adequate mixing is essential for light penetration and CO_2 mass transfer, while the presence of baffles can limit settling and formation of dark zones (Anandraj et al. 2019). Open ponds have been widely adapted for wastewater treatment, and cell density is an essential factor influencing long-term stability and cost-effective algal biomass production. Microalgae growth has been tested to evaluate the potential for wastewater decontamination and contribution of the organic compounds to biomass and lipid accumulation (Abreu et al. 2012; Ji et al. 2013). For instance, growth of *Scenedesmus sp.* was tested in municipal wastewater (McGinn et al. 2012), urban stormwater-ponds (Massimi and Kirkwood 2016), and food wastewater (Ji et al. 2015). Furthermore, enhanced lipid accumulation and fatty acid productivity in *Scenedesmus obliquus* were examined using brewery wastewater (Mata et al. 2013) and rinse water from olive oil extraction industries (Hodaifa et al. 2008).

Hybrid bioreactors that are comprised of both closed and open systems (Fig. 2) have also been deployed for microalgae growth and biomass production (Kothari et al. 2019; Tan et al. 2018). Benefits of such a system include lowering production cost and the risk for contamination while accelerating the production of carbohydrates and lipids (Tan et al. 2018). However, conventional suspended microalgae cultivation in raceway ponds and/or PBRs is energy-intensive and marked by low biomass productivity. These processes also require large amounts of water and are highly prone to contamination with undesired microorganisms.

3.2 Downstream Processing of Microalgal Biomass

Following cultivation, microalgal biomass is harvested then usually dried prior to further processing (Grossmann et al. 2020). Harvesting can be a challenging task since it typically requires the separation of cells with very small diameters (2–20 μm) from large culture volumes. The technique employed depends on the microalgae species, density, culture conditions, and target product(s), and can significantly contribute to the overall cost of downstream processing (Dixon and Wilken 2018; Karemore and Sen 2016). Concentrated cell pellets are then subject to disruption and extraction of products, followed by conversion and/or refining (Dixon and Wilken 2018). Algal biomass has been used to produce biofuel precursors (e.g., lipids and carbohydrates), which may be converted to biodiesel, bioethanol, and jet fuel (Dixon and Wilken 2018). Genera including *Scenedesmus*, *Chlorella*, *Dunaliella*, *Nannochloropsis*, and *Spirulina* offer important biofuel-producing candidates (Dixon and Wilken 2018).

3.2.1 Harvesting and Drying Methods

Harvesting is an essential first step that is required for solid/liquid separation and concentration of algal biomass with up to 20% solid content (Milledge and Heaven 2013). Common harvesting methods currently employed include centrifugation, flotation, flocculation, filtration, and gravity sedimentation (Barros et al. 2015; Sakarika and Kornaros 2019). Harvesting methods may also include collecting algal biomass following growth of biofilms on solid surfaces (Gross et al. 2013). Although centrifugation of microalgae is considered a highly efficient method for biomass recovery, it requires significant equipment and operating costs, and results in high energy consumption and CO_2 emissions (Pacheco et al. 2015). Typically, flotation and flocculation are preferred for harvesting diverse types of microalgae for biofuel production (Yin et al. 2020).

Flotation methods involve dispersed or dissolved air sparging, as well as electro-flotation that requires applying an electrical current to the culture resulting in cell aggregation and biomass accumulation on the surface (Pragya et al. 2013). Harvesting by flotation has several advantages including low costs and ability to reuse the culture media. However, there are also disadvantages such as cell disruption and contamination (Roy and Mohanty 2019). Dissolved air flotation (with 1 mg L^{-1} alum) and electro-flotation (at 24 V) have been utilized in harvesting both *Scenedesmus obliquus* and *Chlorella vulgaris* with over 91% efficiency (Koley et al. 2017). Compared to centrifugation, flocculation requires less energy to concentrate the cells using chemical additives (Chen et al. 2015b). Electro-flocculation, microbial flocculation, and auto-flocculation may also be effective methods for biomass harvesting (Salim et al. 2011). For chemical flocculation, inorganic flocculants (alum and lime) or organic molecules (surfactants and chitosan) are required to harvest microalgae by coagulating the negatively charged cells (Wan et al. 2015). This technique has been efficiently used to flocculate *Nannochloropsis* sp. (Schlesinger et al. 2012), *Chlorococcum* sp. (Karemore and Sen 2016), *Scenedesmus obliquus* (Abomohra et al. 2018). and *Chlorella* sp. (Mubarak et al. 2019). Some drawbacks using this method include costs associated with system upscale, remnants of the chemicals that may be carried in the harvested biomass, as well as environmental impacts of generated effluents. No chemical additives are required for microalgae electro-flocculation. However, cell charges are modified by application of an electric field resulting in biomass coagulation (Pearsall et al. 2011). Microbial flocculation techniques utilize bacteria, fungi, or other microorganisms as drivers for microalgae flocculation, since they can secrete extracellular polymeric substances considered as bio-flocculants that promote cell sedimentation (Barros et al. 2015; Chen et al. 2019). Auto-flocculation of microalgae may also facilitate harvesting since cells naturally respond to stress conditions (e.g., nutrient deprivation and temperature) by adhering to one another forming clumps. pH adjustments can improve microalgal auto-flocculation efficiency (Ahmad et al. 2022). For instance, high pH improved the harvesting of *Dunaliella salina* by flocculation and membrane filtration (Besson and Guiraud 2013; Monte et al. 2018). High efficiency harvesting by flocculation of *Scenedesmus obliquus* was also performed following growth in kitchen wastewater (Wang et al. 2020).

Filtration methods are often applied to further concentrate microalgae biomass and minimize the water content (Mathimani and Mallick 2018). Membrane pore size is essential for filtration efficiency, since cells or clumps may pass through membranes if pores are too large (Barros et al. 2015). Membrane-based cultivation has also been applied for simultaneous nutrients uptake and cell growth, followed by harvesting of microalgae attached as biofilm to a membrane in a specialized photobioreactor (Sun et al. 2020). Following harvesting, the drying process is essential for further concentration of the algal biomass up to ~95% solid content (Xu et al. 2011). In addition to removing extra moisture to prevent contamination and microbial growth, drying reduces shipping, handling, and storage costs (Khoo et al. 2020). It is also required to preserve the chemical integrity of the biomass components prior to further downstream processing (Khoo et al. 2020). Common techniques for drying include solar, hot air, spray, rotary, and freeze drying (Shiratake et al. 2013; Show et al. 2015). Solar drying is slow and may promote contamination by other organisms, whereas high heat generated while drying biomass with hot air or rotary drums may lead to cell disruption and degradation. On the other hand, freeze drying is fast, and reduces contamination, but is a costly process contributing to increasing the overall cost of biomass harvesting (Chen et al. 2015a).

3.2.2 Microalgae Biomass Pretreatment

Following harvesting and drying, biomass pretreatment includes cell disruption and solvent extraction. The pretreatment step is required for industrial-scale extraction of biofuel precursors, since it facilitates the rupture of cell walls, which is crucial for solvent access to lipid and carbohydrate metabolites for high productivity (Khoo et al. 2019). Currently, mechanical and nonmechanical techniques used for biomass pretreatment (Dixon and Wilken 2018) involve energy-intensive equipment and various organic solvents (Harun and Danquah 2011a). Different pretreatment methods previously applied prior to product extraction from some representative microalgae are summarized in Table 2. Mechanical methods can be used with diverse microalgae types by employing physical forces to break down the cellulose and depolymerize the cell wall components. These methods, usually are more expensive when compared to nonmechanical methods (Dahunsi 2019; Lee et al. 2012), include techniques such as milling, homogenization, ultrasonication, and pulsed electric field treatment (Dixon and Wilken 2018). Bead milling and homogenization are more common since these techniques can efficiently disrupt cells using large amounts of biomass (Onumaegbu et al. 2018). Ultrasound and pulsed electric field treatments have also been utilized for cell wall disruption to increase permeability and facilitate the extraction process. An increase in lipid recovery was observed from *Scenedesmus obliquus* biomass following ultrasound treatment (Yoo et al. 2012), and from *Chlorella pyrenoidosa* following pulsed electric field treatment (Han et al. 2019).

Table 2 Pretreatment methods applied to biomass of representative microalgae and the yield of certain target products

Pretreatment technique	Method	Microalgae	Target Product	Yield (%)	References
Mechanical	Bead Milling	*Nannochloropsis gaditana*	Proteins	50	Safi et al. (2017b)
		Tetraselmis suecica	Proteins	43.3	Postma et al. (2017)
		Neochloris oleoabundans	Proteins	47.3	
		Chlorella vulgaris	Proteins	53.1	
	High-Pressure Homogenization	*Chlorella vulgaris*	Proteins	54.01	Carullo et al. (2022)
	Ultrasonication	*Scenedesmus obliquus*	Lipids	26.6	Ido et al. (2018)
		Nannochloropsis sp.	Lipids	18.9	Koberg et al. (2011)
	Pulsed Electric Field	*Chlorella pyrenoidosa*	Lipids	13	Han et al. (2019)
		Chlorella vulgaris	Carbohydrates	24.3	Carullo et al. (2022)
Nonmechanical	Chemical	*Chlorella minutissima*	Carbohydrates	69.2	Margarites and Costa (2014)
	Thermal	*Nannochloropsis oceanica*	Lipids	24.5	Chen et al. (2016)
	Biological	*Chlorella vulgaris*	Lipids	44.3	Bai et al. (2015a)

Nonmechanical disruption methods currently used can be chemical, thermal, or biological methods, and are largely affected by the strain and target product (Dixon and Wilken 2018). Chemical methods use osmotic shock or alkaline and acid solvents combined with high temperatures to solubilize cellulose and disrupt cell walls (Velazquez-Lucio et al. 2018). However, the use of corrosive and/or toxic solvents make chemical methods less suitable for large-scale industrial applications (Khoo et al. 2020; Show et al. 2015). Thermal methods utilize high (e.g., microwaving and autoclaving) or low (e.g., freezing) temperatures to break the cells and release essential metabolites (Dixon and Wilken 2018). As per biological pretreatment methods, microalgal cells can be enzymatically hydrolyzed using specific enzymes (individual or mixtures) or microbes to facilitate the access and recovery of lipids and carbohydrates (Khoo et al. 2019). For instance, protease enzymes used under specific catalytic conditions were able to solubilize cell wall glycoproteins in *Dunaliella salina* (Pirwitz et al. 2016). Algicidal bacteria including

Bacillus thuringiensis and *Bacillus licheniformis* were effectively used to induce cell lysis in *Chlorella* sp. resulting in higher lipid yields (Bai et al. 2015b; Munoz et al. 2014). Although considered safe, biological pretreatment methods still require optimization and are not cost-effective for large-scale applications (Günerken et al. 2015; Vanthoor-Koopmans et al. 2013). Following cell disruption, various types of solvents are currently used to extract the lipid portion including organic solvents (chloroform and methanol), supercritical fluids, and green solvents (ionic liquids/polar covalent molecules) (Dixon and Wilken 2018; Khoo et al. 2020). Supercritical fluids and green solvents are more eco-friendly compared to organic solvents, they are chemically stable, nontoxic, and nonflammable compounds that are highly efficient in separating lipids from other biomass components (Abrahamsson et al. 2018; Choi et al. 2014; Khoo et al. 2019). The release of carbohydrates is usually effective following biomass pretreatment with acids at higher temperature (Karemore and Sen 2016). For bioethanol production, carbohydrates conversion in *Chlorococcum* sp. was accomplished using sulfuric acid and autoclaving that resulted in a mixture of monosaccharides comprised mostly of fermentable glucose (Karemore and Sen 2016).

3.2.3 Products and/or Whole Biomass Conversion

The last step in downstream processing of microalgae is the conversion of extracted biomass to functional products (Dixon and Wilken 2018; Karemore and Sen 2016). In some applications, biochemical conversion using specific enzymes or microbes (e.g., bacteria and yeast) may eliminate some pretreatment steps, allowing direct anaerobic digestion of biomass and fermentation (Khoo et al. 2020). Carbohydrates are converted into monomeric sugars that subsequently undergo anaerobic fermentation to produce bioethanol and biobutanol (Hernández et al. 2015; Karemore and Sen 2016). Extracted microalgal lipids are converted into fatty acid alkyl esters (typically methyl esters, FAME; or ethyl esters, FAEE) or biodiesel and other by-products (e.g., glycerol) *via* transesterification in the presence of acid or alkali catalysts (Zhu et al. 2017). Biofuel and derivatives may also be produced from whole biomass using chemical and thermochemical pathways including pyrolysis, hydrothermal liquefaction (HTL), and gasification (Dixon and Wilken 2018). Pyrolysis, a thermochemical process taking place at high temperatures, results in conversion of dry biomass to bio-oil that is further refined to functional products. During gasification, dry microalgal biomass is readily converted at elevated temperatures into syngas that may be used to produce methanol, ethylene, and propylene (Dixon and Wilken 2018). HTL also uses high temperature and pressure to readily convert harvested wet biomass (20% solid content) into bio-oil and other products (Guo et al. 2015). The latter method, capable of wet biomass processing, can reduce the energy input compared to pyrolysis which requires considerable processing and drying prior to biomass conversion.

3.2.4 Value-Added Products

Cultivation of microalgae and biomass accumulation is not only restricted to biofuel production, since other value-added products including antimicrobials, vaccines, and pigments are also produced. By-products generated following lipid extraction from algal biomass are usually rich in protein content and have been applied as animal feed, fuel, and biosorbents of heavy metals (Ahmad et al. 2022). Prior to extraction of soluble proteins from microalgae, cells are first disrupted using various mechanical methods (Grimi et al. 2014; Halim et al. 2012; Safi et al. 2014). However, high-pressure homogenization has been most efficient when compared to bead milling (Doucha and Livansky 2008; Grossmann et al. 2018), allowing the disruption of about 100% of *Chlorella protothecoides* cells and the release of protein content (Safi et al. 2017a). Protein extracts may then be further subjected to clarification by filtration or centrifugation to separate cell debris, and purification and concentration of the target molecule of interest, especially when used for nutraceutical or pharmaceutical purposes (Dixon and Wilken 2018). For instance, additional processing and purification are required for recombinant antibodies (e.g., Herpes simplex virus glycoprotein D antibody and Immunoglobulin G antibody against anthrax) and vaccines (e.g., viral protein 1-cholera toxin B vaccine) produced from *Chlamydomonas reinhardtii* biomass (Rasala and Mayfield 2015; Tran et al. 2012). Extraction of pigments from microalgal cells is achieved using organic solvents, supercritical fluids, or vegetable oils (Nobre et al. 2013). *Dunaliella salina* usually grown in closed-system PBRs is used to commercially produce β-carotene (Borowitzka and Borowitzka 1990; Kleinegris et al. 2009; Kyriakopoulou et al. 2015), whereas *Haematococcus pluvialis* grown in closed systems or open ponds is used to produce astaxanthin (Dixon and Wilken 2018). Following vegetative growth, cultures usually undergo nutrient stress to promote pigment accumulation at levels about 3% of the dry weight (Shah et al. 2016). *Nannochloropsis* species are a promising source of essential fatty acids including eicosapentaenoic acid (EPA), which can accumulate about 12% of the dry weight, depending on growth and culture conditions (Camacho-Rodríguez et al. 2013; Ma et al. 2016). *Nannochloropsis* cells grown in open ponds are harvested and dried, then subjected to pretreatment to extract EPA using organic solvents, followed by refining and purification (Adarme-Vega et al. 2012). *Spirulina*, usually grown in open raceway ponds, is a highly valuable source of phycobiliproteins and vitamins and is considered a safe food and feed supplement (Harun et al. 2010). Based on their high protein content and composition, *Spirulina* and *Chlorella* have also been used for bioplastic production (Zeller et al. 2013) and are considered a potential alternative to food-derived feedstocks.

4 Economic Viability of Biomass Conversion

Algae offer great potential as a source for biofuel and high-value products (Carriquiry et al. 2011; Cruce and Quinn 2019; Williams et al. 2009). They have fast growth rates, can utilize non-potable water, and do not compete for land or with food and other crop products (Chisti 2007; Dismukes et al. 2008; Dixon and Wilken 2018; Gouveia and Oliveira 2009; Nguyen et al. 2008; Radakovits et al. 2010). However, commercial viability has yet to be attained since large-scale algae cultivation growing photoautotrophically is still very expensive, resulting from low productivity of high-energy chemicals and high costs of downstream processing (harvesting and extraction) (Apt et al. 2011; Cao et al. 2000; Hongjin and Guangce 2009; Nguyen et al. 2008; Volkman et al. 1989). Consequently, the overall cost of microalgal biofuel is still significantly higher than the target set by the U.-S. Department of Energy at 3 USD per gallon of gasoline equivalent (gge^{-1}) (Schwab 2016) and cannot yet compete with petroleum-derived nonrenewable fuels (Benemann 2013; Chauton et al. 2015). Current estimates of the calculated costs of biodiesel production are highly variable depending on the farming system utilized to produce the algal oil for biodiesel synthesis (Nagarajan et al. 2013). Techno-economic analyses and life-cycle assessments have shown that to achieve economic viability, increases in biomass productivity and/or changes in biomass composition will be necessary (Barlow et al. 2016). Exploring cost-effective cultivation techniques that will assure the long-term commercial viability of algal-based biofuel and high-value products is highly desirable.

Some of the major bottlenecks that were cited in the literature have been biomass productivity, harvesting/dewatering, and extraction of target biomolecules (Barsanti and Gualtieri 2018; Beal et al. 2015). These factors are directly influenced by the choice of microalgae species, with a great impact on the economic feasibility of the cultivation systems. For instance, in open raceway ponds, production variability may occur due to changes in environmental conditions which can greatly impact large-scale microalgal growth and biomass accumulation (Grossmann et al. 2020). In contrast, closed systems such as PBRs may be an attractive alternative because the culture conditions are easier to control resulting in higher productivities, less contamination, and easier harvesting. However, these systems are generally accompanied by higher production costs due to high energy and labor demands (Richardson et al. 2014; Slade and Bauen 2013). A previous study highlighted the differences in total cost of algal oil production comparing open raceway ponds to PBRs using techno-economic analysis. In that study, it was shown that to achieve a 10% rate of return, the oils derived from open raceway pond cultivation would need to be sold at 8.52 USD (gge^{-1}), whereas those from the PBR biosynthetic process would require 18.10 USD gge^{-1} price tag (Davis et al. 2011). Biomass harvesting of wet algal cells followed by dewatering account for 20% to 40% of the total production cost, which is highly influenced by the harvesting method and cell density (Barsanti and Gualtieri 2018; Christenson and Sims 2012; Ríos et al. 2013). The costs of harvesting by flocculation or sedimentation are significantly lower than centrifugation (Ríos et al. 2013). Extraction of target products from microalgae is still a major constraint since disruption of extremely small microalgal cells with sizes between 2 μm and 20 μm is energy-intensive (Lee et al. 2012), whereas recovery and purification of biomolecules often require harmful organic solvents that are not environmentally friendly (Barsanti and Gualtieri 2018). One report stated that lipid extraction for biodiesel production accounts for 30% to 40% of the production costs (Kumar et al. 2019), whereas another study compared costs using various extraction methods and concluded that using enzymes for extraction of target products was cheaper than chemical compounds (Acien et al. 2012). Industrial cultivation of microalgae should ideally utilize inexpensive nutrient-rich wastewater and atmospheric CO_2 for the simultaneous production of biomass and biomaterials at lower costs. Coupling of wastewater treatment and CO_2 capture should contribute to energy security and water safety, while simultaneously reducing environmental and ecological footprints. However, studies have shown that using wastewater for microalgae cultivations and biomass production still does not significantly reduce the overall costs (Cruce and Quinn 2019).

Current economic models have shown that the large-scale coproduction of high-value products improves economic efficiency, since producing biomass for the exclusive conversion to biofuel is still not feasible (Cruce and Quinn 2019; Dixon and Wilken 2018; Soratana et al. 2014). Reports have indicated that using microalgae in a biorefinery concept to produce multiple target products improved the overall costs by about 33% (Harun et al. 2011; Karemore and Sen 2016; Laurens et al. 2015). High-value microalgae-derived specialty products (pigments and essential fatty acids) have been marketed by companies worldwide. Moreover, biochar derived from the HTL pathways, used to produce bio-oil and other products, is valuable and can improve the economics of microalgae production systems (Cruce and Quinn 2019). Although biomass production using wastewater is still not economically feasible (Cruce and Quinn 2019), the integration of microalgae biofilm-based cultivation with wastewater treatment has a great potential for cost-effective biomass production. For instance, the revolving algal biofilm (RAB) system utilizes vertically oriented and rotating conveyor belts (Gross et al. 2013; Gross et al. 2015; Gross and Wen 2014; Gross et al. 2016). Attached microalgae biofilms can perform photosynthesis with more efficient light and CO_2 utilization, while simultaneously assimilating mineral and organic elements from the nutrient-rich wastewater, since the lowest parts of the belts are submerged in the medium. In a pilot study, RAB-pond hybrid systems showed significant increases in biomass productivity, outperforming open raceway pond systems, while simultaneously decreasing harvesting costs (Gross et al. 2015; Gross and Wen 2014). Furthermore, integrating microalgae biomass production with wastewater treatment should also have positive environmental impacts by reducing eutrophication risks.

5 Limitations, Research Requirements, and Future Scope

Depleting fossil fuel reserves and ever-rising levels of atmospheric CO_2 are driving the exploration of alternative energy forms that are more sustainable and ecofriendly. Compared to other plant systems, algae permit the highest rates of photosynthetic CO_2 uptake and assimilation capacity, which are key properties for higher biomass productivity. Algae and cyanobacteria have evolved different strategies to maximize the performance of relevant carboxylation enzymes and pathways (Atkinson et al. 2016; Blankenship et al. 2011). For the long-term economic viability of algae as biofuel production feedstocks, it is important to understand the strain morphology and physiology, the effects of stress on the cell metabolism, and the regulation of carbon fixation/allocation and biosynthetic pathways of lipids and carbohydrates.

The widespread cultivation of microalgae and cyanobacteria for biomass production is currently limited by the high capital and operating costs (Cruce and Quinn 2019). Cost reduction and effective management are required at every stage during cultivation and biofuel conversion. Coupling of microalgae grown photoautotrophically with CO_2 pumped into the cultures can significantly boost biomass accumulation and productivity (Acien et al. 2012; Colling Klein et al. 2018). Moreover, cheap inorganic CO_2 can be easily acquired from thermal power plants or other industrial facilities (e.g., breweries), and streamed into PBRs under regulated conditions. Optimization of microalgae cultivations in both closed-system PBRs and open raceway ponds is also required to reduce costs. For instance, improved system design with lower energy requirements for temperature control and illumination can significantly reduce the operating costs (Cruce and Quinn 2019). Similarly, optimization of microalgae biomass flocculation as a harvesting method can potentially reduce the overall costs (Ahmad et al. 2022). The heterotrophic production of biomass may have great potential for decreasing the costs of biofuel production from microalgae and provide a sustainable alternative since several microalgae can utilize inexpensive carbon sources, including glucose, acetate, and glycerol, to produce high-value biomolecules (Morales-Sanchez et al. 2017; Rosenberg et al. 2014). In fact, microalgae grown under heterotrophic conditions can generate as much as three times biomass yield compared to photoautotrophic growth (Perez-Garcia et al. 2011). A recent study using techno-economic analysis showed the industrial-scale heterotrophic cultivation of microalgae resulted in 4.00 € kg^{-1} production cost with the potential to go as low as 1.08 € kg^{-1} (Ruiz et al. 2022). Several microalgae can also switch between photoautotrophic and mixotrophic growth by utilizing organic compounds and other essential nutrients derived from wastewater, and in turn accumulate high biomass (Khoo et al. 2020). In this type of cultivation, microalgae do not compete with food crops requiring fresh water and nutrients for growth, bringing new insights to commercial biofuel production. Biofilm cultures offer an alternative for suspended cultivations. In this method, photosynthetic microalgae, attached on the surface of a supporting material, can capture and sequester CO_2 directly from the atmosphere while simultaneously applied to wastewater treatment

(Gross et al. 2015; Gross and Wen 2014). However, more research is necessary for such technologies to become cost-effective and commercial.

Conventional established approaches to large-scale microalgae biomass and biofuel production have not yet been successful. However, several algal genomes have been sequenced and annotated as prerequisite for any future biotechnology efforts focused on developing feedstock for biomass and biofuel production (Carreres et al. 2017; Dasgupta et al. 2018). Sequenced strains have been specifically chosen for their potential as biofuel producers (e.g., *Chlorella* and *Scenedesmus*), and capacity to grow under stress conditions, particularly high temperature, high light, and high CO_2 levels (Cuaresma et al. 2009; Matsukawa et al. 2000; Sorokin 1959). The availability of complete genome sequences should provide an array of genetic and genomic tools that can be deployed for trait improvement and rerouting of metabolism toward increasing biofuel production (Dixon and Wilken 2018; Singh et al. 2016). Transcriptomic and proteomic analyses under various photoautotrophic and mixotrophic conditions have also allowed for genome-scale metabolic network reconstruction, particularly carbon assimilation and biomass production, as well as carbon partitioning and conversion into lipids (Sirikhachornkit et al. 2018). The availability of complete algal genomes and advances in the basic understanding of gene regulatory networks have allowed the identification of essential genes that function in important metabolic pathways. This may provide potential targets for microalgal strain improvement through biotechnological applications to optimize biofuel and biomaterial productions. However, metabolic engineering with individual lipid biosynthesis genes has resulted in limited increase in TAG accumulation, which emphasizes the importance of a deeper understanding of the intricate metabolic and regulatory pathways controlling biomass productivity and yields.

6 Conclusions

Algae are discussed as an adequate and sustainable source of many bio-based materials. However, commercial viability has yet to be attained since industrial production of microalgal biomass growing photoautotrophically is still expensive, resulting from low productivity of high-energy chemicals and high costs of downstream processing. Nevertheless, the availability of microalgae strains capable of simultaneously producing high-value chemicals and bioenergy molecules, coupled with advancements in cultivation and downstream processing techniques can reduce the overall costs and promote commercialization. Lastly, rational genetic and/or culture manipulations and enhanced CO_2 sequestration should optimize the algal metabolism and allow substantial improvements in microalgal productivity and certain phytochemical accumulation.

References

Abd El-Malek F, Rofeal M, Zabed HM, Nizami A-S, Rehan M, Qi X (2022) Microorganism-mediated algal biomass processing for clean products manufacturing: Current status, challenges and future outlook. Fuel 311:122612

Abeliovich A, Weisman D (1978) Role of heterotrophic nutrition in growth of the alga *Scenedesmus obliquus* in high-rate oxidation ponds. Appl Environ Microbiol 35(1):32–37

Abomohra AE-F, Jin W, Tu R, Han S-F, Eid M, Eladel H (2016) Microalgal biomass production as a sustainable feedstock for biodiesel: current status and perspectives. Renew Sust Energ Rev 64: 596–606

Abomohra AE-F, Jin W, Sagar V, Ismail GA (2018) Optimization of chemical flocculation of *Scenedesmus obliquus* grown on municipal wastewater for improved biodiesel recovery. Renew Energy 115:880–886

Abrahamsson V, Cunico LP, Andersson N, Nilsson B, Turner C (2018) Multicomponent inverse modeling of supercritical fluid extraction of carotenoids, chlorophyll A, ergosterol and lipids from microalgae. J Supercrit Fluids 139:53–61

Abreu AP, Fernandes B, Vicente AA, Teixeira J, Dragone G (2012) Mixotrophic cultivation of *Chlorella vulgaris* using industrial dairy waste as organic carbon source. Bioresour Technol 118:61–66

Acien FG, Fernandez JM, Magan JJ, Molina E (2012) Production cost of a real microalgae production plant and strategies to reduce it. Biotechnol Adv 30(6):1344–1353

Adamczyk M, Lasek J, Skawinska A (2016) CO_2 biofixation and growth kinetics of *Chlorella vulgaris* and *Nannochloropsis gaditana*. Appl Biochem Biotechnol 179(7):1248–1261

Adarme-Vega T, Lim DKY, Timmins M, Vernen F, Li Y, Schenk PM (2012) Microalgal biofactories: a promising approach towards sustainable omega-3 fatty acid production. Microb Cell Factories 11:96

Adegboye MF, Ojuederie OB, Talia PM, Babalola OO (2021) Bioprospecting of microbial strains for biofuel production: metabolic engineering, applications, and challenges. Biotechnol Biofuels 14(1):1–21

Ahmad I, Abdullah N, Koji I, Yuzir A, Muhammad SE (2021) Evolution of photobioreactors: a review based on microalgal perspective. IOP Conf Ser Mater Sci Eng 1142:012004. IOP Publishing

Ahmad S, Iqbal K, Kothari R, Singh HM, Sari A, Tyagi V (2022) A critical overview of upstream cultivation and downstream processing of algae-based biofuels: opportunity, technological barriers and future perspective. J Biotechnol 351:74–98

An JY, Sim SJ, Lee JS, Kim BW (2003) Hydrocarbon production from secondarily treated piggery wastewater by the green alga *Botryococcus braunii*. J Appl Phycol 15:185–191

Anandraj A, White S, Mutanda T (2019) Photosystem I fluorescence as a physiological indicator of hydrogen production in *Chlamydomonas reinhardtii*. Bioresour Technol 273:313–319

Apt KE, Allnutt FCT, Kyle DJ, Lippmeier JC (2011) Trophic conversion of abligate phototrophic algae through metabolic engineering, (Ed.) U.S. Patent, Vol. US 7,939,710 B1, Martek Biosciences Corporation, Columbia, MD (US), USA

Arevalo-Gallegos A, Ahmad Z, Asgher M, Parra-Saldivar R, Iqbal HM (2017) Lignocellulose: a sustainable material to produce value-added products with a zero waste approach—a review. Int J Biol Macromol 99:308–318

Arora K, Kaur P, Kumar P, Singh A, Patel SKS, Li X, Yang Y-H, Bhatia SK, Kulshrestha S (2021) Valorization of wastewater resources into biofuel and value-added products using microalgal system. Front Energy Res 9:646571

Assunçao J, Malcata FX (2020) Enclosed "non-conventional" photobioreactors for microalga production: a review. Algal Res 52:102107

Atkinson N, Feike D, Mackinder LC, Meyer MT, Griffiths H, Jonikas MC, Smith AM, McCormick AJ (2016) Introducing an algal carbon concentrating mechanism into higher plants: location and incorporation of key components. Plant Biotechnol J 14:1302–1315

Bai M-D, Chen C-Y, Lu W-C, Wan H-P, Ho S-H, Chang J-S (2015a) Enhancing the oil extraction efficiency of *Chlorella vulgaris* with cell-disruptive pretreatment using active extracellular substances from Bacillus thuringiensis ITRI-G1. Biochem Eng J 101:185–190

Bai M-D, Chen C-Y, Lu W-C, Wan H-P, Ho S-H, Chang J-S (2015b) Enhancing the oil extraction efficiency of *Chlorella vulgaris* with cell-disruptive pretreatment using active extracellular substances from *Bacillus thuringiensis* ITRI-G1. Biochem Eng 101:185–190

Barlow J, Sims RC, Quinn JC (2016) Techno-economic and life-cycle assessment of an attached growth algal biorefinery. Bioresour Technol 220:360–368

Barros AI, Gonçalves AL, Simões M, Pires JC (2015) Harvesting techniques applied to microalgae: a review. Renew Sust Energ Rev 41:1489–1500

Barsanti L, Gualtieri P (2018) Is exploitation of microalgae economically and energetically sustainable? Algal Res 31:107–115

Beal CM, Gerber LN, Sills DL, Huntley ME, Machesky SC, Walsh MJ, Tester JW, Archibald I, Granados J, Greene CH (2015) Algal biofuel production for fuels and feed in a 100-ha facility: a comprehensive techno-economic analysis and life cycle assessment. Algal Res 11:375–378

Beaumont M, Tran R, Vera G, Niedrist D, Rousset A, Pierre R, Shastri VP, Forget A (2021) Hydrogel-forming algae polysaccharides: from seaweed to biomedical applications. Biomacromolecules 22(3):1027–1052

Behera B, Acharya A, Gargey IA, Aly N, Balasubramanian P (2019) Bioprocess engineering principles of microalgal cultivation for sustainable biofuel production. Bioresour Technol Rep 5:297–316

Belohlav V, Uggetti E, García J, Jirout T, Kratky L, Díez-Montero R (2021) Assessment of hydrodynamics based on Computational Fluid Dynamics to optimize the operation of hybrid tubular photobioreactors. J Environ Chem Eng 9(5):105768

Benemann J (2013) Microalgae for biofuels and animal feeds. Energies 6:5869–5886

Besson A, Guiraud P (2013) High pH induced flocculation flotation of the hypersaline microalga *Dunaliella salina*. Bioresour Technol 147:464–470

Blankenship RE, Tiede DM, Barber J, Brudvig GW, Fleming G, Ghirardi M, Gunner MR, Junge W, Kramer DM, Melis A, Moore TA, Moser CC, Nocera DG, Nozik AJ, Ort DR, Parson WW, Prince RC, Sayre RT (2011) Comparing photosynthetic and photovoltaic efficiencies and recognizing the potential for improvement. Science 332(6031):805–809

Boodhoo K, Harvey A (2013) Process intensification technologies for green chemistry: engineering solutions for sustainable chemical processing. John Wiley & Sons, Hoboken

Boodhoo K, Flickinger M, Woodley J, Emanuelsson E (2022) Bioprocess intensification: A route to efficient and sustainable biocatalytic transformations for the future. Chem Eng Process Intensification 172:108793

Borowitzka MA (1999) Commercial production of microalgae: ponds, tanks, tubes and fermenters. J Biotechnol 70:313–321

Borowitzka MA (2013) High-value products from microalgae-their development and commercialisation. J Appl Phycol 25(3):743–756

Borowitzka LJ, Borowitzka MA (1990) Commercial production of β-carotene by *Dunaliella salina* in open ponds. Bull Mar Sci 47:244–252

Boussiba S, Richmond AE (1980) C-phycocyanin as a storage protein in the blue-green alga *Spirulina platensis*. Arch Microbiol 125(1-2):143–147

Bux F (2013) Biotechnological applications of microalgae: biodiesel and value-added products. CRC Press, Boca Raton

Callegari A, Bolognesi S, Cecconet D, Capodaglio AG (2020) Production technologies current role and future prospects of biofuels feedstocks: a state-of-theart review. Crit Renew Sustain Energy Rev 504:384–436

Camacho-Rodríguez J, Cerón-García M, González-López C, Fernández-Sevilla JM, Contreras-Gómez A, Molina-Grima E (2013) A low-cost culture medium for the production of *Nannochloropsis gaditana* biomass optimized for aquaculture. Bioresour Technol 144:57–66

Cao C, Sun S, Mai K, Liang Y (2000) Fatty acid composition and total lipid content of 30 strains of marine green algae. J Ocean Univ Qingdao 30(3):428–434

Carreres BM, de Jaeger L, Springer J, Barbosa MJ, Breuer G, van den End EJ, Kleinegris DMM, Schäffers I, Wolbert EJH, Zhang H, Lamers PP, Draaisma RB, Martins Dos Santos VA, Wijffels RH, Eggink G, Schaap PJ, Martens DE (2017) Draft genome sequence of the oleaginous green alga *Tetradesmus obliquus* UTEX 393. Genome Announc 5(3):e01449-16

Carriquiry MA, Du X, Timilsina GR (2011) Second generation biofuels: economics and policies. Energy Policy 39:4222–4234

Carullo D, Abera BD, Scognamiglio M, Donsì F, Ferrari G, Pataro G (2022) Application of pulsed electric fields and high-pressure homogenization in biorefinery cascade of *C. vulgaris* microalgae. Foods 11(3):471

Chaudhry B, Akhtar MS, Ahmad M, Munir M, Zafar M, Alhajeri NS, Ala'a H, Ahmad Z, Hasan M, Bokhari A (2022) Membrane based reactors for sustainable treatment of *Coronopus didymus* L. by developing Iodine, doped potassium oxide membrane under Dynamic conditions. Chemosphere 303:135138

Chauton MS, Reitan KI, Norsker NH, Tveterås R, Kleivdal HT (2015) A technoeconomic analysis of industrial production of marine microalgae as a source of EPA and DHA-rich raw material for aquafeed: research challenges and possibilities. Aquaculture 436:95–103

Chen F, Johns MR (1991) Effect of C/N ratio and aeration on the fatty acid composition of heterotrophic *Chlorella sorokiniana*. J Appl Phycol 3:203–209

Chen C-L, Chang J-S, Lee D-J (2015a) Dewatering and drying methods for microalgae. Dry Technol 33(4):443–454

Chen W-H, Lin B-J, Huang M-Y, Chang J-S (2015b) Thermochemical conversion of microalgal biomass into biofuels: a review. Bioresour Technol 184:314–327

Chen L, Li R, Ren X, Liu T (2016) Improved aqueous extraction of microalgal lipid by combined enzymatic and thermal lysis from wet biomass of *Nannochloropsis oceanica*. Bioresour Technol 214:138–143

Chen L, Yu Z, Xu H, Wan K, Liao Y, Ma X (2019) Microwave-assisted co-pyrolysis of *Chlorella vulgaris* and wood sawdust using different additives. Bioresour Technol 273:34–39

Cheng P, Huang J, Song X, Yao T, Jiang J, Zhou C, Yan X, Ruan R (2022) Heterotrophic and mixotrophic cultivation of microalgae to simultaneously achieve furfural wastewater treatment and lipid production. Bioresour Technol 349:126888

Chisti Y (2007) Biodiesel from microalgae beats bioethanol. Trends Biotechnol 26(3):126–131. https://doi.org/10.1016/j.tibtech.2007.12.002

Choi S-A, Lee J-S, Oh Y-K, Jeong M-J, Kim SW, Park J-Y (2014) Lipid extraction from *Chlorella vulgaris* by molten-salt/ionic-liquid mixtures. Algal Res 3:44–48

Christenson LB, Sims RC (2012) Rotating algal biofilm reactor and spool harvester for wastewater treatment with biofuels by-products. Biotechnol Bioeng 109:1674–1684

Colling Klein B, Bonomi A, Maciel Filho R (2018) Integration of microalgae production with industrial biofuel facilities: a critical review. Renew Sust Energ Rev 82:1376–1392

Converti A, Casazza AA, Ortiz EY, Perego P, Del Borghi M (2009) Effect of temperature and nitrogen concentration on the growth and lipid content of *Nannochloropsis oculata* and *Chlorella vulgaris* for biodiesel production. Chem Eng Process Process Intensif 48:1146–1151

Costa JAV, Lucas BF, Alvarenga AGP, Moreira JB, de Morais MG (2021) Microalgae polysaccharides: an overview of production, characterization, and potential applications. Polysaccharides 2(4):759–772

Cruce JR, Quinn JC (2019) Economic viability of multiple algal biorefining pathways and the impact of public policies. Appl Energy 233–234:735–746

Cuaresma M, Janssen M, Vılchez C, Wijffels RH (2009) Productivity of *Chlorella sorokiniana* in a short light-path (SLP) panel photobioreactor under high irradiance. Biotechnol Bioeng 104(2): 352–359

Culaba AB, Mayol AP, San Juan JLG, Vinoya CL, Concepcion II RS, Bandala AA, Vicerra RRP, Ubando AT, Chen W-H, Chang J-S (2022) Smart sustainable biorefineries for lignocellulosic biomass. Bioresour Technol 344:126215

Dahmani S, Zerrouki D, Ramanna L, Rawat I, Bux F (2016) Cultivation of *Chlorella pyrenoidosa* in outdoor open raceway pond using domestic wastewater as medium in arid desert region. Bioresour Technol 219:749–752

Dahunsi SO (2019) Mechanical pretreatment of lignocelluloses for enhanced biogas production: methane yield prediction from biomass structural components. Bioresour Technol 280:18–26

Dasgupta CN, Nayaka S, Toppo K, Singh AK, Deshpande U, Mohapatra A (2018) Draft genome sequence and detailed characterization of biofuel production by oleaginous microalga *Scenedesmus quadricauda* LWG002611. Biotechnol Biofuels 11:308

Davis R, Aden A, Pienkos PT (2011) Techno-economic analysis of autotrophic microalgae for fuel production. Appl Energy 88(10):3524–3531

de Alva MS, Luna-Pabello VM, Cadena E, Ortiz E (2013) Green microalga *Scenedesmus acutus* grown on municipal wastewater to couple nutrient removal with lipid accumulation for biodiesel production. Bioresour Technol 46:744–748

de Alva MS, Pabello VML, Ledesma MTO, Gomez MJC (2018) Carbon, nitrogen, and phosphorus removal, and lipid production by three saline microalgae grown in synthetic wastewater irradiated with different photon fluxes. Algal Res 34:97–103

De Vree JH, Bosma R, Janssen M, Barbosa MJ, Wijffels RH (2015) Comparison of four outdoor pilot-scale photobioreactors. Biotechnol Biofuels 8:1–12

Dismukes GC, Carrieri D, Bennette N, Ananyev GM, Posewitz MC (2008) Aquatic phototrophs: efficient alternatives to land-based crops for biofuels. Curr Opin Biotechnol 19:235–240

Dixon C, Wilken LR (2018) Green microalgae biomolecule separations and recovery. Bioresour Bioprocess 5:14

Do CVT, Dinh CT, Dang MT, Tran TD, Le TG (2022) A novel flat-panel photobioreactor for simultaneous production of lutein and carbon sequestration by Chlorella sorokiniana TH01. Bioresour Technol 345:126552

Doucha J, Livansky K (2008) Influence of processing parameters on disintegration of *Chlorella* cells in various types of homogenizers. Appl Microbiol Biotechnol 81(3):431–440

Ericksen NT (2016) Research trends in the dominating microalgal pigments, β-carotene, astaxanthin, and phycocyanin used in feed, in foods, and in health applications. J Nutr Food Sci. https://doi.org/10.4172/2155-9600.1000507

Goo BG, Baek G, Choi DJ, Park YI, Synytsya A, Bleha R, Seong DH, Lee C-G, Park JK (2013) Characterization of a renewable extracellular polysaccharide from defatted microalgae *Dunaliella tertiolecta*. Bioresour Technol 129:343–350

Gouveia L, Oliveira AC (2009) Microalgae as a raw material for biofuels production. J Ind Microbiol Biotechnol 36:269–274

Grimi N, Dubois A, Marchal L, Jubeau S, Lebovka NI, Vorobiev E (2014) Selective extraction from microalgae *Nannochloropsis* sp. using different methods of cell disruption. Bioresour Technol 153:254–259

Gross M, Wen Z (2014) Yearlong evaluation of performance and durability of a pilot-scale revolving algal biofilm (RAB) cultivation system. Bioresour Technol 171:50–58

Gross M, Henry W, Michael C, Wan Z (2013) Development of a rotating algal biofilm growth system for attached microalgae growth with in situ biomass harvest. Bioresour Technol 150: 195–201

Gross M, Mascarenhas V, Wen Z (2015) Evaluating algal growth performance and water use efficiency of pilot-scale revolving algal biofilm (RAB) culture systems. Biotechnol Bioeng 112: 2040–2050

Gross M, Zhao X, Mascarenhas V, Wen Z (2016) Effects of the surface physicochemical properties and the surface textures on the initial colonization and the attached growth in algal biofilm. Biotechnol Biofuels 9:38

Grossmann L, Ebert S, Hinrichs J, Weiss J (2018) Effect of precipitation, lyophilization, and organic solvent extraction on preparation of protein-rich powders from the microalgae *Chlorella protothecoides*. Algal Res 29:266–267

Grossmann L, Hinrichs J, Weiss J (2020) Cultivation and downstream processing of microalgae and cyanobacteria to generate protein-based technofunctional food ingredients. Crit Rev Food Sci Nutr 60(17):2961–2989

Gu X, Martinez-Fernandez J, Pang N, Fu X, Chen S (2020) Recent development of hydrothermal liquefaction for algal biorefinery. Renew Sust Energ Rev 121:109707

Guan Y, Yan J, Shan Y, Zhou Y, Hang Y, Li R, Liu Y, Liu B, Nie Q, Bruckner B (2023) Burden of the global energy price crisis on households. Nat Energy 8:304–316

Günerken E, Dhondt E, Eppink M, Garcia-Gonzalez L, Elst K, Wijffels RH (2015) Cell disruption for microalgae biorefineries. Biotechnol Adv 33:243–260

Guo Y, Yeh T, Song W, Xu D, Wang S (2015) A review of bio-oil production from hydrothermal liquefaction of algae. Renew Sust Energ Rev 48:776–790

Halim R, Harun R, Danquah MK, Webley PA (2012) Microalgal cell disruption for biofuel development. Appl Energy 91(1):116–121

Han S-F, Jin W, Yang Q, Abomohra AE-F, Zhou X, Tu R, Chen C, Xie G-J, Wang Q (2019) Application of pulse electric field pretreatment for enhancing lipid extraction from *Chlorella pyrenoidosa* grown in wastewater. Renew Energy 133:233–239

Harun R, Danquah MK (2011a) Enzymatic hydrolysis of microalgal biomass for bioethanol production. Chem Eng J 168:1079–1084

Harun R, Danquah MK (2011b) Influence of acid pre-treatment on microalgal biomass for bioerthanol production. Process Biochem 46(1):304–309

Harun R, Danquah MK, Forde GM (2010) Microalgal biomass as a fermentation feedstock for bioethanol production. J Chem Technol Biotechnol 85:199–203

Harun R, Davidson M, Doyle M, Gopiraj R, Danquah M, Forde G (2011) Technoeconomic analysis of an integrated microalgae photobioreactor, biodiesesl and biogas production facility. Biomass Bioenergy 35(1):741–747

Hernández D, Riaño B, Coca M, García-González M (2015) Saccharification of carbohydrates in microalgal biomass by physical, chemical and enzymatic pre-treatments as a previous step for bioethanol production. Chem Eng J 262:939–945

Hodaifa G, Martínez ME, Sánchez S (2008) Use of industrial wastewater from olive-oil extraction for biomass production of *Scenedesmus obliquus*. Bioresour Technol 99:1111–1117

Hongjin Q, Guangce W (2009) Effect of carbon source on growth and lipid accumulation in *Chlorella sorokiniana* GXNN01. Chin J Oceanol Limnol 27(4):762–768

Huang Q, Jiang F, Wang L, Yang C (2017) Design of photobioreactors for mass cultivation of photosynthetic organisms. Engineering 3(3):318–329

Ido AL, de Luna MDG, Capareda SC, Maglinao Jr AL, Nam H (2018) Application of central composite design in the optimization of lipid yield from Scenedesmus obliquus microalgae by ultrasound-assisted solvent extraction. Energy 157:949–956

Ingle A, Antunes F, Paula A, Flumignan D, Terán-Hilares R, Philippini R, Martiniano S, Abdeshahian P, Hernandez-Perez A, Silva G (2022) Nanotechnological interventions for sustainable production of microbial biofuel and bioenergy. In: Microbial biotechnology for renewable and sustainable energy. Springer, Berlin, pp 191–226

Ji M-K, Abou-Shanab RAI, Hwang JH, Timmes TC, Kim HC, Oh YK, Jeon B-H (2013) Removal of nitrogen and phosphorus from piggery wastewater effluent using the green microalga *Scenedesmus obliquus*. J Environ Eng 139:1198–1205

Ji M-K, Yun H-S, Park Y-T, Kabra AN, Oh I-H, Choi J (2015) Mixotrophic cultivation of a microalga *Scenedesmus obliquus* in municipal wastewater supplemented with food wastewater and flue gas CO_2 for biomass production. J Environ Manag 159:115–120

Jin K, Jin C, Wu Y (2022) Synthetic biology-powered microbial co-culture strategy and application of bacterial cellulose-based composite materials. Carbohydr Polym 119171:283

Karemore A, Sen R (2016) Downstream processing of microalgal feedstock for lipid and carbohydrate in a biorefinery concept: a holistic approach for biofuel applications. RSC Adv 6:29486

Karemore A, Pal R, Sen R (2013) Strategic enhancement of algal biomass and lipid in *Chlorococcum infusionum* as bioenergy feedstock. Algal Res 2(2):113–121

Khoo KS, Chew KW, Ooi CW, Ong HC, Ling TC, Show PL (2019) Extraction of natural astaxanthin from *Haematococcus pluvialis* using liquid biphasic flotation system. Bioresour Technol 290:121794

Khoo KS, Chew KW, Yew GY, Leong WH, Chai YH, Show PL, Chen W-H (2020) Recent advances in downstream processing of microalgae lipid recovery for biofuel production. Bioresour Technol 304:122996

Kleinegris DMM, Janssen M, Brandenburg WA, Wijffels RH (2009) The selectivity of milking of *Dunaliella salina*. Mar Biotechnol 12:14–23

Koberg M, Cohen M, Ben-Amotz A, Gedanken A (2011) Bio-diesel production directly from the microalgae biomass of *Nannochloropsis* by microwave and ultrasound radiation. Bioresour Technol 102(5):4265–4269

Koley S, Prasad S, Bagchi SK, Mallick N (2017) Development of a harvesting technique for large-scale microalgal harvesting for biodiesel production. RSC Adv 7(12):7227–7237

Kothari R, Ahmad S, Pathak VV, Pandey A, Kumar A, Shankarayan R, Paul NB, Tyagi VV (2019) Algal-based biofuel generation through flue gas and wastewater utilization: a sustainable prospective approach. Biomass Convers Bioref 11(4):1419–1442

Kruger JS, Wiatrowski M, Davis RE, Dong T, Knoshaug EP, Nagle NJ, Laurens LM, Pienkos PT (2022) Enabling production of algal biofuels by techno-economic optimization of co-product suites. Front Chem Eng 3:83

Kumar G, Shobana S, Chen W-H, Bach Q-V, Kim S-H, Atabani A, Chang J-S (2017) A review of thermochemical conversion of microalgal biomass for biofuels: chemistry and processes. Green Chem 19:44–67

Kumar A, Patel DP, Patel GK, Singh D, Prasad AK, Sarva K (2018) Establishment of rice husk by-product as pharmaceutical excipients. World J Pharmaceu Res 7(7):1790–1821

Kumar V, Arora N, Nanda M, Pruthi V (2019) Different cell disruption and lipid extraction methods from microalgae for biodiesel production. In: Microalgae biotechnology for development of biofuel and wastewater treatment. Springer, Singapore, pp 265–292

Kumar BR, Mathimani T, Sudhakar M, Rajendran K, Nizami A-S, Brindhadevi K, Pugazhendhi A (2021) A state of the art review on the cultivation of algae for energy and other valuable products: application, challenges, and opportunities. Renew Sust Energ Rev 138:110649

Kuravi SD, Mohan SV (2022) Mixotrophic cultivation of Monoraphidium sp. In dairy wastewater using Flat-Panel photobioreactor and photosynthetic performance. Bioresour Technol 348: 126671

Kyriakopoulou K, Papadaki S, Krokida M (2015) Life cycle analysis of β-carotene extraction techniques. J Food Eng 167:51–58

Lafarga T, Acién-Fernández FG, Garcia-Vaquero M (2020) Bioactive peptides and carbohydrates from seaweed for food applications: Natural occurrence, isolation, purification, and identification. Algal Res 48:101909

Laurens LML, Nagle N, Davis R, Sweeney N, Van Wychen S, Lowell A, Pienkos PT (2015) Acid-catalyzed algal biomass pretreatment for integrated lipid and carbohydrate-based biofuels production. Green Chem 17(2):1145–1158

Le Quéré C, Jackson RB, Jones MW, Smith AJ, Abernethy S, Andrew RM, De-Gol AJ, Willis DR, Shan Y, Canadell JG (2020) Temporary reduction in daily global CO2 emissions during the COVID-19 forced confinement. Nat Clim Chang 10(7):647–653

Lee AK, Lewis DM, Ashman PJ (2012) Disruption of microalgal cells for the extraction of lipids for biofuels: processes and specific energy requirements. Biomass Bioenergy 46:89–101

Li S, Li X, Ho S-H (2022) Microalgae as a solution of third world energy crisis for biofuels production from wastewater toward carbon neutrality: an updated review. Chemosphere 291: 132863

Lindblad P, Fuente D, Borbe F, Cicchi B, Conejero JA, Couto N, Čelešnik H, Diano MM, Dolinar M, Esposito S (2019) CyanoFactory, a European consortium to develop technologies needed to advance cyanobacteria as chassis for production of chemicals and fuels. Algal Res 41: 101510

Ma X-N, Chen T-P, Yang B, Li J, Chen F (2016) Lipid production from *Nannochloropsis*. Mar Drugs 14:61

Macías-de la Rosa A, González-Cardoso MÁ, Cerón-García MdC, López-Rosales L, Gallardo-Rodríguez JJ, Seoane S, Sánchez-Mirón A, García-Camacho F (2023) Bioactives overproduction through operational strategies in the ichthyotoxic microalga *Heterosigma akashiwo* culture. Toxins 15(5):349

Margarites ACF, Costa JAV (2014) Increment of carbohydrate concentration of *Chlorella minutissima* microalgae for bioethanol production. J Eng Res Appl 4(2):80–86

Massimi R, Kirkwood AE (2016) Screening microalgae isolated from urban storm- and wastewater systems as feedstock for biofuel. PeerJ 4:e2396

Mata TM, Meloa AC, Meireles S, Mendes AM, Martins AA, Caetano NS (2013) Potential of microalgae *Scenedesmus obliquus* grown in brewery wastewater for biodiesel production. Chem Eng Trans 32:901–906

Mathimani T, Mallick N (2018) A comprehensive review on harvesting of microalgae for biodiesel-Key challenges and future directions. Renew Sust Energ Rev 91:1103–1120

Matsukawa R, Hotta M, Masuda Y, Chihara M, Karube I (2000) Antioxidants from carbon dioxide fixing *Chlorella sorokiniana*. J Appl Phycol 12:263–267

McGinn PJ, Dickinson KE, Park KC, Whitney CG, MacQuarrie SP, Black FJ, Frigon J-C, Guiot SR, O'Leary SJB (2012) Assessment of the bioenergy and bioremediation potentials of the microalga *Scenedesmus* sp. AMDD cultivated in municipal wastewater effluent in batch and continuous mode. Algal Res 1:155–165

Miao X, Wu Q (2006) Biodiesel production from heterotrophic microalgal oil. Bioresour Technol 97(6):841–846

Milledge JJ, Heaven S (2013) A review of the harvesting of micro-algae for biofuel production. Rev Environ Sci Biol 12(2):165–178

Mohler DT, Wilson MH, Fan Z, Groppo JG, Crocker M (2019) Beneficial reuse of industrial CO2 emissions using a microalgae photobioreactor: waste heat utilization assessment. Energies 12(13):2634

Monte J, Sá M, Galinha CF, Costa L, Hoekstra H, Brazinha C, Crespo JG (2018) Harvesting of *Dunaliella salina* by membrane filtration at pilot scale. Sep Purif Technol 190:252–260

Morales-Sanchez D, Martinez-Rodriguez OA, Martinez A (2017) Heterotrophic cultivation of microalgae: production of metabolites of commercial interest. J Chem Technol Biotechnol 92(5):925–936

Mostafa SS (2012) Microalgal biotechnology: prospects and applications. Plant Sci 12:276–314

Mourelle M, Gómez C, Legido J (2017) The potential use of marine microalgae and cyanobacteria in cosmetics and thalassotherapy. Cosmetics 4:46

Mubarak M, Shaija A, Suchithra TV (2019) Flocculation: an effective way to harvest microalgae for biodiesel production. J Environ Chem Eng 7(4):103221

Munir M, Ahmad M, Mubashir M, Asif S, Waseem A, Mukhtar A, Saqib S, Munawaroh HSH, Lam MK, Khoo KS (2021a) A practical approach for synthesis of biodiesel via non-edible seeds oils using trimetallic based montmorillonite nano-catalyst. Bioresour Technol 328:124859

Munir M, Ahmad M, Saeed M, Waseem A, Nizami A-S, Sultana S, Zafar M, Rehan M, Srinivasan GR, Ali AM (2021b) Biodiesel production from novel non-edible caper (*Capparis spinosa* L.) seeds oil employing Cu–Ni doped ZrO2 catalyst. Renew Sust Energ Rev 138:110558

Munir M, Saeed M, Ahmad M, Waseem A, Alsaady M, Asif S, Ahmed A, Khan MS, Bokhari A, Mubashir M (2023) Cleaner production of biodiesel from novel non-edible seed oil (*Carthamus lanatus* L.) via highly reactive and recyclable green nano CoWO3@ rGO composite in context of green energy adaptation. Fuel 332:126265

Munoz C, Hidalgo C, Zapata M, Jeison D, Riquelme C, Rivas M (2014) Use of cellulolytic marine bacteria for enzymatic pretreatment in microalgal biogas production. Appl Environ Microbiol 80:4199–4206

Murphy F, Devlin G, Deverell R, McDonnell K (2013) Biofuel production in Ireland—an approach to 2020 targets with a focus on algal biomass. Energies 6(12):6391–6412

Nagarajan S, Chou SK, Cao S, Wu C, Zhou Z (2013) An updated comprehensive techno-economic analysis of algal biodiesel. Bioresour Technol 145:150–156

Narala RR, Garg S, Sharma KK, Thomas-Hall SR, Deme M, Li Y, Schenk PM (2016) Comparison of microalgae cultivation in photobioreactor, open raceway pond, and a two-stage hybrid system. Front Energy Res 4:29. https://doi.org/10.3389/fenrg.2016.00029

Nguyen MA, Hoang AL (2016) A review on microalgae and cyanobacteria in biofuel production. USTH, Hanoi, Vietnam, pp 1–37

Nguyen AV, Thomas-Hall SR, Malnoë A, Timmins M, Mussgnug JH, Rupprecht J, Kruse O, Hankamer B, Schenk PM (2008) Transcriptome for photobiological hydrogen production induced by sulfur deprivation in the green alga *Chlamydomonas reinhardtii*. Eukaryot Cell 7(11):1965–1979

Nguyen LN, Vu MT, Vu HP, Johir MAH, Labeeuw L, Ralph PJ, Mahlia T, Pandey A, Sirohi R, Nghiem LD (2023) Microalgae-based carbon capture and utilization: A critical review on current system developments and biomass utilization. Crit Rev Environ Sci Technol 53(2): 216–238

Nobre BP, Villalobos F, Barragán BE, Oliveira AC, Batista AP, Marques P, Mendes RL, Sovová H, Palavra AF, Gouveia L (2013) A biorefinery from *Nannochloropsis* sp. microalga-extraction of oils and pigments. Production of biohydrogen from the leftover biomass. Bioresour Technol 135:128–136

Nodooshan KG, Moraga RJ, Chen S-JG, Nguyen C, Wang Z, Mohseni S (2018) Environmental and economic optimization of algal biofuel supply chain with multiple technological pathways. Ind Eng Chem Res 57(20):6910–6925

Onumaegbu C, Mooney J, Alaswad A, Olabi A (2018) Pre-treatment methods for production of biofuel from microalgae biomass. Renew Sust Energ Rev 93:16–26

Pacheco MM, Hoeltz M, Moraes MS, Schneider RC (2015) Microalgae: cultivation techniques and wastewater phycoremediation. J Environ Sci Health Part A 50(6):585–601

Parashar A, Shah N, Rane M, Shastri Y (2023) Biogas-assisted growth of *Chlorella vulgaris* in an open raceway pond: proof of concept and economic assessment. Chem Eng Technol. https://doi.org/10.1002/ceat.202200206

Patel A, Matsakas L, Rova U, Christakopoulos P (2018) Heterotrophic cultivation of *Auxenochlorella protothecoides* using forest biomass as a feedstock for sustainable biodiesel production. Biotechnol Biofuels Bioprod 11:169

Pearsall R, Connelly R, Fountain M, Hearn CS, Werst MD, Hebner RE, Kelley EF (2011) Electrically dewatering microalgae. IEEE Trans Dielectr Electr Insul 18:1578–1583

Perez-Garcia O, Escalante FM, De-Bashan LE, Bashan Y (2011) Heterotrophic cultures of microalgae: metabolism and potential products. Water Res 45(1):11–36

Pirwitz K, Rihko-Struckmann L, Sundmacher K (2016) Valorization of the aqueous phase obtained from hydrothermally treated *Dunaliella salina* remnant biomass. Bioresour Technol 219:64–71

Posten C (2009) Design principles of photo-bioreactors for cultivation of microalgae. Eng Life Sci 9(3):165–177

Postma P, Suarez-Garcia E, Safi C, Yonathan K, Olivieri G, Barbosa M, Wijffels RH, Eppink M (2017) Energy efficient bead milling of microalgae: Effect of bead size on disintegration and release of proteins and carbohydrates. Bioresour Technol 224:670–679

Pragya N, Pandey KK, Sahoo P (2013) A review on harvesting, oil extraction and biofuels production technologies from microalgae. Renew Sust Energ Rev 24:159–171

Quinn JC, Davis R (2015) The potentials and challenges of algae based biofuels: a review of the techno-economic, life cycle, and resource assessment modeling. Bioresour Technol 184:444–452

Radakovits R, Jinkerson RE, Darzins A, Posewitz MC (2010) Biofuels from Eukaryotic Microalgae. Eukaryot Cell. https://doi.org/10.1128/EC.00364-09

Ramaraj R, Unpaprom Y, Dussadee N (2016) Cultivation of green microalga, *Chlorella vulgaris* for biogas purification. Int J New Technol Res 2:117–122

Rao AR, Ravishankar G, Sarada R (2012) Cultivation of green alga *Botryococcus braunii* in raceway, circular ponds under outdoor conditions and its growth, hydrocarbon production. Bioresour Technol 123:528–533

Rasala BA, Mayfield SP (2015) Photosynthetic biomanufacturing in green algae; production of recombinant proteins for industrial, nutritional, and medical uses. Photosynth Res 123:227–239

Rattanapoltee P, Kaewkannetra P (2014) Utilization of agricultural residues of pineapple peels and sugarcane bagasse as cost-saving raw materials in *Scenedesmus acutus* for lipid accumulation and biodiesel production. Appl Biochem Biotechnol 173:1495–1510

Richardson JW, Johnson MD, Zhang X, Zemke P, Chen W, Hu Q (2014) A financial assessment of two alternative cultivation systems and their contributions to algae biofuel economic viability. Algal Res 4:96–104

Ríos SD, Torres CM, Torras C, Salvado J, Mateo-Sanz JM, Jimenez L (2013) Microalgae-based biodiesel: economic analysis of downstream process realistic scenarios. Bioresour Technol 136: 617–625

Rosenberg JN, Kobayashi N, Barnes A, Noel EA, Betenbaugh MJ, Oyler GA (2014) Comparative analyses of three *Chlorella* species in response to light and sugar reveal distinctive lipid accumulation patterns in the microalga *C. sorokiniana*. PLoS One 9(4):e92460

Roy M, Mohanty K (2019) A comprehensive review on microalgal harvesting strategies: Current status and future prospects. Algal Res 44:101683

Ruiz J, Wijffels RH, Dominguez M, Barbosa MJ (2022) Heterotrophic vs autotrophic production of microalgae: Bringing some light into the everlasting cost controversy. Algal Res 64:102698

Safi C, Ursu AV, Laroche C, Zebib B, Merah O, Pontalier P-Y, Vaca-Garcia C (2014) Aqueous extraction of proteins from microalgae: Effect of different cell disruption methods. Algal Res 3: 61–65

Safi C, Olivieri G, Campos RP, Engelen-Smit N, Mulder WJ, van den Broek LAM, Sijtsma L (2017a) Biorefinery of microalgal soluble proteins by sequential processing and membrane filtration. Bioresour Technol 225:151–158

Safi C, Rodriguez LC, Mulder W, Engelen-Smit N, Spekking W, Van den Broek L, Olivieri G, Sijtsma L (2017b) Energy consumption and water-soluble protein release by cell wall disruption of Nannochloropsis gaditana. Bioresour Technol 239:204–210

Sakarika M, Kornaros M (2019) *Chlorella vulgaris* as a green biofuel factory: comparison between biodiesel, biogas and combustible biomass production. Bioresour Technol 273:237–243

Salim S, Bosma R, Vermue MH, Wijffels RH (2011) Harvesting of microalgae by bio-flocculation. J Appl Phycol 23:849–855

Sandmann M, Smetana S, Heinz V, Rohn S (2021) Comparative life cycle assessment of a mesh ultra-thin layer photobioreactor and a tubular glass photobioreactor for the production of bioactive algae extracts. Bioresour Technol 340:125657

Sarker NK (2021) Exploring the potential of wastewater reclamation by means of outdoor cultivation of microalgae in photobioreactors. Energy Ecol Environ 7:473

Sathya AB, Thirunavukkarasu A, Nithya R, Nandan A, Sakthishobana K, Kola AK, Sivashankar R, Tuan HA, Deepanraj B (2023) Microalgal biofuel production: potential challenges and prospective research. Fuel 332:126199

Schlesinger A, Eisenstadt D, Bar-Gil A, Carmely H, Einbinder S, Gressel J (2012) Inexpensive non-toxic flocculation of microalgae contradicts theories; overcoming a major hurdle to bulk algal production. *Biotechnol Adv* 30:1023–1030

Schwab A (2016) Bioenergy technologies office multi-year program plan. (Ed.) D.o. Energy, Bioenergy Technologies Office, Washington, DC.

Scranton MA, Ostrand JT, Fields FJ, Mayfield SP (2015) *Chlamydomonas* as a model for biofuels and bio-products production. Plant J 82:523–531

Seo YH, Sung M, Kim B, Oh Y-K, Kim DK, Han J-I (2015) Ferric chloride based downstream process for microalgae based biodiesel production. Bioresour Technol 181:143–147

Shah MMR, Liang Y, Cheng JJ, Daroch M (2016) Astaxanthin-producing green microalga *Haematococcus pluvialis*: from single cell to high value commercial products. Front Plant Sci 7:531

Sharmila VG, Banu JR, Kumar MD, Kumar SA, Kumar G (2022) Algal biorefinery towards decarbonization: Economic and environmental consideration. Bioresour Technol 364:128103

Shekh A, Sharma A, Schenk PM, Kumar G, Mudliar S (2022) Microalgae cultivation: photobioreactors, CO_2 utilization, and value-added products of industrial importance. J Chem Technol Biotechnol 97(5):1064–1085

Sheldon RA, Brady D (2019) Broadening the scope of biocatalysis in sustainable organic synthesis. ChemSusChem 12(13):2859–2881

Shiratake T, Sato A, Minoda A, Tsuzuki M, Sato N (2013) Air-drying of cells, the novel conditions for stimulated synthesis of triacylglycerol in a green alga, *Chlorella kessleri*. PLoS One. https://doi.org/10.1371/journal.pone.0079630

Show K-Y, Lee D-J, Tay J-H, Lee T-M, Chang J-S (2015) Microalgal drying and cell disruption-recent advances. Bioresour Technol 184:258–266

Siddiki SYA, Mofijur M, Kumar PS, Ahmed SF, Inayat A, Kusumo F, Badruddin IA, Khan TY, Nghiem L, Ong HC (2022) Microalgae biomass as a sustainable source for biofuel, biochemical and biobased value-added products: An integrated biorefinery concept. Fuel 307:121782

Singh P, Baranwal M, Reddy SM (2016) Antioxidant and cytotoxic activity of carotenes produced by *Dunaliella salina* under stress. Pharm Biol 54(10):22269–22275

Sirikhachornkit A, Suttangkakul A, Vuttipongchaikij S, Juntawong P (2018) *De novo* transcriptome analysis and gene expression profiling of an oleaginous microalga *Scenedesmus acutus* TISTR8540 during nitrogen deprivation-induced lipid accumulation. Sci Rep 8:3668

Slade R, Bauen A (2013) Micro-algae cultivation for biofuels: Cost, energy balance, environmental impacts and future prospects. Biomass Bioenergy 53:29–38

Soratana K, Barr WJ, Landis AE (2014) Effects of co-products on the life-cycle impacts of microalgal biodiesel. Bioresour Technol 159:157–166

Sorokin C (1959) Tabular comparative data for the low-temperature and high-temperature strains of *Chlorella*. Nature 184:613–614

Sreekala AGV, Ismail MHB, Nathan VK (2022) Biotechnological interventions in food waste treatment for obtaining value-added compounds to combat pollution. Environ Sci Pollut Res 29(42):62755–62784

Su M, D'Imporzano G, Veronesi D, Afric S, Adani F (2020) *Phaeodactylum tricornutum* cultivation under mixotrophic conditions with glycerol supplied with ultrafiltered digestate: a simple biorefinery approach recovering C and N. J Biotechnol 323:73–81

Sukacová K, Lošák P, Brummer V, Máša V, Vícha D, Zavrel T (2021) Perspective design of algae photobioreactor for greenhouses—a comparative study. Energies 14:1338

Sun Y, Duan D, Chang H, Guo C (2020) Optimizing light distributions in a membrane photobioreactor via optical fibers to enhance CO2 photobiochemical conversion by a *Scenedesmus obliquus* biofilm. Ind Eng Chem Res 59(50):21654–21662

Suparmaniam U, Lam MK, Uemura Y, Lim JW, Lee KT, Shuit SH (2019) Insights into the microalgae cultivation technology and harvesting process for biofuel production: a review. Renew Sust Energ Rev 115:109361

Syrpas M, Venskutonis PR (2020) Algae for the production of bio-based products. In: *Biobased products and industries*. Elsevier, Amsterdam, pp 203–243

Tan CK, Johns MR (1991a) Fatty acid production by heterotrophic *Chlorella saccharophila*. Hydrobiologia 215:13–19

Tan CK, Johns MR (1991b) Fatty avid production by heterotrophic *Chlorella saccharophila*. Hydrobiologia 215:13–19

Tan XB, Lam MK, Uemura Y, Lim JW, Wong CY, Lee KT (2018) Cultivation of microalgae for biodiesel production: a review on upstream and downstream processing. Chin J Chem Eng 26(1):17–30

Teng Z, Zheng L, Yang Z, Li L, Zhang Q, Li L, Chen W, Wang G, Song L (2023) Biomass production and astaxanthin accumulation of *Haematococcus pluvialis* in large-scale outdoor culture based on year-round survey: Influencing factors and physiological response. Algal Res 71:103070

Teo HL, Wahab RA (2020) Towards an eco-friendly deconstruction of agro-industrial biomass and preparation of renewable cellulose nanomaterials: a review. Int J Biol Macromol 161:1414–1430

Tran M, Van C, Barrera DJ, Pettersson PL, Peinado CD, Bui J, Mayfield SP (2012) Production of unique immunotoxin cancer therapeutics in algal chloroplasts. Proc Natl Acad Sci U S A 110: E15–E22

Trautmann A, Watzer B, Wilde A, Forchhammer K, Posten C (2016) Effect of phosphate availability on cyanophycin accumulation in *Synechocystis* sp. PCC 6803 and the production strain BW86. Algal Res 20:189–196

Vanthoor-Koopmans M, Wijffels RH, Barbosa MJ, Eppink MH (2013) Biorefinery of microalgae for food and fuel. Bioresour Technol 135:142–149

Vargas SR, Santos PVD, Giraldi LA, Zaiat M, Calijuri MDC (2018) Anaerobic phototrophic processes of hydrogen production by different strains of microalgae *Chlamydomonas* sp. FEMS Microbiol Lett 365(9):fny073

Vasudevan V, Stratton RW, Pearlson MN, Jersey GR, Beyene AG, Weissman JC, Rubino M, Hileman JI (2012) Environmental performance of algal biofuel technology options. Environ Sci Technol 46:2451–2459

Velazquez-Lucio J, Rodríguez-Jasso RM, Colla LM, Sáenz-Galindo A, Cervantes-Cisneros DE, Aguilar CN, Fernandes BD, Ruiz HA (2018) Microalgal biomass pretreatment for bioethanol production: a review. Biofuel Res J 5:780–791

Volkman JK, Jeffrey SW, Nichds PD (1989) Fatty acid and lipid composition of 10 species of microalgae used in mariculture. J Exp Mar Biol Ecol 128:219–240

Voloshin RA, Rodionova MV, Zharmukhamedov SK, Veziroglu TN, Allakhverdiev SI (2016) Biofuel production from plant and algal biomass. Int J Hydrog Energy 41(39):17257–17273

Walker DA (2009) Biofuels, facts, fantasy, and feasibility. J Appl Phycol 21(5):509–517

Wan C, Alam MA, Zhao X-Q, Zhang X-Y, Guo S-L, Ho S-H, Chang J-S, Bai F-W (2015) Current progress and future prospect of microalgal biomass harvest using various flocculation technologies. Bioresour Technol 184:251–257

Wang H-MD, Chen C-C, Huynh P, Chang J-S (2015) Exploring the potential of using algae in cosmetics. Bioresour Technol 184:355–362

Wang S, Hu S, Shang H, Barati B, Gong X, Hu X, Abomohra AE-F (2020) Study on the co-operative effect of kitchen wastewater for harvest and enhanced pyrolysis of microalgae. Bioresour Technol 317:123983

Wang F, Harindintwali JD, Yuan Z, Wang M, Wang F, Li S, Yin Z, Huang L, Fu Y, Li L (2021a) Technologies and perspectives for achieving carbon neutrality. The Innovation 2(4):100180

Wang L, Wang Q, Zhao R, Tao Y, Ying K-Z, Mao X-Z (2021b) Novel flat-plate photobioreactor with inclined baffles and internal structure optimization to improve light regime performance. ACS Sustain Chem Eng 9:1550–1558

Wang X, Zhang Y, Xia C, Alqahtani A, Sharma A, Pugazhendhi A (2023) A review on optimistic biorefinery products: Biofuel and bioproducts from algae biomass. Fuel 338:127378

Wiatrowski M, Klein BC, Davis RW, Quiroz-Arita C, Tan EC, Hunt RW, Davis RE (2022) Techno-economic assessment for the production of algal fuels and value-added products: opportunities for high-protein microalgae conversion. Biotechnol Biofuels Bioprod 15(1):1–14

Williams PRD, Inman D, Aden A, Heath GA (2009) Environmental and sustainability factors associated with next-generation biofuels in the U.S.: what do we really know? Environ Sci Technol 43:4763–4775

Xu L, Brilman DWF, Withag JAM, Brem G, Kersten S (2011) Assessment of a dry and a wet route for the production of biofuels from microalgae: energy balance analysis. Bioresour Technol 102: 5113–5122

Xu K, Jiang H, Juneau P, Qiu B (2012) Comparative studies on the photosynthetic responses of three freshwater phytoplankton species to temperature and light regimes. J Appl Phycol 24: 1113–1122

Yang F, Xiang W, Li T, Long L (2018) Transcriptome analysis for phosphorus starvation-induced lipid accumulation in *Scenedesmus* sp. Sci Rep 8:16420

Yin Z, Zhu L, Li S, Hu T, Chu R, Mo F, Hu D, Liu C, Li B (2020) A comprehensive review on cultivation and harvesting of microalgae for biodiesel production: environmental pollution control and future directions. Bioresour Technol 301:122804

Yoo G, Park W-K, Kim CW, Choi Y-E, Yang J-W (2012) Direct lipid extraction from wet *Chlamydomonas reinhardtii* biomass using osmotic shock. Bioresour Technol 123:717–722

Yusibov V, Kushnir N, Streatfield SJ (2016) Antibody production in plants and green algae. Annu Rev Plant Biol 67:669–701

Zabed HM, Akter S, Yun J, Zhang G, Zhang Y, Qi X (2020) Biogas from microalgae: technologies, challenges and opportunities. Renew Sust Energ Rev 117:109503

Zeller MA, Hunt R, Jones A, Sharma S (2013) Bioplastics and their thermoplastic blends from *Spirulina* and *Chlorella* microalgae. J Appl Polym Sci 130:3263–3275

Zhang K, Zhang F, Wu Y-R (2021) Emerging technologies for conversion of sustainable algal biomass into value-added products: a state-of-the-art review. Sci Total Environ 784:147024

Zhang D, Zhong D, Ouyang J, He J, Qi Y, Chen W, Zhang X, Tao W, Zhou M (2022) Microalgae-based oral microcarriers for gut microbiota homeostasis and intestinal protection in cancer radiotherapy. Nat Commun 13(1):1413

Zhu L, Nugroho Y, Shakeel S, Li Z, Martinkauppi B, Hiltunen E (2017) Using microalgae to produce liquid transportation biodiesel: what is next? Renew Sust Energ Rev 78:391–400

Overview of Biorefinery Technology

Benjamin Bernard Uzoejinwa, Oji Achuka Nwoke, Felix Uzochukwu Asoiro, and Chukwuemeka Jude Ohagwu

Abstract Biorefinery involves integrated systems or routes/pathways for converting biomass into different targeted valuable products such as energy, fuels, and chemicals by various physical, chemical, thermochemical, and biochemical technologies. It has recently gained increasing attention owing to its relevance. International Energy Agency Bioenergy Task 42 defined biorefinery as "the sustainable processing of biomass into a range of biobased products (chemicals, feed, food, or other useful products) and bioenergy (power, biofuels, and/or heat)," with the aim to optimize the full utilization of biomass, maximize its profitability, and concurrently minimize waste generation. An overview of the technologies/possible routes of algal biomass biorefinery for enhanced biomass utilization, most especially for production of several industrially important phycochemicals and bioenergy, has been provided in this chapter. Besides, the chapter also presented the general concept of biorefinery and the summary of the challenges, opportunities, and recent trends in algal biomass biorefinery.

Keywords Algal biomass · Circular economy · Biochemicals · Bioenergy · Enhanced biomass utilization

1 Introduction

Biorefinery has recently gained wider attention owing to its indispensable relevance in sustainable processing of biomass, with the aid of integrated systems or routes/pathways, into a variety of valuable and commercial bioproducts and bioenergy. The later refers to the renewable energy obtained from biomass, while biofuels refer to liquid, gaseous, or solid fuels produced from biomass or organic matter that can be utilized as a replacement for fossil fuels (Reid et al. 2020). Biorefineries thus include facilities that aim to extract multiple high-value products from renewable biomass

B. B. Uzoejinwa (✉) · O. A. Nwoke · F. U. Asoiro · C. J. Ohagwu
Department of Agricultural and Bio-resources Engineering, University of Nigeria, Enugu, Nigeria
e-mail: benjamin.uzoejinwa@unn.edu.ng

A. Abomohra, S. Ende (eds.), *Value-added Products from Algae*,
https://doi.org/10.1007/978-3-031-42026-9_7

sources such as algae, agricultural residues, urban and wood residues, forest residues and mill residues, as well as herbaceous energy crops and dedicated woody crops (Hingsamer and Jungmeier 2019; Moncada and Aristizábal 2016). These biomass feedstocks can be further grouped into lignocellulosic biomass, starchy biomass, sugar biomass, organic waste biomass, and triglycerides biomass. Biorefineries are thus considered a sustainable solution to replace traditional petrochemical refineries, which are heavily reliant on finite and non-renewable fossil fuels.

Climate change, global warming, and environmental degradation, occasioned by the intensive deployment of fossil resources, are the most severe environmental hitches challenging humanity today (Hassan et al. 2021; Martins et al. 2019; Solarin 2020). These challenges can be ameliorated by employing biofuels of algal origin as an alternative and renewable energy fuel source. Microalgae are ubiquitous organisms that are thought of as traditional substitutes for the viable production of a variety of bioproducts of higher value. However, due to high costs, high-energy requirements, and a lack of readily available biomass and biomolecules, microalgal biorefinery is still very far from being economically feasible (Chew et al. 2017; Chia et al. 2018). Seaweeds (commonly known as the macroalgae) are species of macroscopic multicellular marine photosynthetic organisms. As previously discussed in this book, they are generally categorized into three main groups: brown algae (Phaeophyceae), red algae (Rhodophyta), and green algae (Chlorophyta). The use of seaweeds as feedstock in biorefinery process development is considered as a high potential alternative in the energy, food, pharmaceutical, and medical sectors. However, it is essential to pretreat the feedstocks to make their molecular structure accessible.

With the progressive surge in global population, shortage of fossil fuels, increased energy prices, climate change challenges, and the threat of environmental pollution, mankind are compelled to explore other alternative routes of energy and biochemical sources to meet the increasing energy needs. Currently, huge research efforts are in the areas of identifying new species, maximizing biomass productivity and identifying new and novel strategies to achieving the needs of folder, food, value-added biochemicals, and fuels in the future (Saad et al. 2019). As discussed in previous chapters, energy and fine biochemicals can be produced from the first-, second-, or the third-generation feedstocks (mostly algae) for biorefinery purposes. When compared to other conventional feedstocks, algae have more benefits. For example, the rate of growth of algae has been found to be 5–10 times greater than that of land-based crops in favorable circumstances, indicating a greater rate of production of ideally convertible biomass (Srinivasan and Kulshreshtha 2020; Voloshin et al. 2016; Wang et al. 2023). Besides, some species may contain up to 70–80% by weight of lipids or carbohydrates. Thus, the rising demand for renewable and sustainable resources has led to a growing interest in biorefineries, which convert renewable resources into a variety of bioproducts (fuels, chemicals, and materials) of higher value. One of the most promising resources for biorefineries is algal biomass, which has several advantages over other renewable resources, such as increased lipid yield, greater rate of growth, and the capacity to grow in marginal agricultural land (salty, brackish, and marshy soils) (Sarwer et al. 2022; Ubando et al. 2021). However, biorefineries based on algal biomass development are still in their early stages, and many challenges must be overcome before these systems can be implemented commercially.

This chapter presented the overview of the technologies/possible routes of algal biomass biorefinery for enhanced biomass utilization, most especially, for production of several industrially important phycochemicals and bioenergy. The chapter also gave the general concept of biorefinery and the summary of the challenges, opportunities, and recent trends in algal biomass biorefinery.

2 Concept of Biorefineries

Algal biorefinery is a unified system that integrates the conversion processes of algal biomass and the associated equipment/facility in production of chemicals, power, and fuels in a socially, economically, and environmentally sustainable manner. Every of the refining phase is known as a "cascading phase." For sustainably producing and efficiently utilizing biomass resources, biocascading and biorefining approaches/technologies must be implemented (Balina et al. 2017; Cherubini 2010; Conteratto et al. 2021; de Jong and Jungmeier 2015). The main importance of a biorefinery includes three main aspects: (1) the simultaneous production of both energy (liquid or gaseous biofuels) and value-added materials (feed, food, and chemicals), (2) the use of multiple processes (thermochemical, biological, and/or mechanical processes), and (3) utilization of various raw biomaterials, both from fresh sources and from waste streams (Chew et al. 2017).

Integrating the biorefinery concept into existing industrial parks has the capability to lower the costs of capital and the resulting products. To ensure sustainable growth, strategies for incorporating biorefineries into the future bio-economy must be implemented (Maity 2015). The process of producing bioenergy through biorefinery approach involves numerous steps, including cultivation, harvesting, pretreatment, extraction, and conversion. Pretreatment aims to simplify the complex structure of biomass by, for example, transforming polysaccharides into sugars that can be fermented. There are four methods for pretreatment including physical, chemical, physiochemical, and biological. During the extraction phase, oil is separated from the biomass, resulting in a de-oiled biomass that consists of many components such as carbohydrates, protein, antioxidants, and pigments. The actual composition of these components depends highly on the biomass types and extraction method used (Wiselogel et al. 2018). The extracted lipid is used to produce biodiesel via transesterification, while the remaining components of the biomass are utilized in the synthesis of high-value bioproducts like pharmaceuticals and feed/food ingredients (Gameiro 2016). Figure 1 presents a general overview of biomass biorefinery concept for production of a spectrum of compounds.

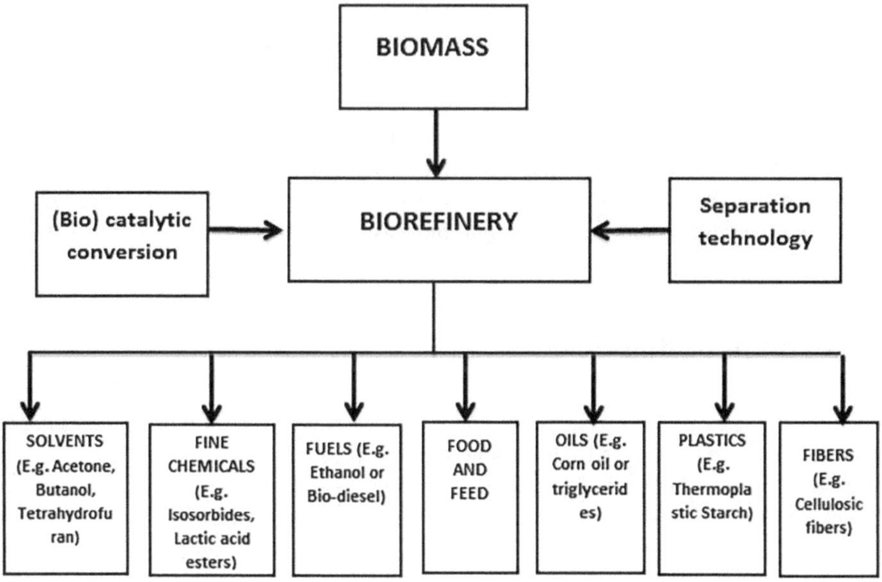

Fig. 1 Typical concept of biorefinery for production of a spectrum of compounds

3 Algal Biorefinery

Algal biorefinery concept is similar to today's oil refineries that produce several fuels and other products from petroleum (Ferreira 2017). Several bioprocessing steps can be employed to extract bioactive compounds, such as carbohydrates, proteins, and lipids from algae. Biorefinery approach has garnered much interest as a process of obtaining fine chemicals, bioactive compounds, and fuels from third-generation feedstocks (Chandra et al. 2019). The idea of using algae as a biomass feedstock for biorefinery looks promising with multiple advantages. It will help in carbon dioxide (CO_2) sequestration, ameliorate the menace of the global warming, and yield components needed for different industrial purposes such as energy, feed, food, pharmaceuticals, cosmetics, fertilizers, nutraceuticals, and other biobased industrial products (Vázquez-Romero et al. 2022). In spite of the huge prospects of algae as biorefinery feedstock, there are, however, some techno-economic challenges. For instance, the present industrial microalgae biomass global productivity and market are still abysmal (15,000–25,000 ton per year) (Fernández et al. 2021; Koyande et al. 2019) and cannot meet up with demand by the industry. The reason for the low rate of production is the inherent high cost of cultivation, harvesting, and extraction (Okoro et al. 2019). As a result of this, algal production is currently directed to the production of high value-added bioproducts. Use of microalgae for energy production can be considered a secondary option due to the fact that it does not command as much price as conventional fossil fuel. Biofuels do not necessarily have to be priced lower than conventional non-renewable fuels. However, biofuel production must be performed with lower energy inputs. Much of the studies only focused on the synthesis of value-added products from algae with relatively few studies on biofuel production, leaving much of the constraints not yet overcome (Halder and Azad 2019). Two main stages have been identified in algae biorefinery processing: (1) upstream and (2) downstream processing. Upstream processing mainly involves algae cultivation. The essential raw materials needed for this stage are space, light, water, nutrients, and CO_2. The algal growth rate and biomass productivity depend largely on many factors, mainly supply of nutrients light intensity, and cultivation system, as discussed in Chapter "Algae Cultivation Systems". Downstream processing of algal biomass includes harvesting, extraction, and purification of phycochemicals. The traditional extraction methods involve physical/mechanical techniques such as manual bead beating, blending, screw press, ultrasound, high-pressure homogenization as well as chemical techniques such as solvent extraction using hexane, acetone, chloroform, and benzene, as well as using supercritical fluid/gas extraction (Aravind et al. 2021). The entire processes are complex, multi-step with huge cost implications. Other methods such as autoclaving and freeze–thawing have also been used which are complex and extremely costly, making the industrial extraction of a single phycochemical from algal biomass an economic challenge. Different products from algal biomass belong to four main cellular components namely carbohydrates, lipids, proteins, and pigments. The resulting by-products or residues can be recycled as nutrients for media during algae cultivation (Abomohra et al. 2018) or used in biorefinery combined heat and power (CHP) plants for electricity generation (Maghzian et al. 2023). Research is presently ongoing to discover and isolate existing or new algal strains with multiple value-added products that can potentially be used as viable inoculum for further biorefinery processing. Many technologies are used to genetically modify and optimize existing algal species to improve strain performance as discussed in Chapter "High Through-put Screening to Accelerate Microalgae-based Phycochemicals". However, the extraction of quality products of higher value from algal biomass on industrial scale is still currently not viable and needs further optimization The need to employ affordable and simple technologies both in the upstream and downstream processing in algal biorefinery was also emphasized (Shahid et al. 2020). Figure 2 presents an overview of integrated algal biorefinery processes, while Fig. 3 provides a typical example of a microalgal biorefinery model.

4 Opportunities and Challenges of Algal Biorefinery

4.1 Opportunities of Algal Biorefinery

Algal biorefineries offer various opportunities for businesses and industries to diversify their products and revenue streams. Algae can serve as a feedstock for production of various high-value-added products including sustainable biofuels, food ingredients, fertilizers, nutraceuticals, and pharmaceuticals. This has the

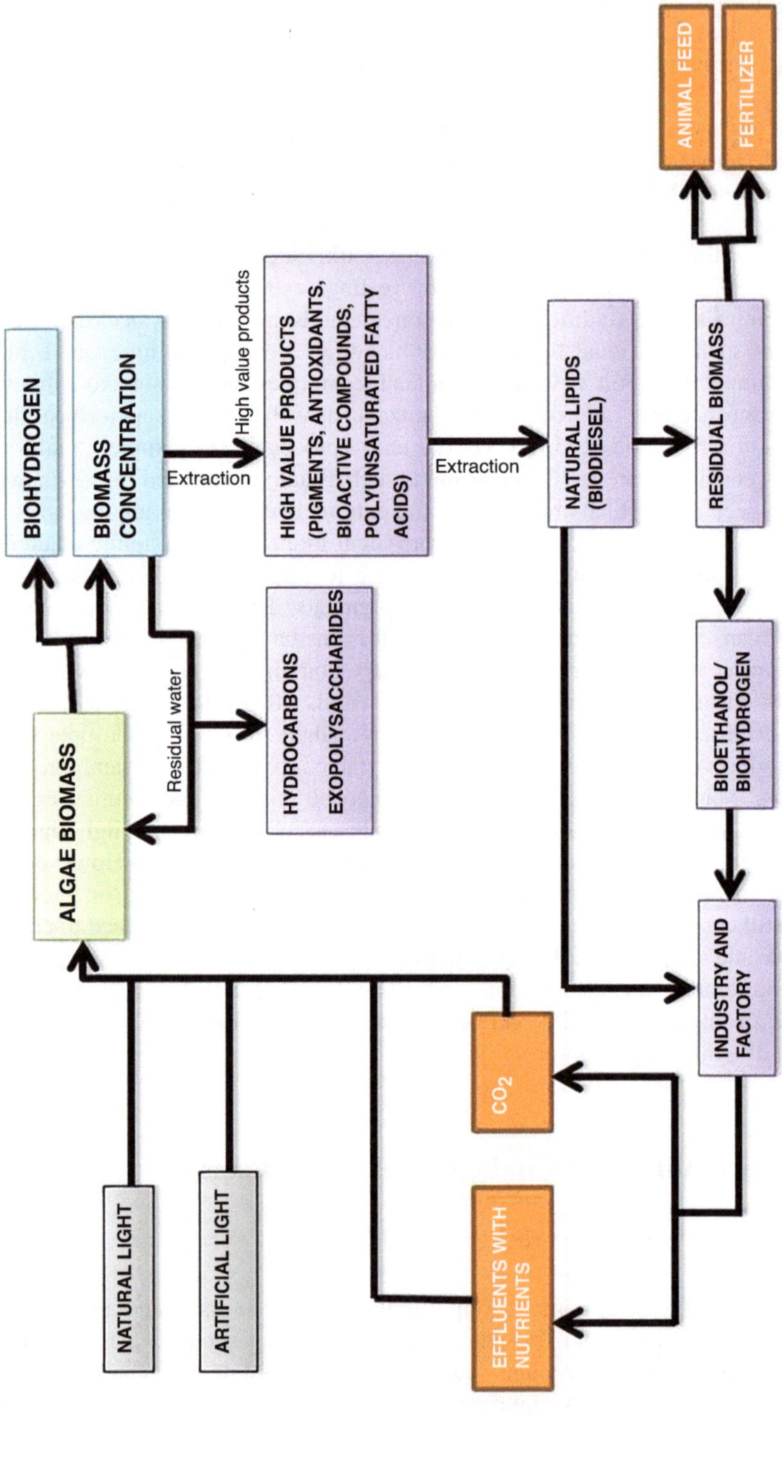

Fig. 2 An overview of integrated algal biorefinery processes showing the closed loop of carbon neutrality

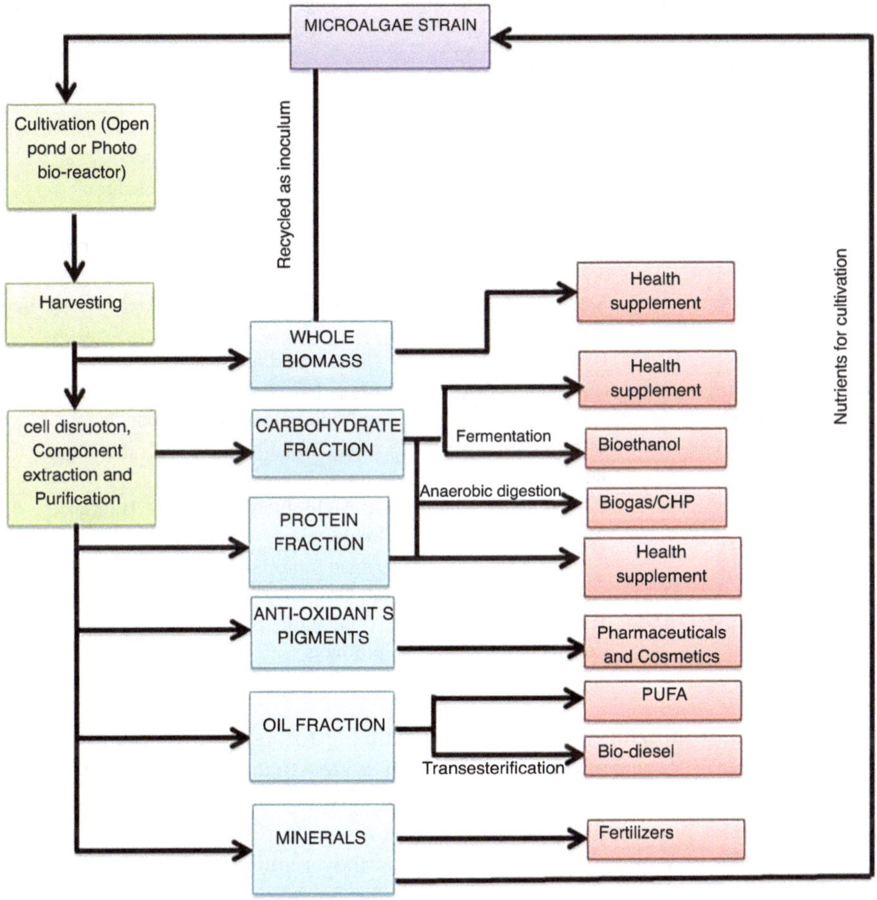

Fig. 3 A typical example of a microalgal biorefinery model

capacity for reduction of dependency on fossil fuels and greenhouse gases emission, increase resource efficiency, and improving the overall economic outcome. In addition, establishment of algal biorefinery could create new jobs and economic opportunities in rural areas. One of the unique characteristics of algae, most especially, the microalgae, is the high lipid content that some strains are capable of producing lipids, especially triacylglycerol (TAG), which is a promising starting material for green gasoline synthesis, green diesel, green jet fuel, and biodiesel (Salman et al. 2023). These biofuels can be produced by either transesterification of the TAG or through combined techniques, such as catalytic cracking and hydroprocessing as discussed in Chapter "Catalyst in Action". Thus, algal biofuels present an opportunity as more sustainable alternatives to bioethanol production from sugarcane and corn or biodiesel from terrestrial oil crops and might even be eco-friendlier than cellulosic ethanol (Bilal and Iqbal 2020). Algae cultivation offers

Table 1 Typical examples of industries where algae are being used as raw materials for commercial biorefinery

Algae-based industry	Location (Headquarter)	Major algal products
Algenol Biofuels	Fort Myers, Florida, USA	Algae-based biofuels (mainly ethanol, gasoline, jet, and biodiesel fuels).
Solix Biosystems	Fort Collins Colorado, USA	Algae-based biofuels and natural ingredients.
Sapphire Energy	Southern New Mexico, USA	Algal biofuels including green crude, biodiesel, and jet fuels.
TerraVia Holdings, Inc. (formerly Solazyme)	South San Francisco, California, USA	Algal oils, ethanol, biodiesel, and jet fuels.
Seambiotic	Israel	Algae-based biofuels.
Aurora Algae (formerly Aurora biofuels)	California, USA	Algae-based biofuels including biodiesel and other algae-based products
Euglena Co., Ltd.	Minato City, Tokyo, Japan	Algae-based biofuels and other algae products.
Biofuels Pty Ltd	Tailem Bend, South Australia	Algae biofuels, Algal oils, BioMax biodiesel.
Algae Farms	Preveza, Greece	Algae bio-oil, biodiesel, algae pellets, briquettes, and other algae products.
Pond Biofuels Inc.	Toronto, Ontario, Canada	Algae biofuels, algal oils, and other algae products

the advantage of utilizing marginal land unsuitable for agriculture and can also be grown in seawater, wastewater, or brackish water that is not in high demand (Debnath and Das 2022). This presents an opportunity for a sustainable source of biofuel that does not compete on the available resources. Microalgae have a significant capability to sequester CO_2 emissions as they require 1.83–2.0 g of CO_2 for every 1 g of biomass produced (Ighalo et al. 2022). Various conversion processes could be employed in the future algae-based biorefinery industries to develop multiple biofuels, such as methane, ethanol, gasoline, aviation fuel, and green biodiesel, as well as valuable by-products such as fatty acids, proteins, and carbohydrates. Table 1 shows typical examples of different global industries where algae are being used as raw materials for commercial biorefinery.

4.2 *Challenges of Algal Biorefinery*

The main challenges of biorefineries currently include market acceptance in the fossil-based economy, availability and quality of the feedstock, quantities required to meet the market demand, and techno-economic feasibility. For instance, the current production of industrial microalgal biomass is limited to approximately 15,000 tons per year (Ali et al. 2017). The main reason for this relatively low production is the high cost associated with cultivation, harvesting, and extraction of microalgae (Chu et al. 2021). Thus, the challenges peculiar to algal biomass biorefinery can be summarized as follows:

1. *Survival of the organisms*: Algal biorefinery systems currently lack the essential capacity for maintenance of the best laboratory organisms under outdoor conditions. Laboratory cultures get contaminated with surrounding organisms (Fawaz et al. 2018). New technologies such as high-throughput screening (Chapter "High Throughput Screening to Accelerate Microalgae-based Phycochemicals") or integrated artificial intelligence (Chapter "Artificial Intelligence in Phytochemicals Recognition") may enable testing and tracking multiple challenges experienced in outdoor conditions, facilitating search for most suitable algal species and identifying the most harmful conditions which need to be controlled.
2. *Carbon dioxide enhancement*: Generally, many algae grow well when "aerated" with carbon dioxide, despite that its levels above 5% reduces the growth rate of higher plants and animals. Carbon dioxide is a cost driver in microalgae production costs, and industrial-scale solutions are needed for substantially reducing the production costs (Nurdiawati et al. 2019).
3. *Light penetration*: Light is an important parameter for autotrophic cultivation of algae. Generally, the intensity of light penetration often reduces by increased water depth. Moreover, the algal species' efficacy of photosynthesis depends largely on the source of light. In open ponds, results of several studies unveiled that shallow depths provided greater surface area with effective mixing and light penetration (Maltsev et al. 2021). However, no industrial solution is on the market to cultivate phototrophic algae in bulk systems with higher water depth, as used in heterotrophic cultivation systems. Such systems would bring benefits of scaling effects and reduce demand for land, hence reducing the overall production costs.
4. *Environmental dependence*: Generally, the high yields of algal biomass achieved in ponds are often seasonal. This is because the growth rate of algae depends largely on the environmental conditions, especially temperature. Previous studies revealed that some species of algae cannot grow well in cold temperatures, while many face serious challenges at high temperatures. Thus, climate independent systems would highly contribute to whole year production on overall increase of algal biomass (Hua et al. 2021).

5 Possible Routes for Integrated Phycochemical Production

Biorefineries aim to convert renewable biological resources into a range of valuable products, including biochemicals, which are every chemical compound found in living things. Biochemical compounds are essential components of the cells and other structures of the organisms and are greatly involved in the performance of the life processes of the organisms. Carbon is the center of all biochemical compounds, and thus, it is very important to life on Earth. Most biochemical compounds form polymers made up of repeating units of smaller monomers. Carbon, hydrogen, and oxygen represent the main elements in biochemicals, as some only contain these basic elements while others may have some additional elements. The vast number of biochemical compounds in algal cells can be categorized into four major groups: lipids, carbohydrates, proteins, and pigments. Thus, carbohydrate-based biochemicals from algal biomass refer to a group of compounds derived from algal carbohydrates, including sugars such as glucose and fructose, which can be converted into various biochemicals such as bioethanol, biobutanol, and biohydrogen (Ak et al. 2022). Lipid-based biochemicals from algal biomass refer to a group of compounds derived from algae that are rich in lipids. These lipids can be converted into a variety of useful products such as biofuels (mainly biodiesel), fatty acid-derived chemicals, bioplastics, lubricants, and cosmetics (Chozhavendhan et al. 2022).

Protein-based biochemicals from algal biomass are a group of compounds derived from algal proteins including amino acids, peptides, and enzymes. Algae are rich in protein, and some species can contain up to 70% by dry weight protein (Wells et al. 2017). This makes algae a potentially valuable source of high-quality protein for use in food, feed, and industrial applications. Algae contain a wide range of visible light-harvesting complexes termed as pigments. Three major categories of pigments present in algae are chlorophylls, carotenoids (carotenes, astaxanthin, and xanthophylls), and phycobilins (phycocyanin and phycoerythrin). Carotenoids are a very popular type of algal pigments which can serve as antioxidants in, e.g., the food and feed industry. They can also be employed for several activities including, among others antitumor, immunoprophylactic, and anti-inflammatory activities. Examples of common carotenoids are astaxanthin, fucoxanthin, lutein, zeaxanthin, canthaxanthin, and β-cryptoxanthin (Duppeti et al. 2017; Patel et al. 2022). Thus, algae represent promising feedstock for biorefinery for production of various phycochemicals, besides bioenergy and other valuable products. Possible routes or technologies for production of such compounds from algal biomass include among others, biochemical conversion, thermal conversion, and chemical conversion (Fig. 4). Biochemical conversion involves the use of enzymes and microorganisms to produce phycochemicals and bioenergy. Thermal conversion involves the use of heat to convert algal biomass into bio-oil, which can be further processed into various biochemicals. Chemical conversion involves the use of chemical reactions to convert algal biomass into biochemicals. Some potential high-valued chemical compounds found in algal biomass include sulfated oligosaccharides, rare sugars (such as D-glucose, D-galactose, D-mannose, D-fructose, D-xylose, D-ribose, L-rhamnose, glucuronic acid, L-fucose, mannitol, and L-arabinose), proteins and amino acids, pigments, and phenolic compounds. However, development of cost-effective and sustainable biorefineries remains a challenge, and further research is needed to optimize algal biorefinery. Various possible routes/technologies for algal biorefinery are described below based on different targeted products.

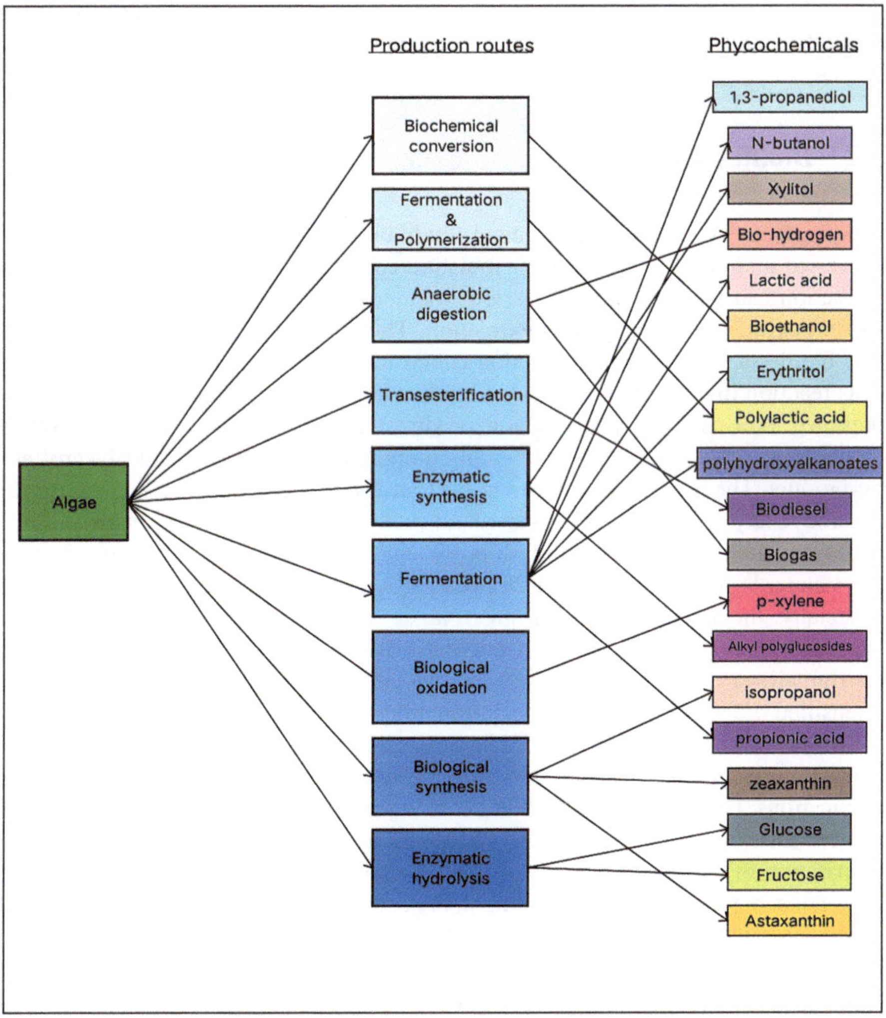

Fig. 4 Algal biomass biorefinery technologies/possible routes for different phycochemicals production

5.1 Biohydrogen

Biohydrogen has the potential to be a clean and renewable energy source, as it produces no greenhouse gases or other harmful emissions when burnt (Benemann and Weissman 1977; Nagarajan et al. 2017). It can be produced by algal cells through a process called photo-biological hydrogen production. This process involves using algae to convert solar energy, water, and carbon dioxide into hydrogen through photosynthesis. The algae used in this process are typically microalgae that have been specifically selected for their high hydrogen production potential. Biohydrogen is mainly produced during the algae cultivation.

5.2 Biodiesel

Biodiesel can be produced from algal lipids by transesterification. First, algae are harvested and dried to reduce their moisture content, then lipids are extracted. The extraction can be done through various methods, including mechanical extraction, solvent extraction, or enzymatic extraction. The extracted lipids are then converted into biodiesel via a chemical reaction called transesterification (Bello et al. 2012). This reaction involves the reaction of lipids with alcohol, typically methanol, in the presence of a catalyst. The reaction produces fatty acid methyl esters (FAMEs) (Nisar et al. 2021), which are the main components of biodiesel and glycerol as a by-product. The biodiesel produced in the transesterification reaction is then purified to remove any residual solvent, water, and catalyst. The final product is tested for its quality, including its viscosity, flash point, and fatty acid profile, to ensure it meets the standards for biodiesel (Obiora 2022). Biodiesel production from algae, most especially, the aquatic unicellular green algae (*Chlorophyceae*) is very economical and easy due to their high lipid content (Belousov et al. 2021). For example, Al-Humairi et al. (2022) employed a homogeneous base catalyst as a part of an intensified process in direct and fast production of biodiesel from algae and obtained a 97% yield of biodiesel in 10 min. The by-products of the biodiesel production from algae include the lipid-free biomass and glycerol, which can be further used in many applications. Glycerol is a vital chemical compound with wide applications and can be used for wounds treatment owing to its antimicrobial and antiviral characteristics, preparation of various pharmaceutical and medical products, preparation of several personal care products, scenes filming in the coastal region in film industries, used as substitute to water owing to its higher and more promising acoustic resistance, and as a raw material in the production of nitroglycerin (Pirzadi and Meshkani 2022). Also, it showed high potential for cultivation of oleaginous fungi that can be used also for biodiesel production (Li et al. 2022). The algal biomass which is free from lipids has substantial quantity of nutrients including carbohydrates, P, N, and other micronutrients. Thus, the demand of the nutrients can be lowered by recycling the nutrients in the residual algal biomass (which is free from lipids) during algae cultivation (Abomohra et al. 2018). In addition, algal lipid-free biomass showed high antioxidant and radical scavenging activity (Abomohra et al. 2022). Apart from being a potential nutrient source, it can also serve as an alternative organic carbon source (Kujawska et al. 2021).

5.3 Fatty Acid-Derived Chemicals

Fatty acid-derived chemicals can be produced from algal lipids through fatty acid hydrolysis. This process involves breaking down the triglycerides in the lipids into their free fatty acids (FFAs). The resulting FFAs can then be converted into various chemicals through different chemical processes, such as esterification, oxidation, and polymerization. After lipid extraction, the lipid-free biomass produced can be used in many applications as described in the previous section.

5.4 Bioethanol

Bioethanol can be produced from algal biomass or lipid-free waste by hydrolysis followed by fermentation. This process involves making the sugars in the algae available for fermentation into alcohol by yeast or other microorganisms. The resulting bioethanol can then be purified and used as a biofuel. Bioethanol yield reports range between 5000–15,000 gal acre^{-1} and 46,760–140,290 L ha^{-1} can be obtained from algae (Kumar and Mukund 2018).

5.5 Butanol

Butanol can be produced from algal biomass or lipid-free waste through bioconversion using microorganisms such as bacteria to convert sugars, starches, and other organic compounds in the algal cells into butanol. The resulting butanol can then be purified and used as a biofuel. Compared to bioethanol, butanol has several advantages including a higher energy content, better stability, and lower vapor pressure. Cheng et al. (2015) evaluated the potential of butanol production from the fermentation of *Clostridium acetobutylicum,* and their results revealed that butanol can be produced at a concentration of 0.2 g g^{-1} glucose, when glucose is considered as the only carbon source, but this may be improved by adding butyrate. However, 3.86 g L^{-1} of butanol can be produced from *C. acetobutylicum* when the substrate is microalgae with 1/3 of carbohydrate still remaining in unused form (Efremenko et al. 2012).

5.6 Enzymes, Peptides, and Amino Acids

Peptides and amino acids are two important components of algal biomass, where peptides are short chains of amino acids. Enzymes are proteins that act as catalysts in biological reactions and are employed in different processes in the industry (i.e., production processes of food and biofuels). Algae are a source of various enzymes, including cellulases, amylases, and lipases, which can be extracted and purified for use in these processes. The extraction and purification of these components from algal biomass or lipid-free biomass can be performed using various methods including chemical hydrolysis, enzymatic hydrolysis, and fermentation (Chronakis and Madsen 2011).

5.7 Biopolymers

Biopolymers are polymers made from renewable resources, such as plant-based materials, rather than petroleum. The extraction of biopolymers from algal biomass can be performed through chemical and enzymatic processes, followed by polymerization and shaping into desired forms (Wells et al. 2017). More details about biopolymers and potential of algae for biopolymers production are discussed in Chapter "Algal-based Biopolymers".

5.8 Methanol

Methanol forms one of the largest volumes of biochemicals produced worldwide, where various other fine chemicals such as formaldehyde, olefins, propylene, and gasoline can be synthesized from it. Methanol can also be converted into some aromatic substances. Besides, methanol has been recognized as proven precursors in several industrial processes and as a potential alternative energy for transportation fuels. Conventionally, methanol is synthesized from coal and methane as feedstock materials by combustion but this process results in emission of CO_2, which is one of the major sources of global warming. However, several eco-friendly alternative processes, such as fermentation, have proven to be effective for production of methanol from biomass materials like algal biomass. The total market demand of fermentation products in 2013 was more than 110 million tons at a monetary value of $207 billion USD, where alcohols accounted for the largest share of 94% (Mahato et al. 2021).

5.9 Sucrose/Hexose

Many sugars such as sucrose can be generated from algal biomass and sugar crops after hydrolysis of starch. Most well-established industries and numerous traditional biorefineries use sugar as a feedstock for synthesis of numerous chemicals. Sugars are rich biochemicals generated from biomass which can be converted catalytically to produce value-added fine biochemicals and liquid alkane compounds. Advanced biorefinery use lignocellulosic and hemicellulosic biomass for the production of fine chemicals, where sugars are produced by pretreatment of biomass accompanied by enzymatic hydrolysis. Glucose is normally produced from cellulose, while a mixture of xylose, arabinose, and galactose can be produced by hydrolysis of hemicellulosic materials like the algal biomass. The utilization of lignocellulosic biorefinery process in the synthesis of glucose and other fine chemical products has outstanding results; however, it is still fraught with many biological, technical, and economic challenges before these opportunities can be fully exploited. Some of these challenges are mainly associated with its complex structure and presence of lignin which offers resistance to biological attacks or enzymatic digestion of the biomass (Singh et al. 2022). Therefore, algae can provide a potential alternative due to their simple structure and lacking of lignin. Through catalytic processes, sugars can be converted into useful chemical products such as 5 hydroxymethylfurfural (HMF), furandicarboxylic, and levulinic acid (van der Waal and de Jong 2016). Sugars can also be subjected to selective dehydration, hydrogenation, and oxidation reactions to produce sorbitol, furfural, glucaric acid, hydroxymethylfurfural, 2,5-furan dicarboxylic acid, levulinic acid, methyl vinyl glycolate, and mono-ethylene glycol (van Putten et al. 2016). The production of sugar alcohols like xylitol and sorbitol for industrial applications currently gains traction across the globe due to the direct use of sorbitol as efficient food ingredients, intermediate in the production of isosorbide, and as a monomer for the production of biobased plastics. The global sorbitol demand in 2018 was put at over two million tons, and researches are currently ongoing in developing novel catalytic processes for conversion of sugars into biochemical products like methyl vinyl glycolate and mono-ethylene glycol which are precursors for the production of biopolymers (Barbaro et al. 2019).

5.10 Levulinic Acid and Gamma-Valerolactone

Among the numerous biochemicals that can be obtained from algal biomass, levulinic acid (LA) has profound importance owing to its wide applications in chemical industries (Kim et al. 2020). Apart from LA, γ-valerolactone (GVL) has attracted huge interest in recent years due to its fantastic performance as a biochemical and can be produced by hydrogenation of LA and the esters (Tong et al. 2021). At present, LA is produced through catalyst-acid reactions involving hemicellulose (C5 sugars) and cellulose (C6 sugars). By carefully selected hydrogenation, furfural is converted into FA. The produced FA is hydrolyzed to LA, which will further be converted to GVL through hydrogenation processes. The production of LA follows some series of careful steps. A vital step in the LA synthesis from furfural is the conversion of furfuryl alcohol into LA; necessary precaution is required otherwise a polymer by-product will be obtained. Efficient large-scale production of LA is still a challenging task due to the complex reaction processes involved and the poor thermal stability of intermediates like furfuryl alcohol, furfural, and 5-hydroxymethyl furfural, which may result in the production of undesirable polymetric material, e.g., humins, if not carefully handled (Kim et al. 2020). Attempts have been made recently on the selective conversion of cellulose and hemicellulose to LA and its derivatives with minimal results. Developing highly efficient catalyst and optimizing the catalytic conversion processes are necessary for the efficient yield of LA and GVL from hemicellulose and cellulose. The process of levulinic acid and formic acid consists of two chemical reaction stages: first, hydrothermal decomposition of cellulose at moderate temperatures (190–270 °C), in the absence of catalysts to produce organic water-soluble compounds (glucose and HMF). The second process involves catalytic (solid acid catalyst) treatment of water-soluble compounds at relatively low temperatures (160 °C) to produce formic and levulinic acid. GVL can easily be extracted in high yield from biomass by employing very moderate catalytic hydrogenation reaction of an aqueous solution of levulinic acid, using a commercial ruthenium supported catalyst together with a heterogeneous acid co-catalyst like the ion exchange resins Amberlyst A70 or A15, niobium oxide or phosphate at 70–50 °C, and at low hydrogen pressure of 3–0.5 MPa (Yu et al. 2020).

5.11 Cyclopentanone and Pentanediol

The selective hydrogenation of furfural produced five-heterocyclic ring chemical compounds with an oxygen atom such as 2-methylfuran (MF), tetrahydrofurfuryl alcohol (THFA), and furfuryl alcohol (FA) cyclopentanol (CPL) and cyclopentanone (CPO). These are fine biochemicals and raw materials with outstanding use in the chemical industry. The simplest member of the class of cyclopetanols with a single hydroxyl group are used for making dyes, chemical solvents, perfumes, pharmaceuticals, and other organic products (Zhang et al. 2016). In general, industrial production of cyclopentanol is done by hydrogenation of cyclopentanone, whereas cyclopentanone is traditionally produced by oxidation of cyclopentene using feedstocks of biomass origin or by decarboxylation of adipic acid. However, the oxidation process is inefficient and often polluting. Naturally, cyclopentanols are only found in few plants and in limited quantity thereby making extraction extremely difficult. However, the production of cycloperanol from furfural in biomass products is a vital and efficient green approach. An efficient and high-quality 1,5-pentanediol is produced by reaction of tetrahydrofurfuryl alcohol with hydrogen. This process requires series of steps. The first step involves the hydrogenolysis reaction of tetrahydrofurfuryl alcohol with hydrogen for production of a crude reaction product usually in the presence of a copper catalyst at a high reaction temperature (200 to 350 °C) and pressure (1 to 40 Mpa), until the conversion rate of tetrahydrofurfuryl alcohol is a little less than 80%. The second step requires the separation of tetrahydrofurfuryl alcohol and crude 1,5-pentanediol from the crude reaction product obtained, followed by supplying recovered tetrahydrofurfuryl alcohol as a raw material in the first step. The final step entails obtaining the high-quality 1,5-pentanediol by the distillation of the crude

1,5-pentanediol. Thus, cyclopentanone is one of the valuable biochemicals synthesized by the furfural rearrangement. The process involves hydrogenation of furfural to furfuryl alcohol in aqueous solution, then followed by further hydrogenation of furfuryl alcohol to cyclopentanone through 2-cyclopentenone (Hronec et al. 2013).

5.12 Furfuryl Alcohol, 2-Methylfuran, and Tetrahydrofurfuryl Alcohol

Furfuryl alcohol (FA) (i.e., 2-furanmethanol) is synthesized by decarbonylation of hydroxymethylfurfural at a temperature of or above 135 ° C in the presence of suitable catalyst (palladium or rhodium) by a liquid-phase process. In this process, the furfuryl alcohol is constantly stripped from the reaction medium. FA production consumes about 65% of the total yearly furfural production. Traditionally, furfural hydrogenation could occur on metal catalyst surface for FA production. Numerous bimetallic and monometallic catalysts have been tested as efficient catalysts for hydrogenation of furfural both in the gaseous and liquid-phase reactions, e.g., Co, Cu, Ni, Pt, Ru, Pd, Cu-Co, Cu–Ni, Pd-Cu, Cu–Zn, and Cu–Cr catalysts (Zhu et al. 2020). The Cu-Cr-based catalyst provides higher activity compared to other catalysts. Copper chromite (Cu–Cr) is a well-established catalyst in the furan industry for FA production due to its high performance with about 35–98% in gaseous phase reaction and 98% in liquid phase. However, Cr element is dangerous to health, thereby limiting its wide-scale application. Furfuryl alcohol (FA) or 2-furanmethanol is produced by hydrogenation of furfural in the presence of suitably selected metal or non-metallic catalysts (Cu–Ni, Ni, Co, Ru, Cu, Pt, Pd, Cu–Co, Cu–Cr, Cu–Zn, and Pd–Cu) in either liquid or gaseous state. The 2-MF can be produced directly from furfural in either by liquid or gaseous phase reactions. Liquid-phase hydrogenation using catalyst under alcohols as hydrogen donors has received global research attention due to safety and economic concerns. Likewise, FA, 2-MF, and tetrahydrofurfuryl alcohol (THFA) can also be produced by hydrogenation of furfural under severe temperature and pressure. This process is usually achieved by first hydrogenating furfural to furfuryl alcohol by addition of hydrogen to the C=C bonds of the furfural, and then further hydrogenation of furfuryl alcohol in the presence of nickel-based catalyst resulted in the production of THFA.

THFA is an important water miscible biochemical solvent with a purity of 98.5%. It is a cheap biodegradable biochemical solvent used majorly as a reactive diluent for epoxy resins. It is also an effective solvent for most of the curatives and catalysts employed in epoxy formulations. Besides, it is also employed in the formulation of biocide and pesticide, electronic cleaner, coatings, dyes, printing ink and epoxy curing agent (Mikucka et al. 2023).

2-Methylfuran (sylvane) is a combustible, water-insoluble liquid with a chocolate odor. It occurs naturally in Myrtle and Dutch Lavender used as a FEMAGRAS flavoring substance. It also possesses the capacity for utilization in alternative fuels.

Industrial manufacture of 2-Methylfuran is by catalytic hydrogenolysis of furfural alcohol or from furfural in the vapor phase by hydrogenation–hydrogenolysis sequential reactions. Quite a number of research focused on the direct production of 2-methylfuran from furfural in both liquid and gaseous phase reactions. Catalyst transfer hydrogenation via liquid phase under alcohol as hydrogen donor has gained more attention in recent times due to the fact that it is much safer than the pure hydrogen system and less costly. Many types of catalysts have been used in furfural hydrodeoxygenation to 2-methylfuran. For example, Mo_2C supported on Al_2O_3 or SiO_2 as well as metallic Ru, Pt, Pd, Rh, Ag, Fe, Co, Ni, and Cu (Martín-Pérez et al. 2019).

5.13 Tetrahydrofuran

Tetrahydrofuran (THF) has a wide range of applications. This includes: precursor in anionic polymerization in fiber manufacture and urethane elastomer solvent for many chromatographic techniques, e.g., gel-phase chromatography (Liu et al. 2019). THF is normally synthesized from the hydrogenation of furan from furfural in the presence of same catalyst used for the decarbonylation of furfural. The Pd catalysts enhanced complete conversion of furan to THF while nickel-based catalysts present more attractive option than noble metal catalyst because of cost consideration. Mixing of Ni and Pd/SiO_2 catalyst has been explored and gave comparable hydrogenation activity with high furan conversion (99%) and high THF selectivity (98%) in acetic acid medium at temperature of 40 °C and pressure of 80 bar hydrogen (Liu et al. 2019). One of the major challenges in the efficient THF production process is the need to suppress the rapid coking rate of the catalysts. This process results in the catalyst deactivation, leading to poor THF yield. The coke formation during the process could be reduced by co-feeding of a hydrogen donor. It is found that a hydrogen pressure of 60 bar will achieve the optimum value for THF production.

5.14 Furan

Furan is a heterocyclic organic compound with five-membered aromatic rings, four carbon atoms, and one oxygen atom. Generally, furans referred to all the chemical compounds with such rings. Furan is a volatile, colorless, and flammable liquid with a strong ethereal chloroform-like odor and a boiling point close to room temperature (Kainulainen et al. 2020). It is slightly soluble in water but highly soluble in common organic solvents (e.g., acetone, alcohol, and ether). It is toxic and may be carcinogenic in humans (Gevrek and Sanyal 2021). Furan serves as a starting point for production of other important chemicals. For example, furan can be converted to tetrahydrofuran by hydrogenation. Tetrahydrofuran is used for production of adipic acid and hexamethylenediamine, the raw materials for nylon-6,6 (Kim et al. 2022). Furan can be produced in the laboratory by the oxidation of furfural to 2-furoic acid, accompanied by decarboxylation reaction or by thermal decomposition of pentose-containing materials, and cellulosic biomass materials. The industrial production of furan can be carried out either in two ways: (1) vapor-phase (2) liquid-phase reaction. Due to ease of operation, possibility of catalyst recycling, and simplicity of operation, the vapor phase by the use of hydrogen is technically feasible and more sustainable (Ricciardi et al. 2022). The presence of hydrogen increases the yield of furan due to the increased rate of reactant-product reaction from the catalyst surface (Nakagawa et al. 2013).

5.15 L-Arabinose

Algal cellulose can be converted to L-arabinose by acid hydrolysis accompanied by multiple purification processes (including ion exchange, neutralization reaction, and other chromatographic separations). Algal biomass materials containing hemicelluloses with high amount of xylose or xylan units in their molecules can also be converted to L-arabinose by acid hydrolysis (Arun et al. 2022). Xylose is a pentose sugar used in the production of xylitol and other sweetening additives for foods.

5.16 Pentanediols (PDO) Production Rote

Good yield of pentanediols (PDO) could be produced by conversion of furfural into THFA, followed by further hydrogenolysis over Rh-supported catalysts like SiO_2. The use of Rh (rhodium) catalysts incorporated with silica or carbon and modified with Re, Mo, or W will support the hydrogenolysis of THFA to 1,5-PDO rather than 1,2-PDO (Nakagawa et al. 2013). The production of 1,5-PDO from furfural and THFA has been conducted through well-known metal catalysts such as *rhodium-based catalysts modified with another metals* (Huang et al. 2017). Figure 4 shows a summary of the algal biomass biorefinery technologies/possible routes for phycochemical production.

6 Different Applications of Phycochemicals

Results of many research studies reported in the literature revealed that algae are promising feedstock for biorefinery and production of various phycochemicals, besides bioenergy and other valuable products (Bhatia et al. 2022). Phycochemicals (including biochemicals and bioenergy) have numerous important applications, and some of those applications are presented in the following subsections below.

6.1 Biochemicals

Previous studies revealed that biochemicals produced from algae have several applications (Rizwan et al. 2018; Russell et al. 2022; Zhou et al. 2022). For example, *galacturonic acid* which is a uronic acid derived from galactose is used as an acidifying agent in foods (Norell 2020). It is an oxidized form of galactose and a major component of pectin. Pectin on the other hand has several applications in numerous industries as it is employed in production of frozen foods, jellies, jams, and more recently as a fat and/or sugar replacer in low-calorie foods. It is also used in the pharmaceutical industry for reduction of blood cholesterol levels and gastrointestinal disorders. Likewise, glucuronic and gluconic acids are essential biochemicals that can also be obtained from algal biomass. They are also the fermentation products in Kombucha tea (Bondar et al. 2022). Glucuronic acid (GA) is a common building block of proteoglycans and glycoglycerolipids and is a cyclic organic compound that can also be isolated from urine. It is also present in different gums such as Arabic gum (18%), xanthan, and kombucha tea. GA is also very vital for the metabolism of animals, plants, and microorganisms. Glucuronic acid is also found in other constituents of the body, such as cartilage and synovial fluid (Martínez-Leal et al. 2020). On the other hand, gluconic acid is an aliphatic organic compound with the molecular formula $C_6H_{12}O_7$ and condensed structural formula $HOCH_2(CHOH)_4COOH$ that can be sustainably produced by oxidation of algal biomass-derived glucose. It is one of the 16 stereoisomers of 2,3,4,5,6-pentahydroxyhexanoic acid. Gluconic acid or gluconate is an electrolyte supplement used in total parenteral nutrition. It is also used for maintenance of the cation–anion balance in electrolyte solutions (Bondar et al. 2022). Likewise, hexoses (mainly fructose and glucose) are algal biochemicals and vital metabolic intermediates employed in the formation of storage pools of carbohydrates (majorly starches) or sucrose and other disaccharides for transportation to the rest of the organism (Zhou et al. 2022).

Pentose, a monosaccharide with five atoms of carbon, is another important biochemical that can be obtained from algae. Pentose has the chemical formula of $C_5H_{10}O_5$ and molecular weight of 150.13 g mol^{-1}. Pentoses are very important biochemicals for synthesis of various important compounds such as lactates (Gómez Millán et al. 2019). Pentoses and hexoses are the most common monosaccharides or higher sugars used for easy production of lactate than the trioses. Besides, almost all carbohydrates in nature are pentoses and hexoses, and they are therefore cheaper and more abundant feedstocks for synthesis of lactates and several other vital compounds. However, in order to employ them, it is essential to have a catalyst that is capable of catalyzing a retro-aldol reaction leading to the shorter C3 sugars, which will form lactates readily at higher reaction temperatures. Besides, minute amounts of pentose sugars (xylose, arabinose, and ribose) are present in wine, naturally. Ribose is a constituent of RNA, and the related molecule, deoxyribose, is a constituent of DNA. Likewise, xylitol is a naturally occurring pentose (C5) sugar alcohol and has important applications in food, cosmetic, confectionary, and pharmaceutical industries (Manishimwe et al. 2022; Sundar and Nampoothiri 2022). For example, it is an excellent artificial sweetener that is used widely by the confectionary industry. Also, phosphorylated pentoses (ribose 5-phosphate and erythrose 4-phosphate) are vital products of the pentose phosphate pathway. Ribose 5-phosphate (R5P) is employed in the production of nucleotides and nucleic acids, while erythrose 4-phosphate is used for aromatic amino acid synthesis (Machelart et al. 2020).

Furfuryl alcohol is another essential organic compound having a furan substituted with a hydroxymethyl group. The primary use of furfuryl alcohol is as a monomer for the synthesis of furan resins (Iroegbu and Hlangothi 2019). These polymers can be employed in cements, thermoset polymer matrix composites, coatings, adhesives, and casting/foundry resins. Furfuryl alcohol is also used in the fabrication of foundry resins which is employed in the synthesis of P-series fuels, which are liquid fuels consisting of blend of methyl tetrahydrofuran (MTHF), ethanol, and hydrocarbon (Gómez Millán and Sixta 2020). It is an essential intermediate in the production of fine biochemicals. It serves as a chemical intermediate for production of vitamin C, lysine, and levulinic acid. It can also serve as a lubricant or a dispersing agent (Gómez Millán et al. 2021). Likewise, 2-methylfuran (C_5H_6O) is another member of furans group where the hydrogen at position 2 is replaced by a methyl group. Generally, the algal biorefinery value chain integrates chemistries in the production of numerous important platform biochemicals and biofuels (such as furfural, levulinic acid, and aromatics), which have several applications (Khoo et al. 2019; Knoshaug et al. 2018).

6.2 Bioenergy

Bioenergy is a renewable energy derived from biomass or organic materials, such as algae, energy crops, forest residues, agricultural wastes, and municipal solid wastes. The applications of bioenergy include

1. *Electricity generation*: Bioenergy can be used to generate electricity using technologies such as the combustion of biomass (like the algae, wood, chips, or agricultural waste), pyrolysis, hydrothermal liquefaction, gasification, fermentation, and anaerobic digestion. Algal biomass power plants produce electricity by burning algal biomass to generate steam, which drives a turbine to produce electricity (Chia et al. 2022; Naina Mohamed et al. 2019).
2. *Transportation fuels*: Bioenergy is also used in the production of transportation fuels such as ethanol and biodiesel. Ethanol is produced by fermenting sugars from algae such as microalgae and seaweeds, while biodiesel is produced by chemically reacting algal oils with alcohol (Raheem et al. 2018).
3. *Heating and cooling*: Algal biomass can be used to heat and cool buildings through technologies such as algal biomass boilers and algal biomass-powered air conditioning. In some cases, bioenergy can be used in combined heat and

power (CHP) systems to provide both heat and electricity to buildings (Beal et al. 2018).

4. *Bioproducts*: Bioenergy can also be used to produce a range of bioproducts, including chemicals, plastics, and materials such as biodegradable plastics and bio-composites. These products can be made from renewable sources, reducing the reliance on fossil fuels (Zhou et al. 2022).
5. *Waste management*: Bioenergy can also be used as a waste management solution. For example, biomass/waste from algal processing, algal cultivation, agriculture, and forestry can be converted into bioenergy, reducing the amount of unwanted biomass/waste going to landfills and reducing greenhouse gas emissions (Zhang et al. 2022).
6. *Biogas production*: Biogas produced through the anaerobic digestion of algae and its residues can be used for heating, cooking, and electricity generation (Perendeci et al. 2019; Torres et al. 2021).
7. *Agriculture*: Algal bioenergy can be used to improve agricultural productivity by providing a source of renewable energy for irrigation, drying crops, and powering farm equipment.
8. Industrial processes: Algal bioenergy can be used in industrial processes, such as pulp and paper production, to provide heat and power (Chandra et al. 2019).

7 Recent Trends, Challenges, and Prospects of Algal Biorefinery

7.1 Recent Trends in Biochemicals

At present, multiple research efforts are in the direction of ways to produce advanced value-added biochemicals and biofuels from furfurals, obtained from algae and other biomass materials, in a sustainable manner (Cesário et al. 2018; Zhou et al. 2022). Various types of routes as well as highly active multifunctional metallic and non-metallic catalysts, which constitute future challenges, have been produced for more efficient synthesis of biochemicals from algal biomass products (Choudhary et al. 2020). One step advanced hydrolysis coupled with thermochemical co-processing for algal biomass valorization to biochemicals and biofuels is the kind of technology required. The development of new algal biorefinery processes for co-production in a single-pot synthesis constitutes extreme challenges. Emphasis is placed on batch and continuous processes using acid hydrolysis in the presence of homocatalyst or heterogenous catalyst, which will help maximize the production of furfural and its derivative products from algae (Martín and Grossmann 2016; Sun et al. 2020). In the future, research concerns will be the development of more active catalysts and co-catalysts for biochemical and biofuel production, advancing chemical reaction routes/pathways, improve the knowledge of kinetic and thermodynamic behavior and improve product yield by integrating reactions and product separation.

The massive production of algal biomass conversion in commercial-scale levels is seen to be the main driver of circular and bio-economy (Chisti 2016). The algal biomass hydrolysis with solid catalysts, gasification, and pyrolysis is the conventional processes which can be employed for wide ranging feedstocks. However, variation in products quality and quantity will be noticed due to different nature and composition of feedstocks. The major hindrances to large-scale implementation of algal biochemical refinery industry are due to competitiveness and economy of scale from petroleum and biochemical refinery (Siddiki et al. 2022; Thanigaivel et al. 2022b). Bulk of the present research is on laboratory scale. The scale-up of research outputs is expected to be a major limitation, as well as cost-effectiveness and quality of fractionated products, before green refinery concept and large-scale production of biochemicals will be achieved. In order to maintain the quality of the raw algal biomass products, standard method for collecting, harvesting, storing, and classification needs to be developed. The use of solid catalysts can produce high product quality with minimal wastes and residue content in the chemical conversion process. The research and development of C5 biochemical produced from algal furfural portends huge future opportunities in terms of design and development; and the control of reaction process, which will directly affect the selectivity of tetrahydrofurfuryl furfuryl alcohol, 2-methylfuran, and other high-valued biochemical products (Abomohra and Elshobary 2019). The use of catalytic liquid-phase hydrogenation under alcohol and hydrogen donor is preferred overusing pure hydrogen due to safety and economic considerations. The key issue for 2-methylfuran, levulinic acid, γ-valerolactone, furan, pentanediol, cyclopentanone, and furfuryl alcohol yields from algal biomass is the development of multifunctional novel catalyst with high activity and selectivity which will aid in improvement of overall process performance and minimize waste generated. More so, a thorough understanding of reaction mechanism, engineering design of active catalyst sites will aid in the research and development of C_5 biochemicals produced from algal furfurals. It can also pave the way to produce other potential high-value biochemicals such as feed/food, active compounds, and pharmaceuticals. Research on furfural production from algal sugar platform has gained global traction since it was first reported. (Ahorsu et al. 2018; Hansen et al. 2017). Several researches have focused on process development, technological system of production, and catalyst design. Catalytic upgrading of the furfural/furan-based compounds into biochemicals and biofuels has attracted significant research attention (Gu et al. 2020; Wu and Chang 2019).

7.2 Recent Trends in Bioenergy

Algae represent a promising source of bioenergy because it can be grown quickly and efficiently. Algal biomass contains high levels of lipids and/or carbohydrates that can be converted into biofuels. Many recent advancements including integration of algae cultivation with other renewable energy systems, genetic engineering, wastewater treatment, and carbon capturing have been suggested. Here are some recent trends in advanced integrated bioenergy production from algal biomass;

1. *Hybrid cultivation systems*: Many researchers are developing hybrid systems that combine algae cultivation with other forms of renewable energy, such as solar or wind power. Hybrid systems combine the photoautotrophic and heterotrophic growth of algae. Advances in cultivation systems and methods such as photobioreactors and open ponds have led to increased biomass productivity and reduced production costs (Kumar et al. 2022; Xiong et al. 2021).
2. *Genetic engineering*: Scientists are exploring ways to genetically modify algae to produce higher levels of lipids and other compounds that can be used for biofuels and also improve their efficiency in converting sunlight and CO_2 into biomass with valuable phycochemical production. This can increase the yield and quality of bioenergy from algal biomass at acceptable production cost. For instance, a study reported the possible protoplast fusion of lipid/astaxanthin-rich microalgae (*Haematococcus pluvialis*) with free fatty acid-secreting microalga (*Ochromonas danica*). The study reported a successful genetic recombination where a hybrid organism was obtained with ability to produce both lipids/astaxanthin and free fatty acid secretion (Abomohra et al. 2016a).
3. *Wastewater treatment*: Algae can be employed in wastewater treatment and removal of pollutants, while also producing biomass for bioenergy and other value-added compounds production. For instance, optimization of nitrogen removal from wastewater coupled with biodiesel production using the green microalga *Chlorella* sp. was evaluated by (Abomohra et al. 2022). The results reported 98.0% nitrogen removal under optimized conditions, with simultaneous increase in the lipid content and dry weight of 20.3% and 31.9% respectively, over the control, which led to a rise of 71.5% in lipid productivity. Besides, biochar-derived from seaweeds showed high potential for wastewater treatment. In that context, Jiang et al. (2023) investigated the mechanism of the adsorption of methylene blue (MB) on carbon derived from a seaweed biomass. In the study, an effective endogenous nitrogen-modified seaweed-activated carbon was prepared from *Enteromorpha* seaweed by pyrolysis and employed it in the adsorption of the MB. Results unveiled that the addition of 0.1 g activated carbon to 100 mg L^{-1} MB solution at pH 5 and 30 °C led to the highest removal efficiency of 100% for MB. It was also observed that the adsorbent still displayed 86% adsorption capability even after four cycles of recovery. Molecular dynamic simulation results revealed that the adsorbent displayed a high adsorption capacity for MB.
4. *Algae-based carbon capture*: Algae can also be used for carbon capture and sequestration. By using algae to capture carbon dioxide emissions from power plants, for example, the resulting biomass can then be used for bioenergy production. A recent study investigated the effect of tobacco smoke on the biochemical compositions, cell growth, and biodiesel characteristics of two strains of model *Chlamydomonas* microalgae, CHL-2220 and CHL-2221 (Barati et al. 2022). Results revealed that the specific growth rate of CHL-2220 remained unaffected (i.e., around 0.5 $days^{-1}$), whereas in CHL-2221, growth reduced

significantly from 0.45 days^{-1} to 0.38 days^{-1} upon exposure to tobacco smoke. Also, a considerable decrease from 15.55% DW to 13.37% DW was noticed in the lipid level of CHL-2221 upon exposure to tobacco smoke. Both strains displayed low-quality biodiesel; however, their fatty acid profiles showed that they are promising nutrient food. Biomass productivity and efficiency of CO_2 sequestration have been observed to increase by improvement in photosynthesis. This improvement in photosynthesis can be achieved by improving the efficiency of enzymes involved in CO_2 fixation, extending the photosynthetically active radiation range to widen the light consumption ability, decreasing the antenna size to prevent energy loss, increasing CO_2 absorption by substituting the existing carbon fixation pathway with more effective pathways and enzymes, and lowering the release of CO_2 captured (Barati et al. 2021).

5. *Extraction methods*: Researchers are developing more efficient methods of extracting lipids and other valuable chemicals from algal biomass. This includes using solvents, supercritical fluid extraction, and ultrasound-assisted extraction. For instance, Abomohra et al. (2016b) investigated an optimized procedure for recovery of esterified fatty acids (EFAs) from the biodiesel promising microalga *Scenedesmus obliquus*. They also examined the effect of diverse solvent blends, pretreatments, time of extraction, and cell-disruption methods on intracellular EFAs and free fatty acids (FFAs) yields. Results revealed that the best solvent blend for the extraction of lipid is the use of chloroform: methanol at the blending ratio of 2:1 for 2 h as it resulted in the highest yield of EFAs. The results also showed that cell disruption is not vital in lipid extraction from *S. obliquus* cells, and besides, it was also noticed that the hot-water pretreatment deactivated the lipases and improved the recovery of EFAs.
6. *Co-production of value-added products*: Biomass from algae can be employed in producing not only biofuels but also other value-added products such as pigments, omega-3 fatty acids, and bioplastics. Co-production of these products can make algal bioenergy more economically viable (Dineshkumar and Sen 2020).

7.3 Challenges in Algal Biorefinery

Algal biorefinery for large-scale production of phycochemicals is faced with numerous obstacles despite its potential benefits. There is therefore a crucial need to address such challenges (as summarized below) to enhance the development and commercialization of the algal biorefinery.

1. *Low conversion efficiency*: The conversion efficiency of algal biomass into biofuels and other algae-based products such as the biochemicals is still relatively low, which limits its commercial viability. Rony et al. (2023) reported that higher yields of microalgae-derived gaseous, solid, and liquid fuels can be obtained by pretreating the microalgal biomass and employment of the appropriate bioconversion processes.

2. *High production costs*: The cost of producing algal biomass is currently high due to the high cost of inputs such as nutrients and energy. (Branco-Vieira et al. 2020) performed the economic analysis of biodiesel production from microalgae in a small-scale facility using a model which assumed 80,000 m^3 of microalgal cultivation, in a set of bubble column photobioreactors installed on 15.247 ha of land, reaching a total of 1811 tons of microalgae biomass and 171,705 L of biodiesel per year. The study results showed that the production cost estimated for microalgae biomass is 2.01 € kg^{-1} and for biodiesel is 0.33 € L^{-1} with biodiesel standing out as the most economic viable option. Thus, the results revealed that despite the project's viability in the medium term, the costs of producing microalgae biomass and biodiesel remain high when compared to fossil fuels. This implies that unless greater technological development is achieved to make the process more economical, it will not be viable in the short term. Therefore, the cost of phycochemicals in general is much higher than the cost of producing conventional synthetic chemicals based on fossil energy.
3. *Nutrient availability*: Algal biomass requires large amounts of nutrients such as nitrogen and phosphorus to grow, which can be expensive and environmentally problematic. To avoid such obstacle, wastewater was suggested to be used for algae cultivation. However, cultivation of algae on wastewater limits the utilization of algal biomass due to possibility of contamination with undesired pollutants or pathogens.
4. *Harvesting and processing challenges*: Harvesting and processing of algal biomass can be energy-intensive and costly. In that regard, Rafa et al. (2021) reported that the conventional harvesting methods for algal biomass such as flocculation, filtration, and centrifugation costs are $*2000, $9884, and $12,500 USD per square hectometer, respectively, which revealed that they are indeed high.*
5. *Scale-up challenges*: Scaling up algal biorefinery from laboratory to commercial scale is difficult due to the complex nature of algal biology, maintaining optimal growth conditions, harvesting and processing of biomass, and managing waste streams. For example, Borowitzka and Vonshak (2017) reported that it is difficult to scale up algal cultures to larger volumes for commercial production since it involves series of complex operations requiring skilled and experienced personnel. First, optimization of the production process and hence the quantity of the inoculum for the large ponds or photobioreactors to reduce the time and cost. Secondly, proper management of the large-scale cultures to prevent substantial contamination or collapse is imperative to minimize the need for re-inoculation. Likewise, further challenges experienced by small-scale laboratory cultures which are not witnessed in the constant environment include among others the maintenance of long-term, stable, high-productivity, large-scale cultures under prevailing outdoor conditions of variable irradiance, temperature, and rainfall.

7.4 *Prospects of Algal Biorefinery*

Algae biorefinery holds great prospect as *a sustainable alternative for the production of biofuels* and a wide range of other high value-added biochemicals. It has the potential to serve the energy, medical, nutraceutical, food and cosmetics industries by using appropriate species and cultivation conditions to produce the desired products. Algal cells can produce different pigments such as chlorophylls, carotenoids, and phycobilins, which can be used to replace synthetic dyes that are derived from fossil resources. In addition, fossil resources may contain lead impurities that is harmful to humans and usually responsible for allergens and irritants (Okeke et al. 2022; Sharma et al. 2022; Thanigaivel et al. 2022a). Chlorophyll, when used to dye wool and derivatives, are degradable and eco-friendly whereas synthetic colorants are hash, recalcitrant and typically not recyclable. In addition, bioplastic was produced from algae that can produce polyhydroxyalkanoates (PHA). In fact, studies conducted with *Chlorella pyrenoidosa* showed a PHA accumulation of 27% (Cinar et al. 2020). Pigment extraction, followed by PHA separation for bioplastics, could reduce the overall production cost and enhances the economic feasibility of algal biomass.

8 Conclusions

The overview of the technologies/possible routes of algal biorefinery for enhanced biomass utilization, most especially for production of several industrially important phycochemicals coupled with bioenergy, has been provided. Thus, in this chapter, several promising routes or pathways for producing bioenergy and phycochemicals from algal biomass, such as fermentation (bioethanol, butanol, etc.,), anaerobic digestion (biogas), transesterification (biodiesel), enzymatic synthesis (1,3-propanediol and alkylpolyglucoside), biological oxidation (P-xylene), and enzymatic hydrolysis (glucose and fructose), have been identified for enhancement of algal biorefinery. It is therefore recommended based on the findings of this work that shifting focus of algae utilization, processing routes, and investigations from biofuels production to biorefinery co-products production that could guarantee more viable and profitable resources. Due to the importance of biofuel production from algal biomass, integrated biofuel routes are discussed further in the next chapter.

References

Abomohra AE-F, Elshobary M (2019) Biodiesel, bioethanol, and biobutanol production from microalgae. In: Microalgae biotechnology for development of biofuel and wastewater treatment. Springer, Singapore, pp 293–321

Abomohra AE-F, El-Sheekh M, Hanelt D (2016a) Protoplast fusion and genetic recombination between Ochromonas danica (Chrysophyta) and Haematococcus pluvialis (Chlorophyta). Phycologia 55(1):65–71

Abomohra AE-F, Jin W, El-Sheekh M (2016b) Enhancement of lipid extraction for improved biodiesel recovery from the biodiesel promising microalga Scenedesmus obliquus. Energy Convers Manag 108:23–29

Abomohra AE-F, Elade H, El-Esawi M, Wang S, Wang Q, He Z, Feng Y, Shang H, Hanelt D (2018) Effect of lipid-free microalgal biomass and waste glycerol on growth and lipid production of Scenedesmus obliquus: innovative waste recycling for extraordinary lipid production. Bioresour Technol 249:992–999

Abomohra AE-F, Eladel H, Mohammed S (2022) Dual use of a local Protosiphon isolate BENHA2020 for biodiesel production and antioxidant activity of lipid-free biomass: A novel biorefinery approach for biomass valorization. Renew Energy 184:1104–1111

Ahorsu R, Medina F, Constantí M (2018) Significance and challenges of biomass as a suitable feedstock for bioenergy and biochemical production: A review. Energies 11(12):3366

Ak İ, Koru E, Türker G, Çankırılıgil EC, Dereli MG (2022) Biochemical compounds of algae: sustainable energy sources for biofuel production. In: Handbook of Algal Biofuels. Elsevier, Amsterdam, pp 57–78

Al-Humairi ST, Lee JG, Harvey AP (2022) Direct and rapid production of biodiesel from algae foamate using a homogeneous base catalyst as part of an intensified process. Energy Convers Manage: X 16:100284

Ali M, Sultana R, Tahir S, Watson IA, Saleem M (2017) Prospects of microalgal biodiesel production in Pakistan–A review. Renew Sust Energ Rev 80:1588–1596

Aravind S, Barik D, Ragupathi P, Vignesh G (2021) Investigation on algae oil extraction from algae Spirogyra by Soxhlet extraction method. Mater Today Proc 43:308–313

Arun J, Vigneshwar SS, Swetha A, Gopinath KP, Basha S, Brindhadevi K, Pugazhendhi A (2022) Bio-based algal (Chlorella vulgaris) refinery on de-oiled algae biomass cake: A study on biopolymer and biodiesel production. Sci Total Environ 816:151579

Balina K, Romagnoli F, Blumberga D (2017) Seaweed biorefinery concept for sustainable use of marine resources. Energy Procedia 128:504–511

Barati B, Zeng K, Baeyens J, Wang S, Addy M, Gan S-Y, Abomohra AE-F (2021) Recent progress in genetically modified microalgae for enhanced carbon dioxide sequestration. Biomass Bioenergy 145:105927

Barati B, Fazeli Zafar F, Amani Babadi A, Hao C, Qian L, Wang S, Abd Abomohra E-F (2022) Microalgae as a natural CO_2 sequester: a study on effect of tobacco smoke on two microalgae biochemical responses. Front Energy Res 10:881758

Barbaro E, Feltracco M, Cesari D, Padoan S, Zangrando R, Contini D, Barbante C, Gambaro A (2019) Characterization of the water soluble fraction in ultrafine, fine, and coarse atmospheric aerosol. Sci Total Environ 658:1423–1439

Beal CM, Archibald I, Huntley ME, Greene CH, Johnson ZI (2018) Integrating algae with bioenergy carbon capture and storage (ABECCS) increases sustainability. Earth's Future 6(3): 524–542

Bello BZ, Nwokoagbara E, Wang M (2012) Comparative techno-economic analysis of biodiesel production from microalgae via transesterification methods. In: Computer aided chemical engineering, vol 30. Elsevier, Amsterdam, pp 132–136

Belousov AS, Esipovich AL, Kanakov EA, Otopkova KV (2021) Recent advances in sustainable production and catalytic transformations of fatty acid methyl esters. Sustain Energy Fuels 5(18): 4512–4545

Benemann JR, Weissman JC (1977) Biophotolysis: problems and prospects. In: Microbial energy conversion. Elsevier, Amsterdam, pp 413–426

Bhatia SK, Ahuja V, Chandel N, Gurav R, Bhatia RK, Govarthanan M, Tyagi VK, Kumar V, Pugazendhi A, Banu JR (2022) Advances in algal biomass pretreatment and its valorisation into biochemical and bioenergy by the microbial processes. Bioresour Technol 358:127437

Bilal M, Iqbal HM (2020) Recent advancements in the life cycle analysis of lignocellulosic biomass. Current Sustainable/Renewable Energy Reports 7:100–107

Bondar M, Pedro F, Oliveira MC, da Fonseca MMR, Cesário MT (2022) Red algae industrial residues as a sustainable carbon platform for the co-production of poly-3-hydroxybutyrate and gluconic acid by Halomonas boliviensis. Front Bioeng Biotechnol 10:934432

Borowitzka MA, Vonshak A (2017) Scaling up microalgal cultures to commercial scale. Eur J Phycol 52(4):407–418

Branco-Vieira M, Mata T, Martins A, Freitas M, Caetano N (2020) Economic analysis of microalgae biodiesel production in a small-scale facility. Energy Rep 6:325–332

Cesário MT, da Fonseca MMR, Marques MM, de Almeida MCM (2018) Marine algal carbohydrates as carbon sources for the production of biochemicals and biomaterials. Biotechnol Adv 36(3):798–817

Chandra R, Iqbal HM, Vishal G, Lee H-S, Nagra S (2019) Algal biorefinery: a sustainable approach to valorize algal-based biomass towards multiple product recovery. Bioresour Technol 278: 346–359

Cheng H-H, Whang L-M, Chan K-C, Chung M-C, Wu S-H, Liu C-P, Tien S-Y, Chen S-Y, Chang J-S, Lee W-J (2015) Biological butanol production from microalgae-based biodiesel residues by Clostridium acetobutylicum. Bioresour Technol 184:379–385

Cherubini F (2010) The biorefinery concept: using biomass instead of oil for producing energy and chemicals. Energy Convers Manag 51(7):1412–1421

Chew KW, Yap JY, Show PL, Suan NH, Juan JC, Ling TC, Lee D-J, Chang J-S (2017) Microalgae biorefinery: high value products perspectives. Bioresour Technol 229:53–62

Chia SR, Chew KW, Show PL, Yap YJ, Ong HC, Ling TC, Chang JS (2018) Analysis of economic and environmental aspects of microalgae biorefinery for biofuels production: a review. Biotechnol J 13(6):1700618

Chia SR, Nomanbhay SBHM, Chew KW, Munawaroh HSH, Show PL (2022) Algae as potential feedstock for various bioenergy production. Chemosphere 287:131944

Chisti Y (2016) Large-scale production of algal biomass: raceway ponds. In: Algae biotechnology: products and processes. Springer, New York, pp 21–40

Choudhary P, Assemany PP, Naaz F, Bhattacharya A, de Siqueira Castro J, do Couto Couto EdA, Calijuri ML, Pant KK, Malik A (2020) A review of biochemical and thermochemical energy conversion routes of wastewater grown algal biomass. Sci Total Environ 726:137961

Chozhavendhan, S., Karthigadevi, G., Praveenkumar, R., Aniskumar, M., Jayakumar, M., Gurunathan, B. 2022. Potentials and challenges in biodiesel production from algae—technological outlook. in: *Biofuels and bioenergy*, Elsevier, Amsterdam, pp. 183–203

Chronakis IS, Madsen M (2011) Algal proteins. In: Handbook of food proteins. Elsevier, Amsterdam, pp 353–394

Chu R, Li S, Zhu L, Yin Z, Hu D, Liu C, Mo F (2021) A review on co-cultivation of microalgae with filamentous fungi: Efficient harvesting, wastewater treatment and biofuel production. Renew Sust Energ Rev 139:110689

Cinar R, Iyer MR, Kunos G (2020) The therapeutic potential of second and third generation CB1R antagonists. Pharmacol Ther 208:107477

Conteratto C, Artuzo FD, Santos OIB, Talamini E (2021) Biorefinery: a comprehensive concept for the sociotechnical transition toward bioeconomy. Renew Sust Energ Rev 151:111527

de Jong E, Jungmeier G (2015) Biorefinery concepts in comparison to petrochemical refineries. In: Industrial biorefineries & white biotechnology. Elsevier, Amsterdam, pp 3–33

Debnath D, Das A (2022) Third-generation biofuels and food security. In: 3rd generation biofuels. Elsevier, Amsterdam, pp 757–768

Dineshkumar R, Sen R (2020) A sustainable perspective of microalgal biorefinery for co-production and recovery of high-value carotenoid and biofuel with CO2 valorization. Biofuels Bioprod Biorefin 14(4):879–897

Duppeti H, Chakraborty S, Das BS, Mallick N, Kotamreddy J (2017) Rapid assessment of algal biomass and pigment contents using diffuse reflectance spectroscopy and chemometrics. Algal Res 27:274–285

Efremenko E, Nikolskaya A, Lyagin I, Senko O, Makhlis T, Stepanov N, Maslova O, Mamedova F, Varfolomeev S (2012) Production of biofuels from pretreated microalgae biomass by anaerobic fermentation with immobilized Clostridium acetobutylicum cells. Bioresour Technol 114:342–348

Fawaz EG, Salam DA, Kamareddine L (2018) Evaluation of copper toxicity using site specific algae and water chemistry: field validation of laboratory bioassays. Ecotoxicol Environ Saf 155:59–65

Fernández FGA, Reis A, Wijffels RH, Barbosa M, Verdelho V, Llamas B (2021) The role of microalgae in the bioeconomy. New Biotechnol 61:99–107

Ferreira AF (2017) Biorefinery concept. Biorefineries: targeting energy, high value products and waste valorisation. Springer, New York, pp 1–20

Gameiro D (2016) *Green waste is Biotech's new black gold: a review of biorefineries*. Retrieved September, 10, 2017

Gevrek TN, Sanyal A (2021) Furan-containing polymeric Materials: Harnessing the Diels-Alder chemistry for biomedical applications. Eur Polym J 153:110514

Gómez Millán G, Sixta H (2020) Towards the green synthesis of furfuryl alcohol in a one-pot system from xylose: a review. Catalysts 10(10):1101

Gómez Millán G, Hellsten S, Llorca J, Luque R, Sixta H, Balu AM (2019) Recent advances in the catalytic production of platform chemicals from holocellulosic biomass. ChemCatChem 11(8): 2022–2042

Gómez Millán G, Bangalore Ashok RP, Oinas P, Llorca J, Sixta H (2021) Furfural production from xylose and birch hydrolysate liquor in a biphasic system and techno-economic analysis. Biomass Convers Biorefin 11:2095–2106

Gu X, Martinez-Fernandez J, Pang N, Fu X, Chen S (2020) Recent development of hydrothermal liquefaction for algal biorefinery. Renew Sust Energ Rev 121:109707

Halder P, Azad A (2019) Recent trends and challenges of algal biofuel conversion technologies. In: Advanced biofuels. Elsevier, Amsterdam, pp 167–179

Hansen ASL, Lennen RM, Sonnenschein N, Herrgård MJ (2017) Systems biology solutions for biochemical production challenges. Curr Opin Biotechnol 45:85–91

Hassan A, Ilyas SZ, Jalil A, Ullah Z (2021) Monetization of the environmental damage caused by fossil fuels. Environ Sci Pollut Res 28:21204–21211

Hingsamer M, Jungmeier G (2019) Biorefineries. In: The role of bioenergy in the bioeconomy. Elsevier, Amsterdam, pp 179–222

Hronec M, Fulajtárova K, Mičušik M (2013) Influence of furanic polymers on selectivity of furfural rearrangement to cyclopentanone. Appl Catal A Gen 468:426–431

Hua L-C, Tsia SR, Ngo DNG, Huang C (2021) Bromide-intrusion into Chlorella sp. and Microcystis aeruginosa growing environments: Its impacts on algal growth and the formation potential of algal-derived DBPs upon chlorination. Sci Total Environ 795:148772

Huang K, Brentzel ZJ, Barnett KJ, Dumesic JA, Huber GW, Maravelias CT (2017) Conversion of furfural to 1, 5-pentanediol: Process synthesis and analysis. ACS Sustain Chem Eng 5(6): 4699–4706

Ighalo JO, Dulta K, Kurniawan SB, Omoarukhe FO, Ewuzie U, Eshiemogie SO, Ojo AU, Abdullah SRS (2022) Progress in microalgae application for CO2 sequestration. *Cleaner*. Chem Eng 100044:100044

Iroegbu AO, Hlangothi SP (2019) Furfuryl alcohol a versatile, eco-sustainable compound in perspective. Chem Afr 2(2):223–239

Jiang D, Li H, Cheng X, Ling Q, Chen H, Barati B, Yao Q, Abomohra A, Hu X, Bartocci P (2023) A mechanism study of methylene blue adsorption on seaweed biomass derived carbon: From macroscopic to microscopic scale. Process Saf Environ Prot 172:1132–1143

Kainulainen TP, Erkkilä P, Hukka TI, Sirviö JA, Heiskanen JP (2020) Application of furan-based dicarboxylic acids in bio-derived dimethacrylate resins. ACS Appl Polym Mater 2(8): 3215–3225

Khoo CG, Dasan YK, Lam MK, Lee KT (2019) Algae biorefinery: Review on a broad spectrum of downstream processes and products. Bioresour Technol 292:121964

Kim M, Yang J, Kim B, Lee JW (2020) Production of levulinic acid from wet microalgae in a biphasic one-pot reaction process. Korean J Chem Eng 37:1933–1941

Kim D, Lee H-J, Shimizu Y, Yoshida J-I, Kim H (2022) Direct C–H metallation of tetrahydrofuran and application in flow. Nat Synth 1(7):558–564

Knoshaug EP, Dong T, Spiller R, Nagle N, Pienkos PT (2018) Pretreatment and fermentation of salt-water grown algal biomass as a feedstock for biofuels and high-value biochemicals. Algal Res 36:239–248

Koyande AK, Show P-L, Guo R, Tang B, Ogino C, Chang J-S (2019) Bio-processing of algal bio-refinery: a review on current advances and future perspectives. Bioengineered 10(1): 574–592

Kujawska N, Talbierz S, Dębowski M, Kazimierowicz J, Zieliński M (2021) Cultivation method effect on Schizochytrium sp. biomass growth and docosahexaenoic acid (DHA) production with the use of waste glycerol as a source of organic carbon. Energies 14(10):2952

Kumar RR, Mukund S (2018) Comprehensive review on lipid extraction methods for the production microalgae biofuel. J Algal Biomass Util 9(1):48–61

Kumar A, Acharya P, Jaiman V (2022) Third-generation hybrid technology for algal biomass production, wastewater treatment, and greenhouse gas mitigation. In: Innovations in environmental biotechnology. Springer, Cham, pp 227–263

Li M, Alotaibi MKH, Li L, Abomohra AE-F (2022) Enhanced waste glycerol recycling by yeast for efficient biodiesel production: Towards waste biorefinery. Biomass Bioenergy 159:106410

Liu Z, Kim J, Lei L, Ning F, Dai S (2019) Tetrahydrofuran hydrate in clayey sediments—laboratory formation, morphology, and wave characterization. J Geophys Res Solid Earth 124(4): 3307–3319

Machelart A, Willemart K, Zúñiga-Ripa A, Godard T, Plovier H, Wittmann C, Moriyón I, De Bolle X, Van Schaftingen E, Letesson J-J (2020) Convergent evolution of zoonotic Brucella species toward the selective use of the pentose phosphate pathway. Proc Natl Acad Sci 117(42): 26374–26381

Maghzian A, Aslani A, Zahedi R, Yaghoubi M (2023) How to effectively produce value-added products from microalgae? Renew Energy 204:262–276

Mahato N, Sharma K, Sinha M, Dhyani A, Pathak B, Jang H, Park S, Pashikanti S, Cho S (2021) Biotransformation of citrus waste-I: Production of biofuel and valuable compounds by fermentation. PRO 9(2):220

Maity SK (2015) Opportunities, recent trends and challenges of integrated biorefinery: Part I. Renew Sust Energ Rev 43:1427–1445

Maltsev Y, Maltseva K, Kulikovskiy M, Maltseva S (2021) Influence of light conditions on microalgae growth and content of lipids, carotenoids, and fatty acid composition. Biology 10(10):1060

Manishimwe C, Feng Y, Sun J, Pan R, Jiang Y, Jiang W, Zhang W, Xin F, Jiang M (2022) Biological production of xylitol by using nonconventional microbial strains. World J Microbiol Biotechnol 38(12):249

Martín M, Grossmann IE (2016) Optimal production of furfural and DMF from algae and switchgrass. Ind Eng Chem Res 55(12):3192–3202

Martínez-Leal J, Ponce-García N, Escalante-Aburto A (2020) Recent evidence of the beneficial effects associated with glucuronic acid contained in kombucha beverages. Curr Nutr Rep 9:163–170

Martín-Pérez C, Navas JF, Perales JC, López-Martín Á, Cordovilla-Guardia S, Portillo M, Maldonado A, Vilar-López R (2019) Brief group-delivered motivational interviewing is equally effective as brief group-delivered cognitive-behavioral therapy at reducing alcohol use in risky college drinkers. PLoS One 14(12):e0226271

Martins F, Felgueiras C, Smitkova M, Caetano N (2019) Analysis of fossil fuel energy consumption and environmental impacts in European countries. Energies 12(6):964

Mikucka W, Witońska I, Zielińska M, Bułkowska K, Binczarski M (2023) Concept for the valorization of cereal processing waste: Recovery of phenolic acids by using waste-derived tetrahydrofurfuryl alcohol and biochar. Chemosphere 313:137457

Moncada J, Aristizábal V (2016) Design strategies for sustainable biorefineries. Biochem Eng J 116:122–134

Nagarajan D, Lee D-J, Kondo A, Chang J-S (2017) Recent insights into biohydrogen production by microalgae–From biophotolysis to dark fermentation. Bioresour Technol 227:373–387

Naina Mohamed S, Jayabalan T, Muthukumar K (2019) Simultaneous bioenergy generation and carbon dioxide sequestration from food wastewater using algae microbial fuel cell. Energy Sources A: Recovery Util Environ Eff 45:1–9

Nakagawa Y, Tamura M, Tomishige K (2013) Catalytic reduction of biomass-derived furanic compounds with hydrogen. ACS Catal 3(12):2655–2668

Nisar S, Hanif MA, Rashid U, Hanif A, Akhtar MN, Ngamcharussrivichai C (2021) Trends in widely used catalysts for fatty acid methyl esters (Fame) production: a review. Catalysts 11(9): 1085

Norell I (2020) Phaeodactylum tricornutum–compositional analysis, carbohydrate-active enzymes and potential applications of residual algal biomass from omega 3 production

Nurdiawati A, Zaini IN, Irhamna AR, Sasongko D, Aziz M (2019) Novel configuration of supercritical water gasification and chemical looping for highly-efficient hydrogen production from microalgae. Renew Sust Energ Rev 112:369–381

Obiora C (2022) Fuel properties comparison of biodiesel produced from Jatropha curcas oil and castor oil: direct transesterification without pretreatment *Available at SSRN* 4225428

Okeke ES, Ejeromedoghene O, Okoye CO, Ezeorba TPC, Nyaruaba R, Ikechukwu CK, Oladipo A, Orege JI (2022) Microalgae biorefinery: an integrated route for the sustainable production of high-value-added products. Energy Convers Manag X:100323

Okoro V, Azimov U, Munoz J, Hernandez HH, Phan AN (2019) Microalgae cultivation and harvesting: Growth performance and use of flocculants-A review. Renew Sust Energ Rev 115:109364

Patel AK, Albarico FPJB, Perumal PK, Vadrale AP, Nian CT, Chau HTB, Anwar C, ud din Wani HM, Pal A, Saini R (2022) Algae as an emerging source of bioactive pigments. Bioresour Technol 351:126910

Perendeci NA, Yılmaz V, Taştan BE, Gökgöl S, Fardinpoor M, Namlı A, Steyer J-P (2019) Correlations between biochemical composition and biogas production during anaerobic digestion of microalgae and cyanobacteria isolated from different sources of Turkey. Bioresour Technol 281:209–216

Pirzadi Z, Meshkani F (2022) From glycerol production to its value-added uses: A critical review. Fuel 329:125044

Rafa N, Ahmed SF, Badruddin IA, Mofijur M, Kamangar S (2021) Strategies to produce cost-effective third-generation biofuel from microalgae. Front Energy Res 9:749968

Raheem A, Prinsen P, Vuppaladadiyam AK, Zhao M, Luque R (2018) A review on sustainable microalgae based biofuel and bioenergy production: Recent developments. J Clean Prod 181: 42–59

Reid WV, Ali MK, Field CB (2020) The future of bioenergy. Glob Chang Biol 26(1):274–286

Ricciardi L, Verboom W, Lange J-P, Huskens J (2022) Production of furans from C 5 and C 6 sugars in the presence of polar organic solvents. Sustain Energy Fuels 6(1):11–28

Rizwan M, Mujtaba G, Memon SA, Lee K, Rashid N (2018) Exploring the potential of microalgae for new biotechnology applications and beyond: a review. Renew Sust Energ Rev 92:394–404

Rony ZI, Mofijur M, Hasan M, Ahmed SF, Badruddin IA, Khan TY (2023) Conversion of algal biomass into renewable fuel: a mini review of chemical and biochemical processes. Front Energy Res 11:1124302

Russell C, Rodriguez C, Yaseen M (2022) High-value biochemical products & applications of freshwater eukaryotic microalgae. Sci Total Environ 809:151111

Saad MG, Dosoky NS, Zoromba MS, Shafik HM (2019) Algal biofuels: current status and key challenges. Energies 12(10):1920

Salman JM, Majrashi N, Hassan FM, Al-Sabri A, Jabar EA-A, Ameen F (2023) Cultivation of blue green algae (Arthrospira platensis Gomont, 1892) in wastewater for biodiesel production. Chemosphere 335:139107

Sarwer A, Hamed SM, Osman AI, Jamil F, Al-Muhtaseb AaH, Alhajeri NS, Rooney DW (2022) Algal biomass valorization for biofuel production and carbon sequestration: a review. Environ Chem Lett 20(5):2797–2851

Shahid A, Khan F, Ahmad N, Farooq M, Mehmood MA (2020) Microalgal carbohydrates and proteins: synthesis, extraction, applications, and challenges. In: Microalgae biotechnology for food, health and high value products. Springer, Singapore, pp 433–468

Sharma P, Gujjala LKS, Varjani S, Kumar S (2022) Emerging microalgae-based technologies in biorefinery and risk assessment issues: bioeconomy for sustainable development. Sci Total Environ 813:152417

Siddiki SYA, Mofijur M, Kumar PS, Ahmed SF, Inayat A, Kusumo F, Badruddin IA, Khan TY, Nghiem L, Ong HC (2022) Microalgae biomass as a sustainable source for biofuel, biochemical and biobased value-added products: an integrated biorefinery concept. Fuel 307:121782

Singh N, Singhania RR, Nigam PS, Dong C-D, Patel AK, Puri M (2022) Global status of lignocellulosic biorefinery: Challenges and perspectives. Bioresour Technol 344:126415

Solarin SA (2020) An environmental impact assessment of fossil fuel subsidies in emerging and developing economies. Environ Impact Assess Rev 85:106443

Srinivasan B, Kulshreshtha G (2020) Algal biomass for biofuels and bioproducts. Bioprocess engineering for bioremediation: valorization and management techniques. Springer, Cham, pp 139–160

Sun C, Liao Q, Xia A, Fu Q, Huang Y, Zhu X, Zhu X, Wang Z (2020) Degradation and transformation of furfural derivatives from hydrothermal pre-treated algae and lignocellulosic biomass during hydrogen fermentation. Renew Sust Energ Rev 131:109983

Sundar ML, Nampoothiri KM (2022) An overview of the metabolically engineered strains and innovative processes used for the value addition of biomass derived xylose to xylitol and xylonic acid. Bioresour Technol 345:126548

Thanigaivel S, Priya A, Kumar PS, Shiong KK, Hoang TK, Rajendran S, Soto-Moscoso M (2022a) Exploration of effective biorefinery approach to obtain the commercial value-added products from algae. Sustain Energy Technol Assess 53:102450

Thanigaivel S, Vickram S, Dey N, Gulothungan G, Subbaiya R, Govarthanan M, Karmegam N, Kim W (2022b) The urge of algal biomass-based fuels for environmental sustainability against a steady tide of biofuel conflict analysis: Is third-generation algal biorefinery a boon? Fuel 317: 123494

Tong KTX, Tan IS, Foo HCY, Tiong ACY, Lam MK, Lee KT (2021) Third-generation L-Lactic acid production by the microwave-assisted hydrolysis of red macroalgae Eucheuma denticulatum extract. Bioresour Technol 342:125880

Torres A, Padrino S, Brito A, Díaz L (2021) Biogas production from anaerobic digestion of solid microalgae residues generated on different processes of microalgae-to-biofuel production. Biomass Convers Biorefin 13:1–14

Ubando AT, Del Rosario AJR, Chen W-H, Culaba AB (2021) A state-of-the-art review of biowaste biorefinery. Environ Pollut 269:116149

van der Waal JC, de Jong E (2016) Avantium Chemicals: the high potential for the levulinic product tree. In: Industrial biorenewables: a practical viewpoint. Wiley, Hoboken, NJ, pp 97–120

van Putten RJ, van der Waal JC, Harmse M, van de Bovenkamp HH, de Jong E, Heeres HJ (2016) A comparative study on the reactivity of various ketohexoses to furanics in methanol. ChemSusChem 9(14):1827–1834

Vázquez-Romero B, Perales JA, Pereira H, Barbosa M, Ruiz J (2022) Techno-economic assessment of microalgae production, harvesting and drying for food, feed, cosmetics, and agriculture. Sci Total Environ 837:155742

Voloshin RA, Rodionova MV, Zharmukhamedov SK, Veziroglu TN, Allakhverdiev SI (2016) Biofuel production from plant and algal biomass. Int J Hydrog Energy 41(39):17257–17273

Wang X, Zhang Y, Xia C, Alqahtani A, Sharma A, Pugazhendhi A (2023) A review on optimistic biorefinery products: Biofuel and bioproducts from algae biomass. Fuel 338:127378

Wells ML, Potin P, Craigie JS, Raven JA, Merchant SS, Helliwell KE, Smith AG, Camire ME, Brawley SH (2017) Algae as nutritional and functional food sources: revisiting our understanding. J Appl Phycol 29:949–982

Wiselogel A, Tyson S, Johnson D (2018) Biomass feedstock resources and composition. In: Handbook on bioethanol. Routledge, Abingdon, pp 105–118

Wu W, Chang J-S (2019) Integrated algal biorefineries from process systems engineering aspects: a review. Bioresour Technol 291:121939

Xiong Q, Hu L-X, Liu Y-S, Zhao J-L, He L-Y, Ying G-G (2021) Microalgae-based technology for antibiotics removal: From mechanisms to application of innovational hybrid systems. Environ Int 155:106594

Yu Z, Lu X, Xiong J, Li X, Bai H, Ji N (2020) Heterogeneous Catalytic Hydrogenation of Levulinic Acid to γ-Valerolactone with Formic Acid as Internal Hydrogen Source. ChemSusChem 13(11): 2916–2930

Zhang G-S, Zhu M-M, Zhang Q, Liu Y-M, He H-Y, Cao Y (2016) Towards quantitative and scalable transformation of furfural to cyclopentanone with supported gold catalysts. Green Chem 18(7):2155–2164

Zhang Y, Ding Z, Hossain MS, Maurya R, Yang Y, Singh V, Kumar D, Salama E-S, Sun X, Sindhu R (2022) Recent advances in lignocellulosic and algal biomass pretreatment and its biorefinery approaches for biochemicals and bioenergy conversion. Bioresour Technol 128281:128281

Zhou Y, Liu L, Li M, Hu C (2022) Algal biomass valorisation to high-value chemicals and bioproducts: Recent advances, opportunities and challenges. Bioresour Technol 344:126371

Zhu X, Cheng X, Luo X, Liu Y, Xu D, Tang X, Gan Z, Yang L, Li G, Liang H (2020) Ultrathin thin-film composite polyamide membranes constructed on hydrophilic poly (vinyl alcohol) decorated support toward enhanced nanofiltration performance. Environ Sci Technol 54(10): 6365–6374

Biofuel-Integrated Routes

Chukwuemeka Samson Ahamefule, Chidimma Osilo, Blessing C. Ahamefule, and James C. Ogbonna

Abstract Despite the enormous potentials of algal biofuels, their costs of production are still too high to compete with fossil fuels. Over the years, applications of algae have spread across various fields. These include applications in the production of high-value and high-volume products and in bioremediation. Making algal biofuels economically feasible requires the adoption of biorefinery concept which properly integrates all/most of the biofuel production with production of pharmaceuticals, nutraceuticals, and cosmetics. Pigments, vitamins, or essential polyunsaturated fatty acids can first be extracted from microalgae before lipid extraction for biodiesel followed by fermentation of defatted biomass for bioethanol. On the other hand, agar, alginate, or carrageenan may be extracted from macroalgae before lipid extraction and the fermentation of lipid-free biomass to bioethanol. To ensure zero waste, spent algal biomass (SAB) can further be fermented or digested anaerobically to produce bio-butanol, dihydroxyacetone, or biogas. Alternatively, SAB may undergo pyrolysis or hydrothermal liquefaction to generate bio-oil, biodiesel, syngas, and biochar. Bioelectricity and algal biomass production may be integrated into wastewater phycoremediation in addition to flue gas sequestration which drastically reduces carbon footprints and addresses the problems of environmental pollution. This chapter holistically presents various integrated routes with respect to economic feasibility of algal biofuels.

C. S. Ahamefule
Department of Microbiology, University of Nigeria, Nsukka, Nigeria

C. Osilo
Department of Microbiology, University of Nigeria, Nsukka, Nigeria

Department of Applied Microbiology, Faculty of Biosciences, Nnamdi Azikiwe University, Awka, Nigeria

B. C. Ahamefule
Department of Microbiology, University of Jos, Jos, Nigeria

J. C. Ogbonna (✉)
Department of Microbiology, University of Nigeria, Nsukka, Nigeria

State University of Medical and Applied Sciences Igbo-Eno, Enugu, Nigeria
e-mail: james.ogbonna@unn.edu.ng

A. Abomohra, S. Ende (eds.), *Value-added Products from Algae*,
https://doi.org/10.1007/978-3-031-42026-9_8

Keywords Algae · Biomass · Phycoremediation · Integrated biorefinery · Zero waste

1 Introduction

Unsustainability of fossil fuels coupled with the enormous environmental havoc caused by their usage has increased the quest for better alternatives. Algal biofuels have gained much attention for some years now because of their many advantages including sustainability and eco-friendliness, among others. However, the costs of producing these biofuels have affected their commercialization and availability. A direct approach to reduce their production costs has been shown to be achieved by numerous integrated algal biofuel approaches, which bring different industries together (Fig. 1). High-value products such as essential pigments, polyunsaturated fatty acids (PUFAs) like ω-3 and ω-6 fatty acids, agar, alginate, several antioxidants, and anticancer compounds have been successfully extracted from algal biomass. These products are known for their enormous medical, pharmaceutical, and nutraceutical properties and are quite expensive. Integration of these products into algal biofuel production will greatly reduce the production costs, thereby making them economically viable. Other high-volume products such as advanced biofuels (acetone and butanol), biochar, and biofertilizers have also been integrated into biofuel production.

This chapter presents various high-value phycochemicals that have been integrated into algal biofuel refineries reported mostly in the last decade. It also outlines how production of these products has been successfully coupled with different biofuels. Recent works in phycoremediation with microalgae that resulted in the production of biomass for assorted biofuels are also reviewed here. Simultaneous bioelectricity generation with wastewater treatment, as well as biofuel production, have also been highlighted. Finally, the challenges restricting the adequate growth of integrated biofuel refinery as well as some adopted solutions to most of them are also presented in this chapter.

2 Integrating Production of Various Algae Biofuels in Biorefinery Systems

Biofuels such as biodiesel, bioethanol, biogas, bio-syngas, bio-oil, acetone, and butanol can be produced from both micro- and macroalgae. Producing only one of these biofuels from algae is not cost-effective and cannot compete favorably with the cost of equivalent products from fossil fuels. Fortunately, these biofuels are usually generated from different components of algal biomass, i.e., biodiesel from lipids, bioethanol mainly from saccharified carbohydrates, and biogas or crude bio-oil from residual components in the defatted biomass and can therefore be integrated in a

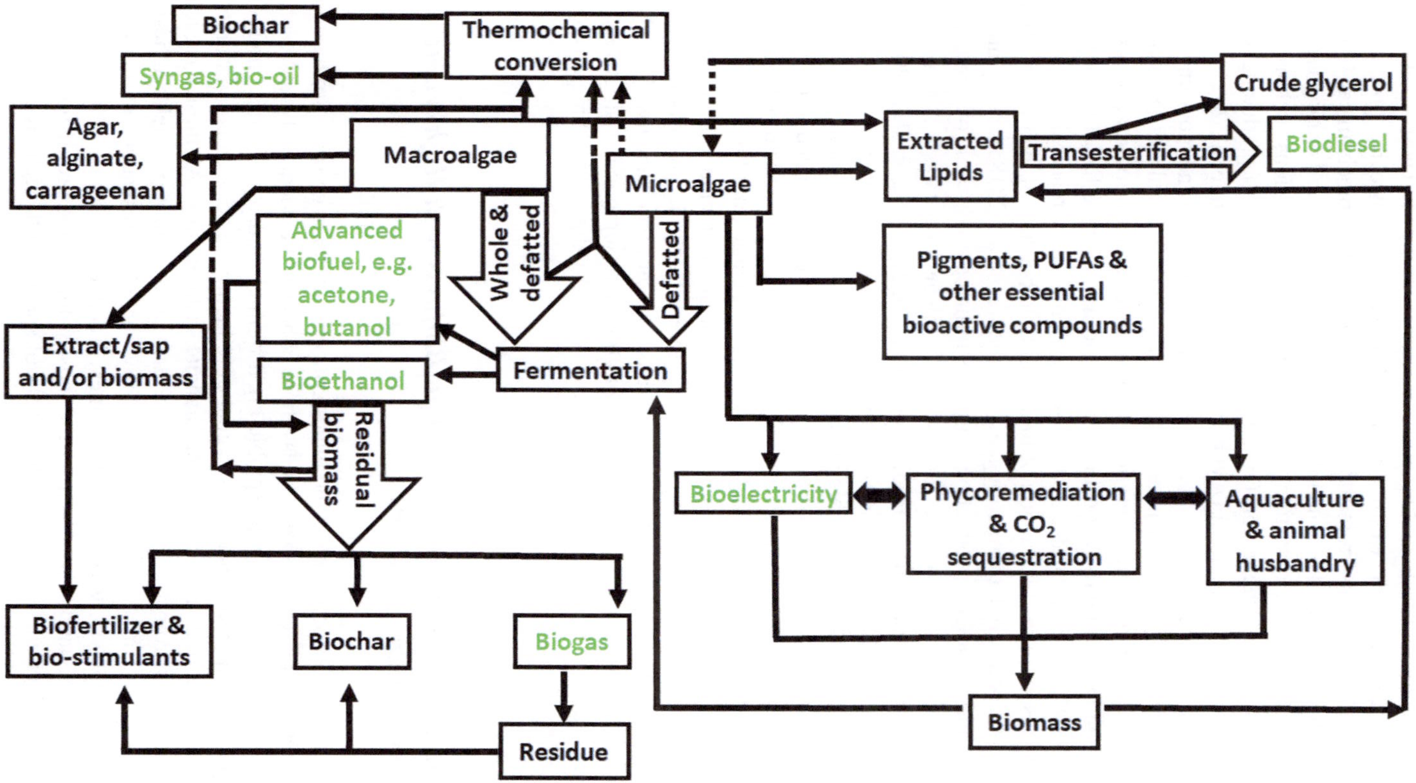

Fig. 1 Various integrated routes for algal biofuels production. Dashed lines were used just to prevent crossing of solid lines

biorefinery system (Karpagam et al. 2021) as shown in Fig. 1. These algal biofuels can be produced through various routes. Biodiesel is basically produced via transesterification reactions from the extracted lipids, while biochemical fermentations lead to the production of ethanol (in anaerobic fermentation) or biogas (in dark fermentation or anaerobic digestion). Thermochemical conversions such as pyrolysis and hydrothermal liquefaction (HTL) are essential routes to producing syngas, bio-oil, biochar, and/or biodiesel (Ayub et al. 2022; Karpagam et al. 2021).

Choice biodiesel lipids [mostly saturated fatty acids (SFAs) and monounsaturated fatty acids (MUFAs)] have been extracted from some macroalgae (such as *Ulva intestinalis* and *Padina tetrastromatica*) for transesterification to produce biodiesel before fermenting the lipid-free/residual algal biomass to bioethanol (Ashokkumar et al. 2017, 2019; Osman et al. 2020). However, only a few macroalgae have been successfully used for the production of biodiesel lipid because of their relatively low lipid contents compared to microalgae (Abomohra et al. 2018a). Integration of bioethanol and/or biogas production with the indirect production of lipids using macroalgal biomass as substrate for growth of oleaginous yeasts has also been reported by several researchers (Hadjkacem et al. 2022; Younes et al. 2020). Some of the integrated biofuel productions by some macroalgae are presented in Table 1.

Many oleaginous microalgae (including *Scenedesmus bijugatus, Chlamydomonas* sp., and *Synechocystis* sp.) have been demonstrated to be adequate for integrated biorefinery to co-produce several biofuels. Good yields of bioethanol have been recorded from such microalgal defatted biomass (after lipid extraction for biodiesel) using mainly *Saccharomyces cerevisiae* for fermentation (Ashokkumar et al. 2015, 2019; Kim et al. 2020) (Table 1). Interestingly, hydrocarbons (another source of bioenergy) have also been reported as a co-biofuel product with biodiesel from the green microalga, *Botryococcus braunii* (Ruangsomboon et al. 2018). The spent algal biomass (SAB) or residual biomass is usually digested to biogas or other high-volume products (Fig. 1).

2.1 Integrating Production of Algal Biofuels with High-Value Low-Volume Products

High cost of production is one of the major concerns in the large-scale production of biofuels among other challenges. Unless these concerns are adequately addressed, a globally strong and generally acceptable algal biofuel industry will not emerge. A viable option is required to achieve the goal of reducing the costs and environmental impact of algal biofuel in order to attain financial sustainability. All these have led to a change in focus from forming a single-centric product to a parallel synthesis of multiple products in a biorefinery model by the algal research community as discussed in Chapter "Overview of Biorefinery Technology".

Algae biomass stands out as one of the most valuable and promising feedstocks for biofuel production. There are wide variety of species with high productivity and

Table 1 Integrating several biofuels production routes from algal biomass

Biofuels	Algae	Other microorganisms involved	Integrated biofuels/lipids production	References
Biodiesel & bioethanol	Macroalga: *Padina tetrastromatica*	*Saccharomyces cerevisiae* was used for bioethanol production from algal residual biomass (ARB)	Maximum lipid yield of 8.15% (*w/w*) was obtained from the algal biomass, while highest bioethanol yield of 83.32% (*w/w* to sugar) was produced from ARB	Ashokkumar et al. (2017)
Biodiesel & bioethanol	Macroalga: *Ulva intestinalis*	*S. cerevisiae* was used for bioethanol production from lipid-free biomass	There was 32.3 mg g^{-1} DW of recovered biodiesel from the whole biomass, while bioethanol yield of 0.081 g g^{-1} DW was obtained from lipid-free algal biomass	Osman et al. (2020)
Bioethanol, biohydrogen & biodiesel	Macroalgae: *Halopteris scoparia*	*S. cerevisiae* was used to produce bioethanol in anaerobic fermentation; bacterial consortium was used to produce biogas through dark fermentation; while *Yarrowia lipolytica* was used to produce lipids from algal biomass	Ethanol production ranging from 0.65–0.72 g L^{-1} was obtained, with maximal yield of 0.35 g-ethanol g^{-1} DW of algal biomass in 72 h. Biohydrogen (1.3 mL H_2 g^{-1} DW substrate) was generated, while lipids (0.04 g g^{-1} DW substrate) were produced by *Y. lipolytica* from raw macroalgal biomass	Hadjkacem et al. (2022)
Biodiesel & bioethanol	10 microalgal species including *Arthrospira platensis, Chlorella marina,* and *Scenedesmus obliquus* 5 macroalgal species including	*S. cerevisiae* was used to produce bioethanol in anaerobic fermentation	*A. platensis* gave the highest ethanol yield and efficiency (45.49% and 89.02%) followed by *C. marina* (23.24% and 45.49%) and	Ismail et al. (2020)

(continued)

Table 1 (continued)

Biofuels	Algae	Other microorganisms involved	Integrated biofuels/lipids production	References
	Ulva linza and *Caulerpa prolifera, etc.*		*U. linza* (12.01% and 37.04%), while their total lipids and biodiesel produced were 8.83% and 8.03%, 15.43% and 12.81%, and 9.82% and 8.32%, respectively. *S. obliquus* gave the highest total lipids and biodiesel of 32.07% and 28.44%, respectively	
Biodiesel & bioethanol	Microalga: *Scenedesmus bijugatus*	*S. cerevisiae* was used for anaerobic fermentation of ARB to bioethanol	Maximal lipid productivity of 63 mg L^{-1} day^{-1} was extracted while maximum biodiesel yield of 0.21 g g^{-1} of algal dry weight was produced. Bioethanol yield of 0.158 g g^{-1} of ARB was also produced	Ashokkumar et al. (2015)
Biodiesel & bioethanol	Microalga: *Chlamydomonas* sp.	*S. cerevisiae* was used for bioethanol fermentation	Highest lipid yield of 0.16 g fatty acid methyl ester (FAME)/g algal biomass was produced, with maximum bioethanol yield of 0.22 g g^{-1} ARM. Total biofuel of ≈300 mg was generated including 156 mg biodiesel and 144 mg bioethanol	Kim et al. (2020)
Biodiesel & bioethanol	Microalgae: *Synechocystis* sp.	*S. cerevisiae* was used to produce bioethanol from lipid extracted	Maximal biodiesel lipid yield of 90.5% was produced, with	Ashokkumar et al. (2019)

(continued)

Table 1 (continued)

Biofuels	Algae	Other microorganisms involved	Integrated biofuels/lipids production	References
		biomass in anaerobic fermentation	bioethanol yield of 0.186 g g^{-1} algal residue	
Biodiesel & hydrocarbon	Microalga: *Botryococcus braunii*	–	Highest proportion of saturated and monounsaturated FAs (for biodiesel) of 43.89% and 45.75%, respectively, were obtained. High hydrocarbon content, yield and productivity of 37.29%, 0.77 g L^{-1} and 90.32 mg L^{-1} day^{-1}, respectively, were also obtained	Ruangsomboon et al. (2018)

potential for carbon sequestration. An integrated approach is used in order to maximize the utilization of biomass by the introduction of other bioproducts into algal biofuel production (Budzianowski 2017). Developing an integrated algal processing system for sequential production of an extensive selection of bioproducts from the same algal biomass promotes technical advancement in sustainability and the generation of high-value products. This also reduces the cost of production substantially, making the industry profitable (Solis et al. 2021).

Essential products with several applications in medical, pharmaceutical, nutraceutical, cosmeceutical, and in food/feed industries are being produced from algal biomass. High-value low-volume products, such as agar and alginate, assorted pigments, essential polyunsaturated fatty acids (FUFAs), proteins, important immunomodulatory carbohydrates (i.e., glucan) or sulfated polysaccharides (i.e., fucoidan), multiple antioxidants, anticancer, and antimicrobial compounds, among others have been extracted from several algal species (Ardalan et al. 2018; Kang et al. 2022; Lee et al. 2018; Singh et al. 2020; Trentin et al. 2022). A good number of these high-value compounds have been reported to be successfully integrated into several biorefinery processes of biofuel production (Table 2), which tremendously reduced the price of algal fuel production (Nagappan and Kumar Verma 2018; Singh et al. 2013; Singh et al. 2020). Essential pigments such as chlorophyll *a* and *b*, as well as assorted carotenoids (including α- and β- carotenes, lutein, and astaxanthin) which have profound medical, pharmaceutical, nutraceutical, and food/feed industrial applications have been extracted from algae (Cao et al. 2022; Singh et al. 2013).

Table 2 Biofuels integrated with some high-value or high-volume products from algae

Biofuel (feedstock)	Integrated product(s)	Algae	Other organisms	Main outcomes	References
Bioethanol	Agar	Macroalga: *Gracilaria verrucosa*	*S. cerevisiae* for fermentation	About 27–33% agar was first extracted from algal biomass before fermentation to produce bioethanol at yield of 0.43 g g^{-1} sugars	Kumar et al. (2013)
Bioethanol	Alginate, glucan	Macroalga: *Sargassum angustifolium*	*S. cerevisiae* for fermentation	Sodium alginate yield of 24.4% and 22.4% were obtained from agal biomass harvested in the winter and summer seasons, respectively, while highest ethanol yield of 70.3% of macroalgal biomass was obtained from winter residual biomass	Ardalan et al. (2018)
Lipids	Carotenoids: Zeaxanthin, β-carotene	Microalga: *Chlorella saccharophila*	–	Highest lipid content of 23.8% was obtained while maximal total carotenoid of 16.39 mg g^{-1} (with 11.32 and 5.07% of zeaxanthin and β-carotene, respectively) was produced	Singh et al. (2013)
Lipids	ω-3 fatty acid [α-linolenic acid, (ALA)]	Microalga: *Desmodesmus* sp.	–	Highest lipid productivity of 15.9 mg L^{-1} day^{-1} with 24% ALA of the total fatty acids (FAs). There was 92% selective removal of ALA from total lipids with the residual lipid having a higher amount of both saturated and monounsaturated FAs for biodiesel	Nagappan and Kumar Verma (2018)
Lipids	Pigment: Fucoxanthin	Microalga: *Nitzschia* sp.	–	Highest productivities of fucoxanthin and lipid of 1.44 and 19.95 mg L^{-1} day^{-1}, respectively, were obtained.	Cao et al. (2022)

				FAME showed excellent lipid quality for biodiesel with 49.97 degree of unsaturation	
Lipids	Pigment: Astaxanthin	Microalga: *Chromochloris zofingiensis*	–	Highest astaxanthin yield and content of 0.318 g L^{-1} and 0.144% DW, respectively, with total FAs of 42% DW were obtained	Chen et al. (2022)
Lipid	Pigment: Astaxanthin	Microalga: *Coelastrum* sp.	–	Highest lipid and astaxanthin productivity of 18.0 g m^{-2} day^{-1} and 168.9 mg m^{-2} day^{-1}, respectively, were co-produced	Liu et al. (2013)
Lipid	Tocopherols	Microalga: *Monoraphidium* sp.	–	High lipid content of 266.6 mg g^{-1} DW and a total tocopherol content of 1450.24 μg g^{-1} DW were produced.	Singh et al. (2020)
Lipid	Pigments: β-carotene, lutein, chlorophylls	Microalga: *Ettlia* sp.	–	Highest lutein, chlorophyll a, β-carotene and lipid contents of 7.9 mg g^{-1}, 27.8 mg g^{-1}, 24.5 mg g^{-1} and 58.2%, respectively, were co-produced	Lee et al. (2018)
Lipid (triacylglycerol, TAG)	Pigments	Microalga: *Nannochloropsis gaditana*	–	Highest pigments, total FAs, and TAG contents of 2.54, 37.34 and 24.24% were obtained at different nitrogen-limiting conditions	Heredia et al. (2021)
Lipids	Carotenoids	Microalga: *Dunaliella tertiolecta*	Gas from integrated yeast fermenter	Carotenoids content and productivity of 4.74 mg g^{-1} DW and 0.86 mg L^{-1} day^{-1}, respectively, were obtained, while 135 mg g^{-1} of lipids was co-produced	Chagas et al. (2015)
Lipid (TAG)	Carotenoids	Microalga: *Neochloris oleoabundans*	–	Highest carotenoids and TAG productivities of 17.9 and 142.4 mg L^{-1} day^{-1} were obtained from cultures	Urreta et al. (2014)

(continued)

Table 2 (continued)

Biofuel (feedstock)	Integrated product(s)	Algae	Other organisms	Main outcomes	References
				with 9- and 3-mM nitrate concentrations, respectively	
Lipid	ω-6 FA (ALA); several pigments	Microalga: *Graesiella emersonii*	–	The amount of extracted chlorophyll a, lutein, and neoxanthin were 19.1, 1.49, and 1.23 mg g^{-1} DW, respectively. There was high amount of choice biodiesel FAs such as palmitic (27.5%), oleic (22.2%), and linoleic (26.3%) acids, while essential ALA was also present in high concentration (22.1%)	Kang et al. (2022)
Lipid	Antioxidants, phenolics	Microalgae: *Tetraselmis marina* (IMA043) & naviculoid diatom (IMA053)	–	Highest total phenolics of 40.58 and 86.14 mg gallic acid equivalent (GAE)/g DW were extracted from IMA053 and IMA043, respectively. Highest radical scavenging activity (RSA) of 99.65% and 103.43% toward 2,2′-azino-bis (3-ethylbenzothiazoline-6-sulfonic acid)diammonium salt (ABTS) and 2,2-diphenyl-1-picrylhydrazyl (DPPH), respectively, were obtained for IMA043, while IMA053 gave copper and chelating activities of 67.48% and 92.05%, respectively. High saturated FAs (38.28% and 45.70%) and monounsaturated FAs (45.41% and 49.81%) were also obtained from IMA043 and IMA053, respectively	

Bioethanol	Bio-stimulant	Macroalgae: *Ulva lactuca Jania rubens* & *Pterocladia capillacea.* The patent extract from these 3 seaweeds gives true-algae-max (TAM®)	*S. cerevisiae* for ethanol fermentation	TAM® 50% concentration gave maximum fresh, plant weight, fruit weight, leaf area, root length and yield of strawberries compared to NPK fertilizer control. There was improved increase of total soluble solids and anthocyanin from 7.58–10.12% and from 23.08–29.42 mg cyanidin-3-glucoside equivalents (CGE) 100 g^{-1}, respectively, compared to NPK. Maximal bioethanol yield of 0.34 g g^{-1} DW was also obtained from algal residual biomass	Ashour et al. (2023)
Bioethanol	Acetone, butanol	Macroalga: *Saccharina latissima*	*Clostridium acetobutylicum* for fermentation	Maximum ethanol, butanol, and acetone of 0.5, 5.1, and 1.1 g L^{-1} were obtained, respectively. Highest acetone, butanol, and ethanol production; and yields of 6.5 g L^{-1} and 0.23 g ABE/g sugar, respectively, were also obtained	Schultze-Jena et al. (2022)
Bioethanol	Acetone, butanol	Algae from wastewater	*Clostridium saccharoperbutylacetonicum* for fermentation	Highest total acetone, butanol, and ethanol (ABE) of 9.74 g L^{-1} was produced which included 1.43 g L^{-1} acetone, 7.79 g L^{-1} butanol, and 0.53 g L^{-1} ethanol	Ellis et al. (2012)

β-Carotene is the very first high-value product to be commercially produced from microalgae. *Dunaliella salina* is the richest source for commercial production of β-carotene, which is capable of expressing up to 98.5% β-carotene in relation to its total carotenoids, representing about 13% of its dry biomass (Molino et al. 2018). Astaxanthin, a red xanthophyll pigment, is the second most industrially exploited carotenoid. *Haematococcus pluvialis* can accumulate up to 81% of astaxanthin of its total carotenoids, representing about 7% of dry weight (Hamed 2016; Molino et al. 2018). It is mostly used in aquaculture feed as dye agent for fish and shellfish due to its high red dyeing power (Hamed 2016).

High-value lipids and fatty acids have several applications across different fields. Essential fatty acids such as ω-3 and ω-6 FUFAs which can also be extracted from algae have proven to have several health benefits with attraction in both food and pharmaceutical industries. These fatty acids have decreased the incidence of several chronic diseases like diabetes, obesity, and cardiovascular diseases (Hamed 2016). Algal phytosterol is also being used as therapeutic steroids, largely as anticancer and anti-cholesterol agents (Francavilla et al. 2012).

In addition, algae are known to be vital sources of protein for both human and animals. Microalgae such as *Arthrospira*, *Chlorella* sp., and *Dunaliella salina* have been mostly used as nutraceuticals or in functional food formulations (Khanra et al. 2018). Protein contents of these algae range from 6% to 70% of their dry weight depending on the species and environmental factors. Microalgal proteins have penetrated the cosmeceutical market, specifically the skin market as they are known to repair early skin aging in the cosmetics industry (Apone et al. 2019).

Typically, microalgae are richly composed of non-essential amino acids (AAs) which includes aspartic and glutamic acids, which accounts for 20% to 30% of protein (8–12% of dry cell weight) as well as high essential amino acid indices (0.9–1.2), rich in leucine, lysine, arginine, and tryptophan (Tibbetts et al. 2015). On the other hand, seaweeds contain 5–43% proteins (dry matter basis) with 1–5% lipids, but are composed primarily of 25–77% carbohydrates (del Río et al. 2020). The high carbohydrate content makes it a promising feedstock for bioethanol or biogas production, with an increased ability for bioremediation of important heavy metals like gold and copper (Abomohra et al. 2021; Lodeiro and Sillanpää 2013).

2.2 Integrating Production of Algal Biofuels with Other High-Volume Low-Value Products

Biofuels such as biodiesel, bioethanol, and biogas are high-volume low-value products as they are usually needed in large quantities at relatively low prices. There are also some other high-volume products that could be integrated into algal biofuel production, even though this is not usual since high-volume low-value products preferably go with low-volume high-value products for a balanced and sustainable production. Some of these high-volume products include acetone, butanol, biofertilizer, and biochar (Ashour et al. 2023; Schultze-Jena et al. 2022). For instance, advanced biofuels (such as acetone and butanol) have been co-produced with bioethanol using either whole or residual algal biomass and several species of *Clostridium* (Ellis et al. 2012; Schultze-Jena et al. 2022). Furthermore, the production of algal stimulants for growing strawberries has also been integrated with bioethanol production. Interestingly, this extracted bio-stimulant (True-Algae-Max, TAM®) was reported to give improved growth parameters compared to the NPK fertilizer as a control, while good bioethanol yield (0.34 g g^{-1} DW) was subsequently generated from the algal residual biomass (Ashour et al. 2023) as shown in Table 2.

Integration of biofuel (especially biodiesel) production with biofertilizer and/or biochar is also quite feasible as defatted algal biomass or SAB can then be converted to manure or biochar. In their study, Khan et al. (2019) reported that using microalgal residual biomass cultivated in wastewater as a manure can save inorganic fertilizer worth about 5584 USD ha^{-1} y^{-1}. This excludes the extra benefits of the biodiesel produced, the phycoremediation, and waste management achieved, and ultimately, the positive impact on the ecosystem.

2.3 *Integrating Production of Algal Biofuels with Both High-Value Low-Volume and High-Volume Low-Value Products*

Just as presented in Sects. 2.1 and 2.2, either high-volume or high-value products can be integrated with algal biofuel production in a biorefinery approach. These have essentially helped to reduce the cost of biofuel production. A step to further reduce the cost of algal biofuels and drastically reduce/eliminate wastes is to integrate both high-value low-volume and high-volume low-value products with production of algal biofuel(s). High-value products such as pigments, essential PUFAs, antioxidants, and anticancer products have been demonstrated to be first extracted from some microalgal species before extracting lipids and some other high-volume products such as carbohydrates. The defatted or residual biomass is then converted to bioethanol, biogas, and/or biofertilizers (Eusébio et al. 2022; Sinha et al. 2021; Zohir et al. 2022), as summarized in Table 3. Some researchers have also demonstrated the feasibility of producing some of the above products in a co-production process (Gao et al. 2022; Ricky et al. 2022; Sisman-Aydin and Simsek 2022).

The microalga, *Tetradesmus obliquus*, was demonstrated to be a promising strain for multiple integrated products coupled with biofuel generation. Antioxidants (e.g., fusarochromanone, benzoquinol, hexadecanoic acid, nigakihemiacetal A, luciferin, convallasaponin A, and tridecanol) and anticancer (e.g., dioscin and cholanoic acid) bioactive compounds were first extracted from the microalgal biomass (Sinha et al. 2021). This was then followed by relatively high fatty acid methyl ester (FAME) and lipid components that were subsequently obtained from the residual biomass. The

Table 3 Algal biofuels integrated with high-value products, low-value products, and wastewater treatment

Algae	Biofuel (feedstocks)	High-value product(s)	Other high-volume product(s)	Waste media and phycoremediation	Integrated/several products production	References
Microalga: *Tetradesmus obliquus*	Lipids (FAME)	Bioactive compounds: antioxidant[a] and anticancer[b] products	Biofertilizer	–	Maximum antioxidant and anticancer IC_{50} values of 137 and 306 μg mL^{-1}, respectively, were obtained from the first extract of algal biomass. Then, lipid content and FAME yield of 35.7–39.1 and 33.1–36.7% w/w, respectively, were obtained from algal post extracted residual biomass (PERB). PERB also gave biofertilizer activity on *Solanum lycopersicum* (tomatoes) seeds with highest germination percentage and germination index of 75–80% and 117.5–118.5, respectively	Sinha et al. (2021)
Macroalgae: *Sargassum muticum*	Biogas	Fucoidan	Biofertilizer / stimulant (sap)	–	Seaweed sap (0.1 L kg^{-1} of fresh algal biomass) used as biofertilizer was first extracted. Then, high fucoidan (>80% after fractionation and concentration) was extracted from the used biomass. Thereafter, biogas (150 mL CH_4 g^{-1}) was obtained from the residual solid biomass	Flórez-Fernández et al. (2021)

Microalgae: *Crypthecodinium cohnii*	Lipids, biogas (methane)	PUFA: Docosahexaenoic acid (DHA)	–	Raw glycerol & corn steep liquor	Lipid content and productivity of 9.2% *w/w* and 38.5 mg L^{-1} h^{-1} were obtained with DHA productivity of 18.5 mg L^{-1} h^{-1} from algal biomass. Extracted algal biomass and waste from both cultivation and lipid extracted processes produced highest biogas yield of 582 L CH_4 kg^{-1} volatile solids (VS) and productivity of 31 L CH_4 kg^{-1} VS d^{-1}	Eusébio et al. (2022)
Microalgae: *Chlorella pyrenoidosa* & *Chlorella vulgaris*	Lipids	Tocopherols, pigments (β-carotene, chlorophylls, lutein), protein	–	Industrial process waste (IPW) at different concentrations (34, 67, and 100%)	*C. vulgaris* and *C. pyrenoidosa* gave the maximum lipid of 17.0% and 17.6% DW and highest protein contents of 55.2% and 65.2% DW, respectively. Maximum α-tocopherol contents of ~200 and >150 μg g^{-1} DW were obtained from *C. vulgaris* and *C. pyrenoidosa,* respectively. *C. vulgaris* produced the highest amount of β-carotene, chlorophyll a, and lutein (1013, 32,444, and 7141 μg g^{-1} DW) compared to	Safafar et al. (2016b)

(continued)

Table 3 (continued)

Algae	Biofuel (feedstocks)	High-value product(s)	Other high-volume product(s)	Waste media and phycoremediation	Integrated/several products production	References
					C. pyrenoidosa (171.8, 5411, and1345 μg g^{-1} DW), respectively	
Microalga: *Chlorella sorokiniana*	Lipids (FAME)	Protein, pigments (-β-carotene, chlorophylls, lutein)	–	Mixed influent of industrial and municipal wastewater in four dilutions (25, 50, 75 and 100%). There was 68.8, 57.5 and 52.1% removal of phosphorus, nitrogen and COD from wastewater, respectively	The average yield of FAME, protein, β-carotene, lutein, and chlorophylls obtained were 62.4, 388.2, 0.44, 1.03 and 11.82 mg g^{-1} DW algal biomass, respectively	De Francisci et al. (2018)
Microalga: *Nannochloropsis Salina*	Lipids	Eicosapentaenoic acid (EPA), protein, pigments: Chlorophyll *a*, several carotenoids (including violaxanthin, vaucheriaxanthin esters, and β-carotene)		IPW plus f/2 standard medium. Large-scale cultivation was done in photobioreactors with twelve hollow chamber sheets (of 2100 mm width, 5600 mm height, and 32 mm depth)	Large-scale cultivation showed increased lipid content from 10.8% to 21.1% and a decrease in protein content from 46.0% to 33.4% DW from day 3 to 21 of cultivation, respectively. EPA reached 44.2% of total fatty acids. There was increase in chlorophyll *a* and total carotenoids obtained from 3,944 to 19,020 μg g^{-1} DW and from 2,038 to 20,797 μg g^{-1} DW, respectively. α-tocopherol also increased from >100 μg g^{-1} DW	Safafar et al. (2016a)

					(in day 3) to 431 μg g^{-1} DW by day 12	
Microalga: *Dysmorphococcus globosus*	Lipids	Proteins, pigments (astaxanthin, carotenes, chlorophylls), unsaturated fatty acids (UFAs)	Carbohydrates	-	Maximum astaxanthin, lipids, carotenoids, chlorophyll *a*, chlorophyll *b* and carbohydrates of 391 mg L^{-1}, 32.5 mg L^{-1}, 1.25 μg mL^{-1}, 2.97 μg mL^{-1}, 1.78 μg mL^{-1} and 135.62 mg L^{-1}, respectively, were obtained. The highest contents of ω3-FAs, ω6-FAs and ω9-FAs of 17.78, 23.11 and 7.06%, respectively, were also obtained	Zohir et al. (2022)
Microalga: *Chroococcus turgidus*	Lipids	Proteins and pigments (chlorophyll *a* and *b*)	Carbohydrates	Municipal wastewater (MWW): Primary effluent (PE), secondary effluent (SE) and final effluent (FE). The removal efficiency of total nitrogen, phosphorus, BOD, and COD in the PE, SE & FE were 71.58, 74.68 & 72.16; 95.29, 92.41 & 89.00; 76.20, 72.23 & 50,00 and 82.56, 58.02 & 73.21%, respectively	The highest lipid content of 28.05% was obtained from SE, while maximum carbohydrate and protein contents of 43.93 and 35.25% were obtained from FE and PE, respectively. The highest chlorophyll *a* (118.05 μg L^{-1}) and chlorophyll *b* (95.86 μg L^{-1}) were obtained from PE and SE, respectively	Sisman-Aydin and Simsek (2022)
Microalgae: *Chlorella vulgaris*	Lipids	Proteins & chlorophyll *a*	Carbohydrates	Antibiotics [ciprofloxacin (CIP) and amoxicillin (AMX)] remediation. There was 76% and 46%	CIP induced stress, caused increase in lipids (50%), proteins (65%), chlorophyll *a* (30%) and carbohydrates	Ricky et al. (2022)

(continued)

Table 3 (continued)

Algae	Biofuel (feedstocks)	High-value product(s)	Other high-volume product(s)	Waste media and phycoremediation	Integrated/several products production	References
				removal of CIP and AMX, respectively	(32%), while AMX caused increase of 46, 49, 22 and 45% for lipids, proteins, chlorophyll *a* and carbohydrates, respectively	
Microalga: *Chlorococcum minutum*	Lipids and bioethanol	Chlorophyll *a* and *b*	–	–	Total lipid content and chlorophyll of 24.5 g 100 g^{-1} and 8.26 mg L^{-1}, respectively, were extracted from algal biomass. The highest ethanol recovery of 32.6 g L^{-1} was obtained from algal biomass with *S. cerevisiae*	Varaprasad et al. (2021)
Microalga: *Barranca yajiagengensis*	Lipids	Pigments: Carotenoids and chlorophylls; proteins	Carbohydrates	–	The maximum total carotenoids, chlorophyll *a* and *b* of 5.57, 13.88 and 6.51 mg g^{-1} DW, respectively, were obtained. Highest proteins, carbohydrates and lipids contents of 20.89, 37.43 and 53.13% DW of algal biomass, respectively, were also obtained	Gao et al. (2022)

[a]The identified most abundant antioxidant compounds were fusarochromanone, benzoquinol, hexadecanoic acid, nigakihemiacetal A, luciferin, convallasaponin A, and tridecanol

[b]The identified most prevalent anticancer compounds were dioscin and cholanoic acid

residual biomass was further used as biofertilizer which gave good yield/performance for the growth of tomatoes. Several studies have also reported different species of *Chlorella* (especially, *C. vulgaris*) to be excellent algal candidates in the production of several high-value and low-value products, alongside the generation of lipids for biodiesel and concomitant/simultaneous phycoremediation (De Francisci et al. 2018; Ricky et al. 2022; Safafar et al. 2016b). High-value antioxidants (i.e., tocopherol) and pigments (including β-carotene, lutein, and chlorophylls) as well as proteins have been extracted from *C. vulgaris* followed by lipids and carbohydrates for generation of different biofuels (Ricky et al. 2022; Safafar et al. 2016b) (Table 3).

In what seems like a reverse process, biofertilizer was reported as the first product from the seaweed *Sargassum muticum* (though using the liquid phase known as "sap extract" and not the biomass) before extracting the high-value product, fucoidan, and then anaerobically digesting the residual biomass to produce biogas (Flórez-Fernández et al. 2021). Interestingly, a number of these integrated biofuel production routes with both high-value and high-volume products have been demonstrated to have used wastes as their main growth medium. Examples of such waste media used include raw glycerol and corn steep liquor (Eusébio et al. 2022; Moniz et al. 2022), industrial process waste (Safafar et al. 2016b), mixed influent of industrial and municipal wastewater (De Francisci et al. 2018), municipal wastewater (Sisman-Aydin and Simsek 2022), and lipid-free algal waste (Abomohra et al. 2018b) (Table 3). In addition, simultaneous bioremediation was recorded in some of these wastes that were adopted as algal media.

2.4 *Integrating Generation of Bioelectricity with Wastewater Treatment*

Bioelectricity includes all the renewable and sustainable forms of electrical energies produced by living organisms or from their biomass. It also refers to the electricity (i.e., electrical potentials, current, and power) generated by microbes at the expense of produced electrons during metabolism (De Souza 2019; Krishnan et al. 2021). Although it is possible to generate bioelectricity from both microalgae and seaweeds, bioelectricity from the later is however very nascent with the first published work by Shlosberg et al. (2022). On the other hand, bioelectricity from microalgae has been studied over a decade and is still evolving. Therefore, this section will focus on the discussion of bioelectricity from microalgae. Microalgae such as *Chlorella vulgaris, Golenkinia* sp., and *Synechococcus* sp. have been largely applied in generation of sustainable electricity using different kinds of systems/devices such as microbial fuel cells (MFC) and various modifications of MFC, photo-bioelectrochemical systems, algal biophotovoltaic (BPV) devices, and microbial carbon capture cells (MCCC) (Table 4). Several works on bioelectricity generation have been done with *C. vulgaris* which have produced power ranging from 78 mW m^{-2} to 110 mW m^{-2} coupled with bioremediating wastes like food remains, azo dyes,

Table 4 Simultaneous bioelectricity generation and wastewater treatment by microalgae

Microalgae	Bioelectricity device and generated electricity	Type of wastewater and remediation	Biodiesel / lipids	References
Chlorella vulgaris	Microbial fuel cell (MFC); Highest power density, working and open circuit voltage (OCV) of 19,151 mW m^{-3}, 170, and 260 mV, respectively, were generated	Food waste; Maximum COD removal efficiency of 44% was obtained	Highest total lipid content of 31% was obtained	Hou et al. (2016b)
	Photo-bioelectrochemical (PBE) system; Maximum power of 110 mW m^{-2} under light was generated	Synthetic high-strength nitrogenous wastewater; There was 100, 86, 83% removal of NH_4-N, NO_3-N, and total nitrogen (TN), respectively	–	Sun et al. (2019)
	Photosynthetic MFC (P-MFC); Highest power of 0.0254 kWh kg^{-1} COD was generated	Synthetic wastewater; Highest COD and NH_4^+-N of 74 and 79%, respectively, were removed	–	Yahampath Arachchige Don and Babel (2021)
	Reversible PBC cell (R-PBEC) with anthraquinone-2,6-disulfonate/MnO_x-doped polypyrrole film electrodes; Maximum power density of 84.58 mW m^{-2} was obtained	Azo dye (Congo red); Enhanced decolorization (>95% efficiency) with over 70% COD removal	–	Sun et al. (2017)
	Cathodic algal-MFC with oxygen-consuming unit; Maximum voltage of 0.39 V was obtained	Landfill leachate; Removal of more than 86, 89.4 and 76.7% of COD, $NH4^+$-N and TN, respectively	–	Elmaadawy et al. (2020)
	pH self-neutralized PBE system; Maximum power density of	Azo dye (Congo red); Highest COD removal of 52% was achieved with	–	Sun et al. (2015)

(continued)

Table 4 (continued)

Microalgae	Bioelectricity device and generated electricity	Type of wastewater and remediation	Biodiesel / lipids	References
	78 mW m^{-2} was produced	90% faster (by 14 h) decolorization		
Golenkinia sp.	Open air cathode MFC; Highest open OCV and power density of 170 mV and 6150 mW m^{-3}, respectively, were generated	Anaerobically digested kitchen waste effluent (ADE-KW); maximum TN, TP, and COD removal of 37.39, 98.00 and 43.59%, respectively, were obtained	Highest lipid content of 38% was obtained	Hou et al. (2016a)
	Dual chamber MFC; Highest OCV and power density of 400 mV and 400 mW, respectively, were generated	ADE-KW; Highest COD removal efficiency of 76% was obtained (in diluted ADE-KW) while >80% of TN was removed	Highest total lipid content of ~55% *w/w* was obtained (undiluted ADE-KW)	Hou et al. (2017)
Chlamydomonas sp.	Algal biomass after remediation (ABAR) electrochemical impedance spectroscopy; Maximum power, power density, and current density of 4.13x10^{-4} W, 1.83 W m^{-2} and 3.6 A m^{-2}, respectively, were obtained	Effluent from textile dyeing mill; Complete decolorization of effluent with 87.54, 83.08, 82.64, 87.15 and 92.36% removal of BOD, COD, total solids, TN, and phosphate, respectively	Enhanced lipid content of 79.1% was obtained from ABAR	Behl et al. (2020)
Leptolyngbya sp.	Algal-MFC photobioreactor (A-MFCP); Maximum power density of 0.008 mW cm^{-2} at 12 mV cell potential was obtained	Raw wastewater; COD and total dissolved solids (TDS) removal of 56.08% and 12.86%, respectively, were obtained	Highest lipid content of 1068.38 mg g^{-1} DW was produced	Maity et al. (2014)
Synechococcus sp.	Double-chamber algal-assisted MFC (AA-MFC); Highest current and	Municipal solid waste leachate; Highest TN and TP removal of 90.2	Highest lipid production of 1079 mg L^{-1} was obtained	Lakshmidevi et al. (2020)

(continued)

Table 4 (continued)

Microalgae	Bioelectricity device and generated electricity	Type of wastewater and remediation	Biodiesel / lipids	References
	power densities of 5.169 A m^{-2} and 110.92 mW m^{-2}, respectively, were generated	and 94.3%, respectively, were obtained		
Synechococcus sp. & *Chlorococcum* sp.	P-MFC; Highest power of 41.5 and 30.2 mW m^{-2} with *Synechococcus* sp. and *Chlorococcum* sp., respectively	Kitchen wastewater; Maximum COD removal of 73.5% was obtained	–	Naina Mohamed et al. (2020)
Chlorella sp.	Algal biophotovoltaic (BPV) device; Highest power of 0.45 mW m^{-2} was obtained	Palm oil mill effluent (POME); There was 80.75, 60.09, 54.92 and 29.41% removal of NH_3-N, COD, NO_3-N and o-PO_4^{3-}, respectively	–	Ng et al. (2021)
C. sorokiniana	Microbial carbon capture cell (MCCC); Highest power density and coulombic efficiency of 3.2 W m^{-3} and 16.53%, respectively, were generated by MCCC with coconut shell (CS) as proton exchange membranes	Synthetic wastewater/pretreated anaerobic sewage sludge; COD removal efficiency of 65.97% was achieved by CS-MCCC, which was lower than 72.14% of Nafion-MCC	–	Neethu et al. (2018)
Microcystis aeruginosa	MFC fed with algae; Maximum power and current densities of 83 mW m^{-2} and 672 mA m^{-2}, respectively, were obtained	Harmful algal biomass; There was 100 and 67.5% removal of microcystin-LR and COD, respectively	–	Ali et al. (2020)
Scenedesmus abundans	H-type MFC; Maximum power	Distillery spent wash diluted with	–	Nayak et al. (2018)

(continued)

Table 4 (continued)

Microalgae	Bioelectricity device and generated electricity	Type of wastewater and remediation	Biodiesel / lipids	References
	density and OCV of 836.81 mW m^{-2} and 745.13 mV, respectively, were generated	sewage wastewater; Removal of TDS and TSS of 39.66 and 97%, respectively, with COD reduction ranging from 66 to 78.66%		
Mixed algal consortia	Oxygenic PBE fuel cell (O-PBEFC); Highest electrogenic activity of 46 mV; O.6 mA was produced during the day	Domestic wastewater; COD removal of 72.6% was achieved	–	Venkata Subhash et al. (2013)
Mixed culture	Up-flow membraneless MFC (ML-MFC) plus photobioreactor (ML-MFC + P); Highest power density of 481 mW m^{-3} was generated	Domestic wastewater; Maximum TP and $NH4^{+}$-N removal of 99.3 and 99%, respectively, in ML-MFC + P while COD removal of 77.9% was obtained in ML-MFC	–	Jiang et al. (2013)
Chlorella pyrenoidosa & *Anabaena ambigua*	Clayware MCCC (C-MCCC); Highest coulombic efficiency, power and current densities of 15.23%, 6.36 W m^{-3} and 44.2 A m^{-3}, respectively, were generated by *C. pyrenoidosa*	Synthetic wastewater; Highest COD removal efficiency of 87.3% was obtained by *C. pyrenoidosa*	–	Jadhav et al. (2017)
Blue green algae	Algae-assisted MFC; Maximum current and power densities of 149.5 mA m^{-2} and 78.12 mW m^{-2}, respectively, were generated	Synthetic wastewater; Total COD removal of 89.23% was achieved	–	Yadav et al. (2015)

landfill leachate (Elmaadawy et al. 2020; Hou et al. 2016b; Sun et al. 2015, 2017, 2019; Yahampath Arachchige Don and Babel 2021).

Such electricity has been attributed to photosynthetic metabolism of some microalgae while the synergistic metabolism/activities of microalgae and bacteria have also been reported to generate bioelectricity (Ng et al. 2021; Sun et al. 2019; Yahampath Arachchige Don and Babel 2021). Microalgal biomass (whole biomass or defatted/residue) has been applied as substrate to generate electricity using MFC (Ali et al. 2020; Behl et al. 2020). The integration of waste treatment with microalgal bioelectricity generation has been an attractive part of modern bioelectricity generation. Efficient phycoremediation of both solid wastes and wastewaters used as substrates for bioelectricity has produced large algal biomass and some other products including lipids for biodiesel production (Behl et al. 2020; Hou et al. 2016a, b; Lakshmidevi et al. 2020) as shown in Table 4. The produced biomass (as well as defatted biomass) may, therefore, be applied in other biofuel production routes. This massive integrated biorefinery is definitely an interesting and cost-effective means of sustainable generation of renewable bioenergy. For example, the high cost of ordinarily treating anaerobic digestate effluents (ADE) due to cost of digester equipment has been positively impacted in the course of simultaneous usage of the waste for bioelectricity and biodiesel production while bioremediating the waste in the process. Several of such ADE (including ADE-kitchen wastes and food wastes) have been successfully used to generate bioelectricity, producing lipids ranging from 31% to 55% with maximum COD removal efficiency of 76% obtained with the aid of microalgae like *C. vulgaris* and *Golenkinia* sp. (Hou et al. 2016a, b, 2017).

3 Biofuels from Phycoremediation

Several water bodies and surfaces have been affected by eutrophication caused by the accumulation of nutrients such as phosphates and nitrates present in wastes, thereby leading to serious environmental problems. Disposal of solid wastes in poorly constructed landfills is associated with soil, surface, and groundwater contamination and, therefore, collection/treatment of these landfills still continues to pose a serious problem. A combination of wastes and water from homes, commercial, and industrial activities make up the municipal wastewater which contains high levels of organic materials, multiple pathogens, nutrients, and toxic compounds (Rawat et al. 2011).

Conventional methods of waste treatment are used widely in various industries currently to reduce the toxicity of metals and nutrients before discharging or reusing (Varjani et al. 2020). Unfortunately, these methods involve physical and chemical processes and, in most cases, not economical and environmentally friendly. However, several biological processes have been used with different microflora with the potential to utilize the nutrients and convert them into energy products (Machineni 2019).

Algae represent a good candidate for biological wastewater remediation since it has been proven to competently remove phosphorus, nitrogen, and toxic metals from several industrial effluents and recover clean water (Cai et al. 2013). The process of using algae for biotransformation or pollutant removal such as nutrients and toxins from wastewater is called phycoremediation (Mulbry et al. 2008). Cultivation of algae using different waste streams is discussed in chapter "The Use of Wastewater for Algal Growth". This will in turn reduce the use of freshwater and chemical nutrients inputs in the commercial cultivation systems.

Comparing to conventional treatment methods, it is beneficial to use algae as they serve a dual role of bioremediation and biomass generation for biofuel production, with concomitant carbon dioxide sequestration while using the wastewater as feed for cellular growth (Mulbry et al. 2008; Ogbonna et al. 2021). Wastewaters from various sources are rich in nitrogen and phosphorus which are needed in large amount for the growth of microalgae since they help to improve biomass as well as the quality and quantity of lipids accumulated (Sarma et al. 2021). The produced lipids serve as raw material in the production of algal biodiesel. Various microalgal species have been investigated for wastewater treatment and biomass accumulation for bioenergy production (Table 5). Algae like *Chlorella* sp., *Ananaena* sp., and *Scenedesmus* sp. have demonstrated high potential to remove nutrients from wastewater while producing high lipid content for the production of various types of biofuels (Table 5). Also, microalgae like *Acutodesmus obliquus* from cassava effluent have exhibited the ability to produce high percentage of biodiesel and bioethanol of 98.75% and 96.83%, respectively (Selvan et al. 2019) (Table 5).

When cultivated in wastewater, some species of *Chlorella* have been reported to accumulate ≥30% lipids while significantly reducing nutrients and pollutants in wastewater substrates used for their cultivation (Ansari et al. 2017; Pandey et al. 2020) (Table 5). Phycoremediation is an eco-friendly process without any secondary pollution as long as there is reuse of biomass produced and efficient nutrient recycling (Mulbry et al. 2008). The process of nutrient removal is simple, beneficial, and sustainable to the environment when algae are used, since it can serve as feedstock for the production of biofuel as well as fertilizer or animal feed (Filippino et al. 2015). Algae assimilate inorganic nitrates and nitrites which are then reduced to ammonium that can be a good source of nitrogen when found in wastewater and can help in the speedy production of biomass. It is also preferred by algae over other inorganic sources due to the less energy required during its assimilation and can also be incorporated into amino acids within the cells (Ahamefule et al. 2019; Sarma et al. 2021). Wastewater having increased level of phosphorus can be a potential source of nutrient for algae cells as it plays a vital role in energy metabolism and forms a major part of nucleotides, proteins, and lipids (Sarma et al. 2021).

In algae, carbon source is taken in the form of CO_2 or soluble carbonates. Carbon has shown to be an essential nutritional element for effective algal growth under photoautotrophic conditions where inorganic atmospheric CO_2 is fixed via photosynthesis (Cai et al. 2013; Sarma et al. 2021). Microelements like calcium, iron as well as heavy metals are also present in wastewater, which showed significant role in lipid accumulation and biomass productivity of algae (Ghafari et al. 2018).

Table 5 Phycoremediation of wastewater coupled with biofuel production

Algae	Wastewater type	Nutrient removal (%)	Lipid content (%)	Lipid Productivity (mg L^{-1} day^{-1})	Biofuels	References
Spirulina platensis	Domestic	23.49 nitrate nitrogen, 17.16 NH_4N, 14.57 orthophosphate, 17.88 COD.	26.65	—	62.38% Biomethane, 165.0 ± 5.39 mL g^{-1} VS biogas	Chavan and Mutnuri (2019)
Scenedesmus obliquus	Aquaculture	77.7 nitrates, 73.83 nitrites, 68.09 total oxidizable nitrogen, 42 COD, 88.71 ammonia	30.85	27.65	–	Ansari et al. (2017)
Chlorella sorokiniana	Aquaculture	75.76 nitrates, 81.79 nitrites, 67.89 total oxidizable nitrogen, 98.91 ammonia, and 69 COD	31.85	34.35	–	Ansari et al. (2017)
Ankistrodesmus falcatus	Aquaculture	80.85 nitrates, 99.73 nitrites, 75.29 total oxidizable nitrogen, 86.4 ammonia and 61 COD.	35.9	57.72	–	Ansari et al. (2017)
Chlamydomonas reinhardtii	Sugarcane biorefinery	63.56 total nitrogen for cl-stress and 83.48 for N, S-stress. 62.02 TC for cl-stress and 26.09 for N, S-stress	10.8 in cl-stress	–	68.3% bioethanol for N, S-stress, 61.9% for cl-stress	Tasic et al. (2020)
Chlorella pyrenoidosa	Fresh cheese whey	68.09 COD, 47.80 turbidity, 73.63 total solids, 74.32–77.91 phosphorus, 90.23–99.05 nitrogen.	31.62	77.41	–	Pandey et al. (2020)
Chlorella pyrenoidosa	Dairy	58.54 BOD, 79.02 phosphate, 88.91 nitrate, 87.50 COD	10.36	–	52% (theoretical BMP)	Brar et al. (2019)
Anabaena ambigua	Dairy	58.54 BOD, 87.83 phosphate, 89.52 nitrate, 81.25 COD	13.13	–	20% (theoretical BMP)	Brar et al. (2019)
Scenedesmus abundans	Dairy	52.44 BOD, 86.51 phosphate, 84.72 nitrate, 62.50 COD.	16.93	–	31% (theoretical BMP)	Brar et al. (2019)
Acutodesmus obliquus	Cassava effluent	96.8 ammonia, 96.89 nitrate, 97.9 potassium, 94.6 inorganic phosphate, 96.3 organic phosphorus.	176.65 mg/mL	—	98.75% biodiesel 96.83% bioethanol	Selvan et al. (2019)

Anabaena variabilis	Piggery	79.90 total nitrogen, 97.97 $NH4^+$-N, 97.14 total phosphorus, 57.11 COD	4.82	0.48	2223.6 mL CH_4 L^{-1} algal culture (theoretical BMP)	Lu et al. (2020a)
Nostoc sp.	Piggery	73.79 total nitrogen, 92.46 NH_4^+-N, 98.46 total phosphorus.	4.91	0.69	3189.1 mL CH_4 L^{-1} algal culture (theoretical BMP)	Lu et al. (2020a)
Haematococcus pluvialis	Cassava processing	60.80 COD, 88.32 BOD, 51.06 total nitrate, 54.68 phosphate.	15.2	0.018 g L^{-1} day^{-1}	–	Sorgatto et al. (2021)
Neochloris (Ettlia) oleoabundans	Cassava processing	69.16 COD, 90.56 BOD, 58.19 total nitrate, 69.84 phosphate.	19.2	0.041 g L^{-1} day^{-1}	–	Sorgatto et al. (2021)

Municipal wastewater, containing nitrogen in form of ammonia and nitrate salts and phosphorus as phosphates, is a strong nutrient source for microalgae growth and can further be channeled into lipids and biofuel production (Sarma et al. 2021). Utilization of wastewater as nutrient source for increased algal biomass production helps make the process cost-effective which makes the biorefinery concept achievable (Nagarajan et al. 2020).

Since it has been identified that both seaweeds and microalgae represent one of the most promising feedstocks for large-scale production of both low-value high-volume products (e.g., biofuels, bioethanol) with low-volume high-value products (e.g., nutraceuticals), and that biomass produced in large quantities is known to improve viability on conversion of biomass to alternate fuel, it is suggested that biomass productivity should be the focus in algal growth rather than just the desired product (Sarwer et al. 2022). Integration of wastewater treatment with CO_2 biofixation remains one of the most attractive approaches in phycoremediation, with production of algal biomass as a feedstock for a variety of biofuels (Phang et al. 2015). The use of effluent from cassava wastewater by Selvan et al. (2019) for cultivating *Acutodesmus obliquus* and producing biodiesel, bioethanol as well as production of other high-value products is an example of such integrated system (Table 5). Asides the obvious benefits, the successful implementation of the combination of algal biomass, biofuel production, and wastewater treatment would lead to minimizing the use of freshwater resources, especially for dry and populous nations (Okpozu et al. 2019; Rawat et al. 2011).

4 Challenges of Algal Integrated Biofuel Routes

Although algal-integrated biofuel production promises to be an avenue for producing assorted green future energy that can compete favorably with or replace fossil fuels, some challenges still restrict the full-scale actualization of this feat. One primary challenge faced is the isolation/selection and/or modification of appropriate algal strains that will give substantial quantities of all or the major integral products of interest coupled with biofuels. Another important feature of such strain should be the ability to utilize available waste substrates without treatment and also adapt to varying culture conditions. So far, many algal isolates have been reported to be a good feedstock for integrated biorefineries as well as a number of genetically modified strains with improved productivity of the products of interest (Ismail et al. 2020; Sisman-Aydin and Simsek 2022). In that context, high-throughput screening techniques and/or OMICs play a significant role in the identification and manipulation of genes expressing these integral products in the algae as discussed in chapters "High Throughput Screening to Accelerate Microalgae-based Phycochemicals" and "Omics Approaches for Algal Applications". Advances in genetical engineering have helped in the identification of essential genes and even the insertion of foreign genes of interest into organisms so that the transformed algae produce all the integral products of choice simultaneously or concurrently. More so, manipulation of algae to over-express genes of interest in integrated biorefinery is also attainable (di Visconte et al. 2019; Kumar et al. 2020).

Another major challenge has always been the high cost of production and net energy requirements of biofuels recovery, arising mainly from cost of the growth media, freshwater, and processing technology (De Bhowmick et al. 2019; Dineshbabu et al. 2019). Since these biofuels are high-volume products, high biomass yield is essentially needed to generate enough products. However, the cost of growth media and freshwater has been adequately addressed in recent times by the adoption of several waste substrates and wastewater or seawater for cultivation (Ahamefule et al. 2021; Ogbonna et al. 2021; Pandey et al. 2020).

Processing technology in this context involves the different operational techniques and equipment used in cultivation, harvesting, and other downstream processes. Though the use of different open ponds to cultivate microalgae is quite cheap as discussed in chapter "Algae Cultivation Systems", several challenges (including low productivity, contamination, etc.) increased the preferential use of more expensive photobioreactors. This tends to substantially increase the cost of algal biomass large-scale production (Alishah Aratboni et al. 2019). However, progress in research has led to the production of several less expensive photobioreactors with high biomass productivity (Roostaei et al. 2018; Zhang et al. 2016). Another bottleneck in the technology-related matter is the variance of techniques applied in different algal integral products in both up- and downstream processes (chapter "Overview of Bioprocess Engineering"). However, to achieve maximum product recovery, mild and non-invasive approaches for biomass fractionation as well as process technologies are necessary. At the same time, integrating and co-optimizing various specific operations needed for the range of integral bioproducts is essential (Gifuni et al. 2019).

Compared to microalgae, seaweeds have a relative advantage due to easy cost-effective harvest. Therefore, the cost of harvesting microalgae is yet another technique that could drastically increase the overall production cost. More details about algae harvest are discussed in chapter "Algae Harvesting". The use of centrifugation and filtration for large-scale biomass production is capital-intensive and time-consuming, respectively. Several cheaper harvesting techniques such as assorted kinds of bio-flocculation (Lu et al. 2020b; Ogbonna and Edeh 2018) as well as chemical flocculation (Abomohra et al. 2018c) have been demonstrated as potential remedies. Generally speaking, most of the integrated biorefinery downstream processes are complex and demand advanced and expensive technologies which further increase the cost of production. Overall, selection of the proper harvesting technology is case dependent and can be decided based on the final targeted product.

The difference in regulatory requirements for production of different integral algal products is another big challenge facing algal biofuel-integrated biorefinery. For example, water quality standard requirement for most pharmaceutical and nutraceutical products is quite different for those needed in biofuels and some other high-volume products. This greatly limits the use of waste streams for algal biomass production. Therefore, simultaneous phycoremediation of waste products and algal biomass production cannot be applied for all integral products from safety aspect (di Visconte et al. 2019; Nethravathy et al. 2019), which drastically reduces the options of these cheap waste nutrient sources. However, some safe waste substrates such as wastewaters from diary and some other food/agricultural wastes that are free from hazardous components might still be applicable in such integrated biorefineries (Fayaz and Honarvar 2022).

Since a number of these products are produced from different parts of the algae biomass, different cultivation conditions could be required for the over-production of one product while also negatively affecting the productivity of another. For example, ample nutrients (i.e., nitrogen, phosphorus, sulfur, etc.) supply increases the biomass (and carbohydrates) production in microalgae, while limiting these nutrients favor lipid production and vice versa. Therefore, trying to get all these integral products from the same culture conditions in an integrated biorefinery will demand a lot of optimization processes, scale-up studies, or even dual culture conditions in some cases (Zhu et al. 2016). Similar effects on algal integral products also apply to most other culture conditions such as light, pH, salinity, and temperature (Morales et al. 2021; Zhu et al. 2016). Potential ways of addressing these challenges and the future prospects of integrated biofuel routes are summarized in Fig. 2. Addressing most of these challenges presented here can be very tasking both in terms of resources, technology, and other aspects, however, the rewards of getting adequate algal species that will meet the demands of future integrated biorefineries proposes to surpass such challenges.

5 Conclusions

Meeting the Paris Agreement on climate change demands immediate shift from fossil fuel usage to reliance on eco-friendly and sustainable energy sources. Renewable and sustainable energy sources like algal biofuels are interesting options as they generate assorted fuels as well as other essential products in several industries. However, the cost of producing algal fuels is still quite high, e.g., previous reports on the economic analysis of algal fuel vs petroleum fuel revealed algal biodiesel production cost to be US \$2.29 kg^{-1} compared to US \$1.08 kg^{-1} for fossil diesel. Fortunately, several kinds of integrated biorefinery studies have demonstrated enormous prospects in reducing the cost of biofuels from algae. For example, the adoption of combined algal processing was shown to reduce the cost of algal biofuel by US \$0.95 per gallon gasoline equivalent which was 9% reduction compared to other biorefinery setup. It has also been reported that the cost of producing microalgal biodiesel can be successfully reduced to US \$0.54 L^{-1} when co-produced with value-added products (such as astaxanthin and polyhydroxyl butyrate), and to US \$0.73 kg^{-1} DW when algae is cultivated in wastewater. Furthermore, previous studies concluded that the joint production of biodiesel with protein extraction reduced the cost of algal biofuel production from US \$17.26 L^{-1} to US \$13.73 L^{-1}. Therefore, integrating algal biofuels to phycoremediation of wastewaters and the production of some high-value products as well as high-volume products will further reduce the cost of production.

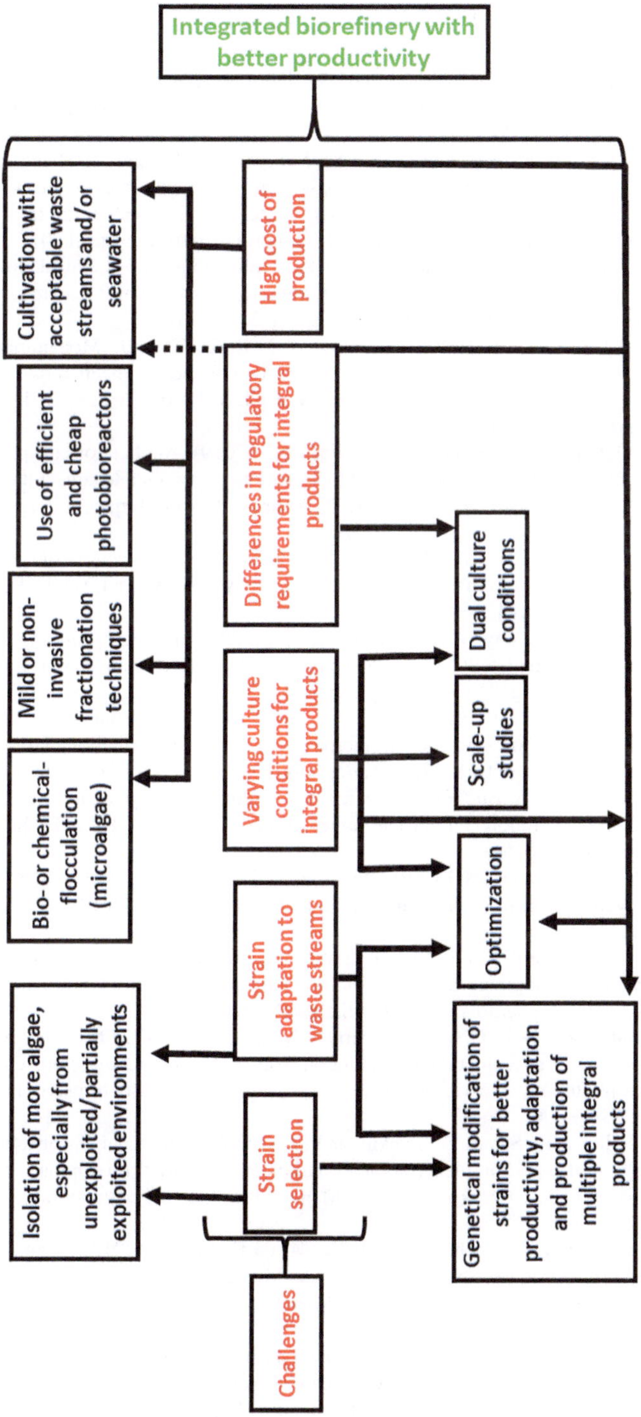

Fig. 2 Challenges and future prospects of integrated biofuel routes

References

Abomohra AEF, El-Naggar AH, Baeshen AA (2018a) Potential of macroalgae for biodiesel production: screening and evaluation studies. J Biosci Bioeng 125:231–237. https://doi.org/10.1016/j.jbiosc.2017.08.020

Abomohra AEF, Eladel H, El-Esawi M, Wang S, Wang Q, He Z, Feng Y, Shang H, Hanelt D (2018b) Effect of lipid-free microalgal biomass and waste glycerol on growth and lipid production of *Scenedesmus obliquus*: innovative waste recycling for extraordinary lipid production. Bioresour Technol 249:992–999. https://doi.org/10.1016/j.biortech.2017.10.102

Abomohra AEF, Jin W, Sagar V, Ismail GA (2018c) Optimization of chemical flocculation of *Scenedesmus obliquus* grown on municipal wastewater for improved biodiesel recovery. Renew Energy 115:880–886. https://doi.org/10.1016/j.renene.2017.09.019

Abomohra AEF, El-Hefnawy ME, Wang Q, Huang J, Li L, Tang J, Mohammed S (2021) Sequential bioethanol and biogas production coupled with heavy metal removal using dry seaweeds: towards enhanced economic feasibility. J Clean Prod 316:128341. https://doi.org/10.1016/j.jclepro.2021.128341

Ahamefule SC, Ogbonna JC, Moneke AN, Ossai NI (2019) Application of photosynthetic microalgae as efficient pH bio-stabilizers and bio-purifiers in sustainable aquaculture of *Clarias gariepinus* (African catfish) fry. Niger J Biotechnol 35:139. https://doi.org/10.4314/njb.v35i2.17

Ahamefule CS, Ugwuodo CJ, Idike PO, Ogbonna JC (2021) Application of photosynthetic microalgae in the direct desalination pretreatment of seawater. Water Environ J 35:657–669. https://doi.org/10.1111/wej.12659

Ali J, Wang L, Waseem H, Song B, Djellabi R, Pan G (2020) Turning harmful algal biomass to electricity by microbial fuel cell: a sustainable approach for waste management. Environ Pollut 266:115373. https://doi.org/10.1016/j.envpol.2020.115373

Alishah Aratboni H, Rafiei N, Garcia-Granados R, Alemzadeh A, Morones-Ramírez JR (2019) Biomass and lipid induction strategies in microalgae for biofuel production and other applications. Microb Cell Factories 18:1–17. https://doi.org/10.1186/s12934-019-1228-4

Ansari FA, Singh P, Guldhe A, Bux F (2017) Microalgal cultivation using aquaculture wastewater: integrated biomass generation and nutrient remediation. Algal Res 21:169–177. https://doi.org/10.1016/j.algal.2016.11.015

Apone F, Barbulova A, Colucci MG (2019) Plant and microalgae derived peptides are advantageously employed as bioactive compounds in cosmetics. Front Plant Sci 10:1–8. https://doi.org/10.3389/fpls.2019.00756

Ardalan Y, Jazini M, Karimi K (2018) Sargassum angustifolium brown macroalga as a high potential substrate for alginate and ethanol production with minimal nutrient requirement. Algal Res 36:29–36. https://doi.org/10.1016/j.algal.2018.10.010

Ashokkumar V, Salam Z, Tiwari ON, Chinnasamy S, Mohammed S, Ani FN (2015) An integrated approach for biodiesel and bioethanol production from *Scenedesmus bijugatus* cultivated in a vertical tubular photobioreactor. Energy Convers Manag 101:778–786. https://doi.org/10.1016/j.enconman.2015.06.006

Ashokkumar V, Salim MR, Salam Z, Sivakumar P, Chong CT, Elumalai S, Suresh V, Ani FN (2017) Production of liquid biofuels (biodiesel and bioethanol) from brown marine macroalgae *Padina tetrastromatica*. Energy Convers Manag 135:351–361. https://doi.org/10.1016/j.enconman.2016.12.054

Ashokkumar V, Chen WH, Ngamcharussrivichai C, Agila E, Ani FN (2019) Potential of sustainable bioenergy production from *Synechocystis* sp. cultivated in wastewater at large scale – a low cost biorefinery approach. Energy Convers Manag 186:188–199. https://doi.org/10.1016/j.enconman.2019.02.056

Ashour M, Al-Souti AS, Hassan SM, Ammar GAG, Goda AMAS, El-Shenody R, Abomohra AEF, El-Haroun E, Elshobary ME (2023) Commercial seaweed liquid extract as strawberry biostimulants and bioethanol production. Life 13(1):85. https://doi.org/10.3390/life13010085

Ayub HMU, Ahmed A, Lam SS, Lee J, Show PL, Park YK (2022) Sustainable valorization of algae biomass via thermochemical processing route: An overview. Bioresour Technol 344:126399. https://doi.org/10.1016/j.biortech.2021.126399

Behl K, SeshaCharan P, Joshi M, Sharma M, Mathur A, Kareya MS, Jutur PP, Bhatnagar A, Nigam S (2020) Multifaceted applications of isolated microalgae *Chlamydomonas* sp. TRC-1 in wastewater remediation, lipid production and bioelectricity generation. Bioresour Technol 304:122993. https://doi.org/10.1016/j.biortech.2020.122993

Brar A, Kumar M, Pareek N (2019) Comparative appraisal of biomass production, remediation, and bioenergy generation potential of microalgae in dairy wastewater. Front Microbiol 10:678. https://doi.org/10.3389/fmicb.2019.00678

Budzianowski WM (2017) High-value low-volume bioproducts coupled to bioenergies with potential to enhance business development of sustainable biorefineries. Renew Sust Energ Rev 70: 793–804. https://doi.org/10.1016/j.rser.2016.11.260

Cai T, Park SY, Li Y (2013) Nutrient recovery from wastewater streams by microalgae: status and prospects. Renew Sust Energ Rev 19:360–369. https://doi.org/10.1016/j.rser.2012.11.030

Cao Z, Shen X, Wang X, Zhu B, Pan K, Li Y (2022) Effect of nitrogen concentration on the Alkalophilic microalga *Nitzschia* sp. NW129-a promising feedstock for the integrated production of lipids and Fucoxanthin in biorefinery. Front Mar Sci 8:830590. https://doi.org/10.3389/fmars.2021.830590

Chagas AL, Rios AO, Jarenkow A, Marcílio NR, Ayub MAZ, Rech R (2015) Production of carotenoids and lipids by *Dunaliella tertiolecta* using CO2 from beer fermentation. Process Biochem 50:981–988. https://doi.org/10.1016/j.procbio.2015.03.012

Chavan R, Mutnuri S (2019) Tertiary treatment of domestic wastewater by *Spirulina platensis* integrated with microalgal biorefinery. Biofuels 10:33–44. https://doi.org/10.1080/17597269.2018.1461509

Chen Q, Chen Y, Xu Q, Jin H, Hu Q, Han D (2022) Effective two-stage heterotrophic cultivation of the unicellular green microalga *Chromochloris zofingiensis* enabled ultrahigh biomass and Astaxanthin production. Front Bioeng Biotechnol 10:834230. https://doi.org/10.3389/fbioe.2022.834230

De Bhowmick G, Sarmah AK, Sen R (2019) Zero-waste algal biorefinery for bioenergy and biochar: a green leap towards achieving energy and environmental sustainability. Sci Total Environ 650:2467. https://doi.org/10.1016/j.scitotenv.2018.10.002

De Francisci D, Su Y, Iital A, Angelidaki I (2018) Evaluation of microalgae production coupled with wastewater treatment. Environ Technol (United Kingdom) 39:581–592. https://doi.org/10.1080/09593330.2017.1308441

De Souza ZJ (2019) Bioelectricity of sugarcane: a case study from Brazil and perspectives, sugarcane biorefinery, technology and perspectives. Elsevier Inc., Amsterdam. https://doi.org/10.1016/B978-0-12-814236-3.00013-5

del Río PG, Gomes-Dias JS, Rocha CMR, Romaní A, Garrote G, Domingues L (2020) Recent trends on seaweed fractionation for liquid biofuels production. Bioresour Technol 299:122613. https://doi.org/10.1016/j.biortech.2019.122613

di Visconte GS, Spicer A, Chuck CJ, Allen MJ (2019) The microalgae biorefinery: a perspective on the current status and future opportunities using genetic modification. Appl Sci 9:4793. https://doi.org/10.3390/app9224793

Dineshbabu G, Goswami G, Kumar R, Sinha A, Das D (2019) Microalgae–nutritious, sustainable aqua- and animal feed source. J Funct Foods 62:103545. https://doi.org/10.1016/j.jff.2019.103545

Ellis JT, Hengge NN, Sims RC, Miller CD (2012) Acetone, butanol, and ethanol production from wastewater algae. Bioresour Technol 111:491–495. https://doi.org/10.1016/j.biortech.2012.02.002

Elmaadawy K, Hu J, Guo S, Hou H, Xu J, Wang D, Liang T, Yang J, Liang S, Xiao K, Liu B (2020) Enhanced treatment of landfill leachate with cathodic algal biofilm and oxygen-consuming unit in a hybrid microbial fuel cell system. Bioresour Technol 310:123420. https://doi.org/10.1016/j.biortech.2020.123420

Eusébio A, Moniz P, Lopes da Silva T, Marques IP (2022) Fermentation wastes from *Chrypthecodinium cohnii* lipid production for energy recovery by anaerobic digestion. PRO 10(11):2463. https://doi.org/10.3390/pr10112463

Fayaz Y, Honarvar M (2022) Using agricultural wastes and food residues for algal cultivation. J Food Sci Nutr 19:39–50. https://doi.org/10.30495/JFTN.2022.62411.11144

Filippino KC, Mulholland MR, Bott CB (2015) Phycoremediation strategies for rapid tertiary nutrient removal in a waste stream. Algal Res 11:125–133. https://doi.org/10.1016/j.algal.2015.06.011

Flórez-Fernández N, Illera M, Sánchez M, Lodeiro P, Torres MD, López-Mosquera ME, Soto M, de Vicente MS, Domínguez H (2021) Integrated valorization of *Sargassum muticum* in biorefineries. Chem Eng J 404:125635. https://doi.org/10.1016/j.cej.2020.125635

Francavilla M, Colaianna M, Zotti M, Morgese MG, Trotta P, Tucci P, Schiavone S, Cuomo V, Trabace L (2012) Extraction, characterization and in vivo Neuromodulatory activity of Phytosterols from microalga *Dunaliella Tertiolecta*. Curr Med Chem 19(18):3058–3067. https://doi.org/10.2174/092986712800672021

Gao B, Dai C, Zhang H, Zhang C (2022) Evaluation of a novel oleaginous filamentous green alga, *Barranca yajiagengensis* (Chlorophyta, Chaetophorales) for biomass, lipids and pigments production. Algal Res 64:102681. https://doi.org/10.1016/j.algal.2022.102681

Ghafari M, Rashidi B, Haznedaroglu BZ (2018) Effects of macro and micronutrients on neutral lipid accumulation in oleaginous microalgae. Biofuels 9:147–156. https://doi.org/10.1080/17597269.2016.1221644

Gifuni I, Pollio A, Safi C, Marzocchella A, Olivieri G (2019) Current bottlenecks and challenges of the microalgal biorefinery. Trends Biotechnol 37:242–252. https://doi.org/10.1016/j.tibtech.2018.09.006

Hadjkacem F, Pierre G, Christophe G, Elleuch J, Fendri I, Boual Z, El Hadj MDO, El Alaoui-Talibi Z, El Modafar C, Dubessay P, Delattre C, Michaud P, Abdelkafi S (2022) Bioconversion of the Brown Tunisian seaweed *Halopteris scoparia*: application to energy. Energies 15:4342. https://doi.org/10.3390/en15124342

Hamed I (2016) The evolution and versatility of microalgal biotechnology: a review. Compr Rev Food Sci Food Saf 15:1104–1123. https://doi.org/10.1111/1541-4337.12227

Heredia V, Gonçalves O, Marchal L, Pruvost J (2021) Producing energy-rich microalgae biomass for liquid biofuels: influence of strain selection and culture conditions. Energies 14(5):1246. https://doi.org/10.3390/en14051246

Hou Q, Nie C, Pei H, Hu W, Jiang L, Yang Z (2016a) The effect of algae species on the bioelectricity and biodiesel generation through open-air cathode microbial fuel cell with kitchen waste anaerobically digested effluent as substrate. Bioresour Technol 218:902–908. https://doi.org/10.1016/j.biortech.2016.07.035

Hou Q, Pei H, Hu W, Jiang L, Yu Z (2016b) Mutual facilitations of food waste treatment, microbial fuel cell bioelectricity generation and *Chlorella vulgaris* lipid production. Bioresour Technol 203:50–55. https://doi.org/10.1016/j.biortech.2015.12.049

Hou Q, Cheng J, Nie C, Pei H, Jiang L, Zhang L, Yang Z (2017) Features of *Golenkinia* sp. and microbial fuel cells used for the treatment of anaerobically digested effluent from kitchen waste at different dilutions. Bioresour Technol 240:130–136. https://doi.org/10.1016/j.biortech.2017.02.092

Ismail MM, Ismail GA, El-Sheekh MM (2020) Potential assessment of some micro- and macroalgal species for bioethanol and biodiesel production. Energy Sources, Part A Recover Util Environ Eff. https://doi.org/10.1080/15567036.2020.1758853

Jadhav DA, Jain SC, Ghangrekar MM (2017) Simultaneous wastewater treatment, algal biomass production and electricity generation in clayware microbial carbon capture cells. Appl Biochem Biotechnol 183:1076–1092. https://doi.org/10.1007/s12010-017-2485-5

Jiang HM, Luo SJ, Shi XS, Dai M, Guo RB (2013) A system combining microbial fuel cell with photobioreactor for continuous domestic wastewater treatment and bioelectricity generation. J Cent South Univ 20:488–494. https://doi.org/10.1007/s11771-013-1510-2

Kang NS, Cho K, An SM, Kim ES, Ki H, Lee CH, Choi G, Hong JW (2022) Taxonomic and biochemical characterization of microalga *Graesiella emersonii* GEGS21 for its potential to become feedstock for biofuels and bioproducts. Energies 15(22):8725. https://doi.org/10.3390/en15228725

Karpagam R, Jawaharraj K, Gnanam R (2021) Review on integrated biofuel production from microalgal biomass through the outset of transesterification route: a cascade approach for sustainable bioenergy. Sci Total Environ 766:144236. https://doi.org/10.1016/j.scitotenv.2020.144236

Khan SA, Sharma GK, Malla FA, Kumar A, Rashmi GN (2019) Microalgae based biofertilizers: a biorefinery approach to phycoremediate wastewater and harvest biodiesel and manure. J Clean Prod 211:1412–1419. https://doi.org/10.1016/j.jclepro.2018.11.281

Khanra S, Mondal M, Halder G, Tiwari ON, Gayen K, Bhowmick TK (2018) Downstream processing of microalgae for pigments, protein and carbohydrate in industrial application: a review. Food Bioprod Process 110:60–84. https://doi.org/10.1016/j.fbp.2018.02.002

Kim EJ, Kim S, Choi HG, Han SJ (2020) Biotechnology for Biofuels Co - production of biodiesel and bioethanol using psychrophilic microalga *Chlamydomonas* sp. KNM0029C isolated from Arctic sea ice. Biotechnol Biofuels 13:1–13. https://doi.org/10.1186/s13068-020-1660-z

Krishnan SK, Kandasamy S, Subbiah K (2021) Fabrication of microbial fuel cells with nanoelectrodes for enhanced bioenergy production. In: Nanomaterials: application in biofuels and bioenergy production systems. Academic Press, Cambridge, pp 677–687. https://doi.org/10.1016/B978-0-12-822401-4.00003-9

Kumar S, Gupta R, Kumar G, Sahoo D, Kuhad RC (2013) Bioethanol production from *Gracilaria verrucosa*, a red alga, in a biorefinery approach. Bioresour Technol 135:150–156. https://doi.org/10.1016/j.biortech.2012.10.120

Kumar G, Shekh A, Jakhu S, Sharma Y, Kapoor R, Sharma TR (2020) Bioengineering of microalgae: recent advances, perspectives, and regulatory challenges for industrial application. Front Bioeng Biotechnol 8:914. https://doi.org/10.3389/fbioe.2020.00914

Lakshmidevi R, Gandhi NN, Muthukumar K (2020) Carbon neutral electricity production from municipal solid waste landfill leachate using algal-assisted microbial fuel cell. Appl Biochem Biotechnol 191:852–866. https://doi.org/10.1007/s12010-019-03160-5

Lee N, Ko SR, Ahn CY, Oh HM (2018) Optimized co-production of lipids and carotenoids from Ettlia sp. by regulating stress conditions. Bioresour Technol 258:234–239. https://doi.org/10.1016/j.biortech.2018.03.006

Liu Z, Liu C, Hou Y, Chen S, Xiao D, Zhang J, Chen F (2013) Isolation and characterization of a marine microalga for biofuel production with astaxanthin as a co-product. Energies 6:2759–2772. https://doi.org/10.3390/en6062759

Lodeiro P, Sillanpää M (2013) Gold recovery from artificial seawater using synthetic materials and seaweed biomass to induce gold nanoparticles formation in batch and column experiments. Mar Chem 152:11–19. https://doi.org/10.1016/j.marchem.2013.03.003

Lu Y, Zhuo C, Li Y, Li H, Yang M, Xu D, He H (2020a) Evaluation of filamentous heterocystous cyanobacteria for integrated pig-farm biogas slurry treatment and bioenergy production. Bioresour Technol 297:122418. https://doi.org/10.1016/j.biortech.2019.122418

Lu W, Asraful Alam M, Liu S, Xu J, Parra Saldivar R (2020b) Critical processes and variables in microalgae biomass production coupled with bioremediation of nutrients and CO2 from livestock farms: a review. Sci Total Environ 716:135247. https://doi.org/10.1016/j.scitotenv.2019.135247

Machineni L (2019) Review on biological wastewater treatment and resources recovery: attached and suspended growth systems. Water Sci Technol 80:2013–2026. https://doi.org/10.2166/wst.2020.034

Maity JP, Hou CP, Majumder D, Bundschuh J, Kulp TR, Chen CY, Te Chuang L, Nathan Chen CN, Jean JS, Yang TC, Chen CC (2014) The production of biofuel and bioelectricity associated with wastewater treatment by green algae. Energy 78:94–103. https://doi.org/10.1016/j.energy.2014.06.023

Molino A, Iovine A, Casella P, Mehariya S, Chianese S, Cerbone A, Rimauro J, Musmarra D (2018) Microalgae characterization for consolidated and new application in human food, animal feed and nutraceuticals. Int J Environ Res Public Health 15:1–21. https://doi.org/10.3390/ijerph15112436

Moniz P, Martins D, Oliveira AC, Reis A, Lopes da Silva T (2022) The biorefinery of the marine microalga *Crypthecodinium cohnii* as a strategy to valorize microalgal oil fractions. Fermentation 8(10):502. https://doi.org/10.3390/fermentation8100502

Morales M, Aflalo C, Bernard O (2021) Microalgal lipids: a review of lipids potential and quantification for 95 phytoplankton species. Biomass Bioenergy 150:106108. https://doi.org/10.1016/j.biombioe.2021.106108

Mulbry W, Kondrad S, Pizarro C, Kebede-Westhead E (2008) Treatment of dairy manure effluent using freshwater algae: algal productivity and recovery of manure nutrients using pilot-scale algal turf scrubbers. Bioresour Technol 99:8137–8142. https://doi.org/10.1016/j.biortech.2008.03.073

Nagappan S, Kumar Verma S (2018) Co-production of biodiesel and alpha-linolenic acid (omega-3 fatty acid) from microalgae, *Desmodesmus* sp. MCC34. Energy Sources Part A Recover Util Environ Eff 40:2933–2940. https://doi.org/10.1080/15567036.2018.1514434

Nagarajan D, Lee DJ, Chen CY, Chang JS (2020) Resource recovery from wastewaters using microalgae-based approaches: a circular bioeconomy perspective. Bioresour Technol 302:122817. https://doi.org/10.1016/j.biortech.2020.122817

Naina Mohamed S, Ajit Hiraman P, Muthukumar K, Jayabalan T (2020) Bioelectricity production from kitchen wastewater using microbial fuel cell with photosynthetic algal cathode. Bioresour Technol 295:122226. https://doi.org/10.1016/j.biortech.2019.122226

Nayak JK, Amit, Ghosh UK (2018) An innovative mixotrophic approach of distillery spent wash with sewage wastewater for biodegradation and bioelectricity generation using microbial fuel cell. J Water Process Eng 23:306–313. https://doi.org/10.1016/j.jwpe.2018.04.003

Neethu B, Bhowmick GD, Ghangrekar MM (2018) Enhancement of bioelectricity generation and algal productivity in microbial carbon-capture cell using low cost coconut shell as membrane separator. Biochem Eng J 133:205–213. https://doi.org/10.1016/j.bej.2018.02.014

Nethravathy MU, Mehar JG, Mudliar SN, Shekh AY (2019) Recent advances in microalgal bioactives for food, feed, and healthcare products: commercial potential, market space, and sustainability. Compr Rev Food Sci Food Saf 18:1882–1897. https://doi.org/10.1111/1541-4337.12500

Ng FL, Phang SM, Thong CH, Periasamy V, Pindah J, Yunus K, Fisher AC (2021) Integration of bioelectricity generation from algal biophotovoltaic (BPV) devices with remediation of palm oil mill effluent (POME) as substrate for algal growth. Environ Technol Innov 21:101280. https://doi.org/10.1016/j.eti.2020.101280

Ogbonna CN, Edeh I (2018) Harvesting Chlorella variabilis biomass using Moringa oleifera seed-induced sedimentation. J Adv Biol Biotechnol 18:1–11. https://doi.org/10.9734/jabb/2018/43608

Ogbonna IO, Ikwebe J, Okpozu OO, Eze CN, Ogbonna JC (2021) Potential of *Dictyosphaerium* sp. LC172264 concomitant remediation of cassava wastewater and accumulation of lipids for biodiesel production. Adv Biosci Biotechnol 12:257–274. https://doi.org/10.4236/abb.2021.128016

Okpozu OO, Ogbonna IO, Ikwebe J, Ogbonna JC (2019) Phycoremediation of cassava wastewater by *Desmodesmus armatus* and the concomitant accumulation of lipids for biodiesel production. Bioresour Technol Reports 7:100255. https://doi.org/10.1016/j.biteb.2019.100255

Osman MEH, Abo-Shady AM, Elshobary ME, Abd El-Ghafar MO, Abomohra AEF (2020) Screening of seaweeds for sustainable biofuel recovery through sequential biodiesel and bioethanol production. Environ Sci Pollut Res 27:32481–32493. https://doi.org/10.1007/s11356-020-09534-1

Pandey A, Srivastava S, Kumar S (2020) Development and cost-benefit analysis of a novel process for biofuel production from microalgae using pre-treated high-strength fresh cheese whey wastewater. Environ Sci Pollut Res 27:23963–23980. https://doi.org/10.1007/s11356-020-08535-4

Phang SM, Chu WL, Rabiei R (2015) Phycoremediation. Springer, Dordrecht. https://doi.org/10.1007/978-94-017-7321-8_13

Rawat I, Ranjith Kumar R, Mutanda T, Bux F (2011) Dual role of microalgae: Phycoremediation of domestic wastewater and biomass production for sustainable biofuels production. Appl Energy 88:3411–3424. https://doi.org/10.1016/j.apenergy.2010.11.025

Ricky R, Chiampo F, Shanthakumar S (2022) Efficacy of ciprofloxacin and amoxicillin removal and the effect on the biochemical composition of *Chlorella vulgaris*. Bioengineering 9(4):134. https://doi.org/10.3390/bioengineering9040134

Roostaei J, Zhang Y, Gopalakrishnan K, Ochocki AJ (2018) Mixotrophic microalgae biofilm: a novel algae cultivation strategy for improved productivity and cost-efficiency of biofuel feedstock production. Sci Rep 8:12528. https://doi.org/10.1038/s41598-018-31016-1

Ruangsomboon S, Sornchai P, Prachom N (2018) Enhanced hydrocarbon production and improved biodiesel qualities of *Botryococcus braunii* KMITL 5 by vitamins thiamine, biotin and cobalamin supplementation. Algal Res 29:159–169. https://doi.org/10.1016/j.algal.2017.11.028

Safafar H, Hass MZ, Møller P, Holdt SL, Jacobsen C (2016a) High-EPA biomass from *Nannochloropsis Salina* cultivated in a flat-panel photo-bioreactor on a process water-enriched growth medium. Mar Drugs 14(8):144. https://doi.org/10.3390/md14080144

Safafar H, Nørregaard PU, Ljubic A, Møller P, Holdt SL, Jacobsen C (2016b) Enhancement of protein and pigment content in two chlorella species cultivated on industrial process water. J Mar Sci Eng 4(4):84–98. https://doi.org/10.3390/jmse4040084

Sarma S, Sharma S, Rudakiya D, Upadhyay J, Rathod V, Patel A, Narra M (2021) Valorization of microalgae biomass into bioproducts promoting circular bioeconomy: a holistic approach of bioremediation and biorefinery. 3 Biotech 11:1–29. https://doi.org/10.1007/s13205-021-02911-8

Sarwer A, Hamed SM, Osman AI, Jamil F, Al-Muhtaseb AH, Alhajeri NS, Rooney DW (2022) Algal biomass valorization for biofuel production and carbon sequestration: a review. Environ Chem Lett 20:2797. https://doi.org/10.1007/s10311-022-01458-1

Schultze-Jena A, Vroon RC, Macleod AKA, Hreggviðsson G, Adalsteinsson BT, Engelen-Smit NPE, de Vrije T, Budde MAW, van der Wal H, López-Contreras AM, Boon MA (2022) Production of acetone, butanol, and ethanol by fermentation of *Saccharina latissima*: cultivation, enzymatic hydrolysis, inhibitor removal, and fermentation. Algal Res 62:102618. https://doi.org/10.1016/j.algal.2021.102618

Selvan ST, Govindasamy B, Muthusamy S, Ramamurthy D (2019) Exploration of green integrated approach for effluent treatment through mass culture and biofuel production from unicellular alga, Acutodesmus obliquus RDS01. Int J Phytoremediation 21:1305–1322. https://doi.org/10.1080/15226514.2019.1633255

Shlosberg Y, Krupnik N, Tóth TN, Eichenbaum B, Meirovich MM, Meiri D, Yehezkeli O, Schuster G, Israel Á, Adir N (2022) Bioelectricity generation from live marine photosynthetic macroalgae. Biosens Bioelectron 198:113824. https://doi.org/10.1016/j.bios.2021.113824

Singh D, Puri M, Wilkens S, Mathur AS, Tuli DK, Barrow CJ (2013) Characterization of a new zeaxanthin producing strain of *Chlorella saccharophila* isolated from New Zealand marine waters. Bioresour Technol 143:308–314. https://doi.org/10.1016/j.biortech.2013.06.006

Singh R, Paliwal C, Nesamma AA, Narula A, Jutur PP (2020) Nutrient deprivation mobilizes the production of unique tocopherols as a stress-promoting response in a new indigenous isolate *Monoraphidium* sp. Front Mar Sci 7:575817. https://doi.org/10.3389/fmars.2020.575817

Sinha A, Goswami G, Kumar R, Das D (2021) A microalgal biorefinery approach for bioactive molecules, biofuel, and biofertilizer using a novel carbon dioxide-tolerant strain *Tetradesmus obliquus* CT02. Biomass Conv Bioref. https://doi.org/10.1007/s13399-021-02098-1

Sisman-Aydin G, Simsek K (2022) Municipal wastewater effects on the performance of nutrient removal, and lipid, carbohydrate, and protein productivity of blue-green algae *Chroococcus turgidus*. Sustain 14(24):17021. https://doi.org/10.3390/su142417021

Solis CMA, Juan JLGS, Mayol AP, Sy CL, Ubando AT, Culaba AB (2021) A multi-objective life cycle optimization model of an integrated algal biorefinery toward a sustainable circular bioeconomy considering resource recirculation. Energies 14(5):1416. https://doi.org/10.3390/en14051416

Sorgatto VG, Soccol CR, Molina-Aulestia DT, de Carvalho MA, de Melo Pereira GV, de Carvalho JC (2021) Mixotrophic cultivation of microalgae in cassava processing wastewater for simultaneous treatment and production of lipid-rich biomass. Fuels 2:521–532. https://doi.org/10.3390/fuels2040030

Sun J, Hu Y, Li W, Zhang Y, Chen J, Deng F (2015) Sequential decolorization of azo dye and mineralization of decolorization liquid coupled with bioelectricity generation using a pH self-neutralized photobioelectrochemical system operated with polarity reversion. J Hazard Mater 289:108–117. https://doi.org/10.1016/j.jhazmat.2015.02.010

Sun J, Cai B, Xu W, Huang Y, Zhang Y, Peng Y, Chang K, Kuo J, Chen K, Ning X, Liu G, Wang Y, Yang Z, Liu J (2017) Enhanced bioelectricity generation and azo dye treatment in a reversible photo-bioelectrochemical cell by using novel anthraquinone-2,6-disulfonate (AQDS)/MnOx-doped polypyrrole film electrodes. Bioresour Technol 225:40–47. https://doi.org/10.1016/j.biortech.2016.11.038

Sun J, Xu W, Cai B, Huang G, Zhang H, Zhang Y, Yuan Y, Chang K, Chen K, Peng Y, Chen K (2019) High-concentration nitrogen removal coupling with bioelectric power generation by a self-sustaining algal-bacterial biocathode photo-bioelectrochemical system under daily light/dark cycle. Chemosphere 222:797–809. https://doi.org/10.1016/j.chemosphere.2019.01.191

Tasic MB, De Jesus A, Maria B, Rocha I, Schiavon B, Colling B, Veljković VB, Filho RM (2020) Cultivation of *Chlamydomonas reinhardtii* in anaerobically digested vinasse for bioethanol production. Waste Biomass Valorization 12:857. https://doi.org/10.1007/s12649-020-01034-0

Tibbetts SM, Milley JE, Lall SP (2015) Chemical composition and nutritional properties of freshwater and marine microalgal biomass cultured in photobioreactors. J Appl Phycol 27:1109–1119. https://doi.org/10.1007/s10811-014-0428-x

Trentin R, Custódio L, Rodrigues MJ, Moschin E, Sciuto K, da Silva JP, Moro I (2022) Total phenolic levels, in vitro antioxidant properties, and fatty acid profile of two microalgae, *Tetraselmis marina* strain IMA043 and Naviculoid diatom strain IMA053, isolated from the North Adriatic Sea. Mar Drugs 20(3):207. https://doi.org/10.3390/md20030207

Urreta I, Ikaran Z, Janices I, Ibañez E, Castro-Puyana M, Castañón S, Suárez-Alvarez S (2014) Revalorization of *Neochloris oleoabundans* biomass as source of biodiesel by concurrent production of lipids and carotenoids. Algal Res 5:16–22. https://doi.org/10.1016/j.algal.2014.05.001

Varaprasad D, Narasimham D, Paramesh K, Sudha NR, Himabindu Y, Keerthi Kumari M, Nazaneen Parveen S, Chandrasekhar T (2021) Improvement of ethanol production using green alga *Chlorococcum minutum*. Environ Technol (United Kingdom) 42:1383–1391. https://doi.org/10.1080/09593330.2019.1669719

Varjani S, Joshi R, Srivastava VK, Ngo HH, Guo W (2020) Treatment of petroleum industry wastewater: current practices and perspectives. Environ Sci Pollut Res 27:27172–27180. https://doi.org/10.1016/B978-0-323-85884-7.00015-1

Venkata Subhash G, Chandra R, Venkata Mohan S (2013) Microalgae mediated bio-electrocatalytic fuel cell facilitates bioelectricity generation through oxygenic photomixotrophic mechanism. Bioresour Technol 136:644–653. https://doi.org/10.1016/j.biortech.2013.02.035

Yadav AK, Nayak SK, Acharya BC, Mishra BK (2015) Algal-assisted microbial fuel cell for wastewater treatment and bioelectricity generation. Energy sources, Part A Recover. Util Environ Eff 37:127–133. https://doi.org/10.1080/15567036.2011.576422

Yahampath Arachchige Don CDY, Babel S (2021) Circulation of anodic effluent to the cathode chamber for subsequent treatment of wastewater in photosynthetic microbial fuel cell with generation of bioelectricity and algal biomass. Chemosphere 278:130455. https://doi.org/10.1016/j.chemosphere.2021.130455

Younes S, Bracharz F, Awad D, Qoura F, Mehlmer N, Brueck T (2020) Microbial lipid production by oleaginous yeasts grown on *Scenedesmus obtusiusculus* microalgae biomass hydrolysate. Bioprocess Biosyst Eng 43:1629–1638. https://doi.org/10.1007/s00449-020-02354-0

Zhang CD, Li W, Shi YH, Li YG, Huang JK, Li HX (2016) A new technology of CO2 supplementary for microalgae cultivation on large scale - a spraying absorption tower coupled with an outdoor open runway pond. Bioresour Technol 209:351–359. https://doi.org/10.1016/j.biortech.2016.03.007

Zhu LD, Li ZH, Hiltunen E (2016) Strategies for lipid production improvement in microalgae as a biodiesel feedstock. Biomed Res Int 2016:7–9. https://doi.org/10.1155/2016/8792548

Zohir WF, Kapase VU, Kumar S (2022) Identification and characterization of a new microalga *Dysmorphococcus globosus*-HI from the Himalayan region as a potential source of natural Astaxanthin. Biology (Basel) 11(6):884. https://doi.org/10.3390/biology11060884

The Use of Wastewater for Algal Growth

Wei Han Foo, Sherlyn Sze Ning Koay, Hooi Ren Lim, and Kit Wayne Chew

Abstract To date, the field of microalgae has been slowly evolving and expanding to accommodate various industries from food and feed to biofuel production coupled with wastewater treatment which subsequently allowed greater access of water for daily usage. This technology is gradually proving its value, despite its limited use in the market. With the depletion of freshwater resources, conventional treatment methods are being replaced by microalgae-mediated wastewater treatment. This method not only has a lower environmental footprint and produces less chemical waste, but also captures nutrients more efficiently than conventional methods which would be subsequently consumed for the development of microalgal biomass. Several factors that influence the recovery of nutrients such as phosphorus and nitrogen that are crucial for algal growth are discussed accordingly in this chapter, including wastewater characteristics, turbidity, concentration of phosphorus and nitrogen as well a chemical oxygen demand and biological oxygen demand. Additionally, currently available microalgae-based wastewater technologies are introduced along with the respective advantages and disadvantages. Sustainability prospect of microalgae-based wastewater technologies is also examined. Lastly, this chapter also includes the potential challenges that hinder the development and commercialization of microalgae-based wastewater technologies and provides adequate future recommendations for resolving the current concerns.

Keywords Microalgae · Nutrient removal · Phycoremediation · Wastewater treatment · Sustainability

---

W. H. Foo · S. S. N. Koay · K. W. Chew (✉)
School of Chemistry, Chemical Engineering and Biotechnology, Nanyang Technological University, Singapore, Singapore
e-mail: kitwayne.chew@ntu.edu.sg

H. R. Lim
Department of Chemical and Environmental Engineering, Faculty of Science and Engineering, University of Nottingham Malaysia, Semenyih, Selangor Darul Ehsan, Malaysia

A. Abomohra, S. Ende (eds.), *Value-added Products from Algae*,
https://doi.org/10.1007/978-3-031-42026-9_9

1 Introduction

As discussed in previous chapters, algae are photosynthetic organisms that use light energy and inorganic carbon sources like carbonate or carbon dioxide to create biomass, releasing oxygen as a by-product. Since both microalgae and cyanobacteria (blue-green algae) are typically present in microalgal-based water treatment systems, the word "microalgae" is broad and frequently comprises both types of algae (Molinuevo-Salces et al. 2019). On the other hand, macroalgae, or seaweeds, have also been considered for wastewater treatment in certain applications. They have the ability to absorb nutrients and remove pollutants from wastewater, similar to microalgae (Wang et al. 2020). However, several reasons have shown that microalgae generally have a higher potential for wastewater treatment compared to macroalgae such as larger surface area-to-volume ratio, broader adaptability, and tolerance to varying environmental conditions, as well as higher nutrient uptake efficiency (Lee et al. 2020). As a result, this chapter mainly discusses about the utilization of microalgae in wastewater treatment systems.

The amount of freshwater in the world is 2.5% of the total water on Earth, and within that only 1.2% of freshwater can be found on the surface of the earth, of which only 20.9% are found in lakes (Ramalingam et al. 2022). Therefore, the amount of freshwater used by humans is recycled and reused, since there are limited supplies. Thus, wastewater treatment has a crucial role to play in sustaining human life. Microalgae have not yet been widely used in large-scale applications of wastewater treatment, but the research done throughout the years shows promising results in meeting the specifications set for urban uses. With the advancement of industrialization and the betterment of human life, the global usage of freshwater for agriculture, industry, and municipal uses has increased sixfold since 1900 to about 4 trillion m^3 (Ritchie and Roser 2017). Consequently, the increase in wastewater has posed a huge conundrum to society. However, in this century's current technologies, the privilege of reusing and recycling wastewater created by human activities is available. Therefore, many efforts have been put in place to treat wastewater.

In general, conventional wastewater treatment involves a series of steps to treat the wastewater, or so-called effluent, and can be divided into four categories: preliminary, primary, secondary, and advanced treatment. Each of these categories include several steps in removing waste or bacteria within wastewater (Michielssen et al. 2016). There are three main stages of the treatment process namely primary, secondary, and tertiary treatment as illustrated in Fig. 1. Before the three main stages of treatment, there is a preliminary stage that involves removing coarse solids and other large substances to improve the efficiency and maintenance of subsequent treatment units. This can include methods like coarse screening, grit removal, and comminution (Sathya et al. 2022). Primary treatment is focused on removing settled organic and inorganic solids to prevent the occurrence of unpleasant conditions in storage or flow-equalizing reservoirs. Methods like sedimentation and skimming can be used for this stage (Rossle 2007). Secondary and tertiary treatment refer to the additional steps taken to treat biodegradable dissolved and colloidal organic matter

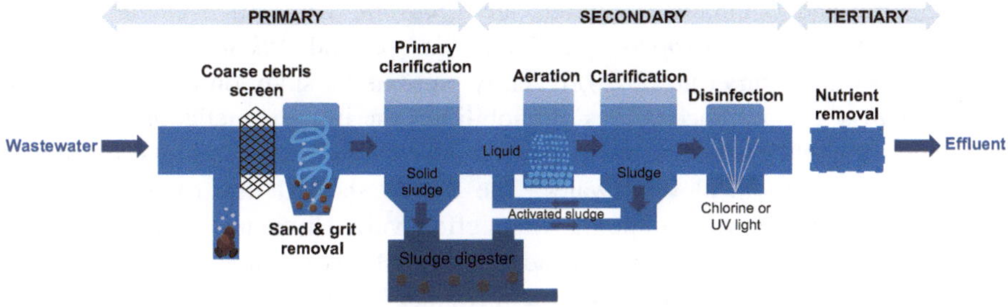

Fig 1 Schematic diagram of three main stages of the conventional wastewater treatment process

in wastewater, as well as removal of nitrogen, phosphorus, suspended solids, refractory organics, heavy metals, and dissolved solids (Sathya et al. 2022). Various methods can be used for these steps, such as activated sludge processes, trickling filters or biofilters, oxidation ditches, and rotating biological contactors (RBC) (Sonthiphand et al. 2022). For water to be safe for human consumption, sterilization is the last step aiming at removing viruses and bacteria, as they can pose a risk to human health. The objective is to remove microbes by injecting a suitable dosage of chlorine, which varies depending on the strength of wastewater and other factors.

Some of the aforementioned steps can be shortened or replaced with the aid of microalgae. The existence of microalgae in wastewater treatment has several advantages, not only for the ecosystem, but also as a substitute for the conventional nutrient recovery method, which is widely known to be much less effective than microalgae (Lim et al. 2021). Afterward, the nutrients from biomass can be converted into useful phycochemicals as previously discussed in this book. Hence, research on microalgae for consumption is conducted with great care for selecting only specific strains. On the other hand, in wastewater treatment plants, any strains that demonstrate positive outcomes can be employed. Therefore, selecting the most efficient microalgal species is crucial for nutrients removal, which is of great importance to prevent eutrophication of rivers and lakes posing a threat to the current supply of clean freshwater source (Wollmann et al. 2019). The common types of microalgae in practice are divided into 3 main categories which are namely municipal, agriculture, and industrial. These main categories can be further split into subcategories depending on what type of contaminants is present in wastewater. For example, municipal wastewater is commonly known to be a combination of domestic wastewater, greywater that primarily comes from non-toilet household activities, and sewage sludge (Prihandrijanti and Firdayati 2011). Agriculture wastewater, in contrast, originates from livestock wastewater and crop irrigation runoff, while industrial wastewater mainly consists of wastewater from food and beverage industry, petrochemical industry, and pharmaceutical industry (Pedersen et al. 2003; Samaei et al. 2018). The microalgal strain used depends on its own nature and also on the properties of wastewater; therefore, some strains of microalgae show more promising results when treating certain wastewater. The main microalgal species that have been extensively studied over years for wastewater treatment in both experimental and commercial settings are *Chlorella vulgaris*, *Scenedesmus* sp., *Spirulina* sp., *Nannochloropsis* sp., *Dunaliella* sp., and *Ankistrodesmus* sp. For instance, Govarthanan et al. (2020) recently presented a study on the use of *Chlorella vulgaris* exopolysaccharides immobilized on iron-magnetic nanoparticles, known as magnetic nano-composite particles (Fe_3O_4@EPS), to remove phosphate and ammonia from local wastewater. The results showed that roughly 91% of phosphorus and 85% of ammonia were effectively eliminated under optimum conditions of 3.5 g L^{-1} Fe_3O_4@EPS and pH 7.0 for 13 h of incubation (Govarthanan et al. 2020). In addition, immobilization of *Nannochloropsis* sp. on sodium alginate beads was found to be able to reduce 71% of chemical oxygen demand with high biomass concentration of 1.27 g L^{-1} during the treatment of palm oil mill effluent (POME) as demonstrated by Emparan et al. (2020). Other than that, microalgal species such as *Chlorella* sp., *Chlorococcus* sp., and *Planktothrix isothrix* are usually cocultured with *Chlorella vulgaris* in the same culture medium are preferred for application in municipal wastewater treatment systems (Min et al. 2011).

Due to the higher potential of microalgae to grow on wastewater than seaweeds, the purpose of this chapter is to provide an overview of the application of microalgae in wastewater treatment systems and highlight its advantages over conventional methods. Various mechanisms of wastewater treatment by microalgae are discussed, including nutrient, phosphorus, and nitrogen recovery, as well as the removal of heavy metals. The chapter also examines the factors that influence wastewater treatment by microalgae, such as wastewater characteristics, turbidity, nitrogen and phosphorus concentrations, and chemical and biological oxygen demand. Two current approaches of microalgae-based wastewater treatment, namely suspended and immobilized microalgae-mediated systems, are presented, along with a discussion on their sustainability. Overall, the chapter addresses the challenges and future perspectives of microalgae-based wastewater treatment systems in a comprehensive manner.

2 Microalgae-Based Wastewater Treatment

Principally, nitrogen, phosphorus, and enzyme-producing metals which are categorized under natural plant nutrients could be found in wastewater that are discharged from various anthropogenic activities including factories, municipalities, and agricultural runoff. Therefore, wastewater serves as a nutrient-rich medium for rapid development and high biomass production of microalgae. Combining wastewater treatment with microalgae cultivation could be an eco-friendly and cost-effective technique in generating sustainable raw materials for further synthesis of renewable algae-based products (Khoo et al. 2021). The primary interest in proposing the application of microalgae valorization in the production of biofuel is due to the high concentration of natural hydrocarbons in microalgae, such as lipids, carbohydrates, and proteins (Chew et al. 2021). Nevertheless, with the intention to obtain optimal microalgal biomass production, physical elements such as nutrient amount and quality, light intensity, carbon dioxide supply, temperature, pH, turbulence, and salinity should be considered and examined (Sukla et al. 2019).

Over last years, scientific community has shown an increasing interest in utilizing microalgae for the treatment of domestic and industrial effluents, specifically in removing newfound pollutants (Yong et al. 2021). Microalgae are recognized as a feasible alternative for wastewater treatment due to their high growth rates and ability to withstand various environmental conditions. Therefore, this has led to the realization of microalgae-based treatment as a sustainable and promising technique for industrial scale (Peter et al. 2021). There are several benefits to use microalgae for bioremediation, including integrated approaches, which aligns with environmental interest and promotes sustainability in the production cycle (Tang et al. 2020). In addition, microalgae play a critical role in carbon dioxide fixation and production of high-value bioproducts. Consequently, microalgae offer a promising solution for the treatment of effluents containing emerging pollutants, overcoming some of the limitations of bacteria and fungi that require specific nutrients and organic carbon sources to achieve their metabolic processes productively (Rempel et al. 2021).

2.1 Benefits of Microalgae in Wastewater Treatment

There exist several types of wastewater treatment plants, including physical, biological, chemical, and sludge treatment, each with unique capabilities and processes for treating wastewater (Saravanan et al. 2021; USEPA 2000). Traditional wastewater treatment (WWT) plants usually focus on removing suspended solids through mechanical means and reducing the biological oxygen demand through the use of activated sludge (Wang et al. 2017). Recently, microalgae have gained enormous attention due to their promising results in carbon fixation, nitrogen and phosphorus removal, as well as their potential to be harvested and turned into value-added products as shown in Table 1. In fact, the use of microalgae in WWT plants has two primary goals; direct effect through transformation of water contaminants, and indirect effect by improving bacterial system purification performance through additional oxygen provided by photosynthesis in symbiotic coculture (Quijano et al. 2017). Generally, different microalgae strains are claimed to be suitable for various types of wastewaters, with *Chlorella vulgaris* being dominant among the most studied for biomass production from diverse wastewater sources due to its ability to grow rapidly and thrive in different environmental conditions.

Another important advantage of using microalgae for wastewater treatment is their ability to valorize carbon dioxide generated by bacteria in wastewater. Since both microalgae and bacteria have a mutualistic relationship, bacteria will need nutrients to respire and metabolize lipids, carbohydrates, and ATP (Khan et al. 2018; Sun et al. 2018). The respiration of bacteria will create CO_2, which in return assists the microalgae to operate with dissolved nutrients, carbon dioxide, light, and water to perform photosynthesis. Dissolved nutrients, such as nitrogen and

Table 1 Products made from microalgae cultivated in wastewater and their benefits

Microalgae	Type of wastewater	Scale and cultivation conditions	Products	Application/ Benefits	References
– *Chlorella* sp. – *Scenedesmus* sp. – *Nostoc muscorum*	– Industrial wastewater	– Laboratory scale – Collected algal culture was inoculated into a tray containing wastewater and exposed to ambient temperature and light, with varying amounts of sunlight during different months	– Biofertilizers	– Improving plant growth – Suppress soil-borne plant pathogens – Repel pests	Khan et al. (2019)
– *Nannochloropsis* sp. – *Phaeodactylum* sp. – *Nitzschia* sp. – *Crypthecodinium* sp. – *Schizochytrium* sp.	– Municipal wastewater	– Laboratory scale – Microalgae was cultured at 23°C and neutral day light, including extra light (12:12 light:dark cycles, 1,380 lumens), 130 rpm agitation until exponential phase	– Health Supplement	– Provides vitamins and antioxidants – Rich in potassium, iron, magnesium, and calcium	Ballesteros-Torres et al. (2019), Priyadarshani and Rath (2012)
– *Spirulina* sp.	– Food processing wastewater and other types of wastewater	– Laboratory scale – Microalgae cultivation in a Zarrouk medium with 1 g L^{-1} NaCl, 5 d at 80 μmol m^{-2} s^{-1} and 30°C; 40 g L^{-1} NaCl, 3 d at 10 μmol m^{-2} s^{-1} and 40°C	– Polysaccharides	– Human nutrition – Pharmaceuticals – Cosmetics	Costa et al. (2021), Li et al. (2019b), Richmond and Hu (2013)
– *Nannochloropsis* sp.	–	– Pilot scale – Microalgae cultivated in raceway ponds of 2000-L capacity at 150 μmol m^{-2} s^{-1}, pH 7.5 ± 0.2, 30 cm water level and 25°C	– Aquaculture – Biodiesel	– Alternative energy source	Moazami et al. (2012)

– *Arthrospira* sp. – *Chlorella* sp. – *Nannochloropsis oculata*	–	– Laboratory scale – Possible for closed (Photobioreactors, PBRs) and open cultivation system (unstirred ponds, raceway ponds and circular ponds) using wastewater	– Cosmetics	– Anti-aging cream, anti-irritant, sun protection	Yarkent et al. (2020)
– *Dunaliella salina* – *D. bardawil*	– Industrial and urban wastewater	– Laboratory scale	– Carotenoids	– Human nutrition – Cosmetics – Food additives	Kalra et al. (2021), Pulz and Gross (2004), Richmond and Hu (2013), Spolaore et al. (2006)
– *Spirulina platensis*	– Municipal wastewater	– Laboratory scale	– Cell mass – Proteins – Phycobiliproteins – Phycocyanin – Vitamins	– Feed additives	Brennan and Owende (2010), Pulz and Gross (2004), Richmond and Hu (2013)

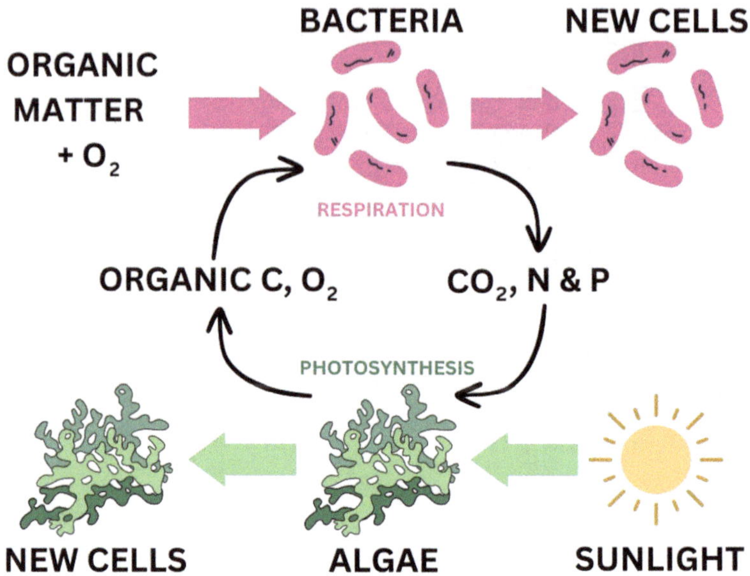

Fig. 2 Symbiotic relationship of microalgae and bacteria cocultured in wastewater medium for bioremediation

phosphorus as major nutrients, are basically produced from organic matter degradation by heterotrophic bacteria that are capable of breaking down complex compounds present in the environment and convert them into simpler forms (Wang et al. 2016). In contrast, microalgae provide benefits to bacteria by generating organic carbon compounds through photosynthesis that can serve as a food source for bacteria while the released oxygen can be utilized for their respiration, as presented in Fig. 2 (Yao et al. 2019). Overall, the mutualistic relationship between microalgae and bacteria in a consortium can enhance nutrient cycling, improve ecosystem stability, and promote growth and productivity of both organisms (Alam et al. 2022).

Conventional methods for removing specific nutrients, such as total nitrogen (TN), phosphorus (TP), ammonia (NH_3-N), and nitrate (NO_3-N), from wastewater can achieve up to 98% removal efficiency. However, studies have demonstrated that the utilization of microalgae for nutrient removal exhibits comparable effectiveness to conventional methods, with efficiencies exceeding 90% depending on the specific microalgal species and wastewater characteristics (Díaz et al. 2022). In some experiments, microalgae have attained higher overall removal efficiency compared to conventional methods (Pham and Bui 2020). For instance, Nguyen et al. (2019) carried out an experiment in pretreating seafood wastewater prior to discharge and successfully proved that involvement of microalgae greatly reduces the chemical oxygen demand (COD) and biological oxygen demand (BOD) by 88% and 81%, respectively (Nguyen et al. 2019). The experiment also explored that microalgae can produce biomass in a range of 0.40–0.54 g L^{-1} in a short period of 9 days, depending on the species used. It can also reduce the concentration of total nitrogen from 64 mg

L^{-1} to as low as 1 mg L^{-1} depending on species (Chen et al. 2018; Delgadillo-Mirquez et al. 2016).

2.2 Conventional Versus Microalgae-Mediated Wastewater Treatment (MMWT)

Regarding the current methods used in WWT plants, most of these methods have the ability to reach high overall COD removal efficiencies ranging from 90% to 95%, relying on the type of wastewater treated which mainly refers to its composition (Rinquest et al. 2019; Zhu et al. 2016). However, these methods are not efficient in nitrogen and phosphorus removal. Generally, the difference in the operation of MMWT plant is the high potential of nitrogen and phosphorus removal by microalgae. The high removal efficiency of nitrogen and phosphorus in MMWT plant is due to the high nutrient uptake capacity in which microalgae have a high affinity for nitrogen and phosphorus as important elements for their growth and metabolism, and they can actively assimilate these nutrients from wastewater (Su 2021). Additionally, as mentioned before, a symbiotic relationship between microalgae and bacteria effectively contribute to nutrient removal by performing additional processes such as denitrification, where they convert nitrate to nitrogen gas and removing nitrogen from the system (Zhang et al. 2021). According to a study done by Matamoros et al. (2015), the results of the study have reported that nitrogen efficiencies ranging from 60% to over 90% using the MMWT system compared with conventional wastewater treatment that has only about 50–80% (Matamoros et al. 2015). In fact, the conventional treatment methods for nitrogen removal is only through ammonification and nitrification, while for phosphorus is phosphate precipitation, which both require additional raw materials to proceed with their respective reaction (Saxenian et al. 2021; Yamashita and Yamamoto-Ikemoto 2014). When compared with conventional WWT plants, MMWT plants can recover the nutrients and reduce the usage of other chemicals with the help of microalgae. Not only does this reduce the chemical usage, but it also reduces the overall carbon dioxide emissions (Mata et al. 2018). However, the removal efficiencies and potential nutrient recovery can vary depending on factors such as the characteristics of wastewater, the microalgae species used, as well as the operational parameters. Since the full genome of most of microalgal species is not fully discovered yet, there might be new strains which can enhance the current microalgae MMWT plants. Therefore, novel technology such as high throughput screening techniques (HTS) coupled with OMICs provide efficient and comprehensive analysis in biological and biomedical research, as discussed in previous chapters, might be useful in allowing researchers to quickly screen and analyze, as well as identify potential strains that could enhance current MMWT systems.

Development of microalgae–bacteria system for wastewater treatment has proven to be efficient in reducing pollutants and pathogens, recovering nutrients in the form of biomass, saving energy, as well as reducing CO_2 emissions (Molinuevo-Salces et al. 2019). Studies have demonstrated that co-immobilization treatment using *C. vulgaris* and *P. putida* might contribute to high COD removal efficiency (97%) compared to other treatments. On the other hand, the culture of *P. putida* without *C. vulgaris* attained the least efficient (92%) COD remediation (Shen et al. 2017). Besides, microalgae and bacteria in the mixed system mutually consume organic carbon through heterotrophic and mixotrophic metabolism (Zeng et al. 2015). Immobilization of microalgae allows the adsorption of nutrients on the surface of the beads, subsequently penetrates through the matrix, and is continuously absorbed into cells (De-Bashan et al. 2002). This environment may promote prompt adaptation of microalgae in a highly concentrated waste environment with respect to nitrogen and phosphorus content, along with interactions that increase survival capabilities of microalgae cells and further increase the nutrient uptake. Moreover, dissolved oxygen and pH levels in the culture are essential for determining pathogen removal in the microalgae–bacteria system (Gonzalez-Fernandez and Muñoz 2017). The advancement of microalgae has resulted in some industrial shift from conventional treatment to microalgae-mediated treatment. So far, more pilot plants are built to study and improve the current methodologies of MMWT. For instance, Algae Production and Research Centre (AlgaePARC) facility located in Netherlands features a pilot-scale wastewater treatment system that valorizes microalgae to remove nutrients and other pollutants from wastewater by using a series of photobioreactors. Other examples include the "Aqualgae" project in Spain and UC Davis Algae Pilot Plant in the United States which also focus on researching and developing innovative approaches for wastewater treatment using microalgae. As a result, microalgae can be considered a step forward in green technology, which can help improve human lifestyles, help the environment, and are self-sustainable.

2.3 Enhanced Water Quality by Algal Treatment

There are various categories of wastewater, such as greywater, aquaculture wastewater, industrial wastewater, and municipal wastewater. Microalgae can react dissimilarly to these constituents. Thus, different algal species synthesize nutrients at a different rate and types depending on the pH, light intensity, temperature, and organics content (Wollmann et al. 2019).

Depending on wastewater, different treatments are employed. For example, greywater is from sinks, showers, baths, and washing machines. Based on the sources, it can be deduced that the contaminants in greywater are mainly soap, food scraps, dead skin, and microplastics from clothes. Due to the presence of food particles and other organic compounds contained in the food scraps and dead skin, respectively, along with nutrients such as nitrogen and phosphorus from soaps, greywater is possible to be treated by microalgae as part of a wastewater treatment system. However, in common cases, greywater from households is mixed with water originated from toilet, which is then needed to be classified and handled like sewage

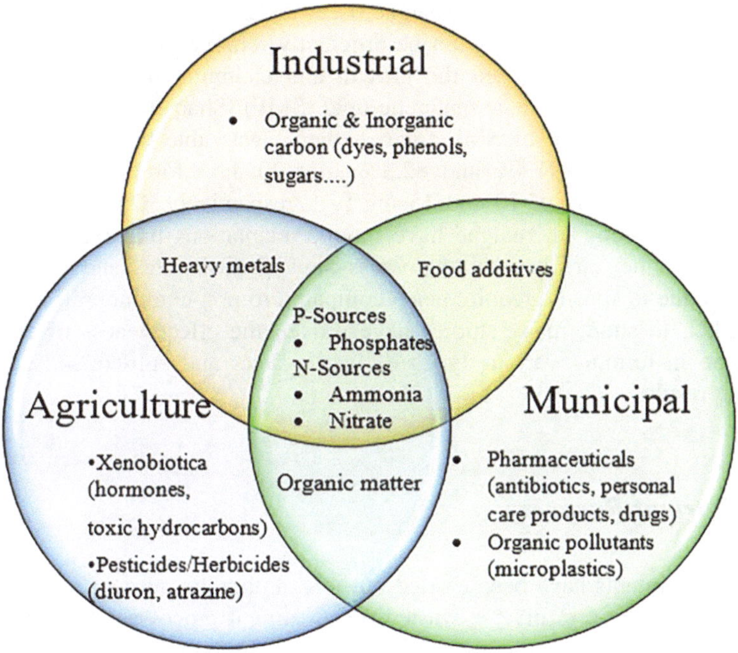

Fig. 3 Wastewater sources and their typical impurities showing the shared nutrients

wastewater (Ferreira et al. 2019). Moreover, since waste components present within municipal wastewater are similar to agricultural wastewater in terms of organic compounds, nutrients, and other contaminants, it can be treated with the same approaches (Liu and Hong 2021). Nevertheless, it is important to note that the characteristics and treatment requirements of municipal and agricultural wastewater can also differ significantly due to variations in composition, source, and regulatory considerations. Typically, both wastewaters contain nitrogen and phosphorus as shown in Figure 3. Therefore, it is suitable to be treated using the MMWT method because nitrogen and phosphorus elimination is claimed to be more efficient utilizing this method under the stimulation of algal–bacterial synergistic interactions, as well as microalgae photosynthetic activity (Sial et al. 2021).

As for industrial wastewater, since it contains heavy metals and organics, as well as inorganic carbon such as dyes, phenols, and sugars, the type of microalgae used should be extremophile. Typical examples of extremophilic microalgae include thermophilic, halophilic, acidophilic or alkaliphilic, and metal-tolerant microalgae. Generally, these microalgae can live in harsh conditions such as pH range of 1.9 to 2.9 and extreme temperatures in the range of 45 °C to 70°C, which is not feasible with the physiology of conventional microalgal species (Abiusi et al. 2022; Patel et al. 2019). Therefore, some adapted microalgae are used in the treatment; however, it is not preferable to use MMWT for removal of high concentrations of heavy metals since certain heavy metals might have an inhibiting effect on the enzyme activity and photosynthesis process of the microalgae (Li et al. 2019a). On the other hand, low concentrations of heavy metals can be efficiently removed by microalgae. In that context, recent study evaluated the growth and cadmium removal efficiency of *Spirulina platensis* under static magnetic field (SMF) (Shao et al. 2018). The study reported that application of SMF for 6 h day^{-1} was able to enhance cadmium removal efficiency by 91.4% and 82.3% after 20 days for cultures with initial cadmium concentration of 10 and 15 mg L^{-1}, respectively. On top of that, even though extremophilic microalgae have unique adaptations under certain extreme environments, they are known to possess slow growth rates and challenges in cultivation due to specific requirements compared to non-extremophiles (Varshney et al. 2015). In short, many studies have proven the effectiveness of employing microalgae in refining various types of wastewaters and subsequently achieved satisfying results.

2.4 Nutrient Recovery

Various experiments have been carried out to examine the nutrient recovery capabilities of microalgae in diverse wastewater treatment processes. These investigations aim to optimize and enhance nutrient recovery through different strategies such as co-metabolism and immobilization of different algal species. The majority of these experiments have demonstrated promising outcomes in terms of nutrient recovery efficiency using various types of wastewater. However, it is necessary to note that these experiments have mostly been conducted on a laboratory scale. For example, Mujtaba et al. (2018) has presented a study on the simultaneous removal of nitrogen, phosphorus, and carbon by coculture of immobilized *Chlorella vulgaris* and suspended activated sludge. Song et al. (2022), on the other hand, reviewed the prospect of nutrient recovery using microalgae via several methods such as microalgae–yeast system, microalgae–bacteria system, microalgae-constructed wetland system, and microalgae–sludge system followed by latest technology, as well as respective challenges. Particularly, in co-metabolism, *C. vulgaris* or *C. sorokiniana* with *A. brasilense* vitally improved the removal of ammonium, although both *Chlorella* species could eliminate most of the ammonium when immobilized alone (De-Bashan et al. 2004). The downside of this method is that this study is the first report of its kind on co-metabolization which is capable of removing content which is under 1 mg L^{-1} only. It is found that this method has the capability of reaching up to 100% ammonium elimination followed by 15% nitrate and 36% phosphorus within 6 days, with varying wastewater sources. However, with the use of sole microalgae, the removal of ammonium, nitrate, and phosphorus were reported to be 75%, 6%, and 19%, respectively (De-Bashan et al. 2004). The conventional methods in capturing these nutrients used are not as efficient and/or eco-friendly compared to microalgae, as conventional methods require the usage of chemicals for the treatment, such as pH neutralizers, anti-foaming agents, coagulants, and flocculants (Udaiyappan et al. 2017). In addition, current conventional methodologies applied

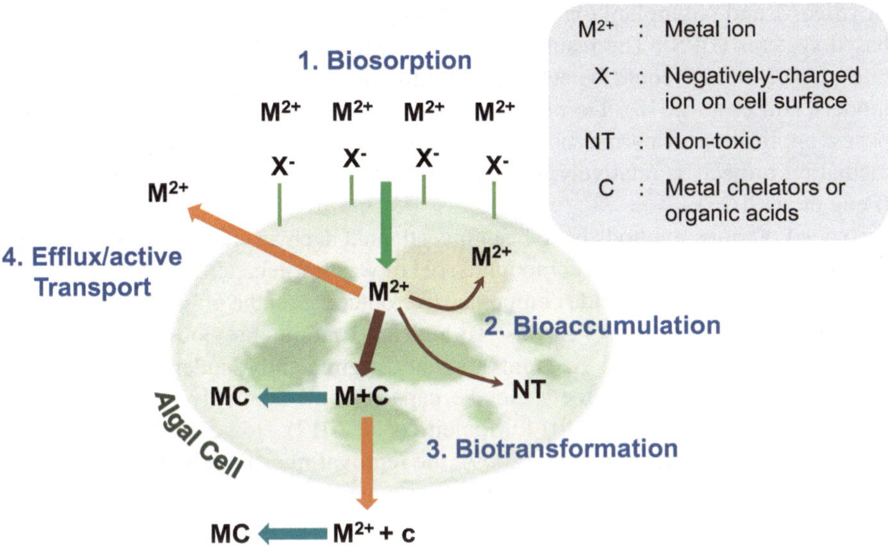

Fig. 4 Schematic diagram of heavy metal detoxification mechanisms employed by microalgae

are known to be complex, expensive, and energy consuming (Ye et al. 2020). As a result, implementation of microalgae can possibly reduce the process cost to a certain extent (Wang et al. 2022).

Some of microalgal technologies such as raceway ponds can be incorporated into current conventional methods depending on the method adapted and integrated into conventional wastewater treatments. Typically, microalgae have different mechanisms to counteract the toxic effects of heavy metals present in industrial waste, which renders toxic heavy metals to innocuous forms. The heavy metal can be taken up by living organisms for metal detoxification and led to the development of metal-binding peptides or proteins such as metallothioneins and phytochelatins (Kaplan 2013). Cells can diffuse metal ions, actively efflux them, and progressively taken up via ion transporters. Subsequently, the ions are transported to the vacuole and kept as either free metal ion or as complexes as illustrated in Fig. 4 (Kaplan 2013).

Furthermore, periphytic algae, which are commonly present in submerged substrates, sand, macrophytes, and rocks, are harvested regularly from the surface of the flow-way using various methods. One common approach is to physically scrape or brush the surface of the substrates to collect the periphytic algae, while another method involves the use of filtration systems or screens that selectively capture the algae when allowing wastewater to pass through. Periphytic algae can efficiently absorb and accumulate heavy metals, as well as capture and valorize excess nutrients from wastewater. They are usually used in phycoremediation projects that aim to remove or reduce the concentration of contaminants in polluted environments. For example, Zhu et al. (2018) investigated on the arsenic removal (As(III)) removal by periphytic biofilm with the presence of microorganisms such as green alga,

C. vulgaris and phytoplankton *C. salina* through the use of biochar and periphyton-based systems (BPS). The results of the study suggested that high As(III) removal rate of 90.2–95.4% could be achieved by utilizing the proposed periphytic biofilm model (Zhu et al. 2018). Despite their extensive research for bioremediation purposes, application of periphytic algae at a commercial-scale and large-scale bioremediation projects is relatively limited due to the need for careful species selection (Peng et al. 2023).

Indeed, various methods have been investigated such as changing the concentration of microalgae and metal compounds, pH of wastewaters, temperature, as well as nutrient availability to provide comprehensive outcome on the effectiveness of using microalgae to remove heavy metals. To take an example, previous studies showed that heavy metal removal rate did not vary significantly under low density of microalgae due to limited metal uptake capacity and relatively low surface area and binding sites (Chan et al. 2014; Monteiro et al. 2012). However, according to the study conducted by Jacinto et al. (2009), the removal rates largely differed with the microalgal strain. It was found that the seaweed *Sargassum* sp. and the dried microalga *Chlorococcum* sp. have the capacity to remove up to 87% and 43–75% of Cu, respectively, under similar treatment systems (Jacinto et al. 2009). Additionally, it was also shown that with strict monitoring of the conditions, increased density of *C. vulgaris* has relatively rapid removal rate of nutrients (Choi and Lee 2012).

2.5 Phosphorus and Nitrogen Recovery

Up to date, there are several approaches for treating wastewater that contain phosphorus including chemical precipitation, high-temperature acid-hydrolysis, biological assimilation, physical–chemical adsorption, as well as ion exchange (Sengupta et al. 2015; Witek-Krowiak et al. 2022). Biological assimilation is a sustainable option, but it has limited use in regions with temperatures that are not tropical or subtropical. This is because of the heat stress induced in high-temperature regions which can negatively impact the metabolic processes of organisms, including their ability to uptake and assimilate nutrients (Liliane and Charles 2020). On the other hand, some plants like water hyacinth, water lettuce, and duckweed can be harvested and sold as a source of nutrients for animal feeds and can remove significant amount of total nitrogen (83–87%) and total phosphorus (70–85%) from wastewater streams (Rezania et al. 2015; Sengupta et al. 2015). In this case, phosphate removal can occur through biotic and abiotic processes via assimilation by biomass, such as bacteria and microalgae, as well as via adsorption and chemical precipitation (Amaro et al. 2023). As discussed earlier in Sect. 2.1, lab-scale recovery of nutrients (phosphorus and nitrogen) through microalgae shows promising results, which indicates that this method is viable for future use in wastewater treatment. In fact, the types of microalgae and its effluent needs to be considered when performing treatment, as some wastewaters are toxic to specific species of microalgae. For example, utilization of immobilized *Phormidium laminosum* in polyurethane and polyvinyl foam in a continuous flow reactor can be used for the removal of nitrate in the water purification process (Garbisu et al. 1991). However, a similar study on nitrate removal from drinking water using immobilized *Scenedesmus obliquus* cells (1 g dry weight L^{-1}) on prepolymers used in polyurethane foams were toxic to the cells (Moreno-Garrido 2008). As a result, proper approaches are essential particularly when considering the carriers used for immobilization techniques as most of the reported synthetic carriers are confirmed to induce certain toxicity toward microalgae which subsequently inhibits the cellular growth.

When it comes to recovering nitrogen from wastewater, NH_4^+ is a major contaminant found in various types of wastewater, including grey water, urine, and wastewater from treatment plants (Besson et al. 2021). To recover nitrogen, various methods were proposed in recent years including ion exchange or adsorption-based methods with the use of zeolites which would lead to the release of nontoxic exchangeable cations (Na^+, K^+, Ca^{2+}, Mg^{2+}) (Cruz et al. 2019). Other than that, bioelectrochemical systems (BES) are also considered, which in this case brings the function of microorganisms which are categorized into 2 types: galvanic and electrolytic cells (Maureira et al. 2023). Basically, BES utilizes the metabolic activity of microorganisms to drive electrochemical reactions which allows the conversion of organic matter and nutrients present in wastewater into useful products or energy. In general, both galvanic and electrolytic cells are reviewed under the pretext of energy production rather than energy efficiency. BES can generate electricity through oxidation of organic matter at the anode, but the overall energy efficiency of the process is relatively low compared to other wastewater treatment methods (Kelly and He 2014). Besides, NH_4^+ could also be extracted from anaerobic digestate by employing a cathode exchange membrane (CEM) as presented in an investigation conducted by Desloover et al. (2012). In the electrochemical cell with a CEM, NH_4^+ can move across the membrane and be stripped by hydrogen that was produced through electrochemical reduction reactions, resulting in significantly lower energy requirements compared to traditional ammonia stripping methods such as air or steam stripping (Desloover et al. 2012). Under optimal process conditions, NH_4^+ charge transfer efficiency up to 96% and NH_4^+ flux of 120 g N m^{-2} d^{-1} could be achieved at a concomitant electricity input of 5 kWh kg^{-1} N removed. In regard to ammonium removal rate, the study reported that ammonium level in the digestate could be reduced from 2.1 to 0.8–1.2 g N L^{-1}. In short, microalgae will in turn assimilate nitrogen from wastewater, converting it into a less energy-intensive form such as ammonia (Delgadillo-Mirquez et al. 2016). Table 2 presents the percentage removal of both nitrogen and phosphorus with incorporation of microalgae. It can be concluded that association of microalgae could indeed significantly facilitate the removal of nitrogen and phosphorus present in wastewater. However, analysis of raw wastewater characteristics in terms of initial concentration for TP and TN, as well as COD/N/P ratio is promising in determining the performance of microalgae in wastewater treatment.

Table 2 Removal rate of nitrogen and phosphorus from different types of wastewater by various microalgae species (Choi and Lee 2012)

Type of wastewater		Microalgae species	Initial concentration		Removal efficiency		References
			TP	TN	TP	TN	
Influent wastewater		*C. vulgaris*	6.23 mg L^{-1}	53.33 mg L^{-1}	32.26–36.26%	81.04–84.81%	Choi and Lee (2012)
Synthetic wastewater	Conditions: 106 h & 10^6 cells mL^{-1}	*Scenedesmus* sp.	6 ppm	44 ppm	8.6 removal proportion of N to P		Kube et al. (2018)
	Conditions: 48 h & 2×10^6 cells mL^{-1}	*S. obliquus*	3.4 ppm	30.1 ppm	5.0 removal proportion of N to P		Ruiz-Marin et al. (2010)
Concentrated wastewater from sludge ozonation		*Scenedesmus* sp.	45 mg L^{-1}	100 mg L^{-1}	52.10–55.70%	60.60–67.20%	Lei et al. (2018)
Municipal wastewater		*C. vulgaris*	3.78 mg L^{-1}	69.05 mg L^{-1}	72.88–96.5%	81.60–95.12%	Molazadeh et al. (2019)

2.6 Heavy Metal Removal

Diverse conventional methods are available for removing heavy metals from industrial wastewater, such as adsorption, chemical precipitation, and membrane filtrations that can be in the forms of micro-, ultra- or nano-filtrations and have shown efficiencies ranging from about 60–99% (Fuerhacker et al. 2012). Table 3 shows the current heavy metal treatment methods along with the associated advantages and disadvantages. However, these methods have limitations such as high costs, maintenance issues, environmental impacts, and process restrictions (Poornima et al. 2022; Vyavahare et al. 2021). As a result, microalgae have been proposed and studied as a better alternative for eliminating heavy metals from wastewater. This is because microalgae can uptake heavy metals through bioaccumulation and biosorption (Abdel-Razek et al. 2019; Fourest and Roux 1992; Salam 2019). In most cases, biosorption is a cost-effective and efficient approach for the removal of heavy metal ions from water solutions (Danouche et al. 2021). It depends on the ability of certain materials, such as biomass or bio-based materials, to bind and remove heavy metal ions through either physiological or physicochemical uptake pathways (Chaudhari 2014; He and Chen 2014; Leong and Chang 2020). In general, physiological uptake refers to some microalgal species (*C. vulgaris*) that have natural physiological pathways to uptake and accumulate heavy metals as part of their metabolic processes, while physicochemical pathway binds heavy metal ions through the formation of physical and chemical interactions (Danouche et al. 2021). Different types of microalgae may exhibit varying degrees of sensitivity or tolerance to heavy metals. Upon absorbing toxic or high concentrations of heavy metal ions, several detrimental effects might occur including inhibition of cellular growth, nutrient uptake inhibition, as well as inducing oxidative stress in microalgae (Kaplan 2013; Leong and Chang 2020). Therefore, a viable option is to use dead cells which are chemically treated or dried, such that they are not affected by the toxicity of the wastewater (Nguyen et al. 2021; Priyadarshanee and Das 2021). Also, the use of dead cells offers the advantage of reduced cost as additional nutrients are not required for growth or maintenance and making the process more economically viable. However, it is important to consider the drawbacks of dead cells such as limited regeneration potential, potential biomass decay and fouling, handling and storage, as well as high processing and preparation costs (Kalra et al. 2021).

Moreover, in terms of its capability to adapt to variations in pH and heavy metal concentrations, biosorption by algae is preferable than conventional methods that rely on precipitation (Kumar et al. 2015). In common practice, precipitation approach involves the formation of insoluble compounds upon the addition of precipitation agent, known as precipitates, by chemical reactions between the heavy metal ions in the wastewater. Nevertheless, precipitation for heavy metal removal is known to be sensitive toward pH and potentially leads to sludge generation (Benalia et al. 2022). Additionally, biosorption exhibits a high tendency for reducing residual heavy metals to levels below 1 ppb in numerous cases (Aksu 1998; Mahamadi 2019). According to the specific concentration limits and guidelines for

Table 3 Advantages and disadvantages of conventional heavy metal treatment methods

Treatment process	Advantages	Disadvantages	References
Biological treatment	• Environmentally friendly compared to chemical-based methods as it relies on natural capabilities of microorganisms • Potential for metal recovery, allowing for the extraction or recycling of heavy metals from the treated water • Use biological materials which are more cost-effective than chemical-based methods • Broad applicability to various water sources such as industrial wastewater and contaminated underground water	• Process optimization and long treatment time with the need to consider factors such as temperature, pH, nutrient availability and the presence of inhibitory substances • Limitations on specific heavy metals that are not readily biodegradable or accessible to the biological materials, affecting the overall efficiency of process • Require careful monitoring and control to maintain optimal conditions for microbial activity	Coccia and Bontempi (2023), Manzoor et al. (2019), Raikova et al. (2019), Singh et al. (2022)
Chemical sedimentation	• Economically viable due to readily available and inexpensive chemicals used for coagulation and flocculation • Wide application for various water sources and treatment scenarios • High removal efficiency under optimized conditions (up to 90% removal efficiency)	• Sludge or sediment generation which requires proper handling and disposal • Potential re-suspension of settled precipitates or sludge • Less effective for wastewater with complex characteristics, such as high organic content or concentrations of dissolved solids	Akcil et al. (2015), Kurt and Özdemir (2022), Xu et al. (2019), Zhang et al. (2019)
Ion exchange	• Lack of sludge production • Cost-effective for potential low-cost materials • Very efficient for low concentration heavy metals (Efficiency > 77% for Cr(VI) removal)	• High investment for system operation and maintenance • High sensitivity toward pH value • Limited applications for certain metals and water conditions • Finite adsorption capacity of heavy metal ions	Fuerhacker et al. (2012), Gunatilake (2015), Homan et al. (2018), Shrestha et al. (2021)

heavy metals by European Union (EU) and World Health Organization (WHO), the maximum contaminant levels for lead, cadmium, mercury are 10 ppb, 5 ppb, 1 ppb and 10 ppb, 3 ppb, 6 ppb, respectively (Izah and Angaye 2016; Tóth et al. 2016).

Overall, research has also shown that microalgal genera like *Chlorella, Ankistrodesmus,* and *Scenedesmus* have the ability to treat wastewater from industries such as olive oil mill and paper (Rawat et al. 2011). For example, in Spain, a pilot-scale project called LIFE ALGAECAN was conducted to explore the use of microalgae for the treatment of olive mill wastewater (Khan et al. 2022). A project in Sweden, on the other hand, called the Algae Pilot aimed to assess the effectiveness of microalgae in treating wastewater from a pulp mill (Ekendahl et al. 2018). Although these examples demonstrate the potential of microalgae-based wastewater treatment in olive oil and paper industries, it is important to note that commercial-scale implementations are still limited for current scenarios.

2.7 *Factors Influencing Wastewater Treatment*

Each wastewater effluent is unique, and its composition and characteristics can fluctuate over time depending on the operational conditions (Nayagam and Prasanna 2022). In the following part, the main characteristics of an effluent that need to be considered for nutrient recovery from effluents, including manure effluent from farms, concentrate generated from anaerobic digestion of manure (liquid fraction separated from solid digestate after the digestion process), as well as sewage effluents, such as raw sewage and activated sludge treated with microalgae will be discussed accordingly.

2.7.1 Wastewater Characteristics

Waste effluents can vary significantly in terms of their physical and chemical characteristics, including nutrient levels, pH, temperature, color, and the presence of hazardous substances like heavy metals, aldehydic and phenolic compounds (Bhat and Gogate 2021). These factors may impose critical impacts on the growth of microalgae, nutrient recovery rates, and biomass yields. As mentioned in Sect. 2.6, high concentrations of heavy metals and toxins in wastewater can negatively impact the microalgal growth and induce toxic effect inhibiting their physiological processes (Leong and Chang 2020). In contrary, microalgal metabolic activities might also be inhibited under extreme pH levels ($pH<4$ and $pH>10$) (Lacroux et al. 2020). Nutrient concentrations which serve as a vital role in microalgal growth could also result in nutrient imbalances, toxicity, and proliferation of undesirable microorganisms under excessively high nutrient concentrations (Tzanakis et al. 2023). Among all types of wastewaters, municipal wastewater and animal wastewater are the most widely studied effluents due to their wide availability and consistent quality. The annual global production of municipal wastewater is greater than 360–380 km^3 and is expected to increase by 24% and 51% by 2030 and 2050, respectively (Kakar et al. 2022). As a result, their wide availability in urban and peri-urban areas makes it readily accessible for research and provides ample opportunities for studying its composition, treatment methods, and reuse options. Since municipal wastewater has standardized sanitation procedures while animal wastewater has similar management practices, it is claimed that their characteristics are relatively consistent and predictable compared to other types of wastewaters (Priya et al. 2021). In addition, other types of effluents, including food processing wastewater from industries such as beverage, dairy, and vegetable oil, as well as waste discharge from ethanol plants, have been explored over the time. Both food processing wastewater and ethanol plant wastewater are rich in organic compounds derived from food materials for the former such as carbohydrates, proteins, and fats or from the fermentation process for the latter which might contribute to high BOD and COD values of these wastewaters (Li et al. 2019b; Nawaz et al. 2021). To put it concisely, the interconnection between wastewater and microalgae is a critical aspect of microalgae-based wastewater treatment systems such as wastewater provides nitrogen and phosphorus that are essential for microalgae growth and metabolism, and in return microalgae help in the removal of pollutants (Li et al. 2019a). The way microalgae interact with wastewater significantly impacts the effectiveness of nutrient removal and overall performance.

In an effort to utilize microalgae in treating diverse wastewater with different profiles, algae require specific conditions for optimal growth, and these conditions are usually achieved through a designed synthetic medium with the right mix of macronutrients and micronutrients (Khan et al. 2018). This is because the direct use of microalgae in wastewater treatment without considering the specific characteristics of wastewater can lead to reduced treatment efficiency, algal growth inhibition, imbalance nutrient removal, as well as high risk of algal blooms (Jiang et al. 2021). However, nutrient profile of wastewater is usually hard to be predicted and not suitable to be directly used for algal growth. With the purpose of addressing this issue, two strategies have been employed. The first involves using acclimatized algae to adapt to the wastewater environment, while the second relates to the modification of wastewater to match particular algal growth conditions. First approach involves an adaptation period where microalgae are continuously exposed to wastewater to enhance their tolerance to specific contaminants, allowing efficiency in treating the specific wastewaters (Acién et al. 2016). Besides, second approach aims to adjust the pH levels, nutrient concentrations, and other parameters to create an environment that is more conducive to the growth and activity of microalgae. (Evans et al. 2017) In practice, both strategies are often applied in conjunction to achieve the desired results of algal biomass production and wastewater treatment (Li et al. 2019a).

Turbidity

The amount of light that can penetrate in a microalgae-based process is essential as cell density, culture depth, as well as optical properties could be the potential factors that might be influenced by the light penetration in the microalgae-based process. In general, light penetration decreases dramatically when the system turbidity increases due to the presence of solids in wastewater. Studies found that microalgae cultures perform poorly when water turbidity exceeds 3000 Nephelometric Turbidity unit (NTU) (Akhiar et al. 2017). Consequently, effluent pretreatment is necessary to address this issue, such as by filtering or diluting the effluent with water (Kadir et al. 2018). However, using the latter approach increases the volume of effluent to be treated, which is not an ideal case. For example, in order to decrease turbidity to an acceptable range, wastewater had to be diluted at several ratios (1:2, 1:3.5, 1:5, 1:10, 1:20) depending on the wastewater turbidity which led to overall high demand for freshwater (Chong et al. 2022). Although suspended solids are eventually hydrolyzed in the microalgae reactor, this process is slow, and solids might persist for a long time. Therefore, reducing the presence of suspended solids in water influent via filtration is recommended to maximize the capacity of the microalgae process (Acién Fernández et al. 2018). By implementing an effective filtration system, larger particles and solids can be physically separated, resulting in a cleaner influent with reduced turbidity and suspended solids which in return allows microalgae to focus their energy and resources on the assimilation of dissolved nutrients (Ghazvini et al. 2022).

Nitrogen and Phosphorus

Nitrogen is an essential element for microalgal growth and is typically added as nitrate at an average concentration of 50 mgN L^{-1} in the synthetic culture media (Lee and Lee 2002). However, nitrogen in effluents is mainly present as ammonia, with the concentration varying between 65 up to more than 9000 mgN L^{-1} (Taziki et al. 2015). In fact, ammonia toxicity starts at concentrations over 100 mg L^{-1}, but different microalgal strains display diverse levels of tolerance to ammonia, which can also be influenced by the culture conditions. For example, *C. vulgaris* which is a commonly studied microalgal species has been found to exhibit tolerance to relatively high concentrations of ammonia in wastewater with concentrations exceeding 100 mg L^{-1} (Cantera et al. 2021). In common case, when microalgae reactors operate continuously, the instantaneous concentration of ammonium at certain time differs from the influent. The actual or real concentration of ammonium within the reactor at a given time is, therefore, determined by several factors including the initial ammonium concentration in the influent, the dilution rate, and the biomass productivity of the culture (Jia and Yuan 2018). In contrary, ammonium concentration inside a batch reactor will change throughout the batch cycle for a fixed duration before the contents are harvested or transferred to another batch (Johnson et al. 2010). The concentration of ammonium inside the reactor can be reduced to less than 100 mg L^{-1} even when the influent ammonium concentrations is greater than 600 mg L^{-1} (Acién Fernández et al. 2018). Primarily, the reduction in ammonium concentration could be achieved through the metabolic activity of the microalgae by converting high concentration ammonium into biomass or other forms of nitrogen (Liu et al. 2019).

Another key ingredient of microalgal growth is phosphorus. In wastewater effluents, phosphorus is found in the form of phosphate or organic compounds.

However, within microalgae reactors, it undergoes a conversion process to be transformed into phosphate, mainly attributed to the prevailing oxidative conditions (Ahmed et al. 2022). There have been no reports on phosphorus toxicity in microalgae cultures, likely because of low doses are always provided to the system as presented in Table 2. Commonly, precipitation of calcium phosphate in alkaline conditions is the primary issue for phosphorus management in the microalgal cultivation system (Abou-Shanab et al. 2013). To alleviate this problem, calcium presence in the growth medium must be restricted, and the pH in the reactor should be lowered while in operation. However, if the phosphorus is effectively harvested along with the microalgal biomass and appropriately utilized in the end application of the biomass, the issue of phosphorus precipitation become less significant and mitigates the concerns associated with phosphorus precipitation in the wastewater treatment processes (Acién Fernández et al. 2018).

Chemical Oxygen Demand (COD)/Biological Oxygen Demand (BOD)

The elimination of organic matter from wastewater is the main interest of the treatment process when aiming for wastewater processing rather than algae biomass production (Abdel-Raouf et al. 2012). Organic matter is composed of biodegradable chemicals that can be quantified using two common measurements, BOD and COD. The first generally measures the amount of biodegradable organic chemicals, while the later encompasses all degradable compounds present in the organic matter (Lv et al. 2022). Although some studies have reported that microalgae can grow heterotrophically and mixotrophically using small organic molecules like glycerol and glucose, it cannot be concluded that they can degrade large organic compounds (Villanova and Spetea 2021). Instead, this function is carried out by heterotrophic bacteria grown together as microalgae–bacteria consortia (Collao et al. 2022). The population of bacteria in the final biomass produced increases with higher organic matter (Acién Fernández et al. 2018). Thus, it is necessary to confirm that the presence of bacteria in microalgae–bacteria consortia is beneficial rather than not harmful or pathogenic as discussed in the previous section (Collao et al. 2022). Furthermore, if the wastewater being treated has high levels of organic matters, it is significant to have longer hydraulic retention times to thoroughly break down these pollutants. For instance, in activated sludge process (biological treatment method), wastewater is mixed with a microbial culture in an aerated tank and remains in the tank for an extended period to promote the degradation of organic matter by microorganisms (Porwal et al. 2015). Similarly, sequencing batch reactor systems, operating by sequentially treating the wastewater in different stages, where the wastewater at one of these stages would be held for an extended duration to facilitate the breakdown of organic compounds (Heidari et al. 2021). In any case, heterotrophic bacteria have a faster rate of oxidizing organic materials into inorganic molecules, typically within hours. Contrastingly, microalgae require a longer period, often days, to uptake and retain the released chemicals resulting from the bacterial activity (Acién Fernández et al. 2018).

3 Current Microalgae-Based Wastewater Treatment Technologies

There are currently two approaches toward MMWT, which include suspended WWT systems and immobilized WWT systems. However, both of these methods require additional technological requirements, particularly with regard to PBR systems (Wollmann et al. 2019). The differentiation of suspended and immobilization of MMWT systems lies in how the microalgae are harvested and processes. In suspended systems, microalgae might need to be harvested manually and further processing is needed to sanitize and clean it before selling to other factories for further processing, which requires high costs (Hoh et al. 2016). Whereas when microalgae are immobilized, the surface-to-volume ratio can be controlled easily and, therefore, growth rates are higher with easier harvest (Lam and Lee 2012).

3.1 Suspended Microalgae-Mediated Wastewater Treatment

Suspended systems are commonly used in MMWT plants, such as treatment ponds and raceway ponds, because of their low construction cost as discussed earlier. However, there are several issues when building an open pond system. One of the challenges is controlling, maintenance, and even distribution of light, ensuring that sunlight reaches the bottom of the pond and providing a sufficient supply of carbon dioxide (Zittelli et al. 2013). Some high-rate algal ponds can address some of these issues by increasing the mixing effectiveness through the use of paddlewheel stirrers and gas introduction (Wollmann et al. 2019). Thus, aeration techniques and the addition of CO_2 were reported to increase the productivity and pollutants removal rates by the algae (Benítez et al. 2019; Wollmann et al. 2019). Generally, the water depths of the channels are maintained between 0.2 and 0.4 m to ensure adequate light penetration and high biomass productivity, but this can result in higher cost for harvesting (Acién et al. 2017). On the other hand, PBR can be more advantageous in terms of offering better control and higher productivity as discussed in Chapter "Algae Cultivation Systems". In PBR, it is reported that $>80\%$ of total nitrogen, phosphorus, and COD can be removed. Also, there are few factors in play to achieve this removal efficiency, which optimize the hydraulic retention time, temperature, and light intensity (Manzoor et al. 2019). With careful tweaking of these 3 main parameters, the efficiency of PBR can be greatly increased (Viruela et al. 2018). Nonetheless, advanced PBRs are mostly used in the production of valuable metabolites at the current state rather than large-scale MMWT due to high capital and operating costs, as well as operational complexity (Wollmann et al. 2019).

3.2 Immobilized Microalgae-Based Wastewater Treatment

Immobilization refers to the use of organic or inorganic, physical or chemical techniques to prevent living free microalgal cells from moving independently in their original position. Microalgal cells which are immobilized in a small area can maintain some of the desired biological activities and be reused in aqueous phase systems (Han et al. 2022). The utilization of this biotechnology has several positive impacts, including protection from aggressive zooplanktons and reduced competition for nutrients with other microbes. Additionally, there are improvements in the metabolism, function, and behavior of microalgae (De-Bashan and Bashan 2010). When compared to free-living cells, the microalgal species *Chlorella sorokiniana* was immobilized in calcium alginate for removal of nitrogen and phosphorus contaminants from synthetic wastewater. From the study, *C. sorokiniana* established higher ammonium (21.84%, 43.59%, and 41.46%) and phosphate removal (87.49%, 88.65%, and 84.84%) than free-living cells (14.35%, 38.57%, and 40.59% for ammonium removal and 20.21%, 42.27%, and 53.52% for phosphate removal) under heterotrophic, mixotrophic, and micro-aerobic conditions, respectively (Liu et al. 2012). Immobilized algal cells have been used in different applications and proven to be highly effective in optimizing metabolite production, allowing for improved control over the culture conditions and maximizing the yield of desired compounds. Additionally, it enables efficient removal of pollutants from aquatic media, providing a sustainable and eco-friendly approach to water remediation and purification, apart from obtaining energy and nutrients, measuring toxicity, and producing co-immobilization systems for different purposes (Han et al. 2022; Kaparapu and Geddada 2016; Moreno-Garrido 2008). Co-immobilization has recently become an area of active research (Kumar et al. 2017). Table 4 summarizes the advantages and disadvantages of different suspended and immobilized designs in MMWT.

4 Sustainability of Microalgae-Based Wastewater Treatment

With fossil fuel depleting drastically and environmental issues on the rise, sustainable clean alternatives must be considered. Therefore, microalgae are considered an alternative, where microalgae-mediated wastewater treatment has the potential to convert waste into valued phytochemicals through biorefinery (Chapter "Overview of Biorefinery Technology"). For microalgae-mediated wastewater treatment to be sustainable, the three criteria namely environmental sustainability, economic sustainability, and social sustainability must be met. When implementing microalgae-based wastewater treatment, generation of high-value products enhances the economy and sustainability of this process (Yadav and Sen 2018). By utilizing the biomass produced by microalgae during wastewater treatment, diverse high-

Table 4 Advantages and disadvantages of suspended and immobilized MMWT designs

Typical designs	Pros	Cons	References
Suspended MMWT			
Open pond systems	Offers advantages such as ease of construction, use, and cleanup, cost-effectiveness post-cultivation, suitability for large-scale cultivation, and the ability to adapt to various shapes.	Challenging aspects include difficulties in achieving optimal light conditions for cultivation, the vulnerability of these systems to evaporation, the loss of CO_2 to the atmosphere, and the substantial land area required for construction.	Cai et al. (2013), Richmond and Hu (2013)
Flat panel photobioreactor	Being simple to scale up, exhibiting low levels of dissolved oxygen accumulation, and benefiting from high light intensity.	Pose challenges related to temperature control, the deposition of algal biofilms, and difficulties in managing strain-specific hydrodynamic stress.	Richmond and Cheng-Wu (2001), Ruiz et al. (2013), Xu et al. (2009), Yan et al. (2016)
Vertical tube photobioreactor	Airlift-designed reactors offer advantages such as high mass transfer efficiency, gentle mixing with low shear stress, low energy consumption, and suitability for immobilizing algae on moving particles.	Limited light penetration to the lower part of reactor, complex and difficulty in scale-up, difficult to control the hydrodynamic conditions which can affect the growth and productivity of microalgae.	Singh and Sharma (2012), Ting et al. (2017)
Horizontal tubular system photobioreactor	Having the ability to utilize natural sunlight effectively due to large surface-to-volume ratio, high biomass productivities, and being a cost-effective option for large-scale production.	Having the potential for fouling which can reduce the efficiency of the system, the occurrence of photoinhibition due to excessive light exposure, and the challenge of maintaining favorable pH conditions for optimal microalgae growth.	Michels et al. (2014), Ting et al. (2017)
Stirred tank photobioreactor	Optimal heat and mass transfer, efficient mixing, easy for monitoring and sampling to facilitate process control due to sufficient agitation.	High energy expenditure due to the need for mechanical agitation, limitation of light penetration throughout the reactor resulting from very low surface area-to-volume ratio, thus reduces the photosynthetic efficiency of microalgae.	Singh and Sharma (2012), Ting et al. (2017)

(continued)

Table 4 (continued)

Typical designs	Pros	Cons	References
Immobilized MMWT			
Microalgae membrane photobioreactor	Offer increased biomass growth and accumulation, high nutrient removal efficiency with minimal environmental impact, and higher biomass productivity compared to suspended cultivation systems.	The cost of membrane fouling in accounts for a large portion of the operation cost and often exhibit low biomass productivity.	Basak and Das (2007), Boonchai and Seo (2015), Praveen and Loh (2016)
Microalgae biofilm	Able to enhance biomass productivity and facilitate easier biomass harvesting.	Challenging in maintaining biofilm stability, controlling and optimizing biofilm thickness and potential sensitivity to toxic components.	Boelee et al. (2014), Gao et al. (2015), Richmond and Cheng-Wu (2001), Xu et al. (2009)
Immobilized beads	High surface area for attachment of microalgae, simplified biomass harvesting, and improved biomass retention by preventing being washout during treatment process.	Potential of bead fouling or clogging due to accumulated organic and inorganic matters, as well as high setup costs in terms of materials and technologies.	De-Bashan and Bashan (2010), Emparan et al. (2020)

value products could be obtained such as biofuels, animal feed, and fertilizers, creating additional revenue streams and reducing the overall costs of the treatment (Hussain et al. 2021). For example, *Chlamydomonas* has been investigated and proven for its capability in producing lipid of 87.5 ± 2.3 mg L^{-1} day^{-1} in simultaneous wastewater treatment and biofuel production system when provided with dairy wastewater as a nutrient medium (Arora et al. 2016). However, there is still room for much improvement in microalgae cultivation to make wastewater treatment a more viable and sustainable process for large-scale application. To achieve this, a thorough environmental analysis using life cycle assessment (LCA) should be conducted (Yadav and Sen 2018). Through LCA, some crucial aspects can be assessed including energy consumption, carbon and water footprint, nutrient balance, as well as waste management in order to identify potential environmental hotspots and areas for improvement in microalgae cultivation for wastewater treatment. The following parts describe some of the common LCA models or tools that could be utilized in determining the environmental impacts of the proposed wastewater treatment.

1. Ecological footprint (EF) method using the following formula (Huijbregts et al. 2017):

$$EF = EF_{direct} + EF_{CO_2} + EF_{nuclear} \tag{1}$$

EF method is sustainability assessment tool which measures the endpoint method and indicates the amount of biologically productive land and water needed to generate the resources for consumption and waste. EF_{direct} represents the direct ecological footprint associated with the activity under consideration where EF_{CO_2} and $EF_{nuclear}$ indicate the ecological footprint associated with CO_2 emissions and nuclear activities, respectively.

2. Centrum voor Milieukunde Leiden (CML 2001) and Tool for the Reduction and Assessment of Chemical and Other Environmental Impacts (TRACI) are both methods used in the impact assessment step of LCA. CML 2001 is a midpoint method that characterizes the potential impacts of a product or process based on certain midpoint indicators. On the other hand, TRACI is a method that characterizes environmental stressors that may have potential effects. Both methods help to provide a comprehensive understanding of the environmental impact of a product or process, allowing better decision-making and sustainable design. Through these methods, various impacts can be examined, including depletion of nonliving resources, acidification, eutrophication, global warming, ozone depletion, human toxicity, and ecotoxicity in freshwater, marine, and terrestrial ecosystems, as well as photochemical oxidation. These impacts are measured in appropriate physical units using physically based formulas that take into account the quantities of various parameters in the PBR system (Bare 2011; Hischier et al. 2010; Pennington et al. 2004).
3. SimaPro is one of the famous LCA models that can quantify environmental impacts and identify areas of concern in a systematic and transparent manner. This software provides results in various impact categories, including midpoint and end-point categories. It allows for the analysis of specific impacts such as carbon footprint and water use. Commonly, SimaPro is widely utilized in industry, consulting firms, universities, and research centers across different countries.

Notwithstanding, in the present climate, there is still lack of comprehensive LCA study of the microalgae-based wastewater treatment, instead many investigations have been done by coupling microalgae wastewater treatment system and value-added products generation in the same scheme. In general, Pessôa et al. (2021) has asserted that with the implementation of so-called 5-steps technology (pretreatment of residual waters, anaerobic bio-digestion, pretreatment of the digestate, microalgal biochemical treatment, and separation of resulting biomass) for the production of biofuels could eventually lead to low ecological footprint (Pessôa et al. 2021). A comparative LCA study has been carried out by Pérez-López et al. (2017) on open raceway ponds (ORP) and horizontal and vertical tubular PBRs at pilot scale. The results of the study suggested that ORP system showed higher environmental impacts than both PBRs because of the PBR systems compensate energy-consuming elements with higher productivity (Pérez-López et al. 2017). On the other hand,

Rothermel et al. (2013) discovered that algae-based wastewater treatment has lower energy use, global warming, eutrophication potential, and other impact categories when associated with conventional type of wastewater treatment (Rothermel et al. 2013). Moreover, high-rate algae pond (HRAP) introduced by Kohlheb et al. (2020) also found to be more beneficial environmentally and economically, contributing to CO_2 sequestration and eutrophication potentials of 146.27×10^{-3} kg CO_2 equiv m^{-3} and 126.14×10^{-6} kg PO_4 equiv m^{-3}, respectively (Kohlheb et al. 2020). To sum up, numerous studies have shown that microalgal wastewater treatment outperforms conventional wastewater treatment in terms of various impact categories.

5 Challenges and Future Perspectives

Over the past few years, there have been pilot-scale plants developed for MMWT, such as membrane filtration system that includes 200-litre feed tank for processing *Scenedesmus* sp. (Gerardo et al. 2014). At Hamburg university, *Scenedesmus obliquus* was cultivated in 320-liters semicontinuously using vertical plastic bags (16 L each) as a cost-effective alternative for biodiesel production followed by application of lipid-free waste for Artemia feed (Abomohra et al. 2014). At Swansea University, both 600-liter and 2000-liter horizontal and vertical tubular PBR were used and harvested periodically during the stationary growth stage, with different culture densities (Gerardo et al. 2015). Despite the promising obtained results for microalgal biomass production, harvesting is a significant challenge because it is costly and can represent 20–30% of the total production cost, which impacts the overall affordability of the process (Fernández et al. 2019; Lv et al. 2017; Ogbonna and Nwoba 2021). Owing to the negatively charged nature of microalgal cell surfaces, the typical method of harvesting microalgae involves dosing a positively charged coagulant to allow the microalgae to flocculate and clump together (Worku and Sahu 2014). Additionally, Plymouth University was working on a project that involves coupling a low-energy 16,000-liter PBR directly to the stack of a modern 15 MW gas turbine power station at Boots Ltd in UK (Gerardo et al. 2015). As for membrane filtration, it shows promising cultures; however, the fouling of membrane is 57.5% of the total cost of the operation (Ye et al. 2018). Also, the relatively small size of microalgae which is about 2–20 microns make the harvesting process a key challenge when adapted in industrial scale (Mennaa et al. 2015). Therefore, when scaling-up, various problems arise which are case-dependent such as the specific species of microalgae being used, the cultivation method employed (e.g., open ponds or PBRs), the characteristics of the wastewater being treated, the infrastructure and equipment available, and the intended application or end-use of the microalgae biomass (Klein et al. 2018). Each case or system may have unique considerations and constraints that need to be addressed in order to overcome the challenges associated with scaling up.

Likewise, the pH of the wastewater must be monitored and controlled at a certain level, as extreme pH values affect the enzyme activity of the microalgae and inhibit its photosynthetic activity. Methods such as immobilization of the microalgae and addition of promoting bacteria are currently not examined when scaled-up, and the combination of different bacteria to achieve a higher efficiency of nutrient recovery is unknown. Therefore, more detailed experiments must be carried out in order to access the feasibility using new techniques to improve MMWT plants. Furthermore, the utilization of extremophiles in the treatment of industrial waste is a viable approach. However, there is a limitation associated with their use. Extremophiles thrive in toxic and harsh environments, which means that additional chemicals or supplements are required to maintain their living conditions. This necessity for supplemental additives can result in increased production or treatment costs, thereby impacting the overall economic feasibility of employing extremophiles for wastewater treatment purposes (Li et al. 2019a). Even if a plausible method has been recognized, implementing it into preexisting conventional wastewater plants is a considerable challenge, since some preexisting plants do not have space to expand or accommodate the PBR/ponds where the microalgae will be cultured. Moreover, concerns arise within society when the topic of algal blooms is raised, particularly when the growth of algae is uncontrolled. Algal blooms can have detrimental effects and lead to a decline in water quality. However, it is important to differentiate between uncontrolled algal growth and carefully studied microalgae. Microalgae, which are unicellular and consist of specific strains, can effectively recover nutrients from wastewater without compromising water quality. These microalgae can be harvested and utilized for the production of biofuels or bioproducts, offering a sustainable and environmentally friendly solution (Khan et al. 2018).

The current market price for microalgae biomass production ranges from a minimum of US \$5.36 kg^{-1} in raceway ponds to US \$46.12 kg^{-1} in tubular photobioreactors depending on the scale of the technologies (Oostlander et al. 2020; Ruiz et al. 2016). However, the estimated cost of production for microalgae oil and biodiesel is US \$10.87 $gallon^{-1}$ and US \$9.84 $gallon^{-1}$, respectively (Valdovinos-García et al. 2022; Zewdie and Ali 2022). In order to establish a financially sustainable microalgae market, the initial price of the value-added products synthesized should exceed the production cost. Correspondingly, to broaden its usage across various sectors, the production of valuable microalgae needs to triple to meet the demand without raising costs, which poses a significant challenge for current scenario. Also, the difficulty of meeting the demand for supply of biofuels, food, feed, and pharmaceuticals is one of the major issues. Thus, most of the recent studies started to focus on the biofuel-integrated biorefinery as one of the possible solutions to produce higher capacity of biofuels with enhanced biomass utilization, as well as reduce the cost (Chapter "Biofuel-Integrated Routes"). Because the technology and design used to accommodate microalgae is not at its peak, the net carbon emission from the overall process is still considered lacking and remunerative (Chanakya et al. 2012). Therefore, additional research and development are required to enhance the existing expertise on cultivating specific microalgae strains which provides higher lipids, carbohydrates, nutrient removal, and enhanced light utilization efficiency. Also, while designing microalgae-mediated wastewater treatment, careful selection of microalgal strain is of great importance where high throughput screening techniques could play a crucial role (Chapter "High Throughput Screening to Accelerate Microalgae-based Phycochemicals"). More detailed implementation schemes can be discussed and presented to the government or proper authorities for a thorough evaluation of MMWT. Therefore, proper authorities should be involved for the implementation and funding of microalgae in current wastewater treatment. In order to get support or help for governments or treatment plants, it is important to establish a method of treatment that presents an enhanced approach compared to current conventional treatments, which does not only solve the water shortage issue and reduce the greenhouse gas and carbon footprint, but also produces usable biomass in order to surpass the conventional methods (Gonçalves et al. 2017).

6 Conclusions

In conclusion, utilization of microalgae in various industries, particularly in wastewater treatment, has shown promising advancements. MMWT offers a more sustainable alternative to conventional methods, as it reduces the environmental impact and efficiently captures nutrients. Municipal wastewater and animal wastewater are the most widely studied effluents due to their availability and consistent quality. The recovery of nutrients such as phosphorus and nitrogen are crucial for the growth of microalgae and, generally, laboratory-scale experiments have shown promising results in terms of nutrient recovery efficiency using various types of wastewaters. Understanding the characteristics of wastewater effluents and their impact on microalgal growth and nutrient recovery is essential for effective implementation of MMWT. Additionally, microalgae have also gained significant attention as a potential alternative for removing heavy metals from industrial wastewater due to their ability to uptake heavy metals through bioaccumulation and biosorption. Some of the LCA analyses have been highlighted in this chapter which successfully demonstrated and proven the environmentally beneficial of employing MMWT. However, challenges still exist in terms of scalability, cost-effectiveness, and technological limitations. To overcome these challenges and fully realize the potential of microalgae-based wastewater technologies, further research and development efforts are needed. With continued advancements and innovative solutions, microalgae-based wastewater treatment coupled with cost-effective phycochemicals production has the potential to revolutionize the industry and contribute to a more sustainable future.

References

Abdel-Raouf N, Al-Homaidan A, Ibraheem I (2012) Microalgae and wastewater treatment. Saudi J Biol Sci 19(3):257–275

Abdel-Razek MA, Abozeid AM, Eltholth MM, Abouelenien FA, El-Midany SA, Moustafa NY, Mohamed RA (2019) Bioremediation of a pesticide and selected heavy metals in wastewater from various sources using a consortium of microalgae and cyanobacteria. Slov Vet 56(Suppl 22):61–73

Abiusi F, Trompetter E, Pollio A, Wijffels RH, Janssen M (2022) Acid tolerant and acidophilic microalgae: an underexplored world of biotechnological opportunities. Front Microbiol 13:51

Abomohra AE-F, El-Sheekh M, Hanelt D (2014) Pilot cultivation of the chlorophyte microalga Scenedesmus obliquus as a promising feedstock for biofuel. Biomass Bioenergy 64:237–244

Abou-Shanab RA, Ji M-K, Kim H-C, Paeng K-J, Jeon B-H (2013) Microalgal species growing on piggery wastewater as a valuable candidate for nutrient removal and biodiesel production. J Environ Manag 115:257–264

Acién Fernández FG, Gómez-Serrano C, Fernández-Sevilla JM (2018) Recovery of nutrients from wastewaters using microalgae. *Frontiers in Sustainable Food Systems* 2:59

Acién FG, Gómez-Serrano C, Morales-Amaral MdM, Fernández-Sevilla JM, Molina-Grima E (2016) Wastewater treatment using microalgae: how realistic a contribution might it be to significant urban wastewater treatment? Appl Microbiol Biotechnol 100:9013–9022

Acién F, Molina E, Reis A, Torzillo G, Zittelli GC, Sepúlveda C, Masojídek J (2017) Photobioreactors for the production of microalgae. In: Microalgae-based biofuels and bioproducts. Elsevier, Amsterdam, pp 1–44

Ahmed SF, Mofijur M, Parisa TA, Islam N, Kusumo F, Inayat A, Badruddin IA, Khan TY, Ong HC (2022) Progress and challenges of contaminate removal from wastewater using microalgae biomass. Chemosphere 286:131656

Akcil A, Erust C, Ozdemiroglu S, Fonti V, Beolchini F (2015) A review of approaches and techniques used in aquatic contaminated sediments: metal removal and stabilization by chemical and biotechnological processes. J Clean Prod 86:24–36

Akhiar A, Battimelli A, Torrijos M, Carrere H (2017) Comprehensive characterization of the liquid fraction of digestates from full-scale anaerobic co-digestion. Waste Manag 59:118–128

Aksu Z (1998) Biosorption of heavy metals by microalgae in batch and continuous systems. In: Wastewater treatment with algae. Springer, Berlin, pp 37–53

Alam MA, Wan C, Tran DT, Mofijur M, Ahmed SF, Mehmood MA, Shaik F, Vo D-VN, Xu J (2022) Microalgae binary culture for higher biomass production, nutrients recycling, and efficient harvesting: a review. Environ Chem Lett 20(2):1153–1168

Amaro HM, Salgado EM, Nunes OC, Pires JC, Esteves AF (2023) Microalgae systems-environmental agents for wastewater treatment and further potential biomass valorisation. J Environ Manag 337:117678

Arora N, Patel A, Sartaj K, Pruthi PA, Pruthi V (2016) Bioremediation of domestic and industrial wastewaters integrated with enhanced biodiesel production using novel oleaginous microalgae. Environ Sci Pollut Res 23:20997–21007

Ballesteros-Torres JM, Samaniego-Moreno L, Gomez-Flores R, Tamez-Guerra RS, Rodríguez-Padilla C, Tamez-Guerra P (2019) Amino acids and acylcarnitine production by *Chlorella vulgaris* and *Chlorella sorokiniana* microalgae from wastewater culture. PeerJ 7:e7977

Bare J (2011) TRACI 2.0: the tool for the reduction and assessment of chemical and other environmental impacts 2.0. Clean Techn Environ Policy 13:687–696

Basak N, Das D (2007) The prospect of purple non-sulfur (PNS) photosynthetic bacteria for hydrogen production: the present state of the art. World J Microbiol Biotechnol 23:31–42

Benalia MC, Youcef L, Bouaziz MG, Achour S, Menasra H (2022) Removal of heavy metals from industrial wastewater by chemical precipitation: mechanisms and sludge characterization. Arab J Sci Eng 47(5):5587–5599

Benítez MB, Champagne P, Ramos A, Torres AF, Ochoa-Herrera V (2019) Wastewater treatment for nutrient removal with Ecuadorian native microalgae. Environ Technol 40(22):2977–2985

Besson M, Berger S, Tiruta-Barna L, Paul E, Spérandio M (2021) Environmental assessment of urine, black and grey water separation for resource recovery in a new district compared to centralized wastewater resources recovery plant. J Clean Prod 301:126868

Bhat AP, Gogate PR (2021) Cavitation-based pre-treatment of wastewater and waste sludge for improvement in the performance of biological processes: a review. J Environ Chem Eng 9(2): 104743

Boelee N, Temmink H, Janssen M, Buisman C, Wijffels R (2014) Balancing the organic load and light supply in symbiotic microalgal–bacterial biofilm reactors treating synthetic municipal wastewater. Ecol Eng 64:213–221

Boonchai R, Seo G (2015) Microalgae membrane photobioreactor for further removal of nitrogen and phosphorus from secondary sewage effluent. Korean J Chem Eng 32:2047–2052

Brennan L, Owende P (2010) Biofuels from microalgae—a review of technologies for production, processing, and extractions of biofuels and co-products. Renew Sust Energ Rev 14(2):557–577

Cai T, Park SY, Li Y (2013) Nutrient recovery from wastewater streams by microalgae: status and prospects. Renew Sust Energ Rev 19:360–369

Cantera S, Fischer PQ, Sánchez-Andrea I, Marín D, Sousa DZ, Muñoz R (2021) Impact of the algal-bacterial community structure, physio-types and biological and environmental interactions on the performance of a high rate algal pond treating biogas and wastewater. Fuel 302:121148

Chan A, Salsali H, McBean E (2014) Heavy metal removal (copper and zinc) in secondary effluent from wastewater treatment plants by microalgae. ACS Sustain Chem Eng 2(2):130–137

Chanakya H, Mahapatra DM, Ravi S, Chauhan V, Abitha R (2012) Sustainability of large-scale algal biofuel production in India. J Indian Inst Sci 92(1):63–98

Chaudhari H (2014) Dielectric properties of soils with inorganic matter at S-band microwave frequency. Int J Chem Physical Sci 3(6):59–66

Chen X, Li Z, He N, Zheng Y, Li H, Wang H, Wang Y, Lu Y, Li Q, Peng Y (2018) Nitrogen and phosphorus removal from anaerobically digested wastewater by microalgae cultured in a novel membrane photobioreactor. Biotechnol Biofuels 11:1–11

Chew KW, Khoo KS, Foo HT, Chia SR, Walvekar R, Lim SS (2021) Algae utilization and its role in the development of green cities. Chemosphere 268:129322

Choi H-J, Lee S-M (2012) Effects of microalgae on the removal of nutrients from wastewater: various concentrations of *Chlorella vulgaris*. Environ Eng Res 17(S1):3–8

Chong CC, Cheng YW, Ishak S, Lam MK, Lim JW, Tan IS, Show PL, Lee KT (2022) Anaerobic digestate as a low-cost nutrient source for sustainable microalgae cultivation: a way forward through waste valorization approach. Sci Total Environ 803:150070

Coccia M, Bontempi E (2023) New trajectories of technologies for the removal of pollutants and emerging contaminants in the environment. Environ Res 229:115938

Collao J, García-Encina PA, Blanco S, Bolado-Rodríguez S, Fernandez-Gonzalez N (2022) Current concentrations of Zn, Cu, and As in piggery wastewater compromise nutrient removals in microalgae–bacteria photobioreactors due to altered microbial communities. Biology 11(8): 1176

Costa JAV, Lucas BF, Alvarenga AGP, Moreira JB, de Morais MG (2021) Microalgae polysaccharides: an overview of production, characterization, and potential applications. Polysaccharides 2(4):759–772

Cruz H, Law YY, Guest JS, Rabaey K, Batstone D, Laycock B, Verstraete W, Pikaar I (2019) Mainstream ammonium recovery to advance sustainable urban wastewater management. Environ Sci Technol 53(19):11066–11079

Danouche M, El Ghachtouli N, El Arroussi H (2021) Phycoremediation mechanisms of heavy metals using living green microalgae: physicochemical and molecular approaches for enhancing selectivity and removal capacity. Heliyon 7(7):e07609

De-Bashan LE, Bashan Y (2010) Immobilized microalgae for removing pollutants: review of practical aspects. Bioresour Technol 101(6):1611–1627

De-Bashan LE, Moreno M, Hernandez J-P, Bashan Y (2002) Removal of ammonium and phosphorus ions from synthetic wastewater by the microalgae *Chlorella vulgaris* coimmobilized in alginate beads with the microalgae growth-promoting bacterium Azospirillum brasilense. Water Res 36(12):2941–2948

De-Bashan LE, Hernandez J-P, Morey T, Bashan Y (2004) Microalgae growth-promoting bacteria as "helpers" for microalgae: a novel approach for removing ammonium and phosphorus from municipal wastewater. Water Res 38(2):466–474

Delgadillo-Mirquez L, Lopes F, Taidi B, Pareau D (2016) Nitrogen and phosphate removal from wastewater with a mixed microalgae and bacteria culture. Biotechnol Rep 11:18–26

Desloover J, Abate Woldeyohannis A, Verstraete W, Boon N, Rabaey K (2012) Electrochemical resource recovery from digestate to prevent ammonia toxicity during anaerobic digestion. Environ Sci Technol 46(21):12209–12216

Díaz V, Leyva-Díaz JC, Almécija MC, Poyatos JM, del Mar Muñío M, Martín-Pascual J (2022) Microalgae bioreactor for nutrient removal and resource recovery from wastewater in the paradigm of circular economy. Bioresour Technol 363:127968

Ekendahl S, Bark M, Engelbrektsson J, Karlsson C-A, Niyitegeka D, Strömberg N (2018) Energy-efficient outdoor cultivation of oleaginous microalgae at northern latitudes using waste heat and flue gas from a pulp and paper mill. Algal Res 31:138–146

Emparan Q, Jye YS, Danquah MK, Harun R (2020) Cultivation of *Nannochloropsis* sp. microalgae in palm oil mill effluent (POME) media for phycoremediation and biomass production: Effect of microalgae cells with and without beads. J Water Process Eng 33:101043

Evans L, Hennige SJ, Willoughby N, Adeloye AJ, Skroblin M, Gutierrez T (2017) Effect of organic carbon enrichment on the treatment efficiency of primary settled wastewater by *Chlorella vulgaris*. Algal Res 24:368–377

Fernández FA, Sevilla JMF, Grima EM (2019) Costs analysis of microalgae production. In: Biofuels from algae. Elsevier, Amsterdam, pp 551–566

Ferreira A, Reis A, Vidovic S, Vladic J, Gkelis S, Melkonyan L, Avetisova G, Congestri R, Acién G, Muñoz R (2019) Combining microalgae-based wastewater treatment with biofuel and bio-based production in the frame of a biorefinery. In: Grand challenges in algae biotechnology. Springer, Cham, pp 319–369

Fourest E, Roux J-C (1992) Heavy metal biosorption by fungal mycelial by-products: mechanisms and influence of pH. Appl Microbiol Biotechnol 37:399–403

Fuerhacker M, Haile TM, Kogelnig D, Stojanovic A, Keppler B (2012) Application of ionic liquids for the removal of heavy metals from wastewater and activated sludge. Water Sci Technol 65(10):1765–1773

Gao F, Yang Z-H, Li C, Zeng G-M, Ma D-H, Zhou L (2015) A novel algal biofilm membrane photobioreactor for attached microalgae growth and nutrients removal from secondary effluent. Bioresour Technol 179:8–12

Garbisu C, Gil JM, Bazin MJ, Hall DO, Serra JL (1991) Removal of nitrate from water by foam-immobilized *Phormidium laminosum* in batch and continuous-flow bioreactors. J Appl Phycol 3:221–234

Gerardo ML, Oatley-Radcliffe DL, Lovitt RW (2014) Minimizing the energy requirement of dewatering *Scenedesmus* sp. by microfiltration: performance, costs, and feasibility. Environ Sci Technol 48(1):845–853

Gerardo ML, Van Den Hende S, Vervaeren H, Coward T, Skill SC (2015) Harvesting of microalgae within a biorefinery approach: A review of the developments and case studies from pilot-plants. Algal Res 11:248–262

Ghazvini M, Kavosi M, Sharma R, Kim M (2022) A review on mechanical-based microalgae harvesting methods for biofuel production. Biomass Bioenergy 158:106348

Gonçalves AL, Pires JC, Simões M (2017) A review on the use of microalgal consortia for wastewater treatment. Algal Res 24:403–415

Gonzalez-Fernandez C, Muñoz R (2017) Microalgae-based biofuels and bioproducts. Woodhead Publishing (Elseiver), Kindlington

Govarthanan M, Jeon C-H, Jeon Y-H, Kwon J-H, Bae H, Kim W (2020) Non-toxic nano approach for wastewater treatment using *Chlorella vulgaris* exopolysaccharides immobilized in iron-magnetic nanoparticles. Int J Biol Macromol 162:1241–1249

Gunatilake S (2015) Methods of removing heavy metals from industrial wastewater. Methods 1(1): 14

Han M, Zhang C, Ho S-H (2022) Immobilized microalgal system: an achievable idea for upgrading current microalgal wastewater treatment. Environ Sci Ecotechnol 14:100227

He J, Chen JP (2014) A comprehensive review on biosorption of heavy metals by algal biomass: materials, performances, chemistry, and modeling simulation tools. Bioresour Technol 160:67–78

Heidari MR, Malakootian M, Boczkaj G, Sun X, Tao Y, Sonawane SH, Mehdizadeh H (2021) Evaluation and start-up of an electro-Fenton-sequencing batch reactor for dairy wastewater treatment. Water Resour Ind 25:100149

Hischier R, Weidema B, Althaus H-J, Bauer C, Doka G, Dones R, Frischknecht R, Hellweg S, Humbert S, Jungbluth N (2010) Implementation of life cycle impact assessment methods. Final Report Ecoinvent 2:2

Hoh D, Watson S, Kan E (2016) Algal biofilm reactors for integrated wastewater treatment and biofuel production: a review. Chem Eng J 287:466–473

Homan NP, Green PG, Young TM (2018) Evaluating ferrous chloride for removal of chromium from ion-exchange waste brines. J Am Water Works Assoc 110(4):E43–E54

Huijbregts MA, Steinmann ZJ, Elshout PM, Stam G, Verones F, Vieira M, Zijp M, Hollander A, Van Zelm R (2017) ReCiPe2016: a harmonised life cycle impact assessment method at midpoint and endpoint level. Int J Life Cycle Assess 22:138–147

Hussain F, Shah SZ, Ahmad H, Abubshait SA, Abubshait HA, Laref A, Manikandan A, Kusuma HS, Iqbal M (2021) Microalgae an ecofriendly and sustainable wastewater treatment option: Biomass application in biofuel and bio-fertilizer production. A review. Renew Sust Energ Rev 137:110603

Izah SC, Angaye TC (2016) Heavy metal concentration in fishes from surface water in Nigeria: potential sources of pollutants and mitigation measures. Sky J Biochem Res 5(4):31–47

Jacinto MLJ, David CPC, Perez TR, De Jesus BR (2009) Comparative efficiency of algal biofilters in the removal of chromium and copper from wastewater. Ecol Eng 35(5):856–860

Jia H, Yuan Q (2018) Ammonium removal using algae–bacteria consortia: the effect of ammonium concentration, algae biomass, and light. Biodegradation 29(2):105–115

Jiang L, Li Y, Pei H (2021) Algal–bacterial consortia for bioproduct generation and wastewater treatment. Renew Sust Energ Rev 149:111395

Johnson K, Kleerebezem R, van Loosdrecht MC (2010) Influence of the C/N ratio on the performance of polyhydroxybutyrate (PHB) producing sequencing batch reactors at short SRTs. Water Res 44(7):2141–2152

Kadir WNA, Lam MK, Uemura Y, Lim JW, Lee KT (2018) Harvesting and pre-treatment of microalgae cultivated in wastewater for biodiesel production: a review. Energy Convers Manag 171:1416–1429

Kakar FL, Tadesse F, Elbeshbishy E (2022) Comprehensive review of hydrothermal pretreatment parameters affecting fermentation and anaerobic digestion of municipal sludge. Processes 10(12):2518

Kalra R, Gaur S, Goel M (2021) Microalgae bioremediation: a perspective towards wastewater treatment along with industrial carotenoids production. J Water Process Eng 40:101794

Kaparapu J, Geddada MNR (2016) Applications of immobilized algae. J Algal Biomass Utilization 7(2):122–128

Kaplan D (2013) Absorption and adsorption of heavy metals by microalgae. In: Handbook of microalgal culture: applied phycology and biotechnology. Wiley, Hoboken, NJ, pp 602–611

Kelly PT, He Z (2014) Nutrients removal and recovery in bioelectrochemical systems: a review. Bioresour Technol 153:351–360

Khan MI, Shin JH, Kim JD (2018) The promising future of microalgae: current status, challenges, and optimization of a sustainable and renewable industry for biofuels, feed, and other products. Microb Cell Factories 17(1):1–21

Khan SA, Sharma GK, Malla FA, Kumar A, Gupta N (2019) Microalgae based biofertilizers: a biorefinery approach to phycoremediate wastewater and harvest biodiesel and manure. J Clean Prod 211:1412–1419

Khan AA, Gul J, Naqvi SR, Ali I, Farooq W, Liaqat R, AlMohamadi H, Štěpanec L, Juchelková D (2022) Recent progress in microalgae-derived biochar for the treatment of textile industry wastewater. Chemosphere 306:135565

Khoo KS, Chia WY, Chew KW, Show PL (2021) Microalgal-bacterial consortia as future prospect in wastewater bioremediation, environmental management and bioenergy production. Indian J Microbiol 61(3):262–269

Klein BC, Bonomi A, Maciel Filho R (2018) Integration of microalgae production with industrial biofuel facilities: a critical review. Renew Sust Energ Rev 82:1376–1392

Kohlheb N, van Afferden M, Lara E, Arbib Z, Conthe M, Poitzsch C, Marquardt T, Becker M-Y (2020) Assessing the life-cycle sustainability of algae and bacteria-based wastewater treatment systems: high-rate algae pond and sequencing batch reactor. J Environ Manag 264:110459

Kube M, Jefferson B, Fan L, Roddick F (2018) The impact of wastewater characteristics, algal species selection and immobilisation on simultaneous nitrogen and phosphorus removal. Algal Res 31:478–488

Kumar KS, Dahms H-U, Won E-J, Lee J-S, Shin K-H (2015) Microalgae–a promising tool for heavy metal remediation. Ecotoxicol Environ Saf 113:329–352

Kumar R, Sharma AK, Ahluwalia SS (2017) Advances in environmental biotechnology. Springer, Singapore

Kurt Z, Özdemir I (2022) Effectiveness of microplastics removal in wastewater treatment plants: A critical analysis of wastewater treatment processes. J Environ Chem Eng 10(3):107831

Lacroux J, Trably E, Bernet N, Steyer J-P, van Lis R (2020) Mixotrophic growth of microalgae on volatile fatty acids is determined by their undissociated form. Algal Res 47:101870

Lam MK, Lee KT (2012) Immobilization as a feasible method to simplify the separation of microalgae from water for biodiesel production. Chem Eng J 191:263–268

Lee C, Lee C-G (2002) Nitrogen removal from wastewaters by microalgae without consuming organic carbon sources. *J Microbiol Biotechnol* 12(6):979–985

Lee XJ, Ong HC, Gan YY, Chen W-H, Mahlia TMI (2020) State of art review on conventional and advanced pyrolysis of macroalgae and microalgae for biochar, bio-oil and bio-syngas production. Energy Convers Manag 210:112707

Lei Y-J, Tian Y, Zhang J, Sun L, Kong X-W, Zuo W, Kong L-C (2018) Microalgae cultivation and nutrients removal from sewage sludge after ozonizing in algal-bacteria system. Ecotoxicol Environ Saf 165:107–114

Leong YK, Chang J-S (2020) Bioremediation of heavy metals using microalgae: recent advances and mechanisms. Bioresour Technol 303:122886

Li K, Liu Q, Fang F, Luo R, Lu Q, Zhou W, Huo S, Cheng P, Liu J, Addy M (2019a) Microalgae-based wastewater treatment for nutrients recovery: a review. Bioresour Technol 291:121934

Li S, Zhao S, Yan S, Qiu Y, Song C, Li Y, Kitamura Y (2019b) Food processing wastewater purification by microalgae cultivation associated with high value-added compounds production—a review. Chin J Chem Eng 27(12):2845–2856

Liliane TN, Charles MS (2020) Factors affecting yield of crops. In: Agronomy-climate change & food security. IntechOpen, London, p 9

Lim HR, Khoo KS, Chew KW, Chang C-K, Munawaroh HSH, Kumar PS, Huy ND, Show PL (2021) Perspective of Spirulina culture with wastewater into a sustainable circular bioeconomy. Environ Pollut 284:117492

Liu X-Y, Hong Y (2021) Microalgae-based wastewater treatment and recovery with biomass and value-added products: a brief review. Current Pollution Rep 7:227–245

Liu K, Li J, Qiao H, Lin A, Wang G (2012) Immobilization of *Chlorella sorokiniana* GXNN 01 in alginate for removal of N and P from synthetic wastewater. Bioresour Technol 114:26–32

Liu X, Wang K, Zhang J, Wang J, Wu J, Peng F (2019) Ammonium removal potential and its conversion pathways by free and immobilized *Scenedesmus obliquus* from wastewater. Bioresour Technol 283:184–190

Lv J, Feng J, Liu Q, Xie S (2017) Microalgal cultivation in secondary effluent: recent developments and future work. Int J Mol Sci 18(1):79

Lv Z, Shan X, Xiao X, Cai R, Zhang Y, Jiao N (2022) Excessive greenhouse gas emissions from wastewater treatment plants by using the chemical oxygen demand standard. Sci China Earth Sci 65:87–95

Mahamadi C (2019) Will nano-biosorbents break the Achilles' heel of biosorption technology? Environ Chem Lett 17(4):1753–1768

Manzoor F, Karbassi A, Golzary A (2019) Removal of heavy metal contaminants from wastewater by using *Chlorella vulgaris* Beijerinck: a review. Current Environ Manage (Formerly: Current Environmental Engineering) 6(3):174–187

Mata TM, Cameira M, Marques F, Santos E, Badenes S, Costa L, Vieira VV, Caetano NS, Martins AA (2018) Carbon footprint of microalgae production in photobioreactor. Energy Procedia 153: 432–437

Matamoros V, Gutiérrez R, Ferrer I, García J, Bayona JM (2015) Capability of microalgae-based wastewater treatment systems to remove emerging organic contaminants: a pilot-scale study. J Hazard Mater 288:34–42

Maureira D, Romero O, Illanes A, Wilson L, Ottone C (2023) Industrial bioelectrochemistry for waste valorization: state of the art and challenges. Biotechnol Adv 64:108123

Mennaa FZ, Arbib Z, Perales JA (2015) Urban wastewater treatment by seven species of microalgae and an algal bloom: biomass production, N and P removal kinetics and harvestability. Water Res 83:42–51

Michels MH, Vaskoska M, Vermuë MH, Wijffels RH (2014) Growth of Tetraselmis suecica in a tubular photobioreactor on wastewater from a fish farm. Water Res 65:290–296

Michielssen MR, Michielssen ER, Ni J, Duhaime MB (2016) Fate of microplastics and other small anthropogenic litter (SAL) in wastewater treatment plants depends on unit processes employed. Environ Sci: Water Res Technol 2(6):1064–1073

Min M, Wang L, Li Y, Mohr MJ, Hu B, Zhou W, Chen P, Ruan R (2011) Cultivating *Chlorella* sp. in a pilot-scale photobioreactor using centrate wastewater for microalgae biomass production and wastewater nutrient removal. Appl Biochem Biotechnol 165:123–137

Moazami N, Ashori A, Ranjbar R, Tangestani M, Eghtesadi R, Nejad AS (2012) Large-scale biodiesel production using microalgae biomass of *Nannochloropsis*. Biomass Bioenergy 39: 449–453

Molazadeh M, Danesh S, Ahmadzadeh H, Pourianfar HR (2019) Influence of CO2 concentration and N: P ratio on *Chlorella vulgaris*-assisted nutrient bioremediation, CO2 biofixation and biomass production in a lagoon treatment plant. J Taiwan Inst Chem Eng 96:114–120

Molinuevo-Salces B, Riaño B, Hernández D, Cruz García-González M (2019) Microalgae and wastewater treatment: advantages and disadvantages. In: Microalgae biotechnology for development of biofuel and wastewater treatment. Springer, Singapore, pp 505–533

Monteiro CM, Castro PM, Malcata FX (2012) Metal uptake by microalgae: underlying mechanisms and practical applications. Biotechnol Prog 28(2):299–311

Moreno-Garrido I (2008) Microalgae immobilization: current techniques and uses. Bioresour Technol 99(10):3949–3964

Mujtaba G, Rizwan M, Kim G, Lee K (2018) Removal of nutrients and COD through co-culturing activated sludge and immobilized *Chlorella vulgaris*. Chem Eng J 343:155–162

Nawaz MZ, Bilal M, Tariq A, Iqbal HM, Alghamdi HA, Cheng H (2021) Bio-purification of sugar industry wastewater and production of high-value industrial products with a zero-waste concept. Crit Rev Food Sci Nutr 61(21):3537–3554

Nayagam JOP, Prasanna K (2022) Utilization of shell-based agricultural waste adsorbents for removing dyes: a review. Chemosphere 291:132737

Nguyen TDP, Tran TNT, Le TVA, Phan TXN, Show P-L, Chia SR (2019) Auto-flocculation through cultivation of *Chlorella vulgaris* in seafood wastewater discharge: Influence of culture conditions on microalgae growth and nutrient removal. J Biosci Bioeng 127(4):492–498

Nguyen HT, Yoon Y, Ngo HH, Jang A (2021) The application of microalgae in removing organic micropollutants in wastewater. Crit Rev Environ Sci Technol 51(12):1187–1220

Ogbonna CN, Nwoba EG (2021) Bio-based flocculants for sustainable harvesting of microalgae for biofuel production. A review. Renew Sust Energ Rev 139:110690

Oostlander P, van Houcke J, Wijffels RH, Barbosa MJ (2020) Microalgae production cost in aquaculture hatcheries. Aquaculture 525:735310

Patel A, Matsakas L, Rova U, Christakopoulos P (2019) A perspective on biotechnological applications of thermophilic microalgae and cyanobacteria. Bioresour Technol 278:424–434

Pedersen JA, Yeager MA, Suffet I (2003) Xenobiotic organic compounds in runoff from fields irrigated with treated wastewater. J Agric Food Chem 51(5):1360–1372

Peng X, Huang S, Yi K, Zhang L, Ge F, Lin Q, Zhang Y, Wu Z, Liu B (2023) Dynamic characteristics of periphytic algae communities on different substrates and the host response in subtropical-urban-landscape lakes. Water 15(4):639

Pennington DW, Potting J, Finnveden G, Lindeijer E, Jolliet O, Rydberg T, Rebitzer G (2004) Life cycle assessment Part 2: Current impact assessment practice. Environ Int 30(5):721–739

Pérez-López P, De Vree JH, Feijoo G, Bosma R, Barbosa MJ, Moreira MT, Wijffels RH, Van Boxtel AJ, Kleinegris DM (2017) Comparative life cycle assessment of real pilot reactors for microalgae cultivation in different seasons. Appl Energy 205:1151–1164

Pessôa LC, Deamici KM, Pontes LAM, Druzian JI, de Jesus Assis D (2021) Technological prospection of microalgae-based biorefinery approach for effluent treatment. Algal Res 60: 102504

Peter AP, Khoo KS, Chew KW, Ling TC, Ho S-H, Chang J-S, Show PL (2021) Microalgae for biofuels, wastewater treatment and environmental monitoring. Environ Chem Lett 19:2891–2904

Pham T-L, Bui MH (2020) Removal of nutrients from fertilizer plant wastewater using Scenedesmus sp.: formation of bioflocculation and enhancement of removal efficiency. J Chemistry 2020:1–9

Poornima S, Manikandan S, Karthik V, Balachandar R, Subbaiya R, Saravanan M, Chi NTL, Pugazhendhi A (2022) Emerging nanotechnology based advanced techniques for wastewater treatment. Chemosphere 303(Pt 3):135050

Porwal H, Mane A, Velhal S (2015) Biodegradation of dairy effluent by using microbial isolates obtained from activated sludge. Water Resour Ind 9:1–15

Praveen P, Loh K-C (2016) Nitrogen and phosphorus removal from tertiary wastewater in an osmotic membrane photobioreactor. Bioresour Technol 206:180–187

Prihandrijanti M, Firdayati M (2011) Current Situation and considerations of domestic waste-water treatment systems for big cities in Indonesia (case study: Surabaya and Bandung). J Water Sustainability 1(2):97–104

Priya A, Pachaiappan R, Kumar PS, Jalil A, Vo D-VN, Rajendran S (2021) The war using microbes: A sustainable approach for wastewater management. Environ Pollut 275:116598

Priyadarshanee M, Das S (2021) Biosorption and removal of toxic heavy metals by metal tolerating bacteria for bioremediation of metal contamination: a comprehensive review. J Environ Chem Eng 9(1):104686

Priyadarshani I, Rath B (2012) Commercial and industrial applications of micro algae–a review. J Algal Biomass Utilization 3(4):89–100

Pulz O, Gross W (2004) Valuable products from biotechnology of microalgae. Appl Microbiol Biotechnol 65:635–648

Quijano G, Arcila JS, Buitrón G (2017) Microalgal-bacterial aggregates: applications and perspectives for wastewater treatment. Biotechnol Adv 35(6):772–781

Raikova S, Piccini M, Surman MK, Allen MJ, Chuck CJ (2019) Making light work of heavy metal contamination: the potential for coupling bioremediation with bioenergy production. J Chem Technol Biotechnol 94(10):3064–3072

Ramalingam G, Perumal N, Priya AK, Rajendran S (2022) A review of graphene-based semiconductors for photocatalytic degradation of pollutants in wastewater. Chemosphere 300: 134391

Rawat I, Kumar RR, Mutanda T, Bux F (2011) Dual role of microalgae: phycoremediation of domestic wastewater and biomass production for sustainable biofuels production. Appl Energy 88(10):3411–3424

Rempel A, Gutkoski JP, Nazari MT, Biolchi GN, Cavanhi VAF, Treichel H, Colla LM (2021) Current advances in microalgae-based bioremediation and other technologies for emerging contaminants treatment. Sci Total Environ 772:144918

Rezania S, Ponraj M, Din MFM, Songip AR, Sairan FM, Chelliapan S (2015) The diverse applications of water hyacinth with main focus on sustainable energy and production for new era: an overview. Renew Sust Energ Rev 41:943–954

Richmond A, Cheng-Wu Z (2001) Optimization of a flat plate glass reactor for mass production of Nannochloropsis sp. outdoors. J Biotechnol 85(3):259–269

Richmond A, Hu Q (2013) Handbook of microalgal culture: applied phycology and biotechnology. John Wiley & Sons, Hoboken

Rinquest Z, Basitere M, Mewa-Ngongang M, Ntwampe SKO, Njoya M (2019) Optimization of the COD removal efficiency for a static granular bed reactor treating poultry slaughterhouse wastewater.

Ritchie H, Roser M (2017) Water use and stress. Our World in Data.

Rossle WH (2007) Liquid and solids management impact on nutrient ratios for in-line wastewater prefermenters. University of Pretoria, Pretoria

Rothermel MC, Landis AE, Barr WJ, Soratana K, Reddington KM, Weschler MK, Witter G, Harper WF (2013) A life cycle assessment based evaluation of a coupled wastewater treatment and biofuel production paradigm. J Environ Prot 4(9):37335. https://doi.org/10.4236/jep.2013.49118

Ruiz J, Álvarez-Díaz P, Arbib Z, Garrido-Pérez C, Barragán J, Perales J (2013) Performance of a flat panel reactor in the continuous culture of microalgae in urban wastewater: prediction from a batch experiment. Bioresour Technol 127:456–463

Ruiz J, Olivieri G, De Vree J, Bosma R, Willems P, Reith JH, Eppink MH, Kleinegris DM, Wijffels RH, Barbosa MJ (2016) Towards industrial products from microalgae. Energy Environ Sci 9(10):3036–3043

Ruiz-Marin A, Mendoza-Espinosa LG, Stephenson T (2010) Growth and nutrient removal in free and immobilized green algae in batch and semi-continuous cultures treating real wastewater. Bioresour Technol 101(1):58–64

Salam KA (2019) Towards sustainable development of microalgal biosorption for treating effluents containing heavy metals. Biofuel Res J 6(2):948

Samaei SM, Gato-Trinidad S, Altaee A (2018) The application of pressure-driven ceramic membrane technology for the treatment of industrial wastewaters–a review. Sep Purif Technol 200: 198–220

Saravanan A, Kumar PS, Jeevanantham S, Karishma S, Tajsabreen B, Yaashikaa P, Reshma B (2021) Effective water/wastewater treatment methodologies for toxic pollutants removal: Processes and applications towards sustainable development. Chemosphere 280:130595

Sathya K, Nagarajan K, Malar CG, G., Rajalakshmi, S., Raja Lakshmi, P. (2022) A comprehensive review on comparison among effluent treatment methods and modern methods of treatment of industrial wastewater effluent from different sources. Appl Water Sci 12(4):70

Saxenian H, Bharali I, Ogbuoji O, Yamey G (2021) A quantitative analysis of sources of changes in government expenditures on health, 2000 to 2015: what can we learn from experience to date? Gates Open Research 3:5

Sengupta S, Nawaz T, Beaudry J (2015) Nitrogen and phosphorus recovery from wastewater. Current Pollution Rep 1:155–166

Shao W, Ebaid R, Abomohra AE-F, Shahen M (2018) Enhancement of Spirulina biomass production and cadmium biosorption using combined static magnetic field. Bioresour Technol 265: 163–169

Shen Y, Gao J, Li L (2017) Municipal wastewater treatment via co-immobilized microalgal-bacterial symbiosis: microorganism growth and nutrients removal. Bioresour Technol 243: 905–913

Shrestha R, Ban S, Devkota S, Sharma S, Joshi R, Tiwari AP, Kim HY, Joshi MK (2021) Technological trends in heavy metals removal from industrial wastewater: a review. J Environ Chem Eng 9(4):105688

Sial A, Zhang B, Zhang A, Liu K, Imtiaz SA, Yashir N (2021) Microalgal–bacterial synergistic interactions and their potential influence in wastewater treatment: a review. BioEnergy Research 14:723–738

Singh R, Sharma S (2012) Development of suitable photobioreactor for algae production–a review. Renew Sust Energ Rev 16(4):2347–2353

Singh A, Pal DB, Mohammad A, Alhazmi A, Haque S, Yoon T, Srivastava N, Gupta VK (2022) Biological remediation technologies for dyes and heavy metals in wastewater treatment: new insight. Bioresour Technol 343:126154

Song Y, Wang L, Qiang X, Gu W, Ma Z, Wang G (2022) The promising way to treat wastewater by microalgae: approaches, mechanisms, applications and challenges. J Water Process Eng 49: 103012

Sonthiphand P, Panthapa C, Mhuantong W, Charanaipayuk N, Homyok P, Limpiyakorn T (2022) Characterization of the nitrogen-transforming microbial community in the biofilms of a full-scale rotating biological contactor system treating wastewater from a fresh market building. Environ Sci: Water Res Technol 8(9):1845–1858

Spolaore P, Joannis-Cassan C, Duran E, Isambert A (2006) Commercial applications of microalgae. J Biosci Bioeng 101(2):87–96

Su Y (2021) Revisiting carbon, nitrogen, and phosphorus metabolisms in microalgae for wastewater treatment. Sci Total Environ 762:144590

Sukla LB, Pradhan D, Subbaiah T (2019) Future prospects of microalgae in wastewater treatment. In: The role of microalgae in wastewater treatment. Springer, Singapore, pp 129–135

Sun H, Zhao W, Mao X, Li Y, Wu T, Chen F (2018) High-value biomass from microalgae production platforms: strategies and progress based on carbon metabolism and energy conversion. Biotechnol Biofuels 11:1–23

Tang DYY, Khoo KS, Chew KW, Tao Y, Ho S-H, Show PL (2020) Potential utilization of bioproducts from microalgae for the quality enhancement of natural products. Bioresour Technol 304:122997

Taziki M, Ahmadzadeh H, Murry MA, Lyon SR (2015) Nitrate and nitrite removal from wastewater using algae. Current Biotechnol 4(4):426–440

Ting H, Haifeng L, Shanshan M, Zhang Y, Zhidan L, Na D (2017) Progress in microalgae cultivation photobioreactors and applications in wastewater treatment: a review. Int J Agric Biol Eng 10(1):1–29

Tóth G, Hermann T, Da Silva M, Montanarella L (2016) Heavy metals in agricultural soils of the European Union with implications for food safety. Environ Int 88:299–309

Tzanakis N, Aravantinou AF, Manariotis ID (2023) Short-term toxicity of ZnO nanoparticles on microalgae at different initial nutrient concentrations. Sustainability 15(10):7853

Udaiyappan AFM, Hasan HA, Takriff MS, Abdullah SRS (2017) A review of the potentials, challenges and current status of microalgae biomass applications in industrial wastewater treatment. J Water Process Eng 20:8–21

USEPA (2000) Wastewater technology fact sheet package plants. United States Environmental Protection Agency, Washington, DC

Valdovinos-García EM, Bravo-Sánchez MG, Olán-Acosta MdlÁ, Barajas-Fernández J, Guzmán-López A, Petriz-Prieto MA (2022) Technoeconomic evaluation of microalgae oil production: effect of cell disruption method. Fermentation 8(7):301

Varshney P, Mikulic P, Vonshak A, Beardall J, Wangikar PP (2015) Extremophilic micro-algae and their potential contribution in biotechnology. Bioresour Technol 184:363–372

Villanova V, Spetea C (2021) Mixotrophy in diatoms: Molecular mechanism and industrial potential. Physiol Plant 173(2):603–611

Viruela A, Robles Á, Durán F, Ruano MV, Barat R, Ferrer J, Seco A (2018) Performance of an outdoor membrane photobioreactor for resource recovery from anaerobically treated sewage. J Clean Prod 178:665–674

Vyavahare G, Sutar S, Gurav R, Patil R, Patil D, Jadhav J (2021) Application of biochar for the treatment of textile dyes and wastewater. In: Biochar and its application in bioremediation. Springer, Berlin, pp 169–191

Wang Y, Ho S-H, Cheng C-L, Guo W-Q, Nagarajan D, Ren N-Q, Lee D-J, Chang J-S (2016) Perspectives on the feasibility of using microalgae for industrial wastewater treatment. Bioresour Technol 222:485–497

Wang Q, Wei W, Gong Y, Yu Q, Li Q, Sun J, Yuan Z (2017) Technologies for reducing sludge production in wastewater treatment plants: State of the art. Sci Total Environ 587:510–521

Wang S, Zhao S, Uzoejinwa BB, Zheng A, Wang Q, Huang J, Abomohra AE-F (2020) A state-of-the-art review on dual purpose seaweeds utilization for wastewater treatment and crude bio-oil production. Energy Convers Manag 222:113253

Wang Q, Li H, Shen Q, Wang J, Chen X, Zhang Z, Lei Z, Yuan T, Shimizu K, Liu Y (2022) Biogranulation process facilitates cost-efficient resources recovery from microalgae-based wastewater treatment systems and the creation of a circular bioeconomy. Sci Total Environ 828:154471

Witek-Krowiak A, Gorazda K, Szopa D, Trzaska K, Moustakas K, Chojnacka K (2022) Phosphorus recovery from wastewater and bio-based waste: an overview. Bioengineered 13(5): 13474–13506

Wollmann F, Dietze S, Ackermann JU, Bley T, Walther T, Steingroewer J, Krujatz F (2019) Microalgae wastewater treatment: Biological and technological approaches. Eng Life Sci 19(12):860–871

Worku A, Sahu O (2014) Reduction of heavy metal and hardness from ground water by algae. J Appl Environ Microbiol 2(3):86–89

Xu L, Weathers PJ, Xiong XR, Liu CZ (2009) Microalgal bioreactors: challenges and opportunities. Eng Life Sci 9(3):178–189

Xu X, Teng Y, Yang Y, Wang G, Zhang S (2019) Washing removal of Pb and Cd from soil using four plant water extracts. J Agro-Environ Sci 38(8):1954–1962

Yadav G, Sen R (2018) Sustainability of microalgal biorefinery: scope, challenges, and opportunities. In: Sustainable energy technology and policies: a transformational journey, vol 1. Springer, Singapore, pp 335–351

Yamashita T, Yamamoto-Ikemoto R (2014) Nitrogen and phosphorus removal from wastewater treatment plant effluent via bacterial sulfate reduction in an anoxic bioreactor packed with wood and iron. Int J Environ Res Public Health 11(9):9835–9853

Yan C, Zhang Q, Xue S, Sun Z, Wu X, Wang Z, Lu Y, Cong W (2016) A novel low-cost thin-film flat plate photobioreactor for microalgae cultivation. Biotechnol Bioprocess Eng 21:103–109

Yao S, Lyu S, An Y, Lu J, Gjermansen C, Schramm A (2019) Microalgae–bacteria symbiosis in microalgal growth and biofuel production: a review. J Appl Microbiol 126(2):359–368

Yarkent Ç, Gürlek C, Oncel SS (2020) Potential of microalgal compounds in trending natural cosmetics: a review. Sustain Chem Pharm 17:100304

Ye J, Zhou Q, Zhang X, Hu Q (2018) Microalgal dewatering using a polyamide thin film composite forward osmosis membrane and fouling mitigation. Algal Res 31:421–429

Ye Y, Ngo HH, Guo W, Chang SW, Nguyen DD, Zhang X, Zhang J, Liang S (2020) Nutrient recovery from wastewater: From technology to economy. Bioresour Technol Rep 11:100425

Yong JJJY, Chew KW, Khoo KS, Show PL, Chang J-S (2021) Prospects and development of algal-bacterial biotechnology in environmental management and protection. Biotechnol Adv 47: 107684

Zeng X, Guo X, Su G, Danquah MK, Zhang S, Lu Y, Sun Y, Lin L (2015) Bioprocess considerations for microalgal-based wastewater treatment and biomass production. Renew Sust Energ Rev 42:1385–1392

Zewdie DT, Ali AY (2022) Techno-economic analysis of microalgal biofuel production coupled with sugarcane processing factories. S Afr J Chem Eng 40(1):70–79

Zhang M, Wang X, Yang L, Chu Y (2019) Research on progress in combined remediation technologies of heavy metal polluted sediment. Int J Environ Res Public Health 16(24):5098

Zhang C, Li S, Ho S-H (2021) Converting nitrogen and phosphorus wastewater into bioenergy using microalgae-bacteria consortia: a critical review. Bioresour Technol 342:126056

Zhu L, He H, Wang C (2016) COD removal efficiency and mechanism of HMBR in high volumetric loading for ship domestic sewage treatment. Water Sci Technol 74(7):1509–1517

Zhu N, Zhang J, Tang J, Zhu Y, Wu Y (2018) Arsenic removal by periphytic biofilm and its application combined with biochar. Bioresour Technol 248:49–55

Zittelli GC, Biondi N, Rodolfi L, Tredici MR (2013) Photobioreactors for mass production of microalgae. In: Handbook of microalgal culture: applied phycology and biotechnology. John Wiley & Sons, Ltd, pp 225–266

High-Throughput Screening to Accelerate Microalgae-Based Phycochemical Production

Juliane Wolf, Robert Chapman, Charu Deepika, Mélanie Pietri, Sakina Bensalem, and Ben Hankamer

Abstract Microalgae biotechnologies tap into the huge energy resource of the sun, absorb CO_2, and are being developed to provide economic solar-driven solutions spanning the production of phytochemicals, foods, biopolymers, renewable fuels and services delivering clean water and bioremediation capabilities. Further, microalgae open up a suite of high-value opportunities in the nutraceutical and pharmaceutical sectors (small molecules and recombinant proteins). Efficient photosynthetic biomass production (upstream processing) represents the first step of the microbiological manufacturing process and requires a systematic optimisation approach to improve resource use efficiency (light, nutrients, CO_2) for individual production strains and growth systems. This chapter looks at general challenges and opportunities in the field of microalgal biotechnology before reviewing how high-throughput screening (HTS) methods can be implemented to improve the understanding and control of critical process parameters. Innovation pathways (miniaturisation, automation), latest advances and bottlenecks (including experimental design and data management), and an outlook of microalgae cell-based HTS assays are discussed for their potential to accelerate the development of new production species and cell lines for the production of molecules of added value. Based on specific application examples for the optimisation of abiotic growth factors, the possibilities of this technology will be presented. HTS data can help to de-risk production scale-up and assist technoeconomic process modelling to accelerate microalgae systems development, technology translation, and commercialisation. Furthermore, HTS can be a useful tool to determine social and environmental impacts of product-specific production processes and help to build resource databases required for technology integration into a circular economy.

J. Wolf (✉) · R. Chapman · C. Deepika · B. Hankamer
Institute for Molecular Bioscience, The University of Queensland, Brisbane, Australia
e-mail: j.wolf@uq.edu.au

M. Pietri · S. Bensalem
Université Paris Saclay, ENS Paris Saclay, CNRS, Institut d'Alembert, LUMIN, Gif-sur-Yvette, France

A. Abomohra, S. Ende (eds.), *Value-added Products from Algae*,
https://doi.org/10.1007/978-3-031-42026-9_10

Keywords Microalgae · Biotechnology · Production scale-up · Process modelling

1 Introduction

As discussed in previous chapters, microalgae have emerged as a promising platform for the production of a wide range of valuable compounds, including biofuels, pharmaceuticals, and nutritional supplements. However, the optimisation of microalgae-based production processes is often challenging, requiring extensive experimentation and optimisation to achieve high yields and productivity. This chapter explores the potential of scale-down approaches for microalgae-based high-throughput screening (HTS). While HTS may also be applicable for macroalgae optimisation, most developments have been made for microalgae-based assays (including cyanobacteria), which is the focus here.

The historical development of solar bio-manufacturing in the context of global challenges and planetary boundaries is reviewed, going from domestication of plants for agriculture (green revolution) to biotechnology and drug discovery supported by automation and machine-learning technologies. The current state of microalgae-based production is briefly introduced, and the challenges/limitations of traditional optimisation approaches are discussed. Examples of scale-down approaches for microalgae-based HTS are also summarised, with a focus on microplate-based assays and microfluidic systems. The advantages and limitations of each approach are summarised, with providing some examples of their successful application in microalgae-based HTS. Finally, the areas for further research and development in the field of microalgae-based HTS using scale-down approaches are highlighted.

1.1 Addressing Planetary Boundaries and Sustainable Development Goals: A Path to Net-Zero Emissions

The international community is facing combined challenges of operating within the planetary boundaries, meeting the United Nations (UN) sustainable development goals (SDGs) 2030 targets, and delivering net-zero emissions by 2050. The cost of delivering CO_2 neutrality alone is estimated to be US$ 130 trillion by 2050 or US$ 4.8 trillion per year (Irena 2022; Mekala et al. 2022). In comparison, global gross domestic product (GDP) in 2022 was ~US$ 100 trillion (Hankamer et al. 2023). Yet, failure to solve these challenges risks food and water security, as well as human health, which combined significantly impact economic, social, political, and climate stability.

Around 48% of the global population in 2019 lives in countries emitting on average more than 6 tCO_2-eq per capita, 35% of the global population live in countries emitting more than 9 tCO_2-eq per capita while another 41% live in countries emitting less than 3 tCO_2-eq per capita (IPCC 2023). Vulnerable communities who historically contributed the least to climate change are disproportionally affected by its impacts (IPCC 2023). Impacts include exposure to acute food insecurity, reduced water security, as well as deaths as a result of floods, droughts, or storms.

Emission reductions in CO_2 from fossil fuels and industrial processes, due to improvements in energy intensity of GDP and carbon intensity of energy, have been less than emission increases due to rising global activity levels in industry, energy supply, transport, agriculture, and buildings (Fig. 1). The 10% of households with the highest per capita emissions contribute 34–45% of global consumption-based household greenhouse gas (GHG) emissions, while the middle 40% contribute 40–53%, and the bottom 50% contribute 13–15% (IPCC 2023). The 2023 IPPC report categorises four key challenges: (1) cut emissions quickly, sharply to create a safer, sustainable world; (2) scale-up practices and infrastructure to enhance resilience; (3) cut the global GHG emissions by nearly half by 2030; and (4) action required along numerous dimensions. Beyond this, CO_2 capture technologies are required to begin the long process of reducing atmospheric and oceanic CO_2 levels.

The current 'linear economies' seem incapable of handling the rapid emergence of global shocks, necessitating the development of robust 'circular economy' solutions (Hankamer et al. 2023). Circular bioeconomy processes are more complex, but promise to provide sustainable and scalable opportunities, integrating high specialisation to deliver implementable, productive, and profitable solutions. However, the high failure rates in translating good science to high-impact scalable industrial solutions must be overcome. The combination of high cost and tight timelines necessitate the use of HTS tools to expedite systems optimisation, de-risk scale-up, and deliver robust business cases for the microalgae science community. Similarly, a mass transition from nondegradable oil-based to degradable plant-based bioplastics (Chapter "Algal-based Biopolymers"), as well as from fossil to renewable fuels (Chapter "Biofuel-Integrated Routes") and direct light-driven industry (Chapter "Diatom Nanostructured Biosilica"), can be made with a similar case.

1.1.1 Decarbonising the Economy: Diversifying Solutions across Sectors

The unit costs of several low-emission technologies, including solar, wind, and lithium-ion batteries, have fallen consistently since 2010. Design and process innovations in combination with the use of digital technologies have led to near-commercial availability of many low- or zero-emission options in buildings, transport, and industry. For almost all basic materials—primary metals, building materials and chemicals—many low- to zero-GHG intensity production processes are at the pilot to near-commercial and in some cases commercial stage, but they are not yet an established industrial practice. Integrated design in construction and retrofit of buildings has led to increasing examples of zero-energy or zero-carbon buildings (e.g. technological innovation enabled the widespread adoption of low-energy LED lighting). Digital technologies including sensors, the Internet of things (IoT), robotics, and artificial intelligence (AI) have further improved energy management in all sectors as they can increase the energy efficiency and promote the adoption of many

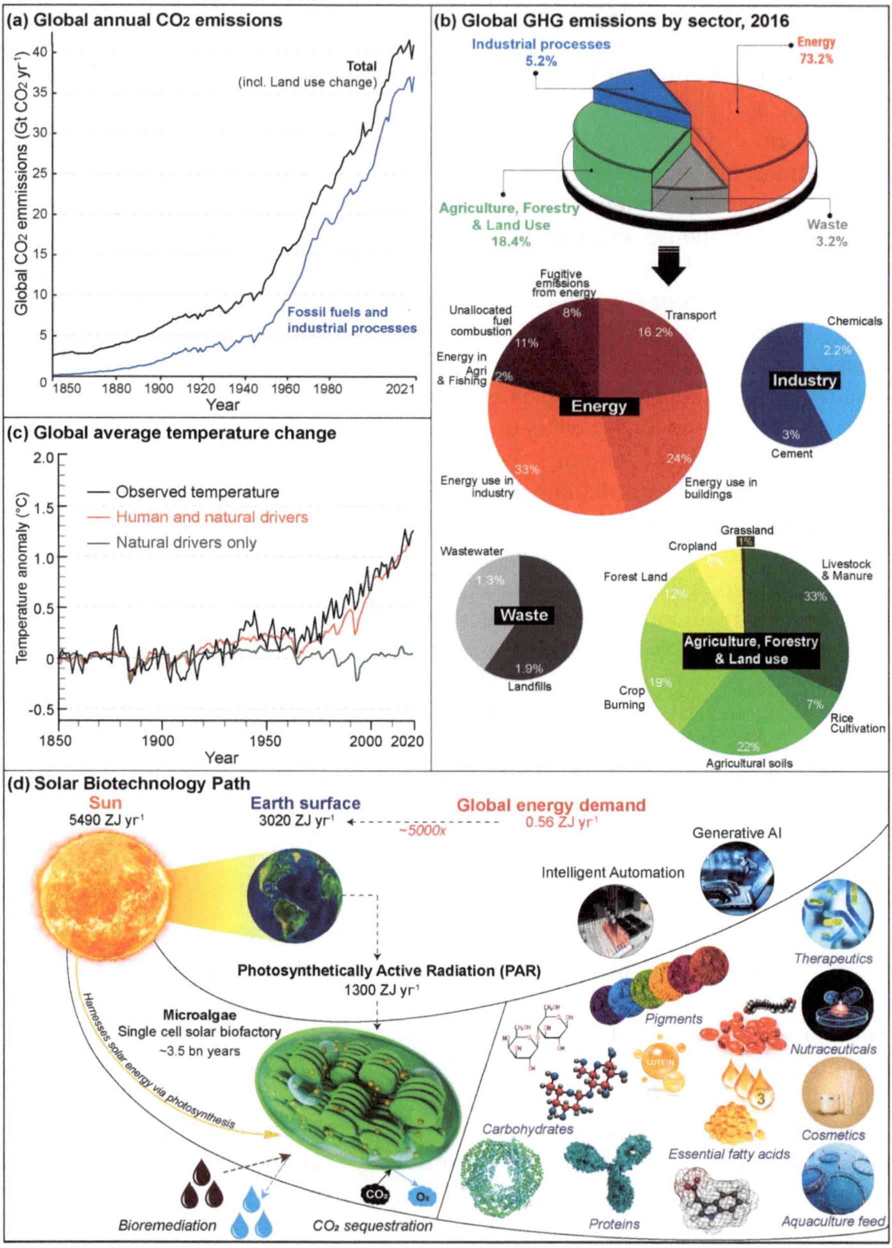

Fig. 1 Global development in GHG emissions and the solar biotechnology path to net zero. (**a**) Annual global net anthropogenic CO_2 emissions since 1850 from fossil fuel combustion and industrial processes (CO_2-FFI) (dark blue); and total (black) which includes net CO_2 emissions from land use, land-use change, and forestry (CO_2- LULUCF); (**b**) global GHG emissions in 2016 by sector (waste, reported as CO_2-equivalent including CH_4; N_2O; and fluorinated gases (HFCs, PFCs, SF_6, NF_3); (**c**) global average temperature change since 1850; (**d**) schematic of the solar

low-emission technologies including decentralised renewable energy. Most importantly innovations create economic opportunities. However, some of these climate change mitigation gains may be reduced or counterbalanced by the growing demand for digital goods and services. Several mitigation options are today technically viable and are becoming increasingly cost-effective, notably solar energy, wind energy, electrification of urban systems, urban green infrastructure, energy efficiency, energy demand management, improved forest- and crop/grassland management, and reduced food waste and loss. Cost-effective mitigation strategies are generally supported by the public, which enables expanded deployment in many regions. GHG emissions come from many sectors (Fig. 1b) (Gütschow et al. 2016; Ritchie 2021; Taylor 2021) stressing that many solutions are needed to decarbonise the economy. To cut emissions quickly, proven resilient infrastructure and practices must be scaled up and implemented across all areas of human life.

1.1.2 IPCC Report 2023: The Power of Adaptation Strategies

Adaptation can generate multiple additional benefits such as improving agricultural productivity, innovation, health and well-being, food security, livelihood, and biodiversity conservation as well as reduction of risks and damages.

Sector-Based Adaptation Strategies: In agriculture, improvements in cultivars, water management, soil conservation, and sustainable land management offer multiple benefits and reduce climate risks. Measures like reducing food loss and waste and promoting balanced diets contribute to nutrition, health, and biodiversity. Ecosystem-based adaptation approaches such as urban greening and wetland restoration mitigate flood risks and provide various co-benefits. Some land-based adaptation options yield immediate benefits, while others, like afforestation and soil restoration, take time. Adaptation and mitigation are synergistic, particularly through sustainable land management. Agroecological principles and practices support food security, livelihoods, biodiversity, and ecosystem services.

Fresh Water: Adaptation to water-related risks and impacts make up the majority (~60%) of all documented adaptation. A large number of these adaptation responses are in the agriculture sector, and these include on-farm water management, water storage, soil moisture conservation, and irrigation. Other adaptations in agriculture include cultivar improvements, agroforestry, community-based adaptation, farm, and landscape diversification. For inland flooding, combinations of non-structural measures like early warning systems, enhancing natural water retention such as by

Fig. 1 (continued) biotechnology path to net zero. Data for panels are a-c retrieved from https://ourworldindata.org/emissions-by-sector (Radiation data from Ringsmuth et al. 2016)

restoring wetlands and rivers, and land-use planning such as no build zones or upstream forest management, can reduce flood risk.

Land and Biodiversity: Some land-related adaptation actions that can have mitigation co-benefits include sustainable food production, improved and sustainable forest management, soil organic carbon management, ecosystem conservation, land restoration, reduced deforestation and degradation, and waste reduction. Adaptation actions that increase the resilience of biodiversity and ecosystem services include minimising additional stresses or disturbances, reducing fragmentation by connecting natural habitat, and increase heterogeneity of protected small-scale refugia where micro-climate conditions can allow species to persist.

Urban: Most innovations in urban adaptation have occurred through advances in disaster risk management, social safety nets, and green/blue infrastructure. Many adaptation measures that benefit health and well-being are found in other sectors (e.g. food, livelihoods, social protection, water, sanitation, and infrastructure).

1.1.3 From Planning to Implementation: Overcoming Fragmented Adaptation Efforts

The current state of adaptation efforts reveals a fragmented, small-scale, and incremental approach that focuses more on planning than actual implementation. Moreover, these observed adaptation measures are unevenly distributed across regions, with the greatest gaps found among lower-income populations. In urban areas, complex risk management projects, such as those addressing the interconnections of food, energy, water, and health, face significant adaptation challenges (IPCC 2023). Insufficient funding, knowledge gaps, and limited practices hinder effective implementation, monitoring, and evaluation, raising concerns about meeting existing adaptation goals. Without significant improvements in adaptation planning and implementation, the adaptation gap will continue to widen.

Examples of maladaptation have been reported in the 2023 IPCC report (IPCC 2023) in various sectors as follows: In urban areas, the development of inflexible and unaffordable urban infrastructure fails to address changing needs. In agriculture, the use of high-cost irrigation in regions projected to face intensified drought conditions proves counterproductive. Ecosystems suffer from misguided practices like fire suppression in naturally fire-adapted ecosystems or reliance on hard defences against flooding. Human settlements experience the detrimental effects of stranded assets and vulnerable communities unable to afford relocation or adaptation, necessitating increased social safety nets. These maladaptations disproportionately impact marginalised and vulnerable groups, including indigenous peoples, ethnic minorities, low-income households, and those living in informal settlements, perpetuating existing inequities. Avoiding maladaptation requires a flexible, multi-sectoral, inclusive, and long-term approach to planning and implementing adaptation actions that benefit various sectors and systems.

2 Microalgae: A Solar Biotechnology Platform Supporting the Path to Net Zero

Microalgal technology is deemed pivotal to deliver climate-resilient solutions efficiently with their inherent ability to capture CO_2, drive complex biochemistry, and utilise light (Sørensen et al. 2022). The photosynthetic processes sustain life on Earth by providing food, feed (Yarnold et al. 2019), fuel (Roles et al. 2021), biomaterials (Karan et al. 2019; Ross et al. 2021), chemicals (Sørensen et al. 2022), biofertilisers (Rupawalla et al. 2021), and other products, including pigment (Deepika et al. 2022c), antioxidants (Gauthier et al. 2020), biostimulants (González-Pérez et al. 2022, Rupawalla et al. 2022), vitamins (Fabregas and Herrero 1990), and polyunsaturated fatty acids (Remize et al. 2021) (Fig. 1d). As discussed in different chapters of this book in more detail, microalgae have the potential to support the path to a net-zero economy and contribute to a circular bioeconomy by:

- *Phycoremediation*: Microalgae can be used to absorb nutrients (e.g. N & P) and other elements (e.g. Cd) to clean up contaminated water, air, and soil, reducing waste and its environmental impacts.
- *Renewable energy*: Microalgae can be used as a feedstock for biofuel production, providing a renewable source of energy.
- *Food and feed*: Microalgae can be used as a source of food and animal/aquaculture feed, reducing dependence on traditional crops and potentially lowering the environmental impact of agriculture. Microalgae-based fertiliser can reduce the environmental impact of agriculture further (Khavari et al. 2021; Rizwan et al. 2018).
- *Bioplastics*: Microalgae can be used as a feedstock for bioplastic production, reducing the need for petroleum-based plastics and promoting a circular economy.
- *Nutritional supplements*: Microalgae can be rich in nutrients and so are used as a natural source of vitamins and minerals, reducing the need for synthetic supplements. Their valuable secondary metabolites include pigments/antioxidants, vitamins, and medicines.
- *Therapeutics and medicines*: Microalgae are capable of producing bioactive small molecules as well as recombinant proteins including monoclonal antibodies, subunit vaccines, hormones, and enzymes, with diagnostic, pharmaceutical and therapeutic applications (Khavari et al. 2021; Rizwan et al. 2018). Microalgae pigments are also reported to possess therapeutic properties such as anti-inflammatory and anti-tumour activities (Eggersdorfer and Wyss 2018; Sumantran et al. 2000).

Additionally, these processes provide ecosystem services, such as the production of atmospheric oxygen and water purification. However, unlocking this immense potential value requires the development of viable commercial processes that deliver economic, social, and environmental benefits. Efficient harnessing of this capability demands accelerated technological innovations highlighting the importance of HTS approaches which together with intelligent automation and generative artificial intelligence approaches offer to fast-track new robust circular economy solutions (Fig. 1d).

2.1 Algae: From Traditional Use to Modern Applications

The synthesis of new biomass from inorganic sources (primary production) is mostly achieved through photosynthesis, only limited by the photosynthetic active radiation (PAR) accounting for ~43% of incident solar energy (350–700 nm) (Kruse et al. 2005; Ringsmuth et al. 2016). Despite representing only 1–2% of the Earth's biomass, microalgae are estimated to be responsible for about 40% of the Earth's photosynthesis (Falkowski 1994), indicating a higher overall photosynthetic efficiency compared to terrestrial plants, providing the net primary production (NPP) representing the resource of 'photosynthates' available to other organisms (Haberl et al. 2007). Some microalgae have extremely high growth rates with a biomass doubling time of 24 hours (Chisti 2007) or less (Wolf et al. 2015). While algae have been used for food, fuel, and medicine in some regions, particularly in Asian cultures, for centuries, modern research into algae is relatively young and started only in the nineteenth century, with the discovery of photosynthesis and the role of algae in producing oxygen (Calvin 1989; Kandler 1950; Powar et al. 2022; Preisig and Andersen 2005,). This led to a better understanding of the biology and potential applications of algae, for example showing that microalgae are an excellent source of proteins (~50% of BDW (Amorim et al. 2021) and lipids (70% (Metting 1996).

Comparatively, domestication of plants and animals started about 10,000 years ago (Özbek 2022), which could be marked as the beginning of agriculture and allowed humans to settle in one place and rely on agriculture for their food supply (Harari 2014).

Mass cultivation of algae for industrial purposes started in the 1950s (Dawiec-Liśniewska et al. 2022), primarily for the production of alginates from seaweeds (Deepika et al. 2022a, 2022b; Tanaka et al. 2022). Algae cultivation has also been used in aquaculture, as a source of food for fish and other aquatic animals (Mishra et al. 2022) and for water treatment to remove excess nutrients and/or microplastic (Abomohra and Hanelt 2022). The use of algae as a source of biofuels has been a major area of research and development since the 1980s, with the potential to provide a sustainable and renewable source of energy (Rodolfi et al. 2009; Brányiková et al. 2011; Schenk et al. 2008). To date, algae are being used to produce high-value non-commodity products (phycochemicals) such as nutritional supplements, cosmetics, and bioplastics, which have a range of applications in food, pharmaceutical, and industrial sectors (Fig. 1d). Advances in cultivation technology, such as closed photobioreactors systems, have improved the efficiency, scalability, and quality control of microalgae cultivation, allowing for more cost-effective and sustainable production.

2.2 *Integrated Solar Biomanufacturing Using Microalgae*

Water-based production systems exist in the form of hydroponics in the agricultural sector or in the form of fermentation in the food and biotechnology sectors. Thus, microalgae production system could be described as an integrated technology of agriculture (solar driven, land plants converting solar to biochemical energy) and fermentation (biochemical driven, microorganisms use energy to improve product shelf-life and nutritional value). Each field has developed in parallel over many centuries where key innovations (revolutions) have accelerated technology deployment by overcoming major bottlenecks at the time that limited product yields, scalability, economic viability, or a mixture thereof. All these technologies, as well as new cross-sector innovations and significant investment, will be required to build a solar biomanufacturing industry on the path to net-zero emissions.

2.2.1 Key Innovations for Enhanced Biomass Production in Agriculture

In traditional agriculture, several techniques and methods have significantly advanced crop production. Irrigation and ploughing were developed around 4000 BCE, which allowed for more efficient cultivation of crops and increased yields. Crop rotation and fertilisation techniques were developed in ancient civilisations such as the Greeks and Romans. Crop rotation helped to maintain soil fertility, while fertilisation improved crop yields. The so-called agricultural revolution started in the eighteenth century with the development of new technologies such as seed drills, which increased agricultural productivity and allowed for more efficient use of land. Industrialisation of agriculture was further revolutionised by the development of chemical fertilisers (Haber Bosch (Ertl 2012)), pesticides, and genetically modified crops in the twentieth century. This allowed for large-scale cultivation of crops for food production, but also raised concerns about environmental (Fig. 1a–c) and health impacts. In the late twentieth century, concerns about the environmental impact of industrial agriculture led to the development of sustainable agriculture practices, which aim to balance economic viability, environmental impact, and social responsibility. Recent advances in technology, such as GPS, drones, and big data analytics, have enabled precision agriculture, which allows farmers to optimise their crop yields and minimise the environmental impact through precise application of inputs like water, fertilisers, and pesticides.

2.2.2 Key Innovations for Enhanced Biomass Production in Fermentation

Fermentation has also been used by humans for thousands of years to produce food (e.g. cheese) and drinks (e.g. wine and beer). The discovery of microorganisms (Antony van Leeuwenhoek) in the seventeenth century and understanding their contributions in fermentation in the nineteenth century by Louis Pasteur was the basis for the industrialisation of the fermentation process in late nineteenth and early twentieth century (Porter 1976; Robertson et al. 2016). This was the beginning of quality control and prevention of contamination (pasteurisation) that enabled large-scale production of fermented products like beer, wine, and cheese, which was further revolutionised through the discovery of antibiotics. Developments in biotechnology and genetic engineering in the late twentieth century expanded the range of fermentation products, including pharmaceuticals, enzymes, and biofuels (e.g. ethanol and biogas).

2.3 *Accelerate Discovery with High-Throughput Screening*

High-throughput screening (HTS) techniques have the potential to accelerate the optimisation process and improve the efficiency and effectiveness of microalgae-based production both on the upstream and downstream processing side. Traditional optimisation approaches involve the screening of large numbers of samples, which even when done in parallel is costly and time-consuming. As a result, there is a growing interest in the development of scale-down approaches for microalgae-based HTS, which can accelerate the optimisation process and reduce the costs while maintaining the biological relevance of the screening results. Scale-down approaches for cell-based assays involve the miniaturisation of cultivation systems, allowing the screening of multiple conditions in parallel. These approaches can include microscale photobioreactors (Bashir et al. 2022; Mehta and Rath 2021), microfluidic systems (Bashir et al. 2022), and microplate-based assays (Radzun et al. 2015; Wolf et al. 2015; Sharma et al. 2022; Sinha et al. 2022).

Despite high-throughput methods and microalgae each attracting increasing research attention, the amount of work applying high-throughput methods to the optimisation of microalgal production remains disproportionately small compared to other applications of high-throughput methods (Fig. 3).

2.3.1 From Compound Screening to Industrial Control: The Future of High-Throughput Screening

High-throughput screening (HTS) is a powerful and automated method that was first used in the discovery of new drugs and therapeutic agents (Fig. 2) (Szymański et al. 2011). HTS aims to quickly test large numbers of compounds or biological entities against a particular target or cellular process to identify the ones with the desired activity (Pereira and Williams 2007). HTS has been used for drug discovery, allowing scientists to screen millions of compounds in a short period of time and rapidly identify potential lead compounds for further optimisation and evaluation. Rapid advancements in molecular biology (assay technologies), equipment, automation, and information technology have all contributed to the rapid growth of HTS

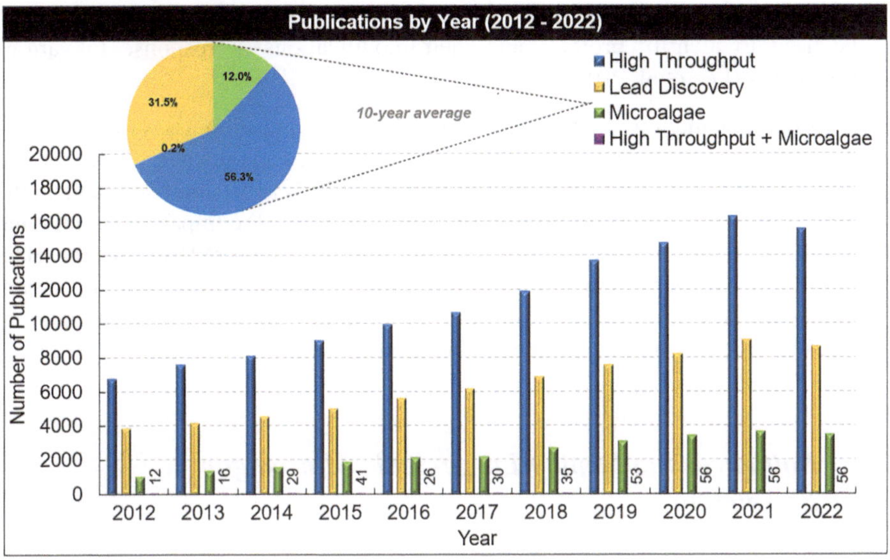

Fig. 2 Publications per year on the topics of "high throughput", "lead discovery", "microalgae", and the combination "high throughput + microalgae" from 2012 to 2022. Data sourced from Web of Science

(Cho et al. 2011; Silva et al. 2022). Many analytical techniques can be incorporated into HTS techniques (e.g. fluorescence-assisted cell sorting (FACS), gas chromatography, gas chromatography–mass spectrometry, high-performance liquid chromatography) if they are capable of processing a large number of samples. Single-cell analytics use nano-tube devices to isolate single cells to characterise them further (e.g. FACS, microfluidics, confocal microscopy, transmission Electron Microscopy). Common HTS techniques include:

- *Cell-based assays*, which involve growing cells in multiwell plates or microfluidic chambers and treating them with various compounds to observe the effects on cellular processes, such as cell growth, cell death, or protein expression (An and Tolliday 2010).
- *Enzyme-linked assays*, which measure the activity of a particular enzyme, such as kinases or proteases, by linking the activity to a change in fluorescence or colour (Lee et al. 2013a).
- *High-content imaging*, which is a type of cell-based assay that uses automated microscopy and image analysis to analyse the morphological changes in cells treated with various compounds (Capus et al. 2016).
- *Luminescence and fluorescence-based assays*, which measure the light emitted by cells or proteins in response to a stimulus, such as changes in pH, ion concentration, or the addition of a specific molecule (Chiaraviglio and Kirby 2014).

- *Mass spectrometry-based assays*, which measure the mass of molecules and can be used to identify proteins and their modifications in response to various treatments (Lee et al. 2013b).

HTS implementation into cell-based production optimisation assays is less common, likely because it is more complex to setup (i.e. time-resolved biomass and compound screening). The automated cultivation of cells at miniaturised scale in the device comes with its own challenges (see Sects. 2.3 and 2.4) compared to traditional analytics of small samples/single cells that have been cultured outside the HTS platform. Finally, the integration of HTS platforms into industrial production stream processes would enable greater control of yields and quality, as well as the integration of machine-learning techniques in the future.

2.4 Bottlenecks to Deploy Large-Scale Microalgae Production

Overcoming the challenges of microalgae production and developing cost-effective systems play a crucial role in realising the full potential of microalgae in a circular economy. Challenges to deploying microalgae production more widely include cost, scalability, variability, harvesting and processing, as well as refinement, also referred to as 'upstream' and 'downstream processing':

- *Cost:* Microalgae cultivation and processing is currently more expensive compared to traditional crops. High energy and resource costs of growing microalgae make it difficult to compete with traditional agricultural crops.
- *Scalability:* The scalability of microalgae production systems is still a major challenge. Large-scale production systems are not yet economically viable, and smaller-scale systems are not yet efficient enough to meet demand.
- *Variability:* Microalgae species can vary significantly in their growth rates, nutritional value, and response to environmental conditions, making it challenging to develop and maintain consistent production processes.
- *Harvesting and Processing:* Harvesting and processing of microalgae can be difficult and expensive, making it challenging to develop cost-effective systems for large-scale production.
- *Refinement:* Conversion to products with the desired level of purity and valorisation of all components extracted from the microalgae biomass are part of the approaches that must be considered to design economically feasible production processes.

Relative to their importance as a transformational platform technology to a circular economy, microalgae are under-studied. Estimates of the number of microalgae species vary from 72,500 (Guiry 2012) to more than 200,000 (Norton et al. 1996), of which approximately 50,000 are currently described in the AlgaeBase database (Guiry and Guiry 2023). Despite their productive potential, only ~40 species of microalgae are under mass cultivation as of 2013 (Richmond and Hu 2013). A key factor to advance the field is finding new productive strains and optimising existing ones. The rapid advances in sequencing, bioinformatics, and genetic engineering have opened up exciting new avenues of research. Some groups are working toward turning microalgae into multi-purpose recombinant protein producers (Rasala and Mayfield 2015; Pacheco et al. 2018). Existing biotechnology approaches tend to be 'top-down', seeking to find and optimise the solutions already developed in nature. The simplicity of microalgae, combined with the depth of knowledge acquired about them and the quality of tools available to researchers, point toward 'bottom-up' synthetic biology approaches becoming viable, allowing fully customised production of desired metabolites (Hlavova et al. 2015). Generally speaking, whether traditional agricultural products or novel products, there is a microalgal strain best suited to the task, along with an optimal set of nutritional and environmental conditions maximisingthat strain's productivity. With tens or possibly hundreds of thousands of yet undiscovered microalgal species, it is incumbent upon researchers to find and characterise new strains and continue the optimisation of known strains.

2.4.1 Developing Scalable Commercial Microalgae Production Systems

To optimise process and design parameters, integrated techno-economic and life-cycle analysis (TELCA) (Roles et al. 2020) can help to simultaneously define the economic (e.g. \$ unit-product^{-1}), social (e.g. jobs or energy efficiency of the industrial facility), and environmental (e.g. greenhouse gas emissions reductions) benefits of an industrial scale facility (e.g. 1–500 ha). Sensitivity analyses of the biomass or compound production processes identify the most important variables (Karan et al. 2023). The adoption of high-performance program languages (e.g. Phython, R, Rust, C++, SQL, Julia) allows for the integration of machine-learning, parallelisation, and data mining techniques bringing decision support systems (DSS) to the next level.

Roles et al. (2021) have simulated complete facilities with their TELCA platform including a 500 ha renewable fuel processes (Roles et al. 2021), multiproduct biorefineries (Karan et al. 2023) through to 1 ha high-value recombinant protein production facilities (Gomez and Andres 2023). This simulation-guided design approach can also be used for process optimisation work including atomic resolution cryo-electron microscopy (Guo et al. 2020; Hankamer et al. 2005; Enami et al. 2008), high-throughput robotic bioprocess optimisation screenings (Radzun et al. 2015; Wolf et al. 2015; Tillich et al. 2014; Rohe et al. 2012), sequential CRISPR gene-editing processes to develop next-generation cell lines with improved light-capture efficiency (Luo et al. 2017; Oey et al. 2016), light-driven recombinant protein (Carrera-Pacheco et al. 2023), and small molecule synthesis (Shah et al. 2022). These combined with pilot-scale trials (Wolf et al. 2016) can fast-track systems optimisation, de-risk scaling-up, and develop robust business cases for the chosen products and services. Furthermore, developing scalable microalgae production systems that are both efficient and economically viable requires a multidisciplinary approach. Some steps that can be taken to accelerate commercialisation include:

- *Research and development:* Investing in research and development to improve the growth rates of microalgae and yields of certain phycochemicals, as well as developing new and more efficient cultivation, harvesting, and processing technologies.
- *Partnership and collaboration:* Encouraging partnerships and collaboration between researchers, industry, and government to leverage resources and expertise and accelerate commercialisation.
- *Pilot projects and demonstration plants:* Developing pilot projects and demonstration plants to validate new technologies and help to build a track record of commercial success.
- *Standardisation and regulation:* Developing industry standards and regulations to promote consistency and quality in microalgae production and processing, as well as to support the development of a sustainable and competitive market.
- *Government incentives and funding:* Encouraging government support for microalgae research and development through funding, tax credits, and other incentives, as well as working to reduce regulatory barriers that may hinder commercialisation.
- *Education and outreach:* Raising awareness of the potential of microalgae and the benefits of circular economy through education and outreach programs, as well as promoting the development of a skilled workforce through training and education initiatives.
- *Investment:* Attracting investment and financing to support the growth and commercialisation of microalgae cultivation systems, including through venture capital, private equity, and public funding.

2.4.2 Enhancing Downstream Processing with Single-Cell Analysis

Traditional techniques used in microalgal research for optimising downstream processing suffer from limitations, including their bulky, labour-intensive, time-consuming nature, and low throughput. However, to achieve economically viable microalgae-based bioproducts, substantial improvements are necessary in the fields of microalgal biology, strain development, and downstream processing. In particular, innovative solutions are needed to reduce energy consumption and enhance the economic viability of the pre-treatment step involved in downstream processes, which aims to extract valuable phytochemicals from microalgae. Among the commonly used cell disruption technologies for microalgae, bead milling and high-pressure homogenisation (HPH) stand out, but both are highly energy-intensive (Günerken et al. 2015; Jaeschke et al. 2019; Scherer et al. 2019). As alternatives, mechanical/physical methods like pulsed electric fields (PEFs) have demonstrated significant promise in extracting intracellular lipids, proteins, and carbohydrates

(Martínez et al. 2020). However, for certain microalgae species, PEF treatment proves ineffective due to the cell wall's structure, which hinders permeability and cell lysis (Stirke et al. 2019). The cell's structural heterogeneity and variations in growth and compound productivity rates have sparked increased interest in individual cell studies. Several approaches for single-cell analysis have been established, including atomic force microscopy (AFM), micropipette aspiration, and optical stretching, although these methods are limited in their analysis rates, typically handling less than 100 cells per hour (Fregin et al. 2019). To overcome the limitations of the aforementioned techniques in microalgal research, the development of high-throughput methods is crucial. These methods enable the rapid study of a substantial number of microalgal samples and facilitate the examination of various species and conditions. This advancement plays a pivotal role in gaining new insights into microalgal process optimisation and holds significant importance for the future of microalgal research.

3 Potential of HTS in Microalgae Research

HTS approaches have been used in microalgae research to rapidly evaluate large numbers of microalgal strains and conditions to screen for specific traits of interest, such as growth rate (Radzun et al. 2015; Wolf et al. 2015), lipid content (Bensalem et al. 2018a; Bensalem et al. 2018b), and tolerance to environmental stressors (Sivakaminathan et al. 2018). Identification of microalgal strains that grow quickly and produce high levels of lipids can be used to identify renewable fuel or omega-3-producing candidates. Here, HTS techniques such as cell-based assays, fluorescence-based assays, and high-content imaging can be used to measure the growth rate and lipid content of microalgal strains under different conditions. HTS has also been used in microalgal research to identify strains that are tolerant to environmental stressors, such as high light intensity (Gouveia et al. 2014; Zhao et al. 2019), low temperature (Arena et al. 2021; Wang et al. 2016), and high salinity (Aburai et al. 2015; Arena et al. 2021). Such data can be used to design cultivation processes at large-scale with improved control algorithms to respond quicker to unforeseen environmental or culture changes. This is particularly important to outdoor cultivations that usually target commodity products such as biofuel production or the production of feedstocks, food, and nutraceuticals. Another application for HTS is the evaluation of different microalgal strains and their pollutant removing efficiency, e.g. from wastewater or the atmosphere (i.e. flue gas). Screening data can be used to develop new strategies for using microalgae to mitigate various environmental problems.

HTS promises to advance the area of microalgae process monitoring and control, which aims to maintain the culture at the optimal growth conditions to maximise productivity for a given bioreactor design. Growth rates and maximum biomass yields vary for different system designs due to differences in factors such as surface area to volume (SA:V) ratio and light supply. These can be simulated in HTS approached to develop more accurate predictive process and control models

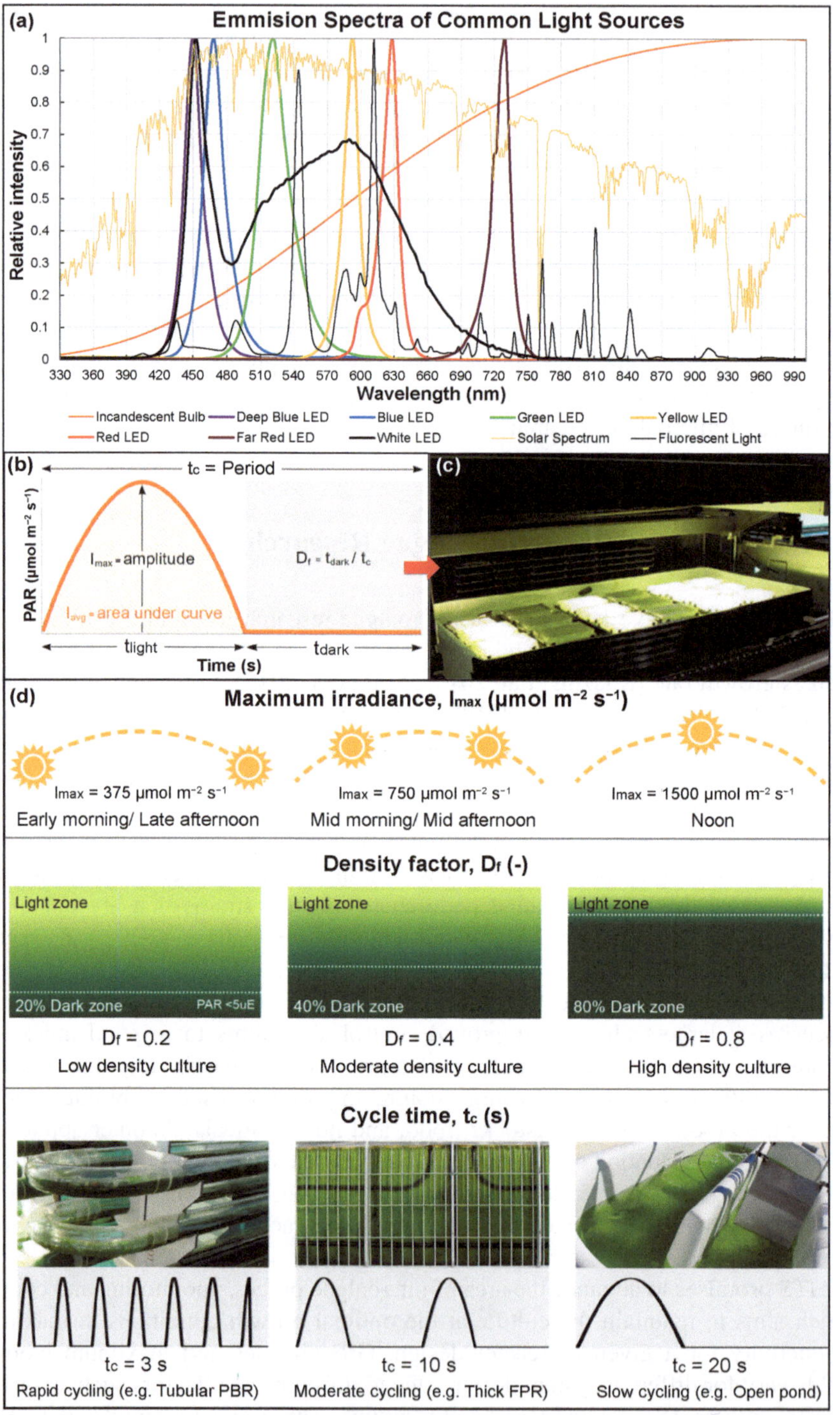

Fig. 3 Integration of light into HTS for microalgae. (**a**) Relative emission spectra profile comparison of the solar spectrum, incandescent bulbs, fluorescent light and various LED systems; (**b**) a

(Fig. 3) (Sivakaminathan et al. 2018; Yarnold et al. 2016). Computational modelling approaches may also be used to predict non-intuitive approaches to optimise metabolic flux. Successful process control is possible with a network of plug-and-play Internet-of-things (IoT) sensors that allow the operators to monitor the algal growth and other parameters in real time (e.g. nutrients, water, CO_2, pH, foam formation). This helps in maintaining a balance and regulation of process parameters at adequately fast time scales resulting in high energy efficiency. The development of reactor-specific computer simulations may enhance process control reducing material wastage and time. Ideally, growth/production models and machine-learning approaches can help to identify which of the 'easy-to-measure' parameters can be used and how they can be implemented to predict culture behaviour and hence optimise process control to reduce costs and increase cultivation robustness.

3.1 Throughput Via Miniaturisation

Increasing throughput via miniaturisation refers to the process of reducing the size of the components used in a HTS assay in order to increase the number of samples that can be screened in a given period of time. This is achieved by miniaturising the sample volume and hence the containers, which are used to contain the samples in HTS assays: fermenters to flasks > flasks to wells >96-multiwell plates to 384-multiwell plates > Microfluidics. The key difference between HTS systems in microalgae research and other fields is the need to illuminate the photosynthetic algal cultures (Fig. 3). This requirement rules out the dense, closely packed configurations characteristic of other HTS systems. Laboratory-scale micro-photobioreactors must be designed carefully in order to ensure that lighting is even across a sample(Bates et al. 2020; Heo et al. 2015; Morschett et al. 2017; Volpe et al. 2021) and that it adequately replicates scaled-up lighting conditions (Leonardi et al. 2021).

In order to replicate production-scale lighting conditions, it is important to mimic the spectral distribution of sunlight as closely as possible. Figure 3a shows the spectral distribution of various light sources compared to that of the Sun. Of the commonly used sources of artificial light, white light LEDs most closely mimic solar

Fig. 3 (continued) combination of light factors can be programmed by changing the light intensity over a cycle time, assuming cell cycling occurs in a sinusoidal trajectory of a microalgae culture. Here, I_{max}, is the amplitude of the sine, simulating the maximum irradiance that a cell would receive when at the 'surface' of a mass culture, D_f, is the proportion of time that PAR is below 5 μmol m^{-2} s^{-1} in one period simulating the fraction of time that a cell spends in the dark (culture density dependant), and t_c is the period of one sine wave that simulates the time required for a cell to cycle through the reactor. I_{avg} is the integration of light received, simulating the average irradiance or light dose received by the cell (Sivakaminathan et al. 2018); (**c**) 18-plate microwell robotic system (Radzun et al. 2015); (**d**) high-throughput light simulations of cells cycling in outdoor microalgae mass cultures (Sivakaminathan et al. 2018)

radiation, especially in the critical photosynthetically active region (PAR) between 400–700 nm.

The effect of different lighting conditions on microalgae growth and composition is complex and still under explored. For example, in natural habitats, the wavelengths of light available to microalgae vary strongly with water depth (Prazeres and Renema 2019). The wide variety of microalgal pigment compositions across species reflects the variety of lighting conditions encountered (Deepika et al. 2022b). The effect of different lighting conditions on plant production has been examined in more detail (Águila Ruiz-Sola et al. 2023; Ozawa et al. 2018). Coloured light has been shown to affect growth rate (e.g. at night (Abomohra et al. 2019)), as well as uptake and storage of major nutrients like potassium (Razzak et al. 2022; Ahammed et al. 2022), and the production of a range of other metabolites (Krzymińska et al. 2022; Zhang et al. 2023).

A variety of light-dependent effects are being reported in microalgae as well. Coloured light reportedly enhances growth for certain species (Das et al. 2011), while other groups have reported wavelength-dependent flocculation behaviours (Lee and Kim 2019), such as in *Spirogyra*. The blue-green algae belong to the genus *Leptolyngbya* have been observed to extensively reconfigure their photosynthetic apparatus in response to changing lighting conditions (Gan et al. 2014), changing their pigment composition entirely in order to maximise their photosynthetic efficiency.

3.1.1 Microwell Plates

By reducing the size of the wells, more wells can be placed on a single plate, allowing more samples to be screened in parallel. This results in a significant increase in the number of samples that can be screened in a given period of time, allowing for more efficient and cost-effective screening. Miniaturisation also has the advantage of reducing the amount of reagents required for each sample, which can lower the cost of the assay and reduce the amount of waste generated. At the same time, this presents a disadvantage, as smaller sample sizes for downstream analytics are not suitable for broad sample characterisation. Other disadvantages are limited availability of suitable online measurements, limited cultivation lengths due to evaporative losses, precipitation, cell aggregation that small volumes experience decrease measurement precision (results in a poor detection limit), signal stability issues (e.g. baseline errors in calibration), and variable signal-to-noise ratios. Furthermore, microwell plates often require harsh shaking methods to achieve even cell distribution per sample and well, which can cause shearing stress to the cells. Therefore, microfluidics have been discussed as one solution to overcome the bottlenecks in downstream analytics.

3.1.2 Microfluidics

In the past decade, microfluidics technologies have been rapidly developed into a powerful approach capable of integrating multiple functions for single-cell analysis (Westerwalbesloh et al. 2019; Zhou et al. 2021). Microfluidic lab-on-a-chip systems offer a promising alternative which can significantly contribute to new information in the microalgae field. Indeed, microfluidic systems can enable high-throughput assays in an automated process (Kim et al. 2018). Microsystems have numerous advantages such as the cost of fabrication and the biocompatible properties of materials used to design them are usually biocompatible polymers. Different fabrication methods have been developed to adapt the microsystem to specific application requirements. Soft lithography is a popular technique involving the creation of a mould by means of photolithography on a master pattern (Whitesides et al. 2001). With an elastomeric material such as polydimethylsiloxane (PDMS), the pattern can be transferred onto a substrate. With this technique, precise patterning with high resolution can be performed as well as the replication of intricate geometries. Another technique involves using a laser to remove material from a substrate in order to create microchannels and other structures (Luo et al. 2007). This technique is well suited for rapid prototyping and useful in case of three-dimensional structures. Mass and low-cost production of microfluidic systems is possible with injection moulding (Attia et al. 2009). 3D printing (Mehta and Rath 2021) and micromilling (Balázs et al. 2021) also create structures with different kinds of materials such as plastics, metals and ceramics.

Using microsystems is a way to massively parallelise analysis over conventional approaches for the downstream process. In comparison with the limitations of analysis rate previously cited as 100 cells per hour (Fregin et al. 2019), using microdroplets in a sorting microfluidic device can achieve 100,000 algal cells per hour (Yu et al. 2021). Various methods can be integrated to the microsystem for sensing or identifying parameters: physical, chemical, or electrical. With these microsystems, it is possible to identify different microalgal cells or classify and separate them according to their properties such as the lipid content (Erickson and Jimenez 2013), thanks to di-electrophoresis as example (Deng et al. 2013). Those microsystems coupled to electrical engineering methods can enhance the single-cell studies. For instance, using electrorotation can provide information about lipid accumulation in a single cell thanks to dielectric properties of cellular compartment and data analysis based on genetic algorithms (Lin et al. 2021). Other techniques such as electroporation and gene insertion with nanowires can be integrated with microsystems allowing a single device gene preparation and transformation processes (Im et al. 2015; Bae et al. 2015). Integrated microfluidic platforms with conventional analytical instrumentation can enhance the analysis on a single-cell level. In particular, studying lipid produced by microalgae can be performed with fluorescent staining methods to quantify them (Koreivienė 2020). A study (Chen et al. 2019) reports the use of single-cell analysis and FACS to identify phenotypic variations in lipid accumulation among a population of the green microalga

Nannochloropsis oceanica. Bensalem et al. (2018a) studied the cells of *Chlamydomonas reinhardtii* during lipid enrichment through their structural changes detected with confocal microscopy. In order to extract microalgae lipids from wet biomass, electro-Fenton-based technique can be used to enhance this process and cell harvest for *N. oceanica* (Zhang et al. 2020).

3.2 Throughput Via Automation

Increasing throughput via automation refers to the use of automated systems and processes to increase the speed and efficiency of HTS assays. Automation in HTS assays is achieved by using robots, software, and other tools to automate various tasks, such as dispensing reagents, mixing solutions, reading results, and transferring samples. Both hardware and software are constantly evolving with new technology, such as smaller sensors and more powerful processors, which are becoming available. Hardware components include readers, robotic manipulators, liquid handlers, mass flow controllers, and a vast diversity of sensors and actuators that can control environmental parameters such as temperature, humidity, pressure, and flow rate. For cell-based microalgae assays, it is critical to enable the control of light and CO_2 supply. The software requires the user to program a sequence of actions customised for a specific sample preparation or handling, data readings, data storage, and sometimes analysis. The goal of automation in HTS assays is to reduce the time and cost required to complete each assay and to increase the number of samples that can be screened in a given period of time. Automation also helps to reduce the risk of human error and improves the consistency and reproducibility of the results.

3.3 The Role of Robotics in HTS Systems

In HTS microalgae assays, robots play a crucial role in automating the process of screening large numbers of compounds or biological entities. They increase the speed, efficiency, and accuracy of a screening process while reducing the risk of human error. There are two main types of robots that are commonly used in HTS assays (or a combination of both) (Radzun et al. 2015);

- *Liquid handling arm (LiHA) robots:* These robots are used to dispense and mix various reagents and compounds into multiwell plates, which are then used for cell-based or enzyme-linked assays, and other types of analytical assays. They may also be used to detect the liquid level of a reagent trough.
- *Robotic manipulator arm (RoMA) robots:* These devices are used to transport samples or objects from position A to position B. For example, it transfers microwell plates (or tubes) from a shaker to a plate reader and back, or it replaces empty tip boxes and reagent troughs with new ones.

A robotic system is usually designed to perform very specific tasks and has different capabilities. The choice of robot for a given HTS assay depends on the type of assay, the specific requirements of the experiment and available funds for the initial capital cost. Some laboratories may use a combination of robots to perform different aspects of an HTS assay, while others may use a single integrated platform that can perform multiple tasks. Robots are then programmed to conduct a sequence of tasks, which includes (micro)environment control of incubation conditions and the use of accessory equipment integrated into the robotics system for sample analysis (Fig. 4). A HTS robotic fitout for analytical assays typically comprises a combination of the following equipments: (1) A plate reader (e.g. fluorescence, luminescence, or absorbance (multiple point reads, area scans), in multiwell plates); (2) a high-content imaging system (e.g. combined automated microscopy and image analysis to perform high-content imaging assays) to evaluate the morphological changes in cells following different treatments; (3) mass spectrometry (MS) and chromatography (GC, HPLC); (4) cell sorting and analysis via flow cytometry to detect and measure physical and chemical characteristics of cell or particle populations (i.e. FACS, or pulse amplitude modulated (PAM) fluorometry). For in-vivo microalgae HTS assays (Radzun et al. 2015; Van Wagenen et al. 2014; Olia et al. 2020; da Silva and Fonseca 2018; Pacheco et al. 2013; Pereira et al. 2011; Podevin et al. 2015; Haire et al. 2018; Tillich et al. 2014), a micro-climate control must be implemented to define cell incubation conditions both in microwell plate assays (mixing via shakers, pipetting of fluidics, temperature control, CO_2 supply, light intensity, spectral quality) and microfluidic setups (cells on a chip, pressure, and flow control) (Sonowal et al. 2019). Lastly, equipment for sample tracking and management is often integrated in robotic platform setups (e.g. barcode readers, RFID, and cameras).

The complexity of developing robotic machines and autonomous systems could be divided into five specialty areas: (1) operator interface, (2) mobility or locomotion, (3) manipulator & effectors, (4) programming, and (5) sensing & perception. Table 1 summarises the multidisciplinary skillsets that are required to develop, build, and design a robot for HTS capabilities. There are several competing companies in the field of laboratory automation and detection solutions aiming to address the needs of research laboratories, diagnostic laboratories, and other scientific institutions. Each competitor has its own strengths and market focus, and the competition in the field drives further innovation and advancements in laboratory automation technology.

- *Tecan* specialises in developing and manufacturing advanced instruments and systems that automate laboratory workflows and enable high-throughput screening and analysis. Tecan's niche lies in the field of laboratory automation, where they offer a wide range of robotic liquid handling platforms, automated sample preparation systems, and detection instruments for applications in life sciences research, diagnostics, and drug discovery.

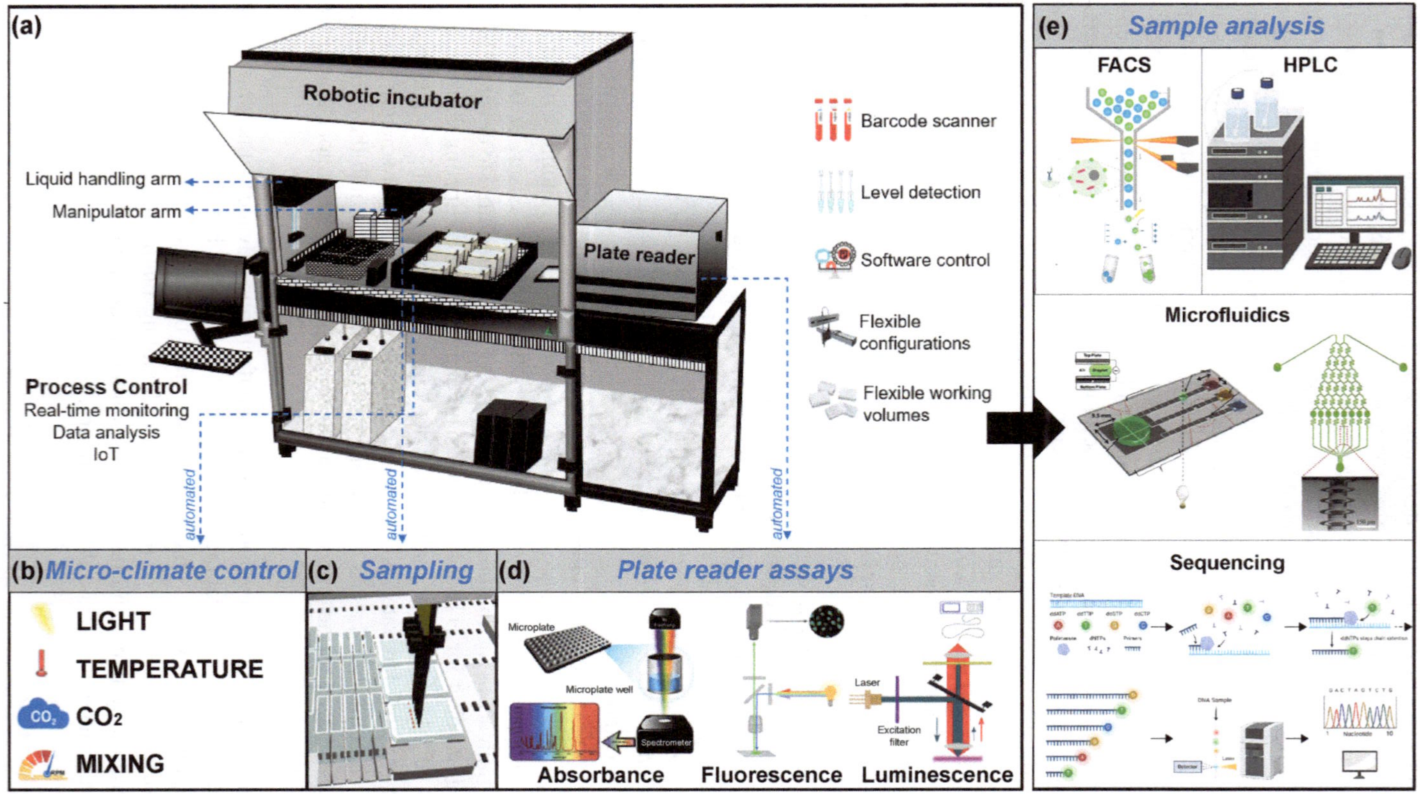

Fig. 4 Schematic of a robotic growth chamber setup for microalgae-based HTS system (**a**), including setup options for micro-climate control (**b**), automated liquid sampling (**c**), integrated plate reader assays (**d**), and further sample analysis (**e**)

Table 1 Capabilities of robotic machines and autonomous systems

Robotic capability	Automation task	Application
Operator interface	• Human workers remain essential; • Human robot interface (HRI) enables effectively communication between robot and human controller • Design of HRIs must prioritise intuitiveness and effectiveness to ensure accurate and efficient task execution.	Gaming controllers, industrial touchscreen computers.
Mobility or locomotion	• Move in its environment (called locomotion in robotics) • Autonomous robotics rely heavily on cameras, lidar, and radar to collect information about their surroundings • Use sensory data to make real-time corrections or adjustments to their movements in order to complete tasks or avoid collisions.	Mimic human movement for assembly; flying robots and drones make use of propellers and other propulsion systems; wheeled rovers deployed on Mars and other celestial bodies.
Manipulator & Effectors	• Interaction with its environment • Pick up objects and move them, or manipulate items that are separate from the system.	Pincers, claws, or pushers which are all uniquely suited to move heavy (or very small) pieces of equipment or materials.
Programming	• Language of the operator to communicate with the robot • Traditionally, any action that an autonomous individual robot was required to perform had to be programmed (usually it is a combination of PLC and user interface software) • Commands are provided by the user in real time for the robot to perform, or the robot is programmed to perform a series of tasks, in sequence, autonomously.	Each robot can be programmed using one (or several) of more than a thousand different programming languages, sending out alarms.
Sensing & Perception	• Use sensors to gather information • Information lets the robot know the physical space it occupies, where it needs to go, and if any obstacles block its path.	Information collection to help the robot decide how to react to objects it encounters, used to maintain a specified environmental parameter (e.g. temperature, light, pressure).
Artificial intelligence	• Advanced programming allows automated robotic systems to learn and adapt to changes within its environment (self-programming).	Autonomous adaption to unforeseen changes.

- *PerkinElmer* is a global leader in scientific instruments and laboratory services. They offer a wide range of solutions for laboratory automation, including liquid handling systems, robotic workstations, and detection instruments.

- *Beckman Coulter* is a provider of biomedical testing instruments and automation solutions. They offer a range of laboratory automation systems, including liquid handling platforms, sample preparation systems, and robotic workstations.
- *Hamilton Company* specialises in laboratory automation technologies, including robotic liquid handling systems, automated sample preparation solutions, and intelligent labware. They cater to a wide range of applications in life sciences research, diagnostics, and biotechnology.
- *Agilent Technologies* is a leading provider of analytical instruments and laboratory solutions. They offer automated liquid handling systems, sample preparation solutions, and detection instruments for various applications in life sciences, pharmaceuticals, and environmental testing.
- *Qiagen* is known for its expertise in molecular biology and diagnostics. They provide automated sample preparation systems, liquid handling platforms, and molecular detection solutions for applications in genomics, proteomics, and clinical diagnostics.

4 Experimental Design and Data Management for HTS Systems

HTS platforms generate large amounts of data, which must be analysed in order to obtain meaningful results. This is where the importance of coupling the experimental activity with rigorous statistically driven data analysis and sorting comes in. Data analysis and cleaning are crucial steps to ensure the accuracy and reliability of the resulting outcomes and data interpretation. This is especially important in HTS assays, where small errors can lead to false-positive or false-negative results, potentially leading to incorrect conclusions. Along the data analytics processing pipeline (i.e. collecting, processing, cleaning, and analysing), the goal is to decrease the content volume while increasing the quality of the content. The use of statistical analysis methods helps to identify trends, systematic patterns, correlations, and interrelations in the data that may not be immediately apparent, leading to a deeper understanding of the results. Furthermore, a rigorous statistical approach can help to quantify the confidence of the results, allowing for the identification of statistically significant findings. This is important because it helps to reduce the risk of false-positive results, which can occur when large numbers of tests are performed.

Key processes used in data analysis pipelines are often iterative and interconnected, including:

- *Data collection:* Gathering and aggregating data from various sources, such as databases, sensors, readers, controllers, and logs.
- *Data ingestion:* Importing and loading data into a data storage or processing system, such as a data lake or a distributed file system.
- *Data cleaning:* Preprocessing and cleansing the data to remove inconsistencies, errors, and missing values.

- *Data transformation:* Converting and restructuring data into a suitable format for analysis, which may involve data normalisation, encoding, or feature extraction.
- *Data storage:* Storing the processed data in a scalable and distributed storage system, such as cloud-based storage solutions.
- *Data processing:* Applying various computational algorithms and techniques to extract insights and derive meaningful patterns from the data.
- *Data analysis:* Performing statistical analysis, machine learning, data mining, or other analytical techniques to gain insights and make predictions from the processed data.
- *Data visualisation:* Presenting the analysed data in visual formats, such as charts, graphs, or dashboards, to facilitate understanding and interpretation.
- *Data interpretation:* Extracting meaningful insights and knowledge from the analysed data to support decision-making and drive business or research objectives.
- *Model Training and Deployment:* Developing and training machine-learning or predictive models using the analysed data and deploying them to make predictions or automate decision-making processes.
- *Monitoring and Maintenance:* Continuously monitoring the data pipeline, ensuring data quality, performance optimisation, and adapting the pipeline to evolving data requirements.

4.1 Design of Experiments (DoE)

DoE approaches can play an important role in reducing the number of experiments required and improving the efficiency of experimental investigations. By allowing the simultaneous testing of multiple factors and optimisation of experimental conditions, DoE approaches can provide more meaningful and accurate results while reducing the time, resources, and cost required to obtain those results. Usually, statistical approaches are employed to minimise the number of experiments required to obtain meaningful results. The goal of DoE is to optimise the conditions of an experiment, maximise the information obtained while minimising the number of experiments required. DoE approaches involve selecting the most appropriate experimental design, such as full-factorial, fractional factorial, or response surface design, and selecting the appropriate number of levels for each factor being tested (Sivakaminathan et al. 2018). Factors can include variables such as temperature, concentration, or time. The experiment is then executed, and the results are analysed to identify the relationship between the studied factors and response of the system.

4.1.1 Experimental Compression with Incomplete Factorial Designs

A significant reduction in the number of experiments is achieved by allowing the simultaneous testing of multiple factors, rather than testing each factor individually. In addition to the identification of optimal conditions for a particular experiment, DoE approaches may also identify the most important factors affecting the response of the system, or interaction of different factors.

Full-Factorial Design A full-factorial design is an experimental design in which all possible combinations of levels of the independent variables are tested. In a full-factorial design, each level of each independent variable is tested in combination with every level of every other independent variable. This allows for a comprehensive understanding of the relationship between the independent variables and the response of the system being studied. However, full-factorial designs require a relatively large number of experiments and might be impractical for systems with a large number of independent variables.

Fractional Factorial Design A fractional factorial design is a reduced version of a full-factorial design that tests only a fraction of the possible combinations of levels of the independent variables. In a fractional factorial design, a smaller number of experiments is performed, but the results may not provide as comprehensive an understanding of the relationship between the independent variables and the response of the system being studied as a full-factorial design.

Response Surface Design A response surface design is an experimental design in which the relationship between the independent variables and the response of the system is modelled as a surface. Response surface designs are used when the relationship between the independent variables and the response of the system is not well understood and when a more comprehensive understanding of the relationship is desired. Response surface designs typically involve a smaller number of experiments than full-factorial designs, but the results provide a more detailed understanding of the relationship between the independent variables and the response of the system.

In conclusion, the choice of experimental design depends on the specific system being studied and the goals of the experimental investigation. Full-factorial designs provide a comprehensive understanding of the relationship between the independent variables and the response of the system but can require a large number of experiments. Fractional factorial designs are a reduced version of a full-factorial design but may not provide as comprehensive an understanding of the relationship. Response surface designs are used to model the relationship between the independent variables and the response of the system and provide a more detailed understanding of the relationship with a smaller number of experiments.

4.2 *Data Fitting Techniques*

Appropriate data analysis tools are required to study different biochemical processes such as growth and biomass production, culture monitoring or quantifying interest components such as lipids, carotenoids, and proteins.

4.2.1 Growth Modelling for Microwell Cultures

For microalgae in a microwell plate, the population is proportional to the optical density (OD). The OD of a growing population of microalgae will follow a logistic function (Eq. 1) as follows:

$$\mathrm{OD}(t) = A + \frac{(B-A)}{1+e^{D(C-t)}} \tag{1}$$

where A is the minimum asymptote of the function, B is the maximum asymptote, C is the midpoint of the curve, D describes the steepness of the curve, and t is time.

A non-linear fitting algorithm such as the Levenberg–Marquardt algorithm (Levenberg 1944) can be used to find these parameters from measured growth data. Libraries capable of calculating these fits are readily available for most modern programming languages. In setting the asymptotes as parameters A and B, the need for background OD subtraction using a blank is obviated, which removes a potential source of systematic error. The midpoint of the curve C is when the population growth rate is maximised. Knowing the midpoint, the maximum growth rate can be found by taking the derivative of the logistic (Eq. 2) with respect to time

$$\frac{d(\mathrm{OD}(t))}{dt} = \frac{D(B-A)e^{D(C-t)}}{\left(e^{D(C-t)}+1\right)^2} \tag{2}$$

Setting $t = C$ in the above equation results in the maximum growth rate, μ_{max}. From the fitting parameters of a measured growth curve, μ_{max} can be calculated (Eq. 3).

$$\mu_{\mathrm{max}} = \frac{D(B-A)}{4} \tag{3}$$

The logistic curve makes a number of assumptions (such as a constant carrying capacity and a growing population) that makes it an unsuitable fit for many measured growth curves. The large amount of data generated with high-throughput techniques makes it infeasible to manually separate these measurements. An effective goodness-of-fit measure that can be used to automatically sort incoming data is therefore required. The coefficient of determination R^2 is traditionally used for this purpose, but R^2 is not suitable for non-linear fits (Spiess and Neumeyer 2010). There are a number of alternative options available; however, one option is to categorise measurements on the basis of specific features. For example, the OD values from the end of a measurement may be below the initial values (indicating population decline), or the measurements may have a limited overall range (indicating population stagnation). In such cases, there is no need to fit these measurements to a logistic function. Another option is to use uncertainty in the fitting parameters as a measure of goodness of fit. For example, a large uncertainty in the 'steepness parameter' D is a strong indication that the fitting algorithm has not found an unambiguously good fit. Finally, statistical measures similar to R^2 can also be used. The standard error of the estimate σ_{est} is one such measure (Eq. 4), defined as

$$\sigma_{\text{est}} = \sqrt{\frac{\sum_i (Y_i - Y_i')^2}{N}} \tag{4}$$

with Y_i being the measured values, Y_i' the fit values, and N the number of data points. This measure is similar to R^2 save that σ_{est} is in the same units as the original measurement, instead of being a unitless coefficient.

4.2.2 Partial Least Squares (PLS) and Principal Component Analysis (PCA)

PLS (Geladi and Kowalski 1986; Wold et al. 2001) and PCA (Wold et al. 1987) are data-driven techniques that are commonly used in various areas of research, including chemistry, biology, and engineering. These techniques are used to analyse and interpret complex datasets and to identify patterns and relationships in the data.

PLS is a regression analysis method that is used to analyse the relationship between two datasets, typically a set of predictor variables and a set of response variables. PLS is commonly used in areas such as chemometrics, where it is used to predict the response of a chemical system based on a set of predictor variables. For example, PLS has been used in the analysis of NMR spectra to predict the content of specific chemical compounds in a sample. PLS has been used to analyse data related to the composition of microalgal and cyanobacterial biomass (Palmer et al. 2021) or to perform metabolomic analysis of biomass (Chen et al. 2019). In (Lin et al. 2021), electrorotation measurements are applied at the single-cell level to evaluate the dielectric properties of microalgae cells during lipid accumulation. PLS regression is then used to develop a calibration model that relates the electrorotation measurements to the total lipid content of the cells. The authors also used a genetic algorithm (GA) to optimise the performance of the PLS regression model. GA is a search algorithm that mimics the process of natural selection and evolution to find optimal solutions to complex problems. In this case, GA was used to select the most informative electrorotation frequencies and to optimise the number of PLS factors used in the calibration model.

PCA is a dimensionality reduction technique that is used to analyse and interpret high-dimensional datasets. PCA is used to identify patterns and relationships in the data by projecting the data onto a lower-dimensional space. For example, PCA has been used in the analysis of gene expression data to identify patterns of gene expression that are associated with a particular disease or condition. Su et al. (2022) used PCA to analyse the data related to the growth and biochemical composition of microalgal consortia. In this study, authors cultivated the microalgae in a mixture of nano-filtered pig slurry and cheese whey under mixotrophic conditions and measured various parameters related to the growth and biochemical composition of the microalgae. The result of PCA analysis showed that the microalgal consortia had a high biomass productivity and a biochemical composition rich in lipids and proteins. The PCA also revealed the relationships among the different parameters, such as the positive correlation between biomass productivity and protein content. In Lin et al. (2021), PCA was also used to study biomass from microalgae culture. Raman spectra were studied with PCA to explore the relationship between the physiological status of microalgae cells and the spectra. In this study, PCA differentiated microalgae cells in the exponential and the stationary phases, indicating these cells had dissimilar cellular compositions, in particular lipids (Abomohra et al. 2021). Indeed, such studies can bring new information in the context of biodiesel production (Abomohra et al. 2021). For example, Jahirul et al. (2021) used PCA to evaluate the impact of chemical composition of biodiesel properties on its individual properties. In this study, variables such as the average number of double bonds (ANDB), average chain length (ACL), mono-unsaturated fatty acids (MUFAs), and polyunsaturated fatty acids (PUFAs) were included. The PCA revealed that the PUFA fraction and an ANDB present in the biodiesels were the most impactful chemical composition components.

Both PLS and PCA have been widely used in various areas of research and have been shown to be effective tools for the analysis and interpretation of complex datasets. However, the choice of the technique depends on the specific dataset being analysed and the goals of the analysis. The tools previously presented are well adapted to perform data fitting and prediction as in Lien et al. (2021) for lipid production (Abomohra et al. 2021). Machine learning algorithms can be interesting to go further and perform predictive modelling from the large amount of data collected in HTS systems.

4.3 Predictive Modelling from HTS Data Based on Machine-Learning Techniques

As discussed in Chapter "Artificial Intelligence in Phytochemicals Recognition", machine learning is one subfield of AI that encompasses a variety of data-driven algorithms and has a long history dating back to the pioneering work of Rosenblatt (1958). Machine-learning techniques can be used to classify different species of algae on images and to study lipid productivity for biofuel production purposes or environmental purification by microalgae. Monitoring the growth of those organisms is also possible with such tools (Ning et al. 2022). Machine learning is also a method to quantify the important role of microalgae in environment protection and pollution prevention (Cruz et al. 2021). In the wide variety of microalgae species, machine learning can identify the properties and characteristics of each species from experimental datasets. This is performed based on various methods available in the field of machine learning such as support vector machine (SVM), decision tree, random forest, neural network, or logistic regression (Ning et al. 2022). However, supervised training in machine learning relies on a large dataset with numerous cell images, which requires time-consuming cell labelling to establish ground truth for evaluating model accuracy. While large datasets such as the Marine Microalgae Detection in Microscopy Image dataset (Zhou et al. 2022) with 937 images and 4201 annotated objects, and WHOI-Plankton (Orenstein et al. 2015) with over 3.5 million images of microscopic marine plankton organised with labels, already exist, conducting specific tasks on personal small datasets for a wide range of algae species can be challenging.

4.3.1 Alternative Machine-Learning Techniques

To address the lack of specific data necessary for extracting biophysical features of individual cells in high-throughput quantitative imaging, deep-learning techniques have been developed for label-free cell classification (Chen et al. 2016). Additionally, data augmentation techniques prove valuable when the original dataset is not wide enough. These techniques involve height adjustment, zooming, or flipping of original images, among others. For instance, in Yadav et al. (2020), a convolutional neural network (CNN) based on ResNeXt was applied to an initial dataset of 100 images, and through data augmentation, the dataset expanded to an impressive 80,000 images. This approach resulted in a high segmentation accuracy of 99.97% (Yadav et al. 2020). To harness the advantages offered by large, labelled datasets, and existing resources, transfer learning becomes a valuable consideration. Transfer learning involves transferring general knowledge from a generalised model to a more specific model, fine-tuning it for the specific task at hand. For instance, in Ward et al. (2022), transfer learning was combined with image augmentation methods to extract rotifer-infested features from microalgae images (Ward et al.).

4.3.2 Coupled Methods with Machine Learning

When machine learning alone may not yield optimal performance, it can be coupled with other methods to enhance results. Genetic algorithms were coupled with extreme learning machine (ELM) in Purnomo et al. (2015) to evaluate the growth rate as a measure of effectiveness in capturing CO_2. ELM, known for its rapid learning speed, enables the training and testing of data collected on diverse microalgal growth behaviours at various pH concentrations. In this study, a genetic algorithm is used to optimised the weight of neuron connector and it is a way to improve the accuracy of the original ELM model (Purnomo et al. 2015). More recent studies can take advantages of combined machine-learning techniques such as in Sonmez et al. (2022). A CNN model was combined with support vector machines (SVM) to classify two microalgae groups, Cyanobacteria and Chlorophyta. This study showed how coupling SVM with AlexNet can improve segmentation accuracy. AlexNet's training speed and simplicity make it an intriguing choice for microalgae classification, although it is less accurate than other models. Without using SVM, AlexNet had the lowest accuracy with 98% and then coupled with SVM this accuracy increased to 99.66%. Data augmentation is also used in this study to increase the classification success of the different CNN models used. This study presents a powerful numerical tool to deal with the challenge of classification of microalgae, usually solved with morphological features and molecular techniques that require a high workload and expert knowledge (Sonmez et al. 2022).

While various aspects of machine learning require further optimisation, these combined methods have already shown promising results in the study of microalgae. Data-driven algorithms and increasing development of machine-learning techniques during the last decade could become an essential tool to optimise processes at different scales to better understand microalgae and exploit them to produce a broad range of microalgae derived bioproducts.

5 HTS Optimisation Applications

Optimisation of culture conditions can be regarded as an initial step for commercial production of many biological products. Microplate-based techniques have been proven as valuable tools that allow the screening of various microalgae and optimisation of targeted compounds much faster and cheaper than ordinary methods. Multifactorial systems optimisation for microalgae (Fig. 5) includes the optimisation of nutrients, CO_2, pH, light, temperature, and mixing. Each of these factors must then be integrated into a suitable process regime at scale.

5.1 Optimising Availability and Interactions of Key Nutrients in Miniaturised HTS

In miniaturised HTS, physical (e.g. solubility), chemical (e.g. ion interaction), and biological (e.g. bioavailability and uptake) availability of key elements can be optimised in a statistically efficient and automated manner to identify limiting parameters/chemical and nutrient interactions, enabling more focused optimisation using nutrient regimes that offers the most practical and robust prospects.

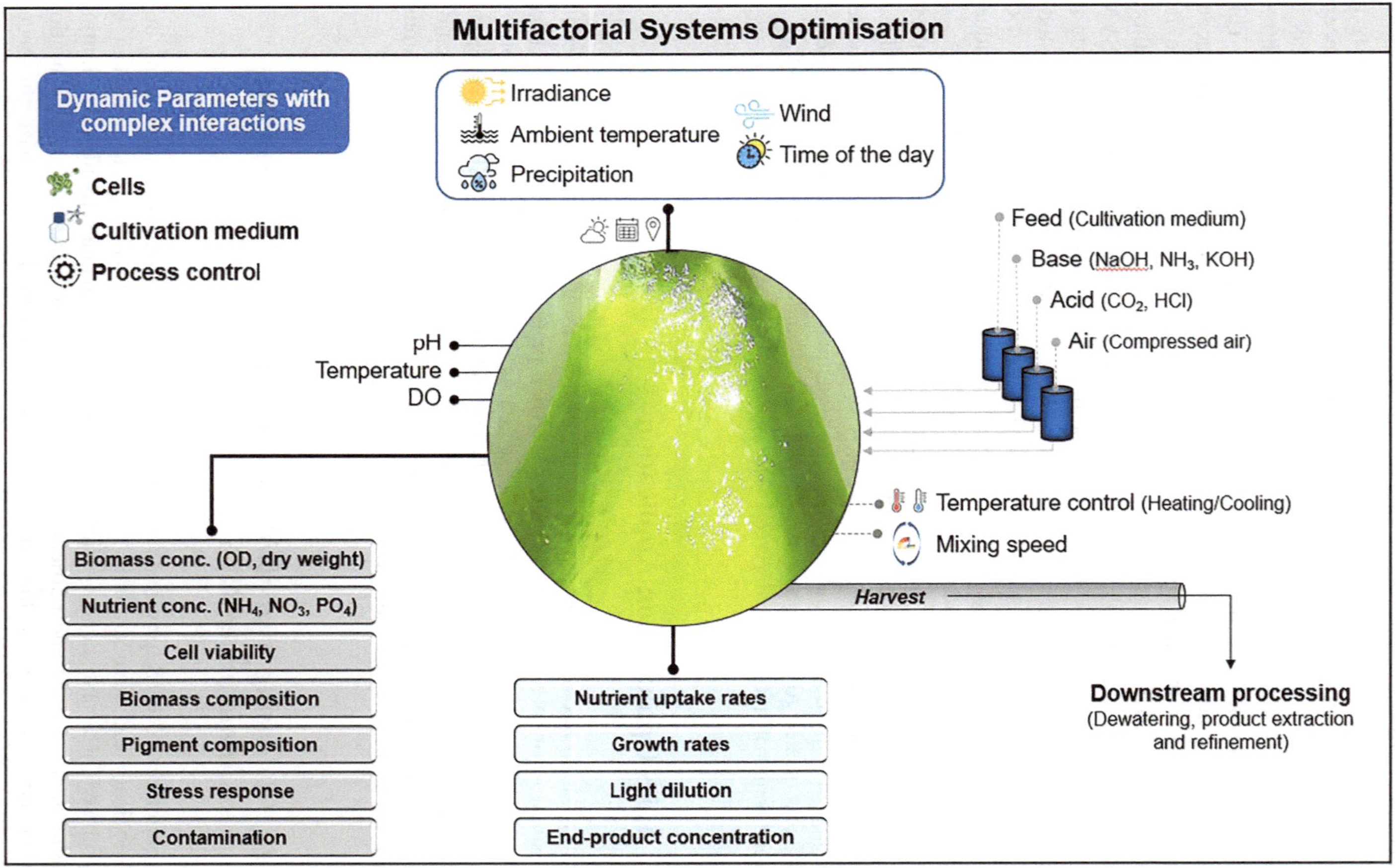

Fig. 5 Multifactorial systems optimisation

Miniaturisation to μL-scale requires suitable process control strategies, such as the use of buffers (e.g. TRIS, Bis Tris Propane, HEPES) as an alternative to dosing-based pH feedback control. Furthermore, vitamins and artificial chelators (e.g. EDTA) may be artificially provided for axenic cultures (Berges and Falkowski 1998; Hui et al. 2023; Tandon and Jin 2017).

The effect and interactions of multiple nutrients at random concentrations on microalgae growth could be described based on 'Liebig's law of the minimum' (Markou et al. 2014), which states that 'growth is dictated not by total resources available, but by the scarcest resource'. In contrast, the 'Droop model' (Droop 1968) states that 'the biomass growth rate (μ)' is dependent on the intracellular concentration of a nutrient and the minimum intracellular concentration of the nutrient below which there is no growth. It is not always the case that once nutrient adequacy has been reached, higher concentrations of a specific nutrient produce the same result as the minimal requirement. Higher nutrient concentrations may reduce uptake of other nutrients by competition, alter metabolic programs, cause toxicity, or produce effects due to solution chemistry, including precipitation. Approximately 21 essential elements are required for microalgae growth, and additional vitamins may also be beneficial for some microalgae. The percentage biomass dry weight contributed by eight of the most abundant elements is estimated for microalgae to be approximately as follows: C (20%–65%), O (12%–29%), N (1%–14%), P (0.05%–3.3%), Ca (0.2%–8%), Mg (0.35%–7.5%), K (1.2%–7.5%), and S (0.15%–1.6%) (Markou et al. 2014). Other elements such as Cu, Mn, Zn, Fe, Co, Mo, Se, Ni, V, B, Na, Cl, and H are generally present in lower amounts but are also essential to mediate a broad range of cellular functions (Bruland et al. 1991; Lommer et al. 2012; Worms et al. 2006). The elemental composition of the biomass of a target microalgae production species, cannot, however, be reliably used to define the media composition required to achieve maximum growth rates. This is because the biomass being analysed may itself have been produced under suboptimal conditions (Kennedy and Krouse 1999).

5.1.1 Streamlining Nutrient Optimisation in HTS: A Combined Complete and Incomplete Factorial Approach

As a starting point for a HTS nutrient screen, a literature review should be conducted to define a robust baseline media composition for a targeted microalgae strain. The most comprehensive approach to optimise the nutrient composition of a given production media is to conduct a full-factorial analysis of all elemental components required (Sect. 4.1.1). However, the analysis of 21 elements even at three different concentrations would require 3^{21} (i.e. 10,460,353,203) experimental runs. A miniaturised microwell approach enables a limited full-factorial analysis (e.g. $3^6 = 729$ experiments), but even 3^{10} conditions (i.e. 59,049) is challenging to conduct. To address this problem, a combination of complete and incomplete factorial-nutrient screens can be employed to optimise media for a diverse set of freshwater microalgae (Radzun et al. 2015; Wolf et al. 2015). This screen combined

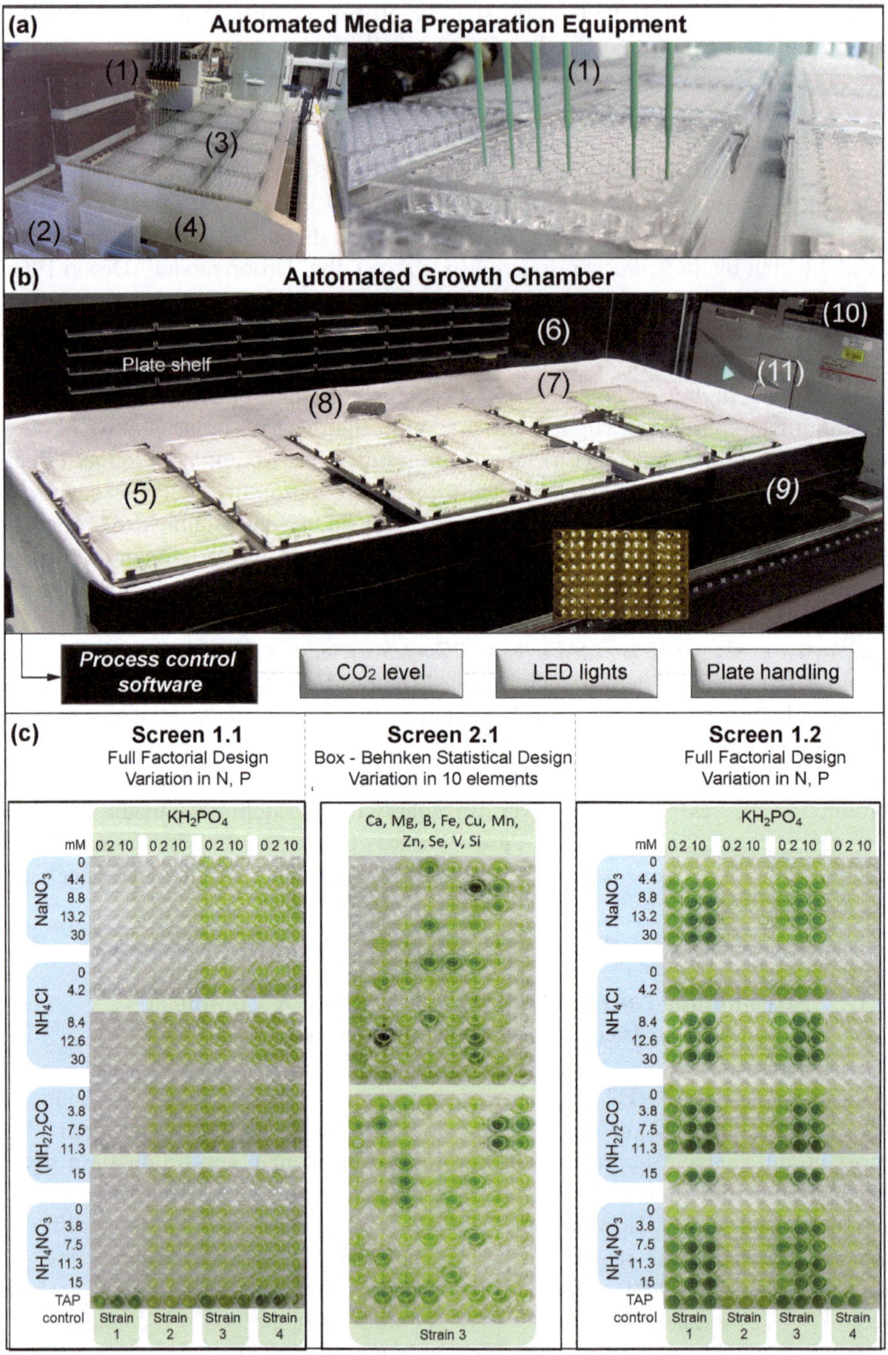

Fig. 6 The robotic screening system used for nutrient optimisation. (**a**) Automated media preparation equipment; and (**b**) automated growth chamber (images from Radzun et al. 2015); (**c**) the multidimensional nutrient HTS assay showing microwell plates post-HTS optimisation runs

complete factorial optimisation of N and P (Screens 1.1 and 1.2), and the refinement of 10 other elements using an incomplete factorial *Box-Behnken* design (Screen 2) (Ferreira et al. 2007) (Fig. 6). Screen 2 essentially reduced a 3^{10} full-factorial multidimensional space analysis (59,049) to 180 experimental conditions (328-fold compression). The effects of the concentrations of 10 elements (i.e. Ca, Mg, K, Fe, Cu, Mn, Zn, Mo, Co, and Se) on growth rate were analysed by varying their concentrations from low (−1), to medium (0), to high (+1). Coupling this sparse matrix approach with response surface analysis allows the identification of elemental *main effects* (e.g. Mg, Ca) and *interaction effects* (e.g. Mg-Ca) using a compressed trial number (Radzun et al. 2015; Wolf et al. 2015). Theoretically, with these twelve elements 'optimised', the remaining 9 elements and/or other factors (e.g. pH, CO_2, temperature) could be analysed next with an incomplete factorial approach.

5.1.2 Navigating the Complexity of Saltwater Media

The optimisation of algal growth in saltwater media adds further complexity. The salinity of natural water sources can vary from hypersaline brine (> 5% up to a maximum of 26–28%), through saline (3–5%; ocean salinity is typically 3.1–3.8%) and brackish water (0.05–3%) to freshwater (<0.05%). While sodium and chloride represent the predominant ions, calcium, magnesium, and potassium ions, other compounds including macro- and microelements (e.g. vitamins) are also important for microalgal growth and can vary in a location and season-specific manner. Moreover, the relatively broad pH range of natural seawater (~7.5 to 8.4) further complicates the chemical diversity of saltwater (Roy and Tim 2012) (Fig. 7, examples for carbonates, phosphorous, silica and boron). For example, light intensity and CO_2 concentrations impact the growth rates of diatoms and coccolithophores differently (Fig. 7), showing that elevated CO_2 stimulates growth in diatoms at low light but inhibits them at high light levels. In contrast, the coccolithophore *Emiliana huxleyi* grows faster under elevated CO_2 regardless of light levels, but calcification decreases (Gao et al. 2019). Marine microalgae cultivation is usually designed around the use of natural seawater that is subsequently enriched with specific nutrients, such as *Walne and Conway* (Lananan et al. 2013; Walne 1970) or *f/2* (Keller et al. 1987) enrichment media. While existing media enrichment compositions (e.g. f/2) support the growth of a wide range of species and are used at pilot, demonstration, and commercial scale (Zittelli et al. 1999), they are usually not fully optimised for a specific production strain or application. Such a standardised formulation may lead to limitations in certain elements (e.g. N, P or K), potential overabundance of some heavy metals which may cause toxicity, high concentration of other salts that cause precipitation, or the limitation of certain vitamins. At production scale, vitamins and chelating agents (e.g. siderophores) are often provided by bacteria (Groussman et al. 2015; Sutak et al. 2012; Vraspir and Butler 2009), while pH can be controlled via CO_2 supply or acid-base dosing. To ensure the achievement of desired product yields and quality, an industrial production line

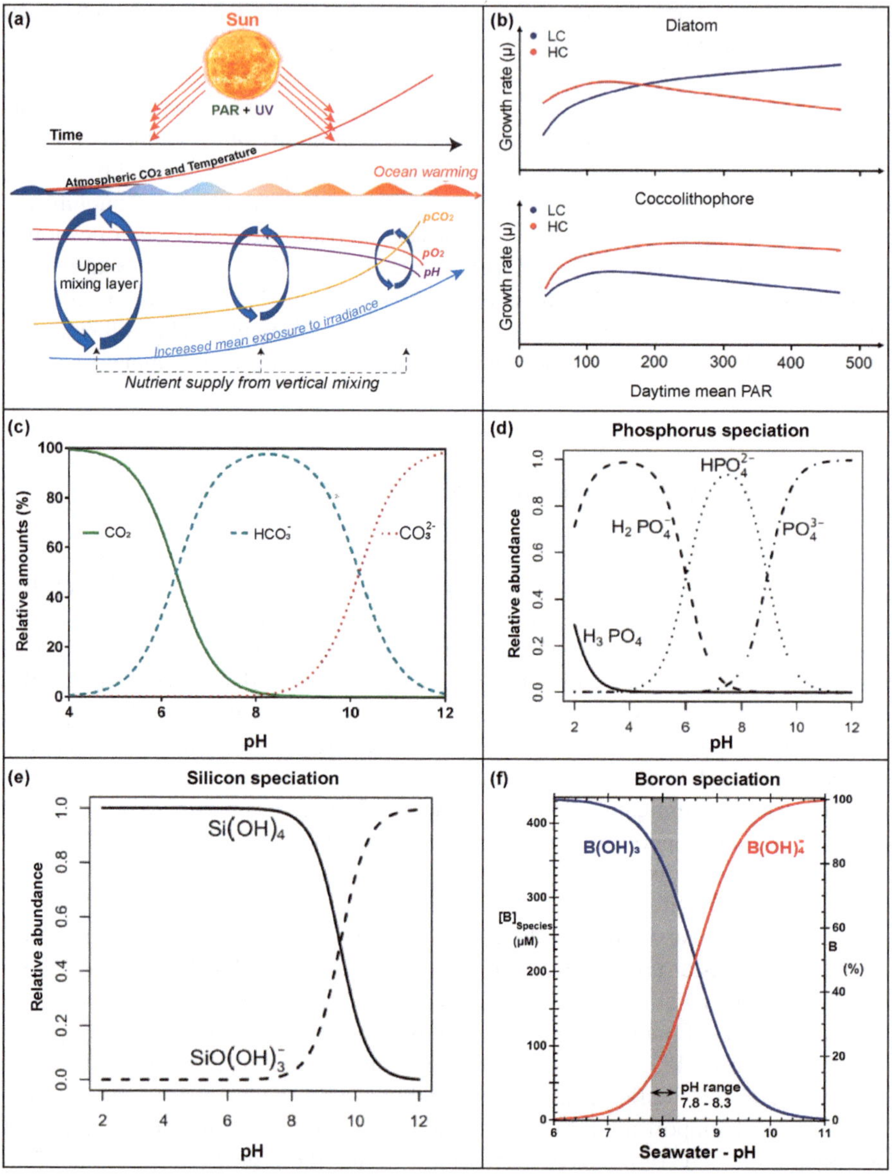

Fig. 7 In marine environment the interaction of light, nutrients and temperature (**a**) defines which species will thrive (**b**) (Gao et al. 2019). Speciation of selected nutrients as a function of pH to visualise complexity of media formulation (Orr et al. 2017). Relative speciation of inorganic carbon (**c**), phosphorous (Orr et al. 2017) (**d**), silicon (Orr et al. 2017) (**e**), and boron (Guillermic et al. 2020) (**f**) in water (or seawater) as a function of pH

relying on the elemental variability of natural and nutrient-enriched seawaters necessitates advanced characterisation knowledge and monitoring procedures. Therefore, HTS nutrient screening can be employed to pin down the influence of single elements in a media background of natural elemental variability. Consequently, to study the effect of individual elements, *artificial seawater* (ASW) media fully formulated from their constituent salts should be used if natural variation is required to be simulated in the HTS research question.

6 Conclusions: Resource Databasing for Circulation of Resources

The increasing demands of a growing population and higher living standards have led to the rapid degradation of Earth's ecosystem. To reverse this trend, it is crucial to recognise and harness the power of photosynthesis, which as yet is underutilised. The transition to a circular bioeconomy, centred around sustainable agricultural systems, necessitates a shift from a lack of stewardship to a strengthened human connection to nature. Photosynthesis-based production (i.e. solar manufacturing) should be prioritised as it not only enables productivity but also captures CO_2 and replenishes atmospheric O_2. Solar manufacturing technology implementation can support the delivery of goals to operate within our planetary boundaries, meet the UN sustainable development goals, and delivering net-zero emissions. The importance of decarbonising the economy and diversifying solutions across sectors would help to connect currently still desperate adaptation efforts.

Algal production systems have the potential to replace fossil fuel-based products, reduce freshwater requirements, and mitigate eutrophication, offering a sustainable alternative for heavy transport, aviation, shipping, and logistics sectors. Careful consideration is required when selecting production species, taking into account growth rates, salinity, temperature and pH tolerance, molecular engineering potential, product storage or export capabilities, and autotoxicity concerns. Technoeconomic models, based on transdisciplinary research and experimental knowledge, facilitate systems optimisation, de-risking scale-up, and the development of robust business models. These models consider location-specific factors such as light, wind, humidity, and temperature to predict biomass productivity. Process inputs, such as taxes, labour costs, carbon pricing, inflation, energy, and material costs, are also considered to identify the most suitable international locations and match processes accordingly. Predictive design approaches support cost-effective and timely implementation of economic processes, which is crucial for large-scale commercial systems needed within the next two decades to mitigate CO_2 emissions effectively.

Despite microalgae having gained prominence as a versatile platform for producing valuable compounds, including biofuels, pharmaceuticals, and nutritional supplements, optimising microalgae-based production processes presents challenges that necessitate extensive experimentation and optimisation to achieve high yields and productivity. Here, the potential of scale-down approaches for microalgae-based HTS, with a particular focus on microplate-based assays and microfluidic systems, provides an opportunity to accelerate knowledge creation. However, further research and development in the field of microalgae-based HTS using scale-down approaches is clearly required.

High-throughput screening (HTS) techniques enable the screening of traits such as growth rate, lipid content, and stress tolerance. It can also determine strains tolerant to environmental stressors, resistance to preditors, supporting the design of cultivation processes, or evaluate pollutant removing efficiency, aiding in environmental problem mitigation. Therefore, HTS advances microalgae process monitoring and control, optimising growth conditions, and maximising productivity. Miniaturisation and automation increase throughput in HTS assays for both microwell plate and microfluidics assays and reduced reagent usage. Automation with robots and software streamlines the screening process, reduces human error, and improves consistency. Therefore, robotics play a crucial role in liquid handling and sample transportation, but require further optimisation in the areas of operator interface, mobility, manipulators, programming, and sensing.

As HTS systems generate large amounts of data, rigorous statistical analysis to obtain meaningful results is required. The data analysis pipeline involves collecting, processing, cleaning, and analysing data, with the goal of decreasing content volume while increasing quality. Statistical methods help identify trends, patterns, correlations, and interrelations in the data, providing a deeper understanding and quantifying confidence in the results. Design of Experiments (DoE) approaches, such as full-factorial, fractional factorial, and response surface design, can optimise experimental conditions, reduce the number of experiments, and identify important factors and interactions. Data fitting techniques, like logistic growth modelling and partial least squares (PLS) and principal component analysis (PCA), aid in analysing complex datasets and identifying patterns and relationships. Additionally, machine-learning techniques, such as support vector machines and deep learning, enable predictive modelling and classification tasks, although they require large datasets or alternative techniques like transfer learning to overcome data limitations.

Microplate-based techniques offer efficient and cost-effective ways to optimise culture conditions for microalgae production. These techniques enable screening and optimisation of targeted compounds in a faster and more affordable manner than traditional methods. Nutrient availability and interactions can be optimised through miniaturised HTS approaches, allowing for focused optimisation of practical and robust nutrient regimes. The effects of multiple nutrients on microalgae growth can be analysed using statistical methods such as factorial analysis and response surface analysis. In saltwater media, the complexity of varying salinity, pH, and nutrient composition adds further challenges, requiring advanced characterisation and monitoring procedures. HTS nutrient screening can help understand the influence of individual elements and simulate natural elemental variability. Artificial seawater media can be used to study the effect of specific elements in HTS experiments.

In conclusion, the selection of suitable algal species for production organisms requires consideration of various properties. These include growth rates, the ability to produce proteins, carbohydrates, or oils in large quantities, the economic value of these products, and the potential for developing solar biorefineries. HTS data can provide a useful tool to build algae strain and resource libraries that enable the match making of location (e.g. climate and weather data), application (e.g. target product or biorefinery concept), strain (e.g. biosecurity limitations, fit-for-purpose, regulatory restraints), and resources (e.g. resource atlas containing available waste streams to supply nutrients or water).

For successful implementation, it is vital to consider environmental, technological, societal, cultural, and economic aspects while actively communicating with the general public. Building a united global effort toward a zero-CO-emission economy requires research outcomes that are acceptable and beneficial to society at large. Lessons can be learned from previous experiences, such as the acceptance of high-yield food crop varieties during the Green Revolution, while considering ecological costs and ethical concerns. Engaging in cross-disciplinary dialogues with social sciences and incorporating art and culture will enhance awareness of sustainable development and ensure ethical considerations are integrated into research project design. Public engagement plays a crucial role in supporting policy-making for future sustainability, even if some decisions may present short-term challenges.

References

Abomohra AE-F, Shang H, El-Sheekh M, Eladel H, Ebaid R, Wang S, Wang Q (2019) Night illumination using monochromatic light-emitting diodes for enhanced microalgal growth and biodiesel production. Bioresour Technol 288:121514

Abomohra A, Hanelt D (2022) Recent advances in micro-/Nanoplastic (MNPs) removal by microalgae and possible integrated routes of energy recovery. Microorganisms 10(12):2400

Abomohra AE-F, Zheng X, Wang Q, Huang J, Ebaid R (2021) Enhancement of biodiesel yield and characteristics through in-situ solvo-thermal co-transesterification of wet microalgae with spent coffee grounds. Bioresour Technol 323:124640

Aburai N, Sumida D, Abe K (2015) Effect of light level and salinity on the composition and accumulation of free and ester-type carotenoids in the aerial microalga Scenedesmus sp. (Chlorophyceae). Algal Res 8:30–36

Águila Ruiz-Sola M, Flori S, Yuan Y, Villain G, Sanz-Luque E, Redekop P, Tokutsu R, Küken A, Tsichla A, Kepesidis G (2023) Light-independent regulation of algal photoprotection by CO2 availability. Nat Commun 14(1):1977

Ahammed GJ, Chen Y, Liu C, Yang Y (2022) Light regulation of potassium in plants. Plant Physiol Biochem 170:316–324

Amorim ML, Soares J, Coimbra JSDR, de Oliveira Leite M, Albino LFT, Martins MA (2021) Microalgae proteins: production, separation, isolation, quantification, and application in food and feed. Crit Rev Food Sci Nutr 61(12):1976–2002

An WF, Tolliday N (2010) Cell-based assays for high-throughput screening. Mol Biotechnol 45: 180–186

Arena R, Lima S, Villanova V, Moukri N, Curcuraci E, Messina C, Santulli A, Scargiali F (2021) Cultivation and biochemical characterization of isolated Sicilian microalgal species in salt and temperature stress conditions. Algal Res 59:102430

Attia UM, Marson S, Alcock JR (2009) Micro-injection moulding of polymer microfluidic devices. Microfluid Nanofluid 7:1–28

Bae S, Park S, Kim J, Choi JS, Kim KH, Kwon D, Jin E, Park I, Kim DH, Seo TS (2015) Exogenous gene integration for microalgal cell transformation using a nanowire-incorporated microdevice. ACS Appl Mater Interfaces 7(49):27554–27561

Balázs BZ, Geier N, Takács M, Davim JP (2021) A review on micro-milling: recent advances and future trends. Int J Adv Manuf Technol 112:655–684

Bashir F, Kovács S, Ábrahám Á, Nagy K, Ayaydin F, Valkony-Kelemen I, Ferenc G, Galajda P, Tóth SZ, Sass L (2022) Viable protoplast formation of the coral endosymbiont alga Symbiodinium spp. in a microfluidics platform. Lab Chip 22(16):2986–2999

Bates H, Zavafer A, Szabó M, Ralph PJ (2020) The Phenobottle, an open-source photobioreactor platform for environmental simulation. Algal Res 52:102105

Bensalem S, Lopes F, Bodénès P, Pareau D, Français O, Le Pioufle B (2018a) Structural changes of Chlamydomonas reinhardtii cells during lipid enrichment and after solvent exposure. Data Brief 17:1283–1287

Bensalem S, Lopes F, Bodénès P, Pareau D, Français O, Le Pioufle B (2018b) Understanding the mechanisms of lipid extraction from microalga Chlamydomonas reinhardtii after electrical field solicitations and mechanical stress within a microfluidic device. Bioresour Technol 257:129–136

Berges JA, Falkowski PG (1998) Physiological stress and cell death in marine phytoplankton: induction of proteases in response to nitrogen or light limitation. Limnol Oceanogr 43(1): 129–135

Brányiková I, Maršálková B, Doucha J, Brányik T, Bišová K, Zachleder V, Vítová M (2011) Microalgae—novel highly efficient starch producers. Biotechnol Bioeng 108(4):766–776

Bruland KW, Donat JR, Hutchins DA (1991) Interactive influences of bioactive trace metals on biological production in oceanic waters. Limnol Oceanogr 36(8):1555–1577

Calvin M (1989) Forty years of photosynthesis and related activities. Photosynth Res 21:3–16

Capus A, Monnerat M, Ribeiro LC, de Souza W, Martins JL, Sant'Anna, C. (2016) Application of high-content image analysis for quantitatively estimating lipid accumulation in oleaginous yeasts with potential for use in biodiesel production. Bioresour Technol 203:309–317

Carrera-Pacheco SE, Hankamer B, Oey M (2023) Environmental and nuclear influences on microalgal chloroplast gene expression. Trends Plant Sci 28:955

Chen CL, Mahjoubfar A, Tai L-C, Blaby IK, Huang A, Niazi KR, Jalali B (2016) Deep learning in label-free cell classification. Sci Rep 6(1):21471

Chen C, Harst A, You W, Xu J, Ning K, Poetsch A (2019) Proteomic study uncovers molecular principles of single-cell-level phenotypic heterogeneity in lipid storage of *Nannochloropsis oceanica*. Biotechnol Biofuels 12:1–14

Chiaraviglio L, Kirby JE (2014) Evaluation of impermeant, DNA-binding dye fluorescence as a real-time readout of eukaryotic cell toxicity in a high throughput screening format. Assay Drug Dev Technol 12(4):219–228

Chisti Y (2007) Biodiesel from microalgae. Biotechnol Adv 25(3):294–306

Cho S-W, Kang D-K, Choo J-B, Demllo AJ, Chang S-I (2011) Recent advances in microfluidic technologies for biochemistry and molecular biology. BMB Rep 44(11):705–712

Cruz RC, Reis Costa P, Vinga S, Krippahl L, Lopes MB (2021) A review of recent machine learning advances for forecasting harmful algal blooms and shellfish contamination. J Mar Sci Eng 9(3):283

da Silva AA, Fonseca GG (2018) Screening of an adapted culture medium composed by different carbon sources for heterotrophic cultivation of Chlorella vulgaris using a microplate assay. Acta Sci Biol Sci 40:1–6

Das P, Lei W, Aziz SS, Obbard JP (2011) Enhanced algae growth in both phototrophic and mixotrophic culture under blue light. Bioresour Technol 102(4):3883–3887

Dawiec-Liśniewska A, Podstawczyk D, Bastrzyk A, Czuba K, Pacyna-Iwanicka K, Okoro OV, Shavandi A (2022) New trends in biotechnological applications of photosynthetic microorganisms. Biotechnol Adv 59:107988

Deepika C, Ravishankar GA, Rao AR (2022a) Potential products from macroalgae: AN overview. In: Sustainable global resources of seaweeds volume 1: bioresources, cultivation, trade and multifarious applications. Springer, Cham, pp 17–44

Deepika C, Wolf J, Roles J, Ross I, Hankamer B (2022b) Sustainable production of pigments from cyanobacteria. Adv Biochem Eng Biotechnol 183:171–251

Deepika C, Wolf J, Moheimani N, Hankamer B, von Herzen B, Rao AR (2022c) Utilisation of seaweeds in the Australian market–commercialisation strategies: current trends and future prospects. In: Sustainable global resources of seaweeds volume 1: bioresources, cultivation, trade and multifarious applications. Springer, pp 265–294

Deng Y-L, Chang J-S, Juang Y-J (2013) Separation of microalgae with different lipid contents by dielectrophoresis. Bioresour Technol 135:137–141

Droop MR (1968) Vitamin B 12 and marine ecology. IV. The kinetics of uptake, growth and inhibition in *Monochrysis lutheri*. J Mar Biol Assoc U K 48(3):689–733

Enami I, Okumura A, Nagao R, Suzuki T, Iwai M, Shen J-R (2008) Structures and functions of the extrinsic proteins of photosystemII from different species. Photosynth Res 98:349–363

Eggersdorfer M, Wyss A (2018) Carotenoids in human nutrition and health. Arch Biochem Biophys 652:18–26

Erickson RA, Jimenez R (2013) Microfluidic cytometer for high-throughput measurement of photosynthetic characteristics and lipid accumulation in individual algal cells. Lab Chip 13(15):2893–2901

Ertl G (2012) The arduous way to the haber–bosch process. Z Anorg Allg Chem 638(3–4):487–489

Fabregas J, Herrero C (1990) Vitamin content of four marine microalgae. Potential use as source of vitamins in nutrition. J Ind Microbiol Biotechnol 5(4):259–263

Falkowski PG (1994) The role of phytoplankton photosynthesis in global biogeochemical cycles. Photosynth Res 39:235–258

Ferreira SC, Bruns R, Ferreira H, Matos G, David J, Brandao G, da Silva EP, Portugal L, Dos Reis P, Souza A (2007) Box-Behnken design: an alternative for the optimization of analytical methods. Anal Chim Acta 597(2):179–186

Fregin B, Czerwinski F, Biedenweg D, Girardo S, Gross S, Aurich K, Otto O (2019) High-throughput single-cell rheology in complex samples by dynamic real-time deformability cytometry. Nat Commun 10(1):415

Gan F, Zhang S, Rockwell NC, Martin SS, Lagarias JC, Bryant DA (2014) Extensive remodeling of a cyanobacterial photosynthetic apparatus in far-red light. Science 345(6202):1312–1317

Gao K, Beardall J, Häder D-P, Hall-Spencer JM, Gao G, Hutchins DA (2019) Effects of ocean acidification on marine photosynthetic organisms under the concurrent influences of warming, UV radiation, and deoxygenation. Front Mar Sci 6:322

Gauthier M, Senhorinho G, Scott J (2020) Microalgae under environmental stress as a source of antioxidants. Algal Res 52:102104

Geladi P, Kowalski BR (1986) Partial least-squares regression: a tutorial. Anal Chim Acta 185:1–17

Gomez G, Andres E (2023) Modelling microalgae based production of recombinant proteins under different scenarios. PhD Thesis, Institute for Molecular Bioscience, The University of Queensland

González-Pérez BK, Rivas-Castillo AM, Valdez-Calderón A, Gayosso-Morales MA (2022) Microalgae as biostimulants: A new approach in agriculture. World J Microbiol Biotechnol 38(1):4

Gouveia L, Neves C, Sebastião D, Nobre BP, Matos CT (2014) Effect of light on the production of bioelectricity and added-value microalgae biomass in a photosynthetic alga microbial fuel cell. Bioresour Technol 154:171–177

Groussman RD, Parker MS, Armbrust EV (2015) Diversity and evolutionary history of iron metabolism genes in diatoms. PLoS ONE 10(6):e0129081

Guillermic M, Misra S, Eagle R, Villa A, Chang F, Tripati A (2020) Seawater pH reconstruction using boron isotopes in multiple planktonic foraminifera species with different depth habitats and their potential to constrain pH and pCO 2 gradients. Biogeosciences 17(13):3487–3510

Guiry MD (2012) How many species of algae are there? J Phycol 48(5):1057–1063

Guiry MD, Guiry GM (2023) AlgaeBase. World-wide electronic publication. National University of Ireland, Galway

Günerken E, D'Hondt E, Eppink M, Garcia-Gonzalez L, Elst K, Wijffels RH (2015) Cell disruption for microalgae biorefineries. Biotechnol Adv 33(2):243–260

Guo W, Feng L, Wang Z, Guo J, Park D, Carroll BL, Zhang X, Liu J, Cheng J (2020) In situ 3D comparison of Chlorella pyrenoidosa with nuclear-irradiated mutagenic strains by using focused ion beam milling and cryo-electron tomography. bioRxiv, 2020.11. 05.369421

Gütschow J, Jeffery ML, Gieseke R, Gebel R, Stevens D, Krapp M, Rocha M (2016) The PRIMAP-hist national historical emissions time series. Earth Syst Sci Data 8(2):571–603

Haberl H, Erb KH, Krausmann F, Gaube V, Bondeau A, Plutzar C, Gingrich S, Lucht W, Fischer-Kowalski M (2007) Quantifying and mapping the human appropriation of net primary production in earth's terrestrial ecosystems. Proc Natl Acad Sci 104(31):12942–12947

Haire TC, Bell C, Cutshaw K, Swiger B, Winkelmann K, Palmer AG (2018) Robust microplate-based methods for culturing and in vivo phenotypic screening of *Chlamydomonas reinhardtii*. Front Plant Sci 9:235

Hankamer B, Barber J, Nield J (2005) Structural analysis of the photosystem II core/antenna holocomplex by electron microscopy. In: Photosystem II: The light-driven water: plastoquinone oxidoreductase. Springer, Dordrecht, pp 403–424

Hankamer B, Pregelj L, O'Kane S, Hussey K, Hine D (2023) Delivering impactful solutions for the bioeconomy. Trends Plant Sci 28(5):583–596

Harari YN (2014) Sapiens: a brief history of humankind by Yuval Noah Harari. The Guardian

Heo J, Cho D-H, Ramanan R, Oh H-M, Kim H-S (2015) PhotoBiobox: A tablet sized, low-cost, high throughput photobioreactor for microalgal screening and culture optimization for growth, lipid content and CO2 sequestration. Biochem Eng J 103:193–197

Hlavova M, Turoczy Z, Bisova K (2015) Improving microalgae for biotechnology—from genetics to synthetic biology. Biotechnol Adv 33(6):1194–1203

Hui C, Schmollinger S, Glaesener AG (2023) Growth techniques. In: The Chlamydomonas sourcebook. Elsevier, Amsterdam, pp 287–314

Im DJ, Jeong S-N, Yoo BS, Kim B, Kim D-P, Jeong W-J, Kang IS (2015) Digital microfluidic approach for efficient electroporation with high productivity: transgene expression of microalgae without cell wall removal. Anal Chem 87(13):6592–6599

IPCC (2023) AR6 synthesis report: climate change 2023

Irena F (2022) World energy transitions outlook 2022:- 1.5°C pathway. International Renewable Energy Agency

Jaeschke DP, Mercali GD, Marczak LDF, Müller G, Frey W, Gusbeth C (2019) Extraction of valuable compounds from Arthrospira platensis using pulsed electric field treatment. Bioresour Technol 283:207–212

Jahirul M, Rasul M, Brown R, Senadeera W, Hosen M, Haque R, Saha S, Mahlia T (2021) Investigation of correlation between chemical composition and properties of biodiesel using principal component analysis (PCA) and artificial neural network (ANN). Renew Energy 168: 632–646

Kandler O (1950) Über die Beziehungen zwischen Phosphathaushalt und Photosynthese I: phosphatspiegelschwankungen bei Chlorella pyrenoidosa als Folge des Licht-Dunkel-Wechsels. Zeitschrift für Naturforschung B 5(8):423–437

Karan H, Funk C, Grabert M, Oey M, Hankamer B (2019) Green bioplastics as part of a circular bioeconomy. Trends Plant Sci 24(3):237–249

Karan H, Roles J, Hankamer B, Ross IL (2023) Targeting greens and yellows: a solar biorefinery analysis for the microalgae-based co-production of pigments, proteins, and fuel. Algal Res 74:103187

Keller MD, Selvin RC, Claus W, Guillard R (1987) Media for the culture of oceanic ultraphytoplankton. J Phycol 23(4):633–638

Kennedy M, Krouse D (1999) Strategies for improving fermentation medium performance: a review. J Ind Microbiol Biotechnol 23(6):456–475

Khavari F, Saidijam M, Taheri M, Nouri F (2021) Microalgae: therapeutic potentials and applications. Mol Biol Rep 48(5):4757–4765

Kim HS, Devarenne TP, Han A (2018) Microfluidic systems for microalgal biotechnology: a review. Algal Res 30:149–161

Koreivienė J (2020) Microalgae lipid staining with fluorescent BODIPY dye. Methods Mol Biol 1980:47–53. Biofuels from algae: methods and protocols

Kruse O, Rupprecht J, Mussgnug JH, Dismukes GC, Hankamer B (2005) Photosynthesis: a blueprint for solar energy capture and biohydrogen production technologies. Photochem Photobiol Sci 4(12):957–970

Krzymińska A, Frąszczak B, Gąsecka M, Magdziak Z, Kleiber T (2022) The content of phenolic compounds and organic acids in two Tagetes patula cultivars flowers and its dependence on light colour and substrate. Molecules 27(2):527

Lananan F, Jusoh A, Lam SS, Endut A (2013) Effect of Conway Medium and f/2 Medium on the growth of six genera of South China Sea marine microalgae. Bioresour Technol 141:75–82

Lee JW, Kim GH (2019) Red and far-red regulation of filament movement correlates with the expression of phytochrome and FHY 1 genes in Spirogyra varians (Zygnematales, Streptophyta). J Phycol 55(3):688–699

Lee TH, Chang JS, Wang HY (2013a) Current developments in high-throughput analysis for microalgae cellular contents. Biotechnol J 8(11):1301–1314

Lee YJ, Leverence RC, Smith EA, Valenstein JS, Kandel K, Trewyn BG (2013b) High-throughput analysis of algal crude oils using high resolution mass spectrometry. Lipids 48(3):297–305

Leonardi RJ, Ibañez MV, Osella EN, Heinrich JM (2021) Laboratory-scale reproduction of lighting conditions for an outdoor vertical column photobioreactor: theoretical fundamentals and operation of a programmable LED module. Algal Res 55:102227

Levenberg K (1944) A method for the solution of certain non-linear problems in least squares. Q Appl Math 2(2):164–168

Lin Y-S, Tsang S, Bensalem S, Tsai C-C, Chen S-J, C.-l S, Lopes F, Le Pioufle B, Wang H-Y (2021) Electrorotation of single microalgae cells during lipid accumulation for assessing cellular dielectric properties and total lipid contents. Biosens Bioelectron 173:112772

Lommer M, Specht M, Roy A-S, Kraemer L, Andreson R, Gutowska MA, Wolf J, Bergner SV, Schilhabel MB, Klostermeier UC (2012) Genome and low-iron response of an oceanic diatom adapted to chronic iron limitation. Genome Biol 13(7):1–21

Luo L, Teo C, Ong W, Tang K, Cheow L, Yobas L (2007) Rapid prototyping of microfluidic systems using a laser-patterned tape. J Micromech Microeng 17(12):N107

Luo L, Bokil NJ, Wall AA, Kapetanovic R, Lansdaal NM, Marceline F, Burgess BJ, Tong SJ, Guo Z, Alexandrov K (2017) SCIMP is a transmembrane non-TIR TLR adaptor that promotes proinflammatory cytokine production from macrophages. Nat Commun 8(1):14133

Markou G, Vandamme D, Muylaert K (2014) Microalgal and cyanobacterial cultivation: the supply of nutrients. Water Res 65:186–202

Martínez JM, Delso C, Álvarez I, Raso J (2020) Pulsed electric field-assisted extraction of valuable compounds from microorganisms. Compr Rev Food Sci Food Saf 19(2):530–552

Mehta V, Rath SN (2021) 3D printed microfluidic devices: a review focused on four fundamental manufacturing approaches and implications on the field of healthcare. Bio-Design Manufacturing 4:311–343

Mekala K, Hamid S, Jonathan W, Sven S, Daniel P, Dickon P, Tomas N, Humayun T, Annabel F, Weige W, Danielle I (2022) The net-zero transition: what it would cost, what it could bring. McKinsey Global Institute

Metting F (1996) Biodiversity and application of microalgae. J Ind Microbiol 17:477–489

Mishra B, Tiwari A, Mahmoud AED (2022) Microalgal potential for sustainable aquaculture applications: bioremediation, biocontrol, aquafeed. Clean Techn Environ Policy 1:1–13

Morschett H, Schiprowski D, Müller C, Mertens K, Felden P, Meyer J, Wiechert W, Oldiges M (2017) Design and validation of a parallelized micro-photobioreactor enabling phototrophic bioprocess development at elevated throughput. Biotechnol Bioeng 114(1):122–131

Ning H, Li R, Zhou T (2022) Machine learning for microalgae detection and utilization. Front Mar Sci 9:947394

Norton TA, Melkonian M, Andersen RA (1996) Algal biodiversity. Phycologia 35(4):308–326

Oey M, Sawyer AL, Ross IL, Hankamer B (2016) Challenges and opportunities for hydrogen production from microalgae. Plant Biotechnol J 14(7):1487–1499

Olia MSJ, Azin M, Sepahi AA, Moazami N (2020) Miniaturized culture method for the statistical study of growth rate and carbohydrate content of Picochlorum sp. D8 isolated from the Persian Gulf. Renew Energy 149:479–488

Orenstein EC, Beijbom O, Peacock EE, Sosik HM (2015) Whoi-plankton-a large scale fine grained visual recognition benchmark dataset for plankton classification. arXiv preprint arXiv:1510.00745

Orr JC, Najjar RG, Aumont O, Bopp L, Bullister JL, Danabasoglu G, Doney SC, Dunne JP, Dutay J-C, Graven H (2017) Biogeochemical protocols and diagnostics for the CMIP6 ocean model Intercomparison project (OMIP). Geosci Model Dev 10(6):2169–2199

Ozawa S-I, Bald T, Onishi T, Xue H, Matsumura T, Kubo R, Takahashi H, Hippler M, Takahashi Y (2018) Configuration of ten light-harvesting chlorophyll a/b complex I subunits in Chlamydomonas reinhardtii photosystem I. Plant Physiol 178(2):583–595

Özbek Ö (2022) Domestication and evolution of ancient wheats. In: Ancient wheats. Springer, Singapore, pp 15–36

Pacheco A, Hernández-Mireles I, García-Martínez C, Álvarez MM (2013) Microplates as a microreactor platform for microalgae research. Biotechnol Prog 29(3):638–644

Pacheco SEC, Hankamer B, Oey M (2018) Optimising light conditions increases recombinant protein production in Chlamydomonasreinhardtii chloroplasts. Algal Res 32:329–340

Palmer JS, Lawton LA, Kindt R, Edwards C (2021) Rapid analytical methods for the microalgal and cyanobacterial biorefinery: application on strains of industrial importance. MicrobiologyOpen 10(1):e1156

Pereira D, Williams J (2007) Origin and evolution of high throughput screening. Br J Pharmacol 152(1):53–61

Pereira H, Barreira L, Mozes A, Florindo C, Polo C, Duarte CV, Custódio L, Varela J (2011) Microplate-based high throughput screening procedure for the isolation of lipid-rich marine microalgae. Biotechnol Biofuels 4:1–12

Podevin M, De Francisci D, Holdt SL, Angelidaki I (2015) Effect of nitrogen source and acclimatization on specific growth rates of microalgae determined by a high-throughput in vivo microplate autofluorescence method. J Appl Phycol 27:1415–1423

Porter J (1976) Antony van Leeuwenhoek: tercentenary of his discovery of bacteria. Bacteriol Rev 40(2):260–269

Powar RS, Yadav AS, Ramakrishna CS, Patel S, Mohan M, Sakharwade SG, Choubey M, Ansu AK, Sharma A (2022) Algae: a potential feedstock for third generation biofuel. Materials Today: Proceedings 63:A27–A33

Prazeres M, Renema W (2019) Evolutionary significance of the microbial assemblages of large benthic foraminifera. Biol Rev 94(3):828–848

Preisig HR, Andersen RA (2005) Historical review of algal culturing techniques. Algal Cult Tech 65:79–82

Purnomo D, Purbarani S, Wibisono A, Hendrayanti D, Bowolaksono A, Mursanto P, Ramdhan D, Jatmiko W (2015) Genetic algorithm optimization for extreme learning machine based microalgal growth forecasting of Chlamydomonas sp. 2015 International Conference on Advanced Computer Science and Information Systems (ICACSIS). IEEE, Piscataway, pp 243–248

Radzun KA, Wolf J, Jakob G, Zhang E, Stephens E, Ross I, Hankamer B (2015) Automated nutrient screening system enables high-throughput optimisation of microalgae production conditions. Biotechnol Biofuels 8(1):65

Rasala BA, Mayfield SP (2015) Photosynthetic biomanufacturing in green algae; production of recombinant proteins for industrial, nutritional, and medical uses. Photosynth Res 123:227–239

Razzak MA, Asaduzzaman M, Tanaka H, Asao T (2022) Effects of supplementing green light to red and blue light on the growth and yield of lettuce in plant factories. Sci Hortic 305:111429

Remize M, Brunel Y, Silva JL, Berthon J-Y, Filaire E (2021) Microalgae n-3 PUFAs production and use in food and feed industries. Mar Drugs 19(2):113

Richmond A, Hu Q (2013) Handbook of microalgal culture. Wiley Online Library, Chichester

Ringsmuth AK, Landsberg MJ, Hankamer B (2016) Can photosynthesis enable a global transition from fossil fuels to solar fuels, to mitigate climate change and fuel-supply limitations? Renew Sust Energ Rev 62:134–163

Ritchie H (2021) Food waste is responsible for 6% of global greenhouse gas emissions. Our world data 2020.

Rizwan M, Mujtaba G, Memon SA, Lee K, Rashid N (2018) Exploring the potential of microalgae for new biotechnology applications and beyond: a review. Renew Sust Energ Rev 92:394–404

Robertson L, Backer J, Biemans C, van Doorn J, Krab K, Reijnders W, Smit H, Willemsen P (2016) Antoni van Leeuwenhoek: Master of the Minuscule. Brill

Rodolfi L, Chini Zittelli G, Bassi N, Padovani G, Biondi N, Bonini G, Tredici MR (2009) Microalgae for oil: strain selection, induction of lipid synthesis and outdoor mass cultivation in a low-cost photobioreactor. Biotechnol Bioeng 102(1):100–112

Rohe P, Venkanna D, Kleine B, Freudl R, Oldiges M (2012) An automated workflow for enhancing microbial bioprocess optimization on a novel microbioreactor platform. Microb Cell Factories 11(1):1–14

Roles J, Yarnold J, Wolf J, Stephens E, Hussey K, Hankamer B (2020) Charting a development path to deliver cost competitive microalgae-based fuels. Algal Res 45:101721

Roles J, Yarnold J, Hussey K, Hankamer B (2021) Techno-economic evaluation of microalgae high-density liquid fuel production at 12 international locations. Biotechnol Biofuels 14(1): 1–19

Rosenblatt F (1958) The perceptron: a probabilistic model for information storage and organization in the brain. Psychol Rev 65(6):386

Ross IL, Shah S, Hankamer B, Amiralian N (2021) Microalgal nanocellulose–opportunities for a circular bioeconomy. Trends Plant Sci 26(9):924–939

Roy C, Tim J (2012) Marine geochemistry. John Wiley & Sons, Hoboken, NJ

Rupawalla Z, Robinson N, Schmidt S, Li S, Carruthers S, Buisset E, Roles J, Hankamer B, Wolf J (2021) Algae biofertilisers promote sustainable food production and a circular nutrient economy–An integrated empirical-modelling study. Sci Total Environ 796:148913

Rupawalla Z, Shaw L, Ross IL, Schmidt S, Hankamer B, Wolf J (2022) Germination screen for microalgae-generated plant growth biostimulants. Algal Res 66:102784

Schenk P, Thomas-Hall S, Stephens E, Marx U, Mussgnug J, Posten C, Kruse O, Hankamer B (2008) Second generation biofuels: high-efficiency microalgae for biodiesel production. Bioenergy Res 1(1):20–43

Scherer D, Krust D, Frey W, Mueller G, Nick P, Gusbeth C (2019) Pulsed electric field (PEF)-assisted protein recovery from Chlorella vulgaris is mediated by an enzymatic process after cell death. Algal Res 41:101536

Schwarzenbach G, Meier J (1958) Formation and investigation of unstable protonation and deprotonation products of complexes in aqueous solution. J Inorg Nucl Chem 8:302–312

Shah S, Li X, Jiang Z, Fahad S, Hassan S (2022) Exploration of the phytohormone regulation of energy storage compound accumulation in microalgae. Food Energy Security 11(4):e418

Sharma K, Nadda AK, Kumar G (2022) Advances in the resources to augment microalgal genetic engineering: omics-based resources, mutant libraries, and high-throughput screening techniques. In: Microalgae for sustainable products: the green synthetic biology platform. The Royal Society of Chemistry, London, pp 85–94

Silva TC, Eppink M, Ottens M (2022) Automation and miniaturization: enabling tools for fast, high-throughput process development in integrated continuous biomanufacturing. J Chem Technol Biotechnol 97(9):2365–2375

Sinha S, Jikare A, Ankulkar R, Mirza Y (2022) Development of miniaturized agar based assays in 96-well microplates applicable to high-throughput screening of industrially valuable microorganisms. J Microbiol Methods 199:106526

Sivakaminathan S, Hankamer B, Wolf J, Yarnold J (2018) High-throughput optimisation of light-driven microalgae biotechnologies. Sci Rep 8(1):1–13

Sonmez ME, Eczacıoglu N, Gumuş NE, Aslan MF, Sabanci K, Aşikkutlu B (2022) Convolutional neural network-support vector machine based approach for classification of cyanobacteria and chlorophyta microalgae groups. Algal Res 61:102568

Sonowal S, Chikkaputtaiah C, Velmurugan N (2019) Role of flow cytometry for the improvement of bioprocessing of oleaginous microorganisms. J Chem Technol Biotechnol 94(6):1712–1726

Sørensen M, Andersen-Ranberg J, Hankamer B, Møller BL (2022) Circular biomanufacturing through harvesting solar energy and CO2. Trends Plant Sci 27:655

Spiess A-N, Neumeyer N (2010) An evaluation of R2 as an inadequate measure for nonlinear models in pharmacological and biochemical research: a Monte Carlo approach. BMC Pharmacol 10(1):1–11

Stirke A, Celiesiute-Germaniene R, Zimkus A, Zurauskiene N, Simonis P, Dervinis A, Ramanavicius A, Balevicius S (2019) The link between yeast cell wall porosity and plasma membrane permeability after PEF treatment. Sci Rep 9(1):14731

Su M, Dell'Orto M, Scaglia B, D'Imporzano G, Bani A, Adani F (2022) Growth performance, biochemical composition and nutrient recovery ability of twelve microalgae consortia isolated from various local organic wastes grown on Nano-filtered pig slurry. Molecules 27(2):422

Sumantran VN, Zhang R, Lee DS, Wicha MS (2000) Differential regulation of apoptosis in normal versus transformed mammary epithelium by lutein and retinoic acid. Cancer Epidemiol Biomark Prev 9(3):257–263

Sutak R, Botebol H, Blaiseau P-L, Léger T, Bouget F-Y, Camadro J-M, Lesuisse E (2012) A comparative study of iron uptake mechanisms in marine microalgae: iron binding at the cell surface is a critical step. Plant Physiol 160(4):2271–2284

Szymański P, Markowicz M, Mikiciuk-Olasik E (2011) Adaptation of high-throughput screening in drug discovery—toxicological screening tests. Int J Mol Sci 13(1):427–452

Tanaka Y, Murase Y, Shibata T, Tanaka R, Mori T, Miyake H (2022) Production of 4-deoxy-L-erythro-5-Hexoseulose Uronic acid using two free and immobilized alginate Lyases from Falsirhodobacter sp. Alg1. Molecules 27(10):3308

Tandon P, Jin Q (2017) Microalgae culture enhancement through key microbial approaches. Renew Sust Energ Rev 80:1089–1099

Taylor RI (2021) Energy efficiency, emissions, tribological challenges and fluid requirements of electrified passenger car vehicles. Lubricants 9(7):66

Tillich UM, Wolter N, Schulze K, Kramer D, Brödel O, Frohme M (2014) High-throughput cultivation and screening platform for unicellular phototrophs. BMC Microbiol 14(1):239

Van Wagenen J, Holdt SL, De Francisci D, Valverde-Pérez B, Plósz BG, Angelidaki I (2014) Microplate-based method for high-throughput screening of microalgae growth potential. Bioresour Technol 169:566–572

Volpe C, Vadstein O, Andersen G, Andersen T (2021) Nanocosm: a well plate photobioreactor for environmental and biotechnological studies. Lab Chip 21(10):2027–2039

Vraspir JM, Butler A (2009) Chemistry of marine ligands and siderophores. Ann Rev Mar Sci 1:43–63

Walne PRJFIS (1970) Studies on the food value of nineteen genera of algae to juvenile bivalves of the genera Ostrea, Crassostrea, Mercenaria and Mytilus. Fish Invest Ser 26(5):62

Wang Y, He B, Sun Z, Chen Y-F (2016) Chemically enhanced lipid production from microalgae under low sub-optimal temperature. Algal Res 16:20–27

Ward A, Deo RC, Sharma E, Soar J (2022) Multi-label classification and deep image analysis methods on marine microalgae and rotifer identification using transfer learning and two-dimensional convolutional neural networks. Available at SSRN 4043821

Westerwalbesloh C, Brehl C, Weber S, Probst C, Widzgowski J, Grünberger A, Pfaff C, Nedbal L, Kohlheyer D (2019) A microfluidic photobioreactor for simultaneous observation and cultivation of single microalgal cells or cell aggregates. PLoS One 14(4):e0216093

Whitesides GM, Ostuni E, Takayama S, Jiang X, Ingber DE (2001) Soft lithography in biology and biochemistry. Annu Rev Biomed Eng 3(1):335–373

Wold S, Esbensen K, Geladi P (1987) Principal component analysis. Chemom Intell Lab Syst 2 (1-3):37–52

Wold S, Sjöström M, Eriksson L (2001) PLS-regression: a basic tool of chemometrics. Chemom Intell Lab Syst 58(2):109–130

Wolf J, Ross IL, Radzun KA, Jakob G, Stephens E, Hankamer B (2015) High-throughput screen for high performance microalgae strain selection and integrated media design. Algal Res 11:313–325

Wolf J, Stephens E, Steinbusch S, Yarnold J, Ross IL, Steinweg C, Doebbe A, Krolovitsch C, Müller S, Jakob G (2016) Multifactorial comparison of photobioreactor geometries in parallel microalgae cultivations. Algal Res 15:187–201

Worms I, Simon DF, Hassler C, Wilkinson KJB (2006) Bioavailability of trace metals to aquatic microorganisms: importance of chemical, biological and physical processes on biouptake. Biochimie 88(11):1721–1731

Yadav D, Jalal A, Garlapati D, Hossain K, Goyal A, Pant G (2020) Deep learning-based ResNeXt model in phycological studies for future. Algal Res 50:102018

Yarnold J, Ross IL, Hankamer B (2016) Photoacclimation and productivity of Chlamydomonas reinhardtii grown in fluctuating light regimes which simulate outdoor algal culture conditions. Algal Res 13:182–194

Yarnold J, Karan H, Oey M, Hankamer B (2019) Microalgal aquafeeds as part of a circular bioeconomy. Trends Plant Sci 24(10):959–970

Yu Z, Geisler K, Leontidou T, Young RE, Vonlanthen SE, Purton S, Abell C, Smith AG (2021) Droplet-based microfluidic screening and sorting of microalgal populations for strain engineering applications. Algal Res 56:102293

Zhang S, Hou Y, Liu Z, Ji X, Wu D, Wang W, Zhang D, Wang W, Chen S, Chen F (2020) Electro-Fenton based technique to enhance cell harvest and lipid extraction from microalgae. Energies 13(15):3813

Zhang R, Liu Y, Pan Q, Khan A, Bai X, Ali M, Yang W, Zhang L, Li B (2023) The effects of short term blue light treatment on promoting nutrition value in Chinese cabbage. Food Chem 412: 135542

Zhao Y, Wang H-P, Han B, Yu X (2019) Coupling of abiotic stresses and phytohormones for the production of lipids and high-value by-products by microalgae: a review. Bioresour Technol 274:549–556

Zhou W-M, Yan Y-Y, Guo Q-R, Ji H, Wang H, Xu T-T, Makabel B, Pilarsky C, He G, Yu X-Y (2021) Microfluidics applications for high-throughput single cell sequencing. J Nanobiotechnol 19:1–21

Zhou S, Jiang J, Hong X, Fang Y, Hong Y, Fu P (2022) Marine microalgae detection in microscopy images: a new dataset. arXiv preprint arXiv:2211.07546

Zittelli GC, Lavista F, Bastianini A, Rodolfi L, Vincenzini M, Tredici MJJoB (1999) Production of eicosapentaenoic acid by Nannochloropsis sp cultures in outdoor tubular photobioreactors. J Biotechnol 70(1–3):299–312

Catalyst in Action

Sivakumar Esakkimuthu, Shuang Wang, and Abdelfatah Abomohra

Abstract Algae exhibit distinguishing potential of producing various products from fuels to wide range of value-added products, described in the present book as phycochemicals. Few decades of research analyzed several pathways of converting microalgal biomass into bioproducts. The product conversion efficiency, sustainability, and economics of the production processes depend on the type of reactions opted and catalyst used for producing the targeted product. Catalyst plays vital role in overall economics and yield of the target product. For biofuel production, chemical and biological catalysts were extensively researched. The inherent disadvantages of homogeneous catalysts include tough separation from the reaction system and thermal instability which increased the heterogeneous catalyst applications. On the other hand, green catalysts are increasingly attractive that are mainly made from biomass, especially enzymes which are effective and environmentally friendly. The term "phycocatalysts" can be identified as the catalysts made from algal biomass and are emerging recently due to their exquisite catalytic potential and acting as another value-added choice for integrated microalgal biorefinery. Moreover, both nanocatalysts and biocatalysts are widely attractive due to plentiful of advantages such as easy synthesis, simple disposal, and high reusability, along with enhanced yield of the desired product. Deep eutectic solvent (DES) and cyanobacteria which are recent intrigue in the field of catalysis were also discussed in the chapter. Thus, catalysts have become indispensable in algal biomass conversion which influences the yield of every algal product synthesis/recovery which have been elucidated elaborately in this chapter.

Keywords Algae · Seaweeds · Phycocatalyst · Biofuels · Enzyme · Nanocatalyst · DES · Energy balance

S. Esakkimuthu · S. Wang (✉)
School of Energy and Power Engineering, Jiangsu University, Jiangsu, China
e-mail: alexjuven@ujs.edu.cn

A. Abomohra
Aquatic Ecophysiology and Phycology, Institute of plant science and Physiology, University of Hamburg, Hamburg, Germany

A. Abomohra, S. Ende (eds.), *Value-added Products from Algae*,
https://doi.org/10.1007/978-3-031-42026-9_11

1 Introduction

Biomass into energy or value-added products conversion could be achieved mainly in three ways including thermochemical conversion, chemical separation, and biochemical processes (Abomohra et al. 2020; Wang et al. 2022a, 2022b). However, the rate of biomass conversion depends upon the path chosen. For instance, though lipid extraction can be achieved through chemical or biological means, the rate of lipid accumulation takes several days to three weeks in most of the algal species (Esakkimuthu et al. 2019, 2020). In case of thermochemical conversion, though time taken for biomass growth is not trivial, the conversion rate and amount of products' production in a single step are the key. On the other hand, the biochemical conversion process involving the production of biogas and bioethanol is significantly taking considerable time. The products from algae can be broadly categorized into two categories including low-value high-volume products and high-value low-volume products. Former includes fuels such as biodiesel, bioethanol, bio-oil, and biogas, whereas the later includes pigments, nutraceutical products, and food supplements.

The rate of biomass conversion is proven to be important irrespective of the type of targeted product and type of pathway chosen for producing it (Mathimani et al. 2019). Although outstanding advancements in industrial scale production of fuels and value-added products have been achieved, the energy cost for the production process is very high as production of microalgal biomass itself remains ten times higher than the cost of fossil fuel (Chisti 2013; Milano et al. 2016). To exemplify, thermochemical conversion of microalgal biomass demands extreme reaction conditions at the expense of high energy. Thermochemical conversion of biomass could be defined as the thermal process applied for decomposition of organic compounds of the biomass resulting in the formation of biofuels and/or value-added compounds. Thermochemical conversion basically includes direct combustion, pyrolysis, gasification, and hydrothermal liquefaction (HTL) (Aliyu et al. 2021). Direct combustion is the process of burning out the biomass at high temperature (800–1000 °C) in furnace for the conversion of stored chemical energy into gases. However, direct combustion requires stringent pretreatment steps including dehydration, cutting, and crushing which are energy demanding and cost-ineffective (Lam et al. 2019). In the process of combustion, the combustible materials of algal biomass are heated to produce gases, whereas in case of gasification, combustion occurs at high temperatures (700–1400 °C) by partial oxidation along with oxidizers (e.g., Air, O_2, CO_2, steam, etc.). In case of pyrolysis, thermal decomposition of biomass occurs in the absence of oxidizers within temperature range of 400–700 °C. Hydrothermal liquefaction involves thermal depolymerization in the presence of water under high pressure within a temperature range of 250–450 °C (Lee et al. 2022). Usually, for the production of hydrogen and power generation, gasification is preferred. In case of multiple applications through multiple products in the form of bio-oil, bio-char, and gas, pyrolysis and hydrothermal liquefaction are preferred. Figure 1 represents

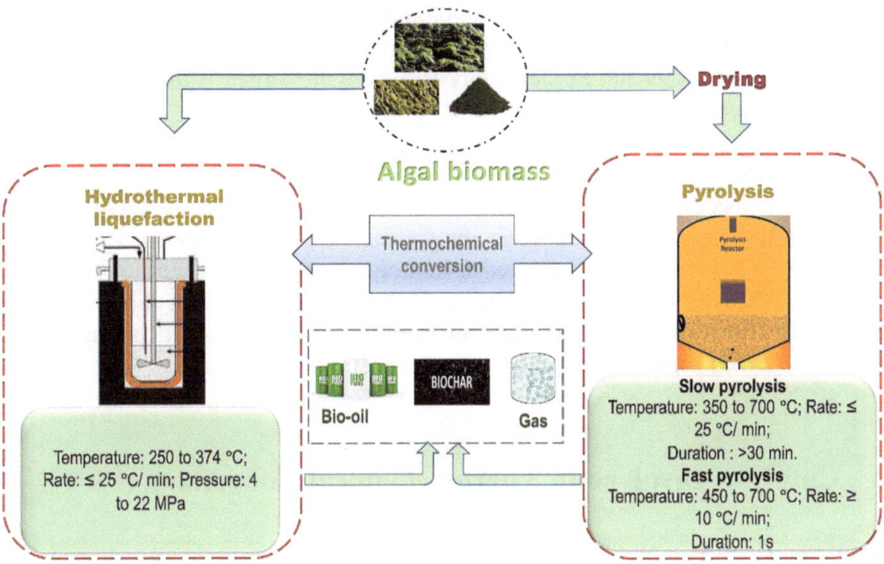

Fig. 1 Hydrothermal liquefaction and pyrolysis as common routes of thermochemical conversion of algal biomass. Data in the figure are extracted from Aliyu et al. (2021)

the above-mentioned two common modes of thermochemical conversion along with the specific reaction conditions.

Notably, the arrival of catalyst to thermochemical conversion processes reduces the energy demand for higher temperature and also influences the yield. In addition, catalyst application enhances the quality of specific targeted product and is proven to exhibit selectivity over the products produced. In recent decades, plenty of research efforts have been made on exploring novel catalysts finding and elucidating their mechanism of action. There are several challenges in catalyst application too, especially due to the complexity of biomass and its elemental composition. To achieve the excellent rate of conversion along with lesser energy consumption and eventual cost-effectiveness, catalysts are widely employed in every bioproduct production from algal biomass. Ample of investigations and widespread knowledge on application of different catalysts have been accumulated so far. Catalyst can be ideally defined as any substance that increases the rate of reaction without undergoing any irreversible chemical changes. Application of catalyst to enhance the reaction rate is not new, because it is believed that thermochemical process involved during fossil fuel formation was enhanced by clay as a catalyst (Castello et al. 2019).

Catalyst application is proven to be effective in several ways. For instance, catalyst application improved the quality of thermochemical products in gas, liquid, and solid forms. Catalysts are proven to reduce the activation energy in tar cracking and increase gas yields (Lee et al. 2022). However, the successful application of catalyst depends upon the wide range of attributes such as selection of appropriate catalyst, compatible reactors in case of thermochemical conversion, optimized proportions, and elemental nature of the biomass (Hu et al. 2021). Catalysts applied for biofuel production, especially thermochemical conversions, are broadly classified as in situ catalysts and ex situ catalysts, where the former is utilized during the process and later is used with the products produced namely gases and liquids (Lee et al. 2022). Catalyst structure plays a key role and it is usually studied with the help of microscopic techniques (Scanning Electron Microscopy, SEM) and X-ray diffraction and spectroscopic studies (X-ray photoelectron spectroscopy). On the other hand, catalyst can be used either in the same phase as like reactants or used in different phases which are categorized as homogeneous and heterogeneous catalysts, respectively (Sharma et al. 2021). With such short note, the present chapter is likely to include basics of catalyst application in producing value-added products from algae along with some needed insights and contemporary glimpses.

2 Catalysts Involved in Thermochemical Conversion

Algal biomass acts as a reservoir for various products and compounds which can be sectioned as fuels and value-added products (Lee et al. 2022). Moreover, "alternative fuels" have been an urging alarm for last few decades and most of the transportation sectors still depend on liquid fuel along with emerging electric vehicles (Bhardwaj et al. 2020). Thus, with the essentiality and prevalent focus on fuels and multiple products conversion, thermochemical conversion using HTL has been on focus and catalysts play a significant role in this process. Due to its advantages over other processes as discussed in Sect. 1, this section will focus on HTL.

2.1 Hydrothermal Liquefaction

For every thermochemical conversion method, catalyst plays a key role in influencing the rate of the reaction and yield of the products. This section of the chapter details the catalysts involved in thermochemical conversion of the process, mainly pyrolysis and HTL. HTL involves the thermal conversion of the wet algal biomass, resulting in the depolymerization of biomass. The major advantage of HTL is skipping the biomass drying as it is energy and cost-intensive process, whereas the other common routes (e.g., pyrolysis) require a drying step (Sharma et al. 2021; Wang et al. 2020). In addition, considering the energy consumption and quality of bio-oil production, HTL is proven to be effective as compared to pyrolysis (Vardon et al. 2012). Indeed, because of biomass drying and higher temperature requirements by pyrolysis, about 1.6 times higher energy consumption than HTL was observed along with 14% lesser energy recovery (Zhang et al. 2017). Although different routes of thermochemical conversion come under one roof, the reaction system is unique and differs based on the process. For instance, HTL has water in the reaction system which slowly converts to nonpolar characteristic due to supercritical conditions and the hot water under elevated pressure exhibits higher reactivity which are mainly catalyzed by acid catalysts (Durak and Aysu 2016a).

The major targeted product of HTL is the crude bio-oil, even though other products are generated during the process with wide range of applications (Almutairi 2022; El-Hefnawy et al. 2023; Li et al. 2021). Microalgal species with higher biomass production and lipid contents are usually preferred for HTL to achieve higher bio-oil yield. The major reaction conditions such as temperature and reaction time are optimized to enhance the yield. Among different factors influencing the optimization of HTL, catalyst application is found to be highly substantial in bio-oil production process. During the process of HTL, at higher temperature and pressure, water exhibits a property such as incremental ionic products and facilitates as medium for acid base catalyst (Déniel et al. 2016). Variety of catalysts such as zeolite, Pt/C, NiO, and HZSM-5 have been widely employed for the enhancement of bio-oil conversion from biomass (Gao et al. 2017). Iron was proven to be the most efficient catalyst for enhancing the bio-oil yield and quality (Durak and Aysu 2016b). Catalyst can influence the targeted product, as Fe catalyst is proven to increase the aqueous phases, whereas Zn as a catalyst improves the gaseous products. Similarly, the quality of the aqueous phase also varies as Fe as a catalyst tends to increase the aromatic compounds, while Zn increases polyaromatic and aliphatic compounds (Durak and Genel 2020).

As compared with non-catalytic reaction, catalytic HTL reduces the energy consumption and introduces cascade of reactions such as cleavage of bonds, transfer of hydrogen, aromatization, decarboxylation, decarbonylation, polymerization, and condensation (Sharma et al. 2021). Thermochemical conversion, including both HTL and pyrolysis, predominantly attracted for bio-oil production, whereas other value-added products could also be recovered from the HTL of algal biomass. HTL of algal biomass results in the formation of bio-oil, aqueous phase, and gaseous products. Recently, emerging interest on effective utilization of aqueous phase for different applications has developed. The waste water resulting from HTL of algal biomass called aqueous phase predominantly contains high concentration of organic compounds (Almutairi 2022; El-Hefnawy et al. 2023). Interestingly, aqueous phase has been applied as reaction medium for HTL and nutritive medium for microalgal growth (Swetha et al. 2021). In addition, the organic compounds in aqueous phase have various applications. For instance, glycolic acid from HTL aqueous phase has demanding reception in cosmetic industries (Vatankhah et al. 2020). Similarly, phenols from aqueous phase have many potential applications and notably in industries dealing with dye, pigment, and resin productions (Holladay et al. 2007). The aqueous phase of HTL also contains N, P, and K which can be used as fertilizer (Li et al. 2019a, b). However, the chemical oxygen demand (COD) is high (84 g L^{-1}) in the HTL aqueous phase of algal biomass load and hence purification of compound and dilution of aqueous phase are inevitably required for successful application (Watson et al. 2020).

2.1.1 Types of Catalyst in HTL

Both homogeneous and heterogeneous catalysts widely used are known to exhibit inherent advantages and disadvantages. Table 1 lists the influence of catalysts on biocrude yield produced from HTL of various feedstocks. Homogeneous catalysts are known to exist in the same phase of the reactants, where mostly salts and acids are widely used. In heterogeneous catalysis, the reactants and catalysts present in different phases and the common heterogeneous catalysts are transition metal oxides and rare metals (Shakya et al. 2017). There are plenty of advantages associated with heterogeneous catalysts over homogeneous catalysts, and hence heterogeneous catalysts are mostly preferred for HTL. Heterogeneous catalysts are advantageous in terms of easy separation, are capable of withstanding wider experimental conditions, and have lesser corrosion to the reactors. In addition, heterogeneous catalysts were proven to increase the bio-oil yield along with reduced heteroatoms which were higher in HTL without catalysts. In the bio-oil produced with heterogeneous catalyzed HTL, 2% increase in higher heating value (HHV) was observed as compared

Table 1 Impact of catalyst on the bio-oil yield under hydrothermal liquefaction process of different algal species

Algal biomass	Catalyst	Catalyst type	Bio-oil yield (wt%)	Percentage of increase over other catalyst	References
Nannochloropsis sp.	Ni/TiO_2	Heterogeneous	48	65 (Fe/TiO_2)	Wang et al. (2018)
Chlorella vulgaris	Pt/Al	Heterogeneous	39	30 (Ni/Al)	Biller and Ross (2011)
Chlorella vulgaris	Na	Homogeneous	48	84 (H_2SO_4)	Ahmad et al. (2013)
Chlorella sorokiniana	KOH, H_2SO_4 Amberlyst-15	Heterogeneous	91.6	–	Dong et al. (2013)
Spirulina platensis	CeO_2	Homogeneous	34	–	Kandasamy et al. (2020)
Chlorella protothecoides	HCl, NaOH in MeOH	Homogeneous	80	90 (Na_2CO_3)	Tran et al. (2010)
Nannochloropsis sp.	H-ZSM5	Homogeneous	55	30 (Na_2CO_3)	Tran et al. (2010)
Ulva prolifera	KOH	Homogeneous	26	22 (Na_2CO_3)	Yan et al. (2019)
Nannochloropsis sp.	Pd/C, Pt,C, Ru/C, Ni/SiO_2-Al_2O_3, $CoMo/Al_2O_3$, Zeolite CaO/Al_2O_3	Heterogeneous	35–37	19	Duan and Savage (2011a)
Spirulina sp.	Kaolin	Heterogeneous	39.71	–	Wang et al. (2017)

wt% - weight percentage to the total biomass

to non-catalytic process and energy recovery was almost 100% which is highly appropriate for combustion (Duan and Savage 2011a). Due to ease of separation, reusing potential of heterogeneous catalysts is comparatively higher than homogeneous catalysts which is beneficial in improving the economy of the process. Both HTL process and upgrading of the produced biocrude were catalyzed by few heterogeneous catalysts in the presence of hydrogen, which revealed that maximum yield was obtained by Ni-Ru/CeO_2 catalysis. It was proven that hydrogen is a critical constituent for the catalysis in order to enhance the bio-oil yield and upgrade the biocrude characteristics (Xu et al. 2018a, b).

A study was carried out to reveal the interaction between temperature and catalysts in which lower temperature (250 °C) along with Na_2CO_3 promoted higher bio-oil yield (Shakya et al. 2015). In another study, sodium bicarbonate proven to increase the bio-oil yield as sodium carbonate effectively reacts with water and produces bicarbonates in which 23% of bio-oil was produced from macroalga *Enteromorpha prolifera* (Zhou et al. 2010). Palladium/carbon as a catalyst was found to be very effective in enhancing the bio-oil yield by 62% as compared to non-catalytic HTL (Duan and Savage 2011a). Similarly, a comparative study on employing different catalysts such as KOH, NaOH, and Na_2CO_3 identified KOH to potentially improve the bio-oil yield by one-fold (Yan et al. 2019). A comparative study on different catalysts such as Ni/C, ZSM-5, Ni/ZSM-5, Ru/C, and Pt/C was made on HTL of *Nannochlorposis* species and found that bio-oil yield was enhanced (61% by weight) under Ni/C, while Pt/C improved the HHV and nitrogen content in the bio-oil (Shakya et al. 2017). *Sargassum tenerrium* was subjected to HTL involving Cao as catalyst supported by cesium oxides, zirconium oxides, and aluminum oxides which resulted in maximum yield of above 30% of bio-oil under CaO supported by zirconium catalysis (Biswas et al. 2020). This study employed water and ethanol as solvent and suggested that solvent is also substantial for bio-oil production processes. It is clearly evident from many studies that catalyst plays a huge role in decomposing pattern of biomolecules, gases (CO_2, methane, and hydrogen), and introduce pronounced effects in all HTL products.

In addition to catalyst, temperature is also proven to be very effective in influencing the HTL reaction performance and products. Temperature, along with catalyst, is proven to influence the characteristics such as quantity and HHV of the bio-oil. From the myriad of reports, it was identified that the optimum range of temperature is between 240 and 370 °C, and minimal variation from these ranges was highly effective for catalytic HTL and products (Biswas et al. 2017; Reddy et al. 2016). Among catalyst types, heterogeneous catalysts are found to be effective in terms of improving the quantity, quality, and other products. However, producing bio-oil by HTL also has elevated levels of heteroatoms such as nitrogen, oxygen, and sulfur which can be alleviated by effective optimization of catalysts and temperature (Wang et al. 2018). The biochemical composition of algae also influences the oil characteristics and various studies dealt with optimization of cultivation conditions for enhanced biomass composition. The other factors such as reaction time, pressure, and feedstock concentration have considerable influences on the yield of HTL products (Ravichandran et al. 2022).

Pressure of the reaction system helps to maintain the reaction phase in a single-medium phase, and as a result, hydrolysis rate has been controlled. Apart from these, under subcritical condition, influence of pressure is negligible on the yield of HTL products. Hence, it is noteworthy that the pressure has negligible effect on the crude yield, whereas it has significant impact on energy consumption (Sangon et al. 2006). Residence time is found to be critical in determining the yield of the products. For instance, if the residence time increased from 10 to 60 mins, the yield of biocrude increased by 2% (wt%) (Valdez et al. 2012). Another important factor is cosolvent addition as water itself acts as catalyst in HTL. For example, when ethanol was added as a cosolvent in HTL of *Nannochloropsis* sp., biocrude yield was increased by 6% (wt%) as compared to HTL without cosolvent addition (Caporgno et al. 2016). Similarly, increment of feedstock (*Nannochloropsis* sp.) from 5% to 35% along with residence time of 60 mins increased the bio-oil yield by 10% (Valdez et al. 2012). However, temperature stands tall in terms of influencing the yield of the products. For example, HTL of *Nannochloropsis* sp. was significantly influenced by temperature even without catalyst and temperature ranging between 300 °C and 350 °C was ideal for increasing biocrude yield (43%) (Brown et al. 2010). Although multiple factors optimization is essential for enhanced production of bio-oil and other value-added products through HTL, catalyst along with temperature is substantial. It is due to the fact that temperature could influence the formation of higher molecular weight which in turn causes deactivation of the catalyst (Zoppi et al. 2021). The phenolic oligomers are found to be responsible for catalyst deactivation due to fouling phenomenon (Zoppi et al. 2021). This catalyst deactivation seems to be one of the major challenges for HTL-mediated bio-oil or products recovery, but with appropriate optimization of temperature and other parameters, the chances of catalyst deactivation can be reduced (Sharma et al. 2021).

2.2 *Catalysts Involved in Pyrolysis*

Pyrolysis is considered as the conventional mode of thermochemical conversion where biomass is decomposed in the absence of oxygen at extremely high temperature. During pyrolysis, sequential processes such as dehydration, decarboxylation, fragmentation, polymerization, and atomics rearrangements occur (Bach and Chen 2017). Algal biomass is mainly composed of carbohydrates, proteins, and lipids, which are decomposed at different temperature ranges. There are potential attentions and insights over bio-oil production through pyrolysis; however, there is a substantial scope for the production of value-added products which can be recovered from algal biomass pyrolysis. This can be exemplified with the recovery of N-containing compounds from algal pyrolysis. Algae contains higher nitrogen content (10%) as compared to other feedstocks (agricultural and woody wastes) in pyrolysis which may result in nitrous oxide emission and nitrogen transformation into NH_3, which could affect the environment as nitrogen-based pollution. Protein concentration in algal biomass attributed to the nitrogen abundancy in pyrolytic products. Pyridine, pyrole, and indole are the major nitrogen-containing compounds in the bio-oil from algal pyrolysis. However, these products, if recovered, can have substantial application in the field of pharmaceuticals, chemicals, and other personal care products. An investigation made by Chen et al. (2017) analyzed the transformation mechanism of nitrogen-containing species by fast pyrolyzing algae such as *Spirulina platensis* and *Enteromorpha prolifera* where amines and amides were decreased under increasing temperature from 400 °C to 800 °C. However, nitriles and N-heterocyclic compounds were increased at the same time, which delineate the transformation and evolution mechanism of nitrogen-containing species.

Another important product of pyrolysis is biochar, which has wider applications. Recent attractive application of biochar includes preparation of catalysts from algal char which was detailed later in Sect. 5. However, biochar has other potential application such as soil fertilizer due to the high nitrogen, phosphorous, and other nutrients (Roberts et al. 2015). Another familiar application of algal biochar is its application in waste water treatment due to its high ion exchange capacity (Michalak et al. 2019). In addition, this algal biochar could also be used as super capacitor material due to its high capacitance and stable nature (Han et al. 2019). Based on the reaction conditions and rate of temperature increment, pyrolysis can be classified as slow and fast pyrolysis. Comparatively, algal biomass pyrolysis is highly efficient in producing the bio-oil yield, whereas dried algal biomass requirement is a costly procedure. Slow pyrolysis promotes biochar formation because of long residence times and eventual secondary reactions towards char accumulation. In case of fast pyrolysis, less residence time prevents such secondary reactions, leading to more bio-oil accumulation (Chen et al. 2015). The major problem with pyrolysis is the formation of oxygen- and nitrogen-containing compounds in the bio-oil due to secondary reactions, which impedes the fluidity of the oil due to high viscosity (Ong et al. 2019). Moreover, it also reduces the HHV under storage which makes it unfit for blending with conventional fuel. In addition, high nitrogen content in the biofuel results in environmental hazards by emitting NO_x on combustion (Abomohra et al. 2021; Ren et al. 2009). Catalytic pyrolysis of algal biomass is helpful in two main ways as it reduces the temperature requirement of the process, and it also controls the oxygen and nitrogen compounds through deoxygenation and denitrogenation, respectively (Dabros et al. 2018). This results in relative increase of hydrocarbons in the bio-oil and thereby exhibiting finer fuel quality. Thus, selecting a suitable catalyst for algal pyrolysis is a very critical step in achieving the bio-oil with finer quality and also for cost-effective processes. So far, there are numerous catalysts that have been studied for algal pyrolysis as discussed below.

2.2.1 Zeolites

Zeolites is one group of familiar catalysts in a wide range of chemical industries due to its characteristic porous structure. Zeolite catalysts can perform catalytic activity by enhancing the rate of decarboxylation, decarbonylation, and deoxygenation of the bio-oil leading to enhanced quality (Hu et al. 2020; Li et al. 2017). It was mainly due

Table 2 Influence of catalysts on the quality of the bio-oil produced from different microalgal species

Feedstock	Sample	Catalyst	Reduced products	Hydrocarbons	References
Microalgae	Crude bio-oil	Ni supported on Zeolite HBeta	NR	78%	Wang et al. (2009)
Nannochloropsis	Crude bio-oil	Sulfided CoMo/F–Al_2O_3	Oxygenates and Nitrogenates	80 to 85%	Elliott et al. (2013)
Scenedesmus sp.	Crude bio-oil	Pt/C, Ru/C, Ni/C,and Co/C	Oxygenates and Nitrogenates	Specific hydrocarbons increased; values not reported	Yang et al. (2016)
Chlorella pyrenoidosa	Bio-oil	Ru/C	NR	Straight chain alkanes enhanced	Xu et al. (2018a, b)
Nannochloropsis sp.	Crude bio-oil	Pt/C, Mo2C, and HZSM-5	Oxygenates and Nitrogenates	Pentadecane increased	Duan et al. (2018)
Nannochloropsis sp.	Crude bio-oil	Pd/C	Oxygenates and Nitrogenates	82%	Duan and Savage (2011b)
Chlorella vulgaris	Crude bio-oil	Zeolite catalyst ZSM-5	NR	24%	Wang and Brown (2013)
Chlorella	Crude bio-oil	Na_2CO_3	Oxygenates and Nitrogenates	3.1%	Babich et al. (2011)
Chlorella pyrenoidosa	Crude bio-oil	Pt/γ-Al_2O_3	NR	n-alkanes increased	Duan et al. (2013)
Chlorella pyrenoidosa	Crude bio-oil	ZSM-5		32%	Xinglong et al. (2013)

NR Not reported

to the presence of acidic sites in the catalysts facilitating carbon-ion mechanism. The acidity of protonic type of Zeolite Socony Mobil-5 catalyst (H-ZSM5) is mainly exhibited due to lower Si/Al ratio which increases aromatic hydrocarbons in the bio-oil (Thangalazhy-Gopakumar et al. 2012). It also involved in breaking larger compounds into smaller compounds which will pass through catalytic pores, resulting in the formation of hydrocarbons such as toluene, anthracene, benzene, naphthalene, and xylene.

Zeolites also involved in reducing the oxygen content through deoxygenation of furans, phenols, and aldehydes which will further enhance the hydrocarbons in the bio-oil. Table 2 shows the impact of various catalysts over hydrocarbon increment in the bio-oil. For instance, hydrocarbon content in the bio-oil from pyrolysis of the green microalga *Chlorella vulgaris* was enhanced up to 25% under H-ZSM5 which was less than 1% under non-catalytic process (Thangalazhy-Gopakumar et al. 2012).

The ratio of catalyst to algal biomass was a substantial factor in improving the hydrocarbon proportion. For instance, increment from 1 to 5 times ratio of the catalyst showed pronounced enrichment of about 24% in aromatic hydrocarbons concentration during pyrolysis of *C. vulgaris*. During these processes, the reduction of nitrogen and oxygen was achieved due to release of ammonia and carbon oxides, respectively (Wang and Brown 2013). Thus, it is well established that zeolites could potentially improve the algal bio-oil quality, and among zeolites, H-ZSM5 was proven to be the most effective form including H-Y and H-B zeolites.

2.2.2 Metal-Loaded Zeolites

The active metals such as copper, nickel, platinum, and palladium are widely loaded with zeolites catalyst and investigated for its influence on pyrolysis. It was observed that aromatic hydrocarbons were magnificently enhanced due to this metal loading as a consequence of enhanced deoxygenation and denitrogenation reactions. On compared with ZSM-5, metal-loaded catalyst has significantly reduced the coke formation (French and Czernik 2010). Recent intrigue on loading H-ZSM5 with noble metals has been widely followed, as bio-oil quality can be improved with the application of different types of catalysts (Gong et al. 2014). At present, rudimentary understanding over metal-loaded zeolite preparation, activity, selectivity, and stability is available, which offers a potential space for exploration.

2.2.3 Metal-Organic Frameworks

Recently, metal organic framework is grabbing attention as it has wider range of applications in the field of drug manufacturing and catalysis. The characteristic microporous structure of metal-organic framework slightly mimics the 3D porous structure of zeolites in catalyzing the pyrolysis (Wang et al. 2009). In addition, the size of porous structure could be altered which was found to be beneficial in improving the catalytic performance. Despite benefits, metal organic frameworks have limitation like catalyst activation along with specific sites seems to be highly challenging. However, further advancements and studies will help in functionalizing these metal-organic frame work along with specific active sites which could potentially overcome the current limitation (Li et al. 2019a, b).

2.2.4 Few Promising Pyrolysis Catalysts

In a row of effective catalysis for algal biomass pyrolysis, silica-supported nickel phosphide is attractive. The catalyst promoted higher amount of hydrocarbon predominantly with alkanes. Compared to non-catalytic reaction, catalytic pyrolysis promoted about 8.42-time in hydrocarbons. Moreover, about 16% increase in HHV was observed as compared to non-catalytic reactions. It was identified that the reusage of catalyst needs to be subjected to calcination to obtain similar effect in the quality of bio-oil as carbon deposits are found to be influential in activity of the catalyst (Wang and Brown 2013). γ-Al_2O_3 was identified as promising catalyst in recent days as it significantly improved the HHV compared to zeolite catalyst and non-catalytic pyrolysis. It is found to be advantageous over zeolites with larger pore structure which is resulting in permitting biomolecules with higher molecular weight and thereby improve the thermal decomposition effectively (Liu et al. 2014).

The nanoporous catalyst Meso-MFI was used to pyrolyze the seaweed *Laminaria japonica* which resulted in significant increment in aromatics and phenolic compounds, which was mainly attributed to the acid sites. These acid sites facilitate the oligomerization of ethylene kind of compounds and simultaneous cyclization is likely to occur to form aromatic compounds. The presence of acid sites is found to be essential in forming hydrocarbons from larger biomolecules (Lee et al. 2011). To date, the major challenge in catalytic pyrolysis prevails in catalyst processing as there are two ways such as in situ and ex situ processes, where the former involves the mixture of catalysts and algae and the later allows the passage of pyrolysis vapor over catalyst (Zainan et al. 2015). Studies prove that ex situ process promoted significant proportions of hydrocarbons as compared to in situ processes. Still significant research is needed in terms of catalytic pyrolysis of algae to identify the suitable catalyst to perform in appropriate mode. With present studies, it is identified that zeolites and metal-loaded zeolites are promising in improving the quality of the bio-oil. To attain refined bio-oil with commendable fuel quality, it is essential to analyze the relationship between catalyst features and catalytic process of deoxygenation and denitrogenation. The studies dealing with catalytic process based on biochemical composition of algal biomass would bring value-added insights for catalytic upgrading (Ahmed et al. 2013). Figure 2 outlines the impact of catalyst type on pyrolysis reaction processes and on the quality of the bio-oil.

As mentioned before, studies on catalytic pyrolysis of algal biomass are still at rudimentary level and need to be explored mainly in deoxygenation and denitrogenation to produce bio-oil with lower viscosity and lower hazard to the environment. So far, zeolites-based catalysts are effective for pyrolysis of microalgal biomass, whereas metal-loaded catalysts were specific for improving deoxygenation reactions. On comparing to lignocellulosic biomass pyrolysis, understandings about catalytic algal pyrolysis are at their infancy level. Thus, it requires reactor design optimization for successful application of catalyst in cost-effective and process-intensive ways (Li et al. 2019a, b). Many modelling and simulation studies are required to serve the cause of achieving algal pyrolysis oil with refinements such as lower viscosity, good stability, and enhanced hydrocarbon contents along with high HHV. In addition, more mechanistic studies on catalytic pyrolysis of microalgae are needed along with insights over designing suitable reactors for microalgal pyrolysis.

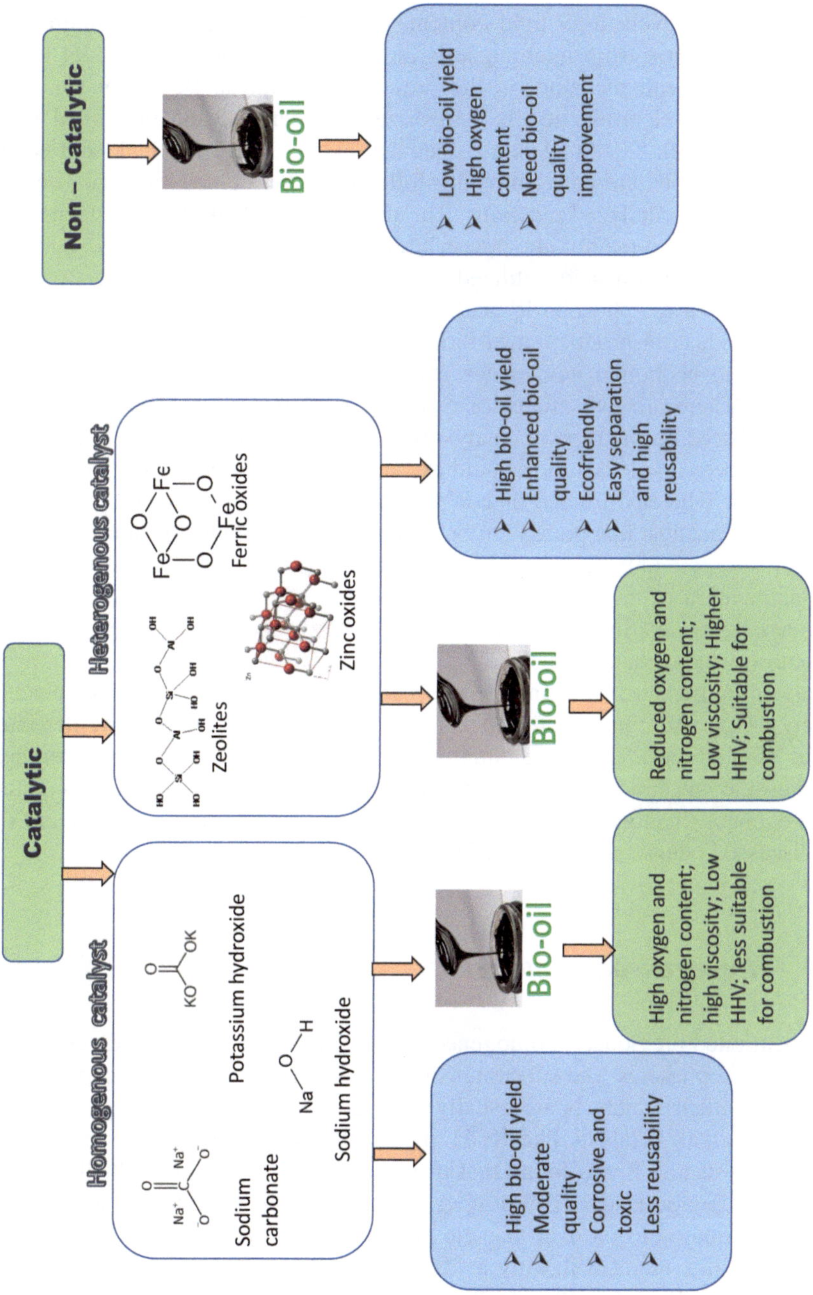

Fig. 2 Influence of catalysts types on the reaction processes and bio-oil quality

3 Catalyst in Biodiesel Production

Due to the relatively high lipid content, the most familiar and common route of biofuel production from microalgae is biodiesel production. Microalgal lipids are transesterified with methanol in the presence of catalyst. Based on the catalysts applied, biodiesel production is categorized as chemical and biological methods (Thangaraj et al. 2014). Among chemical methods, alkali- and acid-catalyzed transesterifications are conventionally followed, where alkali catalysts seem to be faster than acid-catalyzed reactions. The most common alkali catalysts are sodium hydroxide, sodium methoxide, potassium hydroxide, potassium amide, and sodium amide. The common acids utilized as catalysts for biodiesel production include hydrochloric acid, phosphoric acid, and sulfuric acid (Mathew et al. 2021). In addition, industrial production of biodiesel is mostly accomplished by alkali catalysts due to economic advantages and higher yield of fatty acid methyl esters (FAME). There are two modes of reactions for biodiesel conversion in chemical methods based on the type of catalyst used. For instance, the conventional esterification process involves sulfuric acid as catalyst which esterifies the free fatty acids into esters; whereas in case of conventional transesterification process, homogeneous base catalyst like potassium or sodium methylate in the presence of methanol converts the free fatty acids to FAME and glycerol. The biological method involves the application of enzyme (Lipase) for catalyzing the transesterification which is detailed below in Sect. 3.4. This biodiesel conversion and related yield highly depend on many factors such as oil to methanol ratio, type/concentration of the catalyst, temperature, and time of the reaction. Beyond other factors, selection of catalyst type and concentration is found to be crucial in determining the rate and yield of biodiesel (Talha and Sulaiman 2016). The type of catalysts applied for biodiesel conversions includes homogeneous catalysts, heterogeneous catalysts, heterogeneous nanocatalysts, biological catalysts, and deep eutectic solvent (DES) as discussed below and shown in Fig. 3.

3.1 Homogeneous Catalysts

As mentioned previously, homogeneous catalysts are catalysts which are subjected in the same phase as like substrates during the reactions and both of the chemical (acid and alkali) catalysts are usually used as homogeneous catalysts. The above-described alkali catalysts like KOH and NaOH are widely used as homogeneous base catalysts and it was found that instead of alkali hydroxides, alkali methoxides are highly preferred due to the reported higher yields (Mathew et al. 2021). It is mainly attributed to the fact that the production of water by hydroxides results in soap formation along with fatty acids and, thereby, reduces the yield which is not reduced in the case of methoxides catalytic process (Canakci and Van Gerpen 1999). The major advantage with homogeneous base catalysts is their ability to catalyze

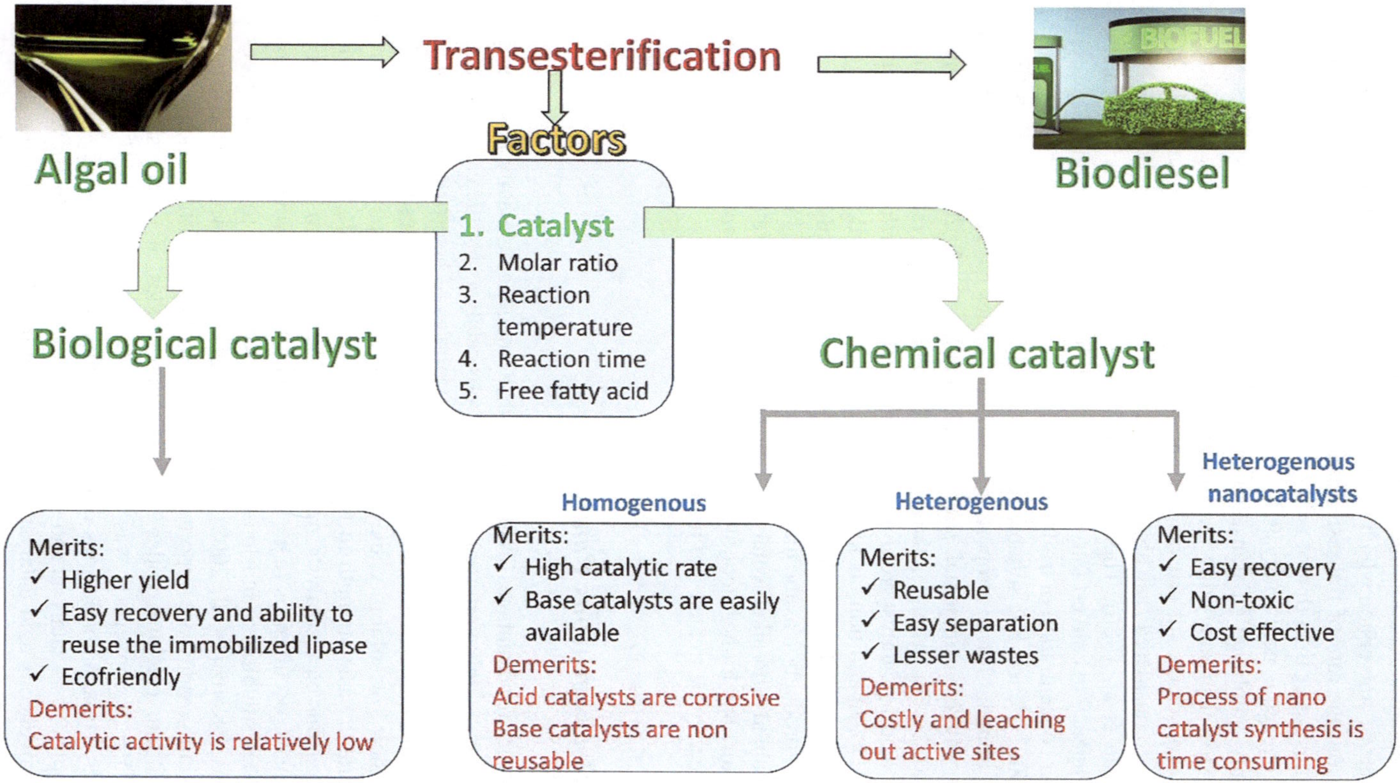

Fig. 3 Types of transesterification catalysts and their influences on biodiesel production

transesterification reaction under low atmospheric temperature and pressure, with a higher rate of conversion which is vital for economy of the process (Lam et al. 2010). In addition, the rate of transesterification under homogeneous catalysts was about 4000 times faster than that of acid catalysts (Fukuda et al. 2001).

Homogeneous acid catalysts like sulfuric acid and hydrochloric acid have been commonly employed in transesterification of microalgal lipids. These acids are usually slower in catalyzing the conversion process, but are able to achieve higher yield. For instance, sulfuric acid-catalyzed transesterification resulted in 99% yield which is considered as highly efficient among homogeneous acid catalysts (Freedman et al. 1986). The major advantages associated with acid catalysts are their efficiency to convert fatty acids to FAME in a single step and more appropriate for lower concentration of triglycerides with free fatty acid content above 5% (Meher et al. 2006). Hence, it is suitable for microalgal feedstock as many of the microalgal species contain lower proportions of triglycerides at normal growth condition without any environmental stress (Esakkimuthu et al. 2016; Esakkimuthu and Wang 2022). For instance, above 92% of biodiesel conversion was obtained using sulfuric acid as catalyst from the green microalga *Chlorella pyrenoidosa* (Hechun et al. 2013). In addition, as its catalytic activity is not hindered by soap formation, yield will not be affected. However, homogeneous acid catalysts are not preferred for industrial production as the major disadvantage is a slower biodiesel conversion rate than obtained with alkali catalysts. Other disadvantages are that they require higher temperature and higher alcohol-oil ratio and lead to corrosiveness due to the effluents which need to be treated and managed effectively.

3.2 Heterogeneous Catalysts

Due to the disadvantages associated with homogeneous catalysts, heterogeneous catalysts are widely preferred. Heterogeneous catalysts are the catalysts that are in a different phase to the reactants. The best examples of heterogeneous base catalysts are calcium oxides, hydrotalcites, and alkaline metal oxides (Widayat et al. 2017). Due to their low solubility, they exhibit higher level of catalytic activity and give a considerable basic strength to the reaction system (Zabeti et al. 2009). Moreover, heterogeneous base catalysts are easily available as calcium is abundant because of its extensive occurrences (e.g., limestone, bone, etc.). However, calcium oxides are sometimes solubilized and loose active sites which need to be heat processed for future use (Kouzu et al. 2008). Although other heterogeneous base catalysts like magnesium oxides are reported, these fail to produce the yield obtained with catalysts like calcium oxides. With calcium oxides of oyster shell as catalyst, about 95% of conversion was achieved in soybean oil at 65 °C (Nakatani et al. 2009). Even the natural forms of calcium oxides like shells of egg and oyster were also employed as catalysts and showed significant effect on biodiesel production.

Heterogeneous acid catalysts are employed mainly to nullify the disadvantages associated with homogeneous acid catalysts (Meher et al. 2006). In this type, solid-based acid catalysts were used instead of acids in liquid form which even prevent the washing step in biodiesel production. In addition, it can also catalyze the transesterification in a single step by simultaneously catalyzing esterification and transesterification between oil and alcohol (Suwannakarn et al. 2009). Although heterogeneous acid and base catalysts are better than homogeneous catalysts, the formation of soap, solubilization, and slower reaction rate with acid catalysts are disadvantageous which are urging the research towards identifying heterogeneous nanocatalysts.

3.3 Heterogeneous Nanocatalysts

Nanocatalysts offer several benefits over conventional catalysts, because nanoforms can provide greater surface area, higher level of catalytic potential, and good rigidity (Wen et al. 2010). Table 3 exemplifies few nanocatalysts and their impact on biodiesel yield, which reveals that nanocatalysts are capable of producing improved

Table 3 Nanocatalysts for biodiesel production from various feedstocks

Feedstocks	Nanocatalysts	Biodiesel yield (%)	Percentage of increase in yield over the other catalyst	References
Simarouba	Ag-ZnO	84.5	5.5 (ZnO)	Nagaraju et al. (2017)
Soybean	Cao	97.6	–	Bharti et al. (2019)
Canola	Zno/$BiFeO_3$	95.4	3.2 (Reused Zno/$BiFeO_3$)	Salimi and Hosseini (2019)
Sunflower	Mg /CeM	94.3	947 (Mg/M)	Dehghani and Haghighi (2019)
Mahua	Mn- ZnO	97	–	Baskar et al. (2017)
Waste cooking oil	MgO-NaOH	97	870 (MgO-KOH)	Rafati et al. (2019)
Chlorella sp.	CaO	92	–	Pandit and Fulekar (2019)
Nannochloropsis oculata	$ZnOMn^{2+}$	87.5	–	Raj et al. (2019)
Neochloris oleoabundans	Fe_2O_3	81	68 (NaOH)	Banerjee et al. (2019)
Mixed microalgal biomass	Fe_2O_3	95.6	–	Kazemifard et al. (2019)
Chlorella vulgaris	CaO	67	–	Davoodbasha et al. (2021)

yield up to 97% and it can effectively replace the conventional alkali and acid catalysts.

The criteria for ideal preparation of nanocatalysts are the need to be an eco-friendly process using a cost-effective feedstock. Some common examples of nanocatalysts include magnesium oxide, zirconium oxide, nanozeolites, and zirconium oxide (Mathew et al. 2021). Titanium oxide nanocatalysts were also reported to achieve a maximum yield of about 92% using palm oil as a feedstock (Madhuvilakku and Piraman 2013). Microalgal biomass was transesterified using calcium oxides nanocatalyst, which resulted in the maximum yield of about 86% (Pandit and Fulekar 2017). The reusability of nanocatalysts is significantly higher compared to other catalysts. For instance, about 5% of Mg/CeM was used to transesterify waste cooking oil, which yielded more than 88% of conversion and was reused up to 7 times without any notable decline in the activity (Dehghani and Haghighi 2019). Similarly, nanohydrotalcites are a group of nanocatalysts used for transesterifying jatropha oil and achieved about 95% biodiesel yield (Deng et al. 2011). Thus, heterogeneous nanocatalysts are advantageous in terms of catalytic efficiency, lower waste generation, nontoxic effluent, and easy to separate from the reaction system.

3.4 Biocatalysts

In order to avoid extreme processing conditions such as high temperature and to avoid toxic effluents and chemicals, biological/greener catalysts have been a research focus for more than two decades. Biodiesel with good quality complying with international standards, such as American Society for Testing and Materials (ASTM), European committee for standardization (CEN), Bureau of Indian Standards (BIS), was produced under enzyme-catalyzed transesterification (Fukuda et al. 2001). One of these catalysts is lipase, which is used in biodiesel production and acts in two ways. This includes, first, direct alcoholysis of triacylglycerol, and second, a two-step reaction involving hydrolysis of triacylglycerol followed by esterification (Marchetti et al. 2007). The maximum catalytic activity is obtained at methanol-oil molar ratio of 3:1 (Kumari et al. 2009). As enzyme activity also depends on temperature, higher temperature above 50 °C inhibits the enzyme activity and, therefore, reaction temperature needs to be under control (Sim et al. 2010). Lipase enzymes are isolated from bacterial and fungal species, e.g., *Pseudomonas* species is the predominant bacterial lipase producer, whereas *Candida* species and *Rhizopus* species are the predominant fungal lipase enzyme producers (Borrelli and Trono 2015).

Novozyme was the familiarly used commercial lipase for laboratory-scale production of biodiesel (Reddy et al. 2018). Lipase production was considered as one of the dominant industrial enzyme productions and Novozyme was found to be one of the leading industrial enzyme producers, especially for lipases (DiCosimo et al. 2013). In general, commercial production of lipases is achieved with the help of fungi and other microorganisms which contain fat degrading enzymes. In addition to transesterification, lipases have been widely used in multiple industries for the production of chemicals, pharmaceuticals, detergents, food, and paper (Navvabi et al. 2018). Solid state cultivation of filamentous fungi using low-cost media has been the most conventional way of producing lipases. The low-cost media for fungal lipase production mainly include agricultural residues and wastes which include, for example, olive pomace (Oliveira et al. 2016), sugarcane bagasse (Vaseghi et al. 2013), and wheat bran (Malilas et al. 2013). Although the industrial production of lipases is prevalent, and utilizing cheap medium for cost-effective solid-state cultivation of fungi has been reported extensively, insights on successful commercial production of lipases in literatures are very rare (Tišma et al. 2019).

There are certain disadvantages for lipase-mediated transesterification such as high enzyme loading for generating more active sites which makes the process as cost-ineffective (Verma et al. 2017). Sometimes, the by-products from biodiesel production block the active sites of the enzyme which reduces the overall enzyme activity, reaction rate, and biodiesel yield. Alternatively, microbes are directly used in the reaction system, but it is not as successful as direct enzyme catalyst. In addition, recovery of enzymes from the reaction system was also found difficult and resulted in eventual loss of enzyme activity which makes the reusability less effective and ultimately make the process expensive (Xu et al. 2011). To overcome the bottlenecks presented in Fig. 3, lipase immobilization was predominantly followed, which enhances the stability and reusability of the enzyme. Lipase enzymes can be immobilized using different strategies such as adsorption, entrapment, covalent bond, and cross-linking (Mathew et al. 2021). In adsorption, weak forces such as hydrophobic interactions are exploited for immobilizing the enzyme to a solid stationary phase. Synthetic polymers, resins, and nanofibers are the commonly used stationary phases to support lipase immobilization using adsorption techniques (Jegannathan et al. 2008).

Entrapment involves the entrapment of enzyme within the matrix/support and forms physical membrane around the enzyme which gives higher stability to the enzymes (Cao 2005). The most familiar and successful carrier used for enzyme entrapment was Kappa-carrageenan (κ-carrageenan) which is cheaper, easily biodegradable, and less toxic to the environment. κ-carrageenan has extensive application in food industry which is a natural polymer extracted from red marine algae. κ-carrageenan is a disaccharide polymer and one of the vital phycochemicals which is very familiar for entrapment application requiring a reaction temperature between 40 °C and 50 °C for entrapment (Raman and Doble 2015). The reusability of enzyme using κ-carrageenan as solid support was significant and the activity retained was more than 82% even after 5 cycles (Jegannathan et al. 2010). Cross-linking involves the formation of three-dimensional network between enzyme and carrier. However, due to the lack of amino group in cross-linking carrier, the enzyme activity seems to be low. In that context, chitosan group was added to the carrier in order to introduce amino group for enhanced enzyme activity (Alamsyah et al. 2017). Covalent bonding seems to be the most stable support, as the linkage is made by covalent bond and not by other weak interactions. This immobilization was stronger among all the strategies, and the enzyme was stable with covalently linked support and no enzyme leakage was observed (Trevan 1988). However, enzyme-mediated transesterifications are not that cost-effective as alkali catalysts; however, the major objective of microalgal biodiesel production was to attain an environmentally friendly fuel without affecting the environment. Hence, greener approach towards biodiesel production is highly important; it requires biological catalyst and lipases have been identified as an appropriate choice. In microalgal biodiesel production chain, the cost-intensive process exists in each stage of the production; the worldwide research has been constantly addressing and identifying each challenge and economic alternative. Similarly, the research focus has been on transesterification processes, and various strategies to tailor the fatty acid profile of the algae have been proposed. The fatty acid chain length reduction was vital to produce biodiesel with finer fuel properties that comply with specific standards and, hence, enriching the short chain fatty acids and medium chain fatty acids could potentially increase the fuel properties of the produced biodiesel (Knothe and Steidley 2005). For instance, medium chain fatty acids such as lauric acid (C12:0) and myristic acid (C14:0) have been increased by seven-fold and four-fold, respectively, by using combinatorial expression platform comprising plant lauric acid-biased TE and MCFA-specific ketoacyl-ACP synthase in *Dunaliella tertiolecta* (Lin et al. 2018).

As a part of biocatalysts, utilizing cyanobacteria as efficient green catalyst has been emerging recently (Śliżewska and Żymańczyk-Duda 2021). Cyanobacteria have been widely used for biocatalysis, especially in reducing carbon-carbon double bonds. Such bioreduction potential has been demonstrated with cyanobacterial species including *Leptolyngbya foveolarum, Synechococcus bigranulatus, Nodularia sphaerocarpa*, and *Arthrospira maxima*. For instance, asymmetric reduction of enones into S ketones was effectively biocatalyzed by *Synechococcus* sp. PCC 7942 (Shimoda et al. 2004). In addition, the above-mentioned cyanobacterial species were potentially known to reduce acetophenone and the resulting chiral alcohols (e.g., 1-(*R*)-phenylethanol) and enantiometric excess were mainly based on biotransformation and strains involved (Żymańczyk-Duda et al. 2019). This biotransformation by cyanobacteria was significantly influenced by light and the process found to be slower in dark conditions. Genetic engineering and molecular approaches have been used to improve the biotransformation potential of cyanobacteria. YqjM enate reductase gene from *Bacillus subtilis* was introduced into *Synechocystis* sp., which resulted in 2-methyl-N-methylmaleimide conversion ability to the cyanobacteria (Köninger et al. 2016). Interestingly, chiral phosphonate synthesis was accomplished with the help of the cyanobacterium *Nodularia sphaerocarpa* which involved diethyl esters reduction in oxophosphonic acids (Górak and Żymańczyk-Duda 2015). Utilizing cyanobacteria for biotransformation and biocatalysts became highly advantageous in terms of applying whole cell instead of isolated enzyme, which could be effective in both energy and economic perspectives.

4 Deep Eutectic Solvents

Deep eutectic solvents (DES) have been recently garnering attention as catalysts for greener biodiesel production. DES are analogues of ionic liquids formed by hydrogen bond donor and hydrogen bond acceptor in a fixed molar ratio (Tao et al. 2010). It was identified as advantageous over other catalysts possessing higher chemical stability, non-corrosiveness, and are less toxic to the environment (Haider et al. 2021). The formed DES have lower melting point than the hydrogen donor and acceptor. This low melting point is because of stronger interaction between the donor and acceptor of the hydrogen bond (Sharma et al. 2022). The other forces such as electrostatic, van der Waals interactions, and lattice energy are also found to be influencing the DES formation (Smith et al. 2014). Based on the hydrogen bond donor and acceptor, DES can be categorized into five types. Type 1 DES are formed from quaternary ammonium salts and metal chlorides, type 2 DES are made of quaternary ammonium salts and metal chloride hydrates, type 3 contains DES made of choline chloride and alcohols, type 4 DES are composed of metal chloride with hydrogen bond donor, and type 5 DES are made of non-ionic molecular hydrogen bond donor and hydrogen bond acceptor (Ijardar et al. 2022). Hydrogen bond formation confers a lower melting point to DES which can also be modified based on the desired requirements through selecting hydrogen donor and acceptor (Maheshwari et al. 2021). DES were identified as efficient catalyst and green solvent for extracting FAME (Table 4). The table displays the significant transesterification yield using DES as catalyst and the maximum is 97%, which is on par with the yield obtained by conventional alkali catalysts. Moreover, DES can also be applied widely to isolate pigments and phenolic products. In such a case, DES could potentially exhibit multifunction (Fig. 4) in algal biotechnology as in transesterification and pigment extraction which could be ideally suitable for integrated algal biorefinery.

Moreover, DES as a cosolvent can improve the catalytic activity of chemical catalyst by reducing the side reactions and eventually facilitate easy separation of the biocatalyst (Škulcová et al. 2016). For example, DES are proven to reduce the side reactions like saponification and also aid in the separation of catalyst from the reaction system. DES are advantageous as transesterification catalysts because they are insensitive to water, not requiring extreme experimental conditions, are simple preparations, and have less toxicity (Soltanmohammadi et al. 2021). Several studies reported the transesterification of palm oil using DES, which revealed about 70% and a maximum of 97% of conversions achieved (Mohebbi et al. 2020). DES can also be easily separated from the reaction system as once the reaction completes, the desired product will be present in the distinct layer, whereas DES mixture will be in the lower phase (Hayyan et al. 2013). The catalytic activity of DES is not considerably reduced even after 4 runs of transesterification as the conversion just reduced to 87% (fourth cycle) from 93% (first cycle), which supports the economy of the process (Hayyan et al. 2013). Thus, DES is an emerging catalyst with numerous advantages over other conventional catalysts for biodiesel production. DES could mitigate economic and environmental challenges by acting as cheaper and less toxic

Table 4 Applications of deep eutectic solvents (DES) as catalysts for biodiesel (FAME) and value-added products recovery from various feedstocks

Feedstocks	Methanol: oil molar ratio	DES	Optimum DES concentration	FAME yield/conversion/ time (value-added products recovery)	References
Soybean oil	10:1–30:1	ChCl: ZnCl (1:2)	10 wt%	54% conversion 72 h^{-1}	Tao et al. (2010)
Rapeseed oil	14.28:1	ChCl: glycerol (1:2)	10.74 wt%	92% yield 3 h^{-1}	Huang et al. (2013)
Miglyol 812	–	ChOAc: glycerol (1:1.5)	1 mL	97% conversion 3 h^{-1}	Gu et al. (2015)
Palm oil	3:1–20:1	ChCl: PTSA (1:3)	0.75 wt%	97% yield 0.5 h^{-1}	Hayyan et al. (2014)
Palm oil	4:1 20:1	DEAC: PTSA (1:3)	0.75 wt%	97% yield 0.5 h^{-1}	Hayyan et al. (2013)
Chlorella pyrenoidosa	1:2	ChCl–$ZnCl_2$	8 wt%	3.6% conversion 7 h^{-1}	Ngatcha et al. (2023)
Lippia citriodora	1:2	ChCl-lactic acid	–	3% (phenols)	Ivanović et al. (2018)
Gnetum gnemom (seeds)	1:1	Bet-lactic acid	–	0.023% (resveratrol)	Aryati and Azka (2020)
Cuminum cyminum L.	1:3	ChCl-lactic acid	–	– (essential oil)	Zhao et al. (2019)
Vitis vinifera (Pomace)	2:1	ChCl-citric acid		0.057% (anthocyanin)	Panić et al. (2019)

wt% - weight percentage to the total biomass

catalyst (Homan et al. 2017). In addition, when added with lipases, DES support higher conversion efficiency for biodiesel production.

Apart from biodiesel production, DES can be potentially involved in other applications like removal of aromatic hydrocarbon. For instance, choline chloride, triethylamine hydrochloride, and ethylamine hydrochloride were used for separating oil in which these ammonium salts were mixed with phenol containing xylene with oil in a tube followed by stirring in hot water bath. The resulting upper phase was analyzed using UV-vis spectroscopy for the detection of DES, where absence of DES confirmed the oil separation (Pang et al. 2012). Similarly, separation of pharmaceutical compounds, proteins, and pesticides was achieved with the help of DES (Pedro et al. 2019; Soltanmohammadi et al. 2021). Another study suggested

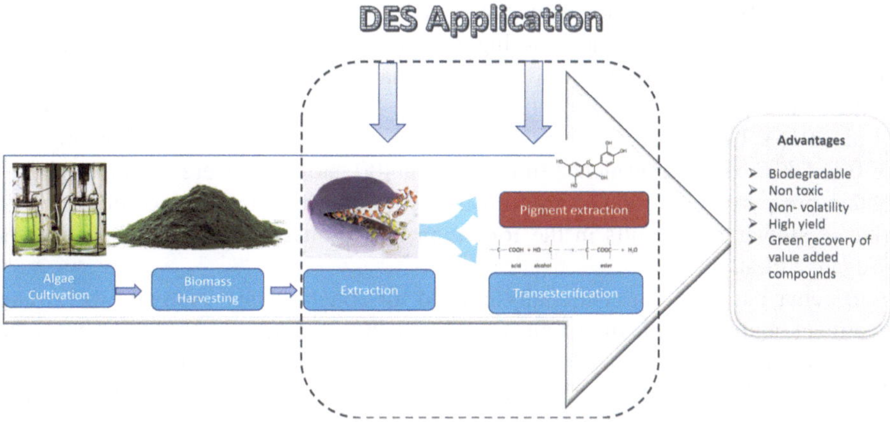

Fig. 4 Applications and advantages of deep eutectic solvents (DES) in the production of value-added products from algae

and demonstrated the application of DES in transdermal drug delivery system (Wang et al. 2022a, b). In addition, DES are efficiently used as gas capture, especially CO_2 and gas absorption efficiency varies with the type of hydrogen bond donor and acceptor used, temperature, time, and molar ratio of the DES (Cui et al. 2019). However, these discussions are from the primitive understanding on DES based on available studies; more studies dealing with molecular mechanism of DES on biodiesel productions need to be done. To ensure economic and environmental sustainability, techno-economic assessment and life cycle assessment of applying DES as catalyst need to be studied. Above all, DES can be tailored to multiple requirements such as the applications as described above which makes it even more attractive (Lu et al. 2016). Thus, with further advancements and exploring insights, DES can be promising as green catalysts for cleaner and sustainable production of algal products.

5 Algae-Based Catalysts (Phycocatalysts)

Biochar from algae can be produced from various thermochemical conversion methods including pyrolysis, hydrothermal carbonization, HTL, and torrefaction, which can be termed as "Phycocatalyst". To the best of our knowledge acquired from the available scientific literature, terming algal-based catalyst as phycocatalyst is appearing for the first time in literature. Interestingly, employing this algal biochar as a catalyst for various value-added products recovery has been incrementally noticed. Algal biochar is widely synthesized through above-mentioned thermochemical conversion and most conventional method of synthesis is through slow pyrolysis (Azizi et al. 2018). It is because of the fact that among other pyrolysis methods (fast and microwave-assisted), slow pyrolysis yields higher proportions of biochar

(Yu et al. 2017). The yield of biochar usually increases with decreased pyrolysis temperature, low heating rate, and higher retention time which are mainly accomplished by slow pyrolysis. In addition, biomass characteristics such as moisture content and particle size also significantly influence the biochar yield (Tripathi et al. 2016). Under slow pyrolysis, the reaction conditions including heating rate of 0.1–1 Ks^{-1} with a residence time between 400 and 500 s were usually set up to obtain pre-pyrolysis followed by further decomposition. Further decomposition of char at slower rate results in the formation of bio-char with high carbon content (Suganya et al. 2016).

Biochar has been conventionally utilized for contaminants removal, wastewater treatment, and anaerobic digestion (Faisal et al. 2023; Jiang et al. 2023), whereas the recent research interests on utilizing algal biochar as catalysts are rapidly emerging. The major advantages of utilizing algal biochar as catalysts are their eco-friendly nature and reusability. In addition, compositional features such as inorganic compounds (potassium and iron) in biochar provide finer catalytic property and functional groups on the surface of the biochar particles that help in metal precursor adsorption for producing biochar-supported metal catalyst (Cheng and Li 2018). Phycocatalyst offers additional advantages such as heterogenicity which are separable from the system, multifunctional (can perform esterification and transesterification), and highly reusable with high porosity (Yu et al. 2011). Phycocatalysts' application has been initiated in wider range of fields. For instance, boron-doped algal biochar obtained from pyrolyzing of the seaweed *Undaria pinnatifida* and boric acid was successfully employed as catalyst for activating peroxymonosulfate towards degradation of diclofenac which is an inevitable toxic pollutant from pharmaceuticals and hospital effluent (Annamalai and Shin 2023). The recyclability of this phycocatalysts was significantly higher (86% of diclofenac removal) even after three cycles. In a recent study, carbon nanotubes (CNT) were produced from algal biochar by microwave irradiation at low energy. The algal species used for producing biochar followed by CNT production were *Macrocystis pyrifera, Sarcothalia crispata,* and *Scenedesmus almeriensis.* The study revealed that algal biochar can be better suitable for the formation of CNT and higher carbon containing char was obtained from *Scenedesmus almeriensis*, which resulted in higher graphitization of wall and contents of the nanotubes (Hidalgo et al. 2023). In addition, phycocatalysts were proven to be attractive catalysts in improving microbial fuel cell performance. A study synthesized biochar catalyst from mixture of algal species collected from high-rate algal ponds in India and employed it as successful electrocatalyst in microbial fuel cell. Three different types of microbial fuel cells such as algal derived biochar cathode catalyst, Pt-C cathode catalyst, and bare carbon black-coated cathode were employed using synthetic wastewater resulting in removal of COD by algal char-based catalysts (79%) that were on par with other two catalysts (Pt-C: 79%; CB: 73%) (Chakraborty et al. 2020). Different metals (Cu, Ce, and Fe) were loaded in biochar derived from the seaweed *Enteromorpha clathrate* and applied as catalyst for pyrolysis of *E. clathrate*, which resulted in high C/H ratio and high heating value of the bio-oil as compared to non-catalytic process. In addition, this photocatalyst significantly reduced the carboxylic acid in the bio-oil (Cao et al. 2021). This study demonstrated the application of photocatalyst in fuel production and it could even act as efficient catalyst towards pyrolyzing algae itself which is highly essential for algal-based biorefinery to attain zero-waste and sustainable production of products and fuels.

6 Energy Balance

Energy return on investment (EROI), calculated based on the product energy output and primary energy inputs, should be greater than 1 to have positive energy balance of the process (Beal et al. 2012). EROI of microalgal HTL revealed that biochemical composition occupies a substantial position in terms of energy recovery. For instance, *Nannochloropsis oceanica* with 30% of lipid content and *Golenkinia* sp. with lipid content of 17.3% have been subjected for HTL along with varying temperature (Yoo et al. 2015). The results showed that *Nannochloropsis oceanica* with higher lipid content over other strains showed maximum EROI of about 4.91 at 200 °C, whereas *Golenkinia* sp. with lower lipid content showed EROI of about 2.7 at 300 °C. It should be noted that increasing temperature in HTL yields more biocrude; however, it also increases the input energy and affects the energy balance. The study concluded that biochemical composition occupies significant role in EROI. Moreover, it also indicates that the choice of HTL is advantageous as there is no requirement of drying the algal biomass which supports EROI.

An energy balance assessment for microalgal bioethanol production has been conducted and revealed a positive net energy ratio (NER) of about 0.45 (Hossain et al. 2019). The study considered bioethanol as the main energy output and the energy inputs were energy required through heat and electricity comprising from cultivation to product purification. Among energy inputs, 16.5 MJ kg^{-1} of energy was expended for drying the algal biomass and considered as maximum in the study. Moreover, solar drying could not be a global solution as it cannot be controlled. Energy consumptions of HTL and pyrolysis of microalgae (*Chlorella* sp.) have been studied and revealed that about 1.6 times higher energy requirement was needed for pyrolysis (Zhang et al. 2017). In addition, energy recovery from HTL was around 90%, whereas pyrolysis recovered around 79%, which suggested HTL as energy efficient way of producing biofuels from algae. Despite many reports on evaluating the energy balance and EROI have been published, most data were based on optimistic assumptions instead of empirical numbers (Ketzer et al. 2018). Very few studies like as exemplified above (Hossain et al. 2019) were obtained data from industrial production of ethanol in Brunei. From literature, the reported EROI values were highly inconsistent which vary between 0 and 3 and mostly based on optimistic assumptions (Ketzer et al. 2018). Two ways have been proposed for improving this EROI to attain positive energy balance. One way is a biological approach to enhance the photo-conversion ability of algae, whereas another way is improvising and adapting less energy-consuming cultivation and harvesting methods to improve the energy balance of algal-based fuel and value-added production. On catalyst application perspective, there are no dedicated reports evaluating its impact on the overall energy balance except the fact that yield improvement could positively push the energy balance.

7 Conclusions

Arrival of catalysts into the production of fuels and value-added products from algae have been undoubtedly advantageous in terms of improving the yield, controlling the product composition, and improving the overall economy of the process. For instance, choosing appropriate catalyst in biodiesel production directly influences the cost, as choosing alkali over acid catalyst could definitely improve the yield and rate of conversion by a distinct margin. Application of novel and interesting catalysts such as phycocatalysts, green catalysts, and DES needs to be considered as a great leap of catalysis that could definitely improve the economy of the process due to its distinguishing impact on the yield as compared to non-catalytic conversion. However, catalysis of algal biomass conversion is a single substantial step, whereas dedicated large-scale production of algal biomass conversion into useful products inevitably requires cost-effectiveness in each step from cultivation to product separation. EROI based on industrial-scale monitoring is required rather than optimistic assumptions, finding cheaper alternative for every energy requirement, breakthroughs in yield improvement, and photo-conversion potential of microalgae which could push the road further towards commercial feasibility of algal-based productions. With all these challenges, application of catalyst is one emerging department which has been effectively moving with novel insights and benefitting in the yield improvisation. Thus, this chapter detailed phycocatalysts as well as other catalysts involved in the most common conversion routes of algal biomass, their contemporary status, challenges, bottlenecks, efficient alternative, and emerging technologies which altogether may aid for enhanced production of phycochemicals.

References

Abomohra A, Elsayed M, Esakkimuthu S, El-Sheekh M, Hanelt D (2020) Potential of fat, oil and grease (FOG) for biodiesel production: a critical review on the recent progress and future perspectives. Prog Energy Combust Sci 81:100868. https://doi.org/10.1016/j.pecs.2020.100868

Abomohra A, Sheikh HMA, El-Naggar AH, Wang Q (2021) Microwave vacuum co-pyrolysis of waste plastic and seaweeds for enhanced crude bio-oil recovery: experimental and feasibility study towards industrialization. Renew Sust Energ Rev 149:111335. https://doi.org/10.1016/J.RSER.2021.111335

Ahmad F, Khan AU, Yasar A (2013) Transesterification of oil extracted from different species of algae for biodiesel production. African J Environ Sci Technol 7:358–364

Ahmed I, Hasan Z, Khan NA, Jhung SH (2013) Adsorptive denitrogenation of model fuels with porous metal-organic frameworks (MOFs): effect of acidity and basicity of MOFs. Appl Catal B Environ 129:123–129

Alamsyah G, Albels VA, Sahlan M, Hermansyah H (2017) Effect of chitosan's amino group in adsorption-crosslinking immobilization of lipase enzyme on resin to catalyze biodiesel synthesis. Energy Procedia 136:47–52

Aliyu A, Lee JGM, Harvey AP (2021) Microalgae for biofuels: a review of thermochemical conversion processes and associated opportunities and challenges. Bioresour Technol Rep 15: 100694

Almutairi AW (2022) Full utilization of marine microalgal hydrothermal liquefaction liquid products through a closed-loop route: towards enhanced bio-oil production and zero-waste approach. 3 Biotech 12:209. https://doi.org/10.1007/s13205-022-03262-8

Annamalai S, Shin WS (2023) Algae-derived metal-free boron-doped biochar acts as a catalyst for the activation of peroxymonosulfate toward the degradation of diclofenac. Environ Pollut 331: 121850

Aryati WD, Azka KM, Mun'im A (2020) Ultrasonic-assisted extraction using a betaine-based natural deep eutectic solvent for resveratrol extraction from melinjo (*Gnetum gnemon*) seeds. Int J Appl Pharm 12(Special Issue 1):26–31

Azizi K, Moraveji MK, Najafabadi HA (2018) A review on bio-fuel production from microalgal biomass by using pyrolysis method. Renew Sust Energ Rev 82:3046–3059

Babich IV, Van der Hulst M, Lefferts L, Moulijn JA, O'Connor P, Seshan K (2011) Catalytic pyrolysis of microalgae to high-quality liquid bio-fuels. Biomass Bioenergy 35:3199–3207

Bach Q-V, Chen W-H (2017) Pyrolysis characteristics and kinetics of microalgae via thermogravimetric analysis (TGA): a state-of-the-art review. Bioresour Technol 246:88–100

Banerjee S, Rout S, Banerjee S, Atta A, Das D (2019) Fe2O3 nanocatalyst aided transesterification for biodiesel production from lipid-intact wet microalgal biomass: a biorefinery approach. Energy Convers Manag 195:844–853

Baskar G, Gurugulladevi A, Nishanthini T, Aiswarya R, Tamilarasan K (2017) Optimization and kinetics of biodiesel production from Mahua oil using manganese doped zinc oxide nanocatalyst. Renew Energy 103:641–646

Beal CM, Stillwell AS, King CW, Cohen SM, Berberoglu H, Bhattarai RP, Connelly RL, Webber ME, Hebner RE (2012) Energy return on investment for algal biofuel production coupled with wastewater treatment. Water Environ Res 84:692–710

Bhardwaj N, Agrawal K, Verma P (2020) Algal biofuels: an economic and effective alternative of fossil fuels. In: Microbial strategies for techno-economic biofuel production. Springer, Berlin, pp 207–227

Bharti P, Singh B, Dey RK (2019) Process optimization of biodiesel production catalyzed by CaO nanocatalyst using response surface methodology. J Nanostructure Chem 9:269–280

Biller P, Ross AB (2011) Potential yields and properties of oil from the hydrothermal liquefaction of microalgae with different biochemical content. Bioresour Technol 102:215–225

Biswas B, Kumar AA, Bisht Y, Singh R, Kumar J, Bhaskar T (2017) Effects of temperature and solvent on hydrothermal liquefaction of *Sargassum tenerrimum* algae. Bioresour Technol 242: 344–350

Biswas B, Kumar A, Fernandes AC, Saini K, Negi S, Muraleedharan UD, Bhaskar T (2020) Solid base catalytic hydrothermal liquefaction of macroalgae: effects of process parameter on product yield and characterization. Bioresour Technol 307:123232

Borrelli GM, Trono D (2015) Recombinant lipases and phospholipases and their use as biocatalysts for industrial applications. Int J Mol Sci 16:20774–20840

Brown TM, Duan P, Savage PE (2010) Hydrothermal liquefaction and gasification of *Nannochloropsis* sp. Energy Fuel 24:3639–3646

Canakci M, Van Gerpen J (1999) Biodiesel production via acid catalysis. Trans ASAE 42:1203

Cao L (2005) Immobilised enzymes: science or art? Curr Opin Chem Biol 9:217–226

Cao B, Yuan J, Jiang D, Wang S, Barati B, Hu Y, Yuan C, Gong X, Wang Q (2021) Seaweed-derived biochar with multiple active sites as a heterogeneous catalyst for converting macroalgae into acid-free biooil containing abundant ester and sugar substances. Fuel 285:119164

Caporgno MP, Pruvost J, Legrand J, Lepine O, Tazerout M, Bengoa C (2016) Hydrothermal liquefaction of *Nannochloropsis oceanica* in different solvents. Bioresour Technol 214:404–410

Castello D, Haider MS, Rosendahl LA (2019) Catalytic upgrading of hydrothermal liquefaction biocrudes: different challenges for different feedstocks. Renew Energy 141:420–430

Chakraborty I, Bhowmick GD, Ghosh D, Dubey BK, Pradhan D, Ghangrekar MM (2020) Novel low-cost activated algal biochar as a cathode catalyst for improving performance of microbial fuel cell. Sustain Energy Technol Assessments 42:100808

Chen Y, Wu Y, Hua D, Li C, Harold MP, Wang J, Yang M (2015) Thermochemical conversion of low-lipid microalgae for the production of liquid fuels: challenges and opportunities. RSC Adv 5:18673–18701

Chen W, Yang H, Chen Y, Xia M, Chen X, Chen H (2017) Transformation of nitrogen and evolution of N-containing species during algae pyrolysis. Environ Sci Technol 51:6570–6579

Cheng F, Li X (2018) Preparation and application of biochar-based catalysts for biofuel production. Catalysts 8:346

Chisti Y (2013) Constraints to commercialization of algal fuels. J Biotechnol 167:201–214

Cui G, Lv M, Yang D (2019) Efficient CO2 absorption by azolide-based deep eutectic solvents. Chem Commun 55:1426–1429

Dabros TMH, Stummann MZ, Høj M, Jensen PA, Grunwaldt J-D, Gabrielsen J, Mortensen PM, Jensen AD (2018) Transportation fuels from biomass fast pyrolysis, catalytic hydrodeoxygenation, and catalytic fast hydropyrolysis. Prog Energy Combust Sci 68:268–309

Davoodbasha M, Pugazhendhi A, Kim J-W, Lee S-Y, Nooruddin T (2021) Biodiesel production through transesterification of *Chlorella vulgaris*: synthesis and characterization of CaO nanocatalyst. Fuel 300:121018

Dehghani S, Haghighi M (2019) Sono-dispersed MgO over cerium-doped MCM-41 nanocatalyst for biodiesel production from acidic sunflower oil: surface evolution by altering Si/Ce molar ratios. Waste Manag 95:584–592

Deng X, Fang Z, Liu Y, Yu C-L (2011) Production of biodiesel from Jatropha oil catalyzed by nanosized solid basic catalyst. Energy 36:777–784

Déniel M, Haarlemmer G, Roubaud A, Weiss-Hortala E, Fages J (2016) Energy valorisation of food processing residues and model compounds by hydrothermal liquefaction. Renew Sust Energ Rev 54:1632–1652

DiCosimo R, McAuliffe J, Poulose AJ, Bohlmann G (2013) Industrial use of immobilized enzymes. Chem Soc Rev 42:6437–6474

Dong J, Lu JJ, Li GB, Song LR (2013) Influences of a submerged macrophyte on colony formation and growth of a green alga. Aquat Biol 19:265–274

Duan P, Savage PE (2011a) Hydrothermal liquefaction of a microalga with heterogeneous catalysts. Ind Eng Chem Res 50:52–61. https://doi.org/10.1021/ie100758s

Duan P, Savage PE (2011b) Catalytic treatment of crude algal bio-oil in supercritical water: optimization studies. Energy Environ Sci 4:1447–1456

Duan P, Bai X, Xu Y, Zhang A, Wang F, Zhang L, Miao J (2013) Catalytic upgrading of crude algal oil using platinum/gamma alumina in supercritical water. Fuel 109:225–233

Duan P-G, Yang S-K, Xu Y-P, Wang F, Zhao D, Weng Y-J, Shi X-L (2018) Integration of hydrothermal liquefaction and supercritical water gasification for improvement of energy recovery from algal biomass. Energy 155:734–745

Durak H, Aysu T (2016a) Structural analysis of bio-oils from subcritical and supercritical hydrothermal liquefaction of *Datura stramonium* L. J Supercrit Fluids 108:123–135

Durak H, Aysu T (2016b) Thermochemical liquefaction of algae for bio-oil production in supercritical acetone/ethanol/isopropanol. J Supercrit Fluids 111:179–198

Durak H, Genel S (2020) Catalytic hydrothermal liquefaction of *Lactuca scariola* with a heterogeneous catalyst: the investigation of temperature, reaction time and synergistic effect of catalysts. Bioresour Technol 309:123375

El-Hefnawy ME, Alhayyani S, Ismail A, El-Sherbiny M, Al-Harbi M, Abomohra A, Sakran M, Zidan N (2023) Integrated approach for enhanced crude bio-oil yield from microalgae cultivated on the aqueous phase of hydrothermal co-liquefaction with agar-free seaweed residues. J Clean Prod 392:136286. https://doi.org/10.1016/j.jclepro.2023.136286

Elliott DC, Hart TR, Schmidt AJ, Neuenschwander GG, Rotness LJ, Olarte MV, Zacher AH, Albrecht KO, Hallen RT, Holladay JE (2013) Process development for hydrothermal liquefaction of algae feedstocks in a continuous-flow reactor. Algal Res 2:445–454

Esakkimuthu S, Wang S (2022) Physical stress for enhanced biofuel production from microalgae. In: Handbook of algal biofuels. Elsevier, Amsterdam, pp 451–475

Esakkimuthu S, Krishnamurthy V, Govindarajan R, Swaminathan K (2016) Augmentation and starvation of calcium, magnesium, phosphate on lipid production of *Scenedesmus obliquus*. Biomass Bioenergy 88:126–134

Esakkimuthu S, Krishnamurthy V, Wang S, Abomohra A, Shanmugam S, Ramakrishnan SG, Subrmaniam S, Swaminathan K (2019) Simultaneous induction of biomass and lipid production in Tetradesmus obliquus BPL16 through polysorbate supplementation. Renew Energy 140:807

Esakkimuthu S, Krishnamurthy V, Wang S, Hu X, Swaminathan K, Abomohra A (2020) Application of p-coumaric acid for extraordinary lipid production in *Tetradesmus obliquus*: a sustainable approach towards enhanced biodiesel production. Renew Energy 157:368–376

Faisal S, Ebaid R, Xiong M, Huang J, Wang Q, El-Hefnawy M, Abomohra A (2023) Maximizing the energy recovery from rice straw through two-step conversion using eggshell-catalytic pyrolysis followed by enhanced anaerobic digestion using calcium-rich biochar. Sci Total Environ 858:159984. https://doi.org/10.1016/J.SCITOTENV.2022.159984

Freedman B, Butterfield RO, Pryde EH (1986) Transesterification kinetics of soybean oil 1. J Am Oil Chem Soc 63:1375–1380

French R, Czernik S (2010) Catalytic pyrolysis of biomass for biofuels production. Fuel Process Technol 91:25–32

Fukuda H, Kondo A, Noda H (2001) Biodiesel fuel production by transesterification of oils. J Biosci Bioeng 92:405. https://doi.org/10.1016/S1389-1723(01)80288-7

Gao L, Sun J, Xu W, Xiao G (2017) Catalytic pyrolysis of natural algae over mg-Al layered double oxides/ZSM-5 (MgAl-LDO/ZSM-5) for producing bio-oil with low nitrogen content. Bioresour Technol 225:293–298

Gong X, Zhang B, Zhang Y, Huang Y, Xu M (2014) Investigation on pyrolysis of low lipid microalgae *Chlorella vulgaris* and *Dunaliella salina*. Energy Fuel 28:95–103

Górak M, Żymańczyk-Duda E (2015) Application of cyanobacteria for chiral phosphonate synthesis. Green Chem 17:4570–4578

Gu L, Huang W, Tang S, Tian S, Zhang X (2015) A novel deep eutectic solvent for biodiesel preparation using a homogeneous base catalyst. Chem Eng J 259:647–652

Haider MB, Maheshwari P, Kumar R (2021) CO2 capture from flue gas using phosphonium based deep eutectic solvents: modeling and simulation approach. J Environ Chem Eng 9:106727

Han J, Lee K, Choi MS, Park HS, Kim W, Roh KC (2019) Chlorella-derived activated carbon with hierarchical pore structure for energy storage materials and adsorbents. Carbon Lett 29:167–175

Hayyan A, Hashim MA, Hayyan M, Mjalli FS, AlNashef IM (2013) A novel ammonium based eutectic solvent for the treatment of free fatty acid and synthesis of biodiesel fuel. Ind Crop Prod 46:392–398

Hayyan A, Hashim MA, Hayyan M, Mjalli FS, AlNashef IM (2014) A new processing route for cleaner production of biodiesel fuel using a choline chloride based deep eutectic solvent. J Clean Prod 65:246–251

Hechun C, Zhang Z, Wu X, Miao X (2013) Direct biodiesel production from wet microalgae biomass of *Chlorella pyrenoidosa* through in situ transesterification. Biomed Res Int 2013: 930686. https://doi.org/10.1155/2013/930686

Hidalgo P, Navia R, Hunter R, Camus C, Buschmann A, Echeverria A (2023) Carbon nanotube production from algal biochar using microwave irradiation technology. J Anal Appl Pyrolysis 172:106017

Holladay JE, White JF, Bozell JJ, Johnson D (2007) Top value-added chemicals from biomass-volume II—results of screening for potential candidates from biorefinery lignin. Pacific Northwest National Lab.(PNNL), Richland, WA

Homan T, Shahbaz K, Farid MM (2017) Improving the production of propyl and butyl ester-based biodiesel by purification using deep eutectic solvents. Sep Purif Technol 174:570–576

Hossain N, Zaini J, Indra Mahlia TM (2019) Life cycle assessment, energy balance and sensitivity analysis of bioethanol production from microalgae in a tropical country. Renew Sust Energ Rev 115:109371. https://doi.org/10.1016/j.rser.2019.109371

Hu Y, Wang H, Lakshmikandan M, Wang S, Wang Q, He Z, Abomohra A (2020) Catalytic co-pyrolysis of seaweeds and cellulose using mixed ZSM-5 and MCM-41 for enhanced crude bio-oil production. J Therm Anal Calorim 143:827. https://doi.org/10.1007/s10973-020-09291-w

Hu Y, Pang K, Cai L, Liu Z (2021) A multi-stage co-gasification system of biomass and municipal solid waste (MSW) for high quality syngas production. Energy 221:119639

Huang W, Tang S, Zhao H, Tian S (2013) Activation of commercial CaO for biodiesel production from rapeseed oil using a novel deep eutectic solvent. Ind Eng Chem Res 52:11943–11947

Ijardar SP, Singh V, Gardas RL (2022) Revisiting the physicochemical properties and applications of deep eutectic solvents. Molecules 27:1368

Ivanović M, Alañón ME, Arráez-Román D, Segura-Carretero A (2018) Enhanced and green extraction of bioactive compounds from *Lippia citriodora* by tailor-made natural deep eutectic solvents. Food Res Int 111:67–76

Jegannathan KR, Abang S, Poncelet D, Chan ES, Ravindra P (2008) Production of biodiesel using immobilized lipase—a critical review. Crit Rev Biotechnol 28:253–264

Jegannathan KR, Jun-Yee L, Chan E-S, Ravindra P (2010) Production of biodiesel from palm oil using liquid core lipase encapsulated in κ-carrageenan. Fuel 89:2272–2277

Jiang D, Li H, Cheng X, Ling Q, Chen H, Barati B, Yao Q, Abomohra A, Hu X, Bartocci P, Wang S (2023) A mechanism study of methylene blue adsorption on seaweed biomass derived carbon: from macroscopic to microscopic scale. Process Saf Environ Prot 172:1132–1143. https://doi.org/10.1016/J.PSEP.2023.02.044

Kandasamy S, Zhang B, He Z, Chen H, Feng H, Wang Q, Wang B, Ashokkumar V, Siva S, Bhuvanendran N (2020) Effect of low-temperature catalytic hydrothermal liquefaction of Spirulina platensis. Energy 190:116236

Kazemifard S, Nayebzadeh H, Saghatoleslami N, Safakish E (2019) Application of magnetic alumina-ferric oxide nanocatalyst supported by KOH for in-situ transesterification of microalgae cultivated in wastewater medium. Biomass Bioenergy 129:105338

Ketzer F, Skarka J, Rösch C (2018) Critical review of microalgae LCA studies for bioenergy production. Bioenergy Res 11:95–105

Knothe G, Steidley KR (2005) Kinematic viscosity of biodiesel fuel components and related compounds. Influence of compound structure and comparison to petrodiesel fuel components. Fuel 84:1059–1065. https://doi.org/10.1016/j.fuel.2005.01.016

Köninger K, Gómez Baraibar Á, Mügge C, Paul CE, Hollmann F, Nowaczyk MM, Kourist R (2016) Recombinant cyanobacteria for the asymmetric reduction of C= C bonds fueled by the biocatalytic oxidation of water. Angew Chemie Int Ed 55:5582–5585

Kouzu M, Kasuno T, Tajika M, Sugimoto Y, Yamanaka S, Hidaka J (2008) Calcium oxide as a solid base catalyst for transesterification of soybean oil and its application to biodiesel production. Fuel 87:2798–2806

Kumari A, Mahapatra P, Garlapati VK, Banerjee R (2009) Enzymatic transesterification of Jatropha oil. Biotechnol Biofuels 2:1–7

Lam MK, Lee KT, Mohamed AR (2010) Homogeneous , heterogeneous and enzymatic catalysis for transesteri fi cation of high free fatty acid oil (waste cooking oil) to biodiesel: a review. Biotechnol Adv 28:500–518. https://doi.org/10.1016/j.biotechadv.2010.03.002

Lam MK, Loy ACM, Yusup S, Lee KT (2019) Biohydrogen production from algae. In: Biohydrogen. Elsevier, Amsterdam, pp 219–245

Lee HW, Jeon J-K, Park SH, Jeong K-E, Chae H-J, Park Y-K (2011) Catalytic pyrolysis of Laminaria japonica over nanoporous catalysts using Py-GC/MS. Nanoscale Res Lett 6:500

Lee D, Nam H, Seo MW, Lee SH, Tokmurzin D, Wang S, Park Y-K (2022) Recent progress in the catalytic thermochemical conversion process of biomass for biofuels. Chem Eng J 447:137501

Li P, Chen X, Wang X, Shao J, Lin G, Yang H, Yang Q, Chen H (2017) Catalytic upgrading of fast pyrolysis products with Fe-, Zr-, and co-modified zeolites based on pyrolyzer–GC/MS analysis. Energy Fuel 31:3979–3986

Li R, Liu D, Zhang Y, Zhou J, Tsang YF, Liu Z, Duan N, Zhang Y (2019a) Improved methane production and energy recovery of post-hydrothermal liquefaction waste water via integration of zeolite adsorption and anaerobic digestion. Sci Total Environ 651:61–69

Li F, Srivatsa SC, Bhattacharya S (2019b) A review on catalytic pyrolysis of microalgae to high-quality bio-oil with low oxygeneous and nitrogenous compounds. Renew Sust Energ Rev 108: 481–497

Li L, Huang J, Almutairi AW, Lan X, Zheng L, Lin Y, Chen L, Fu N, Lin Z, Abomohra A (2021) Integrated approach for enhanced bio-oil recovery from disposed face masks through co-hydrothermal liquefaction with *Spirulina platensis* grown in wastewater. Biomass Convers Biorefin 2021(1):1–12. https://doi.org/10.1007/S13399-021-01891-2

Lin H, Shen H, Lee YK (2018) Cellular and molecular responses of *Dunaliella tertiolecta* by expression of a plant medium chain length fatty acid specific acyl-ACP thioesterase. Front Microbiol 9:619

Liu J, Li X, Guo QJ (2014) Study of catalytic pyrolysis of chlorella with γ-Al2O3 catalyst. Adv Mat Res Trans Tech Publ 1004-1005:562–566

Lu W, Alam MA, Pan Y, Wu J, Wang Z, Yuan Z (2016) A new approach of microalgal biomass pretreatment using deep eutectic solvents for enhanced lipid recovery for biodiesel production. Bioresour Technol 218:123–128

Madhuvilakku R, Piraman S (2013) Biodiesel synthesis by TiO2–ZnO mixed oxide nanocatalyst catalyzed palm oil transesterification process. Bioresour Technol 150:55–59

Maheshwari P, Haider MB, Kumar R (2021) Effect of water addition on CO_2 solubility of glycol-based deep eutectic solvents. In: Proceedings of the 15th Greenhouse Gas Control Technologies Conference, pp 15–18

Malilas W, Kang SW, Kim SB, Yoo HY, Chulalaksananukul W, Kim SW (2013) Lipase from *Penicillium camembertii* KCCM 11268: optimization of solid state fermentation and application to biodiesel production. Korean J Chem Eng 30:405–412

Marchetti JM, Miguel VU, Errazu AF (2007) Possible methods for biodiesel production. Renew Sust Energ Rev 11:1300–1311

Mathew GM, Raina D, Narisetty V, Kumar V, Saran S, Pugazhendi A, Sindhu R, Pandey A, Binod P (2021) Recent advances in biodiesel production: challenges and solutions. Sci Total Environ 794:148751

Mathimani T, Baldinelli A, Rajendran K, Prabakar D, Matheswaran M, van Leeuwen RP, Pugazhendhi A (2019) Review on cultivation and thermochemical conversion of microalgae to fuels and chemicals: process evaluation and knowledge gaps. J Clean Prod 208:1053–1064

Meher LC, Kulkarni MG, Dalai AK, Naik SN (2006) Transesterification of karanja (*Pongamia pinnata*) oil by solid basic catalysts. Eur J Lipid Sci Technol 108:389–397

Michalak I, Baśladyńska S, Mokrzycki J, Rutkowski P (2019) Biochar from a freshwater macroalga as a potential biosorbent for wastewater treatment. Water 11:1390

Milano J, Ong HC, Masjuki HH, Chong WT, Lam MK, Loh PK, Vellayan V (2016) Microalgae biofuels as an alternative to fossil fuel for power generation. Renew Sust Energ Rev 58:180–197

Mohebbi S, Rostamizadeh M, Kahforoushan D (2020) Effect of molybdenum promoter on performance of high silica MoO3/B-ZSM-5 nanocatalyst in biodiesel production. Fuel 266:117063

Nagaraju G, Prashanth SA, Shastri M, Yathish KV, Anupama C, Rangappa D (2017) Electrochemical heavy metal detection, photocatalytic, photoluminescence, biodiesel production and antibacterial activities of Ag–ZnO nanomaterial. Mater Res Bull 94:54–63

Nakatani N, Takamori H, Takeda K, Sakugawa H (2009) Transesterification of soybean oil using combusted oyster shell waste as a catalyst. Bioresour Technol 100:1510–1513

Navvabi A, Razzaghi M, Fernandes P, Karami L, Homaei A (2018) Novel lipases discovery specifically from marine organisms for industrial production and practical applications. Process Biochem 70:61–70

Ngatcha ADP, Zhao A, Zhang S, Xiong W, Sarker M, Xu J, Alam MA (2023) Determination of active sites on the synthesis of novel Lewis acidic deep eutectic solvent catalysts and kinetic studies in microalgal biodiesel production. RSC Adv 13:10110–10122

Oliveira F, Moreira C, Salgado JM, Abrunhosa L, Venâncio A, Belo I (2016) Olive pomace valorization by *Aspergillus* species: lipase production using solid-state fermentation. J Sci Food Agric 96:3583–3589

Ong HC, Chen WH, Farooq A, Gan YY, Lee KT, Ashokkumar V (2019) Catalytic thermochemical conversion of biomass for biofuel production: a comprehensive review. Renew Sustain Energy Rev. https://doi.org/10.1016/j.rser.2019.109266

Pandit PR, Fulekar MH (2017) Egg shell waste as heterogeneous nanocatalyst for biodiesel production: optimized by response surface methodology. J Environ Manag 198:319–329

Pandit PR, Fulekar MH (2019) Biodiesel production from *Scenedesmus armatus* using egg shell waste as nanocatalyst. Mater Today Proc 10:75–86

Pang K, Hou Y, Wu W, Guo W, Peng W, Marsh KN (2012) Efficient separation of phenols from oils via forming deep eutectic solvents. Green Chem 14:2398–2401

Panić M, Gunjević V, Cravotto G, Redovniković IR (2019) Enabling technologies for the extraction of grape-pomace anthocyanins using natural deep eutectic solvents in up-to-half-litre batches extraction of grape-pomace anthocyanins using NADES. Food Chem 300:125185

Pedro SN, Freire MG, Freire CSR, Silvestre AJD (2019) Deep eutectic solvents comprising active pharmaceutical ingredients in the development of drug delivery systems. Expert Opin Drug Deliv 16:497–506

Rafati A, Tahvildari K, Nozari M (2019) Production of biodiesel by electrolysis method from waste cooking oil using heterogeneous MgO-NaOH nano catalyst. Energy Sources, Part A Recover Util Environ Eff 41:1062–1074

Raj J, Bharathiraja B, Vijayakumar B, Arokiyaraj S, Iyyappan J, Praveen Kumar R (2019) Biodiesel production from microalgae *Nannochloropsis oculata* using heterogeneous poly ethylene glycol (PEG) encapsulated ZnOMn 2+ nanocatalyst. Bioresour Technol 282:348–352

Raman M, Doble M (2015) κ-Carrageenan from marine red algae, *Kappaphycus alvarezii*–a functional food to prevent colon carcinogenesis. J Funct Foods 15:354–364

Ravichandran SR, Venkatachalam CD, Sengottian M, Sekar S, Kandasamy S, Subramanian KPR, Purushothaman K, Chandrasekaran AL, Narayanan M (2022) A review on hydrothermal liquefaction of algal biomass on process parameters, purification and applications. Fuel 313: 122679

Reddy HK, Muppaneni T, Ponnusamy S, Sudasinghe N, Pegallapati A, Selvaratnam T, Seger M, Dungan B, Nirmalakhandan N, Schaub T (2016) Temperature effect on hydrothermal liquefaction of *Nannochloropsis gaditana* and *Chlorella* sp. Appl Energy 165:943–951

Reddy SN, Nanda S, Sarangi PK (2018) Applications of supercritical fluids for biodiesel production. In: Recent advancements in biofuels and bioenergy utilization. Springer Nature, Singapore, pp 261–284

Ren Q, Zhao C, Wu X, Liang C, Chen X, Shen J, Tang G, Wang Z (2009) Effect of mineral matter on the formation of NOx precursors during biomass pyrolysis. J Anal Appl Pyrolysis 85:447–453

Roberts DA, Paul NA, Dworjanyn SA, Bird MI, de Nys R (2015) Biochar from commercially cultivated seaweed for soil amelioration. Sci Rep 5:9665

Salimi Z, Hosseini SA (2019) Study and optimization of conditions of biodiesel production from edible oils using ZnO/BiFeO3 nano magnetic catalyst. Fuel 239:1204–1212

Sangon S, Ratanavaraha S, Ngamprasertsith S, Prasassarakich P (2006) Coal liquefaction using supercritical toluene–tetralin mixture in a semi-continuous reactor. Fuel Process Technol 87: 201–207

Shakya R, Whelen J, Adhikari S, Mahadevan R, Neupane S (2015) Effect of temperature and Na2CO3 catalyst on hydrothermal liquefaction of algae. Algal Res 12:80–90

Shakya R, Adhikari S, Mahadevan R, Shanmugam SR, Nam H, Dempster TA (2017) Influence of biochemical composition during hydrothermal liquefaction of algae on product yields and fuel properties. Bioresour Technol 243:1112–1120

Sharma N, Jaiswal KK, Kumar V, Vlaskin MS, Nanda M, Rautela I, Tomar MS, Ahmad W (2021) Effect of catalyst and temperature on the quality and productivity of HTL bio-oil from microalgae: a review. Renew Energy 174:810–822. https://doi.org/10.1016/J.RENENE.2021.04.147

Sharma R, Jisha KJ, Gardas RL, Malek N, Ijardar SP (2022) Insights into experimental and theoretical approach to physicochemical properties of aqueous PEGylated deep eutectic solvents at T=(293.15–323.15) K. J Mol Liq 366:120278

Shimoda K, Kubota N, Hamada H, Kaji M, Hirata T (2004) Asymmetric reduction of enones with Synechococcus sp. PCC 7942. Tetrahedron Asymmetry 15:1677–1679

Sim JH, Kamaruddin AH, Bhatia S (2010) Biodiesel (FAME) productivity, catalytic efficiency and thermal stability of lipozyme TL IM for crude palm oil transesterification with methanol. J Am Oil Chem Soc 87:1027–1034

Škulcová A, Kamenská L, Kalman F, Ház A, Jablonský M, Čížová K, Šurina I (2016) Deep eutectic solvents as medium for pretreatment of biomass. Key Engineering Materials 688:17–24. Trans Tech Publ

Śliżewska A, Żymańczyk-Duda E (2021) Cyanobacteria as valuable tool in biotechnology. Catalysts 11:1259

Smith EL, Abbott AP, Ryder KS (2014) Deep eutectic solvents (DESs) and their applications. Chem Rev 114(21):11060–11082. https://doi.org/10.1021/cr300162p

Soltanmohammadi F, Jouyban A, Shayanfar A (2021) New aspects of deep eutectic solvents: extraction, pharmaceutical applications, as catalyst and gas capture. Chem Pap 75:439–453

Suganya T, Varman M, Masjuki HH, Renganathan S (2016) Macroalgae and microalgae as a potential source for commercial applications along with biofuels production: a biorefinery approach. Renew Sust Energ Rev 55:909–941. https://doi.org/10.1016/j.rser.2015.11.026

Suwannakarn K, Lotero E, Ngaosuwan K, Goodwin JG (2009) Simultaneous free fatty acid esterification and triglyceride transesterification using a solid acid catalyst with in situ removal of water and unreacted methanol. Ind Eng Chem Res 48:2810–2818

Swetha A, ShriVigneshwar S, Gopinath KP, Sivaramakrishnan R, Shanmuganathan R, Arun J (2021) Review on hydrothermal liquefaction aqueous phase as a valuable resource for biofuels, bio-hydrogen and valuable bio-chemicals recovery. Chemosphere 283:131248

Talha NS, Sulaiman S (2016) Overview of catalysts in biodiesel production. ARPN J Eng Appl Sci 11:439–442

Tao L, Yuefeng D, Shucai GAN, Ji C (2010) Application of choline chloride· xZnCl2 ionic liquids for preparation of biodiesel. Chinese J Chem Eng 18:322–327

Thangalazhy-Gopakumar S, Adhikari S, Chattanathan SA, Gupta RB (2012) Catalytic pyrolysis of green algae for hydrocarbon production using H+ ZSM-5 catalyst. Bioresour Technol 118:150–157

Thangaraj B, Ramachandran KB, Raj SP (2014) Homogeneous catalytic transesterification of renewable *Azadirachta indica* (neem) oil and its derivatives to biodiesel fuel via acid/alkaline esterification processes. Int J Renew Energy Biofuels 11:1–16

Tišma M, Tadić T, Budžaki S, Ostojčić M, Šalić A, Zelić B, Tran NN, Ngothai Y, Hessel V (2019) Lipase production by solid-state cultivation of Thermomyces lanuginosus on by-products from cold-pressing oil production. PRO 7:465

Tran NH, Bartlett JR, Kannangara GSK, Milev AS, Volk H, Wilson MA (2010) Catalytic upgrading of biorefinery oil from micro-algae. Fuel 89:265–274

Trevan MD (1988) Enzyme immobilization by covalent bonding. New Protein Tech 3:495–510
Tripathi M, Sahu JN, Ganesan P (2016) Effect of process parameters on production of biochar from biomass waste through pyrolysis: a review. Renew Sust Energ Rev 55:467–481
Valdez PJ, Nelson MC, Wang HY, Lin XN, Savage PE (2012) Hydrothermal liquefaction of *Nannochloropsis* sp.: systematic study of process variables and analysis of the product fractions. Biomass Bioenergy 46:317–331
Vardon DR, Sharma BK, Blazina GV, Rajagopalan K, Strathmann TJ (2012) Thermochemical conversion of raw and defatted algal biomass via hydrothermal liquefaction and slow pyrolysis. Bioresour Technol 109:178–187
Vaseghi Z, Najafpour GD, Mohseni S, Mahjoub S (2013) Production of active lipase by R hizopus oryzae from sugarcane bagasse: solid state fermentation in a tray bioreactor. Int J Food Sci Technol 48:283–289
Vatankhah E, Hamedi S, Ramezani O (2020) Surfactant-assisted incorporation of rosmarinic acid into electrosprayed poly (lactic-co-glycolic acid) microparticles with potential for cosmetic and pharmaceutical applications. Polym Test 81:106180
Verma S, Das LM, Kaushik SC (2017) Effects of varying composition of biogas on performance and emission characteristics of compression ignition engine using exergy analysis. Energy Convers Manag 138:346–359
Wang K, Brown RC (2013) Catalytic pyrolysis of microalgae for production of aromatics and ammonia. Green Chem 15:675–681
Wang Z, Chen G, Ding K (2009) Self-supported catalysts. Chem Rev 109:322–359
Wang W, Zhang S, Yu Q, Lin Y, Yang N, Han W, Zhang J (2017) Hydrothermal liquefaction of high protein microalgae via clay material catalysts. RSC Adv 7:50794–50801
Wang W, Xu Y, Wang X, Zhang B, Tian W, Zhang J (2018) Hydrothermal liquefaction of microalgae over transition metal supported TiO2 catalyst. Bioresour Technol 250:474–480
Wang S, Zhao S, Uzoejinwa BB, Zheng A, Wang Q, Huang J, Abomohra A (2020) A state-of-the-art review on dual purpose seaweeds utilization for wastewater treatment and crude bio-oil production. Energy Convers Manag 222:113253
Wang J, Li M, Duan L, Lin Y, Cui X, Yang Y, Wang C (2022a) Deep eutectic systems as novel vehicles for assisting drug transdermal delivery. Pharmaceutics 14:2265
Wang S, Mukhambet Y, Esakkimuthu S (2022b) Integrated microalgal biorefinery–routes, energy, economic and environmental perspectives. J Clean Prod 131245:131245
Watson J, Wang T, Si B, Chen W-T, Aierzhati A, Zhang Y (2020) Valorization of hydrothermal liquefaction aqueous phase: pathways towards commercial viability. Prog Energy Combust Sci 77:100819
Wen L, Wang Y, Lu D, Hu S, Han H (2010) Preparation of KF/CaO nanocatalyst and its application in biodiesel production from Chinese tallow seed oil. Fuel 89:2267–2271
Widayat W, Darmawan T, Hadiyanto H, Rosyid RA (2017) Preparation of heterogeneous CaO catalysts for biodiesel production. In: Journal of physics: conference series, vol 877. IOP Publishing, Bristol, p 12018
Xinglong D, Zhaoan C, Song X, JInling Z, Jiannan Z, Yapeng L, Yunpeng X, Zhongmin L (2013) Catalytic pyrolysis of microalga *Chlorella pyrennoidosa* for production of ethylene, propylene and butene
Xu Y, Nordblad M, Nielsen PM, Brask J, Woodley JM (2011) In situ visualization and effect of glycerol in lipase-catalyzed ethanolysis of rapeseed oil. J Mol Catal B Enzym 72:213–219
Xu Y-P, Duan P-G, Wang F, Guan Q-Q (2018a) Liquid fuel generation from algal biomass via a two-step process: effect of feedstocks. Biotechnol Biofuels 11:1–16
Xu D, Guo S, Liu L, Hua H, Guo Y, Wang S, Jing Z (2018b) Ni-Ru/CeO2 catalytic hydrothermal upgrading of water-insoluble biocrude from algae hydrothermal liquefaction. Biomed Res Int 2018:1–9
Yan D, Xia P, Song X, Lin T, Cao H (2019) Community structure and functional diversity of epiphytic bacteria and planktonic bacteria on submerged macrophytes in Caohai Lake, southwest of China. Ann Microbiol 69:933–944

Yang C, Li R, Cui C, Liu S, Qiu Q, Ding Y, Wu Y, Zhang B (2016) Catalytic hydroprocessing of microalgae-derived biofuels: a review. Green Chem 18:3684–3699

Yoo G, Park MS, Yang J-W, Choi M (2015) Lipid content in microalgae determines the quality of biocrude and energy return on investment of hydrothermal liquefaction. Appl Energy 156:354–361

Yu JT, Dehkhoda AM, Ellis N (2011) Development of biochar-based catalyst for transesterification of canola oil. Energy Fuel 25:337–344

Yu KL, Lau BF, Show PL, Ong HC, Ling TC, Chen W-H, Ng EP, Chang J-S (2017) Recent developments on algal biochar production and characterization. Bioresour Technol 246:2–11

Zabeti M, Daud WMAW, Aroua MK (2009) Activity of solid catalysts for biodiesel production: a review. Fuel Process Technol 90:770–777

Zainan NH, Srivatsa SC, Bhattacharya S (2015) Catalytic pyrolysis of microalgae *Tetraselmis suecica* and characterization study using in situ synchrotron-based infrared microscopy. Fuel 161:345–354

Zhang B, Wu J, Deng Z, Yang C, Cui C, Ding Y (2017) A comparison of energy consumption in hydrothermal liquefaction and pyrolysis of microalgae. Trends Renew Energy 3:76–85

Zhao Y, Wang P, Zheng W, Yu G, Li Z, She Y, Lee M (2019) Three-stage microwave extraction of cumin (*Cuminum cyminum* L.) seed essential oil with natural deep eutectic solvents. Ind Crop Prod 140:111660

Zhou D, Zhang L, Zhang S, Fu H, Chen J (2010) Hydrothermal liquefaction of macroalgae Enteromorpha prolifera to bio-oil. Energy Fuel 24:4054–4061

Zoppi G, Pipitone G, Galletti C, Rizzo AM, Chiaramonti D, Pirone R, Bensaid S (2021) Aqueous phase reforming of lignin-rich hydrothermal liquefaction by-products: a study on catalyst deactivation. Catal Today 365:206–213

Żymańczyk-Duda E, Głąb A, Gorak M, Klimek-Ochab M, Brzezińska-Rodak M, Strub D, Śliżewska A (2019) Reductive capabilities of different cyanobacterial strains towards acetophenone as a model substrate–Prospect of applications for chiral building blocks synthesis. Bioorg Chem 93:102810

Omics Approaches for Algal Applications

Faisal Shah, Dhara Dixit, Mahdy Elsayed, and Durga Prasad Behera

Abstract Algae grow faster due to their higher photosynthetic efficiency than terrestrial plants, making them more productive. Assimilating carbon dioxide and transforming it into chemical energy/value-added compounds including vitamins, carotenoids, fatty acids, proteins, and nucleic acids is the basis of microalgae and seaweeds (macroalgae) biorefinery, which has applications in bioenergy, health foods, aquaculture feed, pharmaceuticals, and medicine. During the course of the last several decades, a great number of research have been conducted with the goals of boosting the productiveness of algal growth and increasing the value-added substances that can be extracted from various algal species. In particular, genetic engineering, synthetic biology, metabolic design, and regulatory mechanisms have been the focus of these investigations. Among them, "omics" techniques have made significant contributions to improving the understanding on presenting massive data sets on microalgae and seaweeds genomes, transcriptomes, proteomes, and metabolomes. The regulation and network integration for the biosynthesis and degradation of metabolic precursors, intermediates, and end products are disentangled using omics-based strategies, and the networks that control the metabolic flow are identified. This chapter offers an overview of the potential of algae and their involvement in symbiotic association utilizing various omics methodologies, such as genomics, transcriptomics, proteomics, and metabolomics. Additionally, applications in wastewater treatment, phycoremediation, biofuel generation, and medicines are critically examined using algae-based omics and multi-omics methodologies to present an impending prognosis for a variety of ecologically sound and commercially feasible algal applications.

F. Shah · M. Elsayed
Department of Environmental Engineering, School of Architecture and Civil Engineering, Chengdu University, Chengdu, China

D. Dixit (✉) · D. P. Behera
Coastal and Marine Ecology Division, Gujarat Institute of Desert Ecology (GUIDE), Kachchh, Gujarat, India

A. Abomohra, S. Ende (eds.), *Value-added Products from Algae*,
https://doi.org/10.1007/978-3-031-42026-9_12

Keywords Omics · Algae · Proteomics · Transcriptomics · Metabolomics · Genomics

1 Introduction

An 'ome' defines entirety, where term 'omics' refers to the integrative understanding of a collective set of biological entities that include their qualitative and quantitative physiognomies, thereby providing a blueprint of the structure, function, and dynamics of a cell/tissue (Xiaofeng and Li 2022). Thus, the term "omics" defines the strategies used in several disciplines of biology to effectively measure the profiles of metabolites (metabolomics), DNA/genes (genomics), proteins (proteomics), or RNA (transcriptomics) within cells or tissues (Conesa and Beck 2019). Multi-omics/integrated omics is the simultaneous study of multiple molecular 'omes' for obtaining a holistic perspective. Applications of these omics technologies are skyrocketing in manifold fields such as medicine, toxicology, agriculture, cosmetics, health sciences, food, and environmental science. Omics datasets tend to benefit in understanding, analyzing, visualizing, and interpreting the mechanism of a biological process. Nowadays, as the data are reaching wider audience, due to improvisations in sequencing techniques, collection, and distribution of database in public domain and progressive visualization tools, omics studies are progressing fast. However, the escalating numbers of datasets, protocols, tools, and approaches encircling omics may pose confusion to new researchers (Krassowski et al. 2020).

Omics approaches (Fig. 1) are increasingly being used in the field of algal research for their ability to provide a comprehensive understanding of the biological processes and functions. With the advent of next-generation sequencing (NGS)

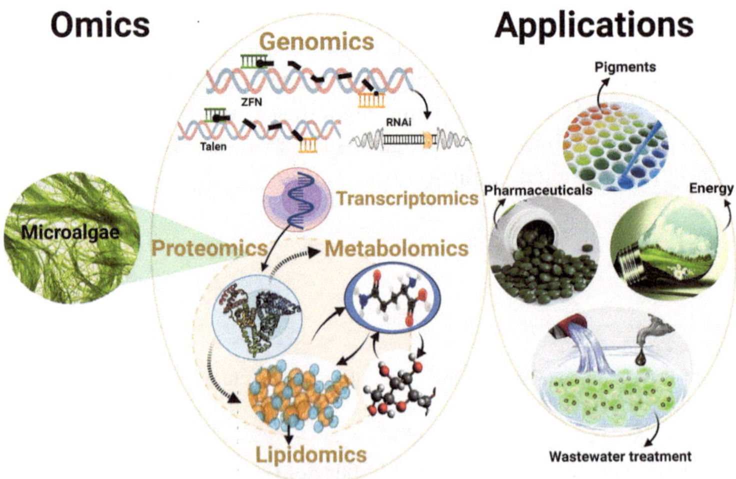

Fig. 1 Conceptual development of microalgae biorefinery through Omics approach

technologies, it is now possible to sequence the entire genome of cells via genomics, which has provided important insights into their biology and evolution (Fort et al. 2018). Metagenomics is a powerful tool for studying the microbial communities associated with algal cells, which are important in marine ecosystems. Transcriptomics involves the analysis of the entire set of transcripts (mRNAs) present in a cell or tissue at a given time. This approach has been used to identify genes involved in various biological processes such as stress responses, development, and reproduction. Proteomics approach has been used to identify proteins involved in various biological processes such as photosynthesis, stress responses, and cell wall biosynthesis. However, metabolomics approach has been used to identify the metabolic pathways involved in various biological processes such as stress responses, growth, and development.

Many studies in the fields of metabolomics, transcriptomics, and proteomics have revealed the abundance or expression of distinctive genes, proteins, and metabolites in diverse microalgae species which are exposed to environmental conditions that indirectly or directly contribute to increased intracellular accumulation of triacylglycerols (TAG). Deep insightful of lipid-metabolism may guide to effective alterations in the abundance and expression of important regulatory components of lipid biosynthesis, leading as a whole to better accumulation of TAG in microalgae (Bellou et al. 2014: Radakovits et al. 2010). Wide range of biochemical analysis of several microalgae in conjunction with complete genome sequence (proteomics, transcriptomics, and metabolomics) has been identified as an extensive method for lipid anabolism and catabolism through multiple growth patterns such as autotrophic, mixotrophic, and heterotrophic stress conditions including chemical, physiological, and operational (Lenka et al. 2016).

Microalgal diversity is enormous as an estimated 40,000 to 50,000 species has been identified and approximately as more as 800,000 species are yet to be recognized (Suganya et al. 2016). An extensive range of microalgal applications has been recognized in animal feed and food items to cosmetics products, bioremediation of municipal wastewater, production of bioplastic, and bioenergy, as well as many other phycochemicals (as discussed in this book). Although a variety of applications have been listed, only few microalgal species have been effectively applied for high value-added products. Marine seaweeds (macroalgae) represent photosynthetic organisms inhabiting the oceans and marine bodies and are crucial primary producers in aquatic ecosystems. They play an impeccable role in marine food webs and provide both habitat and shelter for a variety of marine organisms (García-Poza et al. 2022). Marine algal production has skyrocketed during the last few decades owing to multiple applications, i.e., between 2000 and 2018, it rose from 10.6×10^6 to 32.4×10^6 tons, thereby reaching a primary sale value estimated at 13.3×10^9 USD (Carpena et al. 2021).

A variety of value-added products including (carotenoids, sterols, phycobiliproteins, and various vitamins) as well as lipids, proteins, and carbohydrates can be synthesized by microalgal species and cyanobacteria using CO_2 and sunlight. Carbohydrates and neutral lipids, mainly TAG, are the most important raw materials that can be utilized to produce bioethanol, biodiesel, and other biochemical products. TAG are made up of fatty acids profile containing the glycerol backbone; after transesterification, these fatty acids are converted to fatty acid methyl-esters (FAMEs) (Lohman et al. 2015). TAG facilitate microalgal species to resist the applied stress in order to maintain lipid homeostasis inside the cell, supply of energy, and cellular functions as well (Lenka et al. 2016). However, long-lasting stressed conditions can result in photosynthetic apparatus breakdown, which ultimately results in degradation of chlorophyll called chlorosis, reduction of cell division, and the overall production of TAG is affected (Vonlanthen et al. 2015). Additionally, there is a contrary relation among lipogenesis and active growth; hence, an optimal balance among TAG accumulation and growth rate is important for biofuel production from algae in a commercial level (Davis et al. 2011). To improve cultivation techniques and strain development for maximum biomass and lipid production, microalgal omics tools can be used for maximal achievement and processes decoupling.

The commercial significance of algae needs a thorough comprehension of their biological functions. "Omics" technologies designed for algae, or Algomics, have gained popularity in the biological and biomedical disciplines during the past ten years. These technologies offer thorough descriptions of algae for potential future uses, which could advance our knowledge of algal ecological processes. Using omics, it is possible to look at numerous facets of algal processes using different techniques including genomics, transcriptomics, proteomics, metabolomics, and metagenomics. Algae are showing more and more appealing possibilities for bioremediation, particularly in wastewater treatment. They also have a lot of potential as productive microbial cell factories. It is now feasible to pinpoint the genetic and phenotypic traits that lead to enhance the microalgal growth and nutrient removal in wastewater using omics, microalgal screening, and molecular phylogeny. Researchers are also investigating the use of microalgae in the synthesis of several bioenergy carriers as third-generation biofuel feedstocks, including direct biohydrogen production, lipids for biodiesel, and carbohydrates for alcohols. This chapter focuses on the significant contributions of omics and molecular methodologies to study the diverse applications for algae. It highlights Algomics to provide important insights into cellular components and metabolic processes for enhanced phycochemical production. These technologies open the door for incorporating these components into high-quality output, hence increasing the profitability and sustainability of algal biorefinery systems.

2 Genomics

The aim of genomics is to sequence, analyze, and assemble structure and functions of specific genome (Primrose and Twyman 2003). The tool of DNA sequencing can be used to know what is exactly happening inside an organism if the aim is to express specific gene. Genomics has contributed significantly to understand the biology and evolution of microalgae and seaweeds. For example, the genome of the brown seaweed *Ectocarpus siliculosus* was the first complete genome sequenced in multicellular algae. This genome sequence has been used to study the evolution of multicellularity and the genetic basis of stress responses in algae. This has led to the discovery of several genes involved in the biosynthesis of alginate as one of the important phycochemicals produced by brown algae. In addition, genomic resources have been created for certain microalgae such as *Chlamydomonas*, *Nannochloropsis*, and diatoms. These resources provide insightful information on the genetic makeup and biological functions of these microalgae. To better understand several biological processes, the genome of the green alga *Chlamydomonas reinhardtii* has been fully sequenced and analyzed. Attempts have also been made to identify the lipid synthesis-related genes in *Nannochloropsis*, a green microalga known for having a relatively high lipid content. Diatoms, which have unique cell walls with many industrial applications (Chapter "Diatom Nanostructured Biosilica"), have had their genomes sequenced, allowing researchers to study their biology and potential applications. These genetic resources will be essential for further research and commercial use of these microalgae (Armbrust et al. 2004; Gong et al. 2020; Merchant et al. 2007). The number of microalgae for which genomic resources are accessible continues to grow as a result of developments in next-generation sequencing technology and the resulting cost reductions (Kselíková et al. 2022). With the availability of genomic resources, multi-omics data, and comparative genomics, it is easy to find target genes in certain algal species (Jeong et al. 2023). Genomics have been used so far in many phycological applications as discussed in this section.

2.1 Wastewater Treatment

Genomics has been applied in wastewater treatment to explore the potential of algae in removing pollutants from wastewater. As discussed in Chapter "The Use of Wastewater for Algal Growth", wastewater is an abundant source of nitrogen, phosphorus, and organic carbon which are essential for growth of microbial flora; however, aquatic ecosystem would be harmed by an overabundance of these nutrients and inappropriate handling. Organic pollutants such as petroleum hydrocarbons (PHCs), insecticides, pesticides, heavy metals, antibiotics, phenolics, polyaromatic hydrocarbons (PAHs), and polychlorinated biphenyls (PCBs) contribute to the contamination of water bodies, presenting a serious risk to the environment and public health (Pandey et al. 2019). Microalgae have been shown to be efficient in wastewater treatment, and microalgal biotechnology offers promising avenues for expanding into new environmental fields via applications including bioremediation and the yield of high-value by-products (Patel et al. 2015). Algomics technology has led the way for contemporary effective bioremediation methods and evaluation of dangers related to environmental toxicity. It has also offered a key insight into algal community dynamics in metabolic processes, development, and functions. Several microalgal species have been examined in terms of their efficacy at removing phosphorus and nitrogen from different wastewater resources. These include *C. vulgaris*, *Chlorella* sp., *Scenedesmus* sp., *Cyanobacteria*, *Arthrospira* sp., and *Oscillatoria* sp. The ability of these microalgae to absorb and use nutrients from the water has showed promise in wastewater treatment. It was confirmed that bacterial cofactors such as thiamin pyrophosphate and 4-amino-5-hydroxymethyl-2-methylpyrimidine help to boost *Auxenochlorella protothecoides* metabolic potential (Higgins et al. 2016). The increased biomass output was due in part to the positive effects of these cofactors on processes including lipid accumulation, glucose absorption, and wastewater treatment efficiency. It has been also confirmed that green algae partners are mostly responsible for the removal of phosphate and nitrogen from wastewater. According to metagenomic data, green algae play a substantial role in absorption and assimilation of nitrogen and phosphorus compounds, which dramatically lower their levels in wastewater. This emphasizes how crucial green algae's metabolic capacities are for nitrogen removal and how they raise the overall effectiveness of the wastewater treatment process (Shetty et al. 2019). Even though algae have a lot of promise to clean up wastewater, there are still hurdles and important problems that experts all over the world are trying to solve. Genomics as well as other omics approaches show a detailed report that would be a very useful guide to explore the behavior of wastewater communities and give information about molecular metabolic flows level (Jha et al. 2018).

Seaweeds also have been known to remove organic and inorganic pollutants from wastewater, and genomics can be used to better understand the mechanism of their action and to optimize their use in wastewater treatment. El-Naggar et al. (2018) investigated the potential use of the seaweed *Gelidium amansii* in wastewater treatment. The study showed the high lead (Pb^{2+}) biosorption efficacy of this seaweed from aqueous solutions that can be harnessed in the future for Pb^{2+} removal. Califano et al. (2020) undertook a metagenomic analysis of the microbial communities associated with *Ulva rigida* in wastewater treatment and showed that the seaweed housed different microbes that could potentially eliminate pollutants from wastewater. Another metagenomic investigation of microbial communities in wastewater treatment using *Gracilaria lemaneiformis* was conducted by Pei et al. (2021). This study reported the identification of key microbial taxa responsible for the degradation of pollutants in wastewater. This information can be used to optimize the microbial communities in wastewater treatment systems that use seaweeds.

2.2 Biofuels

Genomics can be used to improve the production of biofuels from algae by identifying the genes and pathways involved in the biosynthesis of biofuel precursors or value-added products in biofuel promising algae and by developing strategies to optimize the expression of these genes. For instance, genomics has been applied to identify and characterize the genes encoding the enzymes involved in the conversion of *Pyropia yezoensis* biomass into biofuels (Zhang et al. 2020b). The genes involved in the biosynthesis of alginate, a polysaccharide found in the biofuel-feedstock Ochrophytes, have been identified and characterized using genomics approaches by Shao et al. (2019). A genomic analysis of the Ochrophyte *Saccharina japonica* revealed genes involved in the biosynthesis of mannitol, a sugar alcohol that can be used as a biofuel precursor. The study led to identification of a mannitol-2- dehydrogenase gene that was upregulated under stress conditions, suggesting that this gene could be targeted to increase mannitol production in seaweeds (Shao et al. 2014), followed by biomass conversion into biofuels. Another study on the same species highlighted the genetic markers associated with its high biomass production and high biofuel production potential (Wargacki et al. 2012). A similar study targeting genomics-assisted breeding and demonstrating the potential application of genomics in developing new seaweed strains with improved biofuel production was used to improve the yield of biofuels from *Undaria pinnatifida*. The identification of a gene involved in the biosynthesis of mannitol was undertaken in order to select high-yielding strains of this seaweed (Shan and Pang 2021). Microorganisms are being employed to produce biomass from seaweeds, where genomics is being used to sequence and analyze the genomes of such microorganisms to understand the underlying genetic mechanisms involved in biofuel production. For example, the genome of the marine bacterium *Saccharophagus degradans*, which can degrade various types of seaweeds and produce biofuels, was sequenced and analyzed by Wayllace et al. (2021). Another metagenomic analysis was undertaken to study the microbial community associated with the Chlorophyte *Ulva ohnoi*, in which several bacterial species that exhibited the potential for producing enzymes involved in the degradation of ulvan were identified. These enzymes have potential biotechnological applications in the production of biofuels and other phycochemicals (Coste et al. 2015).

A technique for locating the regulatory components required for the development of genetic engineering methods in microalgae has been made available using omics. Consequently, the identification of reporter genes, terminators, promoters, and splicing signals has improved the expression of heterologous genes in microalgae. As a result, the development of powerful molecular tools for genome editing, such as RNA interference (RNAi), TALEN (Transcription Activator-Like Effector Nuclease), zinc finger nucleases (ZFN), and clustered regularly interspaced short palindromic repeats (CRISPR-Cas) -related proteins, has gained steam. Researchers have made considerable use of these genomic instruments throughout the last decade (Lin et al. 2019). This approach modifies metabolic mechanisms to upregulate or downregulate genes to boost biomass and product output (Khoo et al. 2023), including lipids for biodiesel production. Kurita et al. (2020) reported two genes in *Nannochloropsis oceanica*, namely nitrate reductase and acyltransferase, which were efficiently mutated after treatment with Platinum TALENs. By disrupting the UDP-glucose pyrophosphorylase gene using TALEN guidance, researchers discovered an aberrant buildup of lipids. Chang et al. (2020) altered the strain's AGP protein, which is responsible for phosphorylating ADP-glucose; this led to an increase in lipid synthesis and successfully extended the approach to the green marine microalga *Tetraselmis* sp. Significant increases of 2.7 and 3.1-fold in lipid content were observed (21.1 and 24.1% dry cell weight, respectively). Gomma et al. (2015) determined the effect of ACC overexpression in *Scenedesmus quadricauda* using *Saccharomyces cerevisiae* to amplify the ACC1 gene. The total fatty acid (TFA) content of the *S. quadricauda* increased by 1.6 times when ACC1 overexpression was compared to the wild strain. Shin et al. (2019) employed CRISPR-Cas9 to make a mutant form of phospholipase-deficient *C. reinhardtii*. One of the mutant strains showed 190.4% increase in TAG contents throughout the development phase. One of the mutant strains had a TAG content increase of 68.21% and a diacylglycerol (DAG) content increase of 117.97%. Despite the promising outcomes of genomics, the price of genetic alteration is costly, and other factors including stability, regulatory problems, safety, and toxicity are actively being studied. Additionally, strict legislations related to outdoor cultivation of genetically modified organisms slow down or prevent the genetic engineering of microalgae for commercial purposes as discussed in Chapter "Legislation and Biosecurity".

2.3 Medical and Therapeutic Applications

Genomics applications in medicine and therapeutics have shown promising results in various areas, including identification of new drug targets, personalized medicine, and disease diagnosis. The application of algal genomics in medicine and therapeutics has revolutionized the diagnosis and treatment of various diseases. Seaweeds, which are a rich source of bioactive compounds, have been studied extensively for their medicinal properties. The role of genomics in the development of new drugs from seaweed-derived compounds has opened new avenues in modern therapeutics. Fucoidan, a sulfated polysaccharide found in Ochrophytes, has proven antitumor, immunomodulatory as well as anticoagulant properties. Genomic analysis has led to the identification of the genes involved in fucoidan biosynthesis, which could further facilitate the development of new drugs (Chi et al. 2018). Genomics approaches have also been used to identify potential drug targets in seaweeds. A recent study by Mikami and Hosokawa (2013) used genomics to study genes-encoding enzymes involved in the biosynthesis of fucoxanthin, a carotenoid found in Ochrophytes. Fucoxanthin has been shown to have anti-inflammatory, anti-obesity, and anticancer properties, and the identification of this enzyme could facilitate the development of drugs targeting fucoxanthin biosynthesis. Genomics approaches have also been used to identify potential drug targets in seaweeds. For example, in *Codium tomentosum*, bioactive compound with anti-inflammatory properties was identified (Rabecca et al. 2022). This compound could be a potential drug for the treatment of inflammatory diseases. Genomics approaches have been used to identify the biosynthetic pathways of bioactive compounds in seaweeds. For example, in the red seaweed *Porphyra umbilicalis*, the biosynthetic pathway for the pigment phycoerythrin was identified using RNA sequencing (RNA-seq) analysis (Collén et al. 2013). Genomics approaches have the potential to facilitate personalized medicine by identifying genetic variants that can affect drug metabolism and response. For example, a study identified genetic variants that affect the metabolism of the anticancer drug capecitabine in patients with colorectal cancer (García-González et al. 2015).

Regarding microalgae, commercialization of the green microalga *Chromochloris zofingiensis* is extremely desirable because of its potential to produce the ketocarotenoid astaxanthin which had nutritional value due to its benefits for human health, as well as lipids for biofuel (Breuer et al. 2012; Liu et al. 2016; Mulders et al. 2014). Astaxanthin is applicable in various industries, including those connected to the manufacturing of pharmaceuticals, cosmetics, nutraceuticals, animal feed, and food (Yuan et al. 2011; Liu et al. 2014). The genome of *Chromochloris zofingiensis* included two genes that encode beta-ketolase (BKT), a crucial enzyme in the synthesis of astaxanthin. High light increased the expression of both genes. Additionally, the adenosine triphosphate (ATP) binding cassette (BC) transporters, cytochromes P450 enzyme, and the acyltransferase were among the potential genes picked up by transcriptome when subjected to intense sunlight that may be involved in crucial, but uncompleted, phases of astaxanthin production. Understanding the green algal ancestry and carotenoid synthesis is made feasible by the high-quality genomic and transcriptomic (Roth et al. 2017). Terpenoids have the potential to cure immunodeficiency syndrome (AIDS) and cancer, with antimicrobial properties. However, terpenoids extraction from their natural plant is costly with limited yield. By genetic engineering, a plant secondary metabolic pathway was introduced into the unicellular alga *P. tricornutum*, which enabled high value terpenoids production (D'Adamo et al. 2019).

2.4 Metal Toxicity and Bioremediation

Genomics is a powerful tool that can be used to understand the mechanisms of metal toxicity and phycoremediation. For example, a study by Ausuri et al. (2022) used genomics to identify microbial genes that are capable of bioremediating heavy metal pollution in polluted water. Genomics can also be used to understand the molecular mechanisms of metal toxicity in algae. Genomic analysis has also been used to identify metal resistance genes in seaweeds. Ritter et al. (2014) used the whole-genome sequencing to identify metal resistance genes in the brown seaweed *Ectocarpus siliculosus*. The study identified several genes involved in metal transport, detoxification, and antioxidant defense. Genome sequencing and analysis of the brown seaweed *Saccharina japonica* led to the identification of genes involved in the biosynthesis of alginate, a commercially important polysaccharide used in food, pharmaceuticals, and industrial applications. They also identified genes responsible for the seaweed's tolerance to environmental stressors such as high salinity, ultraviolet (UV) radiation, and heavy metal toxicity.

Data from several omics approaches were utilized in toxicological studies of microalgae as well as in order to learn how toxic metals exert their effects. When contaminants are given to microalgae, the cell density often decreases initially owing to cell death, but then rises because cells become resistant to the contaminants over time (Ipatova et al. 2012). Microalgae are viable phytoremediation candidates because they are abundant; nevertheless, the challenge of selecting suitable microalgal species for phycoremediation means that they must not only be capable to survive in high metal concentrations, but also need to be able to extract and catch the heavy metals in wastewater (Mishra et al. 2019). Multiple metal systems (copper, cadmium, and lead) were used to treat *Chlorella* sp. and the results were analyzed with a combining metallomics with metabolomics based on nuclear magnetic resonance spectroscopy (NMR). Further studies aimed to comprehend the potential interactions among heavy metals and the destiny of multi-metal systems on the microalgae *C. vulgaris* (Zhang et al. 2014; Zhang et al. 2015). To understand toxicity mechanisms of the contaminants, omics can turnout as an alternative to the conventional methods since this approach is limited by its target analysis character.

2.5 Agriculture, Cosmetics, and Environmental Applications

The critical threshold for anthropogenic CO_2 emissions has been achieved, and it is anticipated that between 2030 and 2050, the global surface temperature would increase by 1.5 °C. To help mitigate the present global warming situation, researchers have been working hard to develop more cost-effective and novel carbon-sequestration techniques (Tarafdar et al. 2023). Algae represent a potential candidate for removal of carbon dioxide (CO_2) from the atmosphere due to their remarkable photosynthetic capacity and rapid growth rate (Sondak et al. 2017). Yong et al. (2022) highlighted the significant role of seaweeds in mitigating climate change. The study reported that seaweeds exhibit characteristics to act as a blue carbon reservoir with an immense carbon sink potential. Another study highlighted the genome sequence of *Saccharina japonica* showing the genes responsible for traits such as stress tolerance and photosynthesis (Ye et al. 2015). The genome sequences of several green algal species have been compared to identify the genes involved in photosynthesis, cell division, and stress responses. This analysis has provided insights into the molecular mechanisms underlying the adaptation of algae to different environments (De Clerck et al. 2018).

Significant carbon-tolerant capability (10–100%) has been shown using different microalgal species including *Spirulina platensis*, *Dunaliella tertiolecta*, *Desmodesmus* sp., *Chlorella* sp., and *Nannochloropsis* sp. Biorefinery techniques can convert microalgal biomass into different value-added products such as biofuels, medicines, and nutraceuticals, lowering the cost of microalgae-based carbon capture (Tarafdar et al. 2023). Several efforts using CRISPR-based technologies have made strides in the past decade towards the goal of increasing the microalgal potential. Tolerance to NH_4Cl in media was achieved, for instance, by deleting the nitrate reductase (NR) gene in *N. oceanica* IMET1 (oleaginous microalgae). CRISPRi (CRISPR interference) was used to regulate the rfp and PEPC1 (genes-encoding protein related to carbon split metabolites which competes with lipid synthesis) in *C. reinhardtii*. Increasing the capture of CO_2 in *C. reinhardtii* with Cia5 gene deletion was achieved (Barati et al. 2021; Asadian et al. 2022). In a fascinating experiment, researchers joined the small subunits of RuBisCO from spinach, sunflower, and *Arabidopsis* with the big algal subunit by transforming a mutant that lacked the rbcS genes from *C. reinhardtii*. Although the velocity remained close to normal, a higher CO_2/O_2 affinity was recorded. The mutated strains also showed a reduced rate of photosynthetic development and no pyrenoid structures. It has been shown that overexpressing of cbbX-homologous, a RuBisCO activase, may increase the biomass production of *N. oceanica* with adequate CO_2 levels. In addition, the mutant growth rate, biomass production, and lipid output all increased by 32%, 46%, and 41%, respectively, in addition to 28% more photosynthetic activity (Wei et al. 2017b). In another study, knockdown of a gene expressing CA (a Beta Type CA; CAH5) in *N. oceanica* increased the growth rate, biomass productivity, and photosynthetic oxygen evolution rate by 30, 45, and 40%, respectively, under 5% CO_2 supplementation (Wei et al. 2019). Despite multiple successful transformations that have increased the overexpression of vital enzymes in the carbon-fixing system, it is still difficult to stabilize the integration of modified genes to the chloroplast genome (Greco et al. 2012).

3 Lipidomics

The branch of metabolomics discipline known as "lipidomics" is used to classify and distinguish different types of lipids as well as the compounds that interact with these lipids (Melo et al. 2015). Identifying lipidomic variations according to various variables in the environment may help to better understand not only the lipid metabolism of algae, but also how to manipulate lipid productivity and profile for the purpose of improving the generation of biofuel metrics (Su et al. 2013). Few investigations have examined how the lipidome changes in responses to diverse types of stress, including high temperatures, high salt levels, and food deficiency. Most studies focused on microalgae as a promising feedstock for sustainable lipid production. These studies showed that various microalgae have different lipid biomarkers that have distinct mechanisms in response to the offered environmental stress, including free fatty acids, digalactosyldiacylglycerol, 1,2diacylglycerol-3-0-4′-(N,N-trimethyl)-homoserine, harderoporphyrin, TAG, sulphoqunovosyldiacylglycerol, phosphatidylglycerol, cholesterol, and lysodigalactosyldiacylglycerol in different microalgae including *Chloromonas*, *Chlamydomonas nivalis*, *C. reinhardtii*, *Nitzschia closterium*, *Fragilaria oceanica*, *Chlorella minutissima*, and *Dunaliella tertiolecta* (Yang et al. 2015a, 2015b, 2015c; Lee et al. 2014; Arora et al. 2018a; Li et al. 2014; Lu et al. 2012; Su et al. 2013).

Researchers have been using lipidomics for years in order to better comprehend how algal lipid metabolism reacts to different stresses in a range of microalgae in recent years. Such results might serve as the foundation for specific genetic engineering. The unicellular red alga *Cyanidioschyzon merolae*, a dweller of sulfur-rich acid hot springs, provided the first microalgal genome, which became accessible in

2004 (Sasso et al. 2012). *C. merolae* has the smallest genome of any photosynthetic eukaryote that has been discovered so far, as well as the only nucleus, mitochondria, and plastids (Matsuzaki et al. 2004). However, there are already accessible entire genome sequences for more than 30 different diatoms and green algae.

3.1 Improving Microalgal Lipids

Triglycerides, phospholipids, fatty acids, and cholesterol are all included in the broad definition of microalgal lipids. The value and appeal of phospholipids and fatty acids are high (Mishra et al. 2019). The composition of algal lipids is significantly influenced not just by genetic and phenotypic factors, but also by external settings and situations. Gene knockout transformants showed an increase in lipid accumulation of 40–55% under nutrient-replete conditions, whereas semicontinuous growth circumstances resulted in a doubling of lipid productivity to 5.0 g m^{-2} day^{-1}. This increase in lipid productivity did not have an adverse effect on the transformants' ability to develop (Ajjawi et al. 2017). Recent advancements in technology, such as lipidomics, have allowed a greater knowledge of the complete genome structure as well as lipid metabolism in a variety of microalgal species. TAG and phosphatidylcholine (PC) molecular species were detected in microalgae, and their accumulation content, fatty acids (FAs) composition, separation, and identification were all made possible by the use of such technology (Arif et al. 2020).

The potential of transgenic microalgae to provide researchers the chance to repair and reconstruct the microalgal lipid biosynthesis (and associated) pathways to study and generate high TAG accumulating phenotypes has led to their increasing appeal. Several genetic techniques, including electroporation, biolistic transformation, homologous recombination (HR), random integrative selection markers, human ribosomal RNA silencing, and HR-mediated DNA incorporation, have been successfully used to alter microalgae in recent years (Ghosh et al. 2016). Depending on what kind of genes or constructs needs to be added or removed, microalgae can undergo transformation whether it be in the mitochondrial, chloroplast, or nuclear genomes (Gimpel et al. 2015). Enzymes that participate in secondary metabolism, for instance, can be modified to express themselves differently in nuclear or plastid genomes. The use of heterologous promoters and untranslated regions is permitted by this method. Nuclear genome transformation benefits from posttranslational modification in addition to simple techniques and flexibility (Gimpel et al. 2015). Despite advances in technology, only few microalgal strains, including *Cyclotella cryptica*, *Chorella minutissimsa*, *P. tricornutum*, *C. reinhardtii*, *C. ellipsoidea*, *T. pseudonana*, and *N. gaditana,* have been metabolically altered for increased TAG deposition (Banerjee et al. 2016; Gimpel et al. 2015; Ho et al. 2014b). Altering carbon absorption pathways and attempting to boost photosynthetic efficiency are two other approaches to enhance TAG productivity and/or yield along with lipid biosynthesis genes. Microalgal capacity for absorbing photons and converting them into energy ATP, nicotinamide adenine dinucleotide phosphate (NADPH), in turn, has an impact on the productivity of lipid synthesis as a whole (Ghosh et al. 2016).

4 Transcriptomics

Transcriptomics is a powerful tool that allows the study of gene expression at the mRNA level, providing valuable insights into the molecular mechanisms underlying biological processes. Transcriptome profiling not only contributed to the identification of genes in response to emulated stresses, but also allowed the discovery of genes involved in secondary metabolite biosynthesis. Global transcriptional profiling of microalgal cells using NGS can provide helpful insights into the molecular mechanisms and regulatory genes associated with a higher production of neutral lipids in microalgae through the identification of significant transcriptional regulators that alter lipid biosynthesis (Rismani-Yazdi et al. 2012). Its absence impairs various cellular processes, particularly the *de novo* synthesis of amino acids, nucleic acids, and other cellular components, which frequently results in a dramatic increase in the amount of specific phycochemicals within cells (Yang et al., 2013). The altered gene expression profiles of numerous microalgae, including *C. reinharditii*, *Chlorella vulgaris*, *Botryococcus braunii*, *Botryosphaerella sudeticus*, *Neochloris oleoabundans*, *Phaeodactylum tricornutum*, *Tetraselmis* sp., *Monoraphidium neglectum*, and *Micractinium pusillum*, have been extensively studied as a result.

These investigations have provided an organized profile of gene expression shifts in a number of metabolic pathways, including ribosome biosynthesis, RNA processing, protein metabolism, photosynthesis, the tricarboxylic acid (TCA) cycle, nitrogen assimilation, energy production, carbon fixation, carbohydrate metabolism, and pentose phosphate metabolism, as well as the elevation in lipid accumulation under nitrogen insufficient scenarios. There has been a substantial enhancement in the transcripts of acyl carrier protein (ACP) gene, diacylglycerol acyltransferase (DGAT) isoforms such as DGAT-1, DGAT-2A, DGAT-2B, and DGAT-2E (which catalyze the final step of TAG synthesis), and biotin carboxylase [which controls ACCase (acetyl CoA carboxylase) activity through carboxylation of biotin moiety of the enzyme]. There have been plentiful studies of thioesterase genes (Fat A and thioesterease oleoyl-ACP hydrolase), acyl-ACP desaturase (AAD), delta 15 saturase, lipases, and saposin under nitrogen deficiency (Li et al. 2016; López García De Lomana et al., 2015; Li et al., 2014; Yang et al., 2013).

Numerous transcriptomics studies in diatom and green microalgae have emphasized the critical role of lipid recycling (as opposed to *de novo* lipid synthesis) in improving the total TAG buildup during nitrogen limitation. The enzymes, phospholipid:diacylglycerol acyltransferase (PDAT), 3-sn-phosphatidate phosphohydrolase (PAP), monogalactosyl diacylglycerol (MGDG) synthase, digalactosyl diacylglycerol synthase (DGD 1n), SQD2 synthase, galactolipase gene, and fatty acid chain modification genes, displayed increased expression in the transformation of lipids from chloroplast glycerolipids, demonstrating that membrane remodeling is in fact utilized in nitrogen deficiency in microalgae (Li et al., 2014; Tanaka et al. 2015; Arora et al. 2018b). Transcriptomics have been used in various algal applications as discussed in the present section.

4.1 Wastewater Treatment

Algae have been studied for wastewater treatment, as they have the ability to remove pollutants from wastewater and provide valuable biomass for various applications (Fig. 2). For instance, the transcriptome of a commonly used seaweed, *G. lemaneiformis*, was analyzed for its potential to remove nutrients from wastewater under nitrogen deficiency (Wei et al. 2017a). A study conducted by Shi et al. (2007) highlighted the removal of nitrogen and phosphorus metabolism of *C. vulgaris and Scenedesmus rubescens*. The findings provide insights into the mechanism underlying the ability of these species to remove nitrogen and phosphorus from wastewater through cell immobilization. A similar study was conducted using *Microcystis aeruginosa* by Zhou et al. (2020) using transcriptomic analysis. The findings revealed that the transcripts of genes associated to nitrogen metabolism were upregulated during nitrogen starvation, while the genes involved in photosynthesis-related gene expression were downregulated. A study conducted by Teo et al. (2009) used transcriptomic analysis to understand the molecular mechanisms of *G. changii* in response to hyper- and hypoosmotic stress in water.

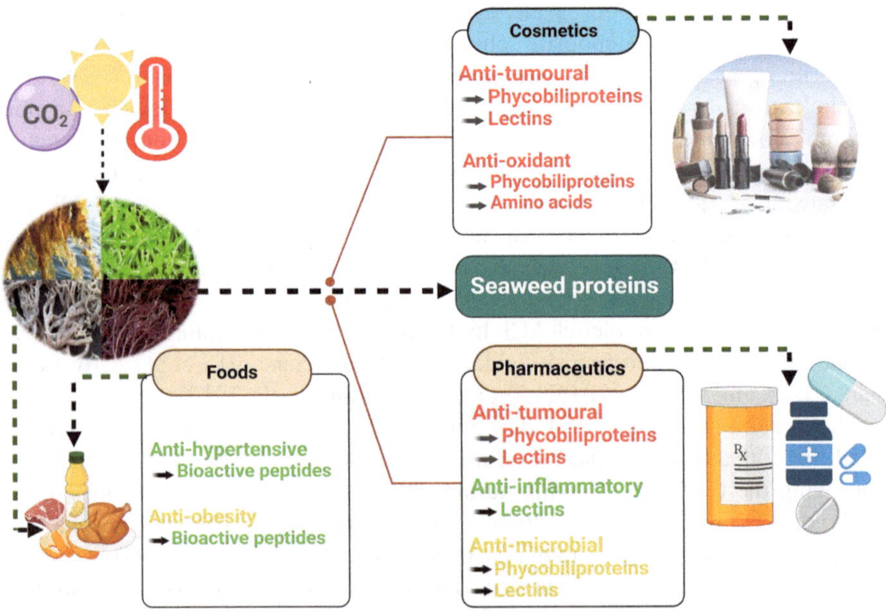

Fig. 2 Potential applications of algal proteins, bioactive peptides, and amino acids

The study revealed the presence of a wide range of genes in the seaweed that were involved in acclimatizing to osmolarity which made it an efficient candidate for seawater treatment. Another study by Mangott et al. (2020) used omics analysis to investigate the response of the seaweed *Ulva lactuca* to wastewater exposure. The study identified differentially expressed genes that were involved in the detoxification and metabolism of pollutants in wastewater. This information can be used to optimize the use of *U. lactuca* in wastewater treatment. A similar attempt for providing insights into the molecular mechanisms underlying the ability of *Ulva lactuca* to tolerate and remove pollutants from wastewater using transcriptomics as an analytical tool was undertaken by Xu et al. (2021) and González et al. (2022). The study reviewed the efficacy of wastewater treatment plants and emerging technologies for the removal of microplastics. Microalgae have been discussed recently as a potential candidate for micro-/nanoplastic removal (Abomohra and Hanelt 2022), where transcriptomics could provide a tool for further understanding the removal process.

Additionally, biocatalytic enzymes have been identified using proteomics and metabolomics, as well as the reactions of microalgae to stressors brought on by pollutants. In reaction to a pharmaceutical pollutant in wastewater, one-half of the coding areas of *Picochlorum* sp. SENEW3 transcriptome study revealed differential expressions. Bacteria and microalgae often coexist during the recycling of microalgal bioresources in wastewater treatment operations. In a previous study, Zhou et al. (2017) examined how chlorophytes responded to the usual quorum sensing (QS) molecules N-acylhomoserine lactones (AHLs) isolated from activated sludge bacteria. After interacting with AHLs, microalgae self-aggregated in 100–200 μm bioflocs by secreting 460–1000 kDa aromatic proteins, and its settling efficiency peaked at 41%. Further transcriptome analysis demonstrated that anthranilate accumulation helped to upregulate the production of aromatic proteins from tyrosine and phenylalanine.

4.2 Biofuel Production

Transcriptomics is a powerful tool to investigate gene expression and regulation in organisms, including algae. Seaweeds are a potential source of biofuel due to their ability to accumulate large amounts of carbohydrates, which can be converted into biofuels such as ethanol and butanol. Other studies evaluated the genes involved in the biosynthesis of lipids and their upregulation under stress using *Pyropia haitanensis* (Lin et al., 2021; Wang et al. 2015). The study highlighted that the changes in gene expression were owing to osmotic stress. It was also revealed that lipid metabolism plays a crucial role in the regulation of conchosporangia maturation and genes underlying such mechanism may serve as important molecular markers in future. A study by Zhang et al. (2020a) used transcriptomics to investigate the gene expression patterns of the brown seaweed *Saccharina japonica* under different stress conditions. The study identified several genes involved in the biosynthesis of carbohydrates, such as cellulose and laminarin, which are potential sources of biofuels. The study also identified genes involved in the regulation of lipid metabolism, which can be converted into biodiesel. Gu et al. (2022) used transcriptomics to investigate the gene expression patterns of the green seaweed *Ulva prolifera* under high light, a stress condition used to induce the accumulation of carbohydrates in seaweeds. They identified several genes involved in the biosynthesis of carbohydrates and lipids metabolism. Likewise, a similar study was also conducted using *Gracilaria tenuistipitata* (Liu and Dong 2001) and concluded that nitrogen demand for supporting growth of *G. tenuistipitata* was less in comparison to *Ulva pertusa*.

Transcriptomics is used in microalgae also to more clearly identify or detect certain metabolic pathways or biological processes, as well as to better understand them. The polar oleaginous microalga *Coccomyxa subellipsoidea* transcriptome analysis revealed the global and cooperative regulation to assimilate carbon and maintain carbon/nitrogen balance under conditions of elevated CO_2 for the support of rapid growth and lipid accumulation by providing an abundant carbon skeleton and abundant metabolic energy (Peng et al., 2016). Positional polymorphism analysis, differential expression levels, selection signature analysis, and the investigation of putative gene function are combined. In the wild-type and modified strains of the microalgae *Tisochrysis lutea* S2M2, 8 genes were discovered to be in charge of the excessive lipid accumulation (Carrier et al., 2014). In a previous study, it was discovered via transcriptomics analysis that putting *N. oleoabundans* under nitrogen stress may increase TAG synthesis. A range of genetic engineering targets and strategies were also supplied by the research in order to increase the rate of synthesis and cellular make-up of potential biofuels (Rismani-Yazdi et al. 2012, b). Recently, a carbonic-anhydrase-deficient *Nannochloropsis* sp. has shown an increase in biomass at 5% CO_2. The whole transcriptomics analysis method indicated that elevated CO_2 tolerance was caused by the inactivation of carbon concentration machinery. This, by itself, may be capable of producing significant amounts of CO_2 which favorably affect oil-based fuel production in the strain (Wei et al., 2019).

4.3 Medical and Therapeutic Applications

Nowadays, transcriptomics has been increasingly used in medicine and therapeutics to investigate the molecular mechanisms underlying diseases and to identify potential drug targets. Algae, which are rich sources of bioactive compounds, have also been studied for their potential therapeutic applications, and transcriptomics has been used to explore the mechanisms of action of these compounds. A recent study by Nishitsuji et al. (2016) used transcriptomics to identify genes involved in the biosynthesis of phlorotannins from *Cladosiphon okamuranus.* The study led to the identification of genes that encoded the enzymes involved in fucoidan synthesis. Fucoidan, a sulfated polysaccharide found in Ochrophytes, possesses anticancer properties. Omics was used to investigate the mechanism of action of fucoidan in human leukemia cells. The study found that fucoidan induced apoptosis in leukemia cells by upregulating the expression of proapoptotic genes and downregulating the expression of antiapoptotic genes (Park et al., 2013). Phlorotannins have been shown to have anti-inflammatory activity. Transcriptomics was used to investigate the mechanisms of action of phlorotannins in human colon cells, and it was reported that phlorotannins inhibited the expression of pro-inflammatory genes and activated the expression of anti-inflammatory genes (Montero et al., 2015). Fucoxanthin, a carotenoid found in brown seaweed, has been shown to have antidiabetic activity. Transcriptomics was used to investigate the mechanism of action of fucoxanthin in diabetic mice (Maeda et al., 2015). The study found that fucoxanthin upregulated the expression of genes involved in insulin signaling and glucose metabolism, leading to improved glucose tolerance. Transcriptomics was also applied to study the antiviral potential of carrageenan against human cells infected with influenza virus (Wang et al., 2012). The study revealed that carrageenan inhibited virus replication. The transcriptomic studies of *Gracilaria changii* revealed the upregulation of genes involved in cell cycle, apoptosis, as well as immune response, highlighting its capability as an antitumor agent (Andriania et al. 2016; Ho et al., 2009). Some insights surrounding the mechanisms involved in the anti-inflammatory action of *Saccharina japonica* have also been highlighted by Islam et al. (2013). Seaweeds have been shown to possess antiviral activity against a range of viruses, including human papillomavirus (HPV), herpes simplex virus (HSV), and influenza virus. Transcriptomic studies of the red seaweed *Laurencia dendroidea* revealed the upregulation of genes involved in the biosynthesis of terpenoids, which play a key role in the innate immune response to microbial infection (de Oliveira et al., 2012). Seaweeds have been used traditionally for wound healing, and transcriptomic studies provided insights into the molecular mechanisms involved. For example, a study of the brown seaweed *Hizikia fusiforme* revealed that the regulation of genes involved in transforming growth factor beta (TGFB), which plays a key role in suppressing metastasis and tumor growth. (Kim et al., 2021).

4.4 Metal Toxicity and Bioremediation

Metal stress can cause significant changes in algal gene expression. Transcriptomic analysis has been applied to understand the molecular mechanisms underlying metal stress response in algae. For instance, a study by Cai et al. (2021) used RNA sequencing to investigate the transcriptomic response of the green alga *Ulva prolifera* to copper stress. The study identified several differentially expressed genes involved in metal detoxification, antioxidant defense, and stress signaling pathways. In another study conducted by Contreras-Porcia et al. (2011), the transcriptome analysis of copper-stressed *Ulva compressa* was performed using RNA sequencing technology. The study identified differentially expressed genes related to metal uptake, transport, and detoxification, including metallothionein, glutathione S-transferase, and ATP-binding cassette (ABC) transporters. A transcriptome analysis of the brown seaweed *Saccharina japonica* in response to copper stress suggested that this brown seaweed has evolved complex mechanisms to cope with metal toxicity, which involve a range of molecular and cellular processes (Zhang et al. 2019). Transcriptome analysis has been used to study the gene expression patterns in the red alga *Porphyra umbilicalis* (Brawley et al., 2017). The study highlighted the novel features of the *Porphyra* genome shared by other Rhodophytes with respect to calcium signaling, cytoskeleton, and augmented stress-tolerance mechanisms. Transcriptomic analysis of the brown seaweed *E. siliculosus* revealed the genes involved in stress responses to high salinity and temperature (Le Bail et al., 2008). *P. yezoensis* was studied to identify genes involved in the biosynthesis of pigments, as well as stress response genes (Sun et al., 2015; Koizumi et al., 2018).

The degree of toxicity in microalgal cells may be determined by the amount of growth inhibition. Numerous studies have shown that the presence of heavy metal ions has a negative impact on the development and biochemical composition of microalgal cells (Arora et al. 2017; Cheng et al., 2019; Samadani et al., 2018; Yang et al. 2015a, b, c). It is believed that heavy metals such as lead and cadmium hinder the normal operation of thylakoid membrane and the manufacture of chlorophyll (Hanikenne et al., 2005; Xu et al., 2020). The precise arsenic detoxification mechanism in *Coccomyxa* sp. was identified by a thorough transcriptomics research combined with untargeted metabolomics (Koechler et al., 2016). In the presence of arsenite (4 mM), RNA sequencing using Illumina followed by *de novo* assembly produced a total of 39,153 contigs. Out of total contigs, the study determined that 329 contigs were induced and 885 were repressed. The transcriptional analysis showed that a gene with a potential role associated to arsenic efflux pump (ACR3) was upregulated in response to arsenic exposure. Thus, microalgae are promising towards persistent emerging contaminants in the ecosystem as well. Overall, transcriptomics as an advanced technology paves way for a better understanding of the potential applications of both seaweeds and microalgae in various fields.

5 Proteomics

The expression of genes is only briefly evaluated by transcriptional profiling of mRNAs, although most of major regulatory mechanisms vary at the posttranscriptional stage. Additionally, it has been shown that fewer than 50% of transcriptomics and proteomic data are correlated, which is likely due to the inadequate turnover of proteins (Choi et al. 2013). Taking lipids as example, integrating transcriptomics and proteomics is eventually necessary for the formation of TAG in microalgal species. In reaction to any environmental stress that increases TAG, quantitative proteomics detects in algae and measures the mechanism of protein loads and its connected activity at the posttranslational and translational stages (Arora et al. 2018b). The deficiency of nitrogen has been used extensively to analyze the modifications of proteome in a variety of algae and diatoms since it is a well-established cause of TAG production in microalgae. Malonyl-CoA:ACP-transacyclase, ACP, ACCase, lipid-droplet surface protein (LDSP), malonyl-CoA acylt-ransferase (MAT), enoyl acyl carrier-protein reductase, trans-2-enoyl-CoA reductase, and the 4-condensed enzymes participating in fatty acids production were the generally listed upregulated proteins that participate in the synthesis of fatty acids, 3-Ketoacyl-ACP synthase (KAS), and enoyl-acyl carrier protein reductase (ENR), DGAT. Similar to oleosins in higher plants, LDSP is a structural element of lipid droplets that provides important function in their expansion and development (Tran et al., 2016). Moreover, to enhance fatty acid biosynthesis in response to nitrogen stress, scientists have frequently observed the decline in the proteins related to photosynthesis that include subunits of RuBisCO, ribosomal proteins, and cytochrome uroporphyrinogen decarboxylase which produce porphyrin, coproporphyrinogen III oxidase (which catalyzes the metabolism of chlorophyll), and geranylgeranyl pyrophosphate (Tran et al., 2016; Longworth et al. 2016; Garnier et al., 2014; Yang et al., 2014; Arora et al. 2018b; Dong et al., 2013).

In addition to the modifications in microalgal proteomes in response to nitrogen stress, some environmental stimuli such as copper deficiency, intensity of the inoculum size, and heterotrophic cultivation have also been studied. Similarly, the effect of salinity on proteomic profile of *C. vulgaris* revealed a boost in Mao-C-like proteins (which helps in regulation of fatty acids production), Spp30-like proteins whose work is to transport lipids to Golgi apparatus (which helps in packaging and profiling to keep cell integrity and structure), while for the proper assembly of cell wall, glycine-rich-like proteins shape a self-regulating structure in the extracellular matrix, resulting in the whole lipogenesis (Li et al., 2015). Overall, proteomics has been applied to a variety of fields in order to decipher the algal potential as mentioned in the following sections.

5.1 Food Industries

Recently, there has been a rising interest in the applications of proteomics in seaweed research. Seaweeds possess a considerable amount of proteins along with their unique carbohydrate content that may be beneficial in feed and food industries (Gregersen et al. 2022). With the increasing global population, food security is alarming day by day, and keeping this in mind the research has been more focused on unconventional protein sources such as algae (Rioux et al., 2017). Seaweeds have comparatively high levels of proteins, which present in many cellular compartments as intracellular components or as enzymes, and also bound to pigments and polysaccharides (Wijesekara et al., 2017). The content of proteins also depends on spatial and temporal scale as well as nutrient supply during the growth phase (Marinho et al., 2015). In Rhodophytes, two main groups of functionally active proteins such as phycobiliproteins (phycoerythrin, phycocyanins, and allophycocyanins) and lectin have been studied more frequently. These proteins are more soluble and functionally active protein (Pangestuti & Kim, 2015) and are used as coloring agents in food industry. The lectins have been extracted from red and green seaweeds, which display an affinity for carbohydrates and glycoproteins as well as participation in many biological processes like intercellular communication (Pangestuti & Kim, 2015). Only few studies have been conducted on edible seaweed proteins, although numerous bioactive peptides have been obtained from characterized following enzymatic hydrolysis of food protein (Harnedy and FitzGerald 2013).

Earlier studies have reported and identified a novel protein in the brown seaweed *Undaria pinnatifida* with antihypertensive activity (Suetsuna and Nakano 2000). The protein, named Ala-lle-Tyr-Lys, was subsequently purified and characterized, with high potential as a functional food ingredient was demonstrated *in vitro* and *in vivo*. A study by Gregersen et al. (2022) used proteomics to characterize protein composition of dried seaweed products and identified several key proteins that were affected by processing conditions. Yotsukura et al. (2012) used proteomics to assess the proteins extracted from *Eklonia cava*. Results showed that the protein quality varies with varying temperatures of seawater. Proteomics can also be used to identify potential allergens in seaweed food products. For example, a study by Polikovskya et al. 2019 used proteomics to identify and characterize a major allergen in dried seaweed products.

5.2 *Medicine and Therapeutic Applications*

One area where proteomics has shown potential in medicine is drug discovery. Proteomics can help to identify new drug targets and also aid in the development of new drugs by providing insights into drug interactions with proteins. A study conducted by Silchenko et al. (2017) used omics to identify proteins in the brown seaweed *Sargassum horneri* that could potentially be used as sequence homologous to fucoidanase. Proteomics has also been applied in disease diagnosis by Niu et al. (2022) in order to identify biomarkers for diagnosis of liver fibrosis. Another area where proteomics has shown promise is in personalized medicine. Proteomics can be used to identify biomarkers for patients stratification and to predict treatment outcomes. Li et al. (2022b) used proteomics to identify biomarkers for predicting the response of patients with advanced gastric cancer to the chemotherapy drug, docetaxel. The study identified a set of proteins that could predict treatment outcomes with a high degree of accuracy. In comparison to most microalgae, seaweeds are a rich source of proteins with potential therapeutic applications. A study conducted by Yim et al. (2021) used omics to identify proteins in the red seaweed *Porphyra tenera* with potential antiviral activity against SARS-CoV-2. The study identified several proteins, including phycobiliproteins and lectins, which could potentially inhibit viral entry and replication. A majority of seaweeds contain around 80% of polysaccharides, of which the cell wall is built with sulfated polysaccharides, which impart flexibility against tidal damage (Ganesan et al., 2019; Freile-Pelegrín and Tasdemir 2019). These sulfated polysaccharides exhibit antioxidant and anti-inflammatory features. Besides this, seaweeds are decent sources of nutraceuticals that provide defense against Alzheimer's disease (Jutur et al., 2016).

Eukaryotic algae contain the phospholipid phosphatidylcholine (PC), which reduces cholesterol and protects the heart. Phosphatidylethanolamine or phosphoethanolamine is methylated three times to create PC. However, some microalgal species like *C. reinhardtii* seldom exhibit it (Hirashima et al., 2018). Sawicki et al. (2017) revealed that magnesium chelatase implied in the production of chlorophyll in *C. reinhardtii* may be stimulated by the 1-N-histidine phosphorylation of ChlD by the AAA (+) ChlI2. Thioredoxins (Trxs), an antioxidant protein that aids in reducing ROS levels in cells, regulate REDOX posttranslational modifications (PTMs). The Trx system has undergone substantial research and is now understood to have a variety of roles in cellular functions and human disorders. When *C. reinhardtii* was exposed to high levels of salt stress, the levels of ROS detoxifying enzymes such as Trxs and ascorbate peroxidase were greater than they were in the control group (Sithtisarn et al. 2017). Some significant housekeeping proteins may also experience PTMs, which activate specific protein activities. By regulating thiol disulfide status of target proteins, Trxs is involved in a number of metabolic pathways, including the processing of genetic information, metabolic activities (including synthesis and degradation of amino acids, lipids, and starch), stress response, and REDOX homeostasis, according to REDOX proteomics investigation of *C. reinhardtii* (Pérez-Pérez et al., 2017). Shrimp farms all across the globe suffer serious damage from the White Spot Syndrome Virus (WSSV). Application of vaccines is designed to target viral protein 28 (VP28), which is the major viral envelop protein of WSSV (Van Hulten et al., 2001). In that context, the nuclear genome of *D. salina* and the chloroplast of *C. reinhardtii* were effectively transformed with the VP28 protein-encoding gene, which plays a significant role to overcome WSSV (Feng et al., 2014; Surzycki et al., 2009).

5.3 Toxicological Studies

Algae act as natural indicators of metal pollution, even at low concentrations of metals. During marine outfall, seaweeds may experience exposure to high metal concentrations. Copper (Cu) unlike other heavy metals, e.g., lead (Pb), cadmium (Cd), and mercury (Hg), does not get bioaccumulated radically, resulting in comparatively lower toxicity to humans and other biota. An interesting study by Wu et al. (2009) highlighted that regulation of mRNA expression related to antioxidant defense mechanisms and redox homeostasis is influenced due to different exposure times to Cu in case of *Ulva fasciata*. Proteomics were also applied for identifying some proteins found in seaweeds. Pathogen-responsive proteins were identified in *P. yezoensis* using proteomics (Khan et al., 2018). Another study analyzed the toxicity effects of cadmium on *Gracilaria tenuistipitata* and reported altered protein expression (Tonon et al., 2018). Such proteins may become potential biomarkers of cadmium toxicity in seaweeds.

5.4 Cosmetic Applications

Proteomics research helped in identifying and understanding the peptides and proteins prevalent in different seaweeds, such as *Porphyra tenera, Ascophyllum nodosum*, and *Ulva lactuca* (Ravanfar et al., 2021). Such data obtained from proteomics studies can be applied to develop novel cosmetics bearing enhanced efficacy and safety profiles. For instance, a recent study on *Ascophyllum nodosum* identified certain phenolic compounds with antioxidant potential and antiaging properties, which could be used in the development of new cosmetic products (Gager et al., 2020). Proteomics can also be used to study the changes in the protein and peptide profiles of algae under different environmental conditions such as salinity, temperature, and light intensity. The data obtained can be employed for studying the growth condition optimization and development of new cultivation techniques for the production of high-quality seaweed biomass for cosmetic applications. Proteomics can also be used to study the interactions between algae-derived bioactive compounds and skin cells (Zheng et al., 2022). This information can be used to evaluate the safety and efficacy of seaweed-derived cosmetic products and to identify the underlying mechanisms of their biological activities.

Lectins are considered a promising seaweed protein group due to their antiviral and antimicrobial characteristics. They offer the promising capability to be incorporated as novel ingredients in contemporary formulations of treatments against several pathogens. The antioxidant potential of algae is directly associated with their cosmetics applications. The mycosporin-like amino acids along with the phycobiliproteins may be considered for certain industrial applications. One such phycobiliprotein named phycoerythrin is currently used as a fluorescent probe. However, the antioxidant, anti-inflammatory, as well as anticancer properties may be harnessed for future applications. The mycosporin-like amino acids may also be used in sunscreen products as they seem to provide protection against UV oxidation in the case of seaweeds (Ryu et al., 2014).

6 Metabolomics

Earlier studies estimated that algal biomass production has risen from 0.56 million tons to 35.82 million tons during 1950–2019 worldwide (FAO 2021). A lot of metabolites (>3,000 compound) have been identified from algae (Leal et al., 2013). Algal metabolites show species specificity and are dependent upon the cultivation conditions and surrounding aquatic ecosystem. Metabolites signify the indicators of variations in proteins, transcripts, and genes. There has been an increase in the application of metabolomics to elucidate complex processes in several organisms, including marine species. The following subsections will discuss some of applications of algal metabolomics.

6.1 Wastewater Treatment

Metabolomics has been increasingly used in water treatment research to identify, quantify, and understand the complex metabolic changes occurring in various water environments. Algae have been reported to be effective in removing pollutants from water. In this context, Poo et al. (2018) used biochar prepared from the seaweed *Sargassum fusiforme* for the removal of heavy metals. Another study by Zou et al. (2014) exhibited the role of Cu in metabolism, wherein the seaweed *Sargassum fusiforme* was evaluated by characterization of metabolic responses using omics techniques. The outcome of the study showed variable toxicological responses on metabolite makeup of *S. fusiforme* exposed to chronic Cu exposures. In addition, metabolomics can be used to optimize water treatment process by identifying the metabolic pathways involved in pollutant removal. For example, a recent study by El Zokm et al. (2022) studied the bioaccumulation capacity of seaweeds in the removal of PAHs. The study highlighted that algal species possess the potential of acting as biomarkers for micropollutants and in the uptake of organic pollutants within an ecosystem. Similarly, Sanchez-Arcos et al. (2022) used metabolomics to observe the changes in the metabolite profile of *Ulva prolifera* in response to changes in water quality. The study reported *Ulva prolifera* as an ideal model to evaluate the alterations in various classes of metabolites due to ocean acidification and to provide crucial information on the impacted metabolic pathways. Pharmaceuticals and personal care products pose a growing concern in water treatment. Algae have been shown to be effective in the removal of these contaminants from water through absorption or secretion of various metabolites. In that regard, researchers have identified several metabolites in seaweeds, such as polysaccharides and polyphenols that can bind to heavy metals and remove them from contaminated water. Metabolomics has been used to identify and quantify these metabolites in algal extracts. For instance, a study by Yaich et al. (2015) used mass spectrometry to quantify the dietary fibers in *Ulva* sp. involved in the removal of cadmium from water. Hardegen et al. (2023) used metabolomics to identify and quantify the metabolites in *Ulva mutabilis* involved in the degradation of organic matter in wastewater.

6.2 Biofuel Production

Metabolomics can be used to identify metabolic pathways involved in the production of biofuels from algae. Wang et al. (2019) used metabolomics to identify the metabolic pathways involved in the production of biofuels from the red seaweed *Pyropia haitanensis*. The key metabolites involved in the biosynthesis of lipids and fatty acids, which are essential for biofuel production, were identified. Bikker et al. (2016) used biorefinery approach to study the production of ethanol from the green seaweed *Ulva lactuca*. The key sugars involved in the biosynthesis of ethanol were identified. Holdt and Kraan (2011) reported that *Ulva lactuca* has several metabolites, including polyphenols, flavonoids, and carotenoids, which have potential applications as antioxidants, antimicrobial agents, and in biofuel production. Metabolomics can also be used to evaluate different seaweed species for their potential as biofuel feedstocks. For example, a study by Wang et al. (2019) reviewed the use of metabolomics for algal characterization and application to biofuel production. Shen et al. (2021) compared the metabolic profiles of three different seaweed species, namely *Saccharina japonica*, *Undaria pinnatifida*, and *Sargassum fusiforme*, and evaluated their potential for polyphenol production. The study identified 12 polyphenolic compounds that comprised flavonoids, phlorotannins, phenolics etc. Jian et al. (2017) used metabolomics to investigate the metabolic changes that occur during the growth of the red seaweed *Pyropia haitanensis*. It was found that the levels of amino acids and fatty acids varied during growth, suggesting that these compounds may be useful for biofuel production. Shao et al. (2019) used metabolomics to identify metabolites that accumulated in *Saccharina japonica* during different growth stages. It was found that the levels of several metabolites, including amino acids and sugars, varied during the growth stages and the genes involved in mannitol and alginate pathways were deduced.

It was found that the levels of several metabolites, including amino acids and sugars, increased during the early growth stages, while the levels of lipids and pigments increased during the later stages of growth. The main focus of research in microalgal metabolomics has been directed on investigation of methods for calculating and detecting secondary metabolites that are valuable for industry in the sectors of pharmaceutical, public health, and food sciences. Research on secondary metabolites in microalgae, including but not limited to fatty acids, polyketides, carotenoids, steroids, lectins, polysaccharides, and toxic chemicals, is the primary focus of the field of microalgal metabolomics (Fernández et al., 2021). The metabolic pathways are regulated to trigger the creation of the desired metabolites by using approaches such as flux balance analysis, changing photosynthetic efficiency, engineering of lipid biogenesis-associated enzymes, mathematical modelling, transcription factor engineering, and so on (Banerjee et al., 2016). NMR-based analysis plays a crucial role in metabolomics studies; according to the findings of a recent NMR-based investigation, an increase in the synthesis of amphidinols and other chemicals from the dinoflagellate *Amphidinium carterae* was achieved by modifying the f/2 medium composition and the daily mean irradiation (Abreu et al., 2019). Recently, Zhang et al. (2021b) inspected the metabolic control of *C. reinhardtii* under increased CO_2 conditions, where combined transcriptomics and metabolomics studies were conducted. The comprehensive approach demonstrated that when compared to the control, the majority of genes and metabolites involved in carbon and nitrogen metabolism were repressed when exposed to high CO_2 levels over extended periods of time, which subsequently led to decreased levels of nutritional components. However, under long-term high CO_2 conditions, cell growth and photosynthesis were greatly boosted, showing the emergence of particular adaptive development by *C. reinhardtii* (Zhang et al.

2021b). These findings confirm that metabolomics is a promising biotechnological tool to evaluate algal metabolites toward enhanced biofuel production.

6.3 Agriculture and Environmental Applications

Based on the Intergovernmental Panel on Climate Change (IPCC) report, an increase in the climate change has been recorded worldwide (Arias et al., 2021) due to which the aquatic environments will be severely impacted (Bernardino et al., 2015; Urrea-Victoria et al., 2020). Algae have been studied for their potential use as biomarkers of environmental pollution and climate change. The response of seaweeds to ocean acidification acts as an indicator for determining the global warming (Hall-Spencer et al., 2008; Martins et al., 2012). Earlier studies reported that the biochemical constitution of seaweed species fluctuates with salinity, temperature, pH, and other environmental factors (Hamid et al., 2019). A study, for instance, reviewed both primary and secondary metabolites in seaweeds along the Brazilian coast (dos Santos et al., 2023). This study highlighted the significance of seaweeds and their bioactive metabolites as potential players for a variety of applications. Metabolomics can be used to identify changes in seaweed metabolites in response to environmental stressors. For example, a study by Kumar et al. (2018) used metabolomics to analyze the metabolite profile of *Sargassum vulgare* in response to ocean acidification. The study found that the seaweed accumulated several metabolites, including fatty acids and phenolic compounds, in response to ocean acidification exposure, suggesting that these compounds could be used as biomarkers for the same. Metabolomics can also be used to analyze the effects of environmental stress on seaweeds. A study by Pilatti et al. (2017) used metabolomics to analyze the metabolic responses of the green seaweed *Ulva lactuca* to different levels of xenobiotic stress. The study found that the metabolism of the seaweed was significantly altered, with changes in the levels of amino acids, organic acids, and sugars. A study by Shen et al. 2021used metabolomics to monitor the polyphenolic content and stress response of the brown seaweed *Undaria pinnatifida* and identified several metabolites involved in the antioxidant mechanism of the seaweed.

In addition, seaweeds have been studied for their potential use as biofertilizers and plant growth promoters. Metabolomics can be used to understand the biochemical composition of seaweeds and their impact on plants. Tran et al. (2023) used metabolomics to compare the metabolite profile of *Arabidopsis thaliana* leaves and roots response mechanisms using two Ochrophytes (*Ascophyllum nodosum* and *Durvillaea potatorum*) and their effects on the growth of the model plant *Arabidopsis thaliana*. The study found that seaweed extracts promoted plant growth and stress tolerance. Rengasamy et al. (2016) studied the metabolites responsible for the growth-promoting effects of the seaweed *Ecklonia maxima*. They studied that the presence and quantification of eckol and phloroglucinol present in liquid seaweed fertilizers would provide useful insights for quality control of such products. Another study showed that seaweed extract from *E. cava* increased the levels of growth and enzyme activity in the maize plants, which in turn contributed to their improved growth (Rengasamy et al., 2015).

Using metabolomics, it is possible to study how different anthropogenic contaminants, such as heavy metals, nanomaterials, pesticides, pharmaceuticals, personal care products, and persistent organic pollutants, affect the metabolism of environmental organisms, including animals (such as rodents, fish, crustaceans, and earthworms) and microorganisms (such as bacteria, yeast, and microalgae) (Zhang et al. 2021a). The earliest definition of environmental metabolomics was "tools that investigates the metabolites of both free-living organisms from the natural environment and laboratory-grown organisms under conditions simulating the natural environment" (Morrison et al., 2007). The practical usage of metabolomics, in common, focuses on understanding how organisms and their environment interact (Bundy et al., 2008). *C. reinhardtii*, a unicellular green alga, was utilized to test the possible toxicity of strictly controlled 4-5 nm ceria nanoparticles (NPs) using unbiased transcriptomics and metabolomics. Results showed that ceria NPs were internalized by *C. reinhardtii* into intracellular vesicles, but had no discernible impact on the algal growth at any exposure dosage (Taylor et al., 2016).

6.4 Medical, Therapeutics, and Health Applications

In medicine and therapeutics, metabolomics has been used to identify biomarkers for disease diagnosis, prognosis, and treatment response, as well as to discover new drugs and understand their mechanisms of action. Algae have been recognized as a rich source of bioactive compounds with potential therapeutic applications. The simultaneous synthesis of many products in a biorefinery approach has lately been promoted by proponents of microalgal biotechnology as a way to offset processing expenses. With uses in cosmetic, culinary, pharmaceutical, and agricultural industries, isoprenoids and other bioactive compounds have significant economic value. To produce bioactive compounds, such as carotene from *Dunaliella salina* and astaxanthin from *Haematococcus pluvialis*, microalgae have been used in industry (Li et al. 2022c). Microalgae provide a range of potential benefits over other biological resources as a natural supply of terpenoids as well, where dynamic metabolomics play an essential role. Combining 13C and 2H labelling techniques may be used to measure isoprenoid exchange among cytosol and plastid (Ladd et al., 2021). The use of carbon that has already been 13C- or 2H-labeled, followed by a metabolomics, enables a complete investigation of the interactions across pathways. A successful example was documented after *Chlamydomonas* was fed a labelled 13C source to track the characteristics of carbon transport to lipids (Ho et al., 2017; Young et al., 2022).

Metabolomics has been used also to identify anti-inflammatory compounds in seaweeds. Kim et al. (2019) identified the anti-inflammatory effects of phlorotannins extracted from *Ecklonia cava* using metabolomics. In a study by Buwono et al. (2018), metabolites from *Sargassum polycystum* were identified and their anti-inflammatory activity was evaluated. In a study by Tanna et al. (2022), metabolites from the red seaweed *Gracilaria corticata* were identified and their anticancer activity was evaluated. Metabolomics has also been used to identify neuroprotective compounds from seaweeds. For example, in a study by Li et al. (2022a), metabolomics highlighted the neuroprotective effects of fucoxanthin from the brown seaweed. The study found that fucoxanthin protected neuronal cells against oxidative stress and inflammation by modulating the expression of genes involved. Metabolomics was used to identify the antioxidant effects of fucosterol from the brown seaweed *Sargassum horneri* (Kirindage et al., 2022). The study found that fucosterol improved glucose metabolism and insulin sensitivity in high-fat diet-induced obese mice by modulating the expression of genes involved in these pathways. In a study by Sulaiman et al. (2019), Ascophyllan, a sulphated polysaccharide from the brown seaweed *Padina tetrastromatica,* was identified and their wound-healing activity was evaluated. The study concluded that this algal polysaccharide could be safely used as a natural alternative for topical application to enhance the wound healing.

6.5 Food and Cosmetic Applications

Algae are a rich source of bioactive compounds such as polysaccharides, lipids, polyphenols, pigments, and vitamins, which have various applications in food industry (Dixit et al., 2018). Algal metabolomics can be used to identify and quantify important metabolites in food applications. These metabolites have been shown to have various nutritional and health benefits, including anti-inflammatory, antioxidant, and antimicrobial properties. For example, fucoxanthin, a carotenoid pigment found in brown seaweed, has been shown to have antiobesity, antidiabetic, and anticancer properties (Pangestuti and Kim 2011). Another example is fucoidan, a sulfated polysaccharide found in brown seaweeds, which has been shown to have immunomodulatory and anticancer effects (Ale and Meyer 2013). Metabolomics has been used to identify and quantify these bioactive compounds in algal cells and to understand their metabolic pathways. UPLC-MS-based metabolomics has been used to analyze the metabolites of an edible seaweed, *Ulva lactuca,* and to study the mechanism responsible for its antiaging and hypoglycemic properties (Chen et al., 2022).

Food industry is continuously seeking new and innovative ways to improve food quality and safety. Metabolomics has been employed in the study of food metabolites, including algae, to identify biomarkers that can be used to assess food quality and safety. A recent study by Oh et al. (2021) used omics to study the metabolic profile of two edible seaweed species, *Saccharina japonica* and *Porphyra tenera*, and reported significant differences in their cholesterol composition. The study identified several compounds that could potentially serve as biomarkers for the authentication and quality control of these seaweeds. Another study highlighted the biochemical, mineral, as well as secondary metabolite makeup of *Iyengaria stellata* using untargeted metabolite profiling for deciphering its potential in contemporary nutraceutical and pharmaceutical compositions (Dixit 2023). A total of 108 putative secondary metabolites were reported which included esters, terpenoids, steroids, anthraquinones, polyphenols, etc., which could account as natural ingredients for nutraceuticals in the future. Furthermore, metabolomics has also been used in food processing and preservation techniques. Gupta et al. (2013) used metabolomics to study the effects of different methods (freeze drying and fresh biomass) on the metabolite detection of seaweeds. They found that the metabolite detection of seaweeds was significantly affected by the drying method employed, highlighting the importance of optimizing detection techniques to preserve the functional properties of seaweeds. Another study investigated the probable inclusion of an Ochrophyte *Dictyota dichotoma* in algal-based functional foods using putative metabolites present in non-targeted metabolomics (Dixit et al., 2020). The study concluded that a total of 16 secondary metabolites could be detected and recommended further exploration of the nutraceutical potential of this species in algal biorefineries.

Metabolomics technology was also used to identify the bioactive compounds in different species of red seaweeds (Norskov et al., 2021). The study found that red seaweeds contain a variety of bioactive compounds, including polyphenols, flavonoids, and carotenoids, which have potential health benefits. A similar study was undertaken to analyze the metabolome of the brown seaweed *Fucus vesiculosus* (Barbosa et al., 2017). The study confirmed the anti-inflammatory potential of phlorotannins that have immense virtues for application in food industries. Catarino et al. (2020) used metabolomics to identify bioactive compounds in the brown seaweed *Fucus vesiculosus*. The study identified several novel compounds with potential anti-inflammatory and antioxidant activities. Fatmawati et al. (2022) used metabolomics to characterize the metabolites present in the red seaweed *Gracilaria verrucosa*. The study identified several metabolites with potential applications in food and pharmaceutical industries. A study by Ye et al. (2014) used metabolomics to evaluate the quality of different seaweed product after processing.

In addition to the food industry, metabolomics technology has also been applied in cosmetic industry to identify and quantify bioactive compounds in seaweeds that have potential applications in cosmetics. For instance, ultra-performance liquid chromatography mass spectrometry (UPLC-MS) -based metabolomics has been used to analyze the metabolites of the *Undaria pinnatifida* and to identify the compounds responsible for its antioxidant properties (Ristivojevic et al. 2021). A recent study by Bhatia et al. (2021) studied the nutraceutical, radical scavenging, and the microbicidal properties of *Pyropia vietnamensis* and found several compounds that exhibited antioxidant and antimicrobial properties. They concluded that these seaweeds could potentially serve as a source of natural antioxidants. Seaweeds have been used in cosmetic industry for their potential to improve skin health and appearance. Metabolomics can be used to identify the bioactive compounds present in seaweeds and their potential benefits for the skin. For example, a study by Xie et al. (2020) analyzed the chemical composition of the red seaweed *Gelidium amansii* and its potential use as a skin-whitening agent. The study determined the composition of 3, 6-anhydrogalactose (AnGal), a monosaccharide, and highlighted it as a potential cosmeceutical ingredient. Seaweeds are also used in cosmetics due to their antioxidant and anti-inflammatory properties. Metabolomics has been used to identify the bioactive compounds responsible for these properties. Dixit and Reddy (2017) investigated the cosmeceutical traits of *Jania rubens* by highlighting its mineral composition, antioxidant potential, and presence of bioactive molecules using non-targeted metabolite profiling. Shen et al. (2021) used metabolomics to identify polyphenols in the brown seaweed *Sargassum fusiforme* that have antioxidant activity. They found that the bioactive compound (phlorotannin) in seaweed extract was able to reduce the oxidative stress. In cosmetics applications, seaweed metabolomics can be used to identify and quantify metabolites such as polyphenols, carotenoids, and fatty acids. These metabolites have been shown to have various benefits for skin health, including antioxidant, antiaging, and moisturizing properties. For example, pigments extracted from red seaweed have been shown to have photoprotective effects against UV radiation (Lalegerie et al., 2019). Additionally, fatty acids extracted from brown seaweeds have been shown to have anti-photoaging and anti-inflammatory effects on the skin (Hwang et al., 2014). Overall, metabolomics analysis of seaweed has shown great potential for the development of new food and cosmetic products.

7 Limitations and Bottlenecks of Algal Omics

Omics technology, which encompasses various high-throughput techniques for analyzing molecular components of cells or organisms, has revolutionized biological research over the past few decades. However, there are still several limitations and bottlenecks associated with algal omics technology which can be summarized as follows;

Sample Processing and Quality Despite the high potential of microalgal bioactive chemicals, there are several technical obstacles, such as limited microalgal culturability, that limit the development of microalgal industrial applications (Mishra et al., 2019). Research to advance the growth of microalgae is required since the high-throughput screening of the microalgal bioactive chemicals needs the use of enormous numbers of samples with repeated large-scale testing. Seaweeds, on the other hand, are complex organisms with unique metabolites and structural components that can affect sample collection and processing. One of the biggest challenges in omics research is obtaining high-quality samples for analysis. Seaweeds are often difficult to collect and preserve, and their composition can vary widely depending on cultivation factors such as season, location, and environmental conditions. The high salt content of seawater can lead to protein degradation, while high carbohydrate content of seaweeds can interfere with RNA extraction. These challenges can affect the quality of omics data generated from seaweed samples. Poor sample quality can lead to inaccurate or inconsistent results, making it challenging to draw meaningful conclusions from the obtained data.

Data Processing and Analysis Omics technologies generate massive amounts of data, which can be difficult to manage and analyze. The computational tools and software needed to process and interpret omics data are constantly evolving, and researchers may lack the expertise needed to effectively analyze complex datasets.

Genome Assembly and Annotation One of the primary challenges in using omics technologies for seaweeds is the lack of a high-quality reference genome assembly. Seaweeds have large genomes with a high degree of heterozygosity and repetitive elements, making genome assembly a time-consuming and challenging process. While some seaweed species have been sequenced, many have not, where assembling and annotating genomes for these organisms remains a significant challenge. As a result, many studies rely on transcriptomics or metagenomics data, which provide a limited view of the genome. This can limit the ability of researchers to identify and study genes and pathways involved in important biological processes. One of the major challenges in metagenomic studies of seaweed-associated microbial communities is the high diversity of the available organisms. Seaweeds are colonized by a wide range of bacteria and other microorganisms, many of which are difficult to cultivate in laboratory. This can make it challenging to accurately identify the different types of organisms present and their functional capabilities.

Transcriptome Annotation Another limitation of seaweeds omics technologies is the lack of well-annotated transcriptomes. Seaweeds have unique gene expression patterns and posttranscriptional modifications, which can make it challenging to accurately annotate transcripts. As a result, many studies rely on homology-based annotation, which can result in misannotation or incomplete annotation of seaweed transcripts.

Functional Validation Even when omics data identify potential targets for further study, functional validation can be difficult. Seaweeds often have complex life cycles and may be difficult to culture in laboratory, making it challenging to test the functional roles of specific genes or pathways.

Data Integration Omics data from different sources and platforms may be difficult to integrate, hindering the ability to draw comprehensive conclusions from multiple datasets. Data integration requires the development of new bioinformatic tools and pipelines, as well as the standardization of data formats and ontologies. This is especially true in case of seaweeds, which are often understudied and lack well-established reference genomes or databases.

Regarding microalgae, omics is a well-established technique for the development of phycochemicals and provides a strong route to strain engineering and hypothesis-driven tactics in microalgal species, especially for improved TAG production. To understand the mechanisms underlying lipid accumulation and the potential targets for strain engineering, detailed comparison omics-based studies have been carried out in microalgae. A variety of omics datasets have revealed several conserved regulatory genes and mechanisms. Significant differences between species still exist though, which emphasizes the need for detailed systems approach that incorporates the insertion of data from physiological coupling omics and genomic-based computational modelings. In fact, addition of these data will be greatly aided by the rapidly developing genome scale modeling capabilities, enabling the predictive capacity for hypothesis driven of strain engineering and formulation of media. On the other hand, biotechnological profile of microalgae has been enhanced by their capacity to produce bioactive compounds. It may be put to use in a variety of ways to enhance the quality of life. Despite the diversity of their biochemistry, only few species are investigated for their metabolites and chemical composition. There are many applications for the concurrent production of valuable bioproducts. It is necessary to take actions during disease treatment and preventions beside to search for extra, unidentified metabolites. The associated difficulties and constraints must be overcome to make microalgal-based metabolites viable for commercialization in pharmaceutical and healthcare sectors. For a vast range of research studies on food and nutritional values, microbiomes analysis, systems microbiology, systems biology, genotype-phenotype interaction, discovery of natural product, disease biology, and significance of integrating omics data have been recognized. However, it is incredibly rare for more than two omics datasets to be successfully implemented.

8 Conclusions

Omics technologies have immense potential in algal research, including microalgae and seaweeds. The use of genomics, transcriptomics, proteomics, and metabolomics can provide insights into genetic makeup, gene expression, protein expression, and metabolic pathways of algae. These technologies can be used for the development of algal strains with enhanced production capabilities for the production of biofuels, pharmaceuticals, and other high-value products. The study of algal communities using omics technologies can also provide insights into their ecological roles in various ecosystems. Genome-level modeling competencies shall pave way for the integration of data, designing media compositions, and hypothesis-based strain engineering. Despite the diverse biochemical configuration of vast numbers of algal strains, the data on metabolite contents along with their chemical makeup are available only for very few species. A contemporary approach towards developing an extensive study and exploration of unknown algal metabolites and their comprehensive applications in healthcare/pharma/food industries needs to be strategized. Though the necessity and significance of integrating omics studies and their associated data in various areas of research have been understood, the limitations and challenges in making them viable for commercialization cannot be overlooked. Therefore, effective application of more than two omics datasets is quite uncommon. Hence, a systematic and continuous approach to enhance the efforts underlying the bioprospecting of algal genetics to engineer novel strains with robust characteristics needs to be adopted. Studies aiming at encompassing targeted efforts towards nurturing, reaping, extracting, and improving algal strains for enhanced stress tolerance, increased photosynthetic potential, higher productivity, as well as better CO_2 utilization capacity should be focused on.

References

Abomohra A, Hanelt D (2022) Recent advances in micro-/nanoplastic (MNPs) removal by microalgae and possible integrated routes of energy recovery. Microorganisms 10:2400

Abreu AC, Molina-Miras A, Aguilera-Saéz LM, López-Rosales L, Cerón-Garciá MDC, Sánchez-Mirón A, Olmo-Garciá L, Carrasco-Pancorbo A, Garciá-Camacho F, Molina-Grima E, Fernández I (2019) Production of amphidinols and other bioproducts of interest by the marine microalga *Amphidinium carterae* unraveled by nuclear magnetic resonance metabolomics approach coupled to multivariate data analysis. J Agric Food Chem 67:9667–9682

Ajjawi I, Verruto J, Aqui M, Soriaga LB, Coppersmith J, Kwok K, Peach L, Orchard E, Kalb R, Xu W, Carlson TJ, Francis K, Konigsfeld K, Bartalis J, Schultz A, Lambert W, Schwartz AS, Brown R, Moellering ER (2017) Lipid production in *Nannochloropsis gaditana* is doubled by decreasing expression of a single transcriptional regulator. Nat Biotechnol 35:647–652

Ale MT, Meyer AS (2013) Fucoidans from brown seaweeds: an update on structures, extraction techniques and use of enzymes as tools for structural elucidation. RSC Adv 3(16):8131–8141

Andriania Y, Syamsumira DF, Yeea TC, Harissonb FS, Herngc GM, Abdullah SA, Oroscoc CA, Alid AM, Latipe J, Kikuzakif H, Mohamada H (2016) Biological activities of isolated compounds from three edible malaysian red seaweeds, *Gracilaria changii*, *G. manilaensis* and *Gracilaria* sp. Nat Prod Commun 11(8):1117–1120

Arias P, Bellouin N, Coppola E, Jones R, Krinner G, Marotzke J, Naik V, Palmer M, Plattner G-K, Rogelj J (2021) Climate change 2021: the physical science basis. Contribution of working group I to the sixth assessment report of the intergovernmental panel on climate change; Technical summary.

Arif M, Bai Y, Usman M, Jalalah M, Harraz FA, Al-Assiri MS, Li X, Salama ES, Zhang C (2020) Highest accumulated microalgal lipids (polar and non-polar) for biodiesel production with advanced wastewater treatment: role of lipidomics. Bioresour Technol 298:122299

Armbrust EV, Berges JA, Bowler C, Green BR, Martinez D, Putnam NH, Zhou S, Allen AE, Apt KE, Bechner M, Brzezinski MA, Chaal BK, Chiovitti A, Davis AK, Demarest MS, Detter JC, Glavina T, Goodstein D, Hadi MZ, Hellsten U, Hildebrand M, Jenkins BD, Jurka J, Kapitonov VV, Kröger N, Lau WWY, Lane TW, Larimer FW, Lippmeier JC, Lucas S, Medina M, Montsant A, Obornik M, Parker MS, Palenik B, Pazour GJ, Richardson PM, Rynearson TA, Saito MA, Schwartz DC, Thamatrakoln K, Valentin K, Vardi A, Wilkerson FP, Rokhsar DS (2004) The genome of the diatom *Thalassiosira Pseudonana*: ecology, evolution, and metabolism. Science 80(306):79–86

Arora N, Gulati K, Patel A, Pruthi PA, Poluri KM, Pruthi V (2017) A hybrid approach integrating arsenic detoxification with biodiesel production using oleaginous microalgae. Algal Res 24:29–39

Arora N, Patel A, Pruthi PA, Poluri KM, Pruthi V (2018a) Utilization of stagnant non-potable pond water for cultivating oleaginous microalga *Chlorella minutissima* for biodiesel production. Renew Energy 126:30–37

Arora N, Pienkos PT, Pruthi V, Poluri KM, Guarnieri MT (2018b) Leveraging algal omics to reveal potential targets for augmenting TAG accumulation. Biotechnol Adv 36(4):1274–1292

Asadian N, Sheikhey MM, Baghban H, Yadipour R (2022) Lasing management using exceptional points in mutually-coupled quantum dot lasers. IEEE J Quantum Electron 58

Ausuri J, Dell'Anno F, Vitale GA, Palma Esposito F, Funari V, Franci G, Galdiero M, Della Sala G, Tedesco P, Coppola D, de Pascale D (2022) Bioremediation of multiple heavy metals mediated by antarctic marine isolated *Dietzia psychralcaliphila* JI1D. J Marine Sci Eng 10(11):1669

Banerjee C, Dubey KK, Shukla P (2016) Metabolic engineering of microalgal based biofuel production: prospects and challenges. Front Microbiol 7:432

Barati B, Zeng K, Baeyens J, Wang S, Addy M, Gan SY, Abomohra A (2021) Recent progress in genetically modified microalgae for enhanced carbon dioxide sequestration. Biomass Bioenergy 145:105927

Barbosa M, Lopes G, Ferreres F, Andrade PB, Pereira DM, Gil-Izquierdo A, Valentão P (2017) Phlorotannin extracts from Fucales: marine polyphenols as bioregulators engaged in inflammation-related mediators and enzymes. Algal Res 28:1–8

Bellou S, Baeshen MN, Elazzazy AM, Aggeli D, Sayegh F, Aggelis G (2014) Microalgal lipids biochemistry and biotechnological perspectives. Biotechnol Adv 32:1476–1493

Bernardino AF, Netto SA, Pagliosa PR, Barros F, Christofoletti RA, Rosa Filho JS, Colling A, Lana PC (2015) Predicting ecological changes on benthic estuarine assemblages through decadal climate trends along Brazilian Marine Ecoregions. Estuar Coast Shelf Sci 166:74–82

Bhatia S, Sharma AK, Cruz DL, Vargas CB, Chaugule B, Ahmed AH (2021) Nutraceutical, antioxidant, antimicrobial properties of *Pyropia vietnamensis* (Tanaka et Pham-Hong Ho) J.E. Sutherl. et Monotilla. Current Bioactive Compounds 17(2):151–164

Bikker P, van Krimpen MM, van Wikselaar P, Houweling-Tan B, Scaccia N, van Hal JW, Huijgen WJ, Cone JW, López-Contreras AM (2016) Biorefinery of the green seaweed *Ulva lactuca* to produce animal feed, chemicals and biofuels. J Appl Phycol 28(6):3511–3525

Brawley SH et al (2017) Insights into the red algae and eukaryotic evolution from the genome of *Porphyra umbilicalis* (Bangiophyceae, Rhodophyta). Proc Natl Acad Sci 114(31):E6361–E6370

Breuer G, Lamers PP, Martens DE, Draaisma RB, Wijffels RH (2012) The impact of nitrogen starvation on the dynamics of triacylglycerol accumulation in nine microalgae strains. Bioresour Technol 124:217–226

Bundy JG, Davey MP, Viant MR (2008) Environmental metabolomics: a critical review and future perspectives. Metabolomics 51(1):3–21

Buwono NR, Risjani Y, Arsad S (2018) Anti-inflammatory and analgesic activity from brown algae *Sargassum polycystum*. J Pharm Sci & Res 10(8):2092–2096

Cai C, Liu X, Zhao H, Jiang T, Jia R, He P (2021) Weakened growth, cell division, and energy metabolism, but enhanced resistance, signaling, and anabolism: responses of *Ulva prolifera* to copper elucidated by omics. J Appl Phycol

Califano G, Kwantes M, Abreu MH, Costa R, Wichard T (2020) Cultivating the macroalgal holobiont: effects of integrated multi-trophic aquaculture on the microbiome of *Ulva rigida* (Chlorophyta). Front Mar Sci 7:52

Carpena M, Caleja C, Pereira E, Pereira C, Ćirić A, Soković M, Soria-Lopez A, Fraga-Corral M, Simal-Gandara J, Ferreira ICFR, Barros L, Prieto MA (2021) Red seaweeds as a source of nutrients and bioactive compounds: optimization of the extraction. Chemosens 9:132

Carrier G, Garnier M, Le Cunff L, Bougaran G, Probert I, De Vargas C, Corre E, Cadoret JP, Saint-Jean B (2014) Comparative transcriptome of wild type and selected strains of the microalgae *Tisochrysis lutea* provides insights into the genetic basis, lipid metabolism and the life cycle. PLoS One 9(1):e86889

Catarino MD, Silva A, Cruz MT, Mateus N, AMS S, Cardoso SM (2020) Phlorotannins from *Fucus vesiculosus*: modulation of inflammatory response by blocking NF-κB signaling pathway. Int J Mol Sci *21*:6897

Chang KS, Kim J, Park H, Hong SJ, Lee CG, Jin ES (2020) Enhanced lipid productivity in AGP knockout marine microalga Tetraselmis sp. using a DNA-free CRISPR-Cas9 RNP method. Bioresour Technol 303:122932

Chen Y, Wu W, Ni X, Farag MA, Capanoglu E, Zhao C (2022) Regulatory mechanisms of the green alga *Ulva lactuca* oligosaccharide via the metabolomics and gut microbiome in diabetic mice. Curr Res Food Sci 5:1127–1139

Cheng SY, Show PL, Lau BF, Chang JS, Ling TC (2019) New prospects for modified algae in heavy metal adsorption. Trends Biotechnol 37:1255–1268

Chi S, Liu T, Wang X, Wang R, Wang S, Wang G, Shan G, Liu C (2018) Functional genomics analysis reveals the biosynthesis pathways of important cellular components (alginate and fucoidan) of *Saccharina*. Curr Genet 64(1):259–273

Choi YE, Hwang H, Kim HS, Ahn JW, Jeong WJ, Yang JW (2013) Comparative proteomics using lipid over-producing or less-producing mutants unravels lipid metabolisms in *Chlamydomonas reinhardtii*. Bioresour Technol 145:108–115

Collén J, Porcel B, Carre W et al (2013) Genome structure and metabolic features in the red seaweed *Porphyra umbilicalis*: a comparative analysis. Curr Biol 23(13):e131–e137

Conesa A, Beck S (2019) Making multi-omics data accessible to researchers. Sci Data 6:251

Contreras-Porcia L, Dennett G, González A, Vergara E, Medina C, Correa JA, Moenne A (2011) Identification of copper-induced genes in the marine alga *Ulva compressa* (Chlorophyta). Mar Biotechnol 13:544–556

Coste O, Malta E-j, López JC, Fernández-Díaz C (2015) Production of sulfated oligosaccharides from the seaweed *Ulva sp.* using a new ulvan-degrading enzymatic bacterial crude extract. Algal Res 10:224–231

D'Adamo S, Schiano di Visconte G, Lowe G, Szaub-Newton J, Beacham T, Landels A, Allen MJ, Spicer A, Matthijs M (2019) Engineering the unicellular alga *Phaeodactylum tricornutum* for high-value plant triterpenoid production. Plant Biotechnol J 17:75–87

Davis R, Aden A, Pienkos PT (2011) Techno-economic analysis of autotrophic microalgae for fuel production. Appl Energy 88:3524–3531

De Clerck O, Kao SM, Bogaert KA, Blomme J, Foflonker F, Kwantes M, Vancaester E, Vanderstraeten L, Aydogdu E, Boesger J, Califano G, Charrier B, Clewes R, Cortona AD, D'Hondt S, Fernandez-Pozo N, Gachon CM, Hanikenne M, Lattermann L, Leliaert F, Liu X, Maggs CA, Popper ZA, Raven JA, Van Bel M, Wilhelmsson PKI, Bhattacharya D, Coates JC, Rensing SA, Van Der Straeten D, Vardi A, Sterck L, Vandepoele K, Van de Peer Y, Wichard T, Bothwell JH (2018) Insights into the evolution of multicellularity from the sea lettuce genome. Curr Biol 28:2921–2933

de Oliveira LS, Gregoracci GB, Silva GGZ, Salgado LT, Filho GA, Ferreira MA, Pereira RC, Thompson FL (2012) Transcriptomic analysis of the red seaweed *Laurencia dendroidea* (Florideophyceae, Rhodophyta) and its microbiome. BMC Genomics 13:487

Dixit D (2023) Untargeted secondary metabolite profiling and putative pathway deduction of a marine Ochrophyte – *Iyengaria stellata* (Børgesen) using liquid chromatography-mass spectrometry. Food Biosci 51(2023):102211

Dixit D, Reddy CRK (2017) Non-targeted secondary metabolite profile study for deciphering the cosmeceutical potential of red marine macro alga *Jania rubens*—an LCMS-based approach. Cosmetics 4(4):45

Dixit DC, Reddy CRK, Balar N, Suthar P, Gajaria T, Gadhavi DK (2018) Assessment of the nutritive, biochemical, antioxidant and antibacterial potential of eight tropical macro algae along kachchh coast, india as human food supplements. J Aquatic Food Prod Technol 27(1):61–79

Dixit D, Reddy CRK, Trivedi MH, Gadhavi DK (2020) Non-targeted metabolomics approach to assess the brown marine macroalga *Dictyota dichotoma* as a functional food using liquid chromatography with mass spectrometry. Sep Sci Plus 3(5):140–149

Dong HP, Williams E, Wang DZ, Xie ZX, Hsia RC, Jenck A, Halden R, Li J, Chen F, Place AR (2013) Responses of *Nannochloropsis oceanica* IMET1 to long-term nitrogen starvation and recovery. Plant Physiol 162:1110–1126

dos Santos GS, de Souza TL, Teixeira TR, Brandão JPC, Santana KA, Barreto LHS, Cunha SS, dos Santos DCMB, Caffrey CR, Pereira NS et al (2023) Seaweeds and corals from the brazilian coast: review on biotechnological potential and environmental aspects. Molecules 28(11):4285

El Zokm GM, Ismail MM, Okbah AE, M. (2022) Seaweed as bioindicators of organic micropollutants polycyclic aromatic hydrocarbons (PAHs) and organochlorine pesticides (OCPs). Environ Sci Pollut Res 29:34738–34748

El-Naggar NEA, Hamouda RA, Mousa IE, Abdel-Hamid MS, Rabei NH (2018) Biosorption optimization, characterization, immobilization and application of *Gelidium amansii* biomass for complete Pb^{2+} removal from aqueous solutions. Sci Rep 8:13456

Fatmawati F, Sibero MT, Trianto A, Wijayanti DP, Sabdono A, Pringgenies D, Radjasa OK (2022) The influence of fermentation using marine yeast *Hortaea werneckii* SUCCY001 on antibacterial and antioxidant activity of *Gracilaria verrucosa*. Biodiversitas 23(10):5258–5266

Feng S, Feng W, Zhao L, Gu H, Li Q, Shi K, Guo S, Zhang N (2014) Preparation of transgenic *Dunaliella salina* for immunization against white spot syndrome virus in crayfish. Arch Virol 159:519–525

Fernández FGA, Reis A, Wijffels RH, Barbosa M, Verdelho V, Llamas B (2021) The role of microalgae in the bioeconomy. New Biotechnol 61:99–107

Food and Agriculture Organization of the United Nations (2021) Global seaweeds and microalgae production, 1950–2019

Fort A, Guiry MD, Sulpice R (2018) Magnetic beads, a particularly effective novel method for extraction of NGS-ready DNA from macroalgae. Algal Res 32:308–313

Freile-Pelegrín Y, Tasdemir D (2019 March) Seaweeds to the rescue of forgotten diseases: a review. Botanica Marina 62:211–226

Gager L, Connan S, Molla M, Couteau C, Arbona JF, Coiffard L, Cerantola S, Pouvreau VS (2020) Active phlorotannins from seven brown seaweeds commercially harvested in Brittany (France) detected by ^{1}H NMR and in vitro assays: temporal variation and potential valorization in cosmetic applications. J Appl Phycol **32**:2375–2386

Ganesan AR, Tiwari U, Rajauria G (2019) Seaweed nutraceuticals and their therapeutic role in disease prevention. Food Sci HumWellness 8:252–263

García-González X, Cortejoso L, García MI, García-Alfonso P, Robles L, Grávalos C, González-Haba E, Marta P, Sanjurjo M, López-Fernández LA (2015) Variants in CDA and ABCB1 are predictors of capecitabine-related adverse reactions in colorectal cancer. Oncotarget 6(8): 6422–6430

García-Poza S, Pacheco D, Cotas J, Marques JC, Pereira L, Gonçalves AMM (2022) Marine macroalgae as a feasible and complete resource to address and promote Sustainable Development Goals (SDGs). Integr Environ Assess Manag *18*:1148–1161

Garnier M, Carrier G, Rogniaux H, Nicolau E, Bougaran G, Saint-Jean B, Cadoret JP (2014) Comparative proteomics reveals proteins impacted by nitrogen deprivation in wild-type and high lipid-accumulating mutant strains of *Tisochrysis lutea*. J Proteome 105:107–120

Ghosh A, Khanra S, Mondal M, Halder G, Tiwari ON, Saini S, Bhowmick TK, Gayen K (2016) Progress toward isolation of strains and genetically engineered strains of microalgae for production of biofuel and other value added chemicals: a review. Energy Convers Manag 113:104–118

Gimpel JA, Henríquez V, Mayfield SP (2015) In metabolic engineering of eukaryotic microalgae: potential and challenges come with great diversity. Front Microbiol 6:1376

Gomma AE, Lee SK, Sun SM, Yang SH, Chung G (2015) Improvement in oil production by increasing malonyl-CoA and glycerol-3-phosphate pools in *Scenedesmus quadricauda*. Indian J Microbiol 55:447–455

Gong Y, Kang NK, Kim YU, Wang Z, Li W, Xin Y, Shen C, Wang Q, You W, Lim J-M, Jeong S-W, Park Y-I, Oh H-M, Pan K, Poliner E, Yang G, Li-Beisson Y, Li Y, Hu Q, Poetsch A, Farre EM, Chang YK, Jeong W-J, Jeong B-R, Xu J (2020) The NanDeSyn database for *Nannochloropsis* systems and synthetic biology. Wiley Online Libr 104:1736–1745

González A, Osorio H, Romero S, Méndez P, Sepúlveda M, Laporte D, Gutierrez-Cutiño M, Santander R, Castro-Nallar E, Moenne A (2022) Transcriptomic analyses reveal increased expression of dioxygenases, monooxygenases, and other metabolizing enzymes involved in anthracene degradation in the marine alga *Ulva lactuca*. *Front Plant Sci* 13:955601

Greco M, Chiappetta A, Bruno L, Bitonti MB (2012) In *Posidonia oceanica* cadmium induces changes in DNA methylation and chromatin patterning. J Exp Bot 63:695–709

Gregersen S, Pertseva M, Marcatili P, Holdt SL, Jacobsen C, García-Moreno PJ, Hansen EB, Overgaard MT (2022) Proteomic characterization of pilot scale hot-water extracts from the industrial carrageenan red seaweed *Eucheuma denticulatum*. Algal Res 62:102619

Gu K, Liu Y, Jiang T, Cai C, Zhao H, Liu X, He P (2022) Molecular response of *Ulva prolifera* to short-term high light stress revealed by a multi-omics approach. Biology (Basel) 11(11):1563

Gupta V, Thakur RS, Reddy CRK, Jha B (2013) Central metabolic processes of marine macrophytic algae revealed from NMR based metabolome analysis. RSC Adv 3(19):7037–7047

Hall-Spencer JM, Rodolfo-Metalpa R, Martin S, Ransome E, Fine M, Turner SM, Rowley SJ, Tedesco D, Buia MC (2008) Volcanic carbon dioxide vents show ecosystem effects of ocean acidification. Nature 454(7200):96–99

Hamid SS, Wakayama M, Ichihara K, Sakurai K, Ashino Y, Kadowaki R, Soga T, Tomita M (2019) Metabolome profiling of various seaweed species discriminates between brown, red, and green algae. Planta 249:1921–1947

Hanikenne M, Krämer U, Demoulin V, Baurain D (2005) A comparative inventory of metal transporters in the green alga *Chlamydomonas reinhardtii* and the red alga cyanidioschizon merolae. Plant Physiol 137:428–446

Hardegen J, Amend G, Wichard T (2023) Lifecycle-dependent toxicity and removal of micropollutants in algal cultures of the green seaweed *Ulva* (Chlorophyta). J Appl Phycol.

Harnedy PA, FitzGerald RJ (2013) Extraction of protein from the macroalga *Palmaria palmata*. LWT Food Sci Technol 51(1):375–382

Higgins BT, Gennity I, Samra S, Kind T, Fiehn O, VanderGheynst JS (2016) Cofactor symbiosis for enhanced algal growth, biofuel production, and wastewater treatment. Algal Res 17:308–315

Hirashima T, Toyoshima M, Moriyama T, Sato N (2018) Evolution of the phosphatidylcholine biosynthesis pathways in green algae: combinatorial diversity of methyltransferases. J Mol Evol 86:68–76

Ho CL, Teoh S, Teo SS, Rahim RA, Phang SM (2009) Profiling the transcriptome of *Gracilaria changii* (Rhodophyta) in response to light deprivation. Mar Biotechnol 11(4):513–519

Ho SH, Ye X, Hasunuma T, Chang JS, Kondo A (2014b) Perspectives on engineering strategies for improving biofuel production from microalgae — a critical review. Biotechnol Adv 32:1448–1459

Ho SH, Nakanishi A, Kato Y, Yamasaki H, Chang JS, Misawa N, Hirose Y, Minagawa J, Hasunuma T, Kondo A (2017) Dynamic metabolic profiling together with transcription analysis reveals salinity-induced starch-to-lipid biosynthesis in alga Chlamydomonas sp. JSC4. Sci Rep 7:45471. https://doi.org/10.1038/srep45471

Holdt SL, Kraan S (2011) Bioactive compounds in seaweed: functional food applications and legislation. J Appl Phycol 23:543–597

Hwang E, Park SY, Sun Z, Shin HS, Lee DG, Yi TH (2014) The protective effects of fucosterol against skin damage in UVB-irradiated human dermal fibroblasts. Mar Biotechnol 16(3): 361–370

Ipatova V, Prokhotskaya V, Dmitrieva A (2012) Structural changes and adaptation of algal population under different regimens of toxic exposure. In: Minai AA, Braha D, Bar-Yam Y (eds) Unifying themes in complex systems VII. Springer, Berlin, pp 149–156

Islam MN, Ishita IJ, Jin SE, Choi RJ, Lee CM, Kim YS, Jung HA, Choi JS (2013) Anti-inflammatory activity of edible brown alga *Saccharina japonica* and its constituents pheophorbide a and pheophytin a in LPS-stimulated RAW 264.7 macrophage cells. Food Chem Toxicol 55:541–548

Jeong BR, Jang J, Jin ES (2023) Genome engineering via gene editing technologies in microalgae. Bioresour Technol 373:128701

Jha V, Puranik Chande S, Purohit HJ (2018) Sequestration options for phosphorus in wastewater. Optim Appl Bioprocesses:115–140

Jian Q, Zhu X, Chen J, Zhu Z, Yang R, Luo Q, Chen H, Yan X (2017) Analysis of global metabolome by gas chromatography-mass spectrometry of *Pyropia haitanensis* stimulated with 1-octen-3-ol. J Appl Phycol 29:2049–2059

Jutur PP, Nesamma AA, Shaikh KM (2016) Algae-derived marine oligosaccharides and their biological applications. Front MarSci 3:83

Khan S, Mao Y, Gao DR, S, Niaz Z, Tang L., Khan S, Wang D. (2018) Identification of proteins responding to pathogen-infection in the red alga *Pyropia yezoensis* using iTRAQ quantitative proteomics. *BMC Genomics* 19:842

Khoo KS, Ahmad I, Chew KW, Iwamoto K, Bhatnagar A, Show PL (2023) Enhanced microalgal lipid production for biofuel using different strategies including genetic modification of microalgae: a review. Prog Energy Combust Sci 96:101071

Kim S, Choi S-I, Kim G-H, Imm J-Y (2019) Anti-Inflammatory Effect of *Ecklonia cava* Extract on *Porphyromonas gingivalis* Lipopolysaccharide-Stimulated Macrophages and a Periodontitis Rat Model. *Nutrients* 11(5):1143

Kim EA, Lee JH, Heo SJ, Jeon YJ (2021) Saringosterol acetate isolated from *Hizikia fusiforme*, an edible brown alga, suppressed hepatocellular carcinoma growth and metastasis in a zebrafish xenograft model. Chem Biol Interact 335:109362

Kirindage KGIS, Jayasinghe AMK, Han EJ, Jee Y, Kim HJ, Do SG, Fernando IPS, Ahn G (2022) Fucosterol isolated from dietary brown alga *Sargassum horneri* protects TNF-α/IFN-γ-stimulated human dermal fibroblasts via regulating Nrf2/HO-1 and NF-κB/MAPK pathways. Antioxidants (Basel) 11(8):1429

Koechler S, Bertin PN, Plewniak F, Baltenweck R, Casiot C, Heipieper HJ, Bouchez O, Arsène-Ploetze F, Hugueney P, Halter D (2016) Arsenite response in *Coccomyxa* sp. Carn explored by transcriptomic and non-targeted metabolomic approaches. Environ Microbiol 18:1289–1300

Koizumi J, Takatani N, Kobayashi N, Mikami K, Miyashita K, Yamano Y, Wada A, Maoka T, Hosokawa M (2018) Carotenoid profiling of a red seaweed *Pyropia yezoensis*: insights into biosynthetic pathways in the order bangiales. Mar Drugs 16(11):426

Krassowski M, Das V, Sahu SK, Misra BB (2020) State of the field in multi-omics research: from computational needs to data mining and sharing. Front Genet 11:610798

Kselíková V, Singh A, Bialevich V, Čížková M, Bišová K (2022) Improving microalgae for biotechnology — From genetics to synthetic biology – Moving forward but not there yet. Biotechnol Adv 58:107885

Kumar A, AbdElgawad H, Castellano I, Selim S, Beemster GTS, Asard H, Buia MC, Palumbo A (2018) Effects of ocean acidification on the levels of primary and secondary metabolites in the brown macroalga *Sargassum vulgare* at different time scales. Sci Total Environ 643:946–956

Kurita T, Moroi K, Iwai M, Okazaki K, Shimizu S, Nomura S, Saito F, Maeda S, Takami A, Sakamoto A, Ohta H, Sakuma T, Yamamoto T (2020) Efficient and multiplexable genome editing using Platinum TALENs in oleaginous microalga, *Nannochloropsis oceanica* NIES-2145. Genes Cells 25:695–702

Ladd SN, Nelson DB, Bamberger I, Daber LE, Kreuzwieser J, Kahmen A, Werner C (2021) Metabolic exchange between pathways for isoprenoid synthesis and implications for biosynthetic hydrogen isotope fractionation. New Phytol 231:1708–1719

Lalegerie F, Lajili S, Bedoux G, Taupin L, Stiger-Pouvreau V, Connan S (2019) Photo-protective compounds in red macroalgae from Brittany: considerable diversity in mycosporine-like amino acids (MAAs). Mar Environ Res 147:37–48

Le Bail A, Dittami SM, de Franco PO, Rousvoal S, Cock MJ, Tonon T, Charrier B (2008) Normalisation genes for expression analyses in the brown alga model *Ectocarpus siliculosus*. BMC Mol Biol 9:75

Leal MC, Munro MHG, Blunt JW, Puga J, Jesus B, Calado R, Rosa R, Madeira C (2013) Biogeography and biodiscovery hotspots of macroalgal marine natural products. Nat Prod Rep 30:1380–1390

Lee B, Choi GG, Choi YE, Sung M, Park MS, Yang JW (2014) Enhancement of lipid productivity by ethyl methane sulfonate-mediated random mutagenesis and proteomic analysis in *Chlamydomonas reinhardtii*. Korean J Chem Eng 31:1036–1042

Lenka SK, Carbonaro N, Park R, Miller SM, Thorpe I, Li Y (2016) Current advances in molecular, biochemical, and computational modeling analysis of microalgal triacylglycerol biosynthesis. Biotechnol Adv 34:1046–1063

Li J, Han D, Wang D, Ning K, Jia J, Wei L, Jing X, Huang S, Chen J, Li Y, Hu Q, Xu J (2014) Choreography of transcriptomes and lipidomes of nannochloropsis reveals the mechanisms of oil synthesis in microalgae. Plant Cell 26:1645–1665

Li Y, Mu J, Chen D, Xu H, Han F, Feng B, Zeng H (2015) Proteomics analysis for enhanced lipid accumulation in oleaginous *Chlorella vulgaris* under a heterotrophic-Na+ induction two-step regime. Biotechnol Lett 37:1021–1030

Li L, Zhang G, Wang Q (2016) De novo transcriptomic analysis of *Chlorella sorokiniana* reveals differential genes expression in photosynthetic carbon fixation and lipid production. BMC Microbiol 16:1–12

Li N, Gao X, Zheng L, Huang Q, Zeng F, Chen H, Farag MA, Zhao C (2022a) Advances in fucoxanthin chemistry and management of neurodegenerative diseases. Phytomedicine 105: 154352

Li Y, Xu C, Wang B, Xu F, Ma F, Qu Y, Jiang D, Li K, Feng J, Tian S, Wu X, Wang Y, Liu Y, Qin Z, Liu Y, Qin J, Song Q, Zhang X, Sujie A, Huang J, Liu T, Shen K, Zhao JY, Hou Y, Ding C (2022b) Proteomic characterization of gastric cancer response to chemotherapy and targeted therapy reveals new therapeutic strategies. Nat Commun 13(1):5723

Li X, Lan C, Li X, Hu Z, Jia B (2022c) A review on design-build-test-learn cycle to potentiate progress in isoprenoid engineering of photosynthetic microalgae. Bioresour Technol 363: 127981

Lin WR, Tan SI, Hsiang CC, Sung PK, Ng IS (2019) Challenges and opportunity of recent genome editing and multi-omics in cyanobacteria and microalgae for biorefinery. Bioresour Technol 291:121932

Lin Y, Xu K, Xu Y, Ji D, Chen C, Wang W, Xie C (2021) Transcriptome co-expression network analysis identifies key genes regulating conchosporangia maturation of *Pyropia haitanensis*. Front Genet 12:680120

Liu J-W, Dong S-L (2001) Comparative studies on utilizing nitrogen capacity between two macroalgae *Gracilaria tenuistipitata* Var. Liui (rhodophyta) and *Ulva Pertusa* (Chlorophyta) I. nitrogen storage under nitrogen enrichment and starvation. J Environ Sci (China) 13(3): 318–322

Liu J, Sun Z, Gerken H, Liu Z, Jiang Y, Chen F (2014) *Chlorella zofingiensis* as an alternative microalgal producer of astaxanthin: biology and industrial potential. Mar Drugs 12:3487–3515

Liu J, Mao X, Zhou W, Guarnieri MT (2016) Simultaneous production of triacylglycerol and high-value carotenoids by the astaxanthin-producing oleaginous green microalga *Chlorella zofingiensis*. Bioresour Technol 214:319–327

Lohman EJ, Gardner RD, Pedersen T, Peyton BM, Cooksey KE, Gerlach R (2015) Optimized inorganic carbon regime for enhanced growth and lipid accumulation in *Chlorella vulgaris* Luisa Gouveia. Biotechnol Biofuels 8:1–13

Longworth J, Wu D, Huete-Ortega M, Wright PC, Vaidyanathan S (2016) Proteome response of *Phaeodactylum tricornutum*, during lipid accumulation induced by nitrogen depletion. Algal Res 18:213–224

López García De Lomana A, Schäuble S, Valenzuela J, Imam S, Carter W, Bilgin DD, Yohn CB, Turkarslan S, Reiss DJ, Orellana MV, Price ND, Baliga NS (2015) Transcriptional program for nitrogen starvation-induced lipid accumulation in *Chlamydomonas reinhardtii*. Biotechnol Biofuels 8:1–18

Lu N, Wei D, Jiang XL, Chen F, Yang ST (2012) Regulation of lipid metabolism in the snow alga *Chlamydomonas nivalis* in response to NaCl stress: an integrated analysis by cytomic and lipidomic approaches. Process Biochem 47:1163–1170

Maeda H, Kanno S, Kodate M, Hosokawa M, Miyashita K (2015) Fucoxanthinol, metabolite of fucoxanthin, improves obesity-induced inflammation in adipocyte cells. Marine Drugs 13(8): 4799–4813

Mangott A, Nappi J, Delli Paoli Carini A, Goncalves P, Hua K, Domingos JA, de Nys R, Thomas T (2020) *Ulva lactuca* as a functional ingredient and water bioremediator positively influences the hepatopancreas and water microbiota in the rearing of *Litopenaeus vannamei*. Algal Res 51: 102040

Marinho GS, Holdt SL, Angelidaki I (2015) Seasonal variations in the amino acid profile and protein nutritional value of *Saccharina latissima* cultivated in a commercial IMTA system. J Appl Phycol 27:1991–2000

Martins CDL, Arantes N, Faveri C, Batista MB, Oliveira EC, Pagliosa PR, Fonseca AL, Nunes JMC, Chow F, Pereira SB, Horta PA (2012) The impact of coastal urbanization on the structure of phytobenthic communities in southern Brazil. Mar Pollut Bull 64:772–778

Matsuzaki M, Misumi O, Shin-I T, Maruyama S, Takahara M, Miyagishima SY, Mori T, Nishida K, Yagisawa F, Nishida K, Yoshida Y, Nishimura Y, Nakao S, Kobayashi T, Momoyama Y, Higashiyama T, Minoda A, Sano M, Nomoto H, Oishi K, Hayashi H, Ohta F, Nishizaka S, Haga S, Miura S, Morishita T, Kabeya Y, Terasawa K, Suzuki Y, Ishii Y, Asakawa S, Takano H, Ohta N, Kuroiwa H, Tanaka K, Shimizu N, Sugano S, Sato N, Nozaki H, Ogasawara N, Kohara Y, Kuroiwa T (2004) Genome sequence of the ultrasmall unicellular red alga *Cyanidioschyzon merolae* 10D. Nature 428:653–657

Melo T, Alves E, Azevedo V, Martins AS, Neves B, Domingues P, Calado R, Abreu MH, Domingues MR (2015) Lipidomics as a new approach for the bioprospecting of marine macroalgae — Unraveling the polar lipid and fatty acid composition of *Chondrus crispus*. Algal Res 8:181–191

Merchant SS, Prochnik SE, Vallon O, Harris EH, Karpowicz SJ, Witman GB, Terry A, Salamov A, Fritz-Laylin LK, Maréchal-Drouard L, Marshall WF, Qu LH, Nelson DR, Sanderfoot AA, Spalding MH, Kapitonov VV, Ren Q, Ferris P, Lindquist E, Shapiro H, Lucas SM, Grimwood J, Schmutz J, Grigoriev IV, Rokhsar DS, Grossman AR, Cardol P, Cerutti H, Chanfreau G, Chen CL, Cognat V, Croft MT, Dent R, Dutcher S, Fernández E, Fukuzawa H, González-Ballester D, González-Halphen D, Hallmann A, Hanikenne M, Hippler M, Inwood W, Jabbari K, Kalanon M, Kuras R, Lefebvre PA, Lemaire SD, Lobanov AV, Lohr M, Manuell A, Meier I, Mets L, Mittag M, Mittelmeier T, Moroney JV, Moseley J, Napoli C, Nedelcu AM, Niyogi K, Novoselov SV, Paulsen IT, Pazour G, Purton S, Ral JP, Riaño-Pachón DM, Riekhof W, Rymarquis L, Schroda M, Stern D, Umen J, Willows R, Wilson N, Zimmer SL, Allmer J, Balk J, Bisova K, Chen CJ, Elias M, Gendler K, Hauser C, Lamb MR, Ledford H, Long JC, Minagawa J, Page MD, Pan J, Pootakham W, Roje S, Rose A, Stahlberg E, Terauchi AM, Yang P, Ball S, Bowler C, Dieckmann CL, Gladyshev VN, Green P, Jorgensen R, Mayfield S, Mueller-Roeber B, Rajamani S, Sayre RT, Brokstein P, Dubchak I, Goodstein D, Hornick L, Huang YW, Jhaveri J, Luo Y, Martínez D, Ngau WCA, Otillar B, Poliakov A, Porter A, Szajkowski L, Werner G, Zhou K (2007) The *Chlamydomonas* genome reveals the evolution of key animal and plant functions. Science (80-) 318:245–251

Mikami K, Hosokawa M (2013) Biosynthetic pathway and health benefits of fucoxanthin, an algae-specific xanthophyll in brown seaweeds. Int J Mol Sci 14(7):13763–13781

Mishra A, Medhi K, Malaviya P, Thakur IS (2019) Omics approaches for microalgal applications: prospects and challenges. Bioresour Technol 291:121890

Montero L, Sánchez-Camargo AP, García-Cañas V, Tanniou A, Stiger-Pouvreau V, Russo M, Rastrelli L, Cifuentes A, Herrero M, Ibáñez E (2015) Anti-proliferative activity and chemical characterization by comprehensive two-dimensional liquid chromatography coupled to mass spectrometry of phlorotannins from the brown macroalga *Sargassum muticum* collected on North-Atlantic coasts. J Chromatogr A 1428:115–125

Morrison N, Bearden D, Bundy JG, Collette T, Currie F, Davey MP, Haigh NS, Hancock D, Jones OAH, Rochfort S, Sansone SA, Štys D, Teng Q, Field D, Viant MRMR (2007) Standard reporting requirements for biological samples in metabolomics experiments: environmental context. Metabolomics 3:203–210

Mulders KJM, Janssen JH, Martens DE, Wijffels RH, Lamers PP (2014) Effect of biomass concentration on secondary carotenoids and triacylglycerol (TAG) accumulation in nitrogen-depleted *Chlorella zofingiensis*. Algal Res 6:8–16

Nishitsuji K, Arimoto A, Iwai K, Sudo Y, Hisata K, Fujie M, Arakaki N, Kushiro T, Konishi T, Shinzato C, Satoh N, Shoguchi E (2016) A draft genome of the brown alga, Cladosiphon okamuranus, S-strain: a platform for future studies of 'mozuku' biology. DNA Res 23(6): 561–570

Niu L, Thiele M, Geyer PE, Rasmussen DN, Webel HE, Santos A, Gupta R, Meier F, Strauss M, Kjaergaard M, Lindvig K, Jacobsen S, Rasmussen S, Hansen T, Krag A, Mann M (2022) Noninvasive proteomic biomarkers for alcohol-related liver disease. Nat Med 28(6):1277–1287

Norskov NP, Bruhn A, Cole A, Nielsen MO (2021) Targeted and untargeted metabolic profiling to discover bioactive compounds in seaweeds and hemp using gas and liquid chromatography-mass spectrometry. Metabolites *11*:259

Oh M-J, So H-J, Hong E-S, Shin J-A, Lee K-T (2021) Presence of cholesterol in non-animal organisms: identification and quantification of cholesterol in crude seed oil from *Perilla frutescens* and dehydrated *Pyropia tenera*. Molecules 26(12):3767

Pandey A, Singh MP, Kumar S, Srivastava S (2019) Phycoremediation of persistent organic pollutants from wastewater: retrospect and prospects. Appl Microalgae Wastewater Treat:207–235

Pangestuti R, Kim SK (2011) Brown algae metabolites for pharmaceutical and cosmetic industries: a review. Fitoterapia 82(2):129–146

Pangestuti R, Kim S-K (2015) Chapter six—Seaweed proteins, peptides, and amino acids. In: Tiwari BK, Troy DJ (eds) Seaweed sustainability food and non-food applications. Academic Press, San Diego, CA, pp 125–140

Park HS, Hwang HJ, Kim G-Y, Cha H-J, Kim W-J, Kim ND, Yoo YH, Choi YH (2013) Induction of apoptosis by fucoidan in human leukemia U937 cells through activation of p38 MAPK and modulation of Bcl-2 family. *Marine Drugs* 11(7):2347–2364

Patel AK, Huang EL, Low-Décarie E, Lefsrud MG (2015) Comparative shotgun proteomic analysis of wastewater-cultured microalgae: nitrogen sensing and carbon fixation for growth and nutrient removal in *Chlamydomonas reinhardtii*. J Proteome Res 14:3051–3067

Pei P, Du H, Aslam M, Liang H, Wang H, Liu X, Chen W (2021) Environmental factors shape the epiphytic bacterial communities of *Gracilariopsis lemaneiformis*. Sci Rep 11(1):8671. https://doi.org/10.21203/rs.3.rs-144624/v1

Peng H, Wei D, Chen G, Chen F (2016) Transcriptome analysis reveals global regulation in response to CO2 supplementation in oleaginous microalga *Coccomyxa subellipsoidea* C-169. Biotechnol Biofuels 9:151

Pérez-Pérez ME, Mauriès A, Maes A, Tourasse NJ, Hamon M, Lemaire SD, Marchand CH (2017) The deep thioredoxome in *Chlamydomonas reinhardtii*: new insights into redox regulation. Mol Plant 10:1107–1125

Pilatti FK, Ramlov F, Schmidt EC, Costa C, de Oliveira ER, Bauer CM, Rocha M, Bouzon ZL, Maraschin M (2017) Metabolomics of *Ulva lactuca* Linnaeus (Chlorophyta) exposed to oil fuels: fourier transform infrared spectroscopy and multivariate analysis as tools for metabolic fingerprint. Mar Pollut Bull 114:831–836

Polikovskya M, Fernanda F, Sack M, Frey W, Müller G, Golberga A (2019) *In silico* food allergenic risk evaluation of proteins extracted from macroalgae *Ulva* sp. with pulsed electric fields. Food Chem 276:735–744

Poo KM, Son EB, Chang JS, Ren X, Choi YJ, Chae KJ (2018) Biochars derived from wasted marine macro-algae (*Saccharina japonica* and *Sargassum fusiforme*) and their potential for heavy metal removal in aqueous solution. J Environ Manag 206:364–372

Primrose SB, Twyman RM (2003) Principals of genome analysis and genomics.3rd edn. Blackwell Publishing, Oxford

Rabecca R, Doss A, Praveen Pole RP, Satheesh S (2022) Phytochemical and anti-inflammatory properties of green macroalga *Codium tomentosum*. Biocatalysis and Agricultural Biotechnology 45:102492

Radakovits R, Jinkerson RE, Darzins A, Posewitz MC (2010) Genetic engineering of algae for enhanced biofuel production. Eukaryot Cell 9:486–501

Ravanfar R et al (2021) Proteomic profiling of *Porphyra tenera* and *Ulva lactuca*: potential sources of bioactive compounds for cosmetic applications. J Appl Phycol 33(2):813–826

Rengasamy KRR, Kulkarni MG, Stirk WA, Van Staden J (2015) Eckol improves growth, enzyme activities, and secondary metabolite content in maize (Zea mayscv. Border King). J Plant Growth Regul 34(2):410–416

Rengasamy KRR, Kulkarni MG, Papenfus HB, Van Staden J (2016) Quantification of plant growth biostimulants, phloroglucinol and eckol, in four commercial seaweed liquid fertilizers and some by-products. Algal Res 20:57–60

Rioux L-E, Beaulieu L, Turgeon SL (2017) Seaweeds: a traditional ingredients for new gastronomic sensation. Food Hydrocoll 68:255–265

Rismani-Yazdi H, Haznedaroglu BZ, Hsin C, Peccia J (2012) Transcriptomic analysis of the oleaginous microalga *Neochloris oleoabundans* reveals metabolic insights into triacylglyceride accumulation. Biotechnol Biofuels 5:1–16

Ristivojevic P, Jovanovic V, Opsenica DM, Park J, Rollinger JM, Velickovic TC (2021) Rapid analytical approach for bioprofiling compounds with radical scavenging and antimicrobial activities from seaweeds. Food Chem 334:127562

Ritter A, Dittami SM, Goulitquer S, Correa JA, Boyen C, Potin P, Tonon T (2014) Transcriptomic and metabolomic analysis of copper stress acclimation in *Ectocarpus siliculosus* highlights signaling and tolerance mechanisms in brown algae. BMC Plant Biol 14:116

Roth MS, Cokus SJ, Gallaher SD, Walter A, Lopez D, Erickson E, Endelman B, Westcott D, Larabell CA, Merchant SS, Pellegrini M, Niyogi KK (2017) Chromosome-level genome assembly and transcriptome of the green alga *Chromochloris zofingiensis* illuminates astaxanthin production. Proc Natl Acad Sci U S A 114:E4296–E4305

Ryu J, Park SJ, Kim IH, Choi YH, Nam TJ (2014) Protective effect of porphyra-334 on UVA-induced photoaging in human skin fibroblasts. Int J Mol Med 34:796–803

Samadani M, Perreault F, Oukarroum A, Dewez D (2018) Effect of cadmium accumulation on green algae *Chlamydomonas reinhardtii* and acid-tolerant Chlamydomonas CPCC 121. Chemosphere 191:174–182

Sanchez-Arcos C, Paris D, Mazzella V, Mutalipassi M, Costantini M, Buia MC, von Elert E, Cutignano A, Zupo V (2022) Responses of the macroalga *Ulva prolifera* Müller to ocean acidification revealed by complementary NMR- and MS-based omics approaches. Marine Drugs 20(12):743

Sasso S, Pohnert G, Lohr M, Mittag M, Hertweck C (2012) Microalgae in the postgenomic era: a blooming reservoir for new natural products. FEMS Microbiol Rev 36(4):761–785

Sawicki A, Zhou S, Kwiatkowski K, Luo M, Willows RD (2017) 1-N-histidine phosphorylation of ChlD by the AAA+ ChlI2 stimulates magnesium chelatase activity in chlorophyll synthesis. Biochem J 474:2095–2105

Shan T, Pang S (2021) Breeding in the economically important brown alga *Undaria pinnatifida*: a concise review and future prospects. Front Genet 12:801937

Shao Z, Zhang P, Li Q, Wang X, Duan D (2014) Characterization of mannitol-2-dehydrogenase in *Saccharina japonica*: evidence for a new polyol-specific long-chain dehydrogenases/reductase. PLoS One 9(5):e97935

Shao Z, Zhang P, Lu C, Li S, Chen Z, Wang X, Duan D (2019) Transcriptome sequencing of *Saccharina japonica* sporophytes during whole developmental periods reveals regulatory networks underlying alginate and mannitol biosynthesis. BMC Genomics 20:975

Shen P, Gu Y, Zhang C, Sun C, Qin L, Yu C, Qi H (2021) Metabolomic approach for characterization of polyphenolic compounds in *Laminaria japonica, Undaria pinnatifida, Sargassum fusiforme* and *Ascophyllum nodosum*. Foods 10(1):192

Shetty P, Boboescu IZ, Pap B, Wirth R, Kovács KL, Bíró T, Futó Z, White RA, Maróti G (2019) Exploitation of algal-bacterial consortia in combined biohydrogen generation and wastewater treatment. Front Energy Res 7:52

Shi J, Podola B, Melkonian M (2007) Removal of nitrogen and phosphorus from wastewater using microalgae immobilized on twin layers: an experimental study. J Appl Phycol 19(5):417–423

Shin YS, Jeong J, Nguyen THT, Kim JYH, Jin ES, Sim SJ (2019) Targeted knockout of phospholipase A 2 to increase lipid productivity in *Chlamydomonas reinhardtii* for biodiesel production. Bioresour Technol 271:368–374

Silchenko AS, Rasin AB, Kusaykin MI, Kalinovsky AI, Zhang M, Changheng L, Malyarenko O, Zueva AO, Zvyagintseva TN, Ermakova SP (2017) Structure, enzymatic transformation, anticancer activity of fucoidan and sulphated fucooligosaccharides from *Sargassum horneri*. Carbohydr Polym 175:654–660

Sithtisarn S, Yokthongwattana K, Mahong B, Roytrakul S, Paemanee A, Phaonakrop N, Yokthongwattana C (2017) Comparative proteomic analysis of *Chlamydomonas reinhardtii* control and a salinity-tolerant strain revealed a differential protein expression pattern. Planta 246:843–856

Sondak CFA, Ang PO, Beardall J, Bellgrove A, Boo SM, Gerung GS, Hepburn CD, Hong DD, Hu Z, Kawai H, Largo D, Lee JA, Lim P-E, Mayakun J, Nelson WA, Oak JH, Phang S-M, Sahoo D, Peerapornpis Y, Yang Y, Chung IK (2017) Carbon dioxide mitigation potential of seaweed aquaculture beds (SABs). J Appl Phycol *29:2363–2373*

Su X, Xu J, Yan X, Zhao P, Chen J, Zhou C, Zhao F, Li S (2013) Lipidomic changes during different growth stages of *Nitzschia closterium* f. minutissima. Metabolomics 9:300–310

Suetsuna K, Nakano T (2000) Identification of an antihypertensive peptide from peptic digest of wakame (*Undaria pinnatifida*). J Nutr Biochem 9:450–454

Suganya T, Varman M, Masjuki HH, Renganathan S (2016) Macroalgae and microalgae as a potential source for commercial applications along with biofuels production: a biorefinery approach. Renew Sust Energ Rev 55:909–941

Sulaiman M, Mahadevan RK, Kurup MG (2019) Effect of ascophyllan from brown algae *Padina tetrastromatica* on cell migration and extracellular matrix stabilisation in burn wounds. Current Bioactive Compounds 15(5):562–572

Sun D, Zhu J, Fang L, Zhang X, Chow Y, Liu J (2013) De novo transcriptome profiling uncovers a drastic downregulation of photosynthesis upon nitrogen deprivation in the nonmodel green alga *Botryosphaerella sudeticus*. BMC Genomics 14:1–18

Sun P, Mao Y, Li G (2015) Comparative transcriptome profiling of *Pyropia yezoensis* (Ueda) M.S. Hwang & H.G. Choi in response to temperature stresses. BMC Genomics 16:463

Surzycki R, Greenham K, Kitayama K, Dibal F, Wagner R, Rochaix JD, Ajam T, Surzycki S (2009) Factors effecting expression of vaccines in microalgae. Biologicals 37:133–138

Tanaka T, Maeda Y, Veluchamy A, Tanaka M, Abida H, Maréchal E, Bowler C, Muto M, Sunaga Y, Tanaka M, Yoshino T, Taniguchi T, Fukuda Y, Nemoto M, Matsumoto M, Wong PS, Aburatani S, Fujibuchi W (2015) Oil accumulation by the oleaginous diatom *Fistulifera solaris* as revealed by the genome and transcriptome. Plant Cell 27:162–176

Tanna B, Choudhary B, Mishra A, Yadav S, Chauhan OP, Elansary HO, Shokralla S, Zin El-Abedin TK, Mahmoud EA (2022) Biochemical and Anti-proliferative activities of seven abundant tropical red seaweeds confirm nutraceutical potential of *Grateloupia indica*. Arab J Chem 15(6):103868

Tarafdar A, Sowmya G, Yogeshwari K, Rattu G, Negi T, Awasthi MK, Hoang AT, Sirohi R (2023) Environmental pollution mitigation through utilization of carbon dioxide by microalgae. Environ Pollut 328:121623

Taylor NS, Merrifield R, Williams TD, Chipman JK, Lead JR, Viant MR (2016) Molecular toxicity of cerium oxide nanoparticles to the freshwater alga *Chlamydomonas reinhardtii* is associated with supra-environmental exposure concentrations. Nanotoxicology 10:32–41

Teo SS, Ho CL, Teoh S, Rahim RA (2009) Transcriptomic analysis of *Gracilaria changii* (rhodophyta) in response to hyper and hypoosmotic stresses. J Phycol 45(5):1093-9

Tonon AP, Zaini PA, dos Reis Falcão V, Oliveira MC, Collen J, Boyen C, Colepicolo P (2018) *Gracilaria tenuistipitata* (Rhodophyta) tolerance to cadmium and copper exposure observed through gene expression and photosynthesis analyses. J Appl Phycol 30:2129–2141

Tran NAT, Padula MP, Evenhuis CR, Commault AS, Ralph PJ, Tamburic B (2016) Proteomic and biophysical analyses reveal a metabolic shift in nitrogen deprived *Nannochloropsis oculata*. Algal Res 19:1–11

Tran TLC, Callahan DL, Islam MT, Wang Y, Arioli T, Cahill D (2023, March 9) Comparative metabolomic profiling of *Arabidopsis thaliana* roots and leaves reveals complex response mechanisms induced by a seaweed extract. Front Plant Sci 14:1114172

Urrea-Victoria V, Nardelli AE, Floh EIS, Chow F (2020) *Sargassum stenophyllum* (Fucales, Ochrophyta) responses to temperature short-term exposure: photosynthesis and chemical composition. Rev Bras Bot 43:733–745

Van Hulten MCW, Witteveldt J, Peters S, Kloosterboer N, Tarchini R, Fiers M, Sandbrink H, Lankhorst RK, Vlak JM (2001) The white spot syndrome virus DNA genome sequence. Virology 286:7–22

Vonlanthen S, Dauvillée D, Purton S (2015) Evaluation of novel starch-deficient mutants of *Chlorella sorokiniana* for hyper-accumulation of lipids. Algal Res 12:109–118

Wang Y, Brahmakshatriya V, Lupiani B, Reddy SM, Soibam B, Benham AL, Gunaratne P, Hc L, Trakooljul N, Ing N, Ron Okimoto R, Zhou H (2012) Integrated analysis of microRNA expression and mRNA transcriptome in lungs of avian influenza virus infected broilers. BMC Genomics 13:278

Wang L, Mao Y, Kong F, Cao M, Sun P (2015) Genome-wide expression profiles of *Pyropia haitanensis* in response to osmotic stress by using deep sequencing technology. BMC Genomics **16**:1012

Wang W, Chang J, Zheng H, Ji D, Xu Y, Chen C, Xie C (2019) Full-length transcriptome sequences obtained by a combination of sequencing platforms applied to heat shock proteins and polyunsaturated fatty acids biosynthesis in *Pyropia haitanensis*. J Appl Phycol 31:1483–1492

Wargacki AJ, Leonard E, Win MN, Regitsky DD, Santos CNS, Kim PB, Cooper SR, Raisner RM, Herman A, Sivitz AB (2012) An engineered microbial platform for direct biofuel production from brown macroalgae. Science 335(6066):308–313

Wayllace NM, Hedín N, Busi MV, Gomez-Casati DF (2021) Characterization of SdGA, a cold-adapted glucoamylase from *Saccharophagus degradans*. Biotechnol Rep 30:e00625

Wei Z, You J, Wu H, Yang F, Long L, Liu Q, Huo Y, He P (2017a) Bioremediation using *Gracilaria lemaneiformis* to manage the nitrogen and phosphorous balance in an integrated multi-trophic aquaculture system in Yantian Bay, China. Marine Pollution Bulletin 121(1–2): 313–319

Wei L, Xin Y, Wang Q, Yang J, Hu H, Xu J (2017b) RNAi-based targeted gene knockdown in the model oleaginous microalgae *Nannochloropsis oceanica*. Plant J 89:1236–1250

Wei L, Shen C, El Hajjami M, You W, Wang Q, Zhang P, Ji Y, Hu H, Hu Q, Poetsch A, Xu J, El M, You W, Wang Q, Zhang P, Ji Y, Hu H, Hu Q, Poetsch A, Xu J, El Hajjami M, You W, Wang Q, Zhang P, Ji Y, Hu H, Hu Q, Poetsch A, Xu J (2019) Knockdown of carbonate anhydrase elevates *Nannochloropsis* productivity at high CO2 level. Metab Eng 54:96–108

Wijesekara I, Lang M, Marty C, Gemin MP, Boulho R, Douzenel P, Wickramasinghe I, Bedoux G, Bourgougnon N (2017) Different extraction procedures and analysis of protein from *Ulva sp.* in Brittany. France J Appl Phycol 29:2503–2511

Wu TM, Hsu YT, Sung MS, Hsu YT, Lee TM (2009) Expression of genes involved in redox homeostasis and antioxidant defense in a marine macroalga *Ulva fasciata* by excess copper. Aquat Toxicol 94:275–285

Xiaofeng D, Li S (2022) Advances and trends in omics technology development. Front Med 9: 911861

Xie XT, Zhang X, Liu Y, Chen XQ, Cheong KL (2020) Quantification of 3,6-anhydro-galactose in red seaweed polysaccharides and their potential skin-whitening activity. 3 Biotech 10:189

Xu S, Ma J, Ji R, Pan K, Miao AJ (2020) Microplastics in aquatic environments: Occurrence, accumulation, and biological effects. Sci Total Environ 703:134699

Xu Z, Bai X, Ye Z (2021) Removal and generation of microplastics in wastewater treatment plants: a review. J Clean Prod 291:125982

Yaich H, Garna H, Bchir B, Besbes S, Paquot M, Richel A, Blecker C, Attia H (2015) Chemical composition and functional properties of dietary fibre extracted by Englyst and Prosky methods from the alga *Ulva lactuca* collected in Tunisia. Algal Res 9:65–73

Yang ZK, Niu YF, Ma YH, Xue J, Zhang MH, Yang WD, Liu JS, Lu SH, Guan Y, Li HY (2013) Molecular and cellular mechanisms of neutral lipid accumulation in diatom following nitrogen deprivation. Biotechnol Biofuels 6:1–14

Yang ZK, Ma YH, Zheng JW, Yang WD, Liu JS, Li HY (2014) Proteomics to reveal metabolic network shifts towards lipid accumulation following nitrogen deprivation in the diatom *Phaeodactylum tricornutum*. J Appl Phycol 26:73–82

Yang D, Song D, Kind T, Ma Y, Hoefkens J, Fiehn O (2015a) Lipidomic analysis of *Chlamydomonas reinhardtii* under nitrogen and sulfur deprivation. PLoS One 10(9):e0137948

Yang JS, Cao J, Xing GL, Yuan HL (2015b) Lipid production combined with biosorption and bioaccumulation of cadmium, copper, manganese and zinc by oleaginous microalgae *Chlorella minutissima* UTEX2341. Bioresour Technol 175:537–544

Yang Y, Feng J, Li T, Ge F, Zhao J (2015c) CyanOmics: an integrated database of omics for the model cyanobacterium *Synechococcus sp.* PCC 7002. Database (Oxford) 2015:bau127. https://doi.org/10.1093/database/bau127

Ye Y, Yang R, Lou Y, Chen J, Yan X, Tang H (2014) Effects of food processing on the nutrient composition of *Pyropia yezoensis* products revealed by NMR-based metabolomic analysis. J Food Nutr Res 2(10):749–756

Ye N, Zhang X, Miao M, Fan X, Zheng Y, Xu D, Wang J, Zhou L, Wang D, Gao Y, Wang Y, Shi W, Ji P, Li D, Guan Z, Shao C, Zhuang Z, Gao Z, Qi J, Zhao F (2015) *Saccharina* genomes provide novel insight into kelp biology. Nat Commun 6:6986

Yim SK, Kim K, Kim IH, Chun SH, Oh TH, Kim JU, Kim JW, Jung WH, Moon HS, Ku BS, Jung KJ (2021) Inhibition of SARS-CoV-2 virus entry by the crude polysaccharides of seaweeds and abalone viscera in vitro. Mar Drugs 19(4):219

Yong WTL, Thien VY, Rupert R, Rodrigues KF (2022) Seaweed: A potential climate change solution. Renew Sust Energ Rev 159:112222

Yotsukura, N., Nagai, K., Tanaka, T. Kimura H, Morimoto K (2012). Temperature stress-induced changes in the proteomic profiles of *Ecklonia cava* (Laminariales, Phaeophyceae). J Appl Phycol 24, 163–171.

Young DY, Pang N, Shachar-Hill Y (2022) 13C-labeling reveals how membrane lipid components contribute to triacylglycerol accumulation in *Chlamydomonas*. Plant Physiol 189:1326–1344

Yuan JP, Peng J, Yin K, Wang JH (2011) Potential health-promoting effects of astaxanthin: A high-value carotenoid mostly from microalgae. Mol Nutr Food Res 55:150–165

Zhang W, Tan NGJ, Li SFY (2014) NMR-based metabolomics and LC-MS/MS quantification reveal metal-specific tolerance and redox homeostasis in *Chlorella vulgaris*. Mol BioSyst 10:149–160

Zhang W, Tan NGJ, Fu B, Li SFY (2015) Metallomics and NMR-based metabolomics of *Chlorella* sp. reveal the synergistic role of copper and cadmium in multi-metal toxicity and oxidative stress. Metallomics 7:426–438

Zhang Y, Wang X, Shan T, Pang S, Xu N (2019) Transcriptome profiling of the meristem tissue of *Saccharina japonica* (Phaeophyceae, Laminariales) under severe stress of copper. Mar Genomics 47:100671

Zhang L, Cao Z, Liang G, Li X, Wu H, Yang G (2020a) Comparative transcriptome analysis reveals candidate genes related to structural and storage carbohydrate biosynthesis in kelp *Saccharina japonica* (Laminariales, Phaeophyceae). J Phycol 56(5):1168–1183

Zhang B, Xie X, Liu X, He L, Sun Y, Wang G (2020b) The carbonate concentration mechanism of *Pyropia yezoensis* (Rhodophyta): evidence from transcriptomics and biochemical data. BMC Plant Biol 20:424

Zhang LJ, Qian L, Ding LY, Wang L, Wong MH, Tao HC (2021a) Ecological and toxicological assessments of anthropogenic contaminants based on environmental metabolomics. Environ Sci Ecotechnol 5:100081

Zhang Y, Gu Z, Ren Y, Wang L, Zhang J, Liang C, Tong S, Wang Y, Xu D, Zhang X, Ye N (2021b) Integrating Transcriptomics and Metabolomics to Characterize Metabolic Regulation to Elevated CO2 in *Chlamydomonas reinhardtii*. Mar Biotechnol 23:255–275

Zheng SY, Hu XM, Huang K, Li ZH, Chen QN, Yang RH, Xiong K (2022) Proteomics as a tool to improve novel insights into skin diseases: what we know and where we should be going. Front Surg 9:1025557

Zhou D, Zhang C, Fu L, Xu L, Cui X, Li Q, Crittenden JC (2017) Responses of the microalga *Chlorophyta* sp. to bacterial quorum sensing molecules (n-acylhomoserine lactones): aromatic protein-induced self-aggregation. Environ Sci Technol 51:3490–3498

Zhou Y, Li X, Xia Q, Dai R (2020) Transcriptomic survey on the microcystins production and growth of *Microcystis aeruginosa* under nitrogen starvation. Sci Total Environ 700:134501

Zou H-X, Pang Q-Y, Lin L-D, Zhang A-Q, Li N, Lin Y-Q et al (2014) Behavior of the edible seaweed *Sargassum fusiforme* to copper pollution: short-term acclimation and long-term adaptation. PLoS One 9(7):e101960

Algal-based Biopolymers

Samer Younes, Nathanael Arnold, Michael Paper, Linda Keil, Dania Awad, and Thomas Brueck

Abstract Worldwide, more than 140 million tons of petroleum-based plastics are consumed yearly, with rising global demand exacerbating the environmental problems. Recycling of plastics solely does not provide a comprehensive solution. Recently, many countries have been investing in eco-friendly solutions, focusing on renewable energies and biodegradable polymers. In this chapter, the potential of algae to fulfill societal needs and mitigate the global demands with value-added chemicals is investigated. Several algal species can valorize waste biomass and remediate wastewater to remove nitrogen, phosphorus, and heavy metals, all while providing bio-based and biodegradable polymeric material in the form of biomass. The major benefit of algae-derived biopolymers over other platforms lies in their autotrophic nature, ultimately geared toward negative carbon footprint biorefinery process and contributing to a sustainable circular economy. Still, the "uneconomical" production, largely attributed to cultivation systems and product titers, remains the major roadblock for market entry of algae-derived biopolymers. Overall, this chapter aims to present and critically analyze the current status and production technologies for algal-derived biopolymers, including polyhydroxybutyrate (PHB) and its equivalents. This work also summarizes the recent process development approaches and optimization opportunities of different algae-derived biopolymers.

Keywords Bioplastic · Polyhydroxy polymers · Polyols · Sustainable biopolymers carbon fibers

1 Introduction

The world of material science has witnessed a significant transformation over the past few decades, driven by the discovery and development of new materials with remarkable properties. Among these materials, polymers have attracted considerable

S. Younes · N. Arnold · M. Paper · L. Keil · D. Awad (✉) · T. Brueck (✉)
Werner Siemens-Chair of Synthetic Biotechnology, TUM School of Natural Sciences, Technical University of Munich (TUM), Garching, Germany
e-mail: dania.awad@tum.de; brueck@tum.de

A. Abomohra, S. Ende (eds.), *Value-added Products from Algae*,
https://doi.org/10.1007/978-3-031-42026-9_13

attention due to their versatility, tunable properties, and potential applications in various industries. Polymers are defined as large molecules composed of repeating structural units called monomers, which are linked together by covalent bonds (Gedde and Hedenqvist 2019). They can be classified into two main categories; synthetic polymers and natural or biopolymers. The former are derived from fossil fuels (such as petroleum or natural gas), while natural polymers are derived from proteins, nucleic acids, and polysaccharides (Sperling 2006).

The current market is dominated by synthetic polymers, including polyethylene, polypropylene, and polyvinyl chloride. Conventional plastics, in particular, have become an essential component of modern society, with an annual global production of about 368 million tons (Dang et al. 2022). Petroleum-based plastics have unique and convenient properties, including durability, flexibility, resistance to water, heat, and electricity, and most importantly ease of production and affordability. However, greenhouse gases (GHGs) emissions pose a major environmental concern, as every 1 ton of plastics requires an input of 1.1 tons of fossil fuel (Suriyamongkol et al. 2007). Furthermore, most plastic products, such as polyethylene (PE), polypropylene (PP), polyethylene terephthalate (PET), and polystyrene (PS), are chemically stable and essentially nonbiodegradable (Dang et al. 2022; Devadas et al. 2021). Globally, plastic waste is accumulating at an annual rate of 25×10^6 tons, posing a threat to the fragile global ecosystem. With 8 million tons of plastics leaking into the oceans annually, it is estimated that by 2050 "White Pollution" will outweigh all the fish in the sea (Balaji et al. 2013; Nanda and Bharadvaja 2022). This ecological problem was greatly exacerbated by the onset of the COVID-19 pandemic, where the demand for single-use-plastics (SUPs), such as masks, gloves, and tests kits, skyrocketed (Nanda and Bharadvaja 2022; Quero and Luna 2017). Microplastics also pose a growing concern, as they have been discovered at the deepest point, the Mariana Trench, and the highest peak, Mount Everest, and recently very alarmingly, in human blood vessels (Rotchell et al. 2023). While recycling plastic waste is undoubtedly valuable, it does not offer a comprehensive stand-alone solution, as only 9% of all plastic is recycled globally (Nanda and Bharadvaja 2022). To truly reduce plastic pollution, the root causes must be tackled, chiefly unsustainable production processes and the recalcitrant nature of the product itself.

Biopolymers are a readily biodegradable class of polymers that are derived from natural sources. In the last decades, biopolymers have gained increasing attention due to their potential to replace traditional polymers derived from nonrenewable sources. Biopolymers can be classified into different groups based on their origin, chemical structure, and properties, including polyesters, polyamides, polyurethanes, and polysaccharides (Jendrossek and Handrick 2002). Biopolymer production processes are very diverse, encompassing direct polymer purification and modifications and indirect chemical and biological (fermentative) polymerization processes of monomers. Currently, commercial production of biopolymers utilizes first- and second-generation feedstocks, including sugarcane, corn, potatoes straw, paper, sawdust, vegetable fats and oils, agricultural food waste, and lignocellulosic biomass (Nanda and Bharadvaja 2022). They could also be derived from processed polysaccharides and proteins, including starch, cellulose, and casein. However, the current approach of using agro-crops to produce biopolymers is not a viable solution in the long term due to competition with human food and animal feed requirements. According to Food and Agriculture Organization (FAO), global demand for food, fodder, and fibers is expected to rise by 70% by 2050 (Coppola et al. 2021; Nanda and Bharadvaja 2022). Additionally, terrestrial crops and plants necessitate large quantities of fresh water, expensive fertilizers and pesticides, as well as arable land, which imposes environmental concerns, i.e., increased deforestation practices.

Algae represent a diverse group of photosynthetic organisms that include seaweeds and microalgae, which have been identified as a viable source of biopolymers due to their ability to synthesize and accumulate various types of intracellular or extracellular biopolymers (Chen and Patel 2012). Various biopolymers can be derived from algae, which can be broadly classified into two groups, polyesters and polysaccharides. Polyesters, such as polyhydroxybutyrate (PHB), polyols, and polyurethanes, are synthesized by algae as intracellular storage compounds under specific environmental conditions, such as nutrient limitation or stress (Geyer et al. 2017). These biopolymers exhibit thermoplastic properties and can be processed into various forms, such as films, fibers, and foams, for different applications in packaging, consumer goods, automotive, construction, and furniture industries (Araújo et al. 2021; Philip et al. 2007). Polysaccharides, such as alginate, carrageenans, and laminarin, are extracellular or cell wall components of seaweeds. These biopolymers exhibit diverse properties, such as gel-forming, thickening, stabilizing, or other functional properties, which allow their incorporation in diverse applications, including food, pharmaceuticals, cosmetics, and biomedical applications (Campo et al. 2009; Ma et al. 2020).

Biopolymer production via algae offers a potentially sustainable alternative route, bypassing the aforementioned limitations. This third-generation platform offers various advantages over the first two generations: (1) higher biomass productivity, (2) superior renewability and sustainability levels, (3) lesser requirement for arable land or scarce resources, (4) usage of waste biomass as raw material, (5) less seasonal, climatic, or geographical limitations, (6) faster growth cycles, (7) smaller carbon footprint, (8) ease of modification (e.g., genetic engineering) to generate enhanced products, and (9) no competition with food or feed (Awad et al. 2020; Younes et al. 2020). To that end, algal-derived biopolymers offer a potentially sustainable solution to preeminent global crises: plastic pollution and CO_2 emissions. However, commercial production of algal biopolymers is still in its infancy, with only few successful examples from seaweeds, such as alginate and carrageenans. The major challenges in algal biopolymer production are the high operation costs, cultivation/harvesting of microalgal biomass, low productivity titers, and complex downstream processes (extraction, purification, and modification of the biopolymers) (Dragone et al. 2010). The development of scalable and cost-effective production processes for algal biopolymers, such as PHB, polyols and polyurethanes, PAN, and laminarin, is a critical challenge that needs to be addressed to unlock their full potential.

In this chapter, the prominent algae-derived biopolymers are explored, together with their production processes, sustainability, as well as their vast commercial applications that span across various industries. The sections include, in the following order, polyhydroxy biopolymers, polyols and polyurethanes (PU), polyacrylonitrile (PAN), alginate, carrageenans, and laminarin. A final paragraph showcases challenges and future directions of algae-based biopolymers.

2 Polyhydroxy Biopolymers

In the last decades, research into biopolymers such as polyhydroxyalkanoates (PHA) and polyhydroxybutyrate (PHB) has been growing in popularity due to their sustainable production, biodegradability, and comparable mechanical and physical properties to petroleum-derived plastics (Dang et al. 2022; Devadas et al. 2021). PHAs are linear polyesters composed of repeating units of hydroxyalkanoates. They can be divided into three groups depending on the carbon atoms in the backbone of the biopolymer: short chain ($\leq$5 carbon atoms), medium chain (6 $\leq$ 14 carbon atoms), and long chain ($\geq$15 carbon atoms) (Raza et al. 2018). Out of the 150 different types of biopolymers from the PHA family, PHB is the preeminent member (Fig. 1). It is an aliphatic polyester consisting of repeating monomeric units of β-hydroxybutyric acid (3HB), which consists of four carbon atoms with one chiral center (Devadas et al. 2021; Raza et al. 2018). PHB-derived plastics are readily biodegradable, whereas synthetic plastics could last thousands of years (Lamberti et al. 2020). Furthermore, due to their sustainable production processes, each ton of synthetic plastics replaced with bio-derived ones would save around 1.8 tons in CO_2 emissions (Nanda and Bharadvaja 2022). Bioplastics have thus received vast scientific, public, and political attention as a potential path towards a sustainable green bio-economy.

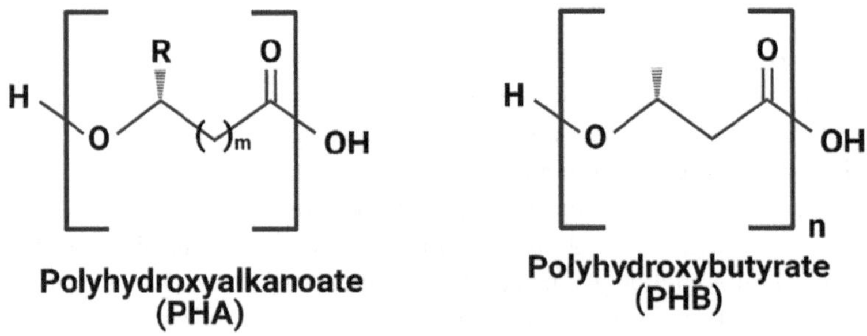

Fig. 1 Chemical structure of polyhydroxyalkanoate (PHA) and polyhydroxybutyrate (PHB), R refers to alkyl functional groups

2.1 Physiochemical Properties

The physiochemical properties of PHB render it a candidate and suitable alternative to conventional petroleum-based plastics. PHB possesses striking similarities or even better material characteristics (e.g., thermal properties, UV resistance) to petroleum-based plastics, such as PP and PS (Markl et al. 2018a; Markl et al. 2018b). This biopolymer is characterized by its favorable thermoplastic processability, hydrophobicity, optical purity, and excellent gas barrier characteristics. A comparison between the physiochemical properties of PHB and PP is presented in Table 1. A major advantage of PHB is its ability to be biodegraded under home-composting conditions, unlike other bioplastics which necessitate industrial decomposition processes, such as polylactic acid (PLA, >50 °C) (Nanda and Bharadvaja 2022; Saratale et al. 2020).

However, PHB-derived biopolymers suffer from several drawbacks, including low elongation at break (EAB) and a high melting point of 175 °C (near degradation temperature) (Nanda and Bharadvaja 2022; Saratale et al. 2020). In addition, the 3HB molecules within the polymeric chain are almost entirely in the R-configuration. This is due to the way microorganisms synthesize the polymer, as the enzymes involved in the polymerization process selectively incorporate the R-3HB enantiomers into the growing chain. The regular packing of the R-3HB units within the chain allows for strong intermolecular interactions, resulting in a tightly ordered and rigid structure. This high stereoregularity leads to high crystallinity, stiffness, and brittleness of the resulting polymer (Madison and Huisman 1999; Younes 2021). Process development and metabolic engineering approaches have recently been employed to generate PHB or copolymers of PHB with improved physiochemical properties. The recombinant expression of various enzymes allowed for the production of Poly-3-Hydroxybutyrate-Co-3-Hydroxyvalerate (P(3HB-co-3HV))—tradename Biopol™—directly from glycerol, bypassing the need for

Table 1 Physiochemical properties of PHB, PP, and other copolymers blends (Balaji et al. 2013; Markl et al. 2018a; Rudnik 2013)

Properties	PHB	PP	PHBV	PHB4B	PHBHx
Melting point °C	175	176	145	150	127
Crystallinity %	60–80	50–70	56	45	34
Average Molecular weight (Da)	2.4×10^5	2×10^5	3.7×10^5	–	8×10^5
Glass transition temperature °C	4	−10	−1	−7	−1
Density (g cm^{-3})	1.250	0.905	1.250	–	–
Flexural modulus (GPa)	3.5	1.7	2.9	–	–
Tensile strength (MPa)	21	38	26	26	21
Elongation at break %	6	400	50	444	850
UV resistance	good	poor	–	–	–
Solvent resistance	poor	good	–	–	–

* PHBV: Poly(3-hydroxybutyrate-co-3-hydroxyvalerate); PHB4B: Poly(3-hydroxybutyrateeco-4-hydroxybutyrate); PHBHx: Poly(3-hydroxyl-hexanoate).

feeding propionate, used in the traditional production process (Aldor and Keasling 2003). Additionally, the limitations of PHB can be overcome by incorporating (biologically or chemically) other PHA-variants, such as polyhydroxyvalerate (PHV) to form blends or copolymers (e.g., PHBV). These copolymers exhibit enhanced thermal and mechanical properties superior to synthetic polymers, such as PP (Table 1) (Devadas et al. 2021; Poltronieri and Kumar 2017).

Blends of synthetic and bio-derived polymers are also actively pursued (Devadas et al. 2021; Onen Cinar et al. 2020). By incorporating polyvinyl acetate (PVA), polypropylene (PP), and polyethylene (PE) with microalgae polymers, the physiochemical properties of the resulting plastic blend can be improved. Furthermore, these polymer blends can be optimized by additives, such as compatibilizers that confer enhanced properties to the individual components (Devadas et al. 2021; Onen Cinar et al. 2020). Examples include diethyl succinate, poly(ethyleneco-glycidyl) methacryloyl carbamate, and maleic anhydride. By binding to both the biopolymer and the synthetic polymer, these compounds can increase the consolidation and improve the mechanical and physiochemical properties of the blend (Devadas et al. 2021; Nanda and Bharadvaja 2022). Moreover, plasticizers as additives, such as octanoic acid, sorbitol, polyethylene glycol (PEG), 1,4-butanediol, and glycerol, can increase the molecular mobility of the polymer chains and reduce the secondary interaction, i.e., hydrogen bonding. Plasticizers render the biopolymer more flexible and malleable, increasing its stretch-ability, processability, biodegradability, and thermoplasticity (Devadas et al. 2021; Nanda and Bharadvaja 2022).

2.2 *Production*

2.2.1 Metabolic Route

Inherent PHB production is well investigated in bacteria like *Alcaligens eutrophus, Azotobacter beijernickii, Pseudomonas, Micrococcus, Bacillus, Rhodococcus, Ralstonia eutropha,* and *Rhdodococcus* sp. as an adaptive response to excess carbon and nutrient-limited stress conditions (Nanda and Bharadvaja 2022). The biosynthetic pathway, depicted in Fig. 2, has been extensively studied. The classical route is catalyzed by three enzymes: (1) acetoacetyl-CoA thiolase (phbA), which catalyzes the condensation of two acetyl-CoA molecules into acetoacetyl-CoA, (2) NADPH-dependent acetoacetyl-CoA reductase (phbB), which reduces acetoacetyl-CoA to (R)-3-hydroxybutyrate-CoA, and (3) PHB synthase (phbC), that polymerizes the (R)-3-hydroxybutyryl-CoA units into PHB (Jarmander et al. 2015; Younes et al. 2021). This polymerase is highly stereospecific that it almost exclusively accepts (R)-3HB monomers as its substrate, resulting in adverse physiochemical properties. Despite an 85% (g g^{-1}) accumulation of PHB in these species, the main disadvantage of adopting heterotrophic bacteria for bioplastic production is the costly fermentative process (Singh Saharan et al. 2014). This is mainly associated with the

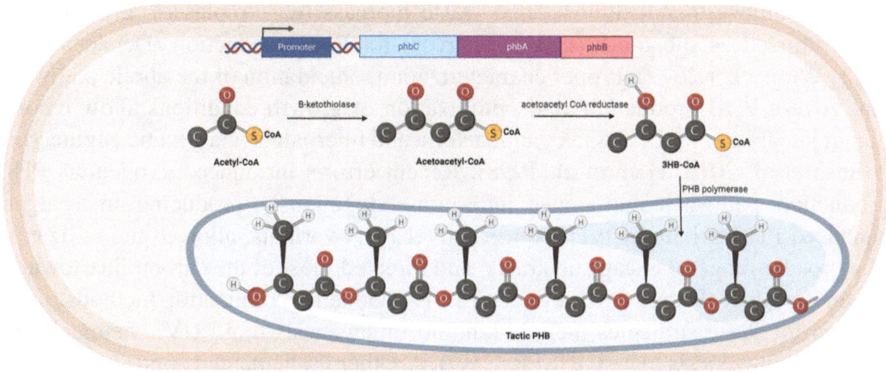

Fig. 2 Illustration of the general PHB metabolic pathway in natural producers. *phbA* acetoacetyl-CoA thiolase, *phbB* acetoacetyl-CoA reductase, *phbC* PHB polymerase, *3HB* β-hydroxybutyric acid, *PHB* polyhydroxybutyrate, *CoA* Coenzyme A

carbon source that accounts for about 48% of the total costs, as well as external aeration and temperature control. Even with the use of waste biomass as carbon sources, a large economic gap remains between synthetic and bioplastic (Afreen et al. 2021; Hempel et al. 2011).

Phototrophic microorganisms, such as microalgae and seaweeds, offer a potentially competitive platform for bioplastic production. The high photosynthetic efficiency, which allows them to convert inorganic nutrients into organic molecules using light as energy source, advances algae as a promising and viable platform for a variety of bioproducts, including bioplastics (Devadas et al. 2021; Nanda and Bharadvaja 2022). In these cells, the organic reserves created by photosynthesis can then be converted to PHB via endogenous or recombinant pathways. The blue-green algae (cyanobacteria) present a preferred host system for algal-PHB production. Strains including *Chlorogloea fritschii*, *Synechocystis* sp., *Arthrospira* sp., *Neochloris oleoabundans*, *Nanocloropsis gaditana*, *Phaeodactylum tricornutum*, *Aphanothece* sp., and *Synechococcus* sp. are well-suited PHB expression systems, owing to their genetic accessibility to the absence of cell differentiation present in higher plants. (Hempel et al. 2011; Nanda and Bharadvaja 2022). Studies showed that certain cyanobacterial strains produce comparable PHB yields from only one-tenth of the bacterial carbon utilization uptake (Sharma and Mallick 2005). Algal-derived PHB would bypass the need for carbon supply, reducing the total cost of this production process.

2.2.2 Optimization Strategies

To achieve economies of scale, research is currently specifically focused on improving product yield, lowering cultivation costs, and upscaling. Cheap renewable (e.g., starch, cellulose, hemicellulose) and waste biomass (e.g., molasses, glycerol) are being utilized as substrates (carbon sources) for PHB production (Devadas et al. 2021; Winnacker 2019). Proper characterization, elucidation of metabolic pathways that govern PHB production, and optimization of growth conditions allow recombinant integration processes in cyanobacteria and microalgae via genetic engineering (Ghosh et al. 2016; Hein et al. 1998). Recent efforts introduced exogenous PHB production pathways and genes into non-endogenously producing microalgae, enhanced PHB productivity, produced novel PHA-variants, allowed the utilization of a broader range of cheap substrates, and directed most of the carbon flux towards the end-product (Kamravamanesh et al. 2018). Genetic modification methods incorporated for these strategies include random mutagenesis (e.g., UV, chemical) and targeted mutagenesis (e.g., CRISPR-Cas9). Another implemented approach involves the use of recycled synthetic polymers (Section 2.1) as additives in the production of algal bioplastic-composites (Nanda and Bharadvaja 2022). This approach would theoretically reduce GHG emissions associated with synthetic plastics by 75% and support the economic acceptance of bioplastics (Posen et al. 2017).

Upscaling microalgae cultivation from lab to commercial scale presents challenges in accurately assessing the variables and limitations. It is widely reported that large-scale cultivation typically generates lower relative yield compared to laboratory experiments, as the theoretical biological potential of microalgae does not meet the achieved productivities (Devadas et al. 2021). Light modulation, temperature, mixing, nutrient provision, contamination, biofilm formation (fouling), and operating costs are major disadvantages for photobioreactor (PBR) scale-up. However, open ponds cultures suffer from low cell densities and productivities, high contamination risk, and difficulties to control culture conditions (Onen Cinar et al. 2020). Considering that cultivation and harvesting operations account for around 62–72% of the total algal-PHB production costs, an open system (ponds) remains favored over PBR cultivation on a commercial scale (Devadas et al. 2021; Onen Cinar et al. 2020; Soroudi and Jakubowicz 2013).

These aforementioned optimization strategies have yet to allow the widespread commercial implementation of algal-based bioplastics. Despite the combined approach of genetic manipulation, optimized cultivation conditions, and calculated upscaling, algal PHB content only reached a peak of 68% ($g\ g^{-1}$), compared to 85% ($g\ g^{-1}$) in bacteria (Devadas et al. 2021; Nanda and Bharadvaja 2022). Furthermore, microalgae-derived plastics remain relatively costly, at around 2.7–5.3 USD per Kg of PHB, compared to 1.3 USD per Kg for PP (Reichert et al. 2020).

2.2.3 Recycling

Recycling of bioplastics feeds into subsequent biopolymer production as a carbon source, thus reducing the overall cost, improving process efficiency, and upholding a cyclic bio-economy approach. PHB is degraded by the environment's microbiome into CO_2, water, and compost within a period of 3 months (Kliem et al. 2020). However, the incorporation of additives, blends, and copolymers aimed at improving the material's physiochemical properties results in low- or nonbiodegradable bioplastic. Without sustainable recycling routes, these materials would pose similar environmental threats as petroleum-derived plastics (Devadas et al. 2021). Therefore, mechanical, chemical, and biological recycling routes are employed for these polymeric materials. Mechanical recycling involves screw extrusion, injection molding, and compression molding to remove contamination. However, the resulting product often suffers from reduced tensile strength, preventing its reshaping or remolding. Therefore, it is typically redirected as a carbon source for fermentative processes, or blended with other bioplastic material (Wojnowska-Baryła et al. 2020). Chemical recycling, such as depolymerization, pyrolysis, thermal cracking, and gasification, breaks down polymers into their monomeric units. These monomers can be re-polymerized to generate the original bioplastic or can be directed for other applications, e.g., the monomer 3HB as a precursor in the synthesis of antibiotics, vitamins, and valuable compounds. Biological or enzymatic recycling, the "greenest" of all three routes, includes composting or fermentative conversion of the polymeric materials into monomers and biogas (Devadas et al. 2021; Wojnowska-Baryła et al. 2020).

2.2.4 Market and Applications

To date, bioplastics industry remains a nascent sector as it accounts for less than 1% of the global plastic production estimated as 368 million tons. Of an annual production capacity of 2.11 million tons, only 55.5% is derived from renewable biodegradable feedstocks (Dang et al. 2022). However, at a cumulative annual growth rate (CAGR) of 22.7%, the global market for bioplastics is predicted to jump from 10.7 billion USD in 2021 to 29.7 billion USD by 2026. PHB market value alone is estimated to reach 284 million USD in the same timeframe (Nanda and Bharadvaja 2022). This growth is largely attributed to the increasing demand for eco-friendly, biodegradable products, as well as the rising awareness of the environmental impact of plastic waste (Devadas et al. 2021). Additional push comes from the European Union (EU), US Food and Drug Administration (FDA), US Environmental Protection Agency (EPA), and other agencies on businesses to use eco-friendly packaging materials in lieu of SUPs in the pharmaceutical, medical, and food sectors.

Besides offering an alternative to plastic containers, packaging materials, and other traditional plastic products, PHB is widely used for a range of applications. In the medical field, it is used for sutures and implants due to its biocompatibility and ability to degrade over a short period of time in the body (Ray and Kalia 2017). PHB bioplastics are also used in the construction industry, for insulation and soundproofing materials, as well as in automotive applications, such as engine covers and interior parts (Balaji et al. 2013; Dahiya et al. 2020; Dang et al. 2022).

3 Polyols and Polyurethanes

When Otto Bayer and his coworkers at IG-Farben sought out a replacement for nylon, owned by the American company DuPont, they discovered and patented the polyisocyanate-polyaddition reaction in 1937, yielding polyurethanes for the first time (Bayer 1947). Polyurethane (PU) is a class of synthetic polymers, generally comprised of a polyol component—a molecule possessing at least two reactive hydroxyl groups—and an organic (poly-)isocyanate linked by a carbamate or polyurethane bond (Fig. 3).

Although deemed useless at first due to shortages of its raw materials, a myriad of applications for the novel organic polymer within the automotive-, aerospace-, adhesives-, construction-, furniture-, textile-, and varnish industries were soon to be found, especially in its nowadays' most commonly propagated form: rigid or flexible foam (GrandViewResearch 2022; IMARC 2022). Although leading industry companies utilize bio-based polyols as chain extenders or cross-linking agents, these account for a very low percentage within the formulation, rendering the resulting PU not as "green" as advertised (Datta et al. 2017; Sardon et al. 2021). To that end, microalgae oil-based polyols present a progressive approach on the sustainability front.

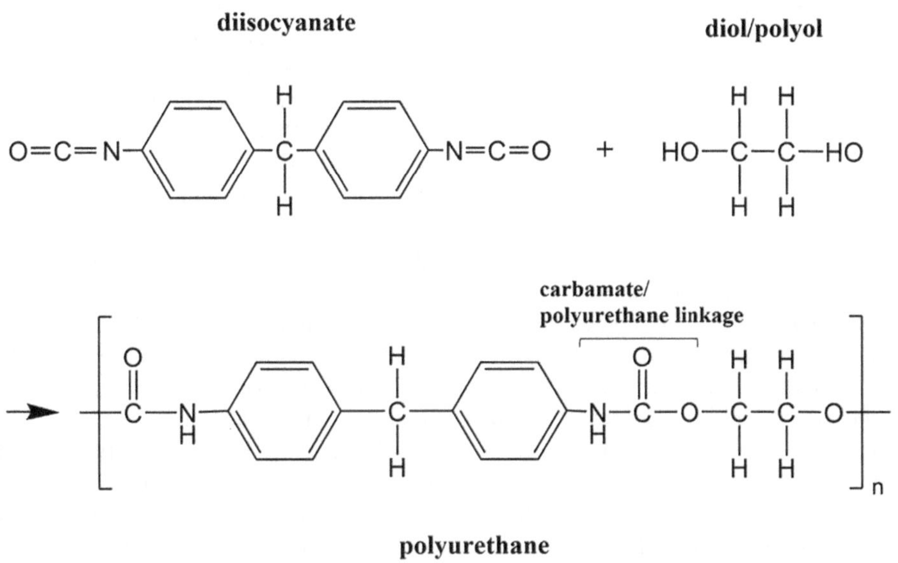

Fig. 3 Schematic reaction mechanism for the synthesis of polyurethane. A polyol, an alcohol with at least two reactive hydroxyl groups, and a diisocyanate or polymeric isocyanate react to produce polyurethane under the presence of suitable catalysts. Additives, e.g., short polyols as chain extenders or blowing agents, e.g., water to obtain foams, could be added to manipulate the properties of the end-products

3.1 *Physiochemical Properties*

The hydroxyl-rich polyol chains of polyurethanes offer high structural flexibility, forging the soft segments, while polyisocyanates and chain extenders, known as hard segments, provide rigidity and durability (Korley et al. 2006). Due to the interchangeability and diversity of polyols combined with isocyanate chemistry, PU versatility is unmatched, which enables tailoring of specific end-use requirements (Cornwall 2021; Gama et al. 2018; Szycher 2012). Commonly deployed polyols exhibit molecular weights in the range of 500–4000 Da, where low MW short chain diols generate in hard plastics, whereas long-chain/high MW diols generate in flexible plastics (Szycher 2012). The extent of intermolecular cross-linking and the overall rigidity of PU is governed by the hydroxyl value, expressed as amount of potassium hydroxide (KOH) required to neutralize one gram of a given acetylated chemical substance with free hydroxyl groups (Akindoyo et al. 2016). Commercial polyols range between 35 and 900 mg KOH g^{-1}, depending on their purpose (BASF 2017).

Although fully polymerized end-products are chemically inert and thus considered harmless, isocyanates are known skin irritants and allergens (Bello et al. 2007). In addition, extreme caution and safety measures during the PU manufacturing process are mandatory due to the evaporation of toluene diisocyanate (TDI) at room temperature (Butcher et al. 1976). Another safety concern involves the highly flammable nature of PU foams, which releases dense, toxic smoke of carbon monoxide, and hydrogen cyanide during rapid combustion (McKenna and Hull 2016).

3.2 *Production*

3.2.1 Conventional Chemical Production

In the chemical synthesis of PUs, isocyanate reacts to gaseous CO_2 and an amine. The former acts as a blowing agent to give rise to foams, while the latter reacts with isocyanate again, resulting in urea groups. This reaction is driven by a catalyst or by exposure to UV in the presence of small amounts of water. To catalyze the reaction between a hydroxyl group and an isocyanate, tertiary amines, organometallic catalysts (predominantly tin compounds), quaternary ammonium salts, and alkali metal carboxylates are most commonly deployed (Furtwengler and Avérous 2018; Sardon et al. 2015). Depending on the fabrication process, duroplasts, elastomers, and thermoplasts can be generated. The melting temperature directly correlates with cross-linking and chain length of the desired PU. If thermoplast material is produced, temperatures of up to 50 °C are required, sufficient for primary reactions of diols and diisocyanates with active hydrogen compounds. However, temperatures of up to

150 °C lead to higher cross-linking and branching by means of secondary reactions (Szycher 2012).

3.2.2 Bio-Based Feedstock and Production

Depending on the raw materials, diverse types of bio-based polyols can be produced. While lipids are mainly explored as starting material for polyester polyols, carbohydrates are primarily utilized for polyether polyol synthesis (Sardon et al. 2021). Bio-derived oils as well as biomass rich in amino acids and carbohydrates can be converted to polyols and subsequently to PU materials (Table 2).

3.2.3 Lipid-Based Polyols

Plant oils, such as castor oil (Ionescu et al. 2016), linseed oil (Calvo-Correas et al. 2015), tung oil (Caideng et al. 2014), soybean oil (Alagi et al. 2018; Xia et al. 2012), sunflower oil (Asare et al. 2022), hemp-, rapeseed- and oilseed radish oil (Polaczek and Kuranska 2022), palm oil (Pawlik and Prociak 2011), corn oil (Ramanujam et al. 2019), canola oil (Kong et al. 2012) as well as cottonseed and karanja oil (Gaikwad et al. 2015), have been the focus of most bio-based feedstock sources for polyol production. These oils are mainly comprised of triglycerides—glycerol moieties esterified with three fatty acids—which differ in structure and composition. The oil's properties vary depending on plant, crop, season, growing conditions, year, and geographical location of crops (Canvin 1965; Kostik et al. 2013; Maisonneueve et al. 2016; Msaada et al. 2009; Samancı and Özkaynak 2003). As a rule of thumb, the more unsaturated the fatty acid is (= lower iodine value), i.e., the more epoxy moieties per molecule (oxirane value) are present, the better an epoxy is suited as plasticizer, lubricant, or polyol precursor (Carlson and Chang 1985; Laurens et al. 2017).

With the exception of castor oil, which contains up to 90% of ricin oleic acid, most vegetable oils do not possess the necessary hydroxyl groups to function as polyols (Maisonneuve et al. 2015). Therefore, chemical modification of the ester groups or the double bonds of unsaturated fatty acid chains should precede polyurethane synthesis. Figure 4 depicts the most common conversion methods, epoxidation (Goud et al. 2006), followed by oxirane ring opening (Hazmi et al. 2013; Pang et al. 2006); but other methods such as oxidation (Petrović et al. 2002), ozonolysis (Petrovic et al. 2005), hydroformylation (Kandanarachchi et al. 2002), olefin metathesis (Chikkali and Mecking 2012), click chemistry (Lligadas et al. 2013), and transesterification (Freedman et al. 1984) have been explored (Laurens et al. 2017; Pawlik and Prociak 2011; Peyrton and Avérous 2021; Singh et al. 2019).

In a similar yet more sustainable approach, these oils can be sourced from algae that do not compete with food sources over arable land and freshwater, while fixating atmospheric carbon with minimal nutrient supply (Fabris et al. 2020; Zappi et al. 2019). These microbial factories contain longer chain fatty acids compared to those

Table 2 Summary of the latest state of the art production routes of polyurethane (PU) materials based on renewable algae- oil or biomass

Organisms	Raw materials	Polyol conversion methods	Products	References
Chlorella sp.	Oil	Ozonolysis/epoxidation/hydroformylation	Rigid foam	Petrović et al. (2013)
Chlorella sp.	Oil	Oxidation with H_2O_2	Foam	Pawar et al. (2015)
Chlorella vulgaris	Oil/palm oil mixture (50:50)	Epoxidation and H_2O_2-guided ring opening	Rigid foam	Chavarro Gomez et al. (2020)
Chlorella sp.	Oil	Chemical conversion	Coating	Patil et al. (2021)
Chlorella sp.	Defatted biomass (10–70%): PEG200	Grinding	Composite film	Saha et al. (2020)
Scenedesmus sp.	Amino acids from biomass	Reaction with 1,2-diaminoethane and ethylene carbonate	Rigid foam	Kumar et al. (2014)
Scenedesmus sp.	Amino acids from biomass	Reaction with ethylene diamine and ethylene carbonate	Coating	Noreen et al. (2022)
Nannochloropsis salina	Oil (3-omega fatty acid production waste stream)	Ozonolysis, condensation with renewable propane diol	Rigid foam/shoe soles	Phung Hai et al. (2021)
Schizochytrium sp.	Oil	Epoxidation and methanol-guided ring opening	PUIR foam	Arbenz et al. (2017)
Schizochytrium sp.	Oil:petrochemical polyols (25:75)	Epoxidation and ring opening guided by acetic acid/diethylamine/ethanol/hydrochloric acid	Foam	Peyrton et al. (2020)
Enteromorpha sp.: *Zostera marina* (10:90)	Dried biomass:petrochemical polyols (up to 30:70)	Solvothermal liquefaction with glycerol:polyethylene glycol (50:50)	Rigid foam	Kosmela et al. (2019)
Enteromorpha sp., *Zostera marina*, and *Chlorella vulgaris*	Biomass:bio-based polyol (up to 15:85)	Solvothermal liquefaction of *Enteromorpha* sp./*Z. marina* (10:90) mixture, grinding of *C. vulgaris*	Thermoplastic elastomer	Glowinska et al. (2023)
Chaetomorpha linum	Dried biomass, activated carbon	Ball milling	Film	Marlina et al. (2020)

(continued)

Table 2 (continued)

Organisms	Raw materials	Polyol conversion methods	Products	References
Gracilaria verrucosa	Dried biomass:castor oil (2:5)	Grinding	Film	Nurman et al. (2021)
N/A	Na-Alginate: PEG:glycerol:1,4-butanediol	N/A	Composite foam	Kim et al. (2012) Kwon et al. (2007)

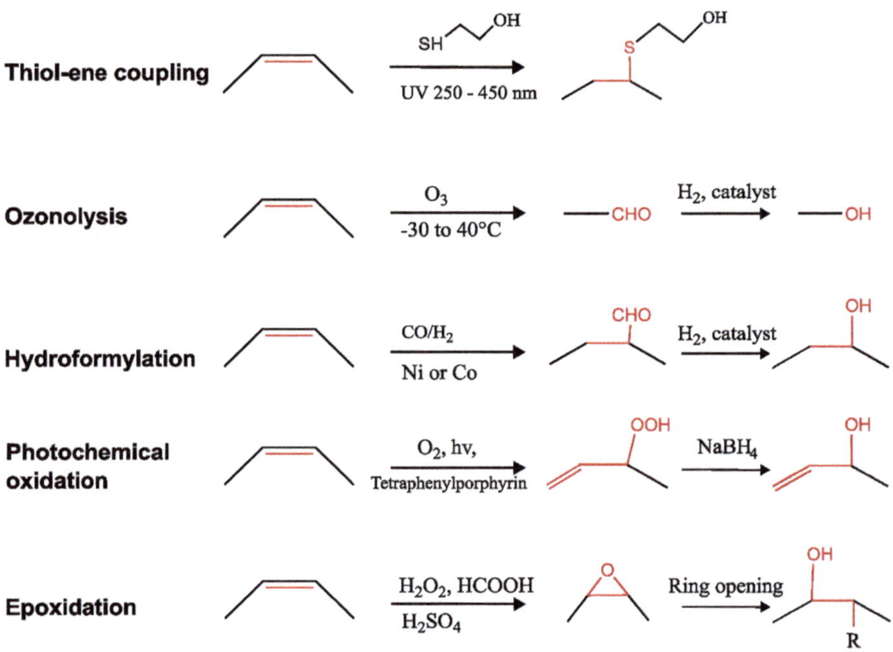

Fig. 4 Reaction mechanisms for the polyol synthesis from plant oils through chemical modification of the ester groups. Adopted from the open access reference (Gadhave et al. 2017)

of vegetable oils, with lower saturation levels, resulting in polyols with OH values of 300 mg per KOH g^{-1} and above (Peyrton et al. 2020). Several algae species have been adopted for the production of bio-based polyols as a component in polyurethane coatings, foams, or composites, most prevalently from the genus of *Chlorella,* but also from *Chaetomorpha, Gracilaria, Nannochloropsis, Scenedesmus, Schizochyztrium,* and *Ulva*; formerly known as *Enteromorpha* according to AlgaeBase. The chemical and structural composition of the oils obtained from algae also varies widely among these species, which inherently necessitates the development of innovative methods of conversion to polyols and subsequently

PUs (Laurens et al. 2017). These methods can be categorized into direct valorization, chemical conversion, and sustainable conversion.

1. Direct valorization: This method accommodates unrefined microalgae oil that is neither refined, bleached, nor deodorized (Petrović et al. 2013). Similar to unrefined plant oils, interfering impurities are present, apparent by the dark green color. The general feasibility and usefulness of the products for PUs could be demonstrated by the example of established reactions in plant oil polyol conversion: (a) Ozonolysis improved the color but was unsuitable for PU applications due to low OH numbers. (b) Epoxidation by peracetic acid- and methanol-guided oxirane ring opening yielded lighter-colored polyol with 150 mg KOH g^{-1}, an OH value similar to that of castor oil. The addition of water and diphenylmethane diisocyanate produced rigid foams with imperfect cell structure, but generally comparable characteristics to petrochemical-based PU foams. (c) Hydroformylation leads to an equally functionalized polyol with 147 mg KOH g^{-1}, albeit black color and higher reactivity, complicating the casting process. This was reflected in PU foams of lower quality compared to the epoxidation/ring opening procedure. For all three examined routes, the inherent impurities required higher catalyst amounts and resulted in lower conversion rates compared to plant oils, clearly indicating the benefits of refined algae oil.
2. Chemical conversion: Patil et al. (2021) explored a chemical two-step conversion route of microalgae oil to polyols, incorporating them into an anticorrosive and antimicrobial PU coating formulation. First, the algae oil was reacted with diethanolamine in presence of sodium methoxide under N_2 atmosphere to produce fatty amide. Second, polyetheramide-polyols were afforded by reacting with either bisphenol-A, 1,4-butanediol, or isosorbide. In a final step, PU coatings were prepared through a reaction with diphenylmethane diisocyanate and cast onto microscope slides. Due to the hydrophobic characteristics of the fatty acid chains, both water resistance and antimicrobial properties were enhanced. Interestingly, in saline aqueous media, the algal oil-based PU coatings outperformed typical poly(tetramethylene ether)glycol-based PU coatings with respect to their corrosion-protective capabilities, protection from an organic solvent (xylene). and an alkaline aqueous solution (5% NaOH).
3. Sustainable conversion: A research group at the University of San Diego developed a method to extract palmitoleic acid (16:1), a waste stream from *Nannochloropsis salina*-based 3-omega fatty acid production and subsequently converted it into azelaic acid (AA) and heptanoic acid via ozonolysis (Phung Hai et al. 2021). This AA was condensed with renewable propane diol (PDO) to form AA-polyester polyols. The same research group developed another method to convert diacids, including AA, through flow chemistry into diisocyanates via Curtius rearrangement. This strategy bypasses the need to isolate explosive acyl azide intermediates, which mitigates safety concerns, especially favorable for process scale up. Applying the aforementioned processes, the renewable AA-polyester polyols and heptametethylene diisocyanates were polycondensated to produce algae-based PU flip-flops.

3.2.4 Amine-Based Polyols

With emphasis on carbon capture and utilization (CCU) of algae, high lipid producers have garnered much interest for their potential as biofuel factories. Kumar et al. (2014) discovered a means to exploit amino acids, often disregarded as side products, for polyol synthesis. Through flash hydrolysis, a method relying on the interplay of subcritical water medium at 280 °C and the different kinetics of polymeric algae biomass contents, lipids and amino acids were separated within 10 seconds of residence time (Garcia-Moscoso et al. 2013). Subsequently filtrated and freeze-dried, the amino acids and peptides were hydrolyzed by HCl and neutralized by NaOH. The hydrolysate was subsequently converted to polyols in a two sequential step process; (1) The addition of ethylene diamine led to amido-amine-terminated intermediate. Excess diamine was removed by distillation. (2) The intermediate was then reacted with ethylene carbonate to yield the hydroxyl-terminated urethane-polyol. Up to 5% of this algae protein-based polyol was blended with plant-based polyols, surfactants, and water. Admixture of isocyanates resulted in self-catalytic rigid foam formation with comparable core density, resiliency, tensile strength, and dry compression set. Although this PU exhibited lesser tensile strength compared to conventional polyurethane foams, higher fire retardancy, highly advantageous in insulating material, was reported.

3.2.5 Cellulose-Based Polyols

Although lipid and amino acid-rich microalgae have been the focal point of polyol and polyurethane-related scientific investigations, seaweeds such as *Enteromorpha* sp., *Chaetomorpha linum*, and *Gracilaria verrucosa* could also be employed. The red alga *Gracilaria* Greville, with inherently elevated carbohydrate contents could also be exploited for polyol production. Two methods of conversion are presented here:

1. Solvothermal liquefaction/acidic hydrolysis: Kosmela et al. (2019) subjected a mixture of macroalgae and seagrass biomass at a 1:9 ratio to a solvothermal liquefaction method, which yielded highly functionalized biopolyols with a hydroxyl number of 650 mg KOH g^{-1}. A 30/70 mixture of these algal polyols and petrochemical polyols, respectively, were subsequently set for rigid PU foam formulation, which resulted in reduced foam rising time and increased glass transition point and compressive strength. In addition, high reaction temperatures improved the reactivity of the biopolyol and the efficiency of the process. Spectroscopic analysis confirmed the presence of PU and isocyanurate rings. The latter contributes to the chemical and thermal stability of the foam.
2. Mechanical comminution: A ball-milling method has been tested on *Chaetomorpha linum,* a hemicellulose-rich alga native to NH_3-N excess aquacultures, for the production of a PU film. To that end, polycondensation of the fine particles dissolved in dioxane with methylene diphenyl diisocyanate (MDI) with

activated carbon as filler was applied without a catalyst. The solution was poured onto a glass plate and dried at room temperature to form the PU film, with promising properties as an immobilizing activated carbon filler matrix (Marlina et al. 2020). The film exhibited a 393.43 μg g^{-1} binding capacity to NH_3-N, a valuable attribute for the decontamination of fishing farms. A phase inversion method was similarly tested on seagrass biomass with toluene diisocyanate to synthesize PU membranes (Nurman et al. 2021). Albeit purely algal biomass-based membranes were brittle, dry, and mechanically unstable, supplementing the mixture with glycerol and castor oil yielded elastic and sturdy membranes with relatively high initial degradation values of 290 °C.

3.3 Recycling

As stated previously, only 9% of all produced plastic was recycled, while 12% was incinerated and 79% was dumped into landfills or natural environments in 2015 (Geyer et al. 2017; Palm and Svensson Myrin 2018). Given that most PUs products are derived from petrochemical building blocks (Furtwengler and Avérous 2018), sustainability and biodegradability remain unattained in this sector (Cozar et al. 2014; Palm and Svensson Myrin 2018). On top of that, PU generates significant amounts of waste during its production and molding, of which only 29.7% is recycled, 39.5% is recovered through energy recovery processes, and 30.8% is deposited in landfills (Gadhave et al. 2019). The slow decay of PU waste and their poor disposal management pose a severe environmental concern (Association of Plastic 2020). Although incineration or energy recovery is widely regarded as the most effective and economical approach currently applied (Zia et al. 2007), a number of other physical, chemical, and biological recycling strategies with higher long-term potential are developed.

3.3.1 Physical Recycling

Mechanical recycling involves shredding of PU waste or waste material from its production process to smaller particle sizes. Depending on the source material, PU can be ground to a fine powder and directly reused for foam production or indirectly utilized as filler material. The most commonly deployed physical recycling method is re-bonding, which entails adding 5–10% adhesive components (e.g., diphenylmethane isocyanate) to PU flakes and compressing the mixture under hot steam. The recycled PU granulates exhibit divergent properties compared to the original material. For example, a higher density renders them a new class of polymers that exhibit desirable specific characteristics during the molding process (Zia et al. 2007). Hot compression molding is another means to reprocess specific types of PU waste, such as reaction injection molded (RIM) polyurethane. To that end, high-quality recycled material for the automotive industry can be obtained through a combination of high pressure, heat, and shear forces (Hulme and Goodhead 2003).

3.3.2 Chemical Recycling

Chemical recycling describes the depolymerization of high molecular weight PU to retrieve smaller oligomers or monomers, ideally allowing for the production of new PU with identical or comparable properties (Yang et al. 2012). Hydrolysis of PU foam with water under high temperatures (superheated steam) is one of the best understood chemical recycling methods (Campbell and Meluch 1975). Both polyols and diamines are breakdown products of isocyanates, which can be retrieved as building blocks for new PU materials. Alternatively, glycolysis denominates the depolymerization of PU waste at high temperatures (above 200 °C) guided by diols, most prominently 1,3-ethylene glycol, due to relatively low product viscosity (Borda et al. 2000). Diethanolamine is commonly used as a co-reactant in this method. Reaction time can be reduced by the deployment of catalysts, alkaline pH conditions, and optimization of diol to PU waste ratios (Bielenia 2021). A split-phase-glycolysis method has been developed for recycling of rigid and flexible foam scrap (preferably MDI-based). First, PU reacts with diethylene glycol over several hours until flexible polyols accumulate in the upper layer, while the heavier phase is transformed to rigid polyol with propene oxide (Scheirs 1998). Accordingly, a catalyst-independent reaction with diethanolamine affords split-phase-alcoholysis, with the apolar polyether polyols in an upper layer and a lower polar phase containing the alcoholizing agent and aromatic side products (Kanaya and Takahashi 1994). Generally, glycolysis is more suited for recycling of PU foam production waste rather than postconsumer waste (Zia et al. 2007).

3.3.3 Biological Recycling

Biodegradation of PU affords higher sustainability standing than its counterparts, all while bypassing the use of toxic and flammable compounds. Fully sustainable PU foams developed by Phung Hai et al. (2021) (Section 3.2.3) were subjected to degradation analysis in a follow-up study. A total mass loss of 70% of microalgae oil-based PU foam during 12 weeks was reported (Gunawan et al. 2020; Gunawan et al. 2022). An ITS and 16S metagenomics sequencing approach revealed an uncharacterized fungus and most notably bacteria of the genera *Pigmentiphaga*, *Roseomonas,* and *Phenylobacterium* as main decomposers. Supported by modeling experiments with commercial enzymes and GC/MS data, biodegradation was mainly accomplished by esterases, depolymerizing the PU into its constitutive diols, diacids, and other PU fragments. Several organisms were identified at six distinct marine sites around San Diego, which were able to depolymerize, uptake, and metabolize bio-based PU as sole carbon source.

3.4 Market and Applications

Sitting at the 6th rank most consumed plastic worldwide with a 7% global market share in 2020, 27 million tons of polyurethanes are produced annually. Projections estimate that this market will grow from 57.34 billion USD in 2021 to 81.74 billion USD in 2028 at a CAGR of 5.1% (Fortunebusinessinsights 2021). Over 80% of the PU foam market is dominated by petrochemically produced polyether-based polyols (IMARC 2021; Markets 2022), due to their superior physiochemical properties (i.e., viscosity, dispersity, mechanical resilience, hydrolytic and aging stability) compared to polyester-based polyols (Furtwengler and Avérous 2018; Ionescu 2005). The global market for polyols, the main component for PU synthesis, is expected to reach 37.9 billion USD by 2028 (IMARC 2022).

The high versatility of PU materials, proceeding from the diverse starting raw materials and the myriad of available conversion methods, allows their deployment in a wide range of industries. Furthermore, formulations can be tailored to the specific end-product needs. Applications include but aren't limited to: shoe soles, tubes, coatings, varnishes, adhesives, preservatives, skis, surfboards, dashboards, sealing compounds, or running tracks (Das and Mahanwar 2020; Gama et al. 2018). Specifically, rigid foams are utilized as building-, sealing- or thermal insulation material in the construction industry, and water-resistant resins as esthetic flooring material (Somarathna et al. 2018). However, flexible foams are installed in furniture, automotive seat, mattress paddings, or carpet cushions (Kausar 2017). Similarly, PU elastomer composites are commonly deployed in textiles, while micro-foams provide breathable components of rain jackets (Lomax 2007). Meanwhile, semiflexible polyurethane foams are deployed as wear-resistant packaging materials, impact-resisting exterior vehicle parts, or jackets for electric leads (Rossio et al. 1993).

4 Polyacrylonitrile (PAN)

Polyacrylonitrile (PAN) was first synthesized in the 1930s by the German chemical conglomerate IG Farben and mass produced in 1946 by American chemical conglomerate DuPont (Houtz 1950; Masson 1995). It is a synthetic semicrystalline polymer that consists of acrylonitrile, a nitrile (CN) functional group attached to a polyethylene backbone as its repeating unit structure (Kobayashi and Müllen 2015). This versatile polymer is a chemical precursor of many widely used materials, including acrylic fibers, resins, plastics, and carbon fibers (CF) (Kaliaguine and Dubois 2020; Karp et al. 2017). Due to the widespread use of PAN, bio-based precursors of its monomer, such as glycerol, are coveted.

4.1 Physiochemical Properties

One of the reasons behind the delayed commercial deployment of PAN was its insolubility in industrial solvents, which was later found to be fusible in ionic liquids (Albert 1948). Due to the strong chemical bonds that form between the nitrile groups, PAN does not possess any of the hazardous properties of its monomer; it is nontoxic and nonflammable. Its insulating property has many benefits in construction and automotive industries, particularly as component of engine exterior. Acrylonitrile (molar mass of 53.06 g mol^{-1}) is a clear flammable liquid that is highly toxic. It is also a known carcinogen, which necessitates strict safety procedures during the polymerization process. The high reactivity of acrylonitrile arises from its two reactive sites: carbon–carbon double bond and nitrile functional group. As this monomer readily polymerizes exothermically, its storage and transportation are strictly regulated (Brazdil 2000).

4.2 Production

4.2.1 Conventional PAN and Acrylonitrile Production

PAN is typically either produced by free radical vinyl polymerization or via anionic polymerization of acrylonitrile. The latter offers control over structure and molecular dimensions of the polymer (Kricheldorf 2013; Matyjaszewski et al. 1997). On an industrial scale, acrylonitrile is synthesized from fossil-based chemicals, namely propylene via ammoxidation—reacting with ammonia in the presence of oxygen (Dubois and Kaliaguine 2020; Ullmann 2003). This process, also known as the SOHIO process, has 80+% conversion efficiency. However, it is responsible for high CO_2 emissions (Davey 2018; Mack et al. 2019). Yearly, more than 7 million tons of acrylonitrile is produced worldwide using advanced Bi-Mo-Ox-based catalysts in this manner (Grasselli and Trifirò 2016). After polymerization, PAN is typically spanned into fibers (Arnold et al. 2018).

4.2.2 Bio-Sourced Acrylonitrile

In the search for sustainable bio-based precursors of acrylonitrile, a number of potential candidates have been identified as products, by-products, or intermediates of fermentative processes, including glycerol, glutamic acid, 3-hydroxypropionic acid, and propionic acid. Considering availability and conversion efficiency, the most promising candidate is glycerol (Dubois and Kaliaguine 2020). This precursor consists of three carbon chains with a hydroxyl group attached to each carbon (Liebig et al. 2013; Tan et al. 2013). Compared to acrylonitrile, this simple diol is nontoxic and nonflammable, with less restriction on its storage and transport (Dams

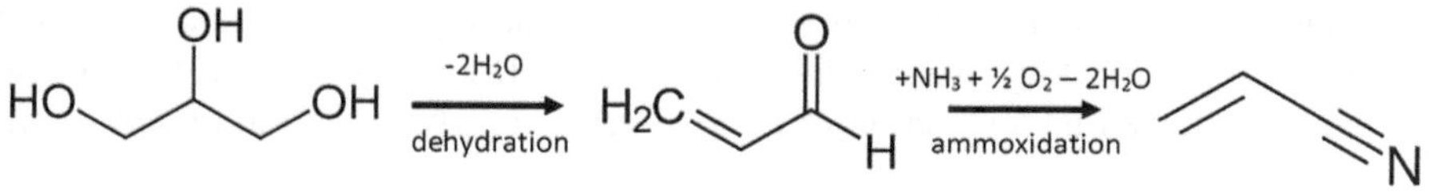

Fig. 5 Chemical reactions involved in the conversion of glycerol into acrylonitrile

et al. 2018). Although glycerol is an achiral molecule, it is prochiral in the two primary hydroxyl groups (Ichikawa et al. 1984). Several strategies for the conversion of glycerol into acrylonitrile have been reported, all of which involve the dehydration of glycerol into an acrolein (propenaldehyde) intermediate, followed by ammoxidation into acrylonitrile (Fig. 5). The different strategies vary in the number of steps involved, often incorporating a purification step of the intermediate as well as implementing distinct catalysts (Grasselli and Trifirò 2016). Conversion efficiencies of up to 60% have been reported. However, further work is needed to overcome technical limitations that include coke formation as well as catalyst and temperature compatibility of both reactions. Glycerol can be sourced from algae via two routes: (1) direct: as a glycerol produced by selection of the algal species and (2) indirect: glycerol produced as a by-product of biofuel production processes.

4.2.3 Direct Route: Glycerol Produced by Select Algal Species

This approach involves sourcing of glycerol directly from algae. Saline microalgae, such as *Dunaliella* spp., are salt-tolerant species that can survive up to 6 M sodium chloride concentrations (Bergstad et al. 2017; Oren 2014). These microalgae lack a rigid polysaccharide cell wall and are only bound by a cytoplasmic membrane. This trait allows them to adapt to hyperosmotic changes through cell size reduction (Hosseini Tafreshi and Shariati 2009). As an immediate response, the cells produce glycerol, benefiting from its nature as an osmolyte to avoid bursting (Chen and Jiang 2009). The level of intracellular glycerol is typically proportional to the extracellular salt concentration. High saline conditions can result in the production of more than 50% (w/w) glycerol of *Dunaliella* dry weight (Ben-Amotz and Avron 1983). *Asteromonas gracilis* also accumulates glycerol as an osmolyte, yet, this strain is poorly characterized (Wegmann et al. 1980). Therefore, most research on direct glycerol production in microalgae is focused on *Dunaliella*. The adaption of this species to a high saline environment as well as the mechanism of glycerol synthesis has been revealed in a number of omics studies (Chen et al. 2012; Fang et al. 2017).

1. Production process: This process involves a two-step fermentation, whereby high biomass yield is generated by cultivation at an optimal salt concentration followed by higher salt concentrations to induce glycerol accumulation (Benamotz and Avron 1990). As the *Dunaliella* cells do not have a cell wall, cell disruption is simply attainable by centrifugation. In this process, water or

ethanol serves as an extraction solvent that is subsequently evaporated (Chow et al. 2013; Mordhay Avron 1978; Xu et al. 2015). Implementing this simple method allowed the extraction of up to 10 g L^{-1} of glycerol from *D. tertiolecta* UTEX LB 99 (Chow et al. 2015).

2. Optimization: Although algal glycerol production has yet to reach industrial scale, *Dunaliella salina* is already cultivated in an open pond system for beta-carotene production (Harvey and Ben-Amotz 2020), which demonstrates the feasibility of this approach. In that respect, the coproduction of glycerol and beta-carotene would increase the value of the process. Several studies have focused on the genetic modification of *Dunaliella* species (Feng et al. 2009). Although enzymes responsible for glycerol production have been identified (Chen et al. 2009), genetic engineering of the pathway has not been employed in *Dunaliella*.

4.2.4 Indirect Route: Glycerol as by-product of Biofuel Production Processes

Algae have been discussed as a promising feedstock for biofuel, where integrated routes have been suggested to enhance the process feasibility, as discussed in Chapter "Biofuel-Integrated Routes". As a component of triglycerides, glycerol comprises the backbone, where three free fatty acids are bound to each hydroxyl group (Cooper 2000; Miles et al. 2004). Thus, glycerol can be found as a by-product of biodiesel production whereby oil is converted via transesterification into biofuel (Kaliaguine and Dubois 2020; Monteiro et al. 2018; Narendra et al. 2010). This process involves mixing the oil with short-chain alcohols, i.e., methanol in the presence of an acid, alkali, or enzyme catalyst (Otera 2002). It is estimated that one kg of glycerol is generated for every 10 kg of the produced biodiesel (Grasselli and Trifirò 2016).

While most of these processes are currently plant-based, microbial platforms, specifically algal-based present another promising and sustainable solution. Several microalgae that can accumulate up to 70% (w/w) lipids of total dry mass, such as *Botryococcus braunii* and *Nannochloropsis salina,* have been identified (Chisti 2007; Narendra et al. 2010). Due to the high lipid accumulation and their ten-fold faster growth compared to terrestrial plants, the overall oil yield of microalgae is 23-fold higher, compared to palm oil (Chisti 2007; Lorenzen et al. 2017).

1. Production process: The production process follows a two-step fermentation (Doan and Obbard 2014), illustrated in Fig. 6. First, algae are grown under optimized conditions to build up biomass rapidly. Subsequently, algal cells are exposed to modified nutritional and/or cultivation conditions (Alishah Aratboni et al. 2019), such as nitrogen- or phosphorous limitation and altered wavelengths or light intensity. These conditions induce lipid accumulation often accompanied by phenotypic changes, e.g., color change from dark green to light green (Alishah Aratboni et al. 2019; Fakhry and El Maghraby 2015; Sharma et al. 2012). By

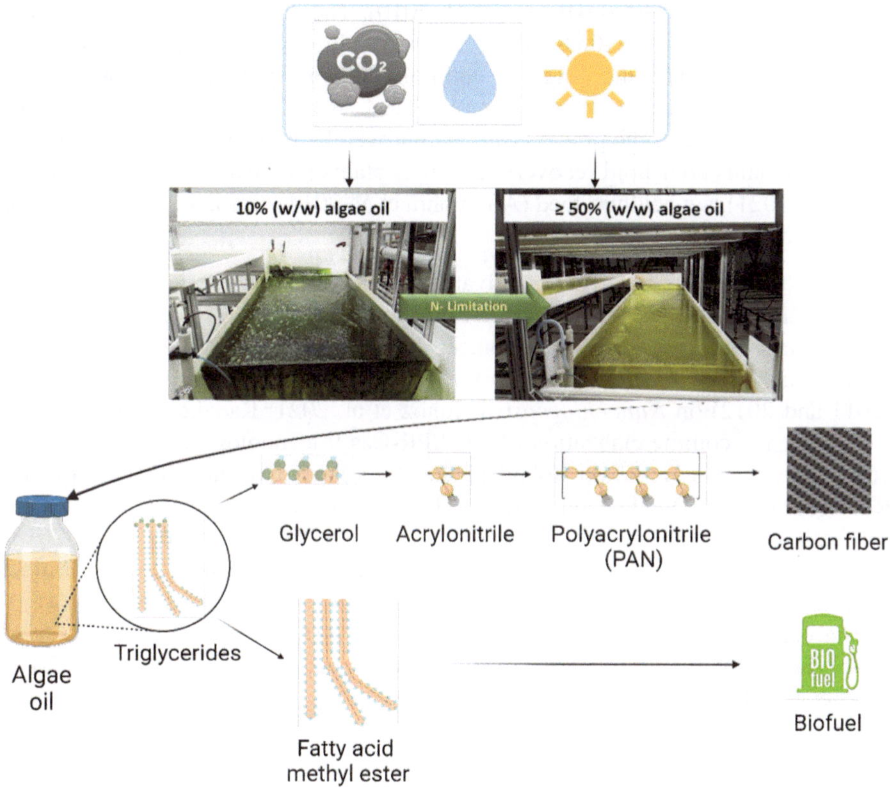

Fig. 6 Illustration of indirect glycerol production as a by-product of algal biofuel production processes

implementing nitrogen limitation, the most commonly applied inducer of lipid accumulation, *Nannochloropsis* can produce 90 kg per hectare per day in open ponds (Rodolfi et al. 2009).

The use of algae offers several advantages, especially with respect to downstream processes. Disruption of algal cells does not require energy-intensive mechanical methods, as commercial hydrolytic enzymes enable solvent-free release of algal oils at relatively low temperatures and a neutral pH (benefit ecological) (Das et al. 2022; Demuez et al. 2015). Although the use of enzymatic cocktails could increase the overall cost of the process, it concurrently reduces device corrosion that often results from thermochemical processes, reducing maintenance costs (Demuez et al. 2015).

2. Process optimization: Despite the amount of research on algae-based biofuels, industrial deployment of the technology has yet to be realized. For algae-based glycerol to compete with propylene as a precursor for acrylonitrile, lipid productivity titers should be increased and production costs should be lowered

(Abomohra et al. 2020; Hannon et al. 2010). The process cost can be rationed when valuable coproducts, such as astaxanthin or exopolysaccharides, are formed along with the lipids. In addition, cultivating the algae in wastewater can further reduce the cost (Abomohra et al. 2016). Furthermore, the sustainability standing of the process can be improved by recycling the waste biomass that remains after lipid recovery in biogas-plants (Achinas et al. 2017; Monlau et al. 2021), aquaculture feed (Abomohra et al. 2014), or microalgae cultivation (Abomohra et al. 2018).

In addition to optimizing cultivation conditions, genetic engineering of microalgae can also improve lipid titers (Munoz et al. 2021). The first report on genetic engineering of algae for enhanced lipid titers emerged in 1996 (Apt et al. 1996; Munoz et al. 2021). Similar work followed a gap of 15 years with two studies (2011 and 2012) on *Nanochloropsis* (Munoz et al. 2021; Radakovits et al. 2012). Even after the commercialization of CRISPR-Cas 9 technology in 2012, less than 100 studies on this topic are found in literature (Munoz et al. 2021; Rodriguez-Rodriguez et al. 2019). Typical targets of these efforts include genes of the fatty acid synthesis pathway (Chen et al. 2017), transcription factors (Kwon et al. 2018), NADPH generation (Koh et al. 2019), and central carbon metabolism (Munoz et al. 2021; Yao et al. 2014). This approach resulted in an increase of up to 129% of triglycerides without growth reduction in *Nannochloropsis oceanica* (Li et al. 2016). Similarly, the lipid content of *Chlorella pyrenoidosa* was increased 3.2-fold (Xue et al. 2020).

4.3 Recycling

Like all plastics or polymers, PAN has a low rate of biodegradability; it pollutes the environment for tens and even hundreds of years. The insolubility of PAN in industrial solvents necessitates the use of ionic solvents which renders its chemical recycling process expensive as well as toxic (IGTPAN 2023). Furthermore, this polymer cannot be repurposed as fuel by incineration in an oven as it requires specific equipment due to the release of highly toxic gases, including hydrocyanic acid (HCN) and ammonia (NH_3) (IGTPAN 2023). Due to the thermosetting properties of the polymer, it cannot be recycled via traditional methods. However, it may be recycled by blending with other polymers to develop copolymers that could be incorporated as a filler strengthening end-products, such as nanoparticle-inclusion bodies of electrically conductive materials. Yet, these approaches present additional problems due to the reduced biodegradation of these copolymers (Adegbola et al. 2020).

4.4 Market and Applications

The global PAN market size is valued at USD 7877.45 million. The expected limited market expansion (CAGR of -0.52) could be attributed to a number of socioeconomic elements and supply chain interruptions as a result of COVID-19 and the Russia-Ukraine war (MarketWatch 2023). This polymer is incorporated in a wide range of industries, including textile, medicine, automobile, aerospace, construction, and packaging (Adegbola et al. 2020). The use of PAN encompasses direct (e.g., filtration membranes) or indirect applications where it serves as precursor to carbon fibers.

4.4.1 Direct Applications

Due to its solvent resistance, surface area, antifouling properties, excellent mechanical properties and high porosity, high electrolyte uptake, and good relative absorption ratio, PAN is directly purposed as membranes for the filtration of air, particularly flash ash particles in industrial plants, water, and chemical treatments (Canalli Bortolassi et al. 2019). As PAN does not bring about undesirable consequences in the human body, it has also been implemented in biomedical materials, such as implants, limbs, and body organs fixation (Adegbola et al. 2020). The flexibility of PAN fibers and its compatibility with other polymers expand its applications to textile industries with knitted clothing and fabrics. Outdoor products, such as tents, especially benefit from the resistant nature of the polymer (Sheng et al. 2017).

4.4.2 Indirect Applications: Precursor to Carbon fibers

Unlike PE, PP, and polyester, PAN is not a thermoplastic polymer that melts when heated. At temperatures above 180 °C, rather than melting, PAN further polymerizes into rigid and dark carbon fiber. This property renders PAN fibers the best known carbon fiber precursor (IGTPAN 2023). PAN serves as a precursor polymer for around 90% of global carbon fiber production, as the resulting carbon fibers exhibit more desirable properties. These properties include average tensile strength, molecular properties, high carbon yield (above 80%), and its ability to decompose to form a char before melting (Kaur et al. 2016). Principally, carbon fiber can be stronger than steel, stiffer than titanium, and lighter than aluminium (Adegbola et al. 2020). Therefore, PAN fibers dominate over other precursor polymers such as pitch (Hawthorne et al. 1970), rayon (Ezekiel and Spain 2007), and polyolefin (Dong 2016). The global market value of carbon fibers is estimated at 2.8 billion USD in 2021 and is expected to reach 4.5 billion USD by 2027, at a CAGR of 8.8% (Lucintel 2022; MaximizeMarketResearch 2022; Peijs et al. 2022). This results in 280,000 tons of demand for PAN precursor (Chemanalyst 2022; IGTPAN 2020). Carbon fibers are used in the majority of current advanced composites (Manocha 2001). In addition to the medical industry, where carbon fibers are designed into ligaments, tendons, cartilages, dental fixtures, skull, artificial bones, and joint replacements, the automotive, aerospace, and construction industries have many prominent applications (Adegbola et al. 2020). Military and sports gear are also often manufactured from carbon fibers (Harussani et al. 2022; Ribeiro et al. 2018). The application of carbon fibers in the automotive, aerospace, and construction industries gained heightened attention;

1. Automotive and aerospace industries: Light and robust machinery components manufactured from carbon fiber composite (CFC) are especially valuable in the transportation sector. Every weight reduction in cars and airplanes benefits the fuel demand and the vehicles' lifetime CO_2 emissions (Lässig et al. 2012; Yuan et al. 2015). Carbon fibers make up around 20–25% of Boeing and Airbus wide-body airframes (Energy.gov 2013). A research group at the Technical University of Munich presented a prototype of an electro-scooter produced from bio-based carbon, highlighting the broad application areas where sustainable carbon could be incorporated (Brueck 2019).
2. Construction industry: The construction industry accounts for 38% of all annual energy-related CO_2 emissions, estimated at 14 gigatons of CO_2 (Mitic 2021). The International Energy Agency (IEA) has estimated that the industry should halve these emissions by 2030 to achieve the climate goals of the Paris Agreement from 2015 (Neill 2020; UnitedNations 2015). Carbon fibers have the potential to replace structural steel in construction materials due to their high strength, excellent fatigue resistance, low thermal coefficient of expansion, and low weight (Liu et al. 2015b). When reinforced with stone, which has high compressive strength, this material would offer a greener solution and a longer service-life. To that end, CarbonFiberStone CFS® has the potential to permanently immobilize CO_2. Since carbon is four times lighter and six times more load-bearing than steel, slimmer and more filigree structures that require less starting material and cost could be built (Arnold et al. 2020).

5 Alginate

In the last decade, the natural biopolymer alginate has gathered attention in the food, cosmetic, biomedical, pharmaceutical as well as construction industries, prompting researchers to explore production processes that could satisfy commercial needs (Doyle 1970; Gheorghita Puscaselu et al. 2020). The wide use of alginate in various applications is mainly due to its biocompatibility, biodegradability, nontoxicity, as well as its favorable physiochemical properties. Consequently, many efforts have been made to investigate potential sources of alginate. Marine brown algae, commonly referred to as seaweeds, are multicellular photosynthetic organisms that can vary in size and shape (Peteiro 2018). Alginate is a linear polysaccharide polymer consisting of alternating β-D-mannuronic acid (M block) and α-L-guluronic acid

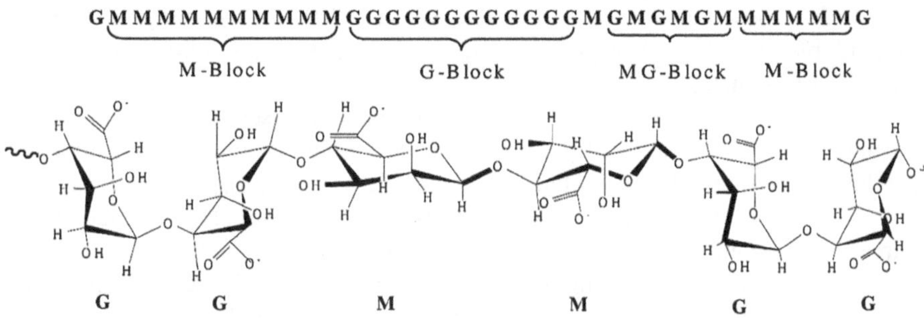

Fig. 7 Structural characteristics of alginate as adopted from Fertah et al. (2017) with permission number 501821048

(G block) joined by 1,4-linkage bonds (Sikorski et al. 2007). The molecular chain is composed of randomly distributed units arranged as either homogeneous blocks (MM or GG) or heterogeneous blocks (MG) (Fig. 7) (Fertah et al. 2017). Alginates are key components of the cell walls and intercellular matrix of these organisms. This biopolymer bestows physiological properties to brown macroalgae similar to those of cellulose in terrestrial plants. It confers strength and flexibility, characteristics essential for the survival of seaweeds in their dynamic aquatic environment (Kleinübing et al. 2013). Under limited nutritional conditions and depending on the species, alginate can make up to 55% of the dry weight of the brown macroalgae (Kleinübing et al. 2013; Lorbeer et al. 2017). Commercial alginate is obtained mainly from the following genera: *Laminaria, Ascophyllum, Durvillaea, Ecklonia, Lessonia, Sargassum, Padina,* and *Macrocystis* (Faidi et al. 2020; Gomez et al. 2009; Łabowska et al. 2019; Sari-Chmayssem et al. 2016).

5.1 Physiochemical Properties

The proportions of M and G blocks, as well as their distribution, influence the physical and chemical properties of the polymer, i.e., alginate with a high concentration of M units provides a flexible molecular structure and greater biocompatibility, whereas alginate with rich G units has a rigid molecular structure (Zhang et al. 2020). M:G ratio also plays a crucial role, as alginates with a high M:G ratios produce elastic gels, while a low M:G polymer generates brittle gels (Zhang et al. 2020). Alginate polymers are rich in carboxyl and hydroxyl groups, making them a great candidate for chemical functionalization and cross-linking. In the presence of divalent or trivalent metal ions, including Fe^{3+}, Al^{3+}, Cr^{3+}, Cu^{2+}, Ba^{2+}, Sr^{2+}, and Ca^{2+}, alginate can form hydrogels (Zhang et al. 2022). Contrary to most gel-forming polymers, alginate gels are cold-setting—forming independently of temperature—if gelling is induced (cations). However, the kinetics of the gelling process as well as the properties of the final product can be affected by the setting temperature (Draget et al. 2005; Łabowska et al. 2019; Lee and Mooney 2012). Designer gels, composites, and blends with other polymers can improve the durability and physical properties of alginate.

Rheological characteristics are critical factors in the selection, manipulation, utilization, and performance of polymers, including alginates and their composites in industrial settings. These include: (1) Intrinsic viscosity [η] is a measure of a polymer's ability to modulate the viscosity in a solvent. The extraction and pretreatment techniques (e.g., bleaching of seaweeds) can impact this property. (2) Gel strength varies depending on the source of alginate. (3) Gelation kinetics is influenced by several variables (e.g., presence of cations favorable for gel formation) and depends on the intended end-applications (e.g., cross-linking, blends) (Draget et al. 2005; Łabowska et al. 2019). Alginate and its derivatives and blends exhibit biologically relevant characteristics, such as phenolic content, antioxidant, radical scavenging activity, and reducing power (Łabowska et al. 2019; Reddy 2021). The biocompatibility of alginate has been extensively explored, both in vitro and in vivo models, with enriched M:G ratio polymers eliciting higher immunogenic cytokine response. Furthermore, cross-linked alginates (mainly Ca^{2+}–Alginate) were found to be nontoxic, with no eyes or mucosal membranes irritation (Łabowska et al. 2019; Reddy 2021).

5.2 *Production*

5.2.1 Biosynthesis

The biosynthetic pathway of alginate in macroalgae and its regulatory mechanisms still suffers from limited understanding (Shao and Duan 2022). Further research would enable the production of alginate with tailored features in various algae strains. Several characterization studies have been conducted on a limited number of algal species, such as *Ectocarpus siliculosus, Saccharina japonica*, and *Laminaria digitate* (Chi et al. 2018; Shao and Duan 2022; Zhang et al. 2021). Due to the striking similarities between algal and bacterial production pathways, studies on the latter provide important insights. Several publications theorize that alginate pathway has complex endosymbiotic gene transfer (EGT) origins (Chi et al. 2018). In recent studies, alginate-specific metabolic steps in *Ectocarpus* sp. were proposed to be acquired by horizontal gene transfer (HGT) from an actinobacterium (Chi et al. 2018; Shao and Duan 2022). Alginate biosynthesis pathway is divided into: (1) synthesis of precursors, (2) polymerization and subsequent cytoplasmic membrane transfer, (3) periplasmic transfer and modification, and (4) export through the outer membrane. Briefly, the process of alginate production begins with the conversion of fructose-6-phosphate to guanosine di-phosphate-mannuronic acid (GDP-ManA) by a number of enzymatic transformations. Mannose-6-phosphate isomerase (MPI), phosphomannomutase (PMM), mannose-1-phosphate guanylyltransferase (MPG), and GDP-mannose/UDP glucose-6-dehydrogenase (GMD/UGD) are sequentially

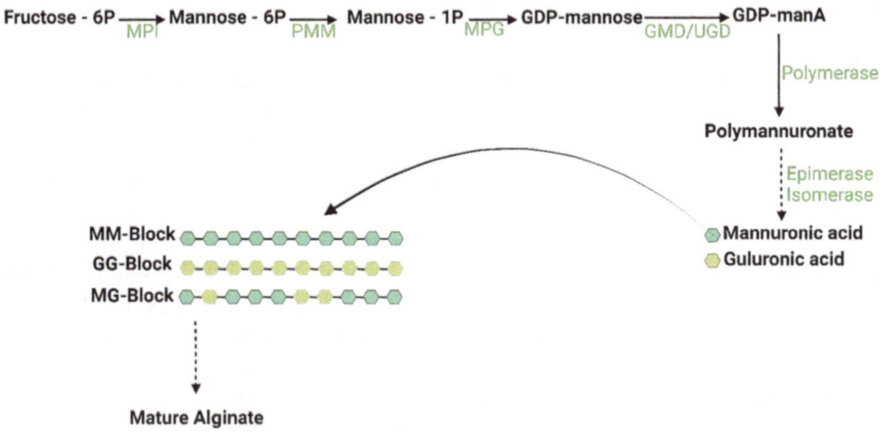

Fig. 8 Illustrative process of alginate production in algae. *MPI* mannose-6-phosphate isomerase, *PMM* phosphomannomutase, *MPG* mannose-1-phosphate guanylyltransferase, *GMD/UGD* GDP-mannose/UDP glucose-6-dehydrogenase

involved in this synthesis. GDP-ManA is then transferred to the cytoplasmic membrane and polymerized to polymannuronate. Epimerization and isomerization of mannuronic acid to guluronic acid result in the mature alginate (Fig. 8).

Bacterial alginate production differs from that of its algal counterpart in that it is O-acetylated, which serves to protect the alginate from degradation. Once the alginate with M and G blocks is formed, it is exported to the outer cellular membrane.

5.2.2 Extraction

The conventional extraction process, which consists of multiple steps aimed at maximizing product yield, is typically preceded by a thermochemical pretreatment step (acidification and/or autoclaving) (Łabowska et al. 2019; Mohammed et al. 2020). The treated biomass is then exposed to a strong alkaline solution to recover sodium alginate, which is then filtered or centrifuged. The extract is then precipitated with either calcium chloride, hydrochloric acid, or sulfuric acid to form alginates either in acid or calcium salt forms. Finally, the alginate is dried and milled for commercial use. Recent interest in green technologies and biorefinery approaches for the extraction of biological compounds. Novel "greener" extraction processes include ultrasound-assisted extraction (UAE) and microwave-assisted extraction (MAE) (Łabowska et al. 2019).

5.2.3 Downstream Processing

The gel-forming properties of alginate allowed the development of various biomaterials such as porous scaffolds, hydrogels, fibers, nonwoven fabrics, and membranes (Li et al. 2022; Vakilian et al. 2021; Zhang et al. 2022). Ionic cross-linking, microfluidic spinning, freeze-drying, wet spinning, and immersive rotary/centrifugal jet spinning methods are often employed to produce these alginate-based materials. Sodium alginates, in particular, are water-soluble and yield highly viscous and homogenous solutions (Vakilian et al. 2021; Zhang et al. 2022). When dropped into calcium salt solutions and then freeze-dried to eliminate water, porous hydrogel scaffolds can be produced. Composite or hydrogel blends can be fabricated by incorporating other polymers such as gelatine, chitosan, and collagen, expanding its applications (Lee and Mooney 2012). The gelling properties of alginate allow the encapsulation of various compounds, including even cell fragments.

5.3 *Recycling*

The linear alginate polymer and alginate-hydrogels can undergo a variety of depolymerization processes. The 1,4-linkage bonds can be cleaved via both acidic and alkaline degradation events, oxidation with free radicals, and autoclaving (Draget et al. 2005). These degradation methods break down the hydrogel structure into smaller fragments, which can be recycled into new hydrogels. However, these methods can result in the loss of some of the polymer's properties. To retain the original polymer properties, alginate lyases and oligomerases can break down the polymer chain, via the β-elimination reactions, into oligomers or monomers (Li et al. 2021). These could be subsequently purified and recycled. Thus, enzymatic degradation is a promising approach due to its mild reaction conditions, high selectivity, low energy consumption, and eco-friendly aspect (Vakilian et al. 2021; Zhang et al. 2022). On the other hand, recycling alginate blends or composites is a more complex process, whereby energy-intense thermochemical methods are required to separate and/or recover the individual components.

5.4 *Market and Applications*

The annual industrial production of alginate is estimated at around 30,000 metric tons, with farmed brown seaweed, primarily *Laminaria* and *Macrocystis* sp., as major source. This number accounts for less than 10% of the global annual production of alginate in the wild (Draget et al. 2005; Hay et al. 2013). The abundance of sourcing and the ease of production suggest that future growth in the alginate market would be of a qualitative rather than quantitative aspect.

5.4.1 Food Applications

FDA have declared alginate to be "generally regarded as safe" (GRAS). Alginate is highly sought after in the food production industry due to its unique properties, i.e., biodegradability, biocompatibility, renewability, and nontoxicity (Yarkent et al. 2021). It is easily tolerated in the human body, and it can be found in a wide range of food products including tinned, baked/frozen foods, meat, poultry, seafood, pet food, cheese, beverages, jams, ice cream, and mayonnaise. Due to its high viscosity and adhesion/cohesion features, alginate is utilized as a thickening agent to improve the texture of products such as yogurt and jelly (Łabowska et al. 2019; Yarkent et al. 2021). This polymer can also be used as a biosurfactant and stabilizer in various products, including beer. Furthermore, alginate is considered a functional food, meaning it has the ability to reduce the risk of chronic diseases and increase quality of life through its anticancer and probiotic benefits. Alginate can be added to dairy liquid products, beer, and drinks consumed by diabetic patients and can help to reduce hunger levels and fight obesity. Alginate also extends the shelf-life of food due to its intrinsic antimicrobial properties. In particular, the encapsulation of important compounds, such as antioxidants and vitamins, within alginate helps protect food products against oxidation, heat, and light degradation (Kontominas 2020; Łabowska et al. 2019; Yarkent et al. 2021).

5.4.2 Cosmetic Applications

With its anticoagulant, antimicrobial, anti-irritating, antioxidant, and anti-inflammatory properties, alginate has been included in high-quality cosmetics that offer numerous skin benefits. These include UV protection, moisture retention, improved skin smoothness, and cell rejuvenation. Alginates are also embedded in face, antiaging, and anti-wrinkle masks. With its high viscosity, it can also form networks that help in lipstick color as well as moisture retention. This biopolymer is also utilized as a thickening agent in shampoos and lotions (Reddy 2021; Yarkent et al. 2021).

5.4.3 Biomedical Applications

Alginate is a popular choice for a variety of pharmaceutical and wound healing applications. It can serve as a thickening, gel forming, and stabilizing agent, as well as a vehicle for controlled drug release (Lee and Mooney 2012; Reddy 2021; Yarkent et al. 2021). When used in oral dosage forms, alginate gels are nanoporous and can release small molecules over a period of time. Alginate gels can also protect protein drugs from denaturation and degradation until their release. For wound healing, alginate dressings are typically produced by ionic cross-linking and freeze-drying, forming a foam or nonwoven material that absorbs wound fluids and helps maintain a moist environment while minimizing the risk of infection. This can help to promote the formation of granulation tissue, rapid epithelialization, and healing (Lee and Mooney 2012). Alginate is also gaining interest in the field of bone tissue engineering due to its biocompatibility and gel-forming characteristics. To date, a variety of composite materials have been studied, such as alginate-polymer (PLGA, PEG, and chitosan) composites, alginate-protein (collagen and gelatine), alginate-ceramic, alginate-bioglass, and alginate-biosilica composites. These scaffolds exhibit enhanced biochemical properties, including increased porosity, improved mechanical strength, increased cell adhesion, increased biocompatibility, increased cell proliferation, increased alkaline phosphatase activity, excellent mineralization, and enhanced osteogenic differentiation (Lee and Mooney 2012; Yarkent et al. 2021).

5.4.4 Textile Applications

Textile-grade alginate is utilized in the imprinting of patterns on fabrics. Sodium alginates specifically are utilized as thickening agents in dyes and color pastes (Reddy 2021; Yarkent et al. 2021). These materials exhibit excellent flame-retardant properties due to the presence of calcium cations and abundant oxygen atoms. When the fiber is exposed to fire, calcium salt is formed, releasing noncombustible carbon dioxide (CO_2) gas and water vapor. This acts like a protective layer and reduces the heat and oxygen penetration, thus effectively slowing down the combustion reaction (Li et al. 2022; Zhang et al. 2022).

5.4.5 Construction Applications

Alginate's physiochemical properties make it an ideal material for use in construction and building materials. This polymer is employed in the production of lightweight and durable roofing materials, including tiles and panels (Mignon et al. 2016). Furthermore, alginate can be utilized to coat walls, floors, and roofs, to waterproof and soundproof the surfaces, and to provide additional insulation (Lacoste et al. 2018; Mignon et al. 2016). It can also improve the adhesion properties of coatings and sealants. Alginate can additionally confer resistance to bacterial and fungal degradation of building materials. This biomaterial can also be used to cast molds for building blocks and bricks, as well as create self-levelling concrete. The strength of concrete, mortar, grout, and other materials can be enhanced by alginate as an admixture (Murugappan and Muthadhi 2022). As for the textile industry, alginate confers significant flame-, fire- and heat-resistance or imperviousness to the building materials. By increasing the thermal stability of products, alginate helps reduce the risk of fire and lower insurance premiums (Xu et al. 2021). Calcium alginate is also employed as admixture for self-healing asphalt (Ruiz-Riancho et al. 2021).

6 Carrageenan

Carrageenans are natural polysaccharides belonging to a class of linear sulphated galactans occurring in the cell wall of red algae (Michel et al. 2003). They are valued for the diversity in their structural attributes, which are linked to a wide range of physicochemical characteristics and biological functions. The backbone structure of carrageenans is based on alternating 3-linked β-D-galactopyranose and 4-linked α-D-galactopyranose (Millane et al. 1988). The polysaccharides are classified into several types based on the structure of the disaccharide repeating units, the sulphation pattern, and the presence of 3,6-anhydrogalactose as a 4-linked residue (Necas and Bartosikova 2013). A standard letter code nomenclature has been proposed by Knutsen et al. (1994) to describe the disaccharide building blocks of the polymer. Carrageenans are identified using Greek prefixes depending on the composition of the two main building blocks (Fig. 9) (Ciancia et al. 2020; Elmarhoum et al. 2023; Ficko-Blean et al. 2015; Pierre et al. 2015).

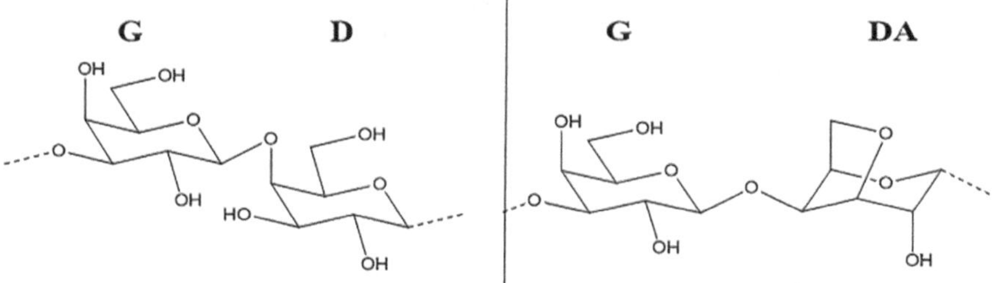

Carrageenan	β-3-linked	α-4-linked
γ (gamma)	G	D6S
λ (lambda)	G2S	D2S,6S
ξ (xi)	G2S	D2S
μ (mu)	G4S	D6S
ν (nu)	G4S	D2S,6S
ω (omega)	G6S	D6S

Carrageenan	β-3-linked	α-4-linked
β (beta)	G	DA
α (alpha)	G	DA2S
θ (theta)	G2S	DA2S
κ (kappa)	G4S	DA
ι (iota)	G4S	DA2S
ψ (psi)	G6S	DA

Letter code*	IUPAC name
G	3-linked β-D-galactopyranose
D	4-linked α-D-galactopyranose
DA	4-linked 3,6-anhydro-α-D-galactopyranose
S	Sulphate ester ($O\text{-}SO_3$)
G2S	3-linked β-D-galactopyranose 2 sulphate
G4S	3-linked β-D-galactopyranose 4 sulphate
G6S	3-linked β-D-galactopyranose 6 sulphate
DA2S	4-linked 3,6-anhydro-α-D-galactopyranose 2 sulphate
D2S,6S	4-linked α-D-galactopyranose 2,6 disulphate
D6S	4-linked α-D-galactopyranose 6 disulphate

Fig. 9 Overview of the repeating units for the idealized structure of distinct carrageenan types and their corresponding letter code nomenclature. *Letter code by Knutsen et al. (1994)

6.1 Physiochemical Properties

The primary differences in chemical properties between various types of carrageenans arise from the number and position of their half-ester sulphate groups. In addition, physical characteristics of carrageenans are governed by the associated cations as well as the conformation of the sugar building blocks in the polymer chain. All carrageenan fractions are soluble in water and insoluble in fats, oils, or organic solvents (Gulrez et al. 2010). The proportion of negatively charged sulphate groups, which are very hydrophilic, and their associated metal ions influence the water solubility of the polymer (Campo et al. 2009). However, some types require higher temperatures for complete solubilization (Bixler 1994). In a similar manner, the amount of sulphate groups and the cation equilibrium in the aqueous solutions also determine the viscosity or strength of gels derived from carrageenans.

The most commonly used carrageenans in commercial and industrial applications are iota (ι-), kappa (κ-), and lambda (λ-) carrageenan (Rupert et al. 2022). λ-carrageenan is primarily used as a thickener additive. At the same time, κ-and ι-carrageenan has gel-forming properties. In comparison with the former, ι-carrageenan-gels are softer and more elastic. The process of gel formation comes about when 3,6-anhydro-d-galactopyranosyl units form helical secondary structures. It has been proposed that intramolecular anhydro bridges play a significant role in this process, which is thermally reversible (Liu et al. 2015a). Gels dissolve when heated to about 80-90 °C and rematerialize after cooling to around 50 °C (Bixler 1994).

6.2 Production

6.2.1 Marine Sources

Carrageenans can be extracted from various marine red algae (*Rhodophyta*), such as *Kappaphycus*, *Chondracanthus, Gigartina, Sarcothalia*, *Mastocarpus,* and *Eucheuma* (Azevedo et al. 2015; Pereira et al. 2009). Generally, most carrageenophyte red algae produce hybrid carrageenans composed of multiple types (Van de Velde 2008; Villanueva et al. 2009). The dominant process for the production of carrageenan is aquaculture-based seaweed cultivation (Rupert et al. 2022). The majority of global production is located in tropical or subtropical regions of South-East Asia, such as Indonesia, the Philippines, and Malaysia (Kumar et al. 2020). κ-carrageenan is predominantly obtained from *Kappaphycus* species, whereas ι-carrageenan is often produced from *Eucheuma*. Together these seaweeds make up roughly 90% of the world's carrageenan production capacity (Campbell and Hotchkiss 2017). λ-carrageenan is mainly extracted from sporophytic plants of *Chondrus crispus,* which are grown in the Northern Atlantic (McCandless et al. 1973; Simpson and Shacklock 1979). In some species, modifications of the biopolymer have been reported, which adds to their chemical complexity. Carrageenan from *Gigartina* spp., for example, have pyruvated β-D-galactopyranose units, whereas carrageen from *Claviclonium ovatum* are highly methylated (Chiovitti et al. 2004; Falshaw and Furneaux 1998). The growth of carrageenan-producing seaweeds is influenced by environmental factors like nitrogen or phosphorus concentration in the water. Studies have shown that the growth under limited nutrient conditions results in higher carrageenan content (Chopin and Wagey 1999). Nevertheless, their overall growth rate is reduced (Ryther and Dunstan 1971).

6.2.2 Extraction

Commercially, carrageenans are available in different purity levels. In industrial processes, the first step generally starts with washing and drying seaweed biomass. Carrageenan is subsequently extracted in an alkaline solution at 70-80 °C using either NaOH or KOH (Bixler 1994). The naturally occurring precursors of κ- and ι-carrageenans, as well as μ- and ν-carrageenans, are converted during alkaline processing. Commercial κ- and ι-carrageenans are therefore typically copolymers of κ-:μ- and ι-:ν-carrageenans. By contrast, λ-carrageenan, as a precursor, is partially converted to θ-carrageenan (Hotchkiss et al. 2016). The length of helices formed during gelation is favored by the relative abundance of μ- and ν-carrageenans, whereas pure κ- and ι-carrageenan blocks cannot form helices. For this reason, alkaline extraction enhances gel strength (Blakemore 2016). Further purification can be achieved by filtration or precipitation with alcohol (Bixler 1994).

6.3 Recycling

Carrageenan polymers are biodegradable and can be hydrolyzed using commercial enzymes, such as α-amylases or acidic treatments (Kamińska-Dwórznicka et al. 2015; Wu 2012). The biodegradability of carrageenan is an important factor in sustainable product design. Innovative composite materials based on carrageenan, that can be used for food packaging, for example, have successfully been tested (Aga et al. 2021). Due to their high biodegradability index, carrageenan-based bio-packaging materials can contribute to reduce environmental concerns.

6.4 Market and Applications

The carrageenan market is the largest global seaweed-derived hydrocolloid market which is predicted to reach 1 billion USD by 2024 (Campbell and Hotchkiss 2017). Commercial carrageenans are available as stable sodium, potassium, or calcium salts. Carrageenans are also implemented in various industry sectors including food, pharmaceutical, cosmetic, and textile (Imeson 2000).

6.4.1 Food

The majority of carrageenan applications are related to the food industry. In the European Union, carrageenan is an approved food additive with the assigned number E407 (Blakemore 2016). Carrageenans are used to improve the texture of cheese, control the viscosity and texture of various dairy products, and utilized as binders and stabilizers in the meat-processing industry (Błaszak et al. 2018; Campbell and Hotchkiss 2017; Hotchkiss et al. 2016). In addition, food coatings based on ι-carrageenan in combination with antimicrobial extracts are proposed as food preservatives (Carocho et al. 2019). Carrageenans can also be used as a clearing agent in the beverage industry, yet other substances like gelatine are more frequently applied (Karim and Bhat 2008).

6.4.2 Pharmaceuticals

Carrageenans exhibit bioactive properties with numerous potential health benefits (Pacheco-Quito et al. 2020; Wijesekara et al. 2011). These include antiviral (Eccles et al. 2015), antitumor (Haijin et al. 2003), and antibacterial activities (Yamashita et al. 2001). For instance, antiviral properties could be attributed to carrageenans' capacity to prevent pathogens from binding or entering the host cells (Loureiro et al. 2017). Nevertheless, the utilization of carrageenan in medicine remains a niche application. Carrageenan has also been explored for new applications in this field. Owing to the high density of negative charges in carrageenans and their ability to form gels, novel formulations for regulated and prolonged drug release have been developed (Li et al. 2014). Carrageenan has also been tested in applications for tissue engineering and wound healing (Tytgat et al. 2019; Yegappan et al. 2018).

6.4.3 Bioplastics

Recently, the incorporation of carrageenans in biodegradable bioplastics, e.g., packaging films, gained increasing attention (Abdou and Sorour 2014; Galus and Kadzińska 2015). These biopolymers are widely regarded as an eco-friendly alternative and potential replacement for petroleum-based plastics (Gironi and Piemonte 2011). Furthermore, the integration of carrageenans can benefit the mechanical and thermal properties of newly developed composite materials (Meng et al. 2018a; Meng et al. 2018b; Tabatabaei et al. 2018). An alternative production process for polyurethane incorporating carrageenan with 2,4 toylulene diisocyanate has been demonstrated in 2010 (Marlina 2010). However, more research and development is required to reach market maturity of carrageenan-derived bioplastics.

7 Laminarin

Laminarin is a long-term storage polysaccharide typically present in brown algae and some diatoms (Biersmith and Benner 1998; Graiff et al. 2016). A significant portion of the global marine carbon cycle is related to the formation of this polymer (Becker et al. 2020). Laminarin is mainly composed of linear chains of β-1-3-linked glucose monomers with partial branches that are formed by β-1-6 glycosidic bonds (Percival and Ross 1951). The reducing end of the polysaccharide backbone can either be linked to glucose or D-mannitol. These variants are referred to as G-type or M-type laminarin, respectively (Fig. 10) (Stark 1976).

7.1 Physiochemical Properties

Laminarin has a relatively low molecular weight, which is determined by the number of building blocks in the polysaccharide. The average molecular weight of this biopolymer is 5 kDa, with a degree of polymerization ranging between 20 and

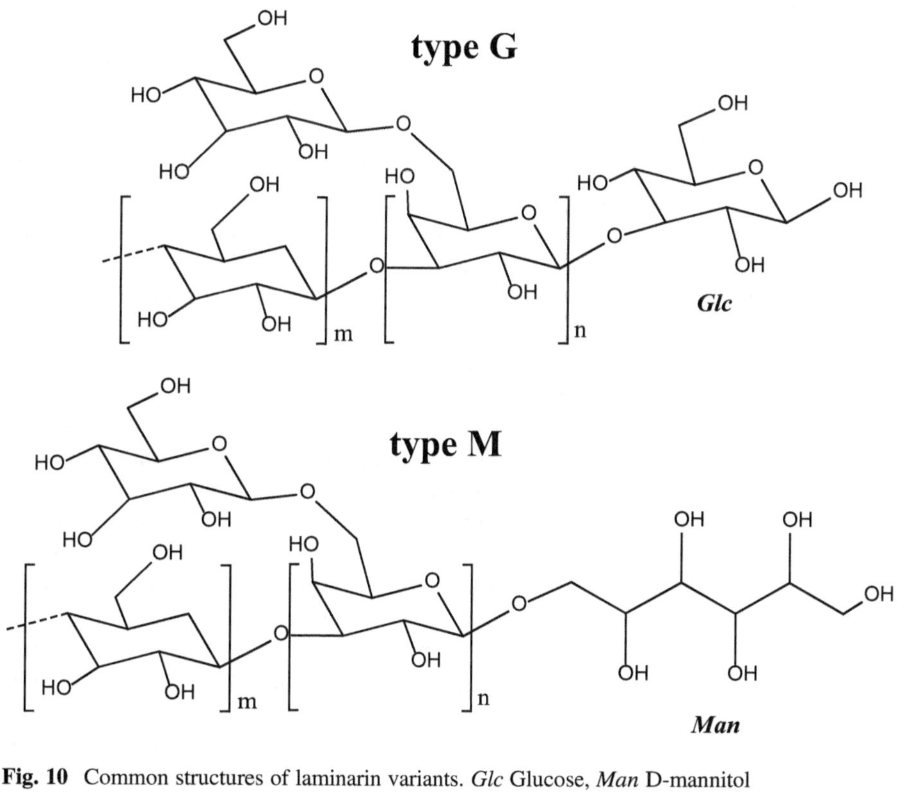

Fig. 10 Common structures of laminarin variants. *Glc* Glucose, *Man* D-mannitol

25 residues (Nelson and Lewis 1974). Recent reports describe laminarin polymers with molecular weights of 2–10 kDa and a polymerization degree of 15–40 building blocks (Hentati et al. 2020). Higher molecular weights of up to 40 kDa have also been reported for some species (Chen et al. 2021; Menshova et al. 2014).

In contrast to many algal-derived polysaccharides, laminarin does not form viscous gels in aqueous solutions (Holdt and Kraan 2011). Depending on the producing organism, laminarin exhibits quantitative as well as chemical variations, which are further influenced by environmental factors, growth conditions, and tissue type (Chizhov et al. 1998; Read et al. 1996; Rioux et al. 2010). In *Laminaria digitate*, for example, the proportion of M-type to G-type chains can reach up to 3: 1 (Read et al. 1996). Due to the presence of both laminarin types, the polysaccharide is polydispersive (Anderson et al. 1958). In contrast, M-types are missing in the laminarin from *Eisenia bicyclis* (Maeda and Nishizawa 1968). In addition, the relative occurrence of branches formed by β-1,6 glycosidic bonds influences the solubility of laminarin in cold water (Annan et al. 1965); higher branching factors increase water solubility. In that respect, water-soluble, as well as water-insoluble, laminarins have been reported in diverse species (Black 1950).

7.2 Production

Traditionally, laminarin is produced by brown algae, such as *Laminaria* and *Saccharina* (Haug and Jensen 1954; Jensen and Haug 1956; Morrissey et al. 2001), but it can also be found in other seaweeds, such as *Ascophyllum, Fucus,* or *Undaria* (Holdt and Kraan 2011). At present, the biosynthetic pathway of laminarin still has to be determined. In that context, only putative pathways for some organisms have been proposed (Chen et al. 2021; Michel et al. 2010). Laminarin biosynthesis starts with the formation of glucose 6-phosphate, which is further converted into glucose-1-phosphate through an isomerization reaction. Then, glucose-1-phosphate is activated by the addition of uridine diphosphate (UDP), forming UDP-glucose. UDP-glucose is the substrate for laminarin synthase which polymerizes glucose molecules to form a growing laminarin chain. The growing chain can undergo branching through the attachment of additional glucose molecules to specific positions on the main chain. The laminarin chain may also undergo modifications after the final chain length is reached.

Laminarin is commonly directly extracted from brown algae following washing, drying, and milling of biomass (Li et al. 2019). Pretreatment in most cases is not required, which reduces the overall cost of the process. Frequently reported methods for the extraction of laminarin employ acidic solutions, e.g., HCl (Black 1950; Devillé et al. 2004; Ermakova et al. 2013) or H_2SO_4 (Voronova et al. 1991). Product yield can be increased by heating at 60 °C during extraction (Zha et al. 2012). Lately, alternative extraction methods have been explored. These include microwave-assisted extraction (MAE) (Graiff et al. 2016), ultrasound-assisted extraction (UAE) (Kadam et al. 2015b), or enzyme-assisted extraction (Charoensiddhi et al.

2016). Following the extraction process, the polysaccharide present in the supernatant can be precipitated using ethanol (Kadam et al. 2015a). For some organisms, $CaCl_2$ or $MnCl_2$ can be added to the solutions to remove alginate, exploiting its gel-forming properties (Cong et al. 2016). Purification of the end-product involves centrifugation, dialysis, filtration, or chromatographic methods (Devillé et al. 2004; Kadam et al. 2015b; Sterner and Gröndahl 2021).

7.3 Recycling

Laminarin is biodegradable and can be hydrolyzed by naturally occurring enzymes, such as β-D-glucanases (Chesters and Bull 1963). Laminarases are well studied in a range of microorganisms, such as *Rasamsonia emersonii*, *Trichoderma viride,* and *Trichoderma reesei*, where they are employed for recycling of laminarin into monomeric sugars and bioethanol (Chesters and Bull 1963; Rocher et al. 2021).

7.4 Market and Applications

Compared to other algal biopolymers, the industrial application of laminarin remains limited. In 2019, the global market size was estimated at approximately 2 million USD (Van Breda 2020). The majority of its common use is in food applications, e.g., dietary fiber with prebiotic effects (Devillé et al. 2004). Laminarin has also been used in cosmetic products, given that beneficial properties for skin cells have been reported (Li et al. 2013; Ozanne et al. 2020). For example, there are indications that laminarin can reduce oxidative stress and inflammation induced by environmental factors (Ozanne et al. 2020). In addition, there likely are positive effects on wound healing (Choi et al. 2013).

New potential uses for laminarin are currently explored, which often involve an alteration of the polymer's backbone structure. The chemical modification of laminarin can open new possibilities for applications in the medical or healthcare sector. For example, sulphated derivatives of laminarin can exhibit an anticoagulant effect on blood (Alban et al. 1992; Hawkins and O'Neill 1955). In 2016, Ren et al. (2016) reported a modification of laminarin, introducing polyethylenimine (PEI) to its surface via N,N′-carbonyldiimidazole, which resulted in cationic PEI-modified laminarin. This modified laminarin was later used as a carrier for a siRNA-based gene therapy against breast cancer. In another study, scientists developed a laminarin-based carrier biomaterial for a more efficient delivery of hydrophobic drugs (Yu et al. 2018). The new polymer was named "Hematin-Laminarin-Dithiodipropionic Acid-MGK". This versatile laminarin-based biomaterial could be applied to develop new nanomedicine carrier drug delivery systems for cancer treatment.

8 Challenges and Future Directions

The shift towards sustainable technologies—such as bio-derived polymers—to mitigate environmental concerns is gaining traction. The Golden Formula can be expressed as follow; $CO_2 \rightarrow [\text{Algae} \rightarrow \text{Bioproduct}_n]_n\, \Delta$ (Gedde and Hedenqvist 2019).

Algae, including macro and micro, are a prominent platform for sustainability. Algal-derived biopolymers, such as alginate and carrageenan, have gained considerable attention due to their sustainable production and widespread commercialization. However, the industrial-scale use of algae for bioplastic production, in particular, remains far from economic feasibility. Extensive efforts are currently carried out in metabolic engineering; however, while these techniques can increase the yield and quality of biopolymers derived from algae, the use of genetically modified organisms (GMOs) in large-scale production has been met with some controversy. Concerns over their potential short- and long-term impact on human health cover in particular their direct applications such as food, cosmetic, and biomedical. Another limiting factor, which spans all sectors from food to construction, is the risk of GMO strains leaking or escaping into the environment. Despite these challenges, the potential benefits of genetically engineered algae biopolymers production cannot be overlooked, and measures could be undertaken to mitigate potential concerns. Studies showed that public acceptance and political support of GMOs can be met if high safety standards are guaranteed. Proper containment methods, closed systems, select production locations, and internal kill switches (biocontainment) could ensure the safe and sustainable commercialization of GMO-algae biopolymer (Cavelius et al. 2023; Varela Villarreal et al. 2020).

Other approaches to improve economical accessibility of algal-biopolymers include improving downstream and modification processes, as well as constructing mathematical or artificial neural network (ANN) models. Research is also conducted on algae cultivation upscaling and growth conditions optimization. As discussed in Chapter "Overview of Biorefinery Technology", a cyclic biorefinery concept could potentially offer a viable economic model for the widespread use of algal-derived products, including biopolymers. This model can help reduce expenditures by creating multiple product streams—e.g., PHB, PU, alginate, carrageenan, laminarin, carbon fiber, oil, and glycerol—that feed into various market segments including bioplastics, biofuels, construction, nutraceuticals, pharmaceuticals, and cosmetics. Bio-derived materials could be recycled back into the biorefinery in a "zero-waste" cyclic approach. Furthermore, renewable energy sources such as solar, wind, and biogas could be utilized, thus leading to a more eco-friendly and cost-effective solution. Finally, in order to promote the use of sustainable algal-biopolymers at the industrial level, a cohesive framework of governmental and international policies as well as funding schemes would rapidly drive biopolymers to consumers.

9 Conclusions

This chapter serves as a comprehensive guide to algal-derived biopolymers, specifically polyhydroxypolymers, polyols, polyurethanes, polyacrylonitrile (PAN), alginate, carrageenan, and laminarin. The highlighted physiochemical properties inherent to each biopolymer provide better allocation of applications based on their corresponding functionality. Production, with respect to common producer strains and biosynthetic pathways, is also detailed. By investigating avenues for their recycling and reprocessing, we offer a roadmap to unlock the full potential of these biomaterials, allowing them to contribute to a circular economy and minimize waste. By understanding the current demand and market trends, we also offer insights into the vast array of opportunities to apply these biopolymers in various industries. Their versatility opens doors to a multitude of industries and sectors, including food, cosmetics, pharmaceuticals, biomedical, and construction. Despite the remarkable potential of algal-derived biopolymers, a number of challenges hinder their widespread implementation. Scaling up production processes, optimizing extraction techniques, and ensuring economic viability, as well as establishing effective circular biorefineries, are crucial areas of focus in order to capitalize on these biopolymers' potential. Continued exploration and innovation in this field will undoubtedly lead to the development of novel biopolymers and applications, paving the way for a more sustainable future.

References

Abdou E, Sorour M (2014) Preparation and characterization of starch/carrageenan edible films. Int Food Res J 21(1):189

Abomohra AE-F, El-Sheekh M, Hanelt D (2014) Pilot cultivation of the chlorophyte microalga Scenedesmus obliquus as a promising feedstock for biofuel. Biomass Bioenergy 64:237–244

Abomohra AE-F, Jin W, Tu R, Han S-F, Eid M, Eladel H (2016) Microalgal biomass production as a sustainable feedstock for biodiesel: current status and perspectives. Renew Sust Energ Rev 64: 596–606

Abomohra AE, Eladel H, El-Esawi M, Wang S, Wang Q, He Z, Feng Y, Shang H, Hanelt D (2018) Effect of lipid-free microalgal biomass and waste glycerol on growth and lipid production of Scenedesmus obliquus: Innovative waste recycling for extraordinary lipid production. Bioresour Technol 249:992–999

Abomohra AE-F, Elsayed M, Esakkimuthu S, El-Sheekh M, Hanelt D (2020) Potential of fat, oil and grease (FOG) for biodiesel production: A critical review on the recent progress and future perspectives. Prog Energy Combust Sci 81:100868

Achinas S, Achinas V, Euverink GJW (2017) A Technological Overview of Biogas Production from Biowaste. Engineering 3(3):299–307

Adegbola TA, Agboola O, Fayomi OSI (2020) Review of polyacrylonitrile blends and application in manufacturing technology: recycling and environmental impact. Results in Engineering 7: 100144

Afreen R, Tyagi S, Singh GP, Singh M (2021) Challenges and perspectives of polyhydroxyalkanoate production from microalgae/cyanobacteria and bacteria as microbial factories: an assessment of hybrid biological system. Frontiers in Bioengineering and Biotechnology 9:624885

Aga MB, Dar AH, Nayik GA, Panesar PS, Allai F, Khan SA, Shams R, Kennedy JF, Altaf A (2021) Recent insights into carrageenan-based bio-nanocomposite polymers in food applications: a review. Int J Biol Macromol 192:197–209

Akindoyo JO, Beg MDH, Ghazali S, Islam MR, Jeyaratnam N, Yuvaraj AR (2016) Polyurethane types, synthesis and applications – a review. RSC Adv 6(115):114453–114482

Alagi P, Ghorpade R, Jang JH, Patil C, Jirimali H, Gite V, Hong SC (2018) Functional soybean oil-based polyols as sustainable feedstocks for polyurethane coatings. Ind Crop Prod 113 (January):249–258

Alban S, Kraus J, Franz G (1992) Synthesis of laminarin sulfates with anticoagulant activity. Arzneim Forsch 42(1008):1005–1008

Albert, J.R. 1948. Polymerization of acrylic acid nitriles., Google Patents.

Aldor IS, Keasling JD (2003) Process design for microbial plastic factories: metabolic engineering of polyhydroxyalkanoates. Curr Opin Biotechnol 14(5):475–483

Alishah Aratboni H, Rafiei N, Garcia-Granados R, Alemzadeh A, Morones-Ramirez JR (2019) Biomass and lipid induction strategies in microalgae for biofuel production and other applications. Microb Cell Factories 18(1):178

Anderson F, Hirst E, Manners D, Ross A (1958) 663. The constitution of laminarin. Part III. The fine structure of insoluble laminarin. J Chemical Soc (Resumed) 1958:3233–3243

Annan W, Hirst E, Manners D (1965) 162. The constitution of laminarin. Part V. The location of 1, 6-glucosidic linkages. J Chem Soc 1:885–891

Apt KE, Kroth-Pancic PG, Grossman AR (1996) Stable nuclear transformation of the diatom Phaeodactylum tricornutum. Mol Gen Genet 252(5):572–579

Araújo R, Vázquez Calderón F, Sánchez López J, Azevedo IC, Bruhn A, Fluch S, Garcia Tasende M, Ghaderiardakani F, Ilmjärv T, Laurans M (2021) Current status of the algae production industry in Europe: an emerging sector of the blue bioeconomy. Front Mar Sci 7: 626389

Arbenz A, Perrin R, Avérous L (2017) Elaboration and properties of innovative biobased PUIR foams from microalgae. J Polym Environ 26(1):254–262

Arnold U, Brück T, De Palmenaer A, Kuse K (2018) Carbon capture and sustainable utilization by algal polyacrylonitrile fiber production: process design, techno-economic analysis, and climate related aspects. Ind Eng Chem Res 57(23):7922–7933

Arnold U, Brück T, Battenberg A, Masri M (2020) PtX-Plus: Synergies through coupling of PtX facilities with a biorefinery. Chemie Ingenieur Technik 92(11):1797–1802

Asare MA, Kote P, Chaudhary S, de Souza FM, Gupta RK (2022) Sunflower oil as a renewable resource for polyurethane foams: effects of flame-retardants. Polymers (Basel) 14(23)

Association of Plastic, M (2020) Plastics – the Facts 2020. PlasticsEurope, 16–16.

Awad D, Younes S, Glemser M, Wagner FM, Schenk G, Mehlmer N, Brueck T (2020) Towards high-throughput optimization of microbial lipid production: from strain development to process monitoring. Sustainable Energy Fuels 4(12):5958–5969

Azevedo G, Torres M, Sousa-Pinto I, Hilliou L (2015) Effect of pre-extraction alkali treatment on the chemical structure and gelling properties of extracted hybrid carrageenan from Chondrus crispus and Ahnfeltiopsis devoniensis. Food Hydrocoll 50:150–158

Balaji S, Gopi K, Muthuvelan B (2013) A review on production of poly β hydroxybutyrates from cyanobacteria for the production of bio plastics. Algal Res 2(3):278–285

BASF. 2017. Isocyanates and Polyols.

Bayer O (1947) Das Di-isocyanat-polyadditionsverfahren (polyurethane). Angew Chem 59(9): 257–272

Becker S, Tebben J, Coffinet S, Wiltshire K, Iversen MH, Harder T, Hinrichs K-U, Hehemann J-H (2020) Laminarin is a major molecule in the marine carbon cycle. Proc Natl Acad Sci 117(12): 6599–6607

Bello D, Herrick CA, Smith TJ, Woskie SR, Streicher RP, Cullen MR, Liu Y, Redlich CA (2007) Skin exposure to isocyanates: reasons for concern. Environ Health Perspect 115(3):328–335
Benamotz A, Avron M (1990) The biotechnology of cultivating the halotolerant algae Dunaliella. Trends Biotechnol 8:121–126
Ben-Amotz A, Avron M (1983) Accumulation of metabolites by halotolerant algae and its industrial potential. Annu Rev Microbiol 37:95–119
Bergstad M, Or D, Withers PJ, Shokri N (2017) Retracted: The influence of NaCl concentration on salt precipitation in heterogeneous porous media. Water Resour Res 53(2):1702–1712
Bielenia M (2021) The role of recyclates in the polyurethane industry : environmental and economic aspects 67(139):1–7
Biersmith A, Benner R (1998) Carbohydrates in phytoplankton and freshly produced dissolved organic matter. Mar Chem 63(1-2):131–144
Bixler H (1994) The carrageenan connection IV. Br Food J 96(3):12–17
Black Wá (1950) The seasonal variation in weight and chemical composition of the common British Laminariaceae. J Mar Biol Assoc U K 29(1):45–72
Blakemore, W. 2016. Polysaccharide ingredients: carrageenan in reference module in food science. https://doi.org/10.1016/B978-0-08-100596-5.03251-0 (Elsevier, 2015).
Błaszak B, Gozdecka G, Shyichuk A (2018) Carrageenan as a functional additive in the production of cheese and cheese-like products. Acta Sci Pol Technol Aliment 17(2):107–116
Borda J, Pásztor G, Zsuga M (2000) Glycolysis of polyurethane foams and elastomers. Polym Degrad Stab 68(3):419–422
Brazdil JF (2000) Acrylonitrile. Ullmann's Encyclopedia of Industrial Chemistry
Brueck T (2019) Innovative materials with carbon fibres made from algae: German Federal Ministry of Education and Research funds TUM's Green Carbon Project. Technical University of Munich
Butcher BT, Salvaggio JE, Weill H, Ziskind MM (1976) Toluene diisocyanate (TDI) pulmonary disease: immunologic and inhalation challenge studies. J Allergy Clin Immunol 58(1 PT 1): 89–100
Caideng Y, Xiaoming Z, Liying S, Kehua T (2014) Preparation and properties of tung oil-based polyurethane. Transactions of Tianjin University 20:273–281
Calvo-Correas T, Mosiewicki MA, Corcuera MA, Eceiza A, Aranguren MI (2015) Linseed oil-based polyurethane rigid foams: synthesis and characterization. Journal of Renewable Materials 3(1):3–13
Campbell R, Hotchkiss S (2017) Carrageenan industry market overview. Tropical seaweed farming trends. Problems and opportunities, Nova Science Publishers, Inc.
Campbell GA, Meluch WC (1975) Polyurethane foam recycling. Environ Sci Technol 10(2): 182–185
Campo VL, Kawano DF, da Silva Jr DB, Carvalho I (2009) Carrageenans: biological properties, chemical modifications and structural analysis–A review. Carbohydr Polym 77(2):167–180
Canalli Bortolassi AC, Guerra VG, Aguiar ML, Soussan L, Cornu D, Miele P, Bechelany M (2019) Composites based on nanoparticle and pan electrospun nanofiber membranes for air filtration and bacterial removal. Nanomaterials (Basel) 9(12):1740
Canvin DT (1965) The effect of temperature on the oil content and fatty acid composition of the oils from several oil seed crops. Can J Bot 43(1):63–69
Carlson KD, Chang SP (1985) Chemical epoxidation of a natural unsaturated epoxy seed oil fromVernonia galamensis and a look at epoxy oil markets. J Am Oil Chem Soc 62(5):934–939
Carocho M, Heleno S, Rodrigues P, Barreiro MF, Barros L, Ferreira IC (2019) A novel natural coating for food preservation: Effectiveness on microbial growth and physicochemical parameters. LWT 104:76–83
Cavelius P, Engelhart-Straub S, Mehlmer N, Lercher J, Awad D, Bruck T (2023) The potential of biofuels from first to fourth generation. PLoS Biol 21(3):e3002063
Charoensiddhi S, Lorbeer AJ, Lahnstein J, Bulone V, Franco CM, Zhang W (2016) Enzyme-assisted extraction of carbohydrates from the brown alga Ecklonia radiata: Effect of enzyme

type, pH and buffer on sugar yield and molecular weight profiles. Process Biochem 51(10): 1503–1510

Chavarro Gomez J, Zakaria R, Aung MM, Mokhtar MN, Yunus RB (2020) Characterization of novel rigid-foam polyurethanes from residual palm oil and algae oil. Journal of Materials Research and Technology 9(6):16303–16316

Chemanalyst (2022) Decode the Future of Carbon Fiber. Chemanalyst

Chen H, Jiang JG (2009) Osmotic responses of Dunaliella to the changes of salinity. J Cell Physiol 219(2):251–258

Chen GQ, Patel MK (2012) Plastics derived from biological sources: present and future: a technical and environmental review. Chem Rev 112(4):2082–2099

Chen H, Jiang JG, Wu GH (2009) Effects of salinity changes on the growth of Dunaliella salina and its isozyme activities of glycerol-3-phosphate dehydrogenase. J Agric Food Chem 57(14): 6178–6182

Chen H, Lu Y, Jiang JG (2012) Comparative analysis on the key enzymes of the glycerol cycle metabolic pathway in Dunaliella salina under osmotic stresses. PLoS One 7(6):e37578

Chen JW, Liu WJ, Hu DX, Wang X, Balamurugan S, Alimujiang A, Yang WD, Liu JS, Li HY (2017) Identification of a malonyl CoA-acyl carrier protein transacylase and its regulatory role in fatty acid biosynthesis in oleaginous microalga Nannochloropsis oceanica. Biotechnol Appl Biochem 64(5):620–626

Chen J, Yang J, Du H, Aslam M, Wang W, Chen W, Li T, Liu Z, Liu X (2021) Laminarin, a major polysaccharide in Stramenopiles. Mar Drugs 19(10):576

Chesters CG, Bull AT (1963) The enzymic degradation of laminarin. 1. The distribution of laminarinase among micro-organisms. Biochem J 86(1):28–31

Chi S, Liu T, Wang X, Wang R, Wang S, Wang G, Shan G, Liu C (2018) Functional genomics analysis reveals the biosynthesis pathways of important cellular components (alginate and fucoidan) of Saccharina. Curr Genet 64(1):259–273

Chikkali S, Mecking S (2012) Refining of plant oils to chemicals by olefin metathesis. Angew Chem Int Ed Eng 51(24):5802–5808

Chiovitti A, Bacic A, Craik DJ, Kraft GT, Liao M-L (2004) A nearly idealized 6′-O-methylated η-carrageenan from the Australian red alga Claviclonium ovatum (Acrotylaceae, Gigartinales). Carbohydr Res 339(8):1459–1466

Chisti Y (2007) Biodiesel from microalgae. Biotechnol Adv 25(3):294–306

Chizhov AO, Dell A, Morris HR, Reason AJ, Haslam SM, McDowell RA, Chizhov OS, Usov AI (1998) Structural analysis of laminarans by MALDI and FAB mass spectrometry. Carbohydr Res 310(3):203–210

Choi JA, Oh TH, Choi JS, Chang DJ, Joo CK (2013) Impact of beta-1,3-glucan isolated from Euglena gracilis on corneal epithelial cell migration and on wound healing in a rat alkali burn model. Curr Eye Res 38(12):1207–1213

Chopin T, Wagey BT (1999) Factorial study of the effects of phosphorus and nitrogen enrichments on nutrient and carrageenan content in Chondrus crispus (Rhodophyceae) and on residual nutrient concentration in seawater 42(1):23–31

Chow YY, Goh SJ, Su Z, Ng DH, Lim CY, Lim NY, Lin H, Fang L, Lee YK (2013) Continual production of glycerol from carbon dioxide by Dunaliella tertiolecta. Bioresour Technol 136: 550–555

Chow Y, Tu WY, Wang D, Ng DH, Lee YK (2015) The role of micronutrients and strategies for optimized continual glycerol production from carbon dioxide by Dunaliella tertiolecta. Biotechnol Bioeng 112(10):2163–2171

Ciancia M, Matulewicz MC, Tuvikene R (2020) Structural Diversity in Galactans From Red Seaweeds and Its Influence on Rheological Properties. Front Plant Sci 11:559986

Cong Q, Chen H, Liao W, Xiao F, Wang P, Qin Y, Dong Q, Ding K (2016) Structural characterization and effect on anti-angiogenic activity of a fucoidan from Sargassum fusiforme. Carbohydr Polym 136:899–907

Cooper GM (2000) The cell: a molecular approach.2nd edn. ASM Press, Washington, DC

Coppola G, Gaudio MT, Lopresto CG, Calabro V, Curcio S, Chakraborty S (2021) Bioplastic from renewable biomass: a facile solution for a greener environment. Earth systems and environment 5:231–251

Cornwall W (2021) The plastic eaters. Science 373(6550):36–39

Cozar A, Echevarria F, Gonzalez-Gordillo JI, Irigoien X, Ubeda B, Hernandez-Leon S, Palma AT, Navarro S, Garcia-de-Lomas J, Ruiz A, Fernandez-de-Puelles ML, Duarte CM (2014) Plastic debris in the open ocean. Proc Natl Acad Sci U S A 111(28):10239–10244

Dahiya S, Katakojwala R, Ramakrishna S, Mohan SV (2020) Biobased products and life cycle assessment in the context of circular economy and sustainability. Materials Circular Economy 2: 1–28

Dams RI, Viana MB, Guilherme AA, Silva CM, dos Santos AB, Angenent LT, Santaella ST, Leitão RC (2018) Production of medium-chain carboxylic acids by anaerobic fermentation of glycerol using a bioaugmented open culture. Biomass Bioenergy 118:1–7

Dang BT, Bui XT, Tran DPH, Hao Ngo H, Nghiem LD, Hoang TK, Nguyen PT, Nguyen HH, Vo TK, Lin C, Lin YA, K., Varjani, S. (2022) Current application of algae derivatives for bioplastic production: A review. Bioresour Technol 347:126698

Das A, Mahanwar P (2020) A brief discussion on advances in polyurethane applications. Advanced Industrial and Engineering Polymer Research 3(3):93–101

Das S, Das S, Ghangrekar MM (2022) Enzymatic cell disruption followed by application of imposed potential for enhanced lipid extraction from wet algal biomass employing photosynthetic microbial fuel cell. Bioresour Technol 363:127924

Datta J, Kasprzyk P, Błażek K, Włoch M (2017) Synthesis, structure and properties of poly(ester-urethane)s obtained using bio-based and petrochemical 1,3-propanediol and 1,4-butanediol. J Therm Anal Calorim 130(1):261–276

Davey SG (2018) Sustainability: Sweet new route to acrylonitrile. Nature Reviews Chemistry 2(1)

Demuez M, Mahdy A, Tomas-Pejo E, Gonzalez-Fernandez C, Ballesteros M (2015) Enzymatic cell disruption of microalgae biomass in biorefinery processes. Biotechnol Bioeng 112(10): 1955–1966

Devadas VV, Khoo KS, Chia WY, Chew KW, Munawaroh HSH, Lam MK, Lim JW, Ho YC, Lee KT, Show PL (2021) Algae biopolymer towards sustainable circular economy. Bioresour Technol 325:124702

Devillé C, Damas J, Forget P, Dandrifosse G, Peulen O (2004) Laminarin in the dietary fibre concept. J Sci Food Agric 84(9):1030–1038

Doan YTT, Obbard JP (2014) Two-stage cultivation of a Nannochloropsis mutant for biodiesel feedstock. J Appl Phycol 27(6):2203–2208

Dong Z (2016) Carbon fibers from oriented polyethylene precursors. J Thermoplast Compos Mater 6(1):38–48

Doyle WT (1970) Nonseed plants: form and function. Springer

Draget KI, Smidsrød O, Skjåk-Bræk G (2005) Alginates from algae. In: Polysaccharides and polyamides in the food industry: properties, production, and patents. Wiley, Weinheim, pp 1–30

Dragone G, Fernandes BD, Vicente AA, Teixeira JA (2010) Third generation biofuels from microalgae. In: Mendez-Vilas A (ed) Current research, technology and education topics in applied microbiology and microbial biotechnology. Formatex, Madrid, pp 1315–1366

Dubois J-L, Kaliaguine S (2020) 2 Alternative routes to more sustainable acrylonitrile: biosourced acrylonitrile. In: Serge K, Jean-Luc D (eds) Industrial Green Chemistry. De Gruyter. Berlin, Boston, pp 31–62

Eccles R, Winther B, Johnston SL, Robinson P, Trampisch M, Koelsch S (2015) Efficacy and safety of iota-carrageenan nasal spray versus placebo in early treatment of the common cold in adults: the ICICC trial. Respir Res 16(1):121

Elmarhoum S, Mathieu S, Ako K, Helbert W (2023) Sulfate groups position determines the ionic selectivity and syneresis properties of carrageenan systems. Carbohydr Polym 299:120166

Energy.gov (2013) Top 9 things you didn't know about carbon fiber | Department of Energy. Department of Energy

Ermakova S, Men'shova R, Vishchuk O, Kim S-M, Um B-H, Isakov V, Zvyagintseva T (2013) Water-soluble polysaccharides from the brown alga Eisenia bicyclis: Structural characteristics and antitumor activity. Algal Res 2(1):51–58

Ezekiel HM, Spain RG (2007) Preparation of graphite fibers from polymeric fibers. Journal of Polymer Science Part C: Polymer Symposia 19(1):249–265

Fabris M, Abbriano RM, Pernice M, Sutherland DL, Commault AS, Hall CC, Labeeuw L, McCauley JI, Kuzhiuparambil U, Ray P, Kahlke T, Ralph PJ (2020) Emerging technologies in algal biotechnology: toward the establishment of a sustainable, algae-based bioeconomy. Front Plant Sci 11(March):279

Faidi A, Farhat F, Boina DA, Touati M, Le-Nouen D, Stumbé JF (2020) Physico-chemical characterization of alginates isolated from a Tunisian Padina pavonica algae as a sustainable biomaterial. Polym Int 69(11):1130–1139

Fakhry EM, El Maghraby DM (2015) Lipid accumulation in response to nitrogen limitation and variation of temperature in Nannochloropsis salina. Bot Stud 56(1):6

Falshaw R, Furneaux RH (1998) Structural analysis of carrageenans from the tetrasporic stages of the red algae, Gigartina lanceata and Gigartina chapmanii (Gigartinaceae, Rhodophyta). Carbohydr Res 307(3-4):325–331

Fang L, Qi S, Xu Z, Wang W, He J, Chen X, Liu J (2017) De novo transcriptomic profiling of Dunaliella salina reveals concordant flows of glycerol metabolic pathways upon reciprocal salinity changes. Algal Res 23:135–149

Feng S, Xue L, Liu H, Lu P (2009) Improvement of efficiency of genetic transformation for Dunaliella salina by glass beads method. Mol Biol Rep 36(6):1433–1439

Fertah M, Belfkira A, Taourirte M, Brouillette F (2017) Extraction and characterization of sodium alginate from Moroccan Laminaria digitata brown seaweed. Arab J Chem 10:S3707–S3714

Ficko-Blean E, Hervé C, Michel G (2015) Sweet and sour sugars from the sea: the biosynthesis and remodeling of sulfated cell wall polysaccharides from marine macroalgae. Perspectives in Phycology 2(1):51–64

Fortunebusinessinsights (2021) The global polyurethane market is projected to grow from $57.34 billion in 2021 to $81.74 billion in 2028 at a CAGR of 5.1% in the forecast period, 2021-2028.

Freedman B, Pryde EH, Mounts TL (1984) Variables affecting the yields of fatty esters from transesterified vegetable oils. JAOCS, J American Oil Chemists' Soc 61(10):1638–1643

Furtwengler P, Avérous L (2018) Renewable polyols for advanced polyurethane foams from diverse biomass resources. Polym Chem 9(32):4258–4287

Gadhave RV, Mahanwar PA, Gadekar PT (2017) Bio-renewable sources for synthesis of eco-friendly polyurethane adhesives 4(7):57–75

Gadhave RV, Srivastava S, Mahanwar PA, Gadekar PT (2019) Recycling and disposal methods for polyurethane wastes: a review. Open Journal of Polymer Chemistry 9(2):39–51

Gaikwad MS, Gite VV, Mahulikar PP, Hundiwale DG, Yemul OS (2015) Eco-friendly polyurethane coatings from cottonseed and karanja oil. Prog Org Coat 86:164–172

Galus S, Kadzińska J (2015) Food applications of emulsion-based edible films and coatings. Trends Food Sci Technol 45(2):273–283

Gama NV, Ferreira A, Barros-Timmons A (2018) Polyurethane foams: Past, present, and future. Materials (Basel) 11(10):1841

Garcia-Moscoso JL, Obeid W, Kumar S, Hatcher PG (2013) Flash hydrolysis of microalgae (*Scenedesmus* sp.) for protein extraction and production of biofuels intermediates. J Supercrit Fluids 82:183–190

Gedde UW, Hedenqvist MS (2019) Fundamental polymer science. Springer

Geyer R, Jambeck JR, Law KL (2017) Production, use, and fate of all plastics ever made. Sci Adv 3(7):3–8

Gheorghita Puscaselu R, Lobiuc A, Dimian M, Covasa M (2020) Alginate: From food industry to biomedical applications and management of metabolic disorders. Polymers 12(10):2417

Ghosh A, Khanra S, Mondal M, Halder G, Tiwari O, Saini S, Bhowmick TK, Gayen K (2016) Progress toward isolation of strains and genetically engineered strains of microalgae for production of biofuel and other value added chemicals: A review. Energy Convers Manag 113: 104–118

Gironi F, Piemonte V (2011) Bioplastics and petroleum-based plastics: strengths and weaknesses. Energy Sources, Part A: Recovery, Utilization, and Environmental Effects 33(21):1949–1959

Glowinska E, Gotkiewicz O, Kosmela P (2023) Sustainable strategy for algae biomass waste management via development of novel bio-based thermoplastic polyurethane elastomers composites. Molecules 28(1):20–20

Gomez CG, Lambrecht MVP, Lozano JE, Rinaudo M, Villar MA (2009) Influence of the extraction–purification conditions on final properties of alginates obtained from brown algae (Macrocystis pyrifera). Int J Biol Macromol 44(4):365–371

Goud VV, Patwardhan AV, Pradhan NC (2006) Studies on the epoxidation of mahua oil (Madhumica indica) by hydrogen peroxide. Bioresour Technol 97(12):1365–1371

Graiff A, Ruth W, Kragl U, Karsten U (2016) Chemical characterization and quantification of the brown algal storage compound laminarin—a new methodological approach. J Appl Phycol 28(1):533–543

GrandViewResearch. 2022. Polyurethane (PU) foam market: Global industry trends, share, size, growth, opportunity and forecast 2022-2027- product image, pp. 105–105

Grasselli RK, Trifirò F (2016) Acrylonitrile from biomass: still far from being a sustainable process. Top Catal 59(17-18):1651–1658

Gulrez S, Al-Assaf S, Phillips G (2010) Characterisation and radiation modification of carrageenan in the solid state. R. K. Khandal & U. Niyogi (Eds.), Radiation processing technology applications (1st ed., pp. 173–186). SRI Publication, New Delhi

Gunawan NR, Tessman M, Schreiman AC, Simkovsky R, Samoylov AA, Neelakantan NK, Bemis TA, Burkart MD, Pomeroy RS, Mayfield SP (2020) Rapid biodegradation of renewable polyurethane foams with identification of associated microorganisms and decomposition products. Bioresour Technol Rep 11:100513–100513

Gunawan NR, Tessman M, Zhen D, Johnson L, Evans P, Clements SM, Pomeroy RS, Burkart MD, Simkovsky R, Mayfield SP (2022) Biodegradation of renewable polyurethane foams in marine environments occurs through depolymerization by marine microorganisms. Sci Total Environ 850(September):158761

Haijin M, Xiaolu J, Huashi G (2003) A κ-carrageenan derived oligosaccharide prepared by enzymatic degradation containing anti-tumor activity. J Appl Phycol 15:297–303

Hannon M, Gimpel J, Tran M, Rasala B, Mayfield S (2010) Biofuels from algae: challenges and potential. Biofuels 1(5):763–784

Harussani MM, Sapuan SM, Nadeem G, Rafin T, Kirubaanand W (2022) Recent applications of carbon-based composites in defence industry: A review. Defence Technology 18(8):1281–1300

Harvey PJ, Ben-Amotz A (2020) Towards a sustainable Dunaliella salina microalgal biorefinery for 9-cis β-carotene production. Algal Res 50:102002

Haug A, Jensen A (1954) Seasonal variations in the chemical composition of Alaria esculenta, Laminaria saccharina. Laminaria hyperborea and Laminaria digitata from northern Norway, Akademisk Trykningssentral

Hawkins WW, O'Neill AN (1955) The anticoagulant action in blood of sulphated derivatives of laminarin. Can J Biochem Physiol 33(4):545–552

Hawthorne HM, Baker C, Bentall RH, Linger KR (1970) High strength, high modulus graphite fibres from pitch. Nature 227(5261):946–947

Hay ID, Ur Rehman Z, Moradali MF, Wang Y, Rehm BH (2013) Microbial alginate production, modification and its applications. Microb Biotechnol 6(6):637–650

Hazmi ASA, Aung MM, Abdullah LC, Salleh MZ, Mahmood MH (2013) Producing Jatropha oil-based polyol via epoxidation and ring opening. Ind Crop Prod 50:563–567

Hein S, Tran H, Steinbüchel A (1998) Synechocystis sp. PCC6803 possesses a two-component polyhydroxyalkanoic acid synthase similar to that of anoxygenic purple sulfur bacteria. Arch Microbiol 170:162–170

Hempel F, Bozarth AS, Lindenkamp N, Klingl A, Zauner S, Linne U, Steinbüchel A, Maier UG (2011) Microalgae as bioreactors for bioplastic production. Microb Cell Factories 10:1–6

Hentati F, Tounsi L, Djomdi D, Pierre G, Delattre C, Ursu AV, Fendri I, Abdelkafi S, Michaud P (2020) Bioactive polysaccharides from seaweeds. Molecules 25(14):3152

Holdt SL, Kraan S (2011) Bioactive compounds in seaweed: functional food applications and legislation. J Appl Phycol 23(3):543–597

Hosseini Tafreshi A, Shariati M (2009) Dunaliella biotechnology: methods and applications. J Appl Microbiol 107(1):14–35

Hotchkiss S, Brooks M, Campbell R, Philp K, Trius A (2016) The use of carrageenan in food. Nova Science

Houtz R (1950) "Orlon" acrylic fiber: Chemistry and properties. Text Res J 20(11):786–801

Hulme AJ, Goodhead TC (2003) Cost effective reprocessing of polyurethane by hot compression moulding. J Mater Process Technol 139(1-3):322–326

Ichikawa J, Asami M, Mukaiyama T (1984) An asymmetric synthesis of glycerol derivatives by the enantioselective acylation of prochiral glycerol. Chem Lett 13(6):949–952

IGTPAN (2020) Carbon fiber market. Grando Institute of Polyacrylonitrile Technology

IGTPAN (2023) Polyacylonitrile recycling method. Grando Institute of Polyacrylonitrile Technology

IMARC. 2021. Polyols market: Global industry trends, share, size, growth, opportunity and forecast 2022-2027.

IMARC. 2022. Global polyurethane market size, share & trends analysis report by product (rigid foam, flexible foam), by application (construction, furniture & interiors), by region, and segment forecasts, 2022-2030.

Imeson, A. 2000. Handbook of hydrocolloids Woodhead Publishing.

Ionescu M (2005) Chemistry and technology of polyols for polyurethanes. Smithers Rapra Publishing

Ionescu M, Radojčić D, Wan X, Shrestha ML, Petrović ZS, Upshaw TA (2016) Highly functional polyols from castor oil for rigid polyurethanes. Eur Polym J 84(620):736–749

Jarmander J, Belotserkovsky J, Sjoberg G, Guevara-Martinez M, Perez-Zabaleta M, Quillaguaman J, Larsson G (2015) Cultivation strategies for production of (R)-3-hydroxybutyric acid from simultaneous consumption of glucose, xylose and arabinose by Escherichia coli. Microb Cell Factories 14:51

Jendrossek D, Handrick R (2002) Microbial degradation of polyhydroxyalkanoates. Annu Rev Microbiol 56(1):403–432

Jensen A, Haug A (1956) Geographical and seasonal variation in the chemical composition of Laminaria hyperborea and Laminaria digitata from the Norwegian coast. Akademisk. trykningssentral

Kadam SU, O'Donnell CP, Rai DK, Hossain MB, Burgess CM, Walsh D, Tiwari BK (2015a) Laminarin from Irish brown seaweeds Ascophyllum nodosum and Laminaria hyperborea: Ultrasound assisted extraction, characterization and bioactivity. Mar Drugs 13(7):4270–4280

Kadam SU, Tiwari BK, O'Donnell CP (2015b) Extraction, structure and biofunctional activities of laminarin from brown algae. Int J Food Sci Technol 50(1):24–31

Kaliaguine S, Dubois J-L (2020) Industrial Green Chemistry. De Gruyter

Kamińska-Dwórznicka A, Antczak A, Samborska K, Lenart A (2015) Acid hydrolysis of kappa-carrageenan as a way of gaining new substances for freezing process modification and protection from excessive recrystallisation of ice. Int J Food Sci Technol 50(8):1799–1806

Kamravamanesh D, Lackner M, Herwig C (2018) Bioprocess engineering aspects of sustainable polyhydroxyalkanoate production in Cyanobacteria. Bioengineering (Basel) 5(4):111

Kanaya K, Takahashi S (1994) Decomposition of polyurethane foams by alkanolamines. J Appl Polym Sci 51(4):675–682

Kandanarachchi P, Guo A, Petrovic Z (2002) The hydroformylation of vegetable oils and model compounds by ligand modified rhodium catalysis. J Mol Catal A Chem 184(1-2):65–71

Karim AA, Bhat R (2008) Gelatin alternatives for the food industry: recent developments, challenges and prospects. Trends Food Sci Technol 19(12):644–656

Karp EM, Eaton TR, Sanchez INV, Vorotnikov V, Biddy MJ, Tan ECD, Brandner DG, Cywar RM, Liu R, Manker LP, Michener WE, Gilhespy M, Skoufa Z, Watson MJ, Fruchey OS, Vardon DR, Gill RT, Bratis AD, Beckham GT (2017) Renewable acrylonitrile production. Science 358(6368):1307–1310

Kaur J, Millington K, Smith S (2016) Producing high-quality precursor polymer and fibers to achieve theoretical strength in carbon fibers: a review. J Appl Polym Sci 133(38)

Kausar A (2017) Polyurethane composite foams in high-performance applications: a review. Polym-Plast Technol Eng 57(4):346–369

Kim J-S, Pathak TS, Yun J-H, Kim K-P, Park T-J, Kim Y, Paeng K-J (2012) Thermal degradation and kinetics of alginate polyurethane hybrid material prepared from alginic acid as a polyol. J Polym Environ 21(1):224–232

Kleinübing SJ, Gai F, Bertagnolli C, Silva MGCd (2013) Extraction of alginate biopolymer present in marine alga Sargassum filipendula and bioadsorption of metallic ions. Mater Res 16:481–488

Kliem S, Kreutzbruck M, Bonten C (2020) Review on the biological degradation of polymers in various environments. Materials 13(20):4586

Knutsen, S., Myslabodski, D., Larsen, B., Usov, A. 1994. A modified system of nomenclature for red algal galactans.

Kobayashi S, Müllen K (2015) Encyclopedia of polymeric nanomaterials. Springer

Koh HG, Kang NK, Jeon S, Shin SE, Jeong BR, Chang YK (2019) Heterologous synthesis of chlorophyll b in Nannochloropsis salina enhances growth and lipid production by increasing photosynthetic efficiency. Biotechnol Biofuels 12:122

Kong X, Liu G, Curtis JM (2012) Novel polyurethane produced from canola oil based poly(ether ester) polyols: Synthesis, characterization and properties. Eur Polym J 48(12):2097–2106

Kontominas MG (2020) Use of alginates as food packaging materials. Foods 9(10):1440

Korley LTJ, Pate BD, Thomas EL, Hammond PT (2006) Effect of the degree of soft and hard segment ordering on the morphology and mechanical behavior of semicrystalline segmented polyurethanes. Polymer 47(9):3073–3082

Kosmela P, Gosz K, Kazimierski P, Hejna A, Haponiuk JT, Piszczyk Ł (2019) Chemical structures, rheological and physical properties of biopolyols prepared via solvothermal liquefaction of Enteromorpha and Zostera marina biomass. Cellulose 26(10):5893–5912

Kostik V, Memeti S, Bauer B (2013) Fatty acid composition of edible oils and fats. Journal of Hygienic Engineering and Design 4:112–116

Kricheldorf H (2013) Handbook of polymer synthesis. John Wiley & Sons

Kumar S, Hablot E, Moscoso JLG, Obeid W, Hatcher PG, DuQuette BM, Graiver D, Narayan R, Balan V (2014) Polyurethanes preparation using proteins obtained from microalgae. J Mater Sci 49(22):7824–7833

Kumar YN, Poong SW, Gachon C, Brodie J, Sade A, Lim PE (2020) Impact of elevated temperature on the physiological and biochemical responses of Kappaphycus alvarezii (Rhodophyta). PLoS One 15(9):e0239097

Kwon O-J, Oh S-T, Lee S-D, Lee N-R, Shin C-H, Park J-S (2007) Hydrophilic and flexible polyurethane foams using sodium alginate as polyol: Effects of PEG molecular weight and cross-linking agent content on water absorbency. Fibers and Polymers 8(4):347–355

Kwon S, Kang NK, Koh HG, Shin SE, Lee B, Jeong BR, Chang YK (2018) Enhancement of biomass and lipid productivity by overexpression of a bZIP transcription factor in Nannochloropsis salina. Biotechnol Bioeng 115(2):331–340

Łabowska MB, Michalak I, Detyna J (2019) Methods of extraction, physicochemical properties of alginates and their applications in biomedical field–a review. Open Chemistry 17(1):738–762

Lacoste C, El Hage R, Bergeret A, Corn S, Lacroix P (2018) Sodium alginate adhesives as binders in wood fibers/textile waste fibers biocomposites for building insulation. Carbohydr Polym 184: 1–8

Lamberti FM, Román-Ramírez LA, Wood J (2020) Recycling of bioplastics: routes and benefits. J Polym Environ 28:2551–2571

Lässig R, Eisenhut M, Arne M, Schulte RT, Peters F, Kühmann T, Waldmann T, Begemann W (2012) Serienproduktion von hochfesten Faserverbundbauteilen: Perspektiven für den deutschen Maschinen- und Anlagenbau

Laurens LML, Markham J, Templeton DW, Christensen ED, Van Wychen S, Vadelius EW, Chen-Glasser M, Dong T, Davis R, Pienkos PT (2017) Development of algae biorefinery concepts for biofuels and bioproducts; a perspective on process-compatible products and their impact on cost-reduction. Energy Environ Sci 10(8):1716–1738

Lee KY, Mooney DJ (2012) Alginate: properties and biomedical applications. Prog Polym Sci 37(1):106–126

Li J, Xie L, Qin Y, Liang WH, Mo MQ, Liu SL, Liang F, Wang Y, Tan W, Liang Y (2013) Effect of laminarin polysaccharide on activity of matrix metalloproteinase in photoaging skin. Zhongguo Zhong Yao Za Zhi 38(14):2370–2373

Li L, Ni R, Shao Y, Mao S (2014) Carrageenan and its applications in drug delivery. Carbohydr Polym 103:1–11

Li DW, Cen SY, Liu YH, Balamurugan S, Zheng XY, Alimujiang A, Yang WD, Liu JS, Li HY (2016) A type 2 diacylglycerol acyltransferase accelerates the triacylglycerol biosynthesis in heterokont oleaginous microalga Nannochloropsis oceanica. J Biotechnol 229:65–71

Li J, Cai C, Yang C, Li J, Sun T, Yu G (2019) Recent advances in pharmaceutical potential of brown algal polysaccharides and their derivatives. Curr Pharm Des 25(11):1290–1311

Li Q, Zheng L, Guo Z, Tang T, Zhu B (2021) Alginate degrading enzymes: an updated comprehensive review of the structure, catalytic mechanism, modification method and applications of alginate lyases. Crit Rev Biotechnol 41(6):953–968

Li S-Q, Tang R-C, Yu C-B (2022) Flame retardant treatment of jute fabric with chitosan and sodium alginate. Polym Degrad Stab 196:109826

Liebig C, Paul S, Katryniok B, Guillon C, Couturier J-L, Dubois J-L, Dumeignil F, Hoelderich WF (2013) Glycerol conversion to acrylonitrile by consecutive dehydration over WO3/TiO2 and ammoxidation over Sb-(Fe,V)-O. Appl Catal B Environ 132–133:170–182

Liu J, Zhan X, Wan J, Wang Y, Wang C (2015a) Review for carrageenan-based pharmaceutical biomaterials: favourable physical features versus adverse biological effects. Carbohydr Polym 121:27–36

Liu Y, Zwingmann B, Schlaich M (2015b) Carbon fiber reinforced polymer for cable structures—A Review. Polymers 7(10):2078–2099

Lligadas G, Ronda JC, Galià M, Cádiz V (2013) Monomers and polymers from plant oils via click chemistry reactions. J Polym Sci A Polym Chem 51(10):2111–2124

Lomax GR (2007) Breathable polyurethane membranes for textile and related industries. J Mater Chem 17(27):2775–2784

Lorbeer AJ, Charoensiddhi S, Lahnstein J, Lars C, Franco CM, Bulone V, Zhang W (2017) Sequential extraction and characterization of fucoidans and alginates from Ecklonia radiata, Macrocystis pyrifera, Durvillaea potatorum, and Seirococcus axillaris. J Appl Phycol 29:1515–1526

Lorenzen J, Igl N, Tippelt M, Stege A, Qoura F, Sohling U, Bruck T (2017) Extraction of microalgae derived lipids with supercritical carbon dioxide in an industrial relevant pilot plant. Bioprocess Biosyst Eng 40(6):911–918

Loureiro RR, Cornish M, Neish IC (2017) Applications of carrageenan: with special reference to iota and kappa forms as derived from the eucheumatoid seaweeds. Focus on Kappaphycus and Eucheuma of Commerce, Tropical Seaweed Farming Trends, Problems and Opportunities, pp 165–171

Lucintel (2022) Market value of carbon fiber worldwide in 2016 and 2021, with a forecast for 2027 (in billion U.S. dollars)

Ma R, Wang B, Chua ET, Zhao X, Lu K, Ho SH, Shi X, Liu L, Xie Y, Lu Y, Chen J (2020) Comprehensive utilization of marine microalgae for enhanced co-production of multiple compounds. Mar Drugs 18(9):467

Mack D, Schatzle S, Traa Y, Klemm E (2019) Synthesis of acrylonitrile from renewable lactic acid. ChemSusChem 12(8):1653–1663

Madison LL, Huisman GW (1999) Metabolic engineering of poly(3-hydroxyalkanoates): from DNA to plastic. Microbiol Mol Biol Rev 63(1):21–53

Maeda M, Nishizawa K (1968) Fine structure of laminaran of Eisenia bicyclis. J Biochem 63(2): 199–206

Maisonneueve L, Chollet G, Grau E, Cramail H (2016) Vegetable oils : a source of polyols for polyurethane materials. OCL Oilseeds and Fats Crops and Lipids 23(5):10–10

Maisonneuve L, Lamarzelle O, Rix E, Grau E, Cramail H (2015) Isocyanate-free routes to polyurethanes and poly(hydroxy Urethane)s. Chem Rev 115(22):12407–12439

Manocha LM (2001) Encyclopedia of materials: science and technology. Elsevier, Amsterdam

Markets MA (2022) Green and biopolyols market by raw material, type, application, end-use industry and region - Global forecast to 2027

MarketWatch (2023) Global polyacrylonitrile (PAN) market

Markl E, Grünbichler H, Lackner M (2018a) Cyanobacteria for PHB bioplastics production: a review. Algae. https://doi.org/10.5772/intechopen.81536

Markl E, Grünbichler H, Lackner M (2018b) PHB-bio based and biodegradable replacement for PP: a review. Nov Tech Nutr Food Sci 2(5):206–209

Marlina R (2010) Sintesis membran poliuretan dari karagenan dan 2, 4 toylulene diisosianat. Jurnal Rekayasa Kimia dan Lingkungan 7(3):138–148

Marlina, Iqhrammullah M, Saleha S, Fathurrahmi, Maulina FP, Idroes R (2020) Polyurethane film prepared from ball-milled algal polyol particle and activated carbon filler for NH(3)-N removal. Heliyon 6(8):e04590

Masson J (1995) Acrylic fiber technology and applications. CRC Press

Matyjaszewski K, Mu Jo S, Paik H-J, Gaynor SG (1997) Synthesis of well-defined polyacrylonitrile by atom transfer radical polymerization. Macromolecules 30(20):6398–6400

MaximizeMarketResearch (2022) Carbon Fiber Market – Global Industry Analysis And Forecast (2022 to 2029), Maximize Market Research

McCandless EL, Craigie JS, Walter JA (1973) Carrageenans in the gametophytic and sporophytic stages of Chondrus crispus. Planta 112(3):201–212

McKenna ST, Hull TR (2016) The fire toxicity of polyurethane foams. Fire Science Reviews 5(1): 1–27

Meng F, Zhang Y, Xiong Z, Wang G, Li F, Zhang L (2018a) Mechanical, hydrophobic and thermal properties of an organic-inorganic hybrid carrageenan-polyvinyl alcohol composite film. Compos Part B 143:1–8

Meng F, Zhou Y, Liu J, Wu J, Wang G, Li R, Zhang Y (2018b) Thermal decomposition behaviors and kinetics of carrageenan-poly vinyl alcohol bio-composite film. Carbohydr Polym 201:96–104

Menshova RV, Ermakova SP, Anastyuk SD, Isakov VV, Dubrovskaya YV, Kusaykin MI, Um BH, Zvyagintseva TN (2014) Structure, enzymatic transformation and anticancer activity of branched high molecular weight laminaran from brown alga Eisenia bicyclis. Carbohydr Polym 99:101–109

Michel G, Helbert W, Kahn R, Dideberg O, Kloareg B (2003) The structural bases of the processive degradation of iota-carrageenan, a main cell wall polysaccharide of red algae. J Mol Biol 334(3): 421–433

Michel G, Tonon T, Scornet D, Cock JM, Kloareg B (2010) Central and storage carbon metabolism of the brown alga Ectocarpus siliculosus: insights into the origin and evolution of storage carbohydrates in Eukaryotes. New Phytol 188(1):67–81

Mignon A, Snoeck D, D'Halluin K, Balcaen L, Vanhaecke F, Dubruel P, Van Vlierberghe S, De Belie N (2016) Alginate biopolymers: Counteracting the impact of superabsorbent polymers on mortar strength. Constr Build Mater 110:169–174

Miles JM, Park YS, Walewicz D, Russell-Lopez C, Windsor S, Isley WL, Coppack SW, Harris WS (2004) Systemic and forearm triglyceride metabolism: fate of lipoprotein lipase-generated glycerol and free fatty acids. Diabetes 53(3):521–527

Millane RP, Chandrasekaran R, Arnott S, Dea IC (1988) The molecular structure of kappa-carrageenan and comparison with iota-carrageenan. Carbohydr Res 182(1):1–17

Mitic, J. 2021. Construction industry needs whole life carbon understanding to hit net zero – new report shows.

Mohammed A, Rivers A, Stuckey DC, Ward K (2020) Alginate extraction from Sargassum seaweed in the Caribbean region: Optimization using response surface methodology. Carbohydr Polym 245:116419

Monlau F, Suarez-Alvarez S, Lallement A, Vaca-Medina G, Giacinti G, Munarriz M, Urreta I, Raynaud C, Ferrer C, Castañón S (2021) A cascade biorefinery for the valorization of microalgal biomass: biodiesel, biogas, fertilizers and high valuable compounds. Algal Res 59:102433

Monteiro MR, Kugelmeier CL, Pinheiro RS, Batalha MO, da Silva César A (2018) Glycerol from biodiesel production: Technological paths for sustainability. Renew Sust Energ Rev 88:109–122

Mordhay Avron, A.B.-A. 1978. Production of glycerol from algae.

Morrissey J, Kraan S, Guiry MD (2001) A guide to commercially important seaweeds on the Irish coast. Irish Bord Iascaigh Mhara/Irish Sea Fisheries Board, Dublin

Msaada K, Hosni K, Taarit MB, Hammami M, Marzouk B (2009) Effects of growing region and maturity stages on oil yield and fatty acid composition of coriander (Coriandrum sativum L.) fruit. Sci Hortic 120(4):525–531

Munoz CF, Sudfeld C, Naduthodi MIS, Weusthuis RA, Barbosa MJ, Wijffels RH, D'Adamo S (2021) Genetic engineering of microalgae for enhanced lipid production. Biotechnol Adv 52: 107836

Murugappan V, Muthadhi A (2022) Studies on the influence of alginate as a natural polymer in mechanical and long-lasting properties of concrete–a review. Mater Today Proc 65:839–845

Nanda N, Bharadvaja N (2022) Algal bioplastics: current market trends and technical aspects. Clean Techn Environ Policy 24(9):2659–2679

Narendra MV, Shakti M, Amitesh S, Bhartendu NM (2010) Prospective of biodiesel production utilizing microalgae as the cell factories: a comprehensive discussion. Afr J Biotechnol 9(10): 1402–1411

Necas J, Bartosikova L (2013) Carrageenan: a review. Vet Med 58(4):187–205

Neill P (2020) Construction industry accounts for 38% of CO2 emissions, environment journal

Nelson TE, Lewis BA (1974) Separation and characterization of the soluble and insoluble components of insoluble laminaran. Carbohydr Res 33(1):63–74

Noreen A, Mahmood S, Khalid A, Takriff S, Anjum M, Riaz L, Ditta A, Mahmood T (2022) Synthesis and characterization of bio-based UV curable polyurethane coatings from algal biomass residue. Biomass Conv Bioref:1–17. https://doi.org/10.1007/s13399-022-03143-3

Nurman S, Saiful S, Ginting B, Rahmi R, Marlina M, Wibisono Y (2021) Synthesis of polyurethane membranes derived from Rred seaweed biomass for ammonia filtration. Membranes (Basel) 11(9):668

Onen Cinar S, Chong ZK, Kucuker MA, Wieczorek N, Cengiz U, Kuchta K (2020) Bioplastic production from microalgae: a review. Int J Environ Res Public Health 17(11):3842

Oren A (2014) The ecology of Dunaliella in high-salt environments. J Biol Res (Thessalon) 21(1): 1–8

Otera J (2002) Transesterification. Chem Rev 93(4):1449–1470

Ozanne H, Toumi H, Roubinet B, Landemarre L, Lespessailles E, Daniellou R, Cesaro A (2020) Laminarin effects, a β-(1, 3)-glucan, on skin cell inflammation and oxidation. Cosmetics 7(3):66

Pacheco-Quito EM, Ruiz-Caro R, Veiga MD (2020) Carrageenan: drug delivery systems and other biomedical applications. Mar Drugs 18(11):583

Palm E, Svensson Myrin E (2018) Mapping the plastics system and its sustainability challenges (IMES/EEES report series; Vol. 108)

Pang K, Kotek R, Tonelli A (2006) Review of conventional and novel polymerization processes for polyesters. Prog Polym Sci 31(11):1009–1037

Patil CK, Jirimali HD, Paradeshi JS, Chaudhari BL, Alagi PK, Mahulikar PP, Hong SC, Gite VV (2021) Chemical transformation of renewable algae oil to polyetheramide polyols for polyurethane coatings. Prog Org Coat 151(August 2020):106084–106084

Pawar MS, Kadam AS, Dawane BS, Yemul OS (2015) Synthesis and characterization of rigid polyurethane foams from algae oil using biobased chain extenders. Polym Bull 73(3):727–741

Pawlik H, Prociak A (2011) Influence of palm oil-based polyol on the properties of flexible polyurethane foams. J Polym Environ 20(2):438–445

Peijs T, Kirschbaum R, Lemstra PJ (2022) Chapter 5: A critical review of carbon fiber and related products from an industrial perspective. Adv Ind Eng Polym Res 5(2):90–106

Percival E, Ross A (1951) 156. The constitution of laminarin. Part II. The soluble laminarin of Laminaria digitata. J Chemical Soc (Resumed) 1951:720–726

Pereira L, Critchley AT, Amado AM, Ribeiro-Claro PJ (2009) A comparative analysis of phycocolloids produced by underutilized versus industrially utilized carrageenophytes (Gigartinales, Rhodophyta). J Appl Phycol 21(5):599–605

Peteiro C (2018) Alginate production from marine macroalgae, with emphasis on kelp farming. In: Alginates and their biomedical applications. Springer, Singapore, pp 27–66

Petrović ZS, Zlatanić A, Lava CC, Sinadinović-Fišer S (2002) Epoxidation of soybean oil in toluene with peroxoacetic and peroxoformic acids— kinetics and side reactions. Eur J Lipid Sci Technol 104(5):293–299

Petrovic ZS, Zhang W, Javni I (2005) Structure and properties of polyurethanes prepared from triglyceride polyols by ozonolysis. Biomacromolecules 6(2):713–719

Petrović ZS, Wan X, Bilić O, Zlatanić A, Hong J, Javni I, Ionescu M, Milić J, Degruson D (2013) Polyols and polyurethanes from crude algal oil. J Am Oil Chem Soc 90(7):1073–1078

Peyrton J, Avérous L (2021) Structure-properties relationships of cellular materials from biobased polyurethane foams. Mater Sci Eng R Rep 145(October 2020):100608

Peyrton J, Chambaretaud C, Sarbu A, Avérous L (2020) Biobased polyurethane foams based on new polyol architectures from microalgae oil. ACS Sustain Chem Eng 8(32):12187–12196

Philip S, Keshavarz T, Roy I (2007) Polyhydroxyalkanoates: biodegradable polymers with a range of applications. J Chem Technol Biotechnol: International Research in Process, Environmental & clean technology 82(3):233–247

Phung Hai TA, Tessman M, Neelakantan N, Samoylov AA, Ito Y, Rajput BS, Pourahmady N, Burkart MD (2021) Renewable polyurethanes from sustainable biological precursors. Biomacromolecules 22(5):1770–1794

Pierre G, Delattre C, Laroche C, Michaud P (2015) Galactans and its applications. In: Polysaccharides. Springer, Berlin, pp 753–794

Polaczek K, Kuranska M (2022) Hemp seed oil and oilseed radish oil as new sources of raw materials for the synthesis of bio-polyols for open-cell polyurethane foams. Materials 15(24): 8891

Poltronieri P, Kumar P (2017) Polyhydroxyalkanoates (PHAs) in industrial applications. Handb Ecomaterials 4:2843–2872

Posen ID, Jaramillo P, Landis AE, Griffin WM (2017) Greenhouse gas mitigation for US plastics production: energy first, feedstocks later. Environ Res Lett 12(3):034024

Quero GM, Luna GM (2017) Surfing and dining on the "plastisphere": microbial life on plastic marine debris. Adv Oceanogr Limnol 8(2):199

Radakovits R, Jinkerson RE, Fuerstenberg SI, Tae H, Settlage RE, Boore JL, Posewitz MC (2012) Draft genome sequence and genetic transformation of the oleaginous alga Nannochloropis gaditana. Nat Commun 3:686

Ramanujam S, Zequine C, Bhoyate S, Neria B, Kahol P, Gupta R (2019) Novel biobased polyol using corn oil for highly flame-retardant polyurethane foams. C 5(1):13–13

Ray S, Kalia VC (2017) Biomedical applications of polyhydroxyalkanoates. Indian J Microbiol 57(3):261–269

Raza ZA, Riaz S, Banat IM (2018) Polyhydroxyalkanoates: properties and chemical modification approaches for their functionalization. Biotechnol Prog 34(1):29–41

Read SM, Currie G, Bacic A (1996) Analysis of the structural heterogeneity of laminarin by electrospray-ionisation-mass spectrometry. Carbohydr Res 281(2):187–201

Reddy SG (2021) Alginates-A seaweed product: its properties and applications. In: Properties and applications of alginates. IntechOpen, London

Reichert CL, Bugnicourt E, Coltelli MB, Cinelli P, Lazzeri A, Canesi I, Braca F, Martinez BM, Alonso R, Agostinis L, Verstichel S, Six L, Mets S, Gomez EC, Issbrucker C, Geerinck R, Nettleton DF, Campos I, Sauter E, Pieczyk P, Schmid M (2020) Bio-based packaging: materials, modifications, industrial applications and sustainability. Polymers 12(7):1558

Ren X, Liu L, Zhou Y, Zhu Y, Zhang H, Zhang Z, Li H (2016) Nanoparticle siRNA against BMI-1 with a polyethylenimine-laminarin conjugate for gene therapy in human breast cancer. Bioconjug Chem 27(1):66–73

Ribeiro I, Kaufmann J, Götze U, Peças P, Henriques E (2018) Fibre reinforced polymers in the sports industry – life cycle engineering methodology applied to a snowboard using anisotropic layer design. Int J Sustain Eng 12(3):201–211

Rioux LE, Turgeon SL, Beaulieu M (2010) Structural characterization of laminaran and galactofucan extracted from the brown seaweed Saccharina longicruris. Phytochemistry 71(13):1586–1595

Rocher DF, Cripwell RA, Viljoen-Bloom M (2021) Engineered yeast for enzymatic hydrolysis of laminarin from brown macroalgae. Algal Res 54:102233

Rodolfi L, Chini Zittelli G, Bassi N, Padovani G, Biondi N, Bonini G, Tredici MR (2009) Microalgae for oil: strain selection, induction of lipid synthesis and outdoor mass cultivation in a low-cost photobioreactor. Biotechnol Bioeng 102(1):100–112

Rodriguez-Rodriguez DR, Ramirez-Solis R, Garza-Elizondo MA, Garza-Rodriguez ML, Barrera-Saldana HA (2019) Genome editing: a perspective on the application of CRISPR/Cas9 to study human diseases (Review). Int J Mol Med 43(4):1559–1574

Rossio RC, Vecchio M, Abramczyk J (1993) Polyurethane energy absorbing foams for automotive applications. J Mater Manuf 102(5)

Rotchell JM, Jenner LC, Chapman E, Bennett RT, Bolanle IO, Loubani M, Sadofsky L, Palmer TM (2023) Detection of microplastics in human saphenous vein tissue using muFTIR: a pilot study. PLoS One 18(2):e0280594

Rudnik E (2013) Compostable polymer properties and packaging applications. In: Plastic films in food packaging. Elsevier, Waltham, MA, pp 217–248

Ruiz-Riancho N, Garcia A, Grossegger D, Saadoon T, Hudson-Griffiths R (2021) Properties of Ca-alginate capsules to maximise asphalt self-healing properties. Constr Build Mater 284: 122728

Rupert R, Rodrigues KF, Thien VY, Yong WTL (2022) Carrageenan from Kappaphycus alvarezii (Rhodophyta, Solieriaceae): metabolism, structure, production, and application. Front Plant Sci 13:859635

Ryther JH, Dunstan WM (1971) Nitrogen, phosphorus, and eutrophication in the coastal marine environment. Science 171(3975):1008–1013

Saha P, Aloui H, Yun JH, Kim HS, Kim BS (2020) Development of a novel composite film based on polyurethane and defatted Chlorella biomass: physical and functional characterization. J Appl Polym Sci 138(14):1–11

Samancı B, Özkaynak E (2003) Effect of planting date on seed yield, oil content and fatty acid composition of safflower (Carthamus tinctorius) cultivars grown in the Mediterranean region of Turkey. J Agron Crop Sci 189(5):359–360

Saratale RG, Cho S-K, Ghodake GS, Shin H-S, Saratale GD, Park Y, Lee H-S, Bharagava RN, Kim D-S (2020) Utilization of noxious weed water hyacinth biomass as a potential feedstock for biopolymers production: a novel approach. Polymers 12(8):1704

Sardon H, Pascual A, Mecerreyes D, Taton D, Cramail H, Hedrick JL (2015) Synthesis of polyurethanes using organocatalysis: a perspective. Macromolecules 48(10):3153–3165

Sardon H, Mecerreyes D, Basterretxea A, Avérous L, Jehanno C (2021) From lab to market: current strategies for the production of biobased polyols. ACS Sustain Chem Eng 9(32):10664–10677

Sari-Chmayssem N, Taha S, Mawlawi H, Guégan J-P, Jeftić J, Benvegnu T (2016) Extracted and depolymerized alginates from brown algae Sargassum vulgare of Lebanese origin: chemical, rheological, and antioxidant properties. J Appl Phycol 28:1915–1929

Scheirs JGM (1998) Polymer recycling: science, technology and applications. J. Wiley & Sons, Chichester

Shao Z, Duan D (2022) The cell wall polysaccharides biosynthesis in seaweeds: a molecular perspective. Front Plant Sci 13:902823

Sharma L, Mallick N (2005) Accumulation of poly-beta-hydroxybutyrate in Nostoc muscorum: regulation by pH, light-dark cycles, N and P status and carbon sources. Bioresour Technol 96(11):1304–1310

Sharma KK, Schuhmann H, Schenk PM (2012) High lipid induction in microalgae for biodiesel production. Energies 5(5):1532–1553

Sheng J, Xu Y, Yu J, Ding B (2017) Robust fluorine-free superhydrophobic amino-silicone oil/SiO2 modification of electrospun polyacrylonitrile membranes for waterproof-breathable application. ACS Appl Mater Interfaces 9(17):15139–15147

Sikorski P, Mo F, Skjak-Braek G, Stokke BT (2007) Evidence for egg-box-compatible interactions in calcium-alginate gels from fiber X-ray diffraction. Biomacromolecules 8(7):2098–2103

Simpson F, Shacklock P (1979) The cultivation of Chondrus crispus. Effect of temperature on growth and carrageenan production.

Singh Saharan B, Grewal A, Kumar P (2014) Biotechnological production of polyhydroxyalkanoates: a review on trends and latest developments. Chin J Biol 42(2022): 102333

Singh I, Samal SK, Mohanty S, Nayak SK (2019) Recent advancement in plant oil derived polyol-based polyurethane foam for future perspective: a review. Eur J Lipid Sci Technol 122(3):1–23

Somarathna HMCC, Raman SN, Mohotti D, Mutalib AA, Badri KH (2018) The use of polyurethane for structural and infrastructural engineering applications: a state-of-the-art review. Constr Build Mater 190:995–1014

Soroudi A, Jakubowicz I (2013) Recycling of bioplastics, their blends and biocomposites: a review. Eur Polym J 49(10):2839–2858

Sperling L (2006) Introduction to physical polymer science. John Willet & Sons, Hoboken

Stark JR (1976) A new method for the analysis of laminarins and for preparative-scale fractionation of their components. Carbohydr Res 47(1):176–178

Sterner M, Gröndahl F (2021) Extraction of laminarin from Saccharina latissima seaweed using cross-flow filtration. J Appl Phycol 33(3):1825–1844

Suriyamongkol P, Weselake R, Narine S, Moloney M, Shah S (2007) Biotechnological approaches for the production of polyhydroxyalkanoates in microorganisms and plants - a review. Biotechnol Adv 25(2):148–175

Szycher M (2012) Structure–property relations in polyurethanes

Tabatabaei RH, Jafari SM, Mirzaei H, Nafchi AM, Dehnad D (2018) Preparation and characterization of nano-SiO2 reinforced gelatin-k-carrageenan biocomposites. Int J Biol Macromol 111: 1091–1099

Tan HW, Abdul Aziz AR, Aroua MK (2013) Glycerol production and its applications as a raw material: a review. Renew Sust Energ Rev 27:118–127

Tytgat L, Van Damme L, Ortega Arevalo MDP, Declercq H, Thienpont H, Otteveare H, Blondeel P, Dubruel P, Van Vlierberghe S (2019) Extrusion-based 3D printing of photo-crosslinkable gelatin and kappa-carrageenan hydrogel blends for adipose tissue regeneration. Int J Biol Macromol 140:929–938

Ullmann F (2003) Ullmann's encyclopedia of industrial chemistry. Wiley-VCH, Weinheim

United Nations (2015) Paris agreement

Vakilian S, Jamshidi-Adegani F, Al Yahmadi A, Al-Broumi M, Ur Rehman N, Anwar MU, Alam K, Al-Wahaibi N, Shalaby A, Alyaqoobi S, Al-Harrasi A, Mustafa K, Al-Hashmi S (2021) A competitive nature-derived multilayered scaffold based on chitosan and alginate, for full-thickness wound healing. Carbohydr Polym 262:117921

Van Breda DJ (2020) Enzymatic extraction of laminarin from brown seaweed Ecklonia maxima. Stellenbosch University, Stellenbosch

Van de Velde F (2008) Structure and function of hybrid carrageenans. Food Hydrocoll 22(5): 727–734

Varela Villarreal J, Burgues C, Rosch C (2020) Acceptability of genetically engineered algae biofuels in Europe: opinions of experts and stakeholders. Biotechnol Biofuels 13:92

Villanueva RD, Hilliou L, Sousa-Pinto I (2009) Postharvest culture in the dark: an eco-friendly alternative to alkali treatment for enhancing the gel quality of kappa/iota-hybrid carrageenan from Chondrus crispus (Gigartinales, Rhodophyta). Bioresour Technol 100(9):2633–2638

Voronova Y, Rekhina N, Nikolaeva T, Tiunova N, Zaikina I, Kobzeva NY, Valiente O (1991) Extraction of carbohydrates fromLaminaria and their utilization. J Appl Phycol 3(3):243–245

Wegmann K, Ben-Amotz A, Avron M (1980) Effect of temperature on glycerol retention in the halotolerant algae dunaliella and asteromonas. Plant Physiol 66(6):1196–1197

Wijesekara I, Pangestuti R, Kim S-K (2011) Biological activities and potential health benefits of sulfated polysaccharides derived from marine algae. Carbohydr Polym 84(1):14–21

Winnacker M (2019) Polyhydroxyalkanoates: recent advances in their synthesis and applications. Eur J Lipid Sci Technol 121(11):1900101

Wojnowska-Baryła I, Kulikowska D, Bernat K (2020) Effect of bio-based products on waste management. Sustainability 12(5):2088

Wu SJ (2012) Degradation of kappa-carrageenan by hydrolysis with commercial alpha-amylase. Carbohydr Polym 89(2):394–396

Xia Y, Zhang Z, Kessler MR, Brehm-Stecher B, Larock RC (2012) Antibacterial soybean-oil-based cationic polyurethane coatings prepared from different amino polyols. ChemSusChem 5(11): 2221–2227

Xu Y, Milledge JJ, Abubakar A, Swamy RAR, Bailey D, Harvey PJ (2015) Effects of centrifugal stress on cell disruption and glycerol leakage from Dunaliella salina. Microalgae Biotechnol 1(1):20–27

Xu Y-J, Qu L-Y, Liu Y, Zhu P (2021) An overview of alginates as flame-retardant materials: pyrolysis behaviors, flame retardancy, and applications. Carbohydr Polym 260:117827

Xue J, Chen TT, Zheng JW, Balamurugan S, Liu YH, Yang WD, Liu JS, Li HY (2020) Glucose-6-phosphate dehydrogenase from the oleaginous microalga nannochloropsis uncovers its potential role in promoting lipogenesis. Biotechnol J 15(2):e1900135

Yamashita S, Suguta-Konishi Y, Shimizu M (2001) *In vitro* bacteriostatic effects of dietary polysaccharides. Food Sci Technol Res 7(3):262–264

Yang W, Dong Q, Liu S, Xie H, Liu L, Li J (2012) Recycling and disposal methods for polyurethane foam wastes. Procedia Environ Sci 16:167–175

Yao Y, Lu Y, Peng K-T, Huang T, Niu Y-F, Xie W-H, Yang W-D, Liu J-S, Li H-Y (2014) Glycerol and neutral lipid production in the oleaginous marine diatom Phaeodactylum tricornutum promoted by overexpression of glycerol-3-phosphate dehydrogenase. Biotechnol Biofuels 7(1):110

Yarkent C, Guler BA, Gurlek C, Sahin Y, Kose A, Oncel SS, Imamoglu E (2021) Algal alginate in biotechnology: biosynthesis and applications. In: Properties and applications of alginates. IntechOpen, London

Yegappan R, Selvaprithiviraj V, Amirthalingam S, Jayakumar R (2018) Carrageenan based hydrogels for drug delivery, tissue engineering and wound healing. Carbohydr Polym 198: 385–400

Younes S (2021) Design and development of microbial platforms for sustainable oleochemicals and biorefinery applications. Technische Universität München, München

Younes S, Bracharz F, Awad D, Qoura F, Mehlmer N, Brueck T (2020) Microbial lipid production by oleaginous yeasts grown on Scenedesmus obtusiusculus microalgae biomass hydrolysate. Bioprocess Biosyst Eng 43(9):1629–1638

Younes S, Awad D, Kassab E, Haack M, Schuler C, Mehlmer N, Brueck T (2021) Systems biology engineering of the pantothenate pathway to enhance 3HB productivity in Escherichia coli. Biotechnol Bioprocess Eng 26:621–629

Yu Y, Zou S, Wang K, Liang R, Fan X, Wang B, Liu M, Fang L, Liu W, Wu Z, Chen D (2018) Synthesis, characterization and *in vitro* evaluation of dual pH/redox sensitive marine laminarin-based nanomedicine carrier biomaterial for cancer therapy. J Biomed Nanotechnol 14(9): 1568–1577

Yuan Z, Eden MR, Gani R (2015) Toward the development and deployment of large-scale carbon dioxide capture and conversion processes. Ind Eng Chem Res 55(12):3383–3419

Zappi ME, Bajpai R, Hernandez R, Mikolajczyk A, Lord Fortela D, Sharp W, Chirdon W, Zappi K, Gang D, Nigam KDP, Revellame ED (2019) Microalgae culturing to produce biobased diesel fuels: an overview of the basics, challenges, and a look toward a true biorefinery future. Ind Eng Chem Res 58(35):15724–15746

Zha X-Q, Xiao J-J, Zhang H-N, Wang J-H, Pan L-H, Yang X-F, Luo J-P (2012) Polysaccharides in Laminaria japonica (LP): extraction, physicochemical properties and their hypolipidemic activities in diet-induced mouse model of atherosclerosis. Food Chem 134(1):244–252

Zhang X, Wang L, Weng L, Deng B (2020) Strontium ion substituted alginate-based hydrogel fibers and its coordination binding model. J Appl Polym Sci 137(16):48571

Zhang L, Li X, Zhang X, Li Y, Wang L (2021) Bacterial alginate metabolism: an important pathway for bioconversion of brown algae. Biotechnol Biofuels 14(1):158

Zhang X, Wang X, Fan W, Liu Y, Wang Q, Weng L (2022) Fabrication, property and application of calcium alginate fiber: a review. Polymers (Basel) 14(15):3227

Zia KM, Bhatti HN, Ahmad Bhatti I (2007) Methods for polyurethane and polyurethane composites, recycling and recovery: a review. React Funct Polym 67(8):675–692

Diatom Nanostructured Biosilica

Mohamed Ghobara, Mostafa El-Sheekh, Adel F. Hamed, Mohamed A. A. Abdelhamid, and Seung P. Pack

Abstract Bacillariophyceae (diatoms) is a unique class of unicellular microalgae of remarkable evolutionary success showing large diversity over a relatively short geologic timescale, leading to exceptional ecological roles in many aquatic ecosystems. Diatoms living cells build outstanding cell walls (frustules) made of amorphous hydrated silica featured with nanometric ultrastructure. Nowadays, diatoms are cultivated on different scales for many purposes and are involved in various industrial and commercial applications. This chapter overviews the wide applicability of diatoms nanostructured biosilica, especially in the fossil form (i.e. diatomaceous earth), as a functional additive and filler in various industries, such as fabrication of building materials, paints, agricultural products, rubbers, and pharmaceuticals. In addition, the chapter will shed light on the biorefinery approach of large-scale cultivated diatoms as a sustainable source of biosilica, besides extracting valuable metabolites. The extracted biosilica can facilitate the industrialization of the recently suggested biotechnological and biomedical applications utilizing diatoms silica as a green alternative to synthetic nanostructured materials.

Keywords Diatoms silica · Diatomaceous earth · Functional additives · Industrial applications · Biorefinery · Biomedical applications

M. Ghobara
Physics Department, Freie Universität Berlin, Berlin, Germany

M. El-Sheekh (✉)
Botany Department, Faculty of Science, Tanta University, Tanta, Egypt
e-mail: mostafaelsheikh@science.tanta.edu.eg

A. F. Hamed
Botany Department, Faculty of Science, Ain Shams University, Cairo, Egypt

M. A. A. Abdelhamid
Department of Botany and Microbiology, Faculty of Science, Minia University, Minia, Egypt

Department of Biotechnology and Bioinformatics, Korea University, Sejong, Republic of Korea

S. P. Pack
Department of Biotechnology and Bioinformatics, Korea University, Sejong, Republic of Korea

A. Abomohra, S. Ende (eds.), *Value-added Products from Algae*,
https://doi.org/10.1007/978-3-031-42026-9_14

1 Introduction

Diatoms are single-celled photosynthetic eukaryotic microalgae, which belong to Class Bacillariophyceae. They are playing crucial roles in the aquatic geochemical cycle while offering about 40% of oceans' primary production and at least 20% of regenerated atmospheric oxygen (Round et al. 1990; Smol and Stoermer 2010). There are possibly up to 200,000 species and varieties spreading across fresh, brackish, and marine aquatic ecosystems (Mann and Droop 1996). Their occurrence also extends to humid terrestrial places. In aquatic environments, they can be found as planktonic, benthic, epiphytic, epizoic, or epilithic (Manoylov and Ghobara 2021; Round et al. 1990). Some species live solitary, while others form colonies, which can be of a simple chain or branching structure (Julius and Theriot 2010; Round et al. 1990), and sometimes have gelatinous layer surrounding them (Aumeier and Menzel 2012). Diatoms are mainly autotrophs; however, some species can survive as mixotrophs or heterotrophs (Cupo et al. 2021; Villanova and Spetea 2021). They have a diplontic life cycle, where living cells mainly reproduce asexually via mitosis (Round et al. 1990). At a certain stage or conditions during diatoms' life, sexual reproduction is triggered forming auxospores and helping diatoms to restore their size (Chepurnov et al. 2004; Poulíčková et al. 2019).

Diatoms are outstanding organisms for many reasons, including their exceptional evolutionary success. They developed a huge diversity within a relatively short evolutionary time compared to other photosynthetic organisms, probably within 200 million years (Benoiston et al. 2017; Hohmann-Marriott and Blankenship 2011). This diversity comes with a wide geographical distribution that tends to cover even extreme environments ranging from polar conditions (Mock et al. 2017) to hot springs and tropical conditions (Fazlutdinova et al. 2020; Pumas et al. 2018). They are found survived under ice sheets (Mock and Kroon 2002), 4 km deep in the ocean (Agusti et al. 2015), and in caves or mountains (Heine-Fuster et al. 2021).

Diatom cells are enclosed within a siliceous cell wall (frustule), which consists of two interlocked halves: epitheca and hypotheca (Fig. 1a). The top half, epitheca, is slightly bigger than the hypotheca, which gives the frustule assembly a Petri-dish-like appearance. Each theca is composed of a valve and one or more girdle bands. This siliceous frustule is the most intriguing structure of diatoms and mostly comes with elaborate ornamentation (Round et al. 1990). Under optical microscope, diatom frustules have two primary views: valve and girdle views. Nevertheless, conventional two-dimensional optical micrographs obtained by optical microscopy are insufficient to describe the morphology of most diatom frustules of three-dimensional structures. Additionally, the optical microscope is unable to disclose the frustules' ultrastructure (Manoylov and Ghobara 2021). Therefore, transmission and scanning electron micrographs obtained by electron microscopy are often used to describe the frustule in detail (Round et al. 1990). Such micrographs reveal the multilayer multiscalar porosity of diatom valves and girdle bands. Moreover, recent research introduced the atomic force microscope to study the topography of the

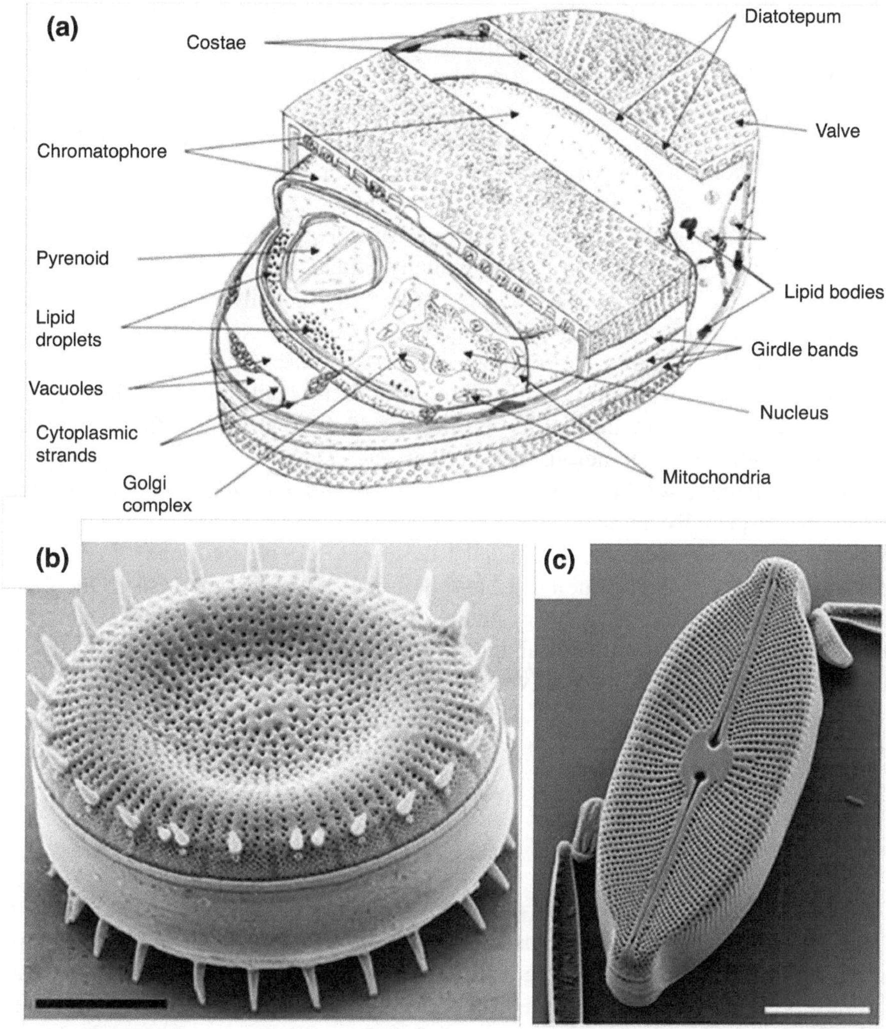

Fig. 1 A schematic diagram of a living diatom cell showing its organelles as well as its frustule (**a**). Scanning electron micrograph (SEM) of a frustule of centric diatom *Stephanodiscus* sp. (**b**), and of biraphid pennate diatom *Cosmioneis* sp. (**c**). The scale bars in (**b**) and (**c**) are 10 μm. (**a**) Reproduced from Yang et al. (2011) after permission (No. 1363360-1). (**c-b**) Reproduced from Mann et al. (2017) after permission (No. 5520840037016)

valves, revealing new features at the nanoscale (Losic et al. 2006). Furthermore, employing focused ion beam scanning electron microscopy enables understanding the complex 3D structure of the frustules (Zgłobicka et al. 2021).

The frustule morphogenesis involves an intricate biomineralization process, which is still partially unrevealed (Babenko et al. 2022). The living cells take up silicic acid from the surrounding environment, then polymerize, condensate, and precipitate it inside the silica deposition vesicle (SDV), where morphogenesis of new valves occurs. The SDV consists during the mitosis by the fusion of vesicles originating either from the Golgi apparatus or the endoplasmic reticulum (Babenko et al. 2022). Inside SDV, many biomolecules are involved in the morphogenesis process, which has been recently reviewed by Kröger (2022). Recent work shows that the cytoskeleton components may also play crucial roles in the morphogenesis process (Babenko et al. 2022).

Nowadays, it is believed that the unique frustules of diatoms could be a major reason for their outstanding success in many aquatic ecosystems. The enhanced mechanical strength, optical properties, and filtration abilities of these frustules reported in literature may help protect and thrive living cells (De Tommasi 2016; Ghobara et al. 2022; Hamm et al. 2003; Losic 2017; Losic et al. 2006). The frustules' design, size, pore size, and pore distribution patterns could affect such properties dramatically. Owing to the unique properties of these frustules, they have been extensively utilized in numerous applications.

In this chapter, we highlight the commercial applications of diatom siliceous frustules—especially in their fossil form—as functional additives and fillers in various fields. In addition, the chapter discusses the recent trends in applying diatoms biosilica in biotechnological and biomedical applications. Given the diversity of diatom frustules morphology and ultrastructure, they can be involved in many nanotechnological and biomedical applications—especially if the challenges facing their biorefinery approach are solved.

2 Multiplicity of Diatom Frustule Morphology and Ultrastructure

The peculiar design of silica frustules served as the basis for the majority of the historical classification of diatoms. Their morphology greatly varies between species, but can be generally distinguished by their radial or bilateral symmetry, and accordingly diatoms can be classified into centrics (e.g. Fig. 1b) or pennates (e.g. Fig. 1c), respectively (Round et al. 1990). Pennate diatoms can be additionally classified, according to the existence of raphe into raphid and araphid. The raphid pennates have a raphe visible on at least one valve (e.g. Fig. 1c), while araphid pennates are without a raphe.

Diatom valves and girdle bands observed via electron microscopy showed an extraordinary decoration on the nanoscale (Round et al. 1990). These structures are generally porous, with pore sizes ranging from few nm to few μm depending on the species (Ghobara et al. 2021a). These pores can be simple (punctate) or complicated (loculate) and can be of periodic to quasi-periodic arrangements with different levels of symmetries and imperfections (Ghobara et al. 2021b). Beside the pores, a large diversity of ultrastructure details is reported in the literature for different taxa, such

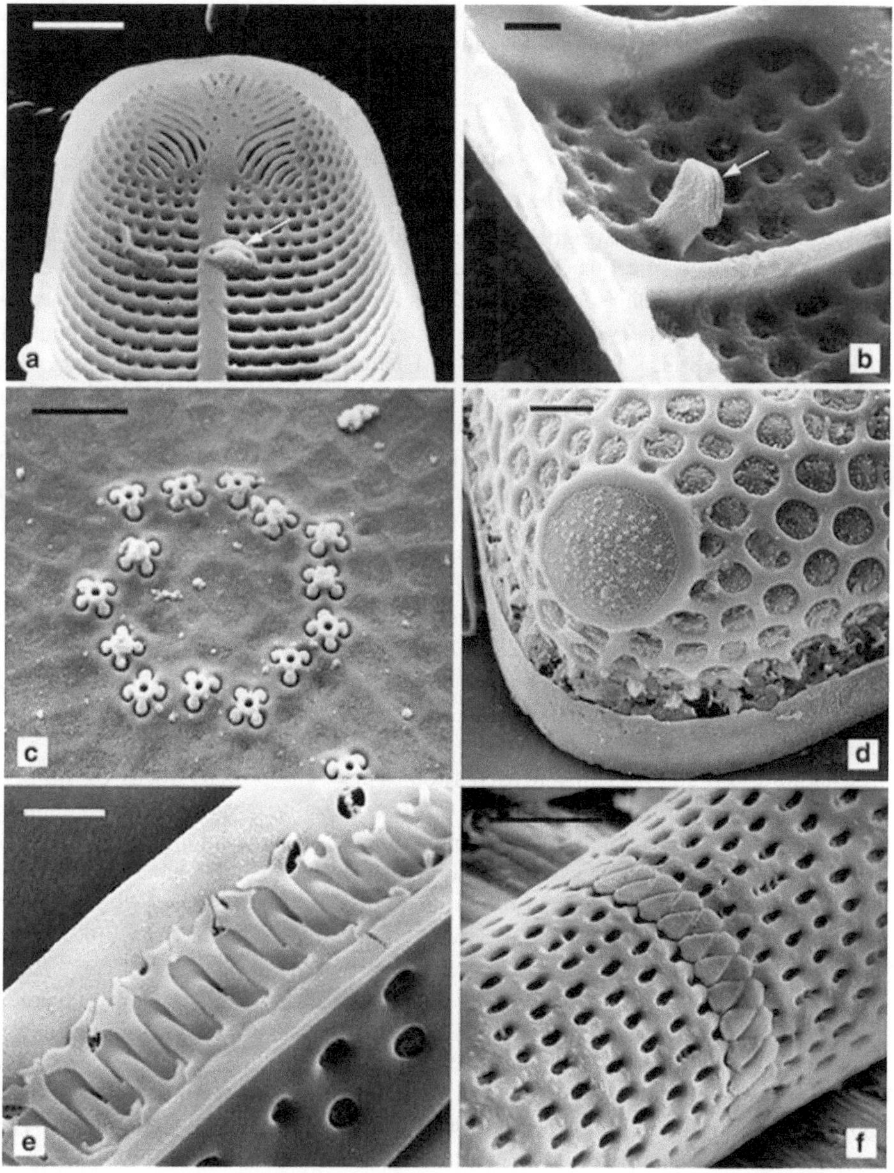

Fig. 2 SEMs show various ultrastructure details on the frustules of different species. Rimoportula in the valves of *Cyclophora* (**a**) and *Triceratium* (**b**), indicated by white arrows. Fultoportulae in the valve of *Thalassiosira* (**c**). Ocellus in the valve of *Odontella* (**d**). Linking spines in a frustule of *Cymatosira* (**e**) and of *Aulacoseira* (**f**). Scale bars are 1 μm. Reproduced from Mann et al. (2017) after required permission (No. 5520840037016)

as spines, central nodules, ocelli, portulae, fasciae, fibulae, and much more (Spaulding et al. 2021). Examples of ultrastructure diversity are illustrated in Fig. 2.

3 Diatomite: A Cheaper Source of Diatoms Silica

Diatomite (DE) is a lightweight biogenic sedimentary rock with a large fraction consisting of diatom siliceous remains, representing more than 50% of its dry weight (Ghobara et al. 2019; Harwood 2010). DE is formed through long-term sedimentation and diagenesis processes over the geologic time, the same as any sedimentary rock (Mackenzie 2005). It is believed that DE has been formed concurrently with thriving living diatoms due to the favourable environmental conditions in paleo-aquatic ecosystems (Flower 2013). Through different sedimentation mechanisms, siliceous frustules were accumulated in the sediments and were preserved from dissolution with a variable quality depending on the conditions of sedimentation and fossilization processes (Flower 2013; Ghobara et al. 2019). Over time, the sediments turned into sedimentary rocks. In general, the quality of a given DE deposit depends on the amount of diatom silica and other debris within, as well as its associated chemical composition. Recently, a classification has been suggested to elucidate the difference between diatomite and other deposits containing diatoms silica to avoid misusing terms in the literature (Zahajská et al. 2020). Further information on the nature of diatomite, its formation, and preservation has been discussed by Flower (2013) and Ghobara et al. (2019).

Diatomite formation mainly involves four steps: The primary focus of the first stage is on the diatom cell's capacity to generate siliceous frustules in ambient conditions. By sophisticated biomineralization processes, diatoms produce their frustules. The Golgi apparatus, microtubules, and a model of other protein nano-structured matrices are thought to play a role in the species-specific process by which the living cells take silicic acid up from the environment, convert it into silica nanospheres, and then arrange them to form various parts of the frustule (Gröger et al. 2016; Kotzsch et al. 2016; Kröger 2007). The ability of the diatom to reach a high enough cell concentration to enter the blooming state, or the expansion of living cells, which necessitates the availability of nutrients (eutrophication) as well as other environmental conditions, is the second step. The deposition of dead diatom remnants, whole frustules, or its fragments is a component of the third phase. The siliceous frustules that settle at the bottom (benthos) create a very fine layer of sediments that typically include other elements. If sedimentation occurs in the ocean one diatom frustule or one valve at a time, it will be too slow (Alldredge and Gotschalk 1989). Nonetheless, there are processes that shorten the time diatom remnants spend settling. The primary mechanism involves grazing zooplankton feeding on living diatoms, which leads to the creation of faecal pellets that contain frustule fragments. The diatom remnants' sedimentation is accelerated by the faecal pellets (Liu and Wu 2016). Another method involves the variation of physicochemical properties during, for example, seasonal shifts, notably in temperate, arctic, and coastal locations, which occurs without the absorption of living cells by grazers. This mechanism might involve the blooms flocculating and the protoplasts rapidly disintegrating (Alldredge and Gotschalk 1989). Once the living diatom cells die, they begin to coagulate with other detritus, bacteria, etc. to create larger coagulated particles that sink more quickly (O'Brien 2002). When phytoplankton cells become more sticky, the flocculation process takes place. In addition to various physico-chemical aspects of the environment, bacteria, and the effective size of algal cells, other factors include the algae's physiological and morphological stickiness. The number of bacteria linked to bloom cells increases quickly when the bloom transitions into its decreasing phase, which can increase the stickiness of the aggregated phytoplankton cells and even include mucus secretion (O'Brien 2002). The final process takes place over a longer period of time, during which time diatom frustules build up layer by layer, getting thicker. These sedimentary layers undergo diagenesis processes and become diatomite.

DE is known for several outstanding characteristics, including its lightweight, high porosity, chemical inertness, absorptive capacity, and thermal insulation (Galal Mors 2010). These are the reasons why DE has been utilized in many applications as a natural material of unique properties, probably since old civilizations even before we discovered its nature (Flower 2013; Ghobara et al. 2019). During the last decades, many DE deposits have been discovered around the globe, and their mining and production have substantially increased. This was associated with increasing the applicability of DE in filtration and many industrial applications (Harwood 2010). In 2018, it was reported that the USA was the world's major DE producer, with annual production reaching about 34% of the market size, estimated to be about 2.84 million metric tons (Crangle 2018). In this report, Denmark, China, Turkey, and the Republic of Korea followed the USA in DE production size and are considered the major producers among 27 countries producing it for commercial purposes (Crangle 2018). However, it should be noted that the annual production size of different countries was changing over time. This can be seen through different subsequent reports published by the United States geological survey. In the market, many commercial grades are available with prices of about a few hundreds of dollars per metric ton (Crangle 2018), which is very cheap compared to commercial porous silica. Interestingly, the estimated amounts of DE deposits available in the already-known mines need approximately 350 years to be consumed under the previously mentioned annual production rate (Crangle 2018). This means that although DE is not sustainable in the short time scale, the world reserves can still be in the market for few hundreds of years. The major application of DE is for filtration purposes, followed by applying DE as a filler in various fields such as the fabrication of building materials, asphalt, rubber, and agriculture products (Crangle 2018; Ghobara et al. 2019), as discussed in the next sections.

3.1 Building Materials

Utilizing DE-based building materials has been reported in ancient civilizations, including ancient Egyptians, Greeks, Romans, and Ottomans (Flower 2013; Ghobara et al. 2019; Siddall 2018). Currently, DE is involved and utilized in fabricating many building materials in the market. Here, we mention some major examples.

3.1.1 Building Bricks

Employing naturally occurring DE rocks as building blocks themselves is not reasonable for the recent complicated constructions that require high mechanical strength that cannot be offered by these rocks. Therefore, recent studies have investigated utilizing DE powder or aggregates as an additive in the fabrication of lightweight building bricks by mixing them with local earth and mud (Elias and Cultrone 2019; Galán-Arboledas et al. 2017; Hasan et al. 2021; Zhang et al. 2005). In the market, there are many DE-based building bricks available, but mainly for heat preservation and as an insulating layer for different types of furnaces due to their excellent thermal insulating properties.

3.1.2 Cement

Cement is an essential binding material extensively used in construction, e.g. in preparing mortar, concrete, and as a binding material for building blocks. In order to enhance the performance of concrete and mortar and increase its durability and compressive strength, pozzolanic materials are often added to the cement, which can be obtained from a natural or synthetic source (Vitola et al. 2017). The effect of blending raw or calcined DE powder with cement—as an example of Pozzolanic additives—has been investigated and shown to improve concrete and mortar properties (Degirmenci and Yilmaz 2009; Ergün 2011; Kipsanai et al. 2022; Liang and Yao 2023; Pavlíková et al. 2022; Sharma et al. 2021; Sun et al. 2020; Yılmaz and Ediz 2008). Interestingly, utilizing DE-blended concrete can affect and control indoor humidity and temperature (Zheng et al. 2017). In recent work, living diatom biofilms were directly deposited onto the surface of concrete to form an adhesive layer, and thus, enhance its properties (Merino-Maldonado et al. 2023).

3.1.3 Paints and Surface Coatings

Some patents and papers have introduced DE powder as an additive and functional filler to paints, often added in significant amounts (Lang and Yongke 2018; Mariusson and Bjarnason 1987; Wang et al. 2014a; Xiaobing and Sen 2013). It has been indicated that adding DE to paints primarily serves as a flatting agent. It gives the paint an irregular texture that reduces its sheen and enhances its durability and adhesion to surfaces (McGonigle and Ciullo 1996). In the market, the utilization of this material in paints is believed to be common; however, to the best of our knowledge, no record offering accurate information regarding the companies adding DE to their paints is available.

Unlike the regular decoration paints, which aim for decoration in the first place, surface coatings are often designed to perform additional functions, e.g. corrosion resistance of metal surfaces, heat and fire resistance, and waterproofing. It has been reported that adding DE enhances the performance of anticorrosion coatings (Veselý et al. 2010). In another study, the addition of DE to Acrylic coating with an adequate amount (4 g L^{-1}) increases the corrosion resistance (Farzaneh et al. 2018). Moreover, adding DE to some coatings showed a significant improvement of their heat and fire resistance properties (Feng 2013; Martinez 2014; Nasirzadeh et al. 2023). Furthermore, modified DE was used as a component to obtain superhydrophobic coating (Perera et al. 2017; Wang et al. 2019), which simultaneously can help increase the coating's robustness (Wang et al. 2019).

3.2 Rubber Industry

Rubbers are elastic polymeric substances that have many natural and synthetic types and are involved in various applications, such as tire fabrication (Simpson 2002). In general, amorphous silica is used as a semi-reinforcing filler in the fabrication of rubber to enhance its strength and longevity. DE has been applied as an additive to silicone rubbers for several decades to fabricate cable insulation layers (Dexter and Servals 1953). Recently, it has been utilized as an additive to reinforce different types of rubbers (Greene et al. 2009; Wu and Chen 2017). Moreover, utilizing DE as a carrier for flame retardant agents while mixing it with silicone rubbers has helped to develop fire-resistant rubbers (Zhang et al. 2019a). Furthermore, adding modified hydrophobic DE to natural rubber latex helps to develop superhydrophobic rubber films (Ambegoda et al. 2021) that can also be employed as coatings.

3.3 Asphalt

The raw material of asphalt basically results as a residue from the petroleum refining process and is used for surfacing car roads, airports, and roofs that have been widely employed in the twentieth century. The asphalt raw material is mixed with different ratios of aggregates and fillers to produce commercially available asphalt mixtures. Blending the asphalt mixture with DE powder gives the asphalt concrete enhanced properties. The DE-modified asphalt mixture showed increased thermal stability, enhanced antiageing (Cong et al. 2012; Cong et al. 2016; Zhang et al. 2019b), and anti-freezing properties (Cheng et al. 2018; Wei et al. 2017). The DE could also enhance the resistance of asphalt to stripping, deformation, and cracking (Yang et al. 2018). In all cases, the best results were obtained when adding DE with 10–20% of the asphalt mixture weight.

3.4 Agriculture Sector

DE has been used in the agriculture sector in the fabrication of insecticides and as additives to animal feed for a long time. It has also been suggested as soil amendments, as explained in (Ghobara et al. 2019). Besides that, today, many products available in the agriculture market are believed to contain DE as a functional filler.

3.5 Insecticides

DE powder considers an effective mechanical insecticide in dry conditions as it injures insects' cuticles with its sharp edges, absorbs their liquids, and eventually leads to their death (Losic and Korunic 2017; Zeni et al. 2021). It can be considered a green and safe insecticide suitable for organic cultivation (Zeni et al. 2021). The applicability of DE as an insecticide has extended from agricultural to urban pests, leading to protecting cultivated crops, stored grains, farm animals, and house pets from insects. For this purpose, DE is often applied as a powder spray or mixed with the stored grains with concentrations exceeding 1000 ppm, much higher than synthetic insecticides, where an outer layer of dusty appearance is formed (Zeni et al. 2021). It can also be applied around the trees and plants to lower the transfer of pests between them in the field. Nevertheless, applying DE powder in that way has some drawbacks. It can cause health problems for farm workers if they breathe the powder or touch their skin for a long time. It can also lead to a loss in stored grain weight and can be harmful to beneficial insects and pests. These drawbacks are not life-threatening as in the case of chemical-based insecticides; however, careful usage is recommended. In the market, DE is also used as an excellent filler or carrier in the formulations of chemical-based insecticides and fungicides, thanks to their high adsorption abilities (Zeni et al. 2021).

3.6 Animal Feed

For decades, food-grade DE powder has been consumed as a food additive by different animals, including beef cattle, sheep, goats, horses, chickens, and house pets. The farmers suggested that adding DE to the animal feed helps to control gastrointestinal parasites and worms. They often used it as an alternative to synthetic drugs and antibiotics in organic livestock farms (Fernandez et al. 1998). It has also been suggested that DE can be used as a mineral supplement as it contains many secondary minerals (Fernandez et al. 1998). Such claims often had the problem that they had little or no supporting scientific evidence (Fernandez et al. 1998).

Nevertheless, during the last two decades, several papers reported the effectiveness of DE in eliminating gastrointestinal parasites and worms as a green alternative to chemical-based treatments and antibiotics (Ikusika et al. 2019). The positive results have been shown in many cases after consuming DE over a long time extending from several weeks to months. As an example, in (Bennett et al. 2011), the effectiveness of DE in controlling internal parasites was tested on hens showing promising results. They also reported that the two different types of hens that consumed food supplemented with 2% DE significantly increased in weight, produced more eggs, and one type of hen produced bigger eggs with more yolk and albumen compared to the control. It has also been suggested that DE can work as a growth promoter for animals, as a minerals supplement, and as a detoxifier against mycotoxins, such as aflatoxins (Ikusika et al. 2019).

3.7 Supplementary Additives for Human Diet

The DE suppliers advocated that food-grade DE powder can be consumed as a supplementary additive for human food showing several benefits similar to its benefits for animals, including eliminating internal parasites, as a source of minerals and silica, and as a detoxifying agent. The food grade DE, consisting of amorphous silica content and permissible limits of heavy metals, is generally safe as recognized by the United States Food and Drug Administration FDA (Terracciano et al. 2018). Therefore, consuming DE directly or as an additive to food or drug formulations should be, in principle, safe. In Wachter et al. (1998), a clinical trial has been conducted by oral administration of 250 mg DE three times daily for two months. The results suggested that utilizing DE in human diet may be able to reduce blood cholesterol, probably through lowering the absorption or digestion of lipids in the intestine. This is unlike utilizing the lipase inhibitors (Heck et al. 2000).

3.8 Pharmaceuticals, Cosmetics, and Dental Materials

DE has been introduced to the pharmaceutical industry since decades mainly for separation and filtration purposes (Graham and Kenner 1973; Hertzog et al. 2015; Milstone 1955). The utilization of DE as a filtration medium demonstrated highly efficient performance in the separation and purification of nucleic acids, proteins, and many biomolecules, sometimes unparallel using other filtration systems (Bleuart et al. 2002; Buyel et al. 2015; Hilbrig et al. 2001; Horn et al. 1996; Tan and Yiap 2009). Recently, many papers and patents have suggested DE as an active ingredient in drug formulations (Lijun 2016; Mikulasik and Albrech 2011), comparable to synthetic porous silica. Moreover, DE is widely utilized in the cosmetics industry (Crangle 2015) as a functional filler for many purposes. It can also work as a natural skin abrasive material that is used in facial masks, and body scrubs, as well as an abrasive in toothpaste formulations (Yeh and Synodis 1984).

Furthermore, DE can be used after suitable surface modifications as a functional additive and filler to the commercial dental composites and ceramics that helped to enhance their mechanical properties (Araújo et al. 2022; Chen and Chang 2023; Lu et al. 2012; Miao et al. 2012; Wang et al. 2011). DE powder can also be added as a filler to alginate powder (up to 60% of its weight), which is widely used as a cheap and efficient material for making mouth casts (Guiraldo et al. 2014). However, DE silica is generally considered to be safe and biocompatible as already mentioned, but cautions should be taken after the report of 4 cases with diatom-induced non-necrotizing granulomatous foreign body reactions (AlHousami et al. 2021).

4 Challenges in Diatom Large-Scale Cultivation

Nowadays, cultivation of many diatom species has become a regular practice in several laboratories and culture collections around the world. Different recipes for growth media are available to cultivate both fresh and marine diatom species (Andersen 2004). It has been found that one of the major limiting factors in their growth is the silicate concentration. The scaling up of diatom cultures to be utilized for commercial purposes is facing many challenges, as is also the case for other microalgae, including the capital and maintenance costs as well as invasion by grazing organisms and competing microalgae (Wang and Seibert 2017). Therefore, only few species have been cultivated on a large-scale so far. The marine chain-forming diatom *Chaetoceros* sp. is a good example of commercially cultivated diatoms on a large scale with an average annual yield reaching 132 MT of dry biomass, which was mainly used to feed shrimps and bivalves in aquacultures (Wang and Seibert 2017). Another successful example of large-scale cultivation of diatoms is *Staurosira* sp., which was cultivated on different scales with green microalgae *Desmodesmus* sp., yielding together up to ~75 MT ha^{-1} $yr.^{-1}$ biomass, and ~ 30 MT ha^{-1} $yr.^{-1}$ lipid in 100-ha cultivation area (Huntley et al. 2015).

As discussed in detail in Chapter " Overview of Biorefinery Technology", biorefinery is the process of converting biomass into value-added products and biomaterials. Similarly, biorefinery of the harvested biomass from diatoms is suggested as a source of value-added metabolites, such as pigments, lipids, and their derivatives, besides the produced biosilica (Bayu et al. 2020; Govindan et al. 2021; Popovich et al. 2020; Prasetiya et al. 2022; Savio et al. 2020). Such metabolites can be extracted and employed in various field, for example, biofuel production from the extracted oil (Dhanker et al. 2022; Popovich et al. 2019). On the other hand, the retrieved porous biosilica from the remaining crude can be employed in biotechnological and nanotechnological applications (Fig. 3). Utilizing freshly prepared diatom biosilica in applications has advantages over DE biosilica, including its

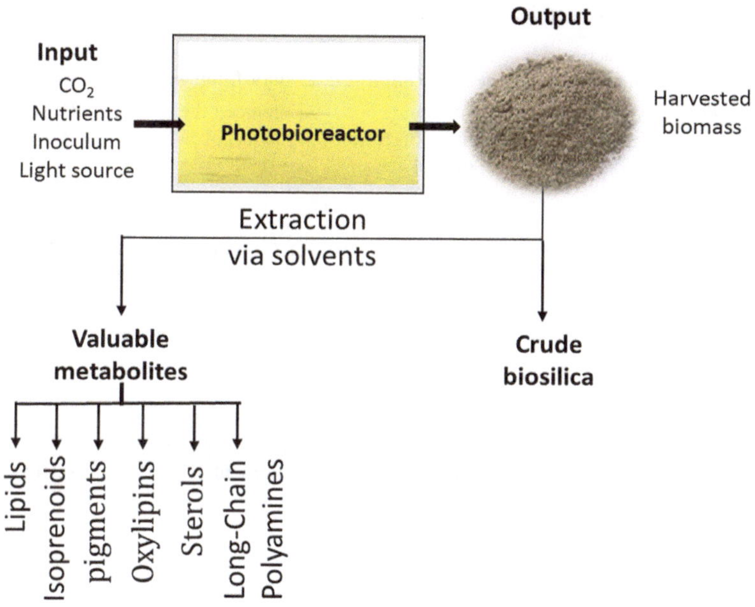

Fig. 3 A diagram illustrates the biorefinery approach of diatoms

sustainability, scalability, uniformity of the valves—a criterion required for some applications, and the ability to modify the biosilica properties during cultivation, besides the future abilities to modify the properties via genetic engineering. On the other hand, there are still limitations facing the commercialization of the cultivated biosilica, including scaling-up the culture, production price, and the ability to retrieve the valves or girdle bands intact, and in parallel extraction of valuable metabolites. This currently cannot compete with the DE from diatomite deposits, which are available in the market with low prices and large amounts. Thus, more efforts are needed in order to establish a commercial approach to utilize freshly cultivated diatom biosilica in applications.

5 Biotechnological and Biomedical Applications

Diatom biosilica is a valuable resource in biotechnology and biomedicine, which is derived from the intricate silica structures produced by diatoms. The unique properties and structural characteristics of diatom biosilica have sparked intense interest among researchers and scientists, paving the way for innovative applications in various biotechnological and biomedical fields. In this section, some examples of the recent trends in utilizing diatom biosilica in biotechnological and biomedical applications are highlighted. An overview of diatom biosilica's applications in biotechnology and biomedicine is presented in Table 1.

Table 1 Diatom biosilica and its surface-functionalization for various applications

Biosilica sources	Surface-functionalization	Applications	Major findings	References
Not specified	APTS-modified surface	Covalent immobilization of tyrosinase enzyme	Enhanced the biodegradation of phenolic compounds	Bayramoglu et al. (2013)
Not specified	EDA-modified p (CEA) graft polymer	Covalent immobilization of lipase enzyme	Enhanced enzyme stability, reusability, and biodiesel conversion efficiency	Bayramoglu et al. (2015)
Thalassiosira weissflogii	–	Physical immobilization of lysozyme and BSA	Effective protein adsorbent	Lim et al. (2015)
Navicula sp.	–	Physical immobilization of lysozyme and BSA	Effective protein adsorbent	Lim et al. (2015)
T. pseudonana	–	GOx immobilization	Effective enzyme support under suspension conditions and flow-through conditions	Begum et al. (2019)
Stephanopyxis turris	–	GOx immobilization	Effective enzyme support under suspension conditions and flow-through conditions	Begum et al. (2019)
Eucampia zodiacus	–	GOx immobilization	Effective enzyme support under suspension conditions and flow-through conditions	Begum et al. (2019)
Diatomite	–	Affinity immobilization of carbonic anhydrase	Stability-controllable immobilization of enzyme	Kim et al. (2020b)
Aulacoseira sp.	–	Affinity immobilization of lipase	Enhanced enzyme stability, reusability, and compatibility to laundry detergents	Abdelhamid et al. (2022)
Thalassiosira pseudonana	LiDSI	GFP immobilization	Specific targeting to diatom frustule	Poulsen et al. (2007)
T. pseudonana	LiDSI	Gox immobilization	Enhanced enzyme stability against denaturation and proteolytic degradation	Sheppard et al. (2012)

(continued)

Table 1 (continued)

Biosilica sources	Surface-functionalization	Applications	Major findings	References
T. pseudonana	LiDSI	HRP immobilization	Specific targeting of expressed heme-proteins to diatom frustule	Sheppard et al. (2012)
T. pseudonana	LiDSI	$sdAb_{EA1}$ immobilization	Engineered diatom biosilica for specific binding of high molecular weight antigen	Ford et al. (2016)
T. pseudonana	LiDSI	Regioselective immobilization of redox enzymes	Specific display of expressed enzyme to girdles and valves of diatom biosilica	Kumari et al. (2020)
Diatomaceous earth	Fe_3O_4 and DOPA-functionalized surface	Drug delivery	Magnetically modified diatom surface-guided delivery of indomethacin drug	Losic et al. (2010)
Diatomaceous earth	Copolymer-grafted surface	Thermo-responsive drug delivery	Controlled delivery of levofloxacin drug from diatom biosilica	Vasani et al. (2015)
Coscinodiscus concinnus	–	Drug delivery	Enhanced in vitro delivery of streptomycin	Gnanamoorthy et al. (2014)
Diatomaceous earth	SNAP-DE	NO delivery	Sustained release of NO	Grommersch et al. (2018)
T. weissflogii	Fe_3O_4 -functionalized surface	Drug delivery microrobots	Engineered biosilica microcarriers for target release of doxorubicin	Li et al. (2022)
Not specified	Fibrous scaffolds with polymer blend solution	Bone regeneration	Developed diatom biosilica-based scaffold to simulate osteogenic activity	Dalgic et al. (2023)
T. weissflogii	In vivo incorporation of sodium alendronate	Bone regeneration	Enhanced osteoblast growth with osteoclast inhibition using alendronate-doped biosilica	Cicco et al. (2019)
Not specified	Composite film with cellulose nanofibril	Biocompatible smart mask	Developed composite surface with more electron-donating and high roughness	Rajabi-Abhari et al. (2021)
Not specified	Hydrogel with catechol-chitosan	Tremor sensor	Developed highly self-healable hydrogel conductor	Kim et al. (2021)
Not specified	Composite film with chitosan	Skin-attachable power generator	Developed skin-attachable motion sensor	Kim et al. (2020a, 2021)

5.1 Enzyme Immobilization

Diatom biosilica is a unique biomaterial with a high surface area and complex three-dimensional structure, making it an ideal substrate for a variety of biocatalytic applications. Interestingly, several industrial-relevant enzymes have been immobilized on diatom biosilica and implemented in biotechnological and biomedicine applications (Bayramoglu et al. 2013; Kim et al. 2020a, b). As discussed below, functional proteins could be loaded on diatom biosilica through in vitro and in vivo immobilization-based approaches (Table 1).

5.1.1 In Vitro Immobilization

In vitro immobilization involves the attachment of proteins to the surfaces of the diatom biosilica using chemical or physical methods (Carlsson et al. 2014). For example, a common method involves physical adsorption or covalent attachment of the protein to the surface of the diatom using a chemical linker (Table 1) (Carlsson et al. 2014). In a recent study, tyrosinase was loaded on diatom biosilica (Bayramoglu et al. 2013). The immobilization process involves modifying diatom frustules microparticles with glutaraldehyde and 3- aminopropyl triethoxysilane (APTES) for covalently loading tyrosinase onto the diatom scaffold. The immobilized tyrosinase was used to degrade three different phenols, and it retained 74% of its initial activity after 10 reuses in a batch system. Similarly, lipase was covalently attached to modified diatom biosilica for biodiesel production and was highly stable and kept 83% of its initial activity after 6 reuses (Bayramoglu et al. 2015).

Several characteristics of diatom biosilica such as morphology, pore volume, surface area, and the active silanol groups on its surface could impact the enzyme's adsorption process. Lim et al. (2015) investigated the effect of the morphology of several diatom frustules from three diatom biosilica samples, diatomite, *Navicula* sp., and *Thalassiosira weissflogii* on the adsorption of bovine serum albumin and lysozyme. The frustules of centric diatom, *T. weissflogii*, were found to have the highest adsorption capability. The adsorption of protein on all tested frustules is monolayer chemisorption. Additionally, the effect of biosilica characteristics on the catalytic activity of immobilized enzymes has been investigated. Interestingly, a study reported that the specific activity of glucose oxidase was strongly dependent on the type of biosilica support, reaching a maximum value on mesoporous synthetic silica (Begum et al. 2019).

Using recombinant technology, a one-step-oriented approach has been developed for enzyme immobilization with high catalytic activity. The genetic fusion of a silica-binding domain to the C-terminus of carbonic anhydrase enzyme from *Hydrogenovibrio marinus* oriented its immobilization on diatom biosilica surfaces (Kim et al. 2020b). Unlike nonspecific adsorption, the ionic interactions between silica-binding tag to diatom biosilica are so strong that the binding is known to be virtually irreversible (Abdelhamid et al. 2020; Kim et al. 2019). The immobilized enzyme showed high reusability and stability even under high temperature conditions, without any enzyme leakage. This research found that there was a relationship between the activity and stability of the immobilized enzyme on biosilica surface, where enzyme's activity decreased as the loading increased, while its stability increased with loading. The study suggested that this phenomenon could be due to the high-density surface of biosilica-loaded enzyme, leading to macromolecular crowding, which affects the enzyme's performance.

Furthermore, the molecular basis of the biomineralization in diatom provides knowledge about the silica-affinity proteins responsible for stabilizing the hierarchical biosilica structure. This knowledge has facilitated the development of methods to load enzymes onto the biosilica structures (Fig. 4) (Abdelhamid and Pack 2021). The molecular understanding of silaffin proteins has been combined with genetic engineering techniques to develop a multivalent-oriented enzyme immobilization approach on the surface of diatom biosilica (Abdelhamid et al. 2022). Silaffins, diatom-driven sequences, are proteins with positively charged sequences rich in the amino acid lysine (Kröger et al. 1999), which have a strong binding affinity to the biosilica structure. This strong binding affinity allows the self-assembly of silaffin polypeptides and diatom biosilica in a site-specific manner, resulting in the stable structure of the frustule. Accordingly, genetic fusion of silaffin-tags to both termini of lipase-A (LipA) from *Bacillus subtilis* impacts its immobilization and stabilization on diatom biosilica surfaces (Abdelhamid et al. 2022). The fusion of LipA with multivalent silaffin Sil3K-tag showed a better capacity for loading the enzyme and recovering its activity compared to the fusion of the enzyme with a single Sil3K-tag. Interestingly, diatom biosilica-immobilized lipase fusion was more stable in the presence of detergents and surfactants, which makes it a suitable candidate for application in detergent formulations. These findings suggest that the multivalent surface immobilization method using Sil3K-tagged lipase enzymes has the potential to be applied in various industrial fields where high-performance biocatalysts are required, such as the production of detergents, biofuels, and pharmaceuticals.

5.1.2 In Vivo Approaches for Enzyme Immobilization

To naturally immobilize a protein on diatom biosilica in vivo, diatom cells are genetically modified to produce biosilica frustule materials containing the desired protein. Employing genetic engineering, diatom-driven silaffins genes and proteins were implemented into in vivo approach for tightly attached functional proteins to biosilica surfaces. In earlier studies, Kröger group developed in vivo biosilica-based approach for enzyme immobilization, named as "living diatom silica immobilization" (LiDSI) (Poulsen et al. 2007; Sheppard et al. 2012). In this study, the fusion of *Thalassiosira pseudonana* silaffin-3 (TpSil3) to the target enzyme was constructed and then genetically transformed into diatom cells. Subsequently, a silaffin-3 tag was used to attach the expressed fusion proteins to diatom frustules. Several expressed proteins, such as green fluorescent protein (GFP) (Poulsen et al. 2007), horseradish

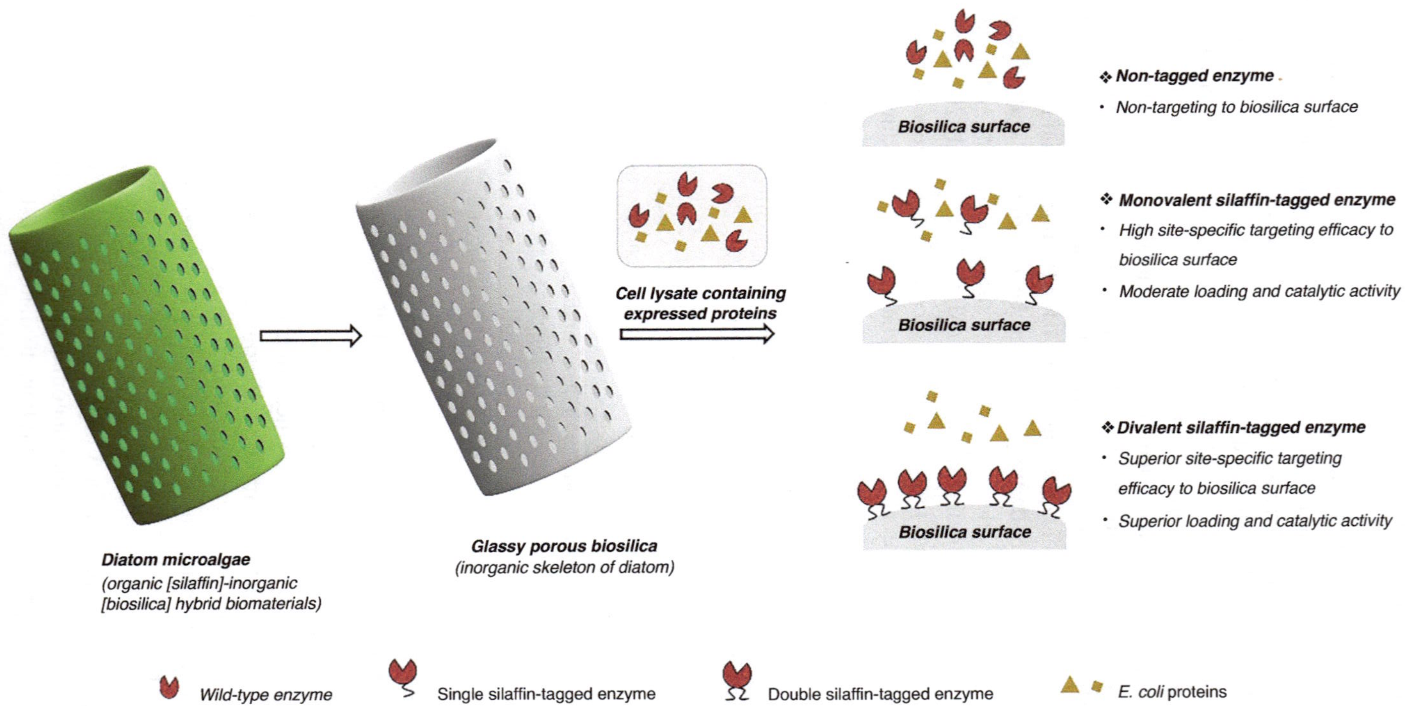

Fig. 4 Schematic illustration showing the affinity immobilization of silaffin-tagged enzymes on diatom biosilica surfaces. Reproduced from Abdelhamid et al. (2022) after required permission (No. 5530541320180)

peroxidase (HRP), and GOx (Sheppard et al. 2012), were directly immobilized on diatom cells, preserving their catalytic activities. Furthermore, single-chain antibodies immobilized on diatom biosilica bind to high molecular antigens, such as layer protein EA1 present in *Bacillus anthracis*, and to small molecular antigens, such as explosive trinitrotoluene (Ford et al. 2016).

For further extension of this repertoire, Kumari et al. (Kumari et al. 2020) designed regioselective affinity immobilization of redox enzymes, HRP, and Gox, on frustules of *T. Pseudonana* (Kumari et al. 2020). While cingulin-derived proteins targeted the enzyme immobilization on the girdle region, silaffin-1 facilitates the localization of its fusion enzymes on the valve region. Interestingly, the fusion enzymes bound to the girdle band had nearly three-fold greater affinity than those bound to the valve region. The authors attributed these findings to the higher negative charge density of the girdle region than the valve region.

The LiDSI immobilization approach of enzymes on diatom biosilica has advantages over in vitro immobilization including skipping required protein purification steps, and surface modifications of the frustule. Additionally, immobilization process occurs under physiological conditions, which preserves the protein's functionality. However, the loading amount of the fusion enzymes is considered as the main disadvantage of this approach. In vivo immobilization of the target enzyme on biosilica of diatom cells resulted in a tenfold reduction in yield efficiency compared to in vitro immobilization on biomimetic silica (Poulsen et al. 2007; Sheppard et al. 2012). This could be explained by the transport capacity of the silaffin fusions to the silica deposition vesicle (SDV). The established system for high loading of functional proteins could be improved by understanding the receptor proteins required silaffin polypeptides targeting to SDV.

5.2 *Drug Delivery*

The unique porous structure and high surface area of diatom biosilica-based materials make them attractive for drug delivery (Delalat et al. 2015; Uthappa et al. 2018). Biosilica demonstrates several advantages over traditional drug delivery systems, including improved drug loading and/or release properties, high biocompatibility, and potential for targeted delivery (Table 1) (Delalat et al. 2015). The nanoporous structure of the diatoms provides high drug loading capacity, while its porous nature ensures controlled drug release (Aw et al. 2011). Furthermore, diatom surfaces can be tailored for improved drug binding and release (Chao et al. 2014). Accordingly, diatoms have become more prevalent due to their advantageous properties (Chao et al. 2014; Uthappa et al. 2018).

In 2010, Losic group was the first to report on the implementation of diatoms in drug delivery (Losic et al. 2010). With the addition of dopamine and iron oxide nanoparticles to diatom biosilica microparticles, drug release profiles could be controlled, and magnetically guided drug delivery systems could be designed. Interestingly, the indomethacin drug was delivered sustainably over two weeks via

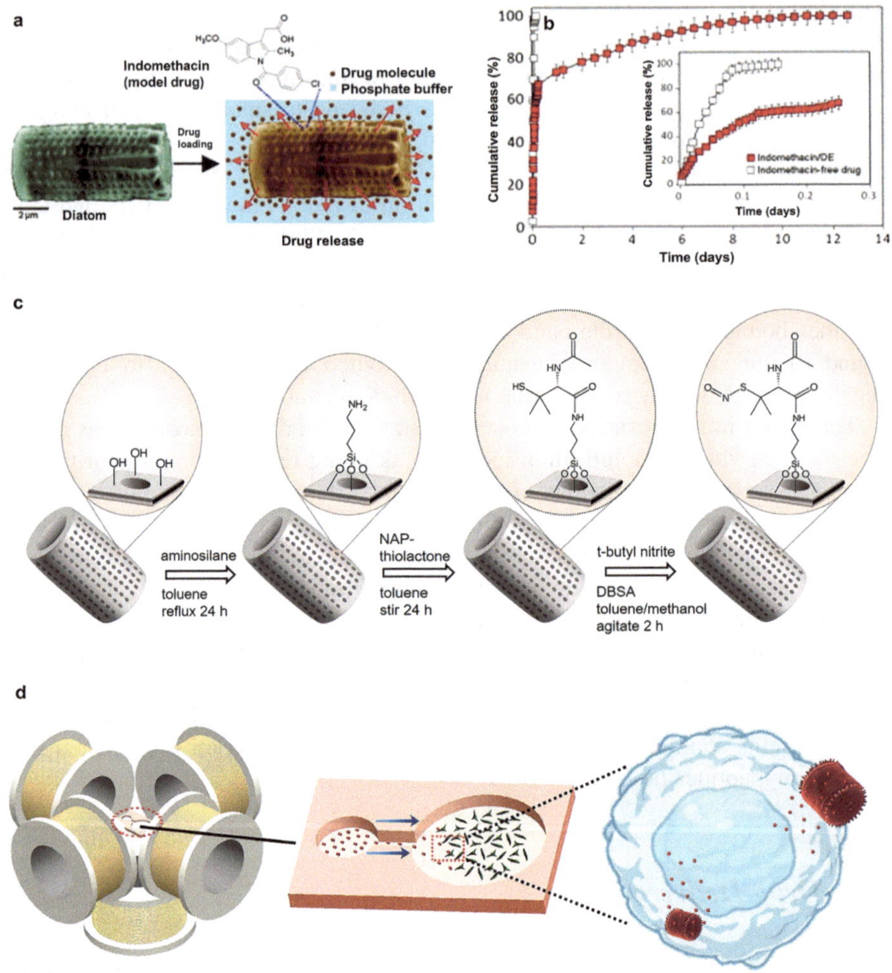

Fig. 5 Targeted drug delivery using diatom biosilica microparticles. (**a**) Scheme presentation showing drug release from biosilica microparticles. (**b**) Indomethacin release curves for biosilica for 14 days and 6 h (inset graph). (**c**) Schematic for derivatized diatom biosilica using APTES as a representative silane for NO release. (**d**) Schematic illustration of targeted delivery of drug-loaded biohybrid magnetic microrobots to MCF-7 cells. (**a** and **b**): Reproduced from Aw et al. (2012) after permission (No. 5530541483033). (**c**) Reproduced from Grommersch et al. (2018) after permission (Copyright 2018 American Chemical Society). (**d**) Reproduced from Li et al. (2022) after permission (No. 5530550107964)

magnetized diatom microstructures. Additionally, the native frustule structure of the *Aulacoseira* diatom has been utilized for the delivery of indomethacin by loading the drug in the pores of diatom microparticles (Fig. 5a and b) (Aw et al. 2012). Interestingly, the delivery profile showed a slow-release phase that lasted for two weeks, suggesting the efficient storage of the drug molecules in the inner structure of diatom biosilica particles (Aw et al. 2012). In the same manner, Vasani et al. (2015) used the atom transfer radical polymerization technique to polymerize thermoresponsive oligo copolymer onto the surfaces of frustule microparticles. The loaded modified matrix with the drug levofloxacin showed thermal-dependent behavior for drug release. Another study explored the potential use of natural biosilica nanoporous biomaterial derived from marine diatoms for drug delivery applications. Specifically, the researchers looked at *Coscinodiscus concinnus*, which has a surface with porous domes. Using streptomycin, a hydrophilic drug, a diatom-based drug delivery model was demonstrated (Gnanamoorthy et al. 2014). Upon adsorption, the drug was found to occupy the inside of pores, as well as the internal hollow structures of diatoms with a maximum loading capacity of 33.33%. The drug release from the diatom foramen had an initial burst followed by sustained release from the cribrum and cribellum, which lasted up to 7 days. According to the findings, diatom biosilica has the potential to be a natural source of siliceous materials that can be used as in vitro drug delivery biomaterials.

Similarly, the use of natural biosilica of *T. weissflogii* as a support for drug delivery has been studied (Saxena et al. 2022). The frustules were able to load 79.05% of curcumin, a model drug. The drug release rate was faster in physiological conditions compared to acidic conditions. The frustules' surface negative charge allowed curcumin to adsorb to the void pores of diatom biosilica particles. Furthermore, the curcumin-loaded biosilica did not show any toxic effects on normal HEK 293 cells, but exhibited toxicity against ACHN cancer cell lines. In addition, researchers have studied the use of nanoporous biosilica derived from diatoms for delivering chemotherapeutic drugs against cancer cells (Delalat et al. 2015). Using genetic engineering to produce a Mg-G-binding domain on the surface of *T. pseudonana* enabled the attachment of cell-targeting antibodies. Engineered biosilica surfaces are coated with antibodies that target and kill neuroblastoma and B-lymphoma cells selectively.

A recent study introduced a novel method for preparing a scaffold that delivers nitric oxide (NO) using diatom biosilica as a biotemplate (Grommersch et al. 2018). Authors evaluated the effectiveness of different amino silanes for maximizing NO loading by attaching N-acetyl-d-penicillamine (NAP) to diatom biosilica (Fig. 5c). The cross-linker APTES was used to achieve the highest degree of NAP tethering to the surface of DE. Through nitrosation, NAP-DE scaffold was converted to S-nitroso-N-acetyl-penicillamine (SNAP)-DE. The biocompatible SNAP-DE scaffold with a total NO loading of 0.0372 μmol mg^{-1} demonstrated high killing efficiency against *Staphylococcus aureus* without any negative impact on mammalian cells. This innovative approach can offer adjustable levels of NO for materials in various fields, such as tissue engineering, drug delivery, polymer chemistry, and wound healing.

Interestingly, the intact diatom frustule microparticles were recently used to prepare biohybrid magnetic microrobots, which were employed to develop drug delivery systems (Li et al. 2022). The authors created a microcarrier that incorporates doxorubicin and magnetic nanoparticles within *T. weissflogii* diatoms (Fig. 5d). The diatoms are designed to release drugs that are pH-responsive, which are activated when they encounter the acidic environment of a tumor region. By applying an external magnetic field, the movement of the microcarrier inside the body can be regulated.

5.3 Bone Regeneration

In the last few decades, there has been growing interest in the potential role of biosilica for bone regeneration (Wang et al. 2014b). Several studies have suggested that silica can enhance bone formation and promote bone healing (Abdelhamid and Pack 2021). Silica is believed to stimulate osteoblasts, the cells responsible for bone formation, by increasing the expression of genes involved in bone formation and mineralization (Wang et al. 2014b). Silica can also enhance the activity of alkaline phosphatase, an enzyme that plays a critical role in bone mineralization (Ki et al. 2023). A recent study created a scaffold by electrospinning that could load and release melatonin in a controlled manner by utilizing diatom silica frustules to enhance osteogenesis (Table 1) (Dalgic et al. 2023). The researchers utilized a blend of polymers, PHBV/PCL, to produce fibrous scaffolds that incorporated diatom frustules loaded with melatonin via wet electrospinning. Frustules were coated with a polymer layer within the 3D fibre matrix, which considerably reduced the release of melatonin and enabled controlled release for seven days. According to this study, alkaline phosphatase (ALP) activity of cells had a dependence on the concentration of melatonin. Scaffolds containing frustules loaded with melatonin demonstrated a significant improvement in ALP activity for Saos-2 cells. By delivering melatonin in a controlled manner, the developed scaffold system stimulated osteogenic activity.

In vivo experiments have demonstrated that sodium alendronate drug can be integrated into biosilica shells of cultured *T. weissflogii* diatoms (Table 1) (Cicco et al. 2019). The resulting biosilica was functionalized with bisphosphonate and characterized for its osteo-inductive properties, showing a strong ability to support tissue regeneration. The researchers estimated the loading percentage of sodium alendronate in the biosilica to be 1.45% *w/w*. When tested, the functionalized biosilica was able to reduce the metabolic activity of osteoclasts-like cells by up to 5% compared to glass control. These findings suggest that diatom cells could be a promising nanoporous material for tissue engineering applications and potential utilization in bone tissue engineering.

5.4 Medical Nanodevices

Currently, there is a growing attention in the scientific community towards using environmentally friendly and safe materials in triboelectric nanogenerators (TENGs) to produce energy from the movements of the human body (Table 1). Researchers have developed a new type of biocomposite film that improves the performance of

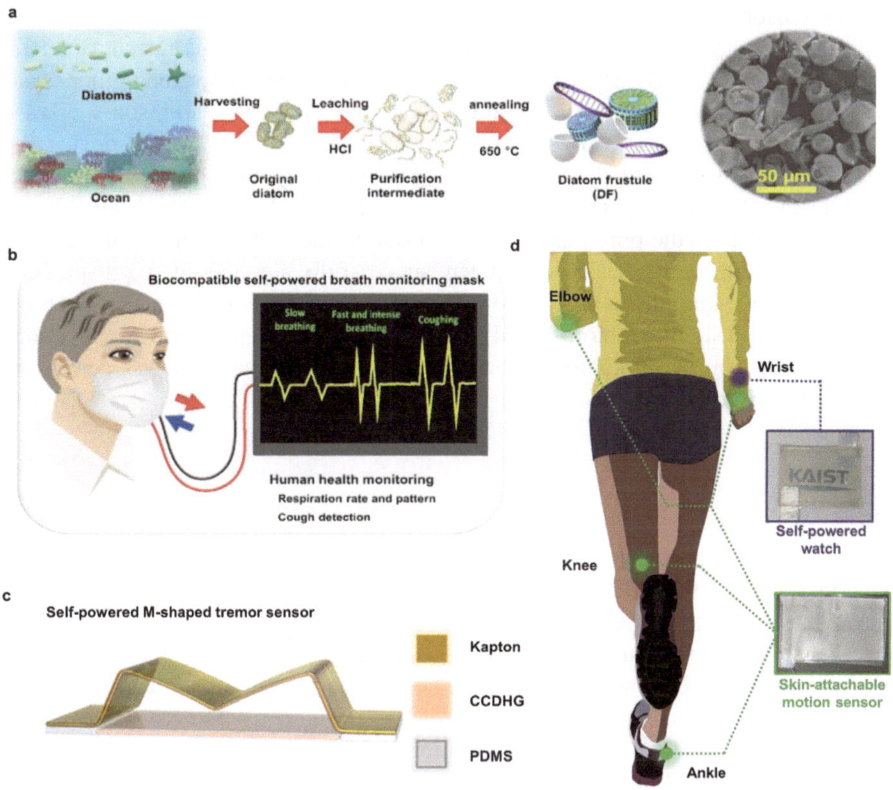

Fig. 6 (**a**) Schematic diagram depicting the steps involved in mass-production of diatom biosilica from cultivated diatom microalgae. Diatom-based TENGs for (**b**) self-powered innovative biocompatible mask, (**c**) tremor sensor, and (**d**) skin-attachable power generator. (**a** and **b**) Reproduced from Rajabi-Abhari et al. (2021) after permission (Copyright 2021 American Chemical Society). (**c**) Reproduced from Kim et al. (2021) after permission (No. 5530541104278). (**d**) Reproduced from Kim et al. (2020a) after permission (No. 5530540948293)

TENGs by combining diatom bio-silica and cellulose nanofibril (CNF) (Fig. 6a and b and Table 1) (Rajabi-Abhari et al. 2021). Diatom biosilica, which has a large surface area and porous three-dimensional structures, was used in the composite film due to their ability to form hydrogen bonds with CNFs, making the surface more electron-donating and rough. The diatom-based TENG demonstrated an output voltage and a time-averaged power of 388 V and 85.5 mW m^{-2}, respectively, with a contact area of 4.9 cm^2, which was sufficient to power 102 light-emitting diodes. Moreover, the biocomposite was deemed biologically safe through cytotoxicity studies and biocompatibility tests on rabbit skin.

A hydrogel conductor called catechol-chitosan-diatom hydrogel (CCDHG) has been developed, which is highly stretchable and self-healable (Fig. 6c) (Kim et al. 2021). This conductor was used to create a stretchable TENG and a self-powered tremor sensor for Parkinson's disease patients. The CCDHG-TENG demonstrated high power output with an open-circuit voltage and short-circuit current of 110 V and 3.8 μA, respectively. By using machine learning, the self-powered tremor sensor was able to identify the health conditions of people with Parkinson's disease. Additionally, Kim et al. (2020a) created a skin-attachable wearable device that addresses biocompatibility issues by utilizing the biocompatible properties of diatom frustule and chitosan (Fig. 6d). The diatom-chitosan-based TENG has a much higher output power than the pure chitosan TENG due to the addition of diatom frustule. Chitosan-diatom-based TENG has been successfully applied in a self-powered watch and a skin-attachable motion sensor. This research represents a novel approach for designing diatom-based TENGs that can have promising applications, such as powering self-monitored health masks, skin-attachable power generators, and tactile feedback systems.

6 Conclusions

Diatom nanostructured biosilica is one of the most outstanding biomaterials offered through evolution. While the morphogenesis of these frustules is still partially a puzzle, there is no doubt that their nanostructuring brings unique mechanical, hydrokinetic, and optical properties that helped them to dominate and flourish in almost all aquatic habitats, including extreme environments. Moreover, the unique features of diatom frustules helped to employ them widely in many applications, especially their fossil form—diatomite. The range of applications is huge, while this chapter focuses more on the applicability of DE as a filler and functional material in several industrial and commercial applications. This gives a brief overview of how DE is really involved in the market. Through diatoms cultivation, it is possible to produce more controllable silica frustules, which can—if all current challenges are solved—provide the recent applications with sufficient amounts required for commercialization. Finally, the recent trends in utilizing diatom biosilica in biotechnological and biomedical applications give a deeper insight into how this silica is promising as an alternative to the widely used synthetic silica for a greener and more sustainable future.

Acknowledgement Seung Pil Pack would like to acknowledge receiving funding from the Korea Ministry of Environment (MOE)(RE202101398) and National Research Foundation of Korea (NRF) grants supported by the Korean government (MSIT) (NRF-2021R1A5A8032895), Republic of Korea.

References

Abdelhamid MAA, Pack SP (2021) Biomimetic and bioinspired silicifications: recent advances for biomaterial design and applications. Acta Biomater 120:38–56. https://doi.org/10.1016/j.actbio.2020.05.017

Abdelhamid MAA, Meligy AMA, Yeo KB, Lee C-S, Pack SP (2020) Silaffin-3-derived pentalysine cluster as a new fusion tag for one-step immobilization and purification of recombinant *Bacillus subtilis* catalase on bare silica particles. Int J Biol Macromol 159:1103–1112. https://doi.org/10.1016/j.ijbiomac.2020.04.172

Abdelhamid MAA, Son RG, Park KS, Pack SP (2022) Oriented multivalent silaffin-affinity immobilization of recombinant lipase on diatom surface: reliable loading and high performance of biocatalyst. Colloids Surf B Biointerfaces 219:112830. https://doi.org/10.1016/j.colsurfb.2022.112830

Agusti S, González-Gordillo JI, Vaqué D, Estrada M, Cerezo MI, Salazar G, Gasol JM, Duarte CM (2015) Ubiquitous healthy diatoms in the deep sea confirm deep carbon injection by the biological pump. Nat Commun 6:7608. https://doi.org/10.1038/ncomms8608

AlHousami T, Stojanov IJ, Deighan P, Rhoads DD, Sundararajan D, Lassetter J, Shamritsky Y, Kabani S, Noonan V (2021) Diatoms: a novel cause of granulomatous inflammation of the head and neck. Oral Surg Oral Med Oral Pathol Oral Radiol 131:565–571. https://doi.org/10.1016/j.oooo.2020.10.014

Alldredge AL, Gotschalk CC (1989) Direct observations of the mass flocculation of diatom blooms: characteristics, settling velocities and formation of diatom aggregates. Deep Sea research Part A. Oceanogr Res Pap 36:159–171. https://doi.org/10.1016/0198-0149(89)90131-3

Ambegoda VT, Egodage SM, Blum FD, Maddumaarachchi M (2021) Enhancement of hydrophobicity of natural rubber latex films using diatomaceous earth. J Appl Polym Sci 138:50047. https://doi.org/10.1002/app.50047

Andersen RA (2004) Algal culturing techniques.1st edn. Elsevier, Amsterdam

Araújo IJ d S, Zanini MM, Guarda MB, Fronza BM, Consani S, Lima IPC, Correr AB (2022) Diatomite filler for resin composites application – a new approach for materials improvement. Res Soc Dev 11:e268111637738. https://doi.org/10.33448/rsd-v11i16.37738

Aumeier C, Menzel D (2012) Secretion in the diatoms. In: Vivanco J, Baluška F (eds) Secretions and exudates in biological systems. signaling and communication in plants, vol 12. Springer, Berlin, pp 221–250. https://doi.org/10.1007/978-3-642-23047-9_10

Aw MS, Simovic S, Addai-Mensah J, Losic D (2011) Silica microcapsules from diatoms as new carrier for delivery of therapeutics. Nanomedicine 6:1159–1173. https://doi.org/10.2217/nnm.11.29

Aw MS, Simovic S, Yu Y, Addai-Mensah J, Losic D (2012) Porous silica microshells from diatoms as biocarrier for drug delivery applications. Powder Technol 223:52–58

Babenko I, Friedrich BM, Kröger N (2022) Structure and morphogenesis of the frustule. In: The molecular life of diatoms. Springer International Publishing, Cham, pp 287–312. https://doi.org/10.1007/978-3-030-92499-7_11

Bayramoglu G, Akbulut A, Yakup Arica M (2013) Immobilization of tyrosinase on modified diatom biosilica: enzymatic removal of phenolic compounds from aqueous solution. J Hazard Mater 244–245:528–536. https://doi.org/10.1016/j.jhazmat.2012.10.041

Bayramoglu G, Akbulut A, Ozalp VC, Arica MY (2015) Immobilized lipase on micro-porous biosilica for enzymatic transesterification of algal oil. Chem Eng Res Des 95:12–21. https://doi.org/10.1016/j.cherd.2014.12.011

Bayu A, Rachman A, Noerdjito DR, Putra MY, Widayatno WB (2020) High-value chemicals from marine diatoms: a biorefinery approach. IOP Conf Ser Earth Environ Sci 460:012012. https://doi.org/10.1088/1755-1315/460/1/012012

Begum G, Oschatz C, Oschatz M, Kaskel S, Brunner E, Kroger N (2019) Influence of silica architecture on the catalytic activity of immobilized glucose oxidase. Bioinspired Biomim Nanobiomater 8:72–80. https://doi.org/10.1680/jbibn.18.00002

Bennett DC, Yee A, Rhee Y-J, Cheng KM (2011) Effect of diatomaceous earth on parasite load, egg production, and egg quality of free-range organic laying hens. Poult Sci 90:1416–1426. https://doi.org/10.3382/ps.2010-01256

Benoiston AS, Ibarbalz FM, Bittner L, Guidi L, Jahn O, Dutkiewicz S, Bowler C (2017) The evolution of diatoms and their biogeochemical functions. Philos Trans R Soc B: Biol Sci. https://doi.org/10.1098/rstb.2016.0397

Bleuart S, Boulis Y, Testud O, Larché-Scrivant L, Mison D, Devillers C (2002) Process for the purification of recombinant proteins from complex media and purified proteins obtained thereby. European Patent office, Munich

Buyel JF, Gruchow HM, Fischer R (2015) Depth filters containing diatomite achieve more efficient particle retention than filters solely containing cellulose fibers. Front Plant Sci 6:1134. https://doi.org/10.3389/fpls.2015.01134

Carlsson N, Gustafsson H, Thörn C, Olsson L, Holmberg K, Åkerman B (2014) Enzymes immobilized in mesoporous silica: a physical–chemical perspective. Adv Colloid Interf Sci 205:339–360. https://doi.org/10.1016/j.cis.2013.08.010

Chao JT, Biggs MJP, Pandit AS, (2014) Diatoms: a biotemplating approach to fabricating drug delivery reservoirs. Expert Opin Drug Deliv 11(11) 1687–1695. Taylor & Francis

Chen P-Y, Chang H-K (2023) Multifunctional bio-ceramic scaffolds and composites fabricated by the freeze casting techniques. In: Najman S et al (eds) Bioceramics, biomimetic and other compatible materials features for medical applications. Engineering materials. Springer, Cham, pp 173–193. https://doi.org/10.1007/978-3-031-17269-4_9

Cheng Y, Yu D, Tan G, Zhu C (2018) Low-temperature performance and damage constitutive model of eco-friendly basalt fiber–diatomite-modified asphalt mixture under freeze–thaw cycles. Materials 11:2148. https://doi.org/10.3390/ma11112148

Chepurnov VA, Mann DG, Sabbe K, Vyverman W (2004) Experimental studies on sexual reproduction in diatoms. Int Rev Cytol 237:91–154. https://doi.org/10.1016/S0074-7696(04)37003-8

Cicco SR, Vona D, Leone G, De Giglio E, Bonifacio MA, Cometa S, Fiore S, Palumbo F, Ragni R, Farinola GM (2019) In vivo functionalization of diatom biosilica with sodium alendronate as osteoactive material. Mater Sci Eng C 104:109897. https://doi.org/10.1016/j.msec.2019.109897

Cong P, Chen S, Chen H (2012) Effects of diatomite on the properties of asphalt binder. Constr Build Mater 30:495–499. https://doi.org/10.1016/j.conbuildmat.2011.11.011

Cong P, Liu N, Tian Y, Zhang Y (2016) Effects of long-term aging on the properties of asphalt binder containing diatoms. Constr Build Mater 123:534–540. https://doi.org/10.1016/j.conbuildmat.2016.07.073

Crangle RD (2015) Diatomite. https://d9-wret.s3-us-west-2.amazonaws.com/assets/palladium/production/mineral-pubs/diatomite/myb1-2015-diato.pdf

Crangle RD (2018) Diatomite. https://d9-wret.s3.us-west-2.amazonaws.com/assets/palladium/production/atoms/files/myb1-2018-diato.pdf

Cupo A, Landi S, Morra S, Nuzzo G, Gallo C, Manzo E, Fontana A, d'Ippolito G (2021) Autotrophic vs. heterotrophic cultivation of the marine diatom *Cyclotella cryptica* for EPA production. Mar Drugs 19:355. https://doi.org/10.3390/md19070355

Dalgic AD, Atila D, Tezcaner A, Gürses S, Keskin D (2023) Diatom silica frustules-doped fibers for controlled release of melatonin for bone regeneration. Eur Polym J 186:111858

De Tommasi E (2016) Light manipulation by single cells: the case of diatoms. J Spectrosc 2016:1–13. https://doi.org/10.1155/2016/2490128

Degirmenci N, Yilmaz A (2009) Use of diatomite as partial replacement for Portland cement in cement mortars. Constr Build Mater 23:284–288. https://doi.org/10.1016/j.conbuildmat.2007.12.008

Delalat B, Sheppard VC, Rasi Ghaemi S, Rao S, Prestidge CA, McPhee G, Rogers M-L, Donoghue JF, Pillay V, Johns TG, Kröger N, Voelcker NH (2015) Targeted drug delivery using genetically engineered diatom biosilica. Nat Commun 6:8791. https://doi.org/10.1038/ncomms9791

Dexter JF, Servals PC (1953) Silicone rubber as cable insulation. In: U. S. Army signal corps wire and cable symposium

Dhanker R, Kumar R, Tiwari A, Kumar V (2022) Diatoms as a biotechnological resource for the sustainable biofuel production: a state-of-the-art review. Biotechnol Genet Eng Rev 38:111–131. https://doi.org/10.1080/02648725.2022.2053319

Elias M, Cultrone G (2019) On the use of sodium chloride and calcined diatomite sludge as additives to improve the engineering properties of bricks made with a clay earth from Jun (Granada, Spain). Fortschr Mineral 9:64. https://doi.org/10.3390/min9010064

Ergün A (2011) Effects of the usage of diatomite and waste marble powder as partial replacement of cement on the mechanical properties of concrete. Constr Build Mater 25:806–812. https://doi.org/10.1016/j.conbuildmat.2010.07.002

Farzaneh A, Erol M, Mermer O, Cobley AJ, Asl SK (2018) Evaluation of the possibility of using diatomite natural mineral as a composite agent in acrylic coating. Prot Met Phys Chem Surfaces 54:99–103. https://doi.org/10.1134/S2070205118010227

Fazlutdinova A, Gabidullin Y, Allaguvatova R, Gaysina L (2020) Diatoms in Kamchatka's hot spring soils. Diversity (Basel) 12:435. https://doi.org/10.3390/d12110435

Feng C (2013) Inner wall body structure with diatom material coatings. CN203284997U

Fernandez MI, Woodward BW, Stromberg BE (1998) Effect of diatomaceous earth as an anthelmintic treatment on internal parasites and feedlot performance of beef steers. Anim Sci 66:635–641. https://doi.org/10.1017/S1357729800009206

Flower RJ (2013) Diatomites: their formation, distribution, and uses. In: Encyclopedia of quaternary science2nd edn. Elsevier Inc., Amsterdam, pp 501–506. https://doi.org/10.1016/B978-0-444-53643-3.00220-X

Ford NR, Hecht KA, Hu D, Orr G, Xiong Y, Squier TC, Rorrer GL, Roesijadi G (2016) Antigen binding and site-directed labeling of biosilica-immobilized fusion proteins expressed in diatoms. ACS Synth Biol 5:193–199

Galal Mors HE (2010) Diatomite: its characterization, modifications and applications. Asian J Mater Sci 2:121–136. https://doi.org/10.3923/ajmskr.2010.121.136

Galán-Arboledas RJ, Cotes-Palomino MT, Bueno S, Martínez-García C (2017) Evaluation of spent diatomite incorporation in clay based materials for lightweight bricks processing. Constr Build Mater 144:327–337. https://doi.org/10.1016/j.conbuildmat.2017.03.202

Ghobara MM, Ghobara MM, Ghobara MM, Mohamed A (2019) Diatomite in use: nature, modifications, commercial applications and prospective trends. In: Diatoms: fundamentals and applications. Wiley, Hoboken, pp 471–509. https://doi.org/10.1002/9781119370741.ch19

Ghobara MM, Gordon R, Reissig L (2021a) The mesopores of Raphid pennate diatoms: toward natural controllable anisotropic mesoporous silica microparticles. In: Diatom morphogenesis. Wiley, Hoboken, pp 383–409. https://doi.org/10.1002/9781119488170.ch16

Ghobara MM, Tiffany MA, Gordon R, Reissig L (2021b) Diatom pore arrays' periodicities and symmetries in the euclidean plane: nature between perfection and imperfection. In: Diatom morphogenesis. Wiley, Hoboken, pp 117–158. https://doi.org/10.1002/9781119488170.ch6

Ghobara M, Oschatz C, Fratzl P, Reissig L (2022) Numerical analysis of the light modulation by the frustule of *Gomphonema parvulum*: the role of integrated optical components. Nano 13:113. https://doi.org/10.3390/nano13010113

Gnanamoorthy P, Anandhan S, Prabu VA (2014) Natural nanoporous silica frustules from marine diatom as a biocarrier for drug delivery. J Porous Mater 21:789–796. https://doi.org/10.1007/s10934-014-9827-2

Govindan N, Maniam G, Ab. Rahim MH, Sulaiman A, Ajit A, Chatsungnoen T, Chisti Y (2021) Production of renewable lipids by the diatom *Amphora copulata*. Fermentation 7:37. https://doi.org/10.3390/fermentation7010037

Graham RE, Kenner CT (1973) Acetonitrile-diatomaceous earth column for separation of steroids and other compounds. J Pharm Sci 62:1845–1849. https://doi.org/10.1002/jps.2600621125

Greene M, Danvers N, Hu Q (2009) Organo-neutralized diatomaceous earth, methods of preparation, and uses thereof. WO2009137394A1

Gröger P, Poulsen N, Klemm J, Kröger N, Schlierf M (2016) Establishing super-resolution imaging for proteins in diatom biosilica. Sci Rep 6:36824. https://doi.org/10.1038/srep36824

Grommersch BM, Pant J, Hopkins SP, Goudie MJ, Handa H (2018) Biotemplated synthesis and characterization of mesoporous nitric oxide-releasing diatomaceous earth silica particles. ACS Appl Mater Interfaces 10:2291–2301. https://doi.org/10.1021/acsami.7b15967

Guiraldo RD, Berger SB, Consani RLX, Consani S, de Carvalho RV, Lopes MB, Meneghel LL, da Silva FB, Sinhoreti MAC (2014) Characterization of morphology and composition of inorganic fillers in dental alginates. Biomed Res Int 2014:1–6. https://doi.org/10.1155/2014/178064

Hamm CE, Merkel R, Springer O, Jurkojc P, Maier C, Prechtel K, Smetacek V (2003) Architecture and material properties of diatom shells provide effective mechanical protection. Nature 421: 841–843. https://doi.org/10.1038/nature01416

Harwood DM (2010) Diatomite. In: The diatoms. Cambridge University Press, Cambridge, pp 570–574. https://doi.org/10.1017/CBO9780511763175.034

Hasan M, Saidi T, Husaini JM, Rhina Z (2021) Characteristics of lightweight bricks composed of clay and diatomite. In: Proceedings of the 2nd international conference on experimental and computational mechanics in engineering. lecture notes in mechanical engineering. Springer, Singapore, pp 9–19. https://doi.org/10.1007/978-981-16-0736-3_2

Heck AM, Yanovski JA, Calis KA (2000) Orlistat, a new lipase inhibitor for the management of obesity. Pharmacotherapy 20:270–279. https://doi.org/10.1592/phco.20.4.270.34882

Heine-Fuster I, López-Allendes C, Aránguiz-Acuña A, Véliz D (2021) Differentiation of diatom guilds in extreme environments in the Andean Altiplano. Front Environ Sci 9:701970. https://doi.org/10.3389/fenvs.2021.701970

Hertzog GI, Soares KL, Caldas SS, Primel EG (2015) Study of vortex-assisted MSPD and LC-MS/MS using alternative solid supports for pharmaceutical extraction from marketed fish. Anal Bioanal Chem 407:4793–4803. https://doi.org/10.1007/s00216-015-8685-3

Hilbrig F, Freitag R, Schumacher I (2001) Method for treating biomass for producing cell lysate containing plasmid DNA. US7378238B2

Hohmann-Marriott MF, Blankenship RE (2011) Evolution of photosynthesis. Annu Rev Plant Biol 62:515–548. https://doi.org/10.1146/annurev-arplant-042110-103811

Horn N, Marquet M, Meek J, Budahazi G (1996) Process for reducing RNA concentration in a mixture of biological material using diatomaceous earth. US5576196A

Huntley ME, Johnson ZI, Brown SL, Sills DL, Gerber L, Archibald I, Machesky SC, Granados J, Beal C, Greene CH (2015) Demonstrated large-scale production of marine microalgae for fuels and feed. Algal Res 10:249–265. https://doi.org/10.1016/j.algal.2015.04.016

Ikusika O, Mpendulo C, Zindove T, Okoh A (2019) Fossil Shell flour in livestock production: a review. Animals 9:70. https://doi.org/10.3390/ani9030070

Julius ML, Theriot EC (2010) The diatoms: a primer. In: The diatoms. Cambridge University Press, Cambridge, pp 8–22. https://doi.org/10.1017/CBO9780511763175.003

Ki M-R, Park KS, Abdelhamid MAA, Pack SP (2023) Novel silicatein-like protein for biosilica production from *Amphimedon queenslandica* and its use in osteogenic composite fabrication. Korean J Chem Eng 40:419–428. https://doi.org/10.1007/s11814-022-1314-x

Kim JK, Abdelhamid MAA, Pack SP (2019) Direct immobilization and recovery of recombinant proteins from cell lysates by using EctP1-peptide as a short fusion tag for silica and titania supports. Int J Biol Macromol 135:969–977. https://doi.org/10.1016/j.ijbiomac.2019.05.105

Kim J-N, Lee J, Go TW, Rajabi-Abhari A, Mahato M, Park JY, Lee H, Oh I-K (2020a) Skin-attachable and biofriendly chitosan-diatom triboelectric nanogenerator. Nano Energy 75: 104904. https://doi.org/10.1016/j.nanoen.2020.104904

Kim S, Joo KI, Jo BH, Cha HJ (2020b) Stability-controllable self-immobilization of carbonic anhydrase fused with a silica-binding tag onto diatom biosilica for enzymatic CO2 capture and utilization. ACS Appl Mater Interfaces 12:27055–27063. https://doi.org/10.1021/acsami.0c03804

Kim J-N, Lee J, Lee H, Oh I-K (2021) Stretchable and self-healable catechol-chitosan-diatom hydrogel for triboelectric generator and self-powered tremor sensor targeting at Parkinson disease. Nano Energy 82:105705. https://doi.org/10.1016/j.nanoen.2020.105705

Kipsanai JJ, Wambua PM, Namango SS, Amziane S (2022) A review on the incorporation of diatomaceous earth as a Geopolymer-based concrete building resource. Materials 15:7130. https://doi.org/10.3390/ma15207130

Kotzsch A, Pawolski D, Milentyev A, Shevchenko A, Scheffel A, Poulsen N, Shevchenko A, Kröger N (2016) Biochemical composition and assembly of biosilica-associated insoluble organic matrices from the diatom *Thalassiosira pseudonana*. J Biol Chem 291:4982–4997. https://doi.org/10.1074/jbc.M115.706440

Kröger N (2007) Prescribing diatom morphology: toward genetic engineering of biological nanomaterials. Curr Opin Chem Biol 11:662–669. https://doi.org/10.1016/j.cbpa.2007.10.009

Kröger N (2022) Biomolecules involved in frustule biogenesis and function, in: the molecular life of diatoms. Springer International Publishing, Cham, pp 313–343. https://doi.org/10.1007/978-3-030-92499-7_12

Kröger N, Deutzmann R, Sumper M (1999) Polycationic peptides from diatom biosilica that direct silica Nanosphere formation. Science 286:1129–1132. https://doi.org/10.1126/science.286.5442.1129

Kumari E, Görlich S, Poulsen N, Kröger N (2020) Genetically programmed Regioselective immobilization of enzymes in biosilica microparticles. Adv Funct Mater 30:2000442. https://doi.org/10.1002/adfm.202000442

Lang Y, Yongke C (2018) A kind of diatom art paint coating containing inorganic salts. CN108359356A

Li M, Wu J, Lin D, Yang J, Jiao N, Wang Y, Liu L (2022) A diatom-based biohybrid microrobot with a high drug-loading capacity and pH-sensitive drug release for target therapy. Acta Biomater 154:443–453. https://doi.org/10.1016/j.actbio.2022.10.019

Liang G, Yao W (2023) Effect of diatomite on the reaction kinetics, early-age chemical shrinkage and microstructure of alkali-activated slag cements. Constr Build Mater 376:131026. https://doi.org/10.1016/j.conbuildmat.2023.131026

Lijun C (2016) Composition and preparation method of effective spurrite medicine. CN106562997A

Lim GW, Lim JK, Ahmad AL, Chan DJC (2015) Influences of diatom frustule morphologies on protein adsorption behavior. J Appl Phycol 27:763–775. https://doi.org/10.1007/s10811-014-0356-9

Liu H, Wu C-J (2016) Effect of the silica content of diatom prey on the production, decomposition and sinking of fecal pellets of the copepod *Calanus sinicus*. Biogeosciences 13:4767–4775. https://doi.org/10.5194/bg-13-4767-2016

Losic D (2017) Diatom nanotechnology. Royal Society of Chemistry, Cambridge. https://doi.org/10.1039/9781788010160

Losic D, Korunic Z (2017) Chapter 10: Diatomaceous earth, a natural insecticide for stored grain protection: recent progress and perspectives. In: Losic D (ed) Diatom nanotechnology: progress and emerging applications. RSC Publishing, Cambridge, pp 219–247. https://doi.org/10.1039/9781788010160-00219

Losic D, Rosengarten G, Mitchell JG, Voelcker NH (2006) Pore architecture of diatom frustules: potential nanostructured membranes for molecular and particle separations. J Nanosci Nanotechnol 6:982–989. https://doi.org/10.1166/jnn.2006.174

Losic D, Yu Y, Aw MS, Simovic S, Thierry B, Addai-Mensah J (2010) Surface functionalisation of diatoms with dopamine modified iron-oxide nanoparticles: toward magnetically guided drug microcarriers with biologically derived morphologies. Chem Commun 46:6323–6325

Lu X, Xia Y, Liu M, Qian Y, Zhou X, Gu N, Zhang F (2012) Improved performance of diatomite-based dental nanocomposite ceramics using layer-by-layer assembly. Int J Nanomedicine 7:2153–2164. https://doi.org/10.2147/IJN.S29851

Mackenzie FT (2005) Sediments, diagenesis, and sedimentary rocks.1st edn. Elsevier, Amsterdam

Mann DG, Droop SJM (1996) 3. Biodiversity, biogeography and conservation of diatoms. Hydrobiologia 336:19–32. https://doi.org/10.1007/BF00010816

Mann DG, Crawford RM, Round FE (2017) Bacillariophyta. In: Handbook of the Protists. Springer International Publishing, Cham, pp 205–266. https://doi.org/10.1007/978-3-319-28149-0_29

Manoylov K, Ghobara M (2021) Introduction for a tutorial on diatom morphology. In: Diatom morphogenesis. Wiley, Hoboken, pp 1–18. https://doi.org/10.1002/9781119488170.ch1

Mariusson O, Bjarnason J (1987) Paint not permeable to water and permeable to aqueous vapour. EP0212790B1

Martinez MF (2014) Fire-resistant, insulating, ecological and corrosion-inhibiting coating. EP3235881A1

McGonigle F, Ciullo PA (1996) Paints & coatings. In: Industrial minerals and their uses. Elsevier, Amsterdam, pp 99–159. https://doi.org/10.1016/B978-081551408-4.50005-3

Merino-Maldonado D, Antolín-Rodríguez A, Serrano-González L, Blanco S, Juan-Valdés A, Morán-del Pozo JM, García-González J (2023) Innovative approach for the protection of recycled concrete by biogenic silica biodeposition. Constr Build Mater 368:130475. https://doi.org/10.1016/j.conbuildmat.2023.130475

Miao X, Zhu M, Li Y, Zhang Q, Wang H (2012) Synthesis of dental resins using diatomite and nano-sized SiO2 and TiO2. Prog Nat Sci Mater Int 22:94–99. https://doi.org/10.1016/j.pnsc.2012.03.006

Mikulasik E, Albrech O (2011) New uses of diatomaceous earth in the pharmaceutical industry. JP2013530153A

Milstone JH (1955) Chromatography of blood-clotting factors and serum proteins on columns of diatomaceous earth. J Gen Physiol 38:743–755. https://doi.org/10.1085/jgp.38.6.743

Mock T, Kroon BMA (2002) Photosynthetic energy conversion under extreme conditions—II: the significance of lipids under light limited growth in Antarctic Sea ice diatoms. Phytochemistry 61:53–60. https://doi.org/10.1016/S0031-9422(02)00215-7

Mock T, Otillar RP, Strauss J, McMullan M, Paajanen P, Schmutz J, Salamov A, Sanges R, Toseland A, Ward BJ, Allen AE, Dupont CL, Frickenhaus S, Maumus F, Veluchamy A, Wu T, Barry KW, Falciatore A, Ferrante MI, Fortunato AE, Glöckner G, Gruber A, Hipkin R, Janech MG, Kroth PG, Leese F, Lindquist EA, Lyon BR, Martin J, Mayer C, Parker M, Quesneville H, Raymond JA, Uhlig C, Valas RE, Valentin KU, Worden AZ, Armbrust EV, Clark MD, Bowler C, Green BR, Moulton V, van Oosterhout C, Grigoriev IV (2017) Evolutionary genomics of the cold-adapted diatom *Fragilariopsis cylindrus*. Nature 541:536–540. https://doi.org/10.1038/nature20803

Nasirzadeh M, Yahyaei H, Mohseni M (2023) Effects of inorganic fillers on the performance of the water-based intumescent fire-retardant coating. Fire Mater 47:51–61. https://doi.org/10.1002/fam.3067

O'Brien NR (2002) Microbial taphonomic processes in the fossilization of insects and plants in the late Eocene Florissant formation, Colorado. Rocky Mountain Geol 37:1–11. https://doi.org/10.2113/gsrocky.37.1.1

Pavlíková M, Rovnaníková P, Záleská M, Pavlík Z (2022) Diatomaceous earth—lightweight Pozzolanic admixtures for repair mortars—complex chemical and physical assessment. Materials 15:6881. https://doi.org/10.3390/ma15196881

Perera HJ, Mortazavian H, Blum FD (2017) Surface properties of Silane-treated diatomaceous earth coatings: effect of alkyl chain length. Langmuir 33:2799–2809. https://doi.org/10.1021/acs.langmuir.7b00015

Popovich CA, Pistonesi M, Hegel P, Constenla D, Bielsa GB, Martín LA, Damiani MC, Leonardi PI (2019) Unconventional alternative biofuels: quality assessment of biodiesel and its blends from marine diatom *Navicula cincta*. Algal Res 39:101438. https://doi.org/10.1016/j.algal.2019.101438

Popovich CA, Faraoni MB, Sequeira A, Daglio Y, Martín LA, Martínez AM, Damiani MC, Matulewicz MC, Leonardi PI (2020) Potential of the marine diatom *Halamphora coffeaeformis* to simultaneously produce omega-3 fatty acids, chrysolaminarin and fucoxanthin in a raceway pond. Algal Res 51:102030. https://doi.org/10.1016/j.algal.2020.102030

Poulíčková A, Mann DG, Mann DG (2019) Diatom sexual reproduction and life cycles. In: Diatoms: fundamentals and applications. Wiley, pp 245–272. https://doi.org/10.1002/9781119370741.ch11

Poulsen N, Berne C, Spain J, Kroeger N (2007) Silica immobilization of an enzyme through genetic engineering of the diatom *Thalassiosira pseudonana*. Angew Chem Int Ed 46:1843–1846

Prasetiya FS, Foret M, Deschênes J-S, Gastineau R, Mouget J-L, Tremblay R (2022) Semi-continuous system for benthic diatom cultivation and marennine production. Algal Res 62: 102633. https://doi.org/10.1016/j.algal.2022.102633

Pumas C, Pruetiworanan S, Peerapornpisal Y (2018) Diatom diversity in some hot springs of northern Thailand. Botanica 24:69–86. https://doi.org/10.2478/botlit-2018-0007

Rajabi-Abhari A, Kim J-N, Lee J, Tabassian R, Mahato M, Youn HJ, Lee H, Oh I-K (2021) Diatom bio-silica and cellulose Nanofibril for bio-triboelectric Nanogenerators and self-powered breath monitoring masks. ACS Appl Mater Interfaces 13:219–232. https://doi.org/10.1021/acsami.0c18227

Round FE, Crawford RM, Mann DG (1990) Diatoms: biology and morphology of the genera. Cambridge University Press, Cambridge

Savio S, Farrotti S, Paris D, Arnaìz E, Díaz I, Bolado S, Muñoz R, Rodolfo C, Congestri R (2020) Value-added co-products from biomass of the diatoms Staurosirella pinnata and *Phaeodactylum tricornutum*. Algal Res 47:101830. https://doi.org/10.1016/j.algal.2020.101830

Saxena A, Dutta A, Kapoor N, Kumar A, Tiwari A (2022) Envisaging marine diatom Thalassiosira weissflogii as a "SMART" drug delivery system for insoluble drugs. J Drug Deliv Sci Technol 68:102983. https://doi.org/10.1016/j.jddst.2021.102983

Sharma N, Sharma P, Verma SK (2021) Influence of diatomite on the properties of mortar and concrete: a review. IOP Conf Ser Mater Sci Eng 1116:012174. https://doi.org/10.1088/1757-899X/1116/1/012174

Sheppard VC, Scheffel A, Poulsen N, Kröger N (2012) Live diatom silica immobilization of multimeric and redox-active enzymes. Appl Environ Microbiol 78:211–218

Siddall R (2018) Mineral pigments in archaeology: their analysis and the range of available materials. Fortschr Mineral 8:201. https://doi.org/10.3390/min8050201

Simpson RB (2002) Rubber basics. iSmithers Rapra Publishing

Smol JP, Stoermer EF (2010) The diatoms: applications for the environmental and earth sciences. Cambridge University Press, Cambridge

Spaulding SA, Potapova MG, Bishop IW, Lee SS, Gasperak TS, Jovanoska E, Furey PC, Edlund MB (2021) *Diatoms.org*: supporting taxonomists, connecting communities. Diatom Res 36: 291–304. https://doi.org/10.1080/0269249X.2021.2006790

Sun M, Zou C, Xin D (2020) Pore structure evolution mechanism of cement mortar containing diatomite subjected to freeze-thaw cycles by multifractal analysis. Cem Concr Compos 114: 103731. https://doi.org/10.1016/j.cemconcomp.2020.103731

Tan SC, Yiap BC (2009) DNA, RNA, and protein extraction: the past and the present. J Biomed Biotechnol 2009:1–10. https://doi.org/10.1155/2009/574398

Terracciano M, De Stefano L, Rea I (2018) Diatoms Green nanotechnology for biosilica-based drug delivery systems. Pharmaceutics 10:242. https://doi.org/10.3390/pharmaceutics10040242

Uthappa UT, Brahmkhatri V, Sriram G, Jung H-Y, Yu J, Kurkuri N, Aminabhavi TM, Altalhi T, Neelgund GM, Kurkuri MD (2018) Nature engineered diatom biosilica as drug delivery systems. J Control Release 281:70–83. https://doi.org/10.1016/j.jconrel.2018.05.013

Vasani RB, Losic D, Cavallaro A, Voelcker NH (2015) Fabrication of stimulus-responsive diatom biosilica microcapsules for antibiotic drug delivery. J Mater Chem B 3:4325–4329. https://doi.org/10.1039/C5TB00648A

Veselý D, Kalendova A, Kalenda P (2010) A study of diatomite and calcined kaoline properties in anticorrosion protective coatings. Prog Org Coat 68:173–179. https://doi.org/10.1016/j.porgcoat.2010.02.007

Villanova V, Spetea C (2021) Mixotrophy in diatoms: molecular mechanism and industrial potential. Physiol Plant 173:603–611. https://doi.org/10.1111/ppl.13471

Vitola L, Sahmenko G, Erdmane D, Bumanis G, Bajare D (2017) The effect of various pozzolanic additives on the concrete strength index. IOP Conf Ser Mater Sci Eng 251:012038. https://doi.org/10.1088/1757-899X/251/1/012038

Wachter H, Lechleitner M, Artner-Dworzak E, Hausen A, Jarosch E, Widner B, Patsch J, Pfeiffer K, Fuchs D (1998) Diatomaceous earth lowers blood cholesterol concentrations. Eur J Med Res 3: 211–215

Wang JK, Seibert M (2017) Prospects for commercial production of diatoms. Biotechnol Biofuels 10:16. https://doi.org/10.1186/s13068-017-0699-y

Wang H, Zhu M, Li Y, Zhang Q, Wang H (2011) Mechanical properties of dental resin composites by co-filling diatomite and nanosized silica particles. Mater Sci Eng C 31:600–605. https://doi.org/10.1016/j.msec.2010.11.023

Wang LY, Xiao XF, Gai GQ (2014a) Preparation and performance of diatomite latex paint for Interior Wall. Adv Mater Res 919–921:1994–1997. https://doi.org/10.4028/www.scientific.net/AMR.919-921.1994

Wang S, Wang X, Draenert FG, Albert O, Schröder HC, Mailänder V, Mitov G, Müller WEG (2014b) Bioactive and biodegradable silica biomaterial for bone regeneration. Bone 67:292–304. https://doi.org/10.1016/j.bone.2014.07.025

Wang P, Yang Y, Wang H, Wang H (2019) Fabrication of super-robust and nonfluorinated superhydrophobic coating based on diatomaceous earth. Surf Coat Technol 362:90–96. https://doi.org/10.1016/j.surfcoat.2019.01.065

Wei H, Li Z, Jiao Y (2017) Effects of diatomite and SBS on freeze-thaw resistance of crumb rubber modified asphalt mixture. Adv Mater Sci Eng 2017:1–14. https://doi.org/10.1155/2017/7802035

Wu W-L, Chen Z (2017) Modified-diatomite reinforced rubbers. Mater Lett 209:159–162. https://doi.org/10.1016/j.matlet.2017.07.133

Xiaobing G, Sen C (2013) Diatom ooze wall paint formula and preparation process thereof. CN103131287A

Yang W, Lopez PJ, Rosengarten G (2011) Diatoms: self assembled silicananostructures, and templates for bio/chemical sensors and biomimetic membranes. Analyst 136:42–53. https://doi.org/10.1039/C0AN00602E

Yang C, Xie J, Zhou X, Liu Q, Pang L (2018) Performance evaluation and improving mechanisms of diatomite-modified asphalt mixture. Materials 11:686. https://doi.org/10.3390/ma11050686

Yeh K-C, Synodis J (1984) Stain removal toothpaste. US4612191A

Yılmaz B, Ediz N (2008) The use of raw and calcined diatomite in cement production. Cem Concr Compos 30:202–211. https://doi.org/10.1016/j.cemconcomp.2007.08.003

Zahajská P, Opfergelt S, Fritz SC, Stadmark J, Conley DJ (2020) What is diatomite? Quat Res 96: 48–52. https://doi.org/10.1017/qua.2020.14

Zeni V, Baliota GV, Benelli G, Canale A, Athanassiou CG (2021) Diatomaceous earth for arthropod Pest control: Back to the future. Molecules 26:7487. https://doi.org/10.3390/molecules26247487

Zgłobicka I, Gluch J, Liao Z, Werner S, Guttmann P, Li Q, Bazarnik P, Plocinski T, Witkowski A, Kurzydlowski KJ (2021) Insight into diatom frustule structures using various imaging techniques. Sci Rep 11:14555. https://doi.org/10.1038/s41598-021-94069-9

Zhang X, Liu X, Meng G (2005) Sintering kinetics of porous ceramics from natural diatomite. J Am Ceram Soc 88:1826–1830. https://doi.org/10.1111/j.1551-2916.2005.00288.x

Zhang C, Wang J, Song S (2019a) Preparation of a novel type of flame retardant diatomite and its application in silicone rubber composites. Adv Powder Technol 30:1567–1575. https://doi.org/10.1016/j.apt.2019.05.002

Zhang P, Guo Q, Tao J, Ma D, Wang Y (2019b) Aging mechanism of a diatomite-modified asphalt binder using Fourier-transform infrared (FTIR) spectroscopy analysis. Materials 12:988. https://doi.org/10.3390/ma12060988

Zheng J, Shi J, Ma Q, Dai X, Chen Z (2017) Experimental study on humidity control performance of diatomite-based building materials. Appl Therm Eng 114:450–456. https://doi.org/10.1016/j.applthermaleng.2016.11.203

Potential of Seaweeds to Mitigate Methane Emissions

Sophie Leonie Steinhausen, Abdelfatah Abomohra, and Dieter Hanelt

Abstract Mitigating greenhouse gas emissions has become a top priority to limit global warming. Methane emissions from the agricultural sector, particularly arising from ruminant animals, are a major concern and a significant contributor to greenhouse gas emissions. Seaweeds have emerged as a promising solution to address this challenge. Some, such as the red alga *Asparagopsis taxiformis*, are rich in bromoform, a volatile halogenated compound that has been found to inhibit methanogenesis in the gastrointestinal tract of ruminant animals. Scientific findings are promising, as trials have demonstrated that incorporating seaweed into animal feed at low levels has shown highly effective methane suppression. The prospect of a seaweed-based feed additive to reduce livestock emissions has sparked excitement, resulting in numerous companies working on solutions using *Asparagopsis*. Ongoing research and development efforts, coupled with the commitment of companies in this field, highlight the potential to make a meaningful contribution to reducing the emissions of livestock. However, overcoming scalability constraints and optimizing the active substances, legislation, and health/ecological concerns are crucial.

Keywords *Asparagopsis* · Bromoform · Halogenated compounds · Methane · Rumen fermentation · Sustainable agriculture

1 Introduction

In the twenty-first century, the challenge of mitigating greenhouse gas emissions has become an urgent priority in the quest to limit global warming. Among the various contributors to global warming is the combustion of fossil fuels within the energy sector, while methane (CH_4) emissions from agricultural sector are also of major concern and a significant contributor to global greenhouse gas (GHG) emissions (Grossi et al. 2019). It is a potent GHG arising particularly from enteric fermentation,

S. L. Steinhausen (✉) · A. Abomohra · D. Hanelt
Aquatic Ecophysiology and Phycology, Institute of Plant Science and Microbiology, Hamburg, Germany
e-mail: sophie.steinhausen@uni-hamburg.de

A. Abomohra, S. Ende (eds.), *Value-added Products from Algae*,
https://doi.org/10.1007/978-3-031-42026-9_15

a process by which methane is produced in the gastrointestinal tract in ruminant animals, mainly cattle. With an estimated number of more than 1.5 billion head of cattle globally (FAO 2021a), enteric emissions from agriculture pose the single largest source of anthropogenic methane (Knapp et al. 2014). To achieve the climate goal outlined in the Paris Agreement of 2015 and SDG 13 "Climate Action" as part of the 2030 Agenda for Sustainable Development, it is imperative to decrease the overall number of farmed animals in the agricultural sector and implement measures to mitigate methane production from livestock. This approach aligns with the United Nations' objective of urgently addressing climate change as a crucial global challenge while promoting collective efforts to reduce greenhouse gas emissions and foster sustainable practices (United Nations 2023).

Considering the projected rise in the demand for meat and dairy products in the coming decades (Beauchemin et al. 2020; Ritchie et al. 2017), a simultaneous reduction of GHG from the same sector is a great obstacle (Ungerfeld et al. 2022). Approaches to reduce emission intensity through elevation of animal productivity through nutrition—dietary modifications (feeds, feeding techniques, and feed management/processing), genetics, breeding programs, microbiology, and physiology—have yielded improvements with only limited success (Gerber et al. 2013; Knapp et al. 2014; Ungerfeld et al. 2022). Over the last decades, a promising solution that lies beneath the ocean's surface has caught the attention of researchers worldwide—seaweeds or macroalgae.

Some seaweeds are rich in volatile halogenated compounds, specifically bromoform $CHBr_3$, which serve as a natural secondary metabolite with various benefits for antifouling, antimicrobial, and antiviral properties (Cabrita et al. 2010; Paul et al. 2006a, 2006b, 2006c). Moreover, bromoform has unique characteristics as an active substance that disrupts key metabolic pathways in ruminant animals (Beauchemin et al. 2020). By inhibiting methanogenesis in microorganisms, specifically methanogenic archaea, bromoform intervenes in a complex series of steps following enteric fermentation, effectively preventing the production of methane. Consequently, this compound has the potential to significantly reduce methane emissions when seaweeds are used as a feed additive, making them a promising biotic anti-methanogen (Abbott et al. 2020; Beauchemin et al. 2020). These scientific findings have sparked excitement within the scientific community and the agricultural industry due to the significant potential impact on reducing greenhouse gas emissions. Over the past decade, extensive research has been conducted on seaweeds that metabolize bromoform, with numerous studies assessing their effectiveness at different concentrations and in various feed formulations (Kinley et al. 2016; Li et al. 2016; Roque et al. 2019a). Among these, the red alga *Asparagopsis taxiformis* (Delile) Trevisan has emerged as the most promising candidate, demonstrating remarkable methane suppression in both laboratory and live animal trials (Kinley et al. 2016, 2020; Machado et al. 2016a; Roque et al. 2021). When incorporated into the diets of ruminant animals at very low inclusion levels, some below 1% of the total dry mass, previous trials have achieved highly effective methane suppression of up to 98% (Kinley et al. 2020; Li et al. 2016; Roque et al. 2019a, 2019b; Stefenoni et al. 2021).

Given that over 70% of the Earth's surface is covered by the ocean, the cultivation potential for marine macroalgae is vast. Already, red algae make up more than 50% of the harvested biomass in seaweed aquaculture, with a significant market value of US $6.3 billion in 2018 (FAO 2020, 2021b). Red algae are consumed as food and find wide application across various industries, including food industry and beyond. The research and development of *A. taxiformis* as a seaweed supplement, including its processing, incorporation in animal feed, and optimization of bromoform as the active substance, has gained considerable attention. Furthermore, several companies are engaged in the R&D large-scale aquaculture production of this species, including land- and sea-based cultivation, hatchery, and nursery. However, research and industry are facing challenges related to scalability and high iodine content as well as legislation and ecological parameters.

The present chapter is a roadmap for a comprehensive understanding of this topic as it explores the potential of seaweeds as a novel solution to mitigate methane emissions of livestock. It explains underlying metabolic pathways and looks at the scientific study background about research on their effectiveness, with a focus on the standout candidate *A. taxiformis*. In addition, it looks at potential alternative seaweeds and provides a state-of-the-art review for companies focused on the industrial implementation. A strong focus is also directed on challenges, foremost scalability of cultivation, optimization of the active substance, and also the potential impact on the environmental and animal health, mostly related to the high iodine content of seaweeds.

2 Overview of Rumen Fermentation and Methane Production

Cattle and other ruminant animals, such as sheep and goats, primarily derive their energy from carbohydrates, which serve as the main dietary source of energy (Beauchemin et al. 2020). They possess a unique digestive system based on their symbiosis with rumen microbes (archaea, bacteria, fungi, and protozoa) in the gastrointestinal tract. The so-called enteric (microbial) fermentation allows them to efficiently digest and break down plant material from a fibrous, herbivorous diet in a specialized fermentation chamber within the digestive system, called rumen, to simple digestible compounds (Knapp et al. 2014) (Fig. 1). During enteric fermentation, plant biomass in the form of carbohydrates (starch, fiber) and other organic matter (OM) from the feed are microbially fermented and broken down in the forestomach (rumen) by a series of enzymes. This series of enzymatic reactions leads to production of carbon dioxide (CO_2) and volatile fatty acids (VFAs)–such as acetate, butyrate, propionate—which serve as essential energy sources for the animal. Besides the major energy supply through VFAs, the ruminant digestion

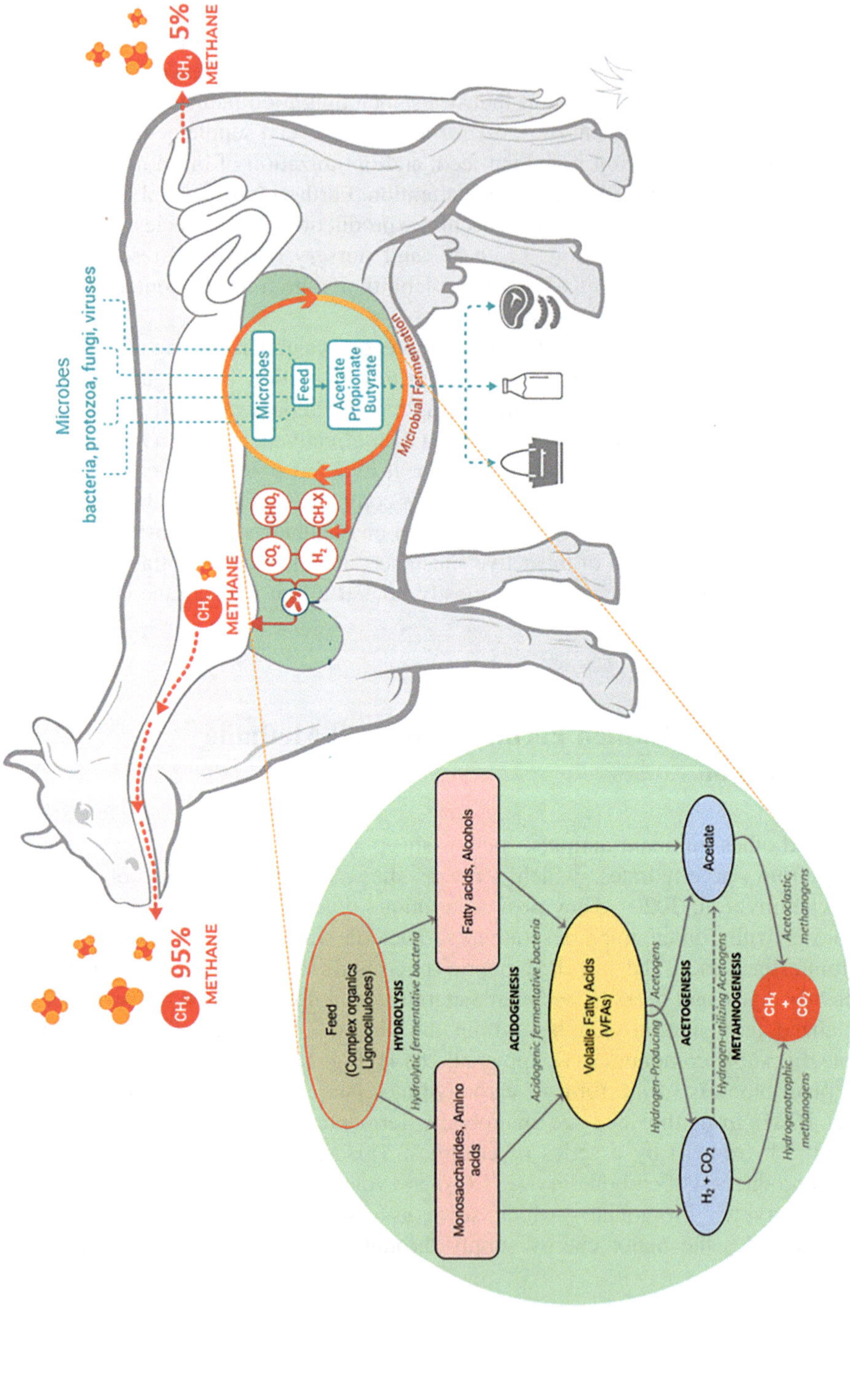

Fig. 1 Ruminant fermentation process and main products of the process showing the microbial methanogenic pathways of plant. *Source*: modified from Glasson et al. (2022) with Permission Number 5565010068410

also results in the production of gaseous waste by-products, such as carbon dioxide and enteric methane (CH_4). The later is formed through methanogenesis by methane-forming archaea, single-celled prokaryotic organisms so called "methanogens".

Thriving within the anaerobic environment of the rumen, these methanogens utilize metabolic hydrogen (H_2) from the enteric fermentation to convert carbon dioxide to methane (Fig. 1), which is then expelled from the animal through belching (FAO 2019; Glasson et al. 2022; Min et al. 2022).

2.1 *Significance of Methane Emissions as GHG*

Methane is a significant component of the Earth's natural environment and carbon cycling. Methane emissions are derived from both natural and anthropogenic sources, which include livestock and agriculture, during production and transport of coal, natural gas, or oil, and the natural decay of organic waste (EPA 2021; Knapp et al. 2014). Even if methane has a shorter lifespan in the atmosphere (~12 years) compared to carbon dioxide where a single life time cannot be given (AR5 IPCC Report 2014, PA, 2021), it is considered a potent greenhouse gas (GHG) with a global warming potential (GWP) estimated to be 28 times greater than that of carbon dioxide over a 100-year timeframe (GWP100). Its direct release to the atmosphere is thus 28 times more efficient at trapping radiation in the atmosphere compared to carbon dioxide (IPCC AR5). This value can vary slightly depending on the specific calculation method and time horizon used. As a result, it contributes more to the greenhouse effect than carbon dioxide (EPA 2021). Additionally, its global temperature potential (GTP), which surpasses GWP by incorporating additional physical processes, is estimated to be 4–11 according to GTP100 and IPCC AR5 (Myhre et al. 2013). At the UN Climate Change Conference of the Parties (COP26) in Glasgow 2021, over 100 countries have committed to a global methane pledge by 2030 aiming to reduce methane gas emissions by 30%. Globally, agriculture stands as the second-largest contributor to the global greenhouse gas emissions, surpassed only by the energy sector. Methane (and nitrous oxide, N_2O) emissions through stored or excreted manure are contributing up to 8% of methane emissions from cattle (FAO 2017; Grossi et al. 2019). Direct enteric fermentation contributes more than 90%, with 2.8 Gt CO_2e in 2019 (ClimateWatch 2020), a substantial contribution to the overall agricultural sector's GHG emissions in the same year estimated at 5.79 Gt CO_2e. Particularly beef and dairy cattle have gained significant attention as major methane emitters, representing 65% of the livestock sector's emissions (Gerber et al. 2013). It is noteworthy to mention that the overall impact of livestock agriculture to global warming in numeric values is frequently debated.

2.2 Mitigation Potential of Various Strategies

The environmental impact of methane emissions from cattle is significant, particularly in relation to the greenhouse effect and the ongoing climate crisis. It has been estimated that with a reduction of 20% by 2030 (from a 2020 baseline), and up to 47% in agricultural methane emissions by 2050 (relative to 2010), it would be necessary to limit the global temperature increase to 1.5 °C (IPCC 2019; Ungerfeld et al. 2022). At the same time, 30% increase in dairy and meat consumption is projected, leading to a corresponding increase in agricultural methane emissions by 2050 (Reisinger et al. 2021). To achieve the desired reduction in enteric methane emissions relative to 2010, emissions per unit of milk (or meat) produced would need to decrease by up to 69% (Ungerfeld et al. 2022). Thus, addressing the global demand for meat and milk production while reducing methane emissions poses a significant challenge for agriculture.

Researchers and farmers are actively exploring strategies to reduce enteric methane production without compromising animal health and the global food supply. The magnitude of methane emissions from cattle and ruminant animals is influenced by various factors, such as feed intake, quality, and digestibility (FAO 2017). Several management strategies to mitigate emissions in ruminant animals are currently being investigated, each with different capacity to reduce methane production. The key strategy to address this issue involves enhancing rumen fermentation efficiency and boosting animal productivity. A range of other strategies encompass selective breeding and animal genetics, such as diet manipulation (composition, dietary interventions, digestibility), size, age, chemical additives, management practices, and even vaccines (Glasson et al. 2022; Grossi et al. 2019; Knapp et al. 2014; Min et al. 2022). In addition, efforts are being made to target the inhibition of specific microbes involved in producing the substrates necessary for methanogenesis. Alternatively, there is a focus on directly addressing the rumen methanogens responsible for methane gas production (FAO 2019). Notably, the latest research with *Asparagopsis* spp. shows a greater potential in inhibiting methanogenesis compared to other approaches (Glasson et al. 2022; Kinley et al. 2020; Li et al. 2016; Stefenoni et al. 2021).

3 *Asparagopsis* Spp.

3.1 Taxonomy and Habitats

Asparagopsis (Delile) Trevisan is a genus of red marine macroalgae within the Bonnemaisonales order, Bonnemaisoniaceae family. It contains two species, *A. armata* Harvey 1855 (Harpoon weed) and *A. taxiformis* Trevisan 1845 (Limu kohu), both present also in the Mediterranean Sea where they are regarded as introduced species (Andreakis et al. 2007a). *Asparagopsis* spp. are commonly found in temperate to tropical and subtropical marine environments (Boni and Hawkes 1987; Martins et al. 2019). They are typically intertidal to shallow subtidal species and prefer rocky substrates, where the thallus can attach itself firmly. The species often grow epiphytic on other seaweeds, widely distributed across different regions, including Atlantic Ocean and Indo-Pacific (Guiry 2018). They also present in the Mediterranean Sea and are regarded as invasive species (Streftaris and Zenetos 2006).

3.2 *Life Cycle*

Like other red algae, the life cycle of *Asparagopsis* spp. follows a complex reproductive pattern known as alternation of generations (Fig. 2) (Lüttge and Kluge 2012). It comprises of haplodiplontic-alternation of generations within a triphasic life cycle: gametophyte (haploid, n), microscopic carposporophyte (diploid, 2n), and the free-living filamentous tetrasporophyte (2n), ensuring both sexual and asexual reproduction (Lüttge and Kluge 2012). The life cycle is also heteromorphic, exhibiting distinct morphological differences among the generations. Haploid tetraspores (n) are formed in the tetrasporangia through meiosis, from which the gametophytes (n) develop, exhibiting a particularly high biomass. After the growth of male and female gametophytes (n), carposporophyte results from fertilization,

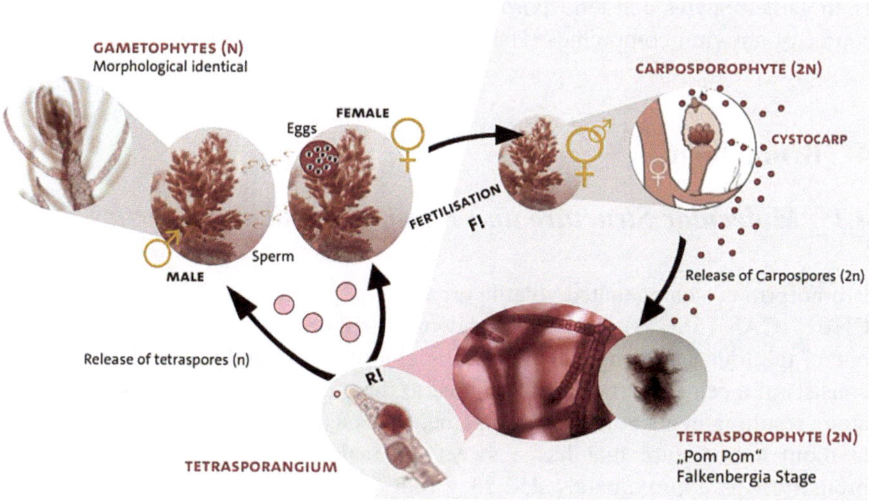

Fig. 2 The life cycle of *A. taxiformis* showing heteromorphic alternation of three generations. Tetrasporophyte: smaller, branched, filamentous morphology. Carposporopyhte: multicellular structure attached within the tissue of the female gametophyte that is formed through the fusion of a male- (sperm) and a female (egg) gamete. Gametophyte: larger stage of the life cycle, characterized by its branching appearance

followed by the tetrasporophytes "Falkenbergia stage" (2n). This diploid phase was formerly known as *Falkenbergia hillebrandii* (Bornet) Falkenberg. The two species of *Asparagopsis* can easily be misidentified, only the gametophyte phenotype enables the differentiation between the species, with *A. armata* having harpoon-like branches (Boni and Hawkes 1987).

Phylogenetic analysis has provided valuable insights into the evolutionary relationships and genetic diversity within this genus. For instance, it has been determined that *A. taxiformis* is a multiphyletic species with several genetic (Indo-Pacific and Atlantic) lineages. They are morphologically cryptic species and can be genetically identified across the globe (Andreakis et al. 2007b; Clark et al. 2018; Greff et al. 2017).

3.3 Nutritional Value and Potential Applications

Asparagopsis spp. are known for their high protein content (Table 1), making them a potential protein source. Furthermore, different species of *Asparagopsis* are abundant in macro- and microelements, including calcium, magnesium, iron and iodine. *A. taxiformis* is traditionally consumed by humans, locally known as *limu kohu* or *līpehe* in Hawai'i (Burreson et al. 1976; McDermid et al. 2019). Applications of this species have great potential in various fields such as food biotechnology, pharmaceutics, and cosmetics, specifically as natural rich source of iodine (Alburquerque et al. 2018; Genovese et al. 2009; Pereira 2018; Ponte et al. 2022). Both gametophytes and tetrasporophytes of *A. armata* also have the potential as a source of antiviral compounds (Haslin et al. 2001).

4 Bromoform

4.1 Molecular Structure and Physio-Chemical Properties

Bromoform is a halogenated volatile organic compound with the molecular formula $CHBr_3$ (CAS 75-22-2) and is also known as tri-bromomethane (WHO 2023). It is one of the four haloforms, also including chloroform, iodoform, and fluoroform. It consists of a central carbon atom bonded to three bromine atoms and one hydrogen atom, resulting in a tetrahedral arrangement. It is a dense, colourless to yellow liquid at room temperature that has a sweet ethereal odor. The molecular weight of bromoform is approximately 252.73 g mol^{-1}. It exhibits a high density of about 2.89 g cm^{-3} at 25 °C, which is significantly higher than that of water (1 g cm^{-3}). In terms of thermodynamics, bromoform is in liquid form at room temperature with a boiling point of approximately 150 °C and a melting point of around 8 °C (PubChem 2023). The cost of bromoform is around 55 € g^{-1} or 691 € L^{-1}, which varies

Table 1 General composition of *Asparagosis armata* and *Asparagosis taxiformis*

Constituents	*A. armata*	*A. taxiformis*	References
Macromolecules (per 100 g)			Ponte et al. (2022)
Water	90.8–91.2 gfw	92.6 g fw	
Polysaccharides	Starch 1.26 g dw Other 72.0 g dw	Starch 8.03 g dw Other 32.47 g dw	
Protein	10.9–14.0 g dw	17.55 g dw	Ponte et al. (2022)
Lipids	2.51 g dw	6.62 g dw	
Ashes	13.36 g dw	23.76 g dw	
Minerals			
Macroelements (g kg^{-1} dw**)			
Ca	14.6–22.3	14.9–17.7	Afonso et al. (2022)
K	13.5–14.2	14.0–14.7	
Mg	9.5–9.8	9.2	
Na	85.6–105.0	97.8–115.3	
P	2.0–2.1	1.7	
S	31.9–36.0	25.3–25.9	
Microelements (mg kg^{-1} dw**)			
Fe	1.19 557.9–1073.0	6.2 917.7–1705.8	Min et al. (2021) Afonso et al. (2022)
Cd	0.5–0.6	0.3	Afonso et al. (2022)
Mn	0.63 21.3–36.9	0.10 27.4–51.3	Min et al. (2021) Afonso et al. (2022)
Cr	2.0–5.8	5.0–13.4	Afonso et al. (2022)
Zn	0.07 36.3–96.6	0.24 11.2–17.4	Min et al. (2021) Afonso et al. (2022)
Cu	2.6–3.5 –	2.1–2.6 0.87	Afonso et al. (2022), Min et al. (2021)
As	16.4–18.5	11.3–12.9	Afonso et al. (2022)
Pb	0.7–1.8	1.3–1.7	Afonso et al. (2022)
Iodine	– 4600–5700** 9387	11,600–33,700* 5100–5700** 8331, 8135, 11,627	Nunes et al. (2018) Afonso et al. (2022) Nunes et al. (2019)
Others (μg g^{-1} dw)			
Bromoform	1320	1723 8080 6550	Min et al. (2021), Machado et al. (2016b), Romanazzi et al. (2021) Kinley et al. (2020)

(continued)

Table 1 (continued)

Constituents	*A. armata*	*A. taxiformis*	References
Bromine content	32,600–59,200	45,500–52,200	Afonso et al. (2022)
Dibromochloromethane	–	15.8 (0.0016%)	Machado et al. (2016b)
Phenolic compounds	– –	5.0–6.0 700–17,100 *	Min et al. (2021) Nunes et al. (2018)
Condensed tannins	5,0	–	Min et al. (2021)
Flavonoid content	–	400–242,500*	Afonso et al. (2022)
Chlorophyll *a*	–	81.0–459.5*	Nunes et al. (2018)
Carotenoid content	–	16.6–362.3	Nunes et al. (2018)

fw: Fresh weight basis; dw: Dry weight basis
– not reported
*Depending on the extraction solvent
**Depending on the drying procedure

depending on the supplier, purity level (>95 or > 99%, stabilized with ethanol), ordered quantity, and market conditions (Sigma-Aldrich 2023).

4.2 *Toxicological Aspects and Safety Considerations*

The reference dose (RfD) is an estimate of the acceptable daily oral exposure to a substance without adverse health effects. It was established for bromoform with 0.02 milligram per kg body weight per day. Higher doses of bromoform can result in adverse impacts to humans and animals. The toxicological properties of bromoform include acute and chronic toxicity to humans as well as other organisms. Bromoform is considered harmful and toxic by ingestion, inhalation, and absorption through skin or eye contact. It can cause adverse health effects, particularly on the liver, kidneys, and nervous system (RAIS 1997). Following the International Agency for Research on Cancer (IARC), bromoform is not classifiable as carcinogenic to humans (group 3), while the US Environmental Protection Agency (EPA) classified it as probable human carcinogen. However, regulatory limits have been established by the EPA.

Bromoform in the environment originates from both natural and human activities. Various species of macroalgae are considered the primary natural source for the synthesis of organic bromides, including bromo- and bromochloro-methanes (Class et al. 1986). Bromoform has low solubility in water, and thus, has the ability to form a separate layer that can persist as a dense phase in water. This characteristic may pose a hazard to aquatic organisms. As a volatile organic compound (VOC), bromoform can volatilize from water to the air due to its physical properties and the process of evaporation, leading to its persistence in the atmosphere. Understanding the volatilization process and the atmospheric fate of bromoform is thus crucial in assessing its environmental behaviour. A baseline concentration of 0.8 ng L^{-1} was found in the surface layer in part of the Atlantic Ocean (Class et al. 1986).

5 Potential of Seaweeds as a Methane Inhibitor in Ruminant Animals

5.1 *Halogenated Bioactive Compounds in Seaweeds*

Halogens are aliphatic compounds that consist of one or two carbon atoms covalently bonded with one or more halogen atoms, such as bromine (Br), chlorine (Cl), fluorine (Fl), or iodine (I) (Pandey et al. 2021). Iodine and bromine are particularly abundant in seawater (Wever and Van Der Horst 2013) and certain seaweeds, specifically red algae of the genus *Asparagopsis* utilize these halogens to produce large amounts of secondary metabolites that play a significant role in their bioactive potential. While numerous studies have highlighted the synthesis of various secondary metabolites, these small halogens are frequently reported and exhibit the highest chemical diversity (Ponte et al. 2022). The compounds are metabolized in specialized structures called the gland cells (Fig. 3), and have attracted scientific interest due to their bioactive properties as potent agents with pharmacological effects (Hutson et al. 2012; Paul et al. 2006c).

Although the precise mechanisms and ecological roles behind the production of halogenated compounds are still under investigation, some hypotheses have been put forward (Machado et al. 2016b). The compounds are believed to play a vital role in the survival and resilience of algae in challenging marine environments. Extracts from algae have exhibited antibacterial properties against both human and animal pathogens (Bansemir et al. 2006; Salvador et al. 2007; Vedhagiri et al. 2009). Halogenated compounds, known for their antimicrobial and antifungal activities, serve as a defense mechanism for algae against predation and grazing by herbivores (Carroll et al. 2021; Paul et al. 2006a, 2006b; Pinteus et al. 2021), by deterring consumption and overgrowth by competing organisms (Paul et al. 2006a, 2006b, 2006c; Pinteus et al. 2015, 2021). Thus, the release of volatile halogenated

Fig. 3 Thallus of *A. taxiformis* gametophyte (**a**), tetrasporophytes (**b**), and enlargement of branched filaments with gland cells within a tetrasporophyte (20×, ZEISS microscope) (**c**). Photographs were captured at the University of Hamburg

metabolites is thought as a natural chemical defense mechanism (Laturnus 1996; Wever and Van Der Horst 2013), and also as a response to mechanical damage or predation (Paul et al. 2006a, 2006b). The production of these metabolites is closely linked to environmental adaptations, especially considering the abundance of *Asparagopsis* spp. in subtropical and tropical waters, where intense seaweed herbivory takes place (Nunes et al. 2018). A noteworthy finding from the study conducted by Greff et al. (2017) is that the production of chemical defenses in species of *Asparagopsis*, as determined through a bioactivity assay, is not closely linked to the genetic lineage of macroalgae, but is predominantly influenced by the environmental factors. This suggests that the environment plays a significant role in shaping the chemical defense mechanisms of *Asparagopsis*. Moreover, the abundance of halogenated metabolites in *Asparagopsis* spp. has been found to contribute to their strong antimicrobial activity, effectively preventing the overgrowth of epiphytic bacteria on the algal surface. This highlights the important ecological role of *Asparagopsis* for reducing bacterial densities on the thalli (Paul et al. 2006b).

5.2 Bromoform in Seaweeds

Among the halogenated compounds in the genus *Asparagopsis*, bromoform stands out as the most abundant, encompassing a diverse array of over 100 brominated metabolites (followed by dibromomethane, dibromochloromethane, bromodichloromethane), chlorinated, and iodinated derivatives including methanogene, acetic acid, acrylic acid, and cyclic compounds (Burreson et al. 1976; Kladi et al. 2004; Ponte et al. 2022). According to Wasson et al. (2022), it was found that the concentration of volatile organic compounds in *Asparagopsis* biomass ranges from 2% to 6% of the dry weight. Mata et al. (2017) reported that the concentration of natural products [sum of bromoform, dibromoacetic acid (DBA), dibromochloromethane (DBCM), and bromochloroacetic acid (BCA)] exhibited significant variation among all isolates and temperatures, spanning a tenfold difference. The range extended from 0.007μg mm^{-2} to 0.07μg mm^{-2} at 20.2 °C. On average, bromoform in *A. armata* constituted approximately 1.54% and 1.67% of the dry weight of tetrasporophyte and gametophyte, respectively (Paul et al. 2006b; Ponte et al. 2022). However, Machado et al. (2016b) detected 1.723 mg g^{-1} dw and Romanazzi et al. (2021) found a mean value of 8.08 mg g^{-1} in freeze-dried biomass and 1.45 mg g^{-1} bromoform in fresh-frozen samples of *A. taxiformis* upon the first extraction, indicating varying and potentially lower levels of the metabolite among the algae.

5.3 Actions in the Rumen

As a natural seaweed-feed additive, bromoform plays a significant role by inhibiting methane biosynthetic pathway of methanogens in ruminant animals (Abbott et al. 2020; Kinley et al. 2016; Li et al. 2016; Machado et al. 2014). Halogenated compounds act as structural analogues of methane and other methanogenic intermediates. The inhibitory effect of bromoform on methane production is attributed to its higher affinity to enzymes, which play a key role in the methanogenesis pathway. As a result, bromoform can competitively inhibit the binding of intermediates or methane substrates in the metabolic pathways in methanogenic archaea, blocking the activity of the enzyme coenzyme-M (CoM) methyltransferase which catalyzes the cobamide-dependent methyl transfer in methanogenesis.

The compounds have structural similarities to CoM, a cofactor produced specifically by methanogens, which is involved in the final reductive step within the methanogenesis process—the Wolfe cycle (Beauchemin et al. 2020; Glasson et al. 2022). The cycle is a sequential reduction of carbon dioxide to methane (see Fig. 1), where methanogenic archaea/bacteria utilize metabolic hydrogen as an electron donor and carbon dioxide as a carbon source (Glasson et al. 2022). If bromoform binds irreversibly to the methyl-CoM reductase or methyl-CoM methyltransferase (with a cobalamin prosthetic group), it influences the activity of methanogens (Glasson et al. 2022). These inhibitory effects of bromoform on multiple enzymatic steps in the methanogenesis pathway disrupt the normal flow of metabolic reactions and ultimately lead to a reduction in methane production. Thus, anti-methanogenic compounds seem to directly influence methane production in ruminants by either decreasing the abundance of rumen methanogens through the disruption of their energy production, or directly on the functional components such as enzymes involved in different stages of methane biosynthesis (Pandey et al. 2021).

The effect of bromoform for its anti-methanogenic properties has evolved through scientific research and experimentation. A comprehensive review of literature between 1960 and 2018 by Beauchemin et al. (2020) revealed a significant number of almost 9000 papers related to rumen methanogenesis in cows, cattle, sheep, and lambs. These trials mainly involved monitoring methane emissions and animal performance, as well as potential side effects to assess the viability of using bromoform-rich seaweeds as a feed additive. Around 20 years ago, in the beginning of the 2000s, research on methane mitigation in ruminant animals started to gain attention due to rising environmental concerns associated with ruminant-produced methane gas (Beauchemin et al. 2020). Research on inhibiting rumen methanogenesis began with both in vitro and in vivo studies aiming to improve energy utilization efficiency in rumen fermentation and to enhance the animal productivity. Initial experiments focused on the energy loss in the form of methane, highlighting factors such as intake level, types of carbohydrates, or lipid supplementation (Beauchemin et al. 2020). These numerous studies have established an important and thorough foundation for understanding rumen methanogenesis, including microbial ecology, carbohydrate fermentation, and associated pathways.

To assess potential anti-methanogenic effects through dietary supplementation, early studies have investigated mainly bromochloromethane (BCM), sometimes as a chemical complex with α-cyclodextrin (Glasson et al. 2022). BCM is derived from the combination of the two halogenated compounds, bromoform and chloroform. It has also shown promising results in suppressing methane production in cattle and goats (Abecia et al. 2012; Denman et al. 2007; Goel et al. 2009).

6 Discovery of Anti-Methanogenic Factors in Seaweeds

With historical evidence suggesting that seaweed was fed to cows by herders in ancient Greece and eighteenth-century Iceland, it seems that cows have been consuming seaweed for a long time (Kinley and Fredeen 2015). Only recently, Joe Dorgan, a farmer from Canada, noticed improved health properties, with higher milk production and stronger reproductive cycles in cows that grazed on seaweed washed up on the shores of Prince Edward Island (Hein 2019). Besides a notable decrease in eructation and flatulence, an indication of improved digestive efficiency, he also found that the cows exhibited accelerated growth rates and reduced incidence of illness. Dorgan captured the interest of agricultural scientist Rob Kinley, who found methane-inhibiting potential in Dorgan's seaweeds of around 18% through experimental trials. Looking for other seaweeds with higher potential to reduce methane emissions in collaboration with scientists from the Commonwealth Scientific and Industrial Research Organization (CSIRO) and the James Cook University, they identified the genus *Asparagopsis* as a standout candidate with superior potential and determined that even small amounts resulted in a nearly complete elimination of methane (Kinley and Fredeen 2015). Today, Dorgan is a part owner of North Atlantic Organics, a company that produces seaweed supplements for livestock, founded in 2007 (NAO 2023).

6.1 In Vitro Studies

These studies involved the incubation of seaweeds with rumen fluid to simulate the rumen environment of cattle. Initial experimental trials were carried out by Dubois et al. (2013) and Machado et al. (2014) to assess the efficiency of more than 20 seaweeds in inhibiting methane production in rumen fluid. Among all tested seaweeds, *A. taxiformis* had the greatest impact on methane reduction, achieving nearly 99% decrease. Subsequent research demonstrated that even a small amount of organic matter (OM) in the daily feed, ≤5%, resulted in methane reductions of 99% (Machado et al. 2016a), 95% (Roque et al. 2019a, or complete elimination of CH_4 (Kinley et al. 2016). *A. taxiformis* emerged as the most suitable "standout candidate" due to its higher bromoform content, which has been shown to effectively reduce methane emissions (Kinley et al. 2020; Machado et al. 2014). Even concentrations of less than 1% in the daily feed can almost completely suppress methane production. The results of these studies consistently demonstrated an inhibitory effect on methane production, with red algae of the genus *Asparagopsis* as the most promising candidates. Subsequent studies have further examined the effect of this red alga with different feed substrates, varying inclusion levels, and an incubation period of up to 4 days (Kinley et al. 2016; Machado et al. 2016a).

6.2 *In Vivo Trials*

Several studies have demonstrated the methane-reducing effects of *A. taxiformis* in vivo (Kinley et al. 2020; Li et al. 2016; Roque et al. 2019a). In sheep, supplementation trials with 0.5% to 3% OM (with a high proportion of fiber) led to a significant reduction in enteric methane emissions by up to 80% over 72 days (Li et al. 2016). Similarly, when the feed of Holstein cattle, a widely and globally distributed dairy breed, was supplemented with a 1% OM inclusion level of *A. armata* (a closely related algal species with lower bromoform content), methane emissions were decreased by 67.2% (Roque et al. 2019b). In beef cattle, the addition of *A. taxiformis* to the diet (0.2% OM) resulted in a 98% reduction in methane emissions over a 90-day period (Kinley et al. 2020). A long-term study was conducted for 21 weeks and confirmed a significant decrease in methanogenesis up to 68% (Roque et al. 2021).

Halogenated compounds are naturally produced by biological sources, including microalgae (marine phytoplankton-diatoms) and other seaweeds (Carpenter and Liss 2000; Quack and Wallace 2003; Wever and Van Der Horst 2013). According to a study by Laturnus (1996), the brown seaweeds *Dictyosyphon foeniculaceus* and *Laminaria saccharina* are considered significant sources of volatile halogenated organic compounds in the Arctic environment. Several seaweeds have been specifically found to contain bromoform (Machado et al. 2016b). Temperate marine seaweeds, including species such as the brown algae *Ascophyllum nodosum* (mean 120 ng g^{-1} dw) and *Fucus vesiculosus* (mean 41 ng g^{-1} dw), the green algae *Enteromorpha linza* (8 ng g^{-1} dw), *Ulva lactuca* (19 and 43 ng g^{-1} dw), and the red alga *Gigartina stellata* (mean 9 ng g^{-1} dw) were found to contain varying levels of bromoform in their tissues. Additionally, these seaweeds were found to release bromoform into seawater, where the measured levels ranged from 8 to 520 ng g^{-1} (Gschwend et al. 1985). In other regions, such as near the Bermuda Islands (Fucales, *Sargassum*) and at the Cape of Good Hope (species of *Laminaria*), seaweeds emitted volatile organohalides into the air. The main components of these emissions were bromoform, bromodichloromethane, and chlorodibromomethane (Class et al. 1986). In an in vitro study by Machado et al. (2014), an inclusion rate of 0.2 OM to 1 g Flinders grass as a basal feed with 17 macroalgae, *Cladophora patentiramea* (green alga), *Dictyota* sp. (brown alga), and *Asparagopsis* sp. (red alga) reduced methane production by 69.7%, 93.1%, and 99.0%, respectively (Machado et al. 2014; Min et al. 2021). Seaweeds contain active compounds with anti-methanogenic potential that are not limited to bromoform alone. These compounds may also include polyphenols (such as phlorotannin), terpenes, and phenolic lipids, which are primarily metabolized in brown seaweeds like *A. nodosum, Cystoseira trinodis, Dictyota bartayresii*, and *Zonaria farlowii* (Pandey et al. 2021). Of the brown seaweeds, *D. bartayresii* and *C. trinodis* showed strong reductions in methane gas generation during in vitro experiments. When 16% of OM was added to the basal substrate Rhodes hay grass, a reduction of 80–90% in methane production was achieved (Dubois et al. 2013; Machado et al. 2014; Pandey et al. 2021). Other brown seaweeds such as *A. nodosum* and *Z. farlowii* have also demonstrated methane reductions of 11–15% in vitro at inclusion rates of 2% and 5% dry matter, respectively (Belanche et al. 2016; Brooke et al. 2020). Maia et al. (2016) and Machado et al. (2014) found that *Ulva* sp. can reduce methane emissions in vitro by 55%, also depending on the feed substrate.

In contrast, Molina-Alcaide et al. (2017) and de la Moneda et al. (2019) assumed an absence of anti-methanogenic compounds in red (*Mastocarpus stellatus, Palmaria palmata* and *Porphyra* sp.), brown (*Alaria esculenta, Laminaria digitata, Pelvetia canaliculata* and *Saccharina latissima),* and green seaweeds (*Acrosiphonia* sp., *Cladophora rupestris*) from their two in vitro studies. They found that methane production is directly related to the amount and different feed ingredients of the fermented substrate (de la Moneda et al. 2019). Belanche et al. (2016) also concluded that the brown seaweeds *L. digitata* or *A. nodosum* have no influence on methane emissions in Rusitec fermenters included in the diet at a 5 % dw inclusion rate.

Red seaweeds, such as *Asparagopsis* spp., are known to have higher levels of crude protein and halogenated compounds compared to brown and green seaweeds (Min et al. 2021). In a previous study, Maia et al. (2016) found that two other red seaweeds, *Gigartina* sp. and *Gracilaria vermiculophylla,* exhibit anti-methanogenic properties in laboratory fermentations (35,8% and 38,2% after 24 hours of incubation). They used a higher inclusion rate of 25% dw seaweed, and the extent of methane reduction varied depending on the substrate used (meadow hay or corn silage), highlighting the importance of the basal diet. A recent study of Mihaila et al. (2022) discovered several new species of red algae demonstrating a dose-dependent reduction through in vitro enteric methane gas production. Different New Zealand red algal species, *Bonnemaisonia hamifera, Euptilota formisissima, Plocamium cirrhosum,* and *Vidalia colensoi,* reduced methane production at certain inclusion levels. Among them, *B. hamifera* performed the best, reducing methane production by up to 98.8%, a nearly complete elimination of enteric methane, at 10% OM inclusion, with increased hydrogen production. *Euptilota formisissima* and *P. cirrhosum* also showed significant methane reduction at higher inclusion levels. Importantly, none of these tested seaweeds contained bromoform.

It is probable that certain species of brown, red, and green macroalgae contain additional anti-methanogenic factors that can effectively reduce methane emissions. These factors can act directly, e.g. through the presence of secondary metabolites like bromoform or dibromochloromethane. Alternatively, they can have indirect impacts on rumen microflora, VFAs composition, or the substrates involved in methane production. Brown seaweeds, in particular, are suspected to possess polyphenols, terpenes, and polysaccharides that contribute to their anti-methanogenic properties, as suggested by Pandey et al. (2021). Therefore, these findings present a potential alternative for anti-methanogenic seaweeds that are less dependent on bromoform (Mihaila et al. 2022). Thus, the quantitative extent to which seaweeds impact methane emissions may vary depending on the feed substrate and type/concentration of the bioactive components metabolized in seaweeds (Pandey et al. 2021). In the search for alternatives, it is desirable to focus on secondary metabolites that are not carcinogenic, stable at ambient temperatures, and do not accumulate in dairy or meat products, ensuring the safety and quality of the food chain. Furthermore, these metabolites should not have adverse effects on animal health and performance. Nonetheless, until a specific compound with proven efficacy is identified, *A. taxiformis* remains the most potent naturally derived feed additive for modifying rumen function and reducing methane emissions in livestock (Wasson et al. 2022).

7 Impacts of Algal Feed Additives

The use of seaweed feed additives requires careful consideration to ensure that it does neither affect animal health, productivity, nor compromise the quality of milk or meat production. FAO (2019) report emphasizes the importance of considering strategies that do not compromise the vital role of the rumen microbiome in animal nutrition when aiming to reduce methane emissions through microbiological approaches. Exploring if a fundamental aspect of animal biology can be modified without compromising the quality of milk or meat production nor animal health is crucial.

7.1 Energy Use, Animal Health, and Performance

It is well understood that the formation of methane gas and its subsequent release into the atmosphere present an energy loss for ruminants (Forbes 1939). Thus, the incorporation of *A. taxiformis* as a feed additive leads to an additional utilization of energy that would otherwise be lost (Kinley et al. 2020). It results in a positive decrease in the acetate:propionate ratio in the rumen (Kinley et al. 2016; Li et al. 2016). Cows increase their production of propionate, a short fatty acid that aids in glucose production during metabolic processes. This allows the animals to grow more efficiently or produce greater quantities of milk, potentially enabling farmers to reduce feed usage and save costs. A positive health effect is plausible due to the profile of unsaturated fatty acids. The high content of docosahexaenoic acid (DHA) promises positive effects similar to those known from salmon, which could reduce the presence of inflammation factors (Weiser et al. 2016).

Incorporating seaweeds into animal diets has been found in certain studies to potentially enhance weight gain and improve feed conversion efficiency, indicating its potential as a beneficial strategy for the livestock industry. Addition of *A. taxiformis* to the feed did not affect meat quality, daily feed intake, feed efficiency, or rumen function (Kinley et al. 2016; Roque et al. 2021). For example, including 0.2% OM basis of *A. taxiformis* in a high grain total mixed ration feed of Brahman-Angus steers increased their average daily weight gain by 42% while feed intake was unaffected (Kinley et al. 2020). Roque et al. (2021) observed a saving of up to 14% in dry feed requirements while maintaining the same weight gain. Notably, both studies also reported a substantial increase in hydrogen production alongside the reduction in methanogenesis. Importantly, these studies demonstrated no adverse effects on other fermentation parameters (Abecia et al. 2012; Roque et al. 2021) or feed intake and fermentation (Kinley et al. 2020; Mitsumori et al. 2012). While low doses of 20 mg g^{-1} (2% OM basis) of seaweed effectively decreased methane emissions, it also decreased the production of VFAs (Dubois et al. 2013; Kinley et al. 2016). On the other hand, feeding *A. taxiformis* at three different levels (67, 133, and 333 g d^{-1}) resulted in decreased dry matter intake (DMI) across all treatment levels (Muizelaar et al. 2021).

It was reported that bromoform does not accumulate in animal tissues, but is mainly excreted through urine (Muizelaar et al. 2021). Studies have shown no residues of bromoform in beef, fat, or organs of cattle (Kinley et al. 2020; Roque et al. 2021) and sheep (Li et al. 2016). The bromoform content remaining in milk was negligible compared to drinking water limits, and it did not differ from the control groups in most studies (Kinley et al. 2020; Roque et al. 2019b; Stefenoni et al. 2021). Experiments with the alga *A. armata* showed no significant residues in milk, with levels of 0.11–0.15μg L^{-1} at both low (0.5% OM) and high (1% OM) seaweed supplementation levels (Roque et al. 2019b).

7.2 Considerations for Dosage and Animal Health

Muizelaar et al. (2021) detected bromoform in the milk of lactating cows fed with *Asparagopsis* (during days 1 to 9 of the feeding period, while it was not consistently detected during days 10 to 22), which was attributed to variations in seaweed intake. However, these were detected only at significantly higher feeding levels showing a slightly increased bromoform content on a single measurement day, which was still within the acceptable limits. The rumen walls of two cows were examined and abnormalities in the rumen wall were reported, suggesting signs of inflammation. Findings from Li et al. (2016) observed mucosal changes in the rumen of sheep supplemented with *A. taxiformis*, characterized by granulomatous and keratotic alterations. However, neither study could definitively establish that damage to the rumen epithelium was a direct consequence of *A. taxiformis* supplementation (Wasson et al. 2022). Both studies solely examined rumen tissue from animals supplemented with *A. taxiformis*, which limits the ability to make direct comparisons with control animals. According to Silva et al. (2021), the exudate of *A. armata* caused oxidative stress and neurotoxicity in the shrimp species *Palaemon elegans*, resulting in altered fatty acid biosynthesis and metabolism. This could indicate a potential risk associated with the incorporation of such algae into animal feed, as highlighted by Ponte et al. (2022). For long-term effects, Stefenoni et al. (2021) observed a loss of the anti-methanogenic effect during the latter half of the experimental period of 28 days. The authors attributed the decreased effectiveness to the reduction of bromoform concentration within the algae over storage time, from approximately 10 mg kg^{-1} to 3 mg kg^{-1} within 120 days of storage (Wasson et al. 2022). In light of the potential risk and diminishing anti-methanogenic effect associated with algae incorporation into animal feed, further research is warranted to explore the long-term consequences and stability of the bromoform effect, emphasizing the need for future studies in this area.

8 Industrial and Commercial Progress

Numerous companies, research organizations, and start-ups are currently engaged in the cultivation and utilization of *A. taxiformis*. They are actively contributing to the advancement of seaweed-based additives and exploring their potential commercial applications in the livestock industry (Table 2). Sea Forest Australia and Greener Grazing™, founded in 2018, are among the recently established companies in the seaweed industry. The majority are based in Australia, taking advantage of *Asparagopsis'* presumed native status to facilitate land-based cultivation. Four (unregistered) trademark products incorporating *Asparagopsis* were introduced: BrominataTM, SeaGrazeTM, Methane TamerTM, and LomeTM Beef. Table 2 presents a list of representative companies actively engaged in research and development related to *A. taxiformis* as seaweed-based additives and its potential applications in the livestock industry.

9 Challenges and Opportunities

9.1 Cultivation

Despite *Asparagopsis* farming is a challenge due to its complex life cycle, it can be invasive in natural habitats. In 2006, both *Asparagopsis* species were listed among the 100 worst invasive species in the Mediterranean by Streftaris and Zenetos (2006), impacting the biodiversity of the infralittoral zone (up to 10 m depth) by forming monospecific, dense coverages, that are avoided by grazers. This reduced the abundance of other macrophytes and the algal richness (Martins et al. 2019). Furthermore, *A. armata* is known to clog up fishing gear. As *Asparagopsis* is non-indigenous in Europe, sea-based cultivation is not advisable. However, with

Table 2 Growing interest into *A. taxiformis* in seaweed-based additives and its potential applications in the livestock industry: a list of representative companies involved in research and development with the red algae

Company	Slogan	Website/Country	Product	Additional information	Value chain
Algal biosciences	We stop cows from burping methane.	alga.bio Berkeley, CA, USA		Founded in 2021 by Caroline McKeon, Daria Balatsky, and Alex Brown Research partners: University of Nebraska-Lincoln, University of Kentucky	Farming & harvesting Applications
Blue Ocean barns	Introducing the most effective burp suppressant for cows: Brominata™	blueoceanbarns.com Hawaii, USA	Brominata™	Cofounder and CEO: Joan Salwen and Matt Rothe Science advisor: Dr. Jennifer Smith, professor of marine ecology and conservation at the Scripps institution of oceanography	Applications, Farming & Harvesting
Fremantle seaweed	The seaweed solution to climate change	fremantleseaweed.com Fremantle, Western Australia	–	*Asparagopsis* cultivation in Western Australia Founders: Chris de Cuyer and Mick Holland Scientific advisors: Dr. Margie rule, phycologist and Dr. Michael rule, marine ecologist (contribution to a sea Forest project)	Farming & Harvesting
SeaExpert Azores	At seaExpert we live the sea	Seaexpert-azores.com Portugal		Seaweed supply: The red alga is abundant in the Azores. Wild harvest from march to may (100% organic)	Consulting Farming & Harvesting
Sea Forest Australia	Cultivating solutions to climate change.	Seaforest.com.au Tasmania, Australia	–	Environmental biotech. Currently, the world's largest producer and commercial supplier of *Asparagopsis* founded in 2018. Focused on sustainable cultivation to create a solution to climate change.	Farming & harvesting

				Actively involved in developing seaweed farming technologies and optimizing the growth of *Asparagopsis*. Sea Forest is a licensee of FutureFeed. Cofounder and CEO: Sam Elsom Chief scientist officer: Dr. rocky de Nys	
SeaStock	SeaStock creates global marine science solutions to produce and extract compounds derived from seaweed and algae.	seastock.com.au Fremantle, Western Australia	–	In cooperation with Flinders University, Adelaide Australia; the pilot plant and commercial production across five large land-based sites along the coastline in Western Australia. Sea Stock is a licensee of FutureFeed.	Farming & Harvesting
Seaweed culture	Sustainability of the future. Through the cultivation of seaweed.	seaweedcultures.com UK		Currently in the process of Research and Development	Farming & Harvesting
Symbrosia	Solving climate change with the world's mightiest seaweed.	symbrosia.com Hawaii, USA	SeaGraze™	Start-up founded and led by Alexia Akbay. Obtained a license from FutureFeed to sell their supplement.	Farming & harvesting
Greener grazing	Scaling up Asparagopsis taxiformis.	greenergrazing.org Vietnam	–	Greener grazing™ is a project by Australis aquaculture, focused on sustainable aquaculture and founded in 2018. Establishment of the world's first seed bank for *A. taxiformis* strains from more than thirty sites across its native range. Project leader: Josh Goldman Chief scientist: Leonardo Mata, PhD	Farming & harvesting Infrastructure and Equipment
CH4 global	Bending the climate curve—Together TM	ch4global.com Australia and New Zealand	Methane tamer™	Developing and commercializing solutions to reduce methane emissions, including *Asparagopsis* cultivation. CEO & co-Founder: Steve Meller, PhD	Farming and Harvesting

(continued)

Table 2 (continued)

Company	Slogan	Website/Country	Product	Additional information	Value chain
				Also partnering with A-culture, Australia a-culture.com.au	
FutureFeed	The world's most effective livestock methane solution: *Asparagopsis* seaweed.	future-feed.com Brisbane, Australia	–	Established in 2020 by CSIRO. Future feed is recruiting potential licensees. A grower and processer based on an intellectual property partnership. Chief scientist: Rob Kinley CEO: Nathaniel Last	Applications
Volta Greentech	Reducing methane emissions from cows.	voltagreentech.com Sweden	Lome™ beef	Indoor vertical farms (bioreactors); Volta Greentech—a tech company—is a licensee of FutureFeed Founder and CEO: Frekrik Åkerman	Applications Farming and Harvesting

Asparagosis as non-indigenous species in the Azores (Martins et al. 2019), harvesting from wild stocks may benefit local biodiversity.

9.2 Meeting the Demands of the Livestock Industry

Scaling up the production to meet commercial demands remains a significant challenge. It has led to an increasing focus on large-scale land and sea-based commercial cultivation of *Asparagopsis* in regions with favourable commercial prospects, considering both genetic and environmental conditions (Mata et al. 2017). In order to estimate the demand for seaweed, it is important to consider the substantial feed requirements of ruminant animals. As there is a growing interest in bromoform-rich seaweed for methane reduction in livestock, efforts are made for scale-up cultivation. While the seaweeds have shown promise in reducing methane emissions in controlled research settings, the scalability and practicality of implementing this strategy on a large commercial scale in livestock production systems need to be considered. However, companies striving to introduce *Asparagopsis* as a bioactive natural product face a challenge due to the limited available biomass. For example, at the University of Hawai'i at Hilo, researchers have received requests for excessively large, unrealistic quantities of Hawaiian *A. taxiformis* for experiments and aquaculture (McDermid et al. 2019). The limited availability has also led to high costs (Hristov et al. 2022).

Considering an average daily intake per animal of approximately 20 kg with a global livestock population of approximately 1.5 billion cattle, the required amount of *A. taxiformis* based on a conservative estimate of 1% of the daily dry seaweed intake supplementation is estimated to be 300 million kg day^{-1}. With a biomass productivity of *A. taxiformis* up to 100 g dry weight per square meter per day (land-based cultivation) (Zanolla et al. 2022), the required cultivation area could be three million square meters. Thus, relying solely on wild harvesting is unlikely to meet the demands of the cattle market. To meet the projected demand for the seaweed, the implementation of large-scale cultivation either along coastal areas or within commercial facilities becomes necessary, which represents a challenge.

9.3 Complicated Lifecycle

The complex triphasic and heteromorph life cycle of red algae is an obstacle for algal cultivation but also enables the utilization of various propagation strategies (Zanolla et al. 2022). On a small scale in the laboratory, e.g. in 3-dimensional vertical containers (reactors) and outdoor-land-based settings (Mata et al. 2006; Zanolla et al. 2022; Schuenhoff et al. 2006), cultures can be maintained and propagated for years. However, the indoor cultivation of the gametophyte remains difficult. An alternative is the use of tetraspores through precultivated tetrasporopyhtes to grow gametophyte seedlings. However, the induction and control of the different sexual life cycles is complex (Liu et al. 2017) and spores require growth attached to sea-based ropes (Zanolla et al. 2022). A further constraint is the seasonal growth cycle, with varying responses to environmental conditions among genetically identical isolates (Mata et al. 2017; Zanolla et al. 2022). For example, the growth of the gametophyte in Northern France is restricted to winter-spring in Brittany, France, allowing only a few harvests per year (Werner et al. 2004).

9.4 Regulatory Considerations

Ensuring that algae-based food products meet regulatory standards and obtaining the necessary approvals for their consumption and commercialization are important milestones to overcome. As bromoform-rich seaweeds as a feed additive gained attention, regulatory bodies began evaluating its safety, environmental impact, and approval processes for commercial use.

In terms of regulations and approvals, using algae as animal feed complies with current German and European laws, which differentiate between feed and feed additives. The latter are subject to strict approval processes governed by the Federal Office for Consumer Protection (Bundesamt für Verbraucherschutz). Since cattle naturally consume *A. taxiformis*, the seaweed falls into the category of general feed (Futtermittelausgangserzeugnis 1996, EU Feed Materials Register EU, F008818-EN) and no separate approval is required for its use as feed. However, suppose *A. taxiformis* is marketed as a feed additive with a specific concentration of the active ingredient bromoform or in pure form (such as tablets) to achieve targeted positive effects. In that case, it must undergo a prior EU administrative approval process.

Also, compliance with the Nagoya Protocol 2010 for the protection of genetic resources 2010 may be required by organizations involved in cultivation, involving prior informed consent from the country providing the genetic resources and negotiating agreed terms for benefit sharing (UNTC 2010).

9.5 Iodine Content

The incorporation of iodine compounds, particularly free and organic iodine, in seaweeds serves as an antioxidant and provides a defence mechanism against predation (Hou et al. 1997). Seaweeds generally contain varying amounts of iodine, with certain species, such as *A. armata* and particularly *A. taxiformis*, being recognized for their high iodine content (Kaliaperumal 2003; Nunes et al. 2018). *A. taxiformis* can have an iodine concentration of up to 33,700 mg kg^{-1} dw (see Table 1), depending on the extraction solvent (Nunes et al. 2018). In comparison, the brown algae *Dictyota dichotoma*, which also exhibits potential anti-methanogenic properties (Nørskov et al. 2021), has an iodine content of 82 mg kg^{-1} dw (Nunes et al. 2018). The maximum recommended level of iodine in the feed of dairy cattle is 0.5 mg kg^{-1} dw (Schöne et al., 2017). A 1% inclusion rate of *A. taxiformis* (260 g dw of algae) would result in an iodine intake of 8762 mg/260 g dw algae (based on iodine content of 33,700 mg kg^{-1}), which is manifold higher than the recommended concentration. Thus, high quantities or regular consumption could result in risks when using the alga as a feed additive for cattle. Therefore, careful consideration must be given to the iodine content in feed formulations.

In summary, while reducing enteric methane emissions from dairy cattle is a crucial step in making dairy products more climate-friendly, it is necessary to consider and address all stages of the production system to achieve significant overall carbon reduction. Further research, transparency, and a comprehensive evaluation of its environmental impact are necessary to determine its sustainability and avoid the pitfalls of greenwashing.

10 Conclusions

The scientific evidence supporting the anti-methanogenic capabilities of seaweeds, particularly within the genus *Asparagopsis*, in reducing methane emissions in livestock has been well-established through numerous studies. The promising effect of the secondary metabolite bromoform within the red algae has sparked scientific and commercial interest, leading to worldwide research efforts and collaborations. While results are promising, there are certain concerns that should be addressed. The invasive nature of *Asparagopsis* spp. in many parts of the world raises environmental and regulatory challenges. Ensuring health safety, particularly halogen concentrations in livestock metabolism as well as dairy and beef products, and obtaining legal approvals are also critical factors. Further research is necessary to obtain a comprehensive understanding of long-term effects, feed additive dosages, and potential side effects of seaweeds with rumen microbiota. Overcoming the scalability constraints of seaweeds associated with the vast number of ruminant animals, specifically the approximately 1.5 billion cattle worldwide, is a challenging endeavour that a single seaweed species cannot solely solve. To address this challenge, maximizing the dosage of bromoform or exploring seaweed alternatives with anti-methanogenic factors are recommended. Given the urgent need to address the global climate crisis and the ambitious target of the Paris Agreement to limit temperature rise to 1.5 °C, it is imperative to find better ways to reduce the environmental impact of livestock, where the seaweed-based solution can become part of the challenge.

Acknowledgments The authors would like to thank Lars Hauck (Brucerius Law School, Hamburg) for his advice and opinion on the legislation of seaweeds as feed and feed additives. Authors are grateful to Hamburg Innovation for funding the project Number 108 005 50, and the UHH Transferfond 2022/23 from the University of Hamburg.

References

Abbott DW, Aasen IM, Beauchemin KA, Grondahl F, Gruninger R, Hayes M, Huws S, Kenny DA, Krizsan SJ, Kirwan SF, Lind V, Meyer U, Ramin M, Theodoridou K, von Soosten D, Walsh PJ, Waters S, Xing X (2020) Seaweed and seaweed bioactives for mitigation of enteric methane: challenges and opportunities. Animals (Basel) 10(12):2432. https://doi.org/10.3390/ANI10122432

Abecia L, Toral PG, Martín-García AI, Martínez G, Tomkins NW, Molina-Alcaide E, Newbold CJ, Yáñez-Ruiz DR (2012) Effect of bromochloromethane on methane emission, rumen fermentation pattern, milk yield, and fatty acid profile in lactating dairy goats. J Dairy Sci 95:2027–2036. https://doi.org/10.3168/JDS.2011-4831

Afonso C, Julião DR, Pinto E, Almeida A, Ferreira IMPLVO, Bandarra NM, Cardoso C (2022) The effect of drying process on undervalued brown and red seaweed species: elemental composition. J Appl Phycol 34:1749–1761. https://doi.org/10.1007/S10811-022-02741-Y/TABLES/4

Alburquerque N, Faize L, Faize M, Nortes MD, Bernardeau J, Fernandez JMR, Burgos L (2018) Towards the valorization of the invasive seaweeds Caulerpa cylindracea and *Asparagopsis taxiformis* in the Mediterranean Sea: applications for in vitro plant regeneration and crop protection. J Appl Phycol 31:1403–1413. https://doi.org/10.1007/S10811-018-1640-X

Andreakis N, Procaccini G, Kooistra WHCF (2007a) *Asparagopsis taxiformis* and *Asparagopsis armata* (Bonnemaisoniales, Rhodophyta): genetic and morphological identification of Mediterranean populations. European J Phycol 39:273–283. https://doi.org/10.1080/0967026042000236436

Andreakis N, Procaccini G, Maggs C, Kooistra WHCF (2007b) Phylogeography of the invasive seaweed *Asparagopsis* (Bonnemaisoniales, Rhodophyta) reveals cryptic diversity. Mol Ecol 16: 2285–2299. https://doi.org/10.1111/J.1365-294X.2007.03306.X

Bansemir A, Blume M, Schröder S, Lindequist U (2006) Screening of cultivated seaweeds for antibacterial activity against fish pathogenic bacteria. Aquaculture 252:79–84. https://doi.org/10.1016/J.AQUACULTURE.2005.11.051

Beauchemin KA, Ungerfeld EM, Eckard RJ, Wang M (2020) Review: fifty years of research on rumen methanogenesis: lessons learned and future challenges for mitigation. Animal 14:S2–S16. https://doi.org/10.1017/S1751731119003100

Belanche A, Ramos-Morales E, Newbold CJ (2016) In vitro screening of natural feed additives from crustaceans, diatoms, seaweeds and plant extracts to manipulate rumen fermentation. J Sci Food Agric 96:3069–3078. https://doi.org/10.1002/JSFA.7481

Boni DR, Hawkes MW (1987) Systematics and life histories of New Zealand bonnemaisoniaceae (bonnemaisoniales, rhodophyta): I. The genus *Asparagopsis*. New Zeal J Bot 25:577–590. https://doi.org/10.1080/0028825X.1987.10410088

Brooke CG, Roque BM, Shaw C, Najafi N, Gonzalez M, Pfefferlen A, De Anda V, Ginsburg DW, Harden MC, Nuzhdin SV, Salwen JK, Kebreab E, Hess M (2020) Methane reduction potential of two Pacific coast macroalgae during in vitro ruminant fermentation. Front Mar Sci 7:522450. https://doi.org/10.3389/FMARS.2020.00561/BIBTEX

Burreson BJ, Moore RE, Roller PP (1976) Volatile halogen compounds in the alga *Asparagopsis taxiformis* (Rhodophyta). J Agric Food Chem 24:856–861. https://doi.org/10.1021/JF60206A040/SUPPL_FILE/JF60206A040_SI_001.PDF

Cabrita MT, Vale C, Rauter AP (2010) Halogenated compounds from marine algae. Mar Drugs 8: 2301. https://doi.org/10.3390/md8082301

Carpenter LJ, Liss PS (2000) On temperate sources of bromoform and other reactive organic bromine gases. J Geophys Res Atmos 105:20539–20547. https://doi.org/10.1029/2000JD900242

Carroll AR, Copp BR, Davis RA, Keyzers RA, Prinsep MR (2021) Marine natural products. Nat Prod Rep 38:362–413. https://doi.org/10.1039/D0NP00089B

Clark BR, Mizobe M, Kaluhiwa JLM, Leong JA, Borris RP (2018) Chemical and genetic differences between Hawaiian lineages of the alga *Asparagopsis taxiformis*. J Appl Phycol 30:2549–2559. https://doi.org/10.1007/S10811-018-1474-6

Class T, Kohnle R, Ballschmiter K (1986) Chemistry of organic traces in air VII: Bromo- and bromochloromethanes in air over the Atlantic Ocean. Chemosphere 15:429–436. https://doi.org/10.1016/0045-6535(86)90536-9

ClimateWatch (2020) Agriculture climate change data | climate watch [WWW document]. World Resour. Institute, Washington, DC. https://www.climatewatchdata.org/sectors/agriculture?emissionType=203&filter=#drivers-of-emissions. Accessed 12 June 2023

de la Moneda A, Carro MD, Weisbjerg MR, Roleda MY, Lind V, Novoa-Garrido M, Molina-Alcaide E (2019) Variability and potential of seaweeds as ingredients of ruminant diets: an in vitro study. Animals (Basel) 9(10):851. https://doi.org/10.3390/ANI9100851

Denman SE, Tomkins NW, McSweeney CS (2007) Quantitation and diversity analysis of ruminal methanogenic populations in response to the antimethanogenic compound bromochloromethane. FEMS Microbiol Ecol 62:313–322. https://doi.org/10.1111/J.1574-6941.2007.00394.X

Dubois B, Tomkins NW, Kinley RD, Bai M, Seymour S, Paul NA, de Nys R, Dubois B, Tomkins NW, Kinley RD, Bai M, Seymour S, Paul NA, de Nys R (2013) Effect of tropical algae as additives on rumen in vitro gas production and fermentation characteristics. Am J Plant Sci 4:34–43. https://doi.org/10.4236/AJPS.2013.412A2005

EPA (2021) Inventory of U.S. Greenhouse Gas Emissions and Sinks: 1990–2019, The United States Environmental Protection Agency

FAO (2017) Global Livestock Environmental Assessment Model (GLEAM) [WWW Document]. Food and Agriculture Organization, Rome. https://www.fao.org/gleam/en/. Accessed 12 June 2023

FAO (2019) The state of the World's biodiversity for food and agriculture [WWW document]. Food and Agriculture Organization, Rome

FAO (2020) The state of world fisheries and aquaculture 2020: sustainability in action. Food and Agriculture Organization, Rome. https://doi.org/10.4060/CA9231EN

FAO (2021a) Crops and livestock products [WWW Document]. https://www.fao.org/faostat/en/#data/QCL. Accessed 12 June 2023

FAO (2021b) Global seaweeds and microalgae production, 1950–2019 WAPI factsheet to facilitate evidence-based policy-making and sector management in aquaculture. Food and Agriculture Organization, Rome

Forbes EB (1939) Nutritional physiology of the adult ruminant 1. Science (80-) 90:270–272. https://doi.org/10.1126/SCIENCE.90.2334.270/ASSET/A28CB978-D996-4D80-91DB-D57C264D1DA1/ASSETS/SCIENCE.90.2334.270.FP.PNG

Genovese G, Tedone L, Hamann MT, Morabito M (2009) The Mediterranean red alga *Asparagopsis*: a source of compounds against Leishmania. Mar Drugs 7:361–366. https://doi.org/10.3390/MD7030361

Gerber PJ, Steinfeld H, Henderson B, Mottet A, Opio C, Dijkman J, Falcucci A, Tempio G (2013) Tackling climate change through livestock: a global assessment of emissions and mitigation opportunities. Food and Agriculture Organization of the United Nations (FAO), Rome

Glasson CRK, Kinley RD, de Nys R, King N, Adams SL, Packer MA, Svenson J, Eason CT, Magnusson M (2022) Benefits and risks of including the bromoform containing seaweed *Asparagopsis* in feed for the reduction of methane production from ruminants. Algal Res 64:102673. https://doi.org/10.1016/J.ALGAL.2022.102673

Goel G, Makkar HPS, Becker K (2009) Inhibition of methanogens by bromochloromethane: effects on microbial communities and rumen fermentation using batch and continuous fermentations. Br J Nutr 101:1484–1492. https://doi.org/10.1017/S0007114508076198

Greff S, Zubia M, Payri C, Thomas OP, Perez T (2017) Chemogeography of the red macroalgae *Asparagopsis*: metabolomics, bioactivity, and relation to invasiveness. Metabolomics 13:1–13. https://doi.org/10.1007/S11306-017-1169-Z/FIGURES/4

Grossi G, Goglio P, Vitali A, Williams AG (2019) Livestock and climate change: impact of livestock on climate and mitigation strategies. Anim Front 9:69–76. https://doi.org/10.1093/AF/VFY034

Gschwend PM, MacFarlane JK, Newman KA (1985) Volatile halogenated organic compounds released to seawater from temperate marine macroalgae. Science 227:1033–1035. https://doi.org/10.1126/SCIENCE.227.4690.1033

Guiry MD (2018) AlgaeBase. World-wide electronic publication, National University of Ireland, Galway [WWW Document]. https://www.algaebase.org. Accessed 19 June 2023

Haslin C, Lahaye M, Pellegrini M, Chermann JC (2001) In vitro anti-HIV activity of sulfated cell-wall polysaccharides from gametic, carposporic and tetrasporic stages of the Mediterranean red alga *Asparagopsis armata*. Planta Med 67:301–305. https://doi.org/10.1055/S-2001-14330

Hein (2019) Rotalgen als Retter [WWW Document]. MARE. https://www.mare.de/rotalgen-als-retter-content-1487. Accessed 12 June 2023

Hou X, Chai C, Qian Q, Yan X, Fan X (1997) Determination of chemical species of iodine in some seaweeds (I). Sci Total Environ 204:215–221. https://doi.org/10.1016/S0048-9697(97)00182-4

Hristov AN, Melgar A, Wasson D, Arndt C (2022) Symposium review: effective nutritional strategies to mitigate enteric methane in dairy cattle. J Dairy Sci 105:8543–8557. https://doi.org/10.3168/JDS.2021-21398

Hutson KS, Mata L, Paul NA, de Nys R (2012) Seaweed extracts as a natural control against the monogenean ectoparasite, Neobenedenia sp., infecting farmed barramundi (Lates calcarifer). Int J Parasitol 42:1135–1141. https://doi.org/10.1016/J.IJPARA.2012.09.007

IPCC (2019) Global warming of 1.5 °C. an IPCC special report on the impacts of global warming of 1.5 °C above pre-industrial levels and related global greenhouse gas emission pathways. In: In the context of strengthening the global response to the threat of climate change, intergovernmental panel on climate change. Cambridge University Press, Cambridge

Kaliaperumal N (2003) Products from seaweeds. SDMRI Res Publ 3:33–42

Kinley RD, Fredeen AH (2015) In vitro evaluation of feeding North Atlantic stormtoss seaweeds on ruminal digestion. J Appl Phycol 27:2387–2393. https://doi.org/10.1007/S10811-014-0487-Z

Kinley RD, de Nys R, Vucko MJ, Machado L, Tomkins NW, Kinley RD, de Nys R, Vucko MJ, Machado L, Tomkins NW (2016) The red macroalgae *Asparagopsis taxiformis* is a potent natural antimethanogenic that reduces methane production during in vitro fermentation with rumen fluid. Anim Prod Sci 56:282–289. https://doi.org/10.1071/AN15576

Kinley RD, Martinez-Fernandez G, Matthews MK, de Nys R, Magnusson M, Tomkins NW (2020) Mitigating the carbon footprint and improving productivity of ruminant livestock agriculture using a red seaweed. J Clean Prod 259:120836. https://doi.org/10.1016/J.JCLEPRO.2020.120836

Kladi M, Vagias C, Roussis V (2004) Volatile halogenated metabolites from marine red algae. Phytochem Rev 3:37–366

Knapp JR, Laur GL, Vadas PA, Weiss WP, Tricarico JM (2014) Invited review: enteric methane in dairy cattle production: quantifying the opportunities and impact of reducing emissions. J Dairy Sci 97:3231–3261. https://doi.org/10.3168/JDS.2013-7234

Laturnus F (1996) Volatile halocarbons released from Arctic macroalgae. Mar Chem 55:359–366. https://doi.org/10.1016/S0304-4203(97)89401-7

Li X, Norman HC, Kinley RD, Laurence M, Wilmot M, Bender H, de Nys R, Tomkins N, Li X, Norman HC, Kinley RD, Laurence M, Wilmot M, Bender H, de Nys R, Tomkins N (2016) *Asparagopsis taxiformis* decreases enteric methane production from sheep. Anim Prod Sci 58: 681–688. https://doi.org/10.1071/AN15883

Liu X, Bogaert K, Engelen AH, Leliaert F, Roleda MY, De Clerck O (2017) Seaweed reproductive biology: environmental and genetic controls. Bot Mar 60:89–108. https://doi.org/10.1515/BOT-2016-0091/ASSET/GRAPHIC/J_BOT-2016-0091_FIG_002.JPG

Lüttge U, Kluge M. (1936-, 2012) Botanik die einführende Biologie der Pflanzen, Wiley-Blackwell; 6. aktualisierte edition (4 April 2012)

Machado L, Magnusson M, Paul NA, De Nys R, Tomkins N (2014) Effects of marine and freshwater macroalgae on in vitro Total gas and methane production. PLoS One 9:e85289. https://doi.org/10.1371/JOURNAL.PONE.0085289

Machado L, Magnusson M, Paul NA, Kinley R, de Nys R, Tomkins N (2016a) Dose-response effects of *Asparagopsis taxiformis* and *Oedogonium* sp. on in vitro fermentation and methane production. J Appl Phycol 28:1443–1452. https://doi.org/10.1007/S10811-015-0639-9/FIGURES/4

Machado L, Magnusson M, Paul NA, Kinley R, de Nys R, Tomkins N (2016b) Identification of bioactives from the red seaweed *Asparagopsis taxiformis* that promote antimethanogenic activity in vitro. J Appl Phycol 28:3117–3126. https://doi.org/10.1007/S10811-016-0830-7

Maia MRG, Fonseca AJM, Oliveira HM, Mendonça C, Cabrita ARJ (2016) The potential role of seaweeds in the natural manipulation of rumen fermentation and methane production. Sci Rep 6: 32321. https://doi.org/10.1038/srep32321

Martins GM, Cacabelos E, Faria J, Álvaro N, Prestes ACL, Neto AI (2019) Patterns of distribution of the invasive alga *Asparagopsis armata* Harvey: a multi-scaled approach. Aquat Invasions 14: 582–593. https://doi.org/10.3391/ai.2019.14.4.02

Mata L, Silva J, Schuenhoff A, Santos R (2006) The effects of light and temperature on the photosynthesis of the *Asparagopsis armata* tetrasporophyte (*Falkenbergia rufolanosa*), cultivated in tanks. Aquaculture 252:12–19. https://doi.org/10.1016/J.AQUACULTURE.2005.11.045

Mata L, Lawton RJ, Magnusson M, Andreakis N, de Nys R, Paul NA (2017) Within-species and temperature-related variation in the growth and natural products of the red alga *Asparagopsis taxiformis*. J Appl Phycol 29:1437–1447. https://doi.org/10.1007/S10811-016-1017-Y/FIGURES/5

McDermid KJ, Martin KJ, Haws MC (2019) Seaweed resources of the Hawaiian islands. Bot Mar 62:443–462. https://doi.org/10.1515/BOT-2018-0091/ASSET/GRAPHIC/J_BOT-2018-0091_CV_003.JPG

Mihaila AA, Glasson CRK, Lawton R, Muetzel S, Molano G, Magnusson M (2022) New temperate seaweed targets for mitigation of ruminant methane emissions: an in vitro assessment. Appl Phycol 3:274–284. https://doi.org/10.1080/26388081.2022.2059700/SUPPL_FILE/TAPY_A_2059700_SM3208.DOCX

Min BR, Parker D, Brauer D, Waldrip H, Lockard C, Hales K, Akbay A, Augyte S (2021) The role of seaweed as a potential dietary supplementation for enteric methane mitigation in ruminants: challenges and opportunities. Anim Nutr 7:1371–1387. https://doi.org/10.1016/J.ANINU.2021.10.003

Min BR, Lee S, Jung H, Miller DN, Chen R (2022) Enteric methane emissions and animal performance in dairy and beef cattle production: strategies, opportunities, and impact of reducing emissions. Animals 12:948. https://doi.org/10.3390/ANI12080948

Mitsumori M, Shinkai T, Takenaka A, Enishi O, Higuchi K, Kobayashi Y, Nonaka I, Asanuma N, Denman SE, McSweeney CS (2012) Responses in digestion, rumen fermentation and microbial populations to inhibition of methane formation by a halogenated methane analogue. Br J Nutr 108:482–491. https://doi.org/10.1017/S0007114511005794

Molina-Alcaide E, Carro MD, Roleda MY, Weisbjerg MR, Lind V, Novoa-Garrido M (2017) In vitro ruminal fermentation and methane production of different seaweed species. Anim Feed Sci Technol 228:1–12. https://doi.org/10.1016/J.ANIFEEDSCI.2017.03.012

Muizelaar W, Groot M, van Duinkerken G, Peters R, Dijkstra J (2021) Safety and transfer study: transfer of Bromoform present in *Asparagopsis taxiformis* to Milk and urine of lactating dairy cows. Foods (Basel, Switzerland) 10(3):584. https://doi.org/10.3390/FOODS10030584

Myhre G, Shindell D, Bréon F, Collins W, Fuglestvedt J, Huang J, Koch D, Lamarque J, Lee D, Mendoza B, Nakajima T, Robock A, Stephens G, Takemura T, Zhang H, Qin D, Plattner G, Tignor M, Allen S, Boschung J, Nauels A, Xia Y, Bex V, Midgley P (2013) Anthropogenic and natural radiative forcing. In: Stocker TF, Qin D, Plattner G-K, Tignor M, Allen SK, Boschung J, Nauels A, Xia Y, Bex V, Midgley PM (eds) Contribution of working group I to the fifth assessment report of the intergovernmental panel on climate change. Cambridge University Press, Cambridge

NAO (2023) Atlantic-Gro organic seaweed supplements for animals and crops - PEI, Canada [WWW document]. North Atl. Org. https://naorganics.com/. Accessed 12 June 2023

Nørskov NP, Bruhn A, Cole A, Nielsen MO (2021) Targeted and untargeted metabolic profiling to discover bioactive compounds in seaweeds and hemp using gas and liquid chromatography-mass spectrometry. Metabolites 11:259. https://doi.org/10.3390/METABO11050259

Nunes N, Valente S, Ferraz S, Barreto MC, Pinheiro de Carvalho MAA (2018) Nutraceutical potential of *Asparagopsis taxiformis* (Delile) Trevisan extracts and assessment of a downstream purification strategy. Heliyon 4:e00957. https://doi.org/10.1016/J.HELIYON.2018.E00957

Nunes N, Valente S, Ferraz S, Barreto MC, de Carvalho MAAP (2019) Validation of a spectrophotometric methodology for a rapid iodine analysis in algae and seaweed casts. Algal Res 42: 101613. https://doi.org/10.1016/J.ALGAL.2019.101613

Pandey D, Mansouryar M, Novoa-Garrido M, Næss G, Kiron V, Hansen HH, Nielsen MO, Khanal P (2021) Nutritional and anti-methanogenic potentials of macroalgae for ruminants. Burleigh Dodds Science Publishing Limited, Cambridge, pp 195–228. https://doi.org/10.19103/AS.2021.0091.14

Paul NA, de Nys R, Steinberg P (2006a) Seaweed-herbivore interactions at a small scale: direct tests of feeding deterrence by filamentous algae. Mar Ecol Prog Ser 323:1–9. https://doi.org/10.3354/meps323001

Paul N, De Nys R, Steinberg PD (2006b) Chemical defence against bacteria in the red alga *Asparagopsis armata*: linking structure with function. Mar Ecol Prog Ser 306:87–101. https://doi.org/10.3354/MEPS306087

Paul NA, Cole L, De Nys R, Steinberg PD (2006c) Ultrastructure of the gland cells of the red alga *Asparagopsis armata* (Bonnemaisoniaceae). J Phycol 42:637–645. https://doi.org/10.1111/J.1529-8817.2006.00226.X

Pereira L (2018) Therapeutic and nutritional uses of algae. Ther Nutr Uses Algae. https://doi.org/10.1201/9781315152844

Pinteus S, Alves C, Monteiro H, Araújo E, Horta A, Pedrosa R (2015) *Asparagopsis armata* and *Sphaerococcus coronopifolius* as a natural source of antimicrobial compounds. World J Microbiol Biotechnol 31:445–451. https://doi.org/10.1007/S11274-015-1797-2/TABLES/2

Pinteus S, Lemos MFL, Alves C, Silva J, Pedrosa R (2021) The marine invasive seaweeds *Asparagopsis armata* and *Sargassum muticum* as targets for greener antifouling solutions. Sci Total Environ 750:141372. https://doi.org/10.1016/J.SCITOTENV.2020.141372

Ponte JMS, Seca AML, Barreto MC (2022) *Asparagopsis* genus: what we really know about its biological activities and chemical composition. Molecules 27:1787. https://doi.org/10.3390/MOLECULES27061787

PubChem (2023) Bromoform | CHBr3 | CID 5558 [WWW Document]. Natl Libr Med. https://pubchem.ncbi.nlm.nih.gov/compound/5558. Accessed 15 June 2023

Quack B, Wallace DWR (2003) Air-sea flux of bromoform: controls, rates, and implications. Global Biogeochem Cycles 17:1023. https://doi.org/10.1029/2002GB001890

RAIS (1997) Formal toxicity summary for BROMOFORM [WWW document]. Risk Assess Inf Syst. https://rais.ornl.gov/tox/profiles/bromofrm.html. Accessed 15 June 2023

Reisinger A, Clark H, Cowie AL, Emmet-Booth J, Gonzalez Fischer C, Herrero M, Howden M, Leahy S (2021) How necessary and feasible are reductions of methane emissions from livestock to support stringent temperature goals? Philos Trans R Soc A 379:20200452. https://doi.org/10.1098/RSTA.2020.0452

Ritchie H, Rosado P, Roser M (2017) Meat and dairy production [WWW document]. Our World Data. https://doi.org/10.1088/1748-9326

Romanazzi D, Sanchez-Garcia C, Svenson J, Mata L, Pes K, Hayman CM, Wheeler TT, Magnusson M (2021) Rapid analytical method for the quantification of Bromoform in the red seaweeds *Asparagopsis armata* and *Asparagopsis taxiformis* using gas chromatography-mass spectrometry. ACS Agric Sci Technol 1:436–442. https://doi.org/10.1021/ACSAGSCITECH.1C00161/ASSET/IMAGES/LARGE/AS1C00161_0003.JPEG

Roque B, Brooke C, Ladau J, Polley T, Marsh LJ, Najafi N, Pandey P, Singh L, Kinley R, Salwen JK, Eloe-Fadrosh E, Kebreab E, Hess M (2019a) Effect of the macroalgae *Asparagopsis taxiformis* on methane production and rumen microbiome assemblage. s 1(1):3. https://doi.org/10.1186/S42523-019-0004-4

Roque BM, Salwen JK, Kinley R, Kebreab E (2019b) Inclusion of *Asparagopsis armata* in lactating dairy cows' diet reduces enteric methane emission by over 50 percent. J Clean Prod 234:132–138. https://doi.org/10.1016/J.JCLEPRO.2019.06.193

Roque BM, Venegas M, Kinley RD, De Nys R, Duarte TL, Yang X, Kebreab E (2021) Red seaweed (*Asparagopsis taxiformis*) supplementation reduces enteric methane by over 80 percent in beef steers. PLoS One 16:e0247820. https://doi.org/10.1371/JOURNAL.PONE.0247820

Salvador N, Gómez Garreta A, Lavelli L, Ribera MA (2007) Antimicrobial activity of Iberian macroalgae. Sci Mar 71:101–114. https://doi.org/10.3989/SCIMAR.2007.71N1101

Schöne F, Spörl K, Leiterer M (2017) Iodine in the feed of cows and in the milk with a view to the consumer's iodine supply. Journal of trace elements in medicine and biology. 39: 202-209.

Schuenhoff A, Mata L, Santos R (2006). The tetrasporophyte of Asparagopsis armata as a novel seaweed biofilter. Aquaculture, 252(1): 3-11.

Sigma-Aldrich (2023) Bromoform [WWW document]. Sigma-Aldrich. https://www.sigmaaldrich.com/DE/de/search/bromoform?focus=products&page=1&perpage=30&sort=relevance&term=bromoform&type=product. Accessed 15 June 2023

Silva CO, Simões T, Félix R, Soares AMVM, Barata C, Novais SC, Lemos MFL (2021) *Asparagopsis armata* exudate cocktail: the quest for the mechanisms of toxic action of an invasive seaweed on marine invertebrates. Biology 10:223. https://doi.org/10.3390/BIOLOGY10030223

Stefenoni HA, Räisänen SE, Cueva SF, Wasson DE, Lage CFA, Melgar A, Fetter ME, Smith P, Hennessy M, Vecchiarelli B, Bender J, Pitta D, Cantrell CL, Yarish C, Hristov AN (2021) Effects of the macroalga *Asparagopsis taxiformis* and oregano leaves on methane emission, rumen fermentation, and lactational performance of dairy cows. J Dairy Sci 104:4157–4173. https://doi.org/10.3168/JDS.2020-19686

Streftaris N, Zenetos A (2006) Alien marine species in the Mediterranean - the 100 "worst invasives" and their impact. Mediterr Mar Sci 7:87–118. https://doi.org/10.12681/mms.180

Ungerfeld EM, Beauchemin KA, Muñoz C (2022) Current perspectives on achieving pronounced enteric methane mitigation from ruminant production. Front Anim Sci 2:795200. https://doi.org/10.3389/FANIM.2021.795200

United Nations (2023) Goal 13 - Take urgent action to combat climate change and its impacts. https://sdgs.un.org/goals/goal13, accessed on 26th of May 2023.

UNTC (2010) Nagoya protocol on access to genetic resources and the fair and equitable sharing of benefits arising from their utilization to the convention on biological diversity [WWW document]. United Nations Treaty Collect. Chapter XXCII Environ. Nagoya. https://treaties.un.org/pages/ViewDetails.aspx?src=IND&mtdsg_no=XXVII-8-b&chapter=27&clang=_en. Accessed 18 June 2023

Vedhagiri K, Manilal A, Valliyammai T, Shanmughapriya S, Sujith S, Selvin J, Natarajaseenivasan K (2009) Antimicrobial potential of a marine seaweed *Asparagopsis taxiformis* against *Leptospira javanica* isolates of rodent reservoirs. Ann Microbiol 59:431–437

Wasson DE, Yarish C, Hristov AN (2022) Enteric methane mitigation through *Asparagopsis taxiformis* supplementation and potential algal alternatives. Front Anim Sci 3:999338. https://doi.org/10.3389/FANIM.2022.999338/BIBTEX

Weiser MJ, Butt CM, Mohajeri MH (2016) Docosahexaenoic acid and cognition throughout the lifespan. Nutrients 8(2):99. https://doi.org/10.3390/NU8020099

Werner A, Clarke D, Kraan S (2004) Strategic review and the feasibility of seaweed aquaculture in Ireland

Wever R, Van Der Horst MA (2013) The role of vanadium haloperoxidases in the formation of volatile brominated compounds and their impact on the environment. Dalt Trans 42:11778–11786. https://doi.org/10.1039/C3DT50525A

WHO (2023) List of classifications – IARC monographs on the identification of carcinogenic hazards to humans [WWW document]. Int. Agency Res. Cancer, World Heal. Organ. https://monographs.iarc.who.int/list-of-classifications/. Accessed 15 June 2023

Zanolla M, Carmona R, Mata L, De la Rosa J, Sherwood A, Barranco CN, Muñoz AR, Altamirano M (2022) Concise review of the genus *Asparagopsis* Montagne, 1840. J Appl Phycol 34:1–17. https://doi.org/10.1007/S10811-021-02665-Z/FIGURES/3

Algae for Aquaculture: Recent Technological Applications

Mostafa Elshobary and Mohamed Ashour

Abstract Algae are essential in aquatic ecosystems as they form the base of food webs, providing bioactive compounds that sustain and improve the growth of commercially important aquatic animals. However, it is critical to provide the required quality and quantity of specific algae strains for each aquatic animal throughout their life cycles on a commercial scale. In sustainable aquaculture, algae, including microalgae and seaweed, could be introduced to aquaculture animals in different forms, such as live, dried, liquid extract, and nanoparticle forms. These forms enhance the benefits of algae bioactive compounds for aquatic animals. The world requires cost-effective, environmentally friendly, and feasible technologies for large-scale production of aquaculture organisms. The integration of algae and animals in sustainable aquaculture offers an intelligent solution to the challenges faced in monoaquaculture. This chapter focuses on different algal forms in aquaculture, including live feeds, biomass concentrates, water conditioners using the "green water technique," aqua-feed additives, co-culturing technologies, and integrated multi-trophic aquaculture (IMTA) to encourage the development of low-cost, highly efficient, and sustainable aquaculture projects in the future. Overall, understanding the role of microalgae in sustainable aquaculture is crucial for improving the growth and health of aquatic animals and for maintaining the ecological balance of aquatic ecosystems.

Keywords Livefeeds · Hatcheries · Nanotechnologies · Algal extract · Seaweeds

M. Elshobary (✉)
Botany Department, Faculty of Science, Tanta University, Tanta, Egypt
e-mail: mostafa_elshobary@science.tanta.edu.eg

M. Ashour
National Institute of Oceanography and Fisheries, NIOF, Cairo, Egypt

A. Abomohra, S. Ende (eds.), *Value-added Products from Algae*,
https://doi.org/10.1007/978-3-031-42026-9_16

1 Introduction

Aquaculture, the cultivation of aquatic organisms for human consumption, has experienced rapid growth in recent years. As the global population continues to increase, it is essential to find sustainable and resource-efficient ways to meet the rising demand for food (Alam et al. 2020). Using plant materials to grow aquatic animals may significantly increase sustainability in this crucial industry, as our knowledge of aquatic ecosystems has advanced alongside the rapid expansion of aquaculture. Science has grown recently, leading to a better international knowledge of the necessity for sustainable management of these resources as well as a much greater grasp of how aquatic ecosystems function (FAO 2020; Mansour et al. 2022a). The population and consumption of the whole world have greatly expanded due to the fast growing change in the world. As a consequence, several international organizations and countries have acknowledged the need for the development and sustainable management of aquatic resources (Shaalan et al. 2018). Many international and national programs that proved the intelligent integration of aquaculture sectors and performance have been conducted by various nations worldwide to address this crucial problem successfully. The global population is growing, and aquaculture enterprises have effectively and quickly expanded in response (Barkia et al. 2019; Limbu et al. 2018; Tocher 2015).

Although aquaculture has advanced greatly and has the potential to be sustainable, there are still many problems facing the sector, according to opponents. For instance, in Nature, Naylor et al. (2000) presented an intriguing assessment that described aquaculture as a viable green approach to reverse the worldwide decline in fisheries stocks before the previous two decades. Since then, around 300 kinds of fish, shellfish, microalgae, and seaweed, have been cultivated globally, from 10 million tonnes (MTs) in 1987 to 29 MTs in 1997. New research (Naylor et al. 2021) by describes how the aquaculture industry has changed over the last 20 years, from playing a minor role to becoming a crucial component of the global food supply. It's interesting to note that the output of fish, shellfish, microalgae, and seaweeds produced by aquaculture has expanded from 29 MT in 1997 to more than 82 MT in 2018. Aquatic animals (crustaceans, finfish, and mollusks) and aquatic plants make up the majority of the species used in commercial aquaculture. These water creatures and plants are the most often cultivated species with significant economic and dietary significance (Alam et al. 2020).

As reported by FAO (2020), 114.5 MTs or 46.0% of the world's fish output came from aquaculture in 2018. Its overall output included 26,000 tonnes (Ts) of pearls and decorative seashells in addition to 82.1 MTs of aquatic plants and 32.4 MTs of microalgae and seaweed. Aquatic animals, such as fish, crustaceans, and molluscs, which are found in marie and/or freshwater environments, are one of the primary sources of animal protein for people. Finfish (54.3 MTs) were the maximum common aquatic animal cultivated in 2018 (47 MTs from inland aquaculture, 7.3 MTs from coastal and marine aquaculture), followed by 17.7 MTs of molluscs including bivalves, 9.4 MTs of crustaceans, 435,400 Ts of marine invertebrates,

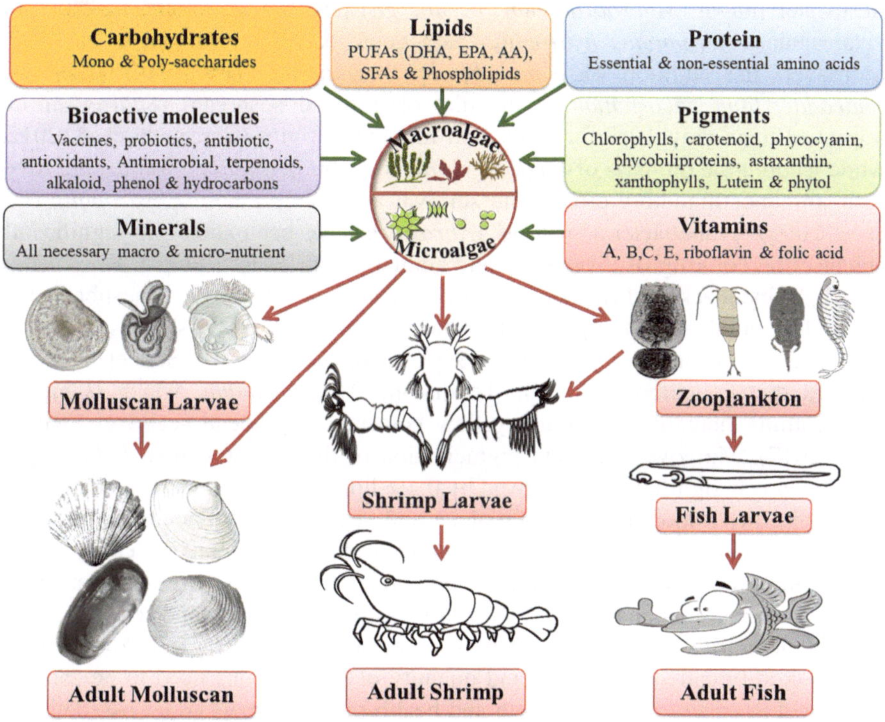

Fig. 1 The role of macro- and microalgae as feed in aquatic systems, as well as in aquaculture

370,000 Ts of aquatic turtles, and 131,300 Ts of frogs (FAO 2020; Naylor et al. 2021).

Algae, a diverse group of photosynthetic organisms, have emerged as a promising solution in various aspects of aquaculture, offering nutritional, environmental, and economic benefits. Algae (Fig. 1) have long played a crucial role in aquatic ecosystems, aiding as a natural food source for many aquatic animals (Naylor et al. 2000; Ruggiero et al. 2015). Moreover, microalgae can contribute to maintaining water quality in aquaculture systems by assimilating excess nutrients and providing oxygen through photosynthesis (Alam et al. 2020; Javed et al. 2019; Muhammad et al. 2020).

Numerous studies have indicated that the addition of microalgae in the diets of various aquatic animals can result in improved growth rate, feed consumption, persistence, better health, gut health, coloration, and immune response (Nagappan et al. 2021). For instance, the incorporation of 1.2% *Schizochytrium* sp. meal in the diet of Nile tilapia (*Oreochromis niloticus*) had a positive impact on the fish's gut microbiota and overall health (Souza et al. 2020). In another study, supplementing the diet of white leg shrimp (*Litopenaeus vannamei*) with 1–2% *D. salina* significantly enhanced their survival and growth rate (Medina-Félix et al. 2014). Replacing fish meals with 6-8% *Chlorella vulgaris* enhanced the immune response of Giant freshwater prawn (*Macrobrachium rosenbergii*) post-larvae, improving their survival against *Aeromonas hydrophila* pathogenic bacteria (Maliwat et al. 2017). Moreover, the addition of *Nannochloropsis gaditana, Tetraselmis chuii*, and *Phaeodactylium tricornutum* in the diet of gilthead seabream (*Sparus aurata*) improved their growth performance and immune activity (Cerezuela et al. 2012), while feeding the biomass of *Euglena viridis* to Rohu fish (*Labeo rohita*) improved their immunostimulatory effects and survival against *Aeromonas hydrophila* (Das et al. 2009). cyanobacteria such as *Arthrospira* have been shown to significantly enhance the coloring of diverse aquatic animals, such as Red tilapia, Koi, Yellow catfish, Striped jack, and Black tiger prawn (Ansarifard et al. 2018; Dineshbabu et al. 2019; Liu et al. 2021). Specifically, the coloration of koi fish was positively enhanced by an inclusion rate of 7.5% *Arthrospira platensis* (Sun et al. 2012), while a diet containing 2.5% of the diatom *P. tricornutum*, which is high in fucoxanthin, increased the bright yellow coloring of gilthead seabream (Ribeiro et al. 2017). Moreover, the skin pigmentation of the yellow catfish *Pelteobagrus fulvidraco* was significantly improved by 0.4% lipid-free Spirulina (Liu et al. 2021).

Despite the importance of microalgae and seaweed in aquaculture, their high production costs continue to pose a significant challenge (Mansour et al. 2022b). One of the primary challenges in the commercial cultivation of these organisms is ensuring a consistent and reliable source of specific microalgae and seaweed strains to meet the nutritional needs of different aquatic animals ranging from zooplanktons to fish. Additionally, the quality and quantity of these strains need to be carefully controlled to ensure optimal growth and health of the animals (Raeisossadati et al. 2019). The variability in environmental conditions, such as temperature, light, and nutrient availability, can also affect the growth and quality of microalgae and seaweeds, making it challenging to maintain a consistent and reliable supply (Nova et al. 2020). Technological advancements are needed to address these challenges and ensure a consistent and reliable source of microalgae and seaweed for commercial aquaculture (Cottier-Cook et al. 2021; Zhang et al. 2022).

Recent technological advancements have made large-scale algae cultivation more feasible and cost-effective, leading to increased interest in their application in aquaculture. Innovations in photobioreactor design, automated monitoring and control systems, genetic engineering, and nutrient recycling have paved the way for more efficient and sustainable algal production (Erbland et al. 2020) Algal biotechnology converges on enhancing the production of high value products including proteins, carbohydrates, polyunsaturated fatty acids (PUFA), pigments, and other from algae via cultivation optimization, carbon flux rate, stress condition manipulation, and metabolic route shifting (Ahmad et al. 2022). For example, the co-production of photo-bioreactors and high value product such as astaxanthin can reduce the cost of algal cultivation four times from $3.90 to $0.54 per litre (Rafa et al. 2021). Fish and microalgae share comparable nutritional characteristics, and the digestibility of feed is an essential element in aquafeed preparation. By using a feed that is highly digestible, it is possible to lower production expenses, decrease feed wastage, and mitigate the possibility of eutrophication. Microalgae-based feed for fish and shrimp has the potential to offer substantial economic advantages due to their low input costs, reduced carbon footprint, wastewater treatment advantages, and ability to generate carbon credits through industrial CO_2 conversion (Ahmad et al. 2022). This chapter aims to provide an overview of the role of algae in aquaculture and explore the recent technological applications that have facilitated their integration into the industry. It will discuss the various types of algae, their cultivation methods, and the processing techniques used to produce algal-based feeds. Furthermore, the nutritional benefits of algal feeds and the environmental and sustainability considerations will be examined. Lastly, the chapter will address the challenges and future prospects for the continued integration of algae into aquaculture.

2 Algae as a Natural Food Source in Aquatic Ecosystems

Algae play a critical role in aquatic ecosystems as primary producers, forming the base of the food web (Alprol et al. 2021b). They convert sunlight, carbon dioxide, and nutrients through the process of photosynthesis into a variety of organic matters including proteins, lipids, carbohydrates, vitamins, minerals, phenols, flavonoids, and pigments. Algae-based products find applications in diverse industries, including food, cosmetics, animal feed, chemicals, and bioenergies. According to industry forecasts, the market for such products is projected to expand from US$ about 1500 million in 2020 to more than US $2800 million by 2028. This represents a compound annual growth rate of 7.9% between 2021 and 2028 (Globenewswire 2021). Zooplanktons, such as small crustaceans and rotifers, feed on algae (phytoplankton) and are, in turn, consumed by larger animals, like fish and shellfish. This transfer of energy via consumption is essential for the proper functioning of aquatic ecosystems (Elshobary et al. 2020).

Recent technological advancements have allowed for the large-scale cultivation of algae for use in aquaculture. These innovations include novel cultivation systems, such as photobioreactors and raceway ponds, which allow for the controlled growth of algae in high densities. Additionally, advances in processing techniques have made it possible to produce high-quality algal products, like dried biomass, extracts, and oils, for use in aquafeeds (Meitei et al. 2022; Sirohi et al. 2023). As the aquaculture industry persists to grow, the use of algae as an ecological and nutritious food source will become increasingly important. By harnessing the natural abilities of algae, we can help minimize the environmental impacts of aquaculture and support the development of a more sustainable food system (Hamdi 2022).

The commonly recognized definition of aquaculture includes the production of microalgae and macroalgae. On the other hand, worldwide maintenance and regulation of algae production occur independently of aquaculture (Barra et al. 2014; FAO 2020; Ferdouse et al. 2018). Several research in the field of aquaculture has acknowledged that microalgae are the "superfood" for all aquatic creatures (Shields and Lupatsch 2012). In this regard, Fig. 2 compares the nutritional contents

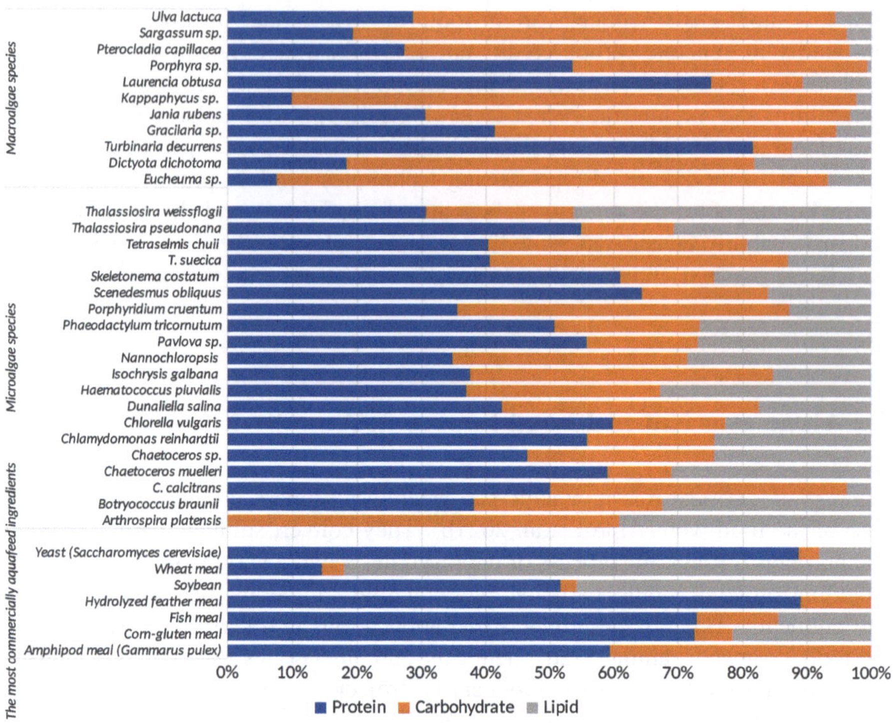

Fig. 2 The usual nutritional composition of the most widely used aquafeed ingredients, as compared to microalgae and seaweed species (protein, lipid, and carbohydrate, depending on % of dry mass). Data in the figure are adopted from Mansour et al. (2022b)

(carbohydrate, fat, and protein) of micro- and macroalgae with the most widely used aquafeed ingredients.

Several studies have concentrated on the chemical and nutrient composition of microalgae to determine their suitability as a feedstock for different organisms. Proximate analysis of microalgae provides information on the total amount of protein, carbohydrate, and lipid content (Kaparapu 2018; Wikfors and Ohno 2001). Microalgae offer promise as a fish feed ingredient primarily because of their protein, lipid, and carbohydrate content, which are all essential for fish health. Compared to other alternative ingredients like yeast and bacteria, microalgae are particularly rich in protein and lipid (Fig. 2) and have a balanced amino acid content that eliminates the need for costly amino acid supplements in fish diets. Some species of microalgae, such as *Chlorella, Porphyridium, Chlamydomonas, Porphyridium, Nannochloropsis,* and *Isochrysis*, are high in methionine, an amino acid that is often insufficient in plant-based ingredients (Wan et al. 2019). The type of carbohydrate is also a crucial feed property. The content of easily digestible starch in microalgal species ranges from 7% to 45%, with *Chlamydomonas rheinhardtii, Tetraselmis subcordiformis*, and *C. vulgaris* having relatively higher starch content (30–49%) than other microalgae (Dragone et al. 2011; Yao et al. 2012) The fiber content in microalgae ranges from 5% to 18%, and unlike plant fiber, microalgal fiber lacks lignin and has low hemicellulose content, which suggests better digestibility (Matos et al. 2016; Niccolai et al. 2019). Some microalgae species, such as *Spirulina* sp. and *C. vulgaris*, have low fiber content (8.5% and 5.6%, respectively), while others, such as those in the genus *Nannochloropis, Tisochrysis, Tetraselmis*, and *Phaeodactylum*, have higher ash content (Niccolai et al. 2019). Additionally, nutrient limitation has been shown to increase the carbohydrate and lipid content in microalgae (Chen et al. 2017; Nagappan et al. 2019; Nagappan and Kumar 2021). Exposure to different stress levels has been found to have a positive effect on the proximate composition of various micro- and macroalgae, making them more suitable for use in aquaculture. Reducing nitrogen and phosphorus levels in *Micractinium reisseri* grown in 75% wastewater resulted in increased lipid content and productivity. A 50% decrease in nitrogen levels led to a 2-fold increase in lipid content, while a similar decrease in phosphorus levels resulted in a 2.5-fold increase. Lipid productivity also showed significant improvements, nearly tripling in the case of nitrogen reduction. These findings highlight the importance of nitrogen and phosphorus stress conditions in lipid metabolism and suggest potential strategies for enhancing lipid production in microalgae cultures (Elshobary et al. 2019). An additional research study delves into how different ocean acidification levels affect the growth and biochemical composition of *Ulva fasciata*, a type of green seaweed, and its potential use in the European sea bass (*Dicentrarchus labrax*) aquaculture. The study found that ocean acidification at 550 μatm increased the growth rate, protein and pigment content of *U. fasciata*, which could be a good source of food for fish farming of *Dicentrarchus labrax* (El-Sayed et al. 2022).

The amount of PUFA in microalgae is one of the most proximates that strongly influence their nutritional value (Gressler et al. 2010; Molinoa et al. 2020). EPA (eicosapentaenoic acid) and DHA (docosahexaenoic acid) are produced by a variety of auto- and heterotrophic microalgae species from different groups, but alpha-linolenic acid (ALA) is often found in small levels (Remize et al. 2021). Recent reviews of total lipid extracts suggest that Bacillariophyceae (diatoms) and Chrysophyceae species are rich sources of EPA and DHA. Cryptophyceae, Eustigmatophyceae, Rhodophyceae, Xanthophyceae, Glaucophyceae, and Prasinophyceae can be interesting sources of EPA, whereas DHA is primarily found in significant amounts in Dinophyceae (Hu et al. 2008; Lang et al. 2011; Sharma et al. 2020). The nutritional characteristics of several microalgae, including *Arthrospira* (Spirulina), *Chlorella*, and *Dunaliella*, are well known, and they have a long history of usage as sources of animal feed. Recent researches have shown the potential of microalgal biomass as a replacement or supplement for fishmeal or fish oil in aquafeed.

Microalgae possess various pigments that have antioxidant properties, and certain microalgae can produce high levels of immunostimulants and vitamins, which improve the health of aquatic species (Awasthi et al. 2020; Prabha et al. 2020; Zhou et al. 2019). The pigments found in microalgae, such as astaxanthin, can also provide attractive colors to fish, enhancing their marketability (Posten and Schaub 2009). In addition, microalgae are rich in organic minerals, as they have unique structural characteristics that enable them to bind metals with high affinity and have a large surface-to-volume ratio. The use of minerals-rich microalgal biomass in fish feed can help overcome the drawback of mineral leaching before pellets are consumed by fish (Nagappan et al. 2021).

Algal cells may generally be used throughout the whole aquaculture process, which is primarily divided into three phases: (1) hatchery inoculum and/or cultivation; (2) early raising and/or nurseries; and (3) on-growing and/or production. At hatcheries, microalgae were fed to juveniles, larvae, post-larvae, bivalve adults, mollusk, and crustacean species, as well as the early developmental stages of several fish species. It also serves as the primary source of food for certain zooplankton (Mansour et al. 2022b).

Diverse aquaculture operations use microalgae in a variety of ways, including in fresh feed (as live feed), biomass (pastes), dried biomass, spray-dried, defatted biomass, freeze-dried, defatted biomass (as a byproduct for biodiesel), and residuals (wastes) from bioindustries that use microalgae. Aquatic plants are utilized as nutritional supplements or as replacements for dietary components, which are mostly fishmeal or fish oil, in the aquafeed business because of their distinct biochemical makeup (Cardinaletti et al. 2018; Delgado et al. 2021). The incorporation of seaweed into the culture of many aquatic species, on the other hand, may have a good effect on the environment, the economy, and sustainability (Yang et al. 2020).

3 Forms of Algae for Sustainable Aquaculture

There are various ways in which microalgae can be utilized in the aquaculture, including live food, pastes, dried biomass, spray-dried, freeze-dried, residual biomass, and waste products from the bioindustries that use microalgae (Cardinaletti et al. 2018; Delgado et al. 2021). Due to their unique biochemical composition, microalgae are often used in aquafeed as a dietary supplement or as a substitute for traditional dietary ingredients like fishmeal or fish oil. Additionally, seaweed can be integrated into the cultivation of aquatic animals, which can have beneficial effects on sustainability, the environment, and the economy (Yang et al. 2020).

3.1 Live Feed

Unfortunately, only a few microalgae species have been extensively employed in aquaculture, and these species must have special features in order to be helpful in aquaculture. Acceptable ingestion size, high digestibility, quick growth rate, simplicity of mass production, adaptability to ecological conditions, cultivability under nutrient constraints, and adequate nutritional content are all characteristics that make a microalgae species appropriate for aquaculture (Duy et al. 2017; Spolaore et al. 2006). These properties are essential for microalgae to be a viable food source for aquatic animals. Many microalgae species (*Nannochloris sp., Nannochloropsis* sp., *Chlorella* sp.,*C. salina, Dunaleilla sp, Chaetoceros* sp., *Coscinodiscus* sp. *Isochrysis* sp.*;* Tetraselmis, *Pseudo-nitzschia* sp., *Prorocentrum* sp., *Rhodomonas* sp., and *Navicula* sp.) (Mansour et al. 2022b) and macroalgae species (*Ulva* sp., *Undaria* sp., *Gracilaria* sp., *Sargassum* sp. *Porphyra* sp. *Ascophyllum* sp., and *Macrocystis* sp) (Rajauria 2015) have been investigated as live food in marine hatcheries, but only around 10 are regularly employed in aquaculture. As a result, microalgae for aquaculture must be chosen based on their compatibility with the target aquatic animal species as well as their properties (Brown and Robert 2002).

3.1.1 Zooplankton

Microalgae are the primary sources of nutrition for several wild zooplankton species in aquatic habitats (Alprol et al. 2021b). These zooplankton species, including rotifers, amphipods, copepods, and daphnia, are sequentially consumed as live feeds by crustaceans, fish, and their larvae (Alprol et al. 2021b). In saline hatcheries, microalgae are cultivated to feed zooplankton such as artemia, rotifers, and copepods, which are then used as food for fish and shellfish. Artemia, rotifers, and copepods are the most worthy zooplankton live feed species used in aquatic hatcheries (El-Gamal et al. 2020), with *Brachionus rotundiformis* and *B. plicatilis* being the two global species of rotifer (Navarro and Sarasquete 1998). The most widely and commercially used species of artemia for the rearing of marine larvae are *Artemia franciscana* and *A. salina* [96], while several copepod species, such as *Oithona nana, O. rigida, Acartia* spp., *Temora stylifera Bestiolina* sp., *Paracalanus pas*, and *Nannocalanus minor*, are commercially produced (Altaff and Vijayaraj 2021; Ananth and Santhanam 2019; Camus and Zeng 2010; Ianora et al. 2011; Jothiraj and Santhanam 2019; Magouz et al. 2021; Santhanam et al. 2019; Schipp et al. 1999). Although all single cell microalgae species are suitable for the feeding and rearing of artemia, copepods, and rotifers, microalgae strains vary in their importance for aquatic animal feed due to various characteristics, such as their morphological charachters, cell wall composition, cell size, digestibleness, shapes, and nutritious values (Ferreira et al. 2009). Another study used four microalgae as live feed for artemia to enhance the quality of *Solea aegyptiaca*, with *Nannochloropsis salina* found to be the most effective in improving the artemia's EPA and DHA content. This, in turn, resulted in an improvement in the quality of *S. aegyptiaca*, which showed significant increases in various parameters and had the highest ARA, EPA, and DHA content (El-Khodary et al. 2021). Additionally, various feeding protocols can be used to enhance the biochemical composition, such as increasing the content of fatty acids especially, EPA, and DHA, of zooplanktonic species by microalgae before using them as prey (Navarro and Sarasquete 1998; Palmtag et al. 2006; Seychelles et al. 2009).

3.1.2 Shrimp

According to recent reports, there has been a global increase in the demand for shrimp due to their nutritional value. The Pacific *Litopenaeus vannamei* is the most widely hatched and cultivated shrimp species (producing around 52.91×10^3 tons in 2018) followed by *Penaeus monodon* (producing around 7.99×10^3 tons in 2018) (FAO 2020). During larval stages, all shrimp species still rely on live diatoms and microalgae as essential live feeds. Among the microalgae groups, diatoms are the most appropriate for shrimp larval feeding due to their physiological and nutritional aspects, such as small cell size, fast growth, and high content of EPA, DHA, and PUFA (Sivakumar et al. 2011). Some of the extensively used diatom species in shrimp larval culture include *Chaetoceros* sp. (Rohani-Ghadikolaei et al. 2015), *Thalassiosira weissflogii* (El-Khodary et al. 2021), *Detonula* sp., *Phaeodactylum sp., Navicula* sp. (Sivakumar et al. 2011), *Tisochrysis Isochrysis* (Rohani-Ghadikolaei et al. 2015), and *Tetraselmis*. In one study (Rohani-Ghadikolaei et al. 2015), a mixture of *Isochrysis galbana* and *Chaetoceros muelleri* was found to be more effective for *Penaeus* indicus larvae growth and development than either species alone. Additionally, six species of macroalgae were utilized in the form of liquid extract as supplement to F/2 medium, and the results suggested that *I. galbana* and *C. muelleri* cultured with the seaweed extract supplement preserve as an substitution cost-effective method for F/2 media preparation in the manufacture of live foods for the rearing of shrimp larvae in marine hatcheries (Rohani-Ghadikolaei et al. 2015).

3.1.3 Sea Cucumbers

Sea cucumbers are highly valuable worldwide, largely due to their nutritional and medicinal properties (Bordbar et al. 2011). There is global interest in developing feeding protocols that can enhance the growth, survival rate, settlement, metamorphosis, and nutritional value of sea cucumber larvae, juveniles, and adults of various commercially important species. To achieve this goal, several microalgae species have been used as live feed diets, including *Chaetoceros* sp., *Isochrysis* sp., *Tetraselmis* sp., *Dunaliella* sp., *Phaeodactilum* sp., *Dicrateria* sp., *Arthrospira* sp., *Rhodomonas* sp., *Nitzschia sp.*, and *Cylindrotheca sp*. The use of these species has been shown to significantly increase the survival rate, enhance metamorphosis, and improve the nutritious value and caloritic content during all stages of sea cucumber species, including red sea cucumber (*Parastichopus tremulus*), sandfish (*Holothuria scabra*), California sea cucumber (*P. californicus*), and Selenka (*Apostichopus japonicus*) (Abdelaty et al. 2021; Ren et al. 2016; Shi et al. 2013; Sibonga et al. 2021). However, among all the microalgae species studied, the marine diatom *Chaetoceros calcitrans* was found to be the most effective in achieving significant larval development, survival, growth, and metamorphosis (Ren et al. 2016; Schagerström et al. 2021; Sibonga et al. 2021).

3.1.4 Sea Urchins and Seahorses

Sea urchins and seahorses are not only important as food products, but also possess various nutritional, pharmacologic, and therapeutic properties, and play a significant role in environmental ecosystems. However, overfishing has caused a decline in their populations, highlighting the need for their hatching, larval rearing, and aquaculture. Various microalgal species, including *Isochrysis* sp., *Chaetoceros gracilis*, *C. muelleri*, and *Pavlova lutheri*, have been identified as suitable live feeds for numerous sea urchin species, including *Paracentrotus lividus* and *Tripneustes gratilla* (Carboni et al. 2012; Scholtz et al. 2013; Shigai et al. 2020). Moreover, studies have reported that feeding Longsnout seahorse (*Hippocampus reidi*) juveniles with *Nanochlorpsis oculata* or *Isochrysis galbana* resulted in higher ingestion and survival rates, as well as growth performances. Thus, these findings emphasize the importance of utilizing suitable feeding protocols to enhance the growth and survival of these valuable aquatic animals (Kasimala et al. 2020; Koldewey 2005; Olivotto et al. 2008; Pham and Lin 2013; Willadino et al. 2012).

3.1.5 Shellfish

Juvenile shellfish, such as oysters, mussels, and clams, are grown in a controlled environment before being transferred to the open sea for further growth and maturation. Microalgae play important roles in spat production by providing essential nutrients and enhancing water quality. The fatty acid profile of the spat is highly dependent on the microalgal diet supplied, indicating that microalgae can be used to manipulate the fatty acid profile of shellfish *Venerupis pullastra* (Albentosa et al. 1994). Major differences were observed in several fatty acids. The absence of 22:6n-3 in *Tetraselmis suecica* cells and the spat fed on this microalga may explain the lower growth rate obtained with this diet. This suggests that the fatty acid composition of microalgae plays a crucial role in the growth and development of shellfish (Albentosa et al. 1994). Another study investigated the effects of two microalgae species, *Isochrysis galbana* and *Tetraselmis suecica*, on the growth and biochemical composition of clam spat *Venerupis pullastra*. Clam spat fed on mixed or *I. galbana* diets had higher growth rates than those fed on *T. suecica* diet due to differences in ingestion rates, digestibility, and dietary protein and lipid content. The study found that the spat's biochemical composition was correlated with growth indices. The study concluded that the nutritional value of microalgae to bivalves is not only determined by biochemical composition but also correlated to the ingestion and digestibility of microalgae by larvae (Albentosa et al. 1993).

Aranda-Burgos et al. (2014) examined the nutritional requirements and developmental stages of the clam *Ruditapes decussatus* by studying its survival, growth, and biochemical changes. The study finds that the type of microalgal diet, specifically *Isochrysis galbana*, *Tetraselmis suecica*, and *Chaetoceros calcitrans*, affects larval survival and growth. The study also shows that the protein content is positively correlated with larval growth, while the lipid content is negatively correlated. In addition, the fatty acid composition of larvae reflected that of the microalgae, with higher levels of EPA and DHA in larvae fed on *I. galbana* and *T. suecica*.

Another study examined the effects of broodstock and larval diets on the hatchery production of European flat oyster. Feeding broodstock with mixed diets of *Rhodomonas salina* and *Chaetoceros neogracile* or *R. salina* and *Thalassiosira weissflogii* produced more and faster-growing larvae than unfed broodstock. Larval growth, survival, and settlement were influenced by both broodstock and larval diets. The best results were obtained by larvae from fed broodstock receiving a bi-specific diet of *Tisochrysis lutea* and *C. neogracile*, while the worst results were obtained by larvae from any broodstock receiving a single diet of *Diacronema lutheri*.

3.1.6 Fish

Farmers are reducing the use of fish-based feed and supplements for fish, and instead turning to seaweed-based feed which is cost-effective and protein-rich. Seaweed is not only attractive to fish, but also helps increase triglyceride and protein deposition in their muscles, resulting in weight gain. Many studies have been conducted on the use of seaweed as a fish feed additive, with various species including *Ulva* sp., *Undaria* sp., *Ascophyllum* sp., *Laminaria* sp., *Sargassum* sp., *Porphyra* sp., and *Gracilaria* sp. being widely researched for their potential use in fish diets (Rajauria 2015). According to Van Alstyne et al. (2001), feeding *Ulva fasciata, Sargassum wightii*, and *Spyridia insignis* seaweed to rohu fish resulted in increased protein efficiency ratio, feed conversion ratios, food absorption efficiency, and nutrient digestibility. This improvement is believed to be due to the abundance of the dimethyl sulfonyl propionate and dimethyl-beta-propionthein compound in green seaweed, particularly in the Ulvales category.

3.2 Dried Microalgae

In the field of aquaculture, algae have been long applied as living feeds for various aquatic organisms, and improving the quality and quantity of cultured zooplankton is vital. However, microalgae production is unpredictable, which makes it a critical point in marine hatcheries. Live feeds account for $> 50\%$ of the total production costs (Oostlander et al. 2020), prompting researchers to look for alternative food sources such as micro-particulate diets (Medina-Reyna et al. 2005), and micro-encapsulated or inert food, as well as microalgae paste and biomass concentrates or using lipid-free biomass after biodiesel production (Raja et al. 2018). Dried microalgae can be incorporated into aquafeeds as powders or pellets (Amjad and Jones 1992), providing a concentrated source of nutrients and bioactive compounds.

Different preservation techniques are used for microalgae concentrates (Brown and Robert 2002). Drying techniques and feed extrusion are both commonly used methods in the production of aquaculture feeds. Each method has its pros and cons, and the choice of method will depend on the type of feed, the ingredients used, and the desired end product (Delgado and Reyes-Jaquez 2018). Drying techniques include air-drying, sun-drying, and freeze-drying. Air-drying and sun-drying are the most traditional methods and are often used for low-value feeds such as tilapia or carp. They are relatively low-cost and simple, but they may not be suitable for all types of feed, as they can result in nutrient loss and variable quality. Freeze-drying, on the other hand, is a more expensive method but results in better nutrient retention and quality (Ansari et al. 2021; Buitrago 1992).

Feed extrusion is a process that involves heating and pressing a mixture of ingredients through a die to form pellets. This method is commonly used in the production of high-value aquaculture feeds as it results in a consistent and uniform product. It also improves feed digestibility and can increase nutrient retention. However, the high temperatures and pressures used in extrusion can result in the denaturation of bioactive compounds such as pigments and enzymes, which can reduce their efficacy (Maehre et al. 2016; Shi et al. 2016).

In the aquaculture feed processing market, hot extrusion is still the most widely used method due to its ability to disrupt algae cell walls to produce a high-quality, stable product with good nutrient retention to consume more nutrients (Gong et al. 2018). A study conducted on the salmon *Salmo salar* showed that the extrusion processed feed had higher digestibility of ash and dry matter when *Nannochloropsis* sp. biomass was used as feed compared to non-extrusion processed feed (Gong et al. 2018). Similarly, in a study conducted on gibel carp (*Carassius gibelio*), extruded feeds were found to have higher protein and dry matter digestibility compared to pelleted feeds (Shi et al. 2016). However, research has shown that some bioactive compounds in microalgae may be lost during hot extrusion. Lipid-rich microalgae may act as a lubricant in the extruder barrel, reducing viscous heat dissipation and lowering pellet quality (Samuelsen et al. 2018). Thus, the recommended lipid content for making fish feed pellets through an extruder is 12% dryweight (Rokey 1994). If high-lipid feed (>30%) is required, such as in the case of salmon feed, then oil coating on the dried pellet using a vacuum coating process is necessary (Rokey 1994). However, the high pressure and temperature in the extrusion process may degrade the functional compounds of microalgae, so some manufacturers are exploring alternative processing methods such as cold extrusion or cold pelleting supplementation to improve the retention of bioactive compounds. Overall, the choice of processing method will depend on the specific requirements of the feed and the desired end product. While hot extrusion remains the dominant method in the aquaculture feed market, there is increasing interest in exploring alternative methods to better preserve the bioactive compounds in microalgae and other feed ingredients.

When collected, dried, and stored appropriately, dried microalgae can effectively substitute for live microalgae in raising aquatic larval and juvenile stages, molluscs, abalone, adult bivalves, shrimp larvae, and various zooplankton species in marine hatcheries, either partially or entirely (Knuckey et al. 2006; Nunes et al. 2009). Microalgae concentrates have shown promising results as a replacement for traditional live feed supply in marine hatcheries. The main advantage of dried microalgae is their longer shelf life compared to live microalgae, which makes them easier to transport and store. Moreover, converting microalgae from live feeds to microalgae-dried biomass eliminates ciliated protozoa, which can contaminate and harm various marine larvae (Raja et al. 2018). In recent times, different forms of microalgae including fresh biomass, dried biomass, and lipid-free biomass have been found to have potential applications for various zooplankton species including rotifer (*B. plicatilis*), copepod species such as *Oithona nana, Acartia sinjiensis* and artemia (*A. franciscana*) (Knuckey et al. 2005), *Pseudodiaptomus euryhalinus* (Puello-Cruz et al. 2009), or *Cyclopina kasignete* (Alprol et al. 2021a; Ashour et al. 2021). Several microalgal species, such as *I. galbana, P. lutheri, C. calcitrans*, *Tetraselmis* sp., and *C. muelleri*, have been collected and stored for 40-60 days, and studies have shown that their preserved biomass achieved growth and survival rates comparable to those of the same algal species in live feeds for the larvae and juvenile of *S. glomerata* (Heasman et al. 2000). Similarly, Ponis et al. (2008) found no substantial differences in the development or persistence between Pacific oyster larvae reared on fresh or dried biomass of *P. lutheri*. Marketable dried feed of algal species, such as *Isochrysis* 1800®, Shellfish Diet 1800® composed of different *Isochrysis* and *Tetraselmis* sp. that have been approved to provide analogous results to live feeds usually used in the commercial manufacture of larvae and juveniles of sandfish (*H. scabra*).

Dobberfuhl and Elser (1999) conducted a study where they used dried *Scenedesmus acutus* as a food source for *Daphnia magna* and found that the growth of *D. magna* was the same on both fresh and dried food. They suggested that using dried algae could be a useful method for studying food quality and nutrient release in situations where it may not be feasible to maintain active algal cultures. Yousef and Hegab (2017) conducted a study where they used dried *Chlorella vulgaris* as a food source for brine shrimp *(Heterocypris salina)* and discovered that they had similar survival rates, growth rates, and biochemical composition as those fed with live algae. They suggested that dried algae may be a viable alternative for aquaculture applications. A book chapter by Guedes et al. (2015) reviewed the use of microalgae protein in aquafeed and noted that microalgae are an important food source for zooplankton, as critical microalgal nutrients are transferred to higher trophic levels via zooplankton intermediates.

On the other hand, seaweeds also could be used as dried feed. Recently, a study was conducted on the potential of Shellfish Diet 1800® and the dried powder of macroalgae species including *Ascophyllum nodosum* and *Saccharina latissima* as dried feed for sea cucumber (*Cucumaria frondosa*). Their findings suggested that the physiological traits of *C. frondosa* were comparable when fed dried microalgae and macroalgae and that dried macroalgae is a favorable feed source for aquaculture of *C. frondosa. Sargassum* sp. is an abundant source of vitamins, minerals, carbohydrates, and essential amino acids and lacks antineutrino substances, making it a good alternative feed source. Livestock can digest the dry matter of seaweed between 28 and 67%, while protein digestibility can reach up to 95%, according to some reports (Casas-Valdez et al. 2006).

Collecting (dewatering) microalgae remains a tricky process due to various factors (Muhammad et al. 2020). Since microalgae cells are very tiny and carry a negative charge, they are not easily separated from the culture medium by gravity or filtration (Chen et al. 2015). The process of harvesting and dewatering microalgae necessitates a significant amount of time and energy, which raises the production costs and environmental impact of microalgae-based products (Musa et al. 2019). Furthermore, the methods employed to harvest and dewater microalgae may impact the quality and viability of the microalgae biomass, reducing its worth and applicability for various applications (Musa et al. 2020; Vrasna et al. 2022). So, more research needs to be conducted to develop effective and sustainable microalgae harvesting and dewatering techniques that are both cost-effective and environmentally friendly. On the other hand, seaweed has a lower surface-to-volume ratio, which makes dewatering relatively easier. Seaweed can be air-dried or sun-dried, which is a simpler and less energy-intensive process compared to microalgae dewatering.

3.3 Combined Algae Diets

One of the challenges of using microalgae in aquaculture is to preserve their nutritional quality after dewatering and drying processes, which can cause degradation of essential fatty acids and other compounds. Another challenge is to match the nutritional composition of microalgae with the specific requirements of different aquatic species, which may vary in their preferences and needs. Therefore, combining algae diet may be a promising strategy to overcome these challenges and enhance the performance and health of aquaculture organisms. In this context, Sarker et al. (2020) conducted an experiment that used two types of microalgae, *Nannochloropsis oculata* and *Schizochytrium* sp., to replace fishmeal and fish oil in tilapia (*Oreochromis niloticus*) feed. The results showed that the microalgae-based feed improved the growth, weight gain, protein digestibility, and DHA content of tilapia compared to a conventional feed.

The study investigated the use of seaweeds *Gracilaria gracilis* and microalga *Nannochloropsis oceanica*, either alone or blended, in the diet of European seabass (*Dicentrarchus labrax*). The fish were fed with four different diets for 106 days, including a control diet, a diet with 8% *G. gracilis*, a diet with 8% *N. oceanica*, and a blend of 4% of each algae. The results showed that all fish had similar growth and feed intake (Batista et al. 2020), while the inclusion of *N. oceanica* did not affect nutrient utilization. The use of *G. gracilis* resulted in higher nitrogen and energy retention efficiency.

White sea urchin (*T. gratilla*) juveniles were raised for 3 weeks and fed with different seaweed species including *Ulva pertusa, Undaria pinnatifida,* and *Gloiopeltis furcata* and a mixture of them (1:1:1). The study revealed that specific growth rates, feed conversion efficiencies, and fatty acid profiles varied among the different diets, which could be attributed to the varying FAs profiles of the seaweeds used (Floreto et al. 1996). The study by Lyons and Scheibling (2007) reported that a combined diet consisting of *Codium fragile* and *Laminaria longicruris* may be the optimal feeding approach in terms of food utilization and feeding rate for the green sea urchin *Strongylocentrotus droebachiensis*.

3.4 Algae as a Substitute for Traditional Diet Ingredients in Aquafeed

Aquatic feed manufacturers are looking to decrease their use of food fish by partly replacing with plant ingredients due to the drawbacks of fishmeal and fish oil. Fishmeal costs have indeed increased due to competition from other industries for its use in animal feed and human nutritional supplements (Mabrouk et al. 2022; Sarker et al. 2018). The form and level of inclusion are the most limiting factors for incorporating aquatic plants, such as microalgae and seaweeds, into aquatic animal diets (Ashour et al. 2020; Sharawy et al. 2022). Microalgae have superior potential for aquaculture owing to their higher nutritious value, high growth rate, and contents of antimicrobial antioxidants, probiotics, and coloring compounds, all of which are vital for the growth and health of aquatic animals (Nagappan et al. 2021; Niccolai et al. 2019). However, the consumption of microalgae as feed additives and/or fishmeal replacements is restricted by several factors such as availability, digestability, cost, the mass production, and overall food value (Nagappan et al. 2021; Skrede et al. 2011).

Table 1 shows the effect of different microalgae as a portion of feed on the growth of various fish species at different stages of development. Substitute fish protein hydrolysate of *Carassius auratus* with 25% *Isocrysis galbana* showed almost 100% survival rates similar to control with no significant differences between them. However, the survival rate of aquatic species fed with diets where microalgal biomass replaced 100% of fish protein hydrolysate (MP100) or vitamin premix (MV) was 78% and 66%, respectively (Coutinho et al. 2006). A 90-day culture trial was conducted to study the effects of *Spirulina platensis* meal on the growth and body composition of two Indian major carps, catla (*Catla catla*) and rohu (*Labeo rohita*). *Spirulina* meal was used to replace fish meal protein in four different diets at 25%, 50%, 75%, and 100% levels. The final weight of catla was not affected by *Spirulina* meal at any level, but rohu's growth was significantly better when fish meal was replaced by more than 25% *Spirulina* meal (Coutinho et al. 2006). Juvenile Atlantic cod *Gadus morhua* were fed with isonitrogenous, isocaloric diets that contained a mixture of dried *Nannochloropsis* sp. and *Isochrysis* sp. as a substitute for fish meal protein. Three diets (55% protein, 16.5% fat, and calculated gross energy = 5328 kcal/kg) were prepared to replace 0, 15, or 30 % of fish meal protein in the diets. After the feeding period, the survival, feed conversion ratios, viscerosomatic indices, and n-3 and n-6 fatty acids in the muscle of the fish did not differ among the groups that received different diets (Walker and Berlinsky 2011).

Table 1 Different microalgae as a portion of feed on the growth of various fish species at different stages of development

Fish species	Scientific name	Stage of development	Microalgae	Portion of feed (%)	Growth	References
Goldfish	*Carassius auratus*	Larvae	*Isocrysis galbana*	25	sign. Neg.	Coutinho et al. (2006)
Goldfish	*Carassius auratus*	Larvae	*Isocrysis galbana*	100	sign. Neg.	Coutinho et al. (2006)
Carp	*Catla*	Juvenile	*Spirulina platensis*	25-100%	no significant	Nandeesha et al. (2001)
Carp	*Labeo rohita*	Juvenile	*Spirulina platensis*	25-100%	sign. Pos.	Nandeesha et al. (2001)
Siberian sturgeon	*Acipenser baeri*	Juvenile	*Spirulina platensis*	50%	sign. Pos.	Palmegiano et al. (2005)
Tilapia	*Oreochromis mossambicus*	Juvenile	*Spirulina maxima*	20-40%	no significant	Olvera-Novoa et al. (1998)
Tilapia	*Oreochromis mossambicus*	Juvenile	*Spirulina maxima*	>40%	sign. Neg.	Olvera-Novoa et al. (1998)
Atl. Cod	*Gadus morhua*	Juvenile	*Nannochloropsis+I.* sp.	15-30%	sign. Neg.	Walker and Berlinsky (2011)
Tilapia	*Oreochromis niloticus*	Juvenile	*Desmochloris* sp.	> 50%	sign. Pos.	García-Ortega et al. (2015)
Atl. Salmon	*Salmo salar* L.	Adult	*Nanofrustulum+ Tetraselmis*	≤ 10%	no significant	Kiron et al. (2012)
Carp	*Cyprinus carpio* L.	Juvenile	*Nanofrustulum+ Tetraselmis*	≤ 10%	no significant	Kiron et al. (2012)
Rainbow trout	*Oncorhynchus mykiss*	Juvenile	*Spirulina* sp	10%	sign. Pos.	Sirakov et al. (2012)
Siberian sturgeon	Acipenser baeri	Juvenile	*Spirulina* sp	50%	sign. Pos.	Palmegiano et al. (2005)
Shrimp	*Litopenaeus vannamei* Boone	Juvenile	*Nanofrustulum+ Tetraselmis*	≤ 10%	no significant	Kiron et al. (2012)

The other researches summerized in Table 1 indicate that some microalgae have a significant positive effect on growth, such as *Spirulina platensis* on *Labeo rohita* (rohu) and *Acipenser baeri* (Siberian sturgeon). On the other hand, some microalgae have a significant negative effect on growth, such as *Isocrysis galbana* on goldfish and *Spirulina maxima* on *Oreochromis mossambicus*. Some microalgae showed no significant effect on growth, such as nanofrustulum and tetraselmis on Atlantic salmon and *Litopenaeus vannamei* Boone. Accordingly, algae are considered a promising nutritional substitute for fish meal due to their ability to improve nutrification value, survival rates, and growth performance, or to provide similar results as fish meal under less favorable conditions.

Earlier studies have found significant increases in growth criteria, immune responses, and zooplankton community in Nile tilapia *O. niloticus* braved with *A. hydrophila* when utilizing seaweed liquid extract as an aqua-feed additive. Due to the presence of heavy metals and anti-nutritional components such as lectins, phlorotannins, and phytic acids, utilizing large amounts of seaweed as feed additions can, however, result in reduced digestibility (Ashour et al. 2020). In another study, several seaweed species, including *S. linearifolium, Gracilaria sp., Ecklonia radiata, U. lactuca,* and *Lophocladia kuetzingii* were used as increasing dry feed supplements for the sea urchin *T. gratilla*. These seaweeds additives substantially improved the growth criteria, feed intake, and protein utilization of *T. gratilla* (Dworjanyn et al. 2007). Additionally, dried seaweed, such as *Solieria robusta, U. lactuca*, and *Sargassum* spp., were used as single or combined feed additives at different protein and lipid contents to improve the gonad indices of sea urchin *Heliocidaris erythrogramma* (Warren-Myers et al. 2021).

3.5 Algae Assisted Aquaculture

Algae as an aquaculture 'conditioner' is a method that employs microalgae, seaweeds, or their extracts and bioactive compounds as water conditioner, resulting in better growth parameters, bacterial pathogens control, elevated disease resistance, enhanced feed, and stimulus of the immunogenicity of cultured aquatic species (Mansour et al. 2022b). The practice of growing aquatic animals and its larvae combined microalgae is known as "microalgae assisted aquaculture", and it has been linked to greater persistence and growth rates than larvae maintained in clear water (Navarro and Sarasquete 1998; Navarro et al. 2001). Microalgae assisted aquaculture is used to regulate the aquaculture growing environment (Yang et al. 2020). Several microalgae, yeast, bacteria, and zooplankton were cultivated in ponds where fish larvae were reared in this system. This approach may be based on natural microalgal populations that are encouraged to develop using fertilizer, or cultivated microalgae species can be introduced into culture ponds if the water system has been treated previously to eliminate competing microorganisms (Shields and Lupatsch 2012). Numerous researchers have stated that the improved growth and survival rates in this technique are primarily due to improved direct and indirect feeding of larvae, lower stress levels, improved feeding environmental conditions by light, increasing turbidity, improved oxygenation rates, boosted visual contrast, and increased antiseptic properties in rearing tanks (Yang et al. 2020).

There are many processes linked with green water's lucrative and therapeutic effects, such as the creation of bioactive chemicals by microalgal cells, which include antibacterial and antioxidant molecules that inhibit virulence genes (Kokou et al. 2012; Natrah et al. 2014). Some of the most important microalgal species applied for this application are *Isochrysis* sp., *I. galbana*, *Nannochloropsis* sp., *Chlorella* sp., and *Tetraselmis sp.* (Kaparapu 2018; Tendencia, and dela Peña 2003).

Artemia franciscana metanauplii fed on different microalgae species had different growth rates and fatty acid compositions, and that these affected the survival and growth of Pikeperch (*Sander lucioperca*) (Turcihan et al. (2021). The best results were obtained by using a mixed diet of *Tisochrysis lutea* and *Chaetoceros neogracile* for both *Artemia* and Pikeperch larvae (Turcihan et al. 2021). A study by Chen et al. (2020) showed that adding *Nannochloropsis oculata* to the rearing water of clownfish (*Amphiprion ocellaris*) larvae increased their survival rate from 30% to 70%, and their growth rate from 0.13 mm day^{-1} to 0.18 mm day^{-1}. The authors suggested that microalgae improved the water quality, enhanced the nutritional value of rotifers, and stimulated the feeding behavior of clownfish larvae (Chen et al. 2020). Adding *Isochrysis galbana* to the rearing water of Atlantic cod (*Gadus morhua*) larvae increased their survival rate from 10% to 40%, and their growth rate from 0.08 mm day^{-1} to 0.14 mm day^{-1} (Barakat et al. 2021). The author suggested that microalgae provided visual cues for feeding, reduced bacterial load, and increased dissolved oxygen in the water. Recent paper observed that *U. fasciata* methanol extract as a daily water additive reduced the pathogenic bacterial community and increased the survival and growth rate of sea bass larvae (*Dicentrarchus labrax*) due to its total phenoloic, flavonoids, and fatty acid content (Barakat et al. 2021).

Green water is a low-cost approach in general. Shrimp farmed in "green water," for example, costs US $1–3 per kg, but shrimp fed on typical diets costs US $4–8 per kg (Biao and Kaijin 2007; Neori 2011). Several commercial aquatic animal species have increased their survival rates and improved their fry-rearing conditions as a result of this technology, including Nile tilapia (*O. niloticus*) (Suárez-Puerto et al. 2021), hybrid tilapia (Jimenez et al. 2016), whiteleg shrimp (*P. vannamei*) (Chithambaran et al. 2017), banana prawn (*F. merguiensis*) (Palmer et al. 2007), Tiger shrimp (*P. monodon*) (Tendencia et al. 2012), dusky flathead (*P. fuscus*), (Palmer et al. 2007), Australian bass (*M. novemaculeata*), red sea bream (*P. auratus*), sand whiting fish (*S. ciliata*), and gilthead seabream (*S. aurata*) (Navarro and Sarasquete 1998; Navarro et al. 2001).

Seaweed extract, on the other hand, has lately shown remarkable promise as a water conditioner. There have been few investigations on this critical topic. Interestingly, previous research concluded that employing methanol extracts of *U. fasciata* as a daily water supplement for sea bass larvae, gave the largest decrease in harmful bacterial population and the best survival rate for sea bass larvae. Sea bass larval development was likewise greatest, probably due to the greater phenolic, flavonoid, and polyunsaturated fatty acid contents. Such nutritional advantages may have been obtained by sea bass larvae by direct absorption or via secondary transfer, by which live feed (*Brachionus* or *Artemia*) ingested by larvae collected more nutrients (El-Sayed et al. 2022).

3.6 Co-Cultivation of Algae and Integration with Aquatic Animals

Land-based aquaculture systems, like recirculating aquaculture systems (RAS), generate substantial amounts of wastewater that may contain high concentrations of nitrogen (N) and phosphate (P). If released untreated, these excess nutrients can harm the environment by causing eutrophication in nearby water bodies, leading to harmful algal blooms, oxygen depletion, and fish deaths. To address this issue, many countries have implemented regulations that mandate land-based aquaculture facilities to lower the levels of N and P in their wastewater before releasing it into the environment. Therefore, reducing N and P levels in aquaculture wastewater before release using the coculture system can help protect the environment and surrounding ecosystems from the negative impacts of nutrient pollution.

'Blue agriculture', which is the use of freshwater, marine, or brackish water resources for a sustainable economy, has the added benefits of increased production and less environmental impact (Agarwal et al. 2020). As a result of the algae's ability to consume nutrients from the wastewater effectively, waste produced in the aquaculture system may be cleaned up, and biomass can be produced to make goods with an additional value. A closed loop may be created by replacing the typical feed in the aquaculture system with the gathered biomass. Yet, an unbalanced C:N ratio and possible ammonia toxicity may be obstacles to the development of microalgae in aquaculture effluent, necessitating careful selection of the microalgal species (Agarwal et al. 2020). Aquaculturists have seen a variety of advantages from the commercial integration of algae with aquatic animals, which also allows for the production of significant quantities of additional valuable aquaculture products and has many positive effects on the aquaculture environment. Aquaculture may increase the commercial outputs of aquatic animals by minimizing environmental effects without putting further environmental pressure on already-stretched coastal resources and places (Kalasariya et al. 2016). The growth and survival rates for co-cultured aquatic animals and seaweed were consistently higher than those for commercial monoculture (Anh et al. 2020). Black tiger shrimp mortality was much higher in monoculture (75.6%) than when integrated with *Gracilaria tenuistipitata* (17.8–31.1%). Moreover, the shrimp in tanks without seaweed showed a lower growth rate compared to those in coculture tanks containing seaweed at densities of 1.0, 1.5, and 2.0 kg m^{-3} (Anh et al. 2020). By eliminating up to 80% or 90% of inorganic wastes and nitrogen, the use of seaweed in polyculture systems may help to increase the value of aquaculture products (Anh et al. 2020; Kang et al. 2011).

In co-culture, waste reduction can increase the product value in several ways. Firstly, reducing waste can improve the water quality in which the fish or shellfish are grown, which can have positive impacts on their health, growth, and overall quality. Better water quality can also result in lower mortality rates and reduced need for medication, which can increase the market value of the final product. Secondly, waste reduction can lead to more sustainable and environmentally friendly aquaculture practices, which can be attractive to consumers who are increasingly concerned about the environmental impact of the food they consume. This can result in a premium price for products that are produced using sustainable and eco-friendly methods. Thirdly, waste reduction can lead to more efficient use of resources, such as water and feed, which can result in cost savings for the aquaculture producer. This can translate into lower production costs, which can in turn increase profit margins and make the product more competitive in the market. Overall, waste reduction in aquaculture can have numerous benefits, including improved product quality, increased market value, more sustainable production practices, and cost savings. This has a favorable effect on resource usage efficiency (Table 2).

Seaweeds may be used into integrated multi-trophic aquaculture (IMTA) to solve several environmental problems in aquaculture, instead of just producing seaweed as a single product (Chávez-Crooker and Obreque-Contreras 2010; Granada et al. 2016). Breeding animals from different trophic levels that are near to one another may involve the IMTA model. Because of this, the waste materials (both organic and inorganic) from one grown species are reused as food inputs for the others (Namvar et al. 2012). In this type of aquaculture, known as integrated multi-trophic aquaculture (IMTA), chemical fertilizers are not necessary to promote seaweed growth. This leads to a more sustainable and profitable system, as there is no additional cost for fertilizers and no excess nutrients are released into the environment (García-Poza et al. 2020). Although IMTA has been touted as a promising solution for reducing the environmental impact of aquaculture, there are several challenges that have hindered its widespread commercial adoption. For example, there are technical and operational challenges associated with managing multiple species and their interactions in the same system. Additionally, there may be regulatory and permitting challenges associated with the co-culture of different species in the same location. The economics of IMTA can also be complex, as the costs and benefits of different species must be carefully balanced to ensure profitability. Finally, there may be a lack of knowledge and experience among aquaculture operators regarding the management of IMTA systems. While there are certainly hurdles to overcome, the potential benefits of IMTA make it an area of ongoing research and development in the aquaculture industry. The nutrients in IMTA systems mainly come from the fecal waste and urine of the aquatic animals rather than from their metabolic waste, such as ammonia and phosphates. These nutrients are then utilized by lower trophic organisms, such as seaweed or microalgae, to produce biomass. The conversion of nutrients into biomass by lower trophic organisms helps to reduce the environmental impact of aquaculture by removing excess nutrients from the water and preventing

Table 2 Macro- and microalgal remediation of aquaculture effluents

Algae	Mode of cultivation	Remediation	Application	References
Microalgae				
Chlorella vulgaris	Aquaculture hatcheries	Phosphate in the effluent is completely consumed; after 23 days	Algae is used as a protein supply for Coho salmon (fish)	Saejung and Ektasaeng (2022)
Parachlorella kessleri	Aquaculture hatcheries	96.2% of ammonium, 94.4% of COD, 94.3% of nitrate, 99% of nitrite, and 95.6% of phosphate were removed after 3 days of incubation using 100 mg L^{-1} of algal material	Using bioflocculation as a green approach for treating aquaculture effluent	Liu et al. (2019)
Spirulina sp.	Aquaculture hatcheries	99.8% of the phosphate was removed, 89.34% of the COD, and 81.10% of the nitrate were removed. The algal biomass had a significant lipid (12.8%) and carbohydrate (69.77%) content	The consumption of biomass as food and feed is made possible by the high level of omega-3 fatty acids and the lack of heavy metals	Cardoso et al. (2020)
Macroalgae				
Gracilaria verrucosa	Open water systems	Remove nutrients from Mediterranean mussels *Mytilus galloprovincialis*	Development and survival, and serve as a natural food supply for Mediterranean mussels *Mytilus galloprovincialis*	Alfaro et al. (2004)
Ulva lactuca	Experimental tanks (black-plastic 0.64 m^{-2} (0.15 m^{-3}) tanks)	Improved settleable solids by 34.2 %. Decreased total suspended particles by 12.9%, nitrite-nitrogen by 72.8%, phosphate by 24.6%, and total ammonia nitrogen by 25.9% in the water	Gross mass of Pacific white shrimp *Litopenaeus vannamei* by 6.9 %	Brito et al. (2014)
Kappaphycus striatum	Tropical lagoon	Not detected	Improved the growth performance for *Holothuria scabra*	Kang et al. (2011).

eutrophication, while maintaining steady levels of oxygen, pH, and CO_2 (Abreu et al. 2011; Tanaka et al. 2020; Zheng et al. 2019). According to the most recent research, the IMTA system generated more seaweed biomass than either wild or monoculture seaweed. Moreover, IMTA-based seaweeds are described as a source of enhanced protein profiles, medicinal chemicals, and technologically useful elements (Machado et al. 2020).

3.7 Nanoparticle Forms

Algal nanoparticles have recently gained attention as an innovative and sustainable technology for encapsulating bioactive compounds derived from microalgae in nanoparticles for targeted delivery in aquaculture systems (Becker 2013). When using microalgae as a carrier for encapsulating different compounds, the composition of the microalgae can be important because it can affect the stability, release rate, and bioavailability of the encapsulated compounds. The physical and chemical properties of the microalgae, such as the surface charge and hydrophobicity, can also impact the encapsulation efficiency and release kinetics of the encapsulated compounds (Garti and McClements 2012). Furthermore, targeted delivery can be achieved by modifying the surface of the nanoparticles with specific ligands or receptors that recognize and bind to target cells or tissues in aquatic animals (Abu-Dief et al. 2022).

Several methods have been developed to synthesize algal nanoparticles, including emulsification, nanoprecipitation, and electrospraying. These methods vary in terms of their complexity, cost, and scalability (Abualnaja et al. 2021; Shera and Banik 2022).

- Emulsification: This method involves the formation of oil-in-water or water-in-oil emulsions using algal extracts and metal salts as the dispersed and continuous phases, respectively. The emulsions are then subjected to high shear forces or ultrasonication to form nanodroplets that act as microreactors for the reduction of metal ions and the formation of nanoparticles. This method is simple, fast, and versatile, but it requires the use of organic solvents and surfactants that may affect the stability and purity of the nanoparticles (Shera and Banik 2022).
- Nanoprecipitation: This method involves the addition of a water-miscible organic solvent containing algal extracts and metal salts to an aqueous solution under stirring. The solvent diffuses into the aqueous phase, creating a supersaturation condition that leads to the nucleation and growth of nanoparticles. This method is easy, rapid, and scalable, but it also requires the use of organic solvents that may pose environmental and health risks (Shera and Banik 2022).
- Electrospraying: This method involves the application of a high voltage to a liquid jet containing algal extracts and metal salts that is sprayed through a nozzle. The jet breaks up into fine droplets that are charged and evaporate in a heated chamber. The remaining solid particles are collected on a grounded substrate.

This method is precise, controllable, and energy-efficient, but it requires sophisticated equipment and parameters that may affect the size and morphology of the nanoparticles (Shera and Banik 2022).

Based on these methods, electrospraying seems to be the most promising in terms of yield, cost, and effectiveness, as it can produce uniform and stable nanoparticles with high purity and low energy consumption. However, they all share the common goal of producing nanoparticles with optimal particle size, shape, and surface properties for targeted delivery (Khan et al. 2022). There is a growing body of evidence supporting the efficacy of algal nanoparticles in aquaculture systems. For example, microalgal-derived astaxanthin nanoparticles were shown to enhance the growth, survival, and antioxidant capacity of the white shrimp (*L. vannamei*) (Wang et al. 2022). Despite the promising potential of algal nanoparticles in aquaculture, there are several challenges that need to be addressed. These include the optimization of nanoparticle synthesis methods, assessment of the potential ecological risks associated with nanoparticle release, and evaluation of the long-term effects of nanoparticle exposure on aquatic organisms. Moreover, research is needed to identify and validate the most effective bioactive compounds from microalgae and develop strategies for their targeted delivery and controlled release.

4 Challenges and Future Prospects of Using Algae in Aquaculture

Aquaculture plays a crucial role in providing sustainable seafood for the growing global population. The use of algae as a feed source in aquaculture has garnered significant attention due to its potential to reduce environmental impacts and decrease reliance on fishmeal and fish oil. However, there are challenges to overcome and prospects to consider for the successful integration of algae in aquaculture systems. This book chapter discusses the economic viability and scale-up challenges, regulatory and market acceptance issues, and future research and development directions.

The economic viability of using algae as a feedstock in aquaculture depends on the cost of production and the ability to scale up production to meet industry demand. Currently, the cost of producing microalgae is higher than traditional feed sources, such as fishmeal and soybean meal (Venkata Subhash et al. 2020). As shown in the table, the production cost of using algae in aquaculture can be reach to \$50.5 kg^{-1} using Photobioreactors greenhouse compared to \$0.55 kg^{-1} fishmeal (Ceyhan and Emir 2015). To improve economic viability, there is a need to develop cost-effective and energy-efficient cultivation, harvesting, and processing techniques (Saravana et al. 2022). Table 3 provides a comparison of different types of microalgae cultivation systems based on their plant size, culture medium, and production costs per kilogram of microalgae. The cultivation systems compared include open raceway systems, thin layer cascade systems, and photobioreactors in

Table 3 Production costs of microalgal biomass by system, plant size, and location

System Type	Plant size	Culture medium	Production costs	References
Open raceway	1 ha	Conventional	12.9 \$ kg^{-1}	Sui et al. (2020)
Open raceway	5 ha	Conventional	5.3 \$ kg^{-1}	Fernández et al. (2019)
Open raceway	5 ha	Wastestream	1,38 \$ kg^{-1}	Fernández et al. (2019)
Open raceway	405 ha	Conventional	0.65 \$ kg^{-1}	Hoffman et al. (2017)
Open raceway turf system	405 ha	Conventional	0.49 \$ kg^{-1}	Hoffman et al. (2017)
Thin layer cascade	5 ha	Conventional	2,65 \$ kg^{-1}	Fernández et al. (2019)
Thin layer cascade	5 ha	Wastestream	0,69 \$ kg^{-1}	Fernández et al. (2019)
Photobioreactors greenhouse	1500 m^2	Conventional	50.5 \$ kg^{-1}	Oostlander et al. (2020)

greenhouses. The table shows that the use of wastestream as a culture medium can significantly reduce production costs in open raceway and thin layer cascade systems. Scaling-up algae production is another challenge that must be addressed to meet the demand in aquaculture. Large-scale cultivation systems, such as open ponds and photobioreactors, need to be optimized for various factors, including light, temperature, and nutrient availability, to maximize algae productivity (El-Seesy et al. 2022). According to Table 3, larger cultivation sizes appear to lead to lower production costs per kilogram of microalgae, as seen in the open raceway turf system and the 405 ha conventional open raceway systems. Additionally, the selection of suitable algal strains with high biomass productivity and desired nutrient profiles is essential for the successful scale-up of algae production (Wijffels et al. 2010).

The regulatory framework for using algae in aquaculture feeds is still evolving. The safety and quality of algal biomass must be ensured to gain regulatory approval and market acceptance (Vizcaíno et al. 2014). For example, the presence of toxins, heavy metals, and other contaminants in algae could pose risks to human health or the environment (Falaise et al. 2016). To address these concerns, standardized cultivation and processing methods must be developed to ensure the safety and quality of algae-derived products. Market acceptance is another challenge for algae-based aquaculture feeds. Consumers may have concerns about the taste, texture, and nutritional value of seafood products raised on algae-based diets (Emerenciano et al. 2017). To promote market acceptance, it is essential to communicate the benefits of using algae, such as improved sustainability and reduced reliance on fishmeal and fish oil, to consumers and the aquaculture industry.

To overcome the challenges of using algae in aquaculture, future research should focus mainly on strain selection and optimization. It could achieved to identify and develop algal strains with high biomass productivity, desired nutrient profiles, and tolerance to various environmental conditions (Wijffels et al. 2010). The following studies are suggested in that regard:

1. Optimization of cultivation systems: Develop cost-effective and energy-efficient cultivation, harvesting, and processing techniques for large-scale algae production (Chauton et al. 2015).
2. Nutrient recycling and integrated systems: Explore opportunities to recycle nutrients from wastestreams, such as aquaculture effluents and agricultural byproducts, to enhance algal growth and reduce production costs (Molino et al. 2018).
3. Safety and quality assurance: Establish standardized methods for the cultivation and processing of algae to ensure the safety and quality of algae-derived products (Vizcaíno et al. 2014).
4. Consumer acceptance studies: Conduct studies on consumer preferences and perceptions to better understand the potential market for algae-based aquaculture products (Emerenciano et al. 2017).

5 Conclusions

The use of algae in aquaculture has the potential to address many of the sustainability concerns associated with traditional feed sources. However, several challenges need to be addressed to ensure the economic viability, regulatory approval, and market acceptance of algae-based aquaculture systems. Future research and development efforts should focus on overcoming these challenges to pave the way for the successful integration of algae in aquaculture. Different forms of algae, such as live, dried, and extracts, can enhance the benefits for aquaculture animals. Sustainable aquaculture practices integrate algae with animals using techniques like aquafeed additives, co-culturing, and integrated multi-trophic aquaculture (IMTA). These approaches aim to develop cost-effective and efficient aquaculture projects. Future research should focus on overcoming challenges and advancing algae integration in aquaculture for economic viability and market acceptance. Understanding the role of microalgae in sustainable aquaculture is crucial for improving animal growth, health, and ecological balance in aquatic ecosystems.

References

Abdelaty M, Al-Solami L, Al-Harbi M, Abu El-Regal M (2021) Utilization of different types of microalgae to improve hatcheries production of the sea cucumber *Holothuria scabra* Jaeger, 1833 in the Red Sea, Egypt. Egyptian J Aquatic Biol Fish 25(2):193–204

Abreu MH, Pereira R, Yarish C, Buschmann AH, Sousa-Pinto I (2011) IMTA with *Gracilaria vermiculophylla*: productivity and nutrient removal performance of the seaweed in a land-based pilot scale system. Aquaculture 312(1-4):77–87

Abualnaja KM, Alprol AE, Abu-Saied MA, Ashour M, Mansour AT (2021) Removing of anionic dye from aqueous solutions by adsorption using of multiwalled carbon nanotubes and poly (Acrylonitrile-styrene) impregnated with activated carbon. Sustainability 13(13):7077

Abu-Dief AM, Alsehli M, Al-Enizi A, Nafady A (2022) Recent advances in mesoporous silica nanoparticles for targeted drug delivery applications. Current Drug Delivery 19(4):436–450

Agarwal A, Mhatre A, Pandit R, Lali AM (2020) Synergistic biorefinery of Scenedesmus obliquus and Ulva lactuca in poultry manure towards sustainable bioproduct generation. Bioresour Technol 297:122462

Ahmad A, Hassan W, S., Banat, F. (2022) An overview of microalgae biomass as a sustainable aquaculture feed ingredient: food security and circular economy. Bioengineered 13(4): 9521–9547

Alam MA, Xu J-L, Wang Z (2020) Microalgae biotechnology for food, health and high value products. Springer, Berlin

Albentosa M, Perez-Camacho A, Labarta U, Beiras R, Fernàndez-Reiriz MJ (1993) Nutritional value of algal diets to clam spat Venerupis pullastra. Mar Ecol Prog Ser 97:261–269

Albentosa M, Labarta U, Pérez-Camacho A, Fernández-Reiriz MJ, Beiras R (1994) Fatty acid composition of Venerupis pullastra spat fed on different microalgae diets. Comp Biochem Physiol A Physiol 108(4):639–648

Alfaro AC, Jeffs AG, Creese RG (2004) Bottom-drifting algal/mussel spat associations along a sandy coastal region in northern New Zealand. Aquaculture 241(1–4):269–290

Alprol AE, Heneash AMM, Ashour M, Abualnaja KM, Alhashmialameer D, Mansour AT, Sharawy ZZ, Abu-Saied MA, Abomohra AE (2021a) Potential applications of *Arthrospira platensis* lipid-free biomass in bioremediation of organic dye from industrial textile effluents and its influence on marine rotifer (*Brachionus plicatilis*). Materials (Basel) 14(16):4446

Alprol AE, Heneash AMM, Soliman AM, Ashour M, Alsanie WF, Gaber A, Mansour AT (2021b) Assessment of water quality, eutrophication, and zooplankton community in lake burullus, Egypt. Diversity 13(6):268

Altaff K, Vijayaraj R (2021) Micro-algal diet for copepod culture with reference to their nutritive value–a review. Int J Cur Res Revl 13(7):86

Amjad S, Jones D (1992) An evaluation of artificial larval diets used in the culture of penaeid shrimp larvae *Penaeus monodon* (Fabricius). Pakistan J Zool 24:135–135

Ananth S, Santhanam P (2019) Intensive culture, biochemical composition analysis, and use of zooplankton tisbe sp. (Copepoda: Harpacticoida) as an alternative live feed for shrimp larviculture. In: Basic and applied zooplankton biology. Springer, Berlin, pp 329–362

Anh NTN, Vinh NH, An BNT, Lan LM, Hai TN (2020) Polyculture culture of black tiger shrimp *Penaeus monodon* and red seaweed *Gracilaria tenuistipitata* under different densities: effects on water quality, post-larvae performance and their resistance against *Vibrio parahaemolyticus*. J Appl Phycol 32(6):4333–4345

Ansari FA, Guldhe A, Gupta SK, Rawat I, Bux F (2021) Improving the feasibility of aquaculture feed by using microalgae. Environ Sci Pollut Res 28(32):43234–43257

Ansarifard F, Rajabi Islami H, Shamsaie Mehrjan M, Soltani M (2018) Effects of *Arthrospira platensis* on growth, skin color and digestive enzymes of Koi, *Cyprinus carpio*. Iran J Fish Sci 17(2):381–393

Aranda-Burgos JA, da Costa F, Nóvoa S, Ojea J, Martínez-Patiño D (2014) Effects of microalgal diet on growth, survival, biochemical and fatty acid composition of Ruditapes decussatus larvae. Aquaculture 420:38–48

Ashour M, Mabrouk MM, Ayoub HF, El-Feky MMMM, Zaki SZ, Hoseinifar SH, Rossi W, Van Doan H, El-Haroun E, Goda AMAS (2020) Effect of dietary seaweed extract supplementation on growth, feed utilization, hematological indices, and non-specific immunity of Nile Tilapia, *Oreochromis niloticus* challenged with *Aeromonas hydrophila*. J Appl Phycol 32(5):3467–3479

Ashour M, Alprol AE, Heneash AMM, Saleh H, Abualnaja KM, Alhashmialameer D, Mansour AT (2021) Ammonia bioremediation from aquaculture wastewater effluents using *Arthrospira platensis* NIOF17/003: impact of biodiesel residue and potential of ammonia-loaded biomass as rotifer feed. Materials (Basel) 14(18):5460

Awasthi MK, Sarsaiya S, Patel A, Juneja A, Singh RP, Yan B, Awasthi SK, Jain A, Liu T, Duan Y, Pandey A, Zhang Z, Taherzadeh MJ (2020) Refining biomass residues for sustainable energy and bio-products: An assessment of technology, its importance, and strategic applications in circular bio-economy. Renew Sust Energ Rev 127:109876

Barakat KM, El-Sayed HS, Khairy HM, El-Sheikh MA, Al-Rashed SA, Arif IA, Elshobary ME (2021) Effects of ocean acidification on the growth and biochemical composition of a green alga (Ulva fasciata) and its associated microbiota. Saudi J Biological Sci 28(9):5106–5114

Barkia I, Saari N, Manning SR (2019) Microalgae for high-value products towards human health and nutrition. Marine drugs 17(5):304

Barra L, Chandrasekaran R, Corato F, Brunet C (2014) The challenge of ecophysiological biodiversity for biotechnological applications of marine microalgae. Marine drugs 12(3):1641–1675

Batista S, Pereira R, Oliveira B, Baião LF, Jessen F, Tulli F, Messina M, Silva JL, Abreu H, Valente LM (2020) Exploring the potential of seaweed Gracilaria gracilis and microalga Nannochloropsis oceanica, single or blended, as natural dietary ingredients for European seabass Dicentrarchus labrax. J Appl Phycol 32:2041–2059

Becker EW (2013) Microalgae for aquaculture: nutritional aspects. In: Handbook of microalgal culture: applied phycology and biotechnology. John Wiley and Sons, Hoboken, pp 671–691

Biao X, Kaijin Y (2007) Shrimp farming in China: operating characteristics, environmental impact and perspectives. Ocean Coast Manag 50(7):538–550

Bordbar S, Anwar F, Saari N (2011) High-value components and bioactives from sea cucumbers for functional foods—a review. Marine drugs 9(10):1761–1805

Brito LO, Arantes R, Magnotti C, Derner R, Pchara F, Olivera A, Vinatea L (2014) Water quality and growth of Pacific white shrimp *Litopenaeus vannamei* (Boone) in co-culture with green seaweed *Ulva lactuca* (Linaeus) in intensive system. Aquac Int 22(2):497–508

Brown M, Robert R (2002) Preparation and assessment of microalgal concentrates as feeds for larval and juvenile Pacific oyster (*Crassostrea gigas*). Aquaculture 207(3-4):289–309

Buitrago E (1992) Concentración y preservación de microalgas como reserva de alimento de organismos marinos cultivados. Prog Iberoam Cienc Tec Subprog II—Acuicultura, CYTED-D 1:47–53

Camus T, Zeng C (2010) Roles of microalgae on total egg production over female lifespan and egg incubation time, naupliar and copepodite survival, sex ratio and female life expectancy of the copepod Bestiolina similis. Aquac Res 41(11):1717–1726

Carboni S, Vignier J, Chiantore M, Tocher DR, Migaud H (2012) Effects of dietary microalgae on growth, survival and fatty acid composition of sea urchin *Paracentrotus lividus* throughout larval development. Aquaculture 324:250–258

Cardinaletti G, Messina M, Bruno M, Tulli F, Poli B, Giorgi G, Chini-Zittelli G, Tredici M, Tibaldi E (2018) Effects of graded levels of a blend of Tisochrysis lutea and Tetraselmis suecica dried biomass on growth and muscle tissue composition of European sea bass (Dicentrarchus labrax) fed diets low in fish meal and oil. Aquaculture 485:173–182

Cardoso LG, Duarte JH, Andrade BB, Lemos PVF, Costa JAV, Druzian JI, Chinalia FA (2020) *Spirulina* sp. LEB 18 cultivation in outdoor pilot scale using aquaculture wastewater: High biomass, carotenoid, lipid and carbohydrate production. Aquaculture 525:735272

Casas-Valdez M, Hernández-Contreras H, Marin-Alvarez A, Aguila-Ramirez R, Hernández-Guerrero C, Sánchez-Rodríguez I, Carrillo-Domínguez S (2006) The seaweed Sargassum (Sargassaceae) as tropical alternative for goats' feeding. Rev Biol Trop 54(1):83–92

Cerezuela R, Guardiola FA, Meseguer J, Esteban MA (2012) Enrichment of gilthead seabream (*Sparus aurata* L.) diet with microalgae: effects on the immune system. Fish Physiol Biochem 38(6):1729–1739

Ceyhan V, Emir M (2015) Structural and economic analysis of Turkish fishmeal and fish oil. Turk J Fish Aquat Sci 15(4):841–850

Chauton MS, Reitan KI, Norsker NH, Tveterås R, Kleivdal HT (2015) A techno-economic analysis of industrial production of marine microalgae as a source of EPA and DHA-rich raw material for aquafeed: research challenges and possibilities. Aquaculture 436:95–103

Chávez-Crooker P, Obreque-Contreras J (2010) Bioremediation of aquaculture wastes. Curr Opin Biotechnol 21(3):313–317

Chen C-L, Chang J-S, Lee D-J (2015) Dewatering and drying methods for microalgae. Dry Technol 33(4):443–454

Chen B, Wan C, Mehmood MA, Chang J-S, Bai F, Zhao X (2017) Manipulating environmental stresses and stress tolerance of microalgae for enhanced production of lipids and value-added products–a review. Bioresour Technol 244:1198–1206

Chen JY, Zeng C, Jerry DR, Cobcroft JM (2020) Recent advances of marine ornamental fish larviculture: broodstock reproduction, live prey and feeding regimes, and comparison between demersal and pelagic spawners. Rev Aquac 12(3):1518–1541

Chithambaran S, Harbi M, Broom M, Khobrani K, Ahmad O, Fattani H, Sofyani A, Ayaril N (2017) Green water technology for the production of Pacific white shrimp *Penaeus vannamei* (Boone, 1931). Indian J Fish 64(3):43–49

Cottier-Cook EJ, Nagabhatla N, Asri A, Beveridge M, Bianchi P, Bolton J, Bondad-Reantaso MG, Brodie J, Buschmann A, Cabarubias J (2021) Ensuring the sustainable future of the rapidly expanding global seaweed aquaculture industry–a vision. UNU Institute on Comparative Regional Integration Studies.

Coutinho P, Rema P, Otero A, Pereira O, Fábregas J (2006) Use of biomass of the marine microalga Isochrysis galbana in the nutrition of goldfish (Carassius auratus) larvae as source of protein and vitamins. Aquac Res 37(8):793–798

Das BK, Pradhan J, Sahu S (2009) The effect of Euglena viridis on immune response of rohu, *Labeo rohita* (Ham.). Fish Shellfish Immunol 26(6):871–876

Delgado E, Reyes-Jaquez D (2018) Extruded aquaculture feed: a review. In: Extrusion of metals, polymers and food products. InTechOpen, London, pp 145–163

Delgado E, Valles-Rosales DJ, Flores NC, Reyes-Jáquez D (2021) Evaluation of fish oil content and cottonseed meal with ultralow gossypol content on the functional properties of an extruded shrimp feed. Aquac Rep 19:100588

Dineshbabu G, Goswami G, Kumar R, Sinha A, Das D (2019) Microalgae–nutritious, sustainable aqua-and animal feed source. J Funct Foods 62:103545

Dobberfuhl DR, Elser JJ (1999) Use of dried algae as a food source for zooplankton growth and nutrient release experiments. J Plan Res 21(5)

Dragone G, Fernandes BD, Abreu AP, Vicente AA, Teixeira JA (2011) Nutrient limitation as a strategy for increasing starch accumulation in microalgae. Appl Energy 88(10):3331–3335

Duy NDQ, Francis DS, Southgate PC (2017) The nutritional value of live and concentrated micro-algae for early juveniles of sandfish, *Holothuria scabra*. Aquaculture 473:97–104

Dworjanyn SA, Pirozzi I, Liu W (2007) The effect of the addition of algae feeding stimulants to artificial diets for the sea urchin *Tripneustes gratilla*. Aquaculture 273(4):624–633

El-Gamal MM, Othman SI, Abdel-Rahim MM, Mansour AT, Alsaqufi AS, El Atafy MM, Mona MH, Allam AA (2020) Palaemon and artemia supplemented diet enhances sea bass, *Dicentrarchus labrax*, broodstock reproductive performance and egg quality. Aquacult Rep 16:100290

El-Khodary GM, El-Sayed HS, Khairy HM, El-Sheikh MA, Qi X, Elshobary ME (2021) Comparative study on growth, survival and pigmentation of Solea aegyptiaca larvae by using four different microalgal species with emphasize on water quality and nutritional value. Aquac Nutr 27(2):615–629

El-Sayed HS, Elshobary ME, Barakat KM, Khairy HM, El-Sheikh MA, Czaja R, Allam B, Senousy HH (2022) Ocean acidification induced changes in Ulva fasciata biochemistry may improve Dicentrarchus labrax aquaculture via enhanced antimicrobial activity. Aquaculture 560:738474

El-Seesy AI, Elshobary ME, He Z (2022) Biofuel versus fossil fuel. In: Handbook of algal biofuels. Elsevier, Amsterdam, pp 181–193

Elshobary ME, Abo-Shady AM, Khairy HM, Essa D, Zabed HM, Qi X, Abomohra AE-F (2019) Influence of nutrient supplementation and starvation conditions on the biomass and lipid productivities of Micractinium reisseri grown in wastewater for biodiesel production. J Environ Manag 250:109529

Elshobary ME, Essa DI, Attiah AM, Salem ZE, Qi X (2020) Algal community and pollution indicators for the assessment of water quality of Ismailia canal, Egypt. Stoch Environ Res Risk Assess 34:1089–1103

Emerenciano MGC, Martínez-Córdova LR, Martínez-Porchas M, Miranda-Baeza A (2017) Biofloc technology (BFT): a tool for water quality management in aquaculture. Water quality 5:92–109

Erbland P, Caron S, Peterson M, Alyokhin A (2020) Design and performance of a low-cost, automated, large-scale photobioreactor for microalgae production. Aquac Eng 90:102103

Falaise C, François C, Travers M-A, Morga B, Haure J, Tremblay R, Turcotte F, Pasetto P, Gastineau R, Hardivillier Y (2016) Antimicrobial compounds from eukaryotic microalgae against human pathogens and diseases in aquaculture. Marine drugs 14(9):159

FAO (2020) The state of world fisheries and aquaculture 2020. Sustainability in action. FAO, Rome

Ferdouse F, Holdt SL, Smith R, Murúa P, Yang Z (2018) The global status of seaweed production, trade and utilization. Globefish Res Programme 124:I

Fernández FA, Sevilla JMF, Grima EM (2019) Costs analysis of microalgae production. In: Biofuels from algae. Elsevier, Amsterdam, pp 551–566

Ferreira M, Coutinho P, Seixas P, Fábregas J, Otero A (2009) Enriching rotifers with "premium" microalgae. Nannochloropsis gaditana. Mar Biotechnol 11(5):585–595

Floreto EA, Teshima S-I, Ishikawa M (1996) The Effects of seaweed diets on the growth, lipid and acids of juveniles of the white sea Urchin *Tripneustes gratilla*. Fish Sci 62(4):589–593

García-Ortega A, Martínez-Steele L, Gonsalves D, Wall MM, Sarnoski PJ (2015) Use of biofuel by-product from the green algae Desmochloris sp. and diatom Nanofrustulum sp. meal in diets for nile tilapia Oreochromis niloticus. J Aquac Eng Fish Res 1(4):144–154

García-Poza S, Leandro A, Cotas C, Cotas J, Marques JC, Pereira L, Gonçalves AM (2020) The evolution road of seaweed aquaculture: Cultivation technologies and the industry 4.0. Int J Environ Res Public Health 17(18):6528

Garti N, McClements DJ (2012) Encapsulation technologies and delivery systems for food ingredients and nutraceuticals. Elsevier, Amsterdam

Globenewswire (2021) Microalgae-based products market forecast to 2028. https://www.globenewswire.com/.

Gong Y, Guterres H, Huntley M, Sørensen M, Kiron V (2018) Digestibility of the defatted microalgae N annochloropsis sp. and D esmodesmus sp. when fed to A tlantic salmon, S almo salar. Aquac Nutr 24(1):56–64

Granada L, Sousa N, Lopes S, Lemos MF (2016) Is integrated multitrophic aquaculture the solution to the sectors' major challenges?–a review. Rev Aquac 8(3):283–300

Gressler V, Yokoya NS, Fujii MT, Colepicolo P, Mancini Filho J, Torres RP, Pinto E (2010) Lipid, fatty acid, protein, amino acid and ash contents in four Brazilian red algae species. Food Chem 120(2):585–590

Guedes AC, Sousa-Pinto I, Malcata FX (2015) Application of microalgae protein to aquafeed. In: Handbook of marine microalgae. Elsevier, Amsterdam, pp 93–125

Hamdi M (2022) Wastewaters treatment and algal biomass production by using innovative multitrophic airlift raceway reactor (MA2R) without O2 supply and CO2 release (ICSEWEN19). In: Sustainable energy-water-environment nexus in deserts: proceeding of the first international conference on sustainable energy-water-environment nexus in desert climates. Springer, Berlin, pp 591–596

Heasman M, Diemar J, O'connor W, Sushames T, Foulkes L (2000) Development of extended shelf-life microalgae concentrate diets harvested by centrifugation for bivalve molluscs–a summary. Aquac Res 31(8-9):637–659

Hoffman J, Pate RC, Drennen T, Quinn JC (2017) Techno-economic assessment of open microalgae production systems. Algal Res 23:51–57

Hu Q, Sommerfeld M, Jarvis E, Ghirardi M, Posewitz M, Seibert M, Darzins A (2008) Microalgal triacylglycerols as feedstocks for biofuel production: perspectives and advances. Plant J 54(4): 621–639

Ianora A, Romano G, Carotenuto Y, Esposito F, Roncalli V, Buttino I, Miralto A (2011) Impact of the diatom oxylipin 15S-HEPE on the reproductive success of the copepod *Temora stylifera*. Hydrobiologia 666(1):265–275

Javed MR, Bilal MJ, Ashraf MUF, Waqar A, Mehmood MA, Saeed M, Nashat N (2019) Microalgae as a feedstock for biofuel production: current status and future prospects. In: Top 5 contributions energy research and development. Springer, Cham, pp 1–39

Jimenez BD, Gaid RD, Jimenez CR (2016) Microalgal profile and anti-Vibrio activity of crude extracts from green water reservoir stocked with tilapia. Int J Adv Agric Environ Eng 3(2): 259–262

Jothiraj K, Santhanam P (2019) Optimisation of the culture conditions of *Nannocalanus minor* (Copepoda: Calanoida). In: Basic and applied zooplankton biology. Springer, Berlin, pp 225–246

Kalasariya HS, Patel RV, Pandya KY, Jasrai R, Brahmbhatt NH (2016) A review on nutritional facets of seaweeds. Int J Chem Sci Technol 1:27–32

Kang YH, Park SR, Chung IK (2011) Biofiltration efficiency and biochemical composition of three seaweed species cultivated in a fish-seaweed integrated culture. Algae 26(1):97–108

Kaparapu J (2018) Application of microalgae in aquaculture. Phykos 48(1):21–26

Kasimala M, Mogos G, Negasi K, Bereket G, Abdu M, Melake H (2020) Biochemical composition of selected seaweeds from intertidal shallow waters of Southern Red Sea, Eritrea. IJMS 49(7): 1153–1157

Khan F, Shahid A, Zhu H, Wang N, Javed MR, Ahmad N, Xu J, Alam MA, Mehmood MA (2022) Prospects of algae-based green synthesis of nanoparticles for environmental applications. Chemosphere 293:133571

Kiron V, Phromkunthong W, Huntley M, Archibald I, De Scheemaker G (2012) Marine microalgae from biorefinery as a potential feed protein source for Atlantic salmon, common carp and whiteleg shrimp. Aquac Nutr 18(5):521–531

Knuckey RM, Semmens GL, Mayer RJ, Rimmer MA (2005) Development of an optimal microalgal diet for the culture of the calanoid copepod *Acartia sinjiensis*: effect of algal species and feed concentration on copepod development. Aquaculture 249(1-4):339–351

Knuckey RM, Brown MR, Robert R, Frampton DM (2006) Production of microalgal concentrates by flocculation and their assessment as aquaculture feeds. Aquac Eng 35(3):300–313

Kokou F, Makridis P, Kentouri M, Divanach P (2012) Antibacterial activity in microalgae cultures. Aquac Res 43(10):1520–1527

Koldewey H (2005) Syngnathid husbandry in public aquariums 2005 manual. Project Seahorse, London, UK

Lang I, Hodac L, Friedl T, Feussner I (2011) Fatty acid profiles and their distribution patterns in microalgae: a comprehensive analysis of more than 2000 strains from the SAG culture collection. BMC Plant Biol 11:1–16

Limbu SM, Zhou L, Sun S-X, Zhang M-L, Du Z-Y (2018) Chronic exposure to low environmental concentrations and legal aquaculture doses of antibiotics cause systemic adverse effects in Nile tilapia and provoke differential human health risk. Environ Int 115:205–219

Liu Y, Lv J, Feng J, Liu Q, Nan F, Xie S (2019) Treatment of real aquaculture wastewater from a fishery utilizing phytoremediation with microalgae. J Chem Technol Biotechnol 94(3):900–910

Liu C, Li Y, Chen Z, Yuan L, Liu H, Han D, Jin J, Yang Y, Hu Q, Zhu X (2021) Effects of dietary whole and defatted *Arthrospira platensis* (Cyanobacterium) on growth, body composition and pigmentation of the yellow catfish *Pelteobagrus fulvidraco*. J Appl Phycol 33(4):2251–2259

Lyons DA, Scheibling RE (2007) Effect of dietary history and algal traits on feeding rate and food preference in the green sea urchin *Strongylocentrotus droebachiensis*. J Exp Mar Biol Ecol 349(1):194–204

Mabrouk MM, Ashour M, Labena A, Zaki MAA, Abdelhamid AF, Gewaily MS, Dawood MAO, Abualnaja KM, Ayoub HF (2022) Nanoparticles of *Arthrospira platensis* improves growth, antioxidative and immunological responses of Nile tilapia (*Oreochromis niloticus*) and its resistance to *Aeromonas hydrophila*. Aquac Res 53(1):125–135

Machado M, Machado S, Pimentel FB, Freitas V, Alves RC, Oliveira M (2020) Amino acid profile and protein quality assessment of macroalgae produced in an integrated multi-trophic aquaculture system. Foods 9(10):1382

Maehre HK, Edvinsen GK, Eilertsen K-E, Elvevoll EO (2016) Heat treatment increases the protein bioaccessibility in the red seaweed dulse (Palmaria palmata), but not in the brown seaweed winged kelp (Alaria esculenta). J Appl Phycol 28:581–590

Magouz FI, Essa MA, Matter M, Tageldein Mansour A, Alkafafy M, Ashour M (2021) Population dynamics, fecundity and fatty acid composition of *Oithona nana* (Cyclopoida, Copepoda), fed on different diets. Animals (Basel) 11(5):118

Maliwat GC, Velasquez S, Robil JL, Chan M, Traifalgar RF, Tayamen M, Ragaza JA (2017) Growth and immune response of giant freshwater prawn *Macrobrachium rosenbergii* (De Man) postlarvae fed diets containing *Chlorella vulgaris* (Beijerinck). Aquac Res 48(4):1666–1676

Mansour AT, Alprol AE, Abualnaja KM, El-Beltagi HS, Ramadan KMA, Ashour M (2022a) Dried brown seaweed's phytoremediation potential for methylene blue dye removal from aquatic environments. Polymers 14(7):1375

Mansour AT, Ashour M, Alprol AE, Alsaqufi AS (2022b) Aquatic plants and aquatic animals in the context of sustainability: cultivation techniques, integration, and blue revolution. Sustainability 14(6):3257

Matos ÂP, Feller R, Moecke EHS, de Oliveira JV, Junior AF, Derner RB, Sant'Anna, E.S. (2016) Chemical characterization of six microalgae with potential utility for food application. J Am Oil Chem Soc 93:963–972

Medina-Félix D, López-Elías JA, Martínez-Córdova LR, López-Torres MA, Hernández-López J, Rivas-Vega ME, Mendoza-Cano F (2014) Evaluation of the productive and physiological responses of *Litopenaeus vannamei* infected with WSSV and fed diets enriched with *Dunaliella* sp. J Invertebr Pathol 117:9–12

Medina-Reyna CE, Ronson-Paulin JA, Hernandez-Rojas F, Santiago-Morales I, Pedroza-Islas R, Vernon-Carter EJ (2005) Dual benefits of whey protein concentrate in a microencapsulated diet for larval white shrimp, *Litopenaeus vannamei*. J World Aquacult Soc 36(3):401–410

Meitei MM, Singh SK, Mangang YA, Meena DK, Debbarma R, Biswas P, Waikhom G, Patel AB, Ngasotter S, Newmei T (2022) Effective valorization of precision output of algaquaculture towards eco-sustainability and bioeconomy concomitant with biotechnological advances: an innovative concept. Cleaner Waste Syst 3:100026

Molino A, Larocca V, Chianese S, Musmarra D (2018) Biofuels production by biomass gasification: a review. Energies 11(4):811

Molinoa A, Iovine A, Leonec G, Di Sanzod G, Palazzod S, Martinod M, Sangiorgiod P, Marinob T, Musmarrab D (2020) Microalgae as alternative source of nutraceutical polyunsaturated fatty acids. Chem Eng 79:277–282

Muhammad G, Alam MA, Xiong W, Lv Y, Xu J-L (2020) Microalgae biomass production: an overview of dynamic operational methods. In: Microalgae biotechnology for food, health and high value products. Springer, Singapore, pp 415–432

Musa M, Doshi A, Brown R, Rainey TJ (2019) Microalgae dewatering for biofuels: A comparative techno-economic assessment using single and two-stage technologies. J Clean Prod 229:325–336

Musa M, Ward A, Ayoko GA, Rösch C, Brown R, Rainey TJ (2020) Single-step dynamic dewatering of microalgae from dilute suspensions using flocculant assisted filtration. Microb Cell Factories 19:1–18

Nagappan S, Kumar G (2021) Investigation of four microalgae in nitrogen deficient synthetic wastewater for biorefinery based biofuel production. Environ Technol Innov 23:101572

Nagappan S, Devendran S, Tsai P-C, Dahms H-U, Ponnusamy VK (2019) Potential of two-stage cultivation in microalgae biofuel production. Fuel 252:339–349

Nagappan S, Das P, AbdulQuadir M, Thaher M, Khan S, Mahata C, Al-Jabri H, Vatland AK, Kumar G (2021) Potential of microalgae as a sustainable feed ingredient for aquaculture. J Biotechnol 341:1–20

Namvar F, Mohamed S, Fard SG, Behravan J, Mustapha NM, Alitheen NBM, Othman F (2012) Polyphenol-rich seaweed (*Eucheuma cottonii*) extract suppresses breast tumour via hormone modulation and apoptosis induction. Food Chem 130(2):376–382

Nandeesha M, Gangadhara B, Manissery J, Venkataraman L (2001) Growth performance of two Indian major carps, catla (Catlacatla) and rohu (Labeorohita) fed diets containing different levels of Spirulina platensis. Bioresour Technol 80(2):117–120

Natrah FM, Bossier P, Sorgeloos P, Yusoff FM, Defoirdt T (2014) Significance of microalgal–bacterial interactions for aquaculture. Rev Aquac 6(1):48–61

Navarro N, Sarasquete C (1998) Use of freeze-dried microalgae for rearing gilthead seabream, *Sparus aurata*, larvae: I. Growth, histology and water quality. Aquaculture 167(3–4):179–193

Navarro N, Yúfera M, García-Gallego M (2001) Use of freeze-dried microalgae for rearing gilthead seabream, *Sparus aurata* L., larvae. II. Biochemical composition. Hydrobiologia 452(1):69–77

Naylor RL, Goldburg RJ, Primavera JH, Kautsky N, Beveridge MC, Clay J, Folke C, Lubchenco J, Mooney H, Troell M (2000) Effect of aquaculture on world fish supplies. Nature 405(6790): 1017–1024

Naylor RL, Hardy RW, Buschmann AH, Bush SR, Cao L, Klinger DH, Little DC, Lubchenco J, Shumway SE, Troell M (2021) A 20-year retrospective review of global aquaculture. Nature 591(7851):551–563

Neori A (2011) "Green water" microalgae: the leading sector in world aquaculture. J Appl Phycol 23(1):143–149

Niccolai A, Zittelli GC, Rodolfi L, Biondi N, Tredici MR (2019) Microalgae of interest as food source: Biochemical composition and digestibility. Algal Res 42:101617

Nova P, Martins AP, Teixeira C, Abreu H, Silva JG, Silva AM, Freitas AC, Gomes AM (2020) Foods with microalgae and seaweeds fostering consumers health: A review on scientific and market innovations. J Appl Phycol 32:1789–1802

Nunes M, Pereira A, Ferreira JF, Yasumaru F (2009) Evaluation of the microalgae paste viability produced in a mollusk hatchery in Southern Brazil. J World Aquacult Soc 40(1):87–94

Olivotto I, Avella MA, Sampaolesi G, Piccinetti C, Ruiz PN, Carnevali O (2008) Breeding and rearing the longsnout seahorse *Hippocampus reidi*: rearing and feeding studies. Aquaculture 283(1-4):92–96

Olvera-Novoa M, Dominguez-Cen L, Olivera-Castillo L, Martínez-Palacios CA (1998) Effect of the use of the microalga Spirulina maxima as fish meal replacement in diets for tilapia, Oreochromis mossambicus (Peters), fry. Aquac Res 29(10):709–715

Oostlander P, van Houcke J, Wijffels RH, Barbosa MJ (2020) Microalgae production cost in aquaculture hatcheries. Aquaculture 525:735310

Palmegiano GB, Agradi E, Forneris G, Gai F, Gasco L, Rigamonti E, Sicuro B, Zoccarato I (2005) Spirulina as a nutrient source in diets for growing sturgeon (Acipenser baeri). Aquac Res 36(2): 188–195

Palmer PJ, Burke MJ, Palmer CJ, Burke JB (2007) Developments in controlled green-water larval culture technologies for estuarine fishes in Queensland, Australia and elsewhere. Aquaculture 272(1-4):1–21

Palmtag MR, Faulk CK, Holt G (2006) Highly unsaturated fatty acid composition of rotifers (*Brachionus plicatilis*) and Artemia fed various enrichments. J World Aquacult Soc 37(1): 126–131

Pham NK, Lin J (2013) The effects of different feed enrichments on survivorship and growth of early juvenile longsnout seahorse, *Hippocampus reidi*. J World Aquacult Soc 44(3):435–446

Ponis E, Parisi G, Zittelli GC, Lavista F, Robert R, Tredici M (2008) *Pavlova lutheri*: Production, preservation and use as food for *Crassostrea gigas* larvae. Aquaculture 282(1-4):97–103

Posten C, Schaub G (2009) Microalgae and terrestrial biomass as source for fuels—a process view. J Biotechnol 142(1):64–69

Prabha SP, Nagappan S, Rathna R, Viveka R, Nakkeeran E (2020) Blue biotechnology: a vision for future marine biorefineries. In: Refining biomass residues for sustainable energy and bioproducts. Elsevier, Amsterdam, pp 463–480

Puello-Cruz A, Mezo-Villalobos S, Gonzalez-Rodriguez B, Voltolina D (2009) Culture of the calanoid copepod *Pseudodiaptomus euryhalinus* (Johnson 1939) with different microalgal diets. Aquaculture 290(3–4):317–319

Raeisossadati M, Moheimani NR, Parlevliet D (2019) Luminescent solar concentrator panels for increasing the efficiency of mass microalgal production. Renew Sust Energ Rev 101:47–59

Rafa N, Ahmed SF, Badruddin IA, Mofijur M, Kamangar S (2021) Strategies to produce cost-effective third-generation biofuel from microalgae. Front Energy Res 9:749968

Raja R, Coelho A, Hemaiswarya S, Kumar P, Carvalho IS, Alagarsamy A (2018) Applications of microalgal paste and powder as food and feed: an update using text mining tool. Beni-Suef Univ J Basic Appl Sci 7(4):740–747

Rajauria G (2015) Seaweeds: a sustainable feed source for livestock and aquaculture. In: Seaweed sustainability. Elsevier, Amsterdam, pp 389–420

Remize M, Brunel Y, Silva JL, Berthon J-Y, Filaire E (2021) Microalgae n-3 PUFAs production and use in food and feed industries. Marine drugs 19(2):113

Ren Y, Liu W, Dong S, Pearce C (2016) Effects of mono-species and bi-species microalgal diets on the growth, survival and metamorphosis of auricularia larvae of the California sea cucumber, *Parastichopus californicus* (Stimpson, 1857). Aquac Nutr 22(2):304–314

Ribeiro AR, Gonçalves A, Barbeiro M, Bandarra N, Nunes ML, Carvalho ML, Silva J, Navalho J, Dinis MT, Silva T (2017) *Phaeodactylum tricornutum* in finishing diets for gilthead seabream: effects on skin pigmentation, sensory properties and nutritional value. J Appl Phycol 29(4): 1945–1956

Rohani-Ghadikolaei K, Abdolalian E, Hojatollah F, Masoud G, Ng WK (2015) The nutritional effect of *Isochrysis galbana* and *Chaetoceros muelleri* cultured with different seaweed extracts on the larval development, growth and survival of the marine shrimp, *Penaeus indicus*. Aquac Res 46(6):1444–1454

Rokey G (1994) Petfood and fishfood extrusion. In: The technology of extrusion cooking. Springer, Boston, pp 144–189

Ruggiero MA, Gordon DP, Orrell TM, Bailly N, Bourgoin T, Brusca RC, Cavalier-Smith T, Guiry MD, Kirk PM (2015) A higher level classification of all living organisms. PLoS One 10(4): e0119248

Saejung C, Ektasaeng T (2022) Evaluation of Chlorella vulgaris grown in sugar industry wastewater for use as aquaculture feed. Int J Environ Sci Technol 20:5957

Samuelsen TA, Oterhals Å, Kousoulaki K (2018) High lipid microalgae (*Schizochytrium* sp.) inclusion as a sustainable source of n-3 long-chain PUFA in fish feed—Effects on the extrusion process and physical pellet quality. Anim Feed Sci Technol 236:14–28

Santhanam P, Jeyaraj N, Jothiraj K, Ananth S, Kumar SD, Pachiappan P (2019) Assessing the efficacy of marine copepods as an alternative first feed for larval production of tiger shrimp *Penaeus monodon*. In: Basic and applied zooplankton biology. Springer, Berlin, pp 293–303

Saravana PS, Ummat V, Bourke P, Tiwari BK (2022) Emerging green cell disruption techniques to obtain valuable compounds from macro and microalgae: a review. Crit Rev Biotechnol 43(6): 904–919. https://doi.org/10.1080/07388551.2022.2089869

Sarker PK, Kapuscinski AR, Bae AY, Donaldson E, Sitek AJ, Fitzgerald DS, Edelson OF (2018) Towards sustainable aquafeeds: Evaluating substitution of fishmeal with lipid-extracted microalgal co-product (Nannochloropsis oculata) in diets of juvenile Nile tilapia (Oreochromis niloticus). PLoS One 13(7):e0201315

Sarker PK, Kapuscinski AR, McKuin B, Fitzgerald DS, Nash HM, Greenwood C (2020) Microalgae-blend tilapia feed eliminates fishmeal and fish oil, improves growth, and is cost viable. Sci Rep 10(1):19328

Schagerström E, Christophersen G, Sunde J, Bakke S, Matusse NR, Dupont S, Sundell KS (2021) Controlled spawning and rearing of the sea cucumber, *Parastichopus tremulus*. J World Aquacult Soc 53(1):224–240

Schipp GR, Bosmans JM, Marshall AJ (1999) A method for hatchery culture of tropical calanoid copepods, *Acartia* spp. Aquaculture 174(1-2):81–88

Scholtz R, Bolton J, Macey B (2013) Effects of different microalgal feeds and their influence on larval development in the white-spined sea urchin *Tripneustes gratilla*. Afr J Mar Sci 35(1): 25–34

Seychelles L, Audet C, Tremblay R, Fournier R, Pernet F (2009) Essential fatty acid enrichment of cultured rotifers (*Brachionus plicatilis*, Müller) using frozen-concentrated microalgae. Aquac Nutr 15(4):431–439

Shaalan M, El-Mahdy M, Saleh M, El-Matbouli M (2018) Aquaculture in Egypt: insights on the current trends and future perspectives for sustainable development. Rev Fish Sci Aquac 26(1): 99–110

Sharawy ZZ, Ashour M, Labena A, Alsaqufi AS, Mansour AT, Abbas EM (2022) Effects of dietary *Arthrospira platensis* nanoparticles on growth performance, feed utilization, and growth-related gene expression of Pacific white shrimp, *Litopenaeus vannamei*. Aquaculture 551:737905

Sharma J, Sarmah P, Bishnoi NR (2020) Market perspective of EPA and DHA production from microalgae. In: Nutraceutical fatty acids from oleaginous microalgae: a human health perspective. Scrivener Publishing LLC, Beverly, MA, pp 281–297

Shera SS, Banik RM (2022) Algal nanoparticles: synthesis and characterization. In: Bioprospecting algae for nanosized materials. Springer, Berlin, pp 25–69

Shi C, Dong S, Wang F, Gao Q, Tian X (2013) Effects of four fresh microalgae in diet on growth and energy budget of juvenile sea cucumber *Apostichopus japonicus* (Selenka). Aquaculture 416:296–301

Shi Z, Li X-Q, Chowdhury MK, Chen J-N, Leng X-J (2016) Effects of protease supplementation in low fish meal pelleted and extruded diets on growth, nutrient retention and digestibility of gibel carp, Carassius auratus gibelio. Aquaculture 460:37–44

Shields R, Lupatsch I (2012) 5 Algae for aquaculture and animal feeds. In: Microalgal biotechnology: integration and economy. De Gruyter, Berlin, pp 79–100

Shigai X, Chuanxin Q, Zhenhua M, Gang Y, Jinhui S, Wanni P, Tao Z, Hongmei M, Wentao Z (2020) Effects of dietary microalgae on growth and survival of larval development of sea urchin (*Anthocidaris crassispina*). 南方水产科学 16(2):115–120

Sibonga RC, Laureta LV, Lebata-Ramos MJH, Nievales MFJ, Pedroso FL (2021) Single and mixed species of microalgae as larval food for the tropical sea cucumber *Holothuria scabra*. J Appl Phycol 33:3103–3112

Sirakov I, Velichkova K, Nikolov G (2012) The effect of algae meal (Spirulina) on the growth performance and carcass parameters of rainbow trout (Oncorhynchus mykiss). J BioSci Biotech:151–156

Sirohi R, Sim SJ, Pandey A (2023) Photobioreactors: an introduction. In: Current developments in biotechnology and bioengineering. Elsevier, Amsterdam, pp 3–10

Sivakumar N, Sundararaman M, Selvakumar G (2011) Efficacy of micro algae and cyanobacteria as a live feed for juveniles of shrimp *Penaeus monodon*. Afr J Biotechnol 10(55):11594–11599

Skrede A, Mydland L, Ahlstrøm Ø, Reitan K, Gislerød H, Øverland M (2011) Evaluation of microalgae as sources of digestible nutrients for monogastric animals. J Anim Feed Sci 20(1): 131–142

Souza FPd, Lima ECSd, Urrea-Rojas AM, Suphoronski SA, Facimoto CT, Bezerra Júnior JdS, Oliveira TESd, Pereira UdP, Santis GWD, Oliveira CALd (2020) Effects of dietary supplementation with a microalga (*Schizochytrium* sp.) on the hemato-immunological, and intestinal histological parameters and gut microbiota of Nile tilapia in net cages. PLoS One 15(1): e0226977

Spolaore P, Joannis-Cassan C, Duran E, Isambert A (2006) Commercial applications of microalgae. J Biosci Bioeng 101(2):87–96

Suárez-Puerto B, Delgadillo-Díaz M, Sánchez-Solís MJ, Gullian-Klanian M (2021) Analysis of the cost-effectiveness and growth of Nile tilapia (*Oreochromis niloticus*) in biofloc and green water technologies during two seasons. Aquaculture 538:736534

Sui Y, Jiang Y, Moretti M, Vlaeminck SE (2020) Harvesting time and biomass composition affect the economics of microalgae production. J Clean Prod 259:120782

Sun X, Chang Y, Ye Y, Ma Z, Liang Y, Li T, Jiang N, Xing W, Luo L (2012) The effect of dietary pigments on the coloration of Japanese ornamental carp (koi, *Cyprinus carpio* L.). Aquaculture 342:62–68

Tanaka Y, Ashaari A, Mohamad FS, Lamit N (2020) Bioremediation potential of tropical seaweeds in aquaculture: Low-salinity tolerance, phosphorus content, and production of UV-absorbing compounds. Aquaculture 518:734853

Tendencia EA, dela Peña M (2003) Investigation of some components of the greenwater system which makes it effective in the initial control of luminous bacteria. Aquaculture 218(1-4): 115–119

Tendencia EA, Bosma RH, Sorio LR (2012) Effect of three innovative culture systems on water quality and whitespot syndrome virus (WSSV) viral load in WSSV-fed *Penaeus monodon* cultured in indoor tanks. Aquaculture 350:169–174

Tocher DR (2015) Omega-3 long-chain polyunsaturated fatty acids and aquaculture in perspective. Aquaculture 449:94–107

Turcihan G, Turgay E, Yardımcı RE, Eryalçın KM (2021) The effect of feeding with different microalgae on survival, growth, and fatty acid composition of Artemia franciscana metanauplii and on predominant bacterial species of the rearing water. Aquac Int 29(5):2223–2241

Van Alstyne KL, Wolfe GV, Freidenburg TL, Neill A, Hicken C (2001) Activated defense systems in marine macroalgae: evidence for an ecological role for DMSP cleavage. Mar Ecol Prog Ser 213:53–65

Venkata Subhash G, Chugh N, Iyer S, Waghmare A, Musale AS, Nandru R, Dixit RB, Gaikwad MS, Menon D, Thorat R (2020) Application of in vitro protein solubility for selection of microalgae biomass as protein ingredient in animal and aquafeed. J Appl Phycol 32:3955–3970

Vizcaíno A, López G, Sáez M, Jiménez J, Barros A, Hidalgo L, Camacho-Rodríguez J, Martínez T, Cerón-García M, Alarcón F (2014) Effects of the microalga Scenedesmus almeriensis as fishmeal alternative in diets for gilthead sea bream, Sparus aurata, juveniles. Aquaculture 431: 34–43

Vrasna DK, Goh PS, Lau WJ, Ismail AF, Matsuyama H, Gonzales RR (2022) Microalgae dewatering using forward osmosis membrane: a review. Mater Today Proc 65:3073–3080

Walker AB, Berlinsky DL (2011) Effects of partial replacement of fish meal protein by microalgae on growth, feed intake, and body composition of Atlantic cod. N Am J Aquac 73(1):76–83

Wan AH, Davies SJ, Soler-Vila A, Fitzgerald R, Johnson MP (2019) Macroalgae as a sustainable aquafeed ingredient. Rev Aquac 11(3):458–492

Wang Y, Sun M, Tang Y, Xu A, Tang J, Song Z (2022) Effects of Haematococcus pluvialis on the water quality and performance of Litopenaeus vannamei using artificial substrates and water exchange systems. Aquac Int 30(4):1779–1797

Warren-Myers F, Turchini G, Swearer SE, Francis D, Dempster T (2021) The balancing act: Protein, lipid and seaweed dietary levels to maximize gonad quantity in a wild-caught sea urchin. Aquac Nutr 27(4):1019–1030

Wijffels RH, Barbosa MJ, Eppink MH (2010) Microalgae for the production of bulk chemicals and biofuels. Biofuels, Bioproducts and Biorefining: Innovation for a sustainable economy 4(3): 287–295

Wikfors GH, Ohno M (2001) Impact of algal research in aquaculture. J Phycol 37(6):968–974

Willadino L, Souza-Santos LP, Mélo RC, Brito AP, Barros NC, Araújo-Castro CM, Galvão DB, Gouveia A, Regis CG, Cavalli RO (2012) Ingestion rate, survival and growth of newly released seahorse *Hippocampus reidi* fed exclusively on cultured live food items. Aquaculture 360:10–16

Yang W, Zheng Z, Lu K, Zheng C, Du Y, Wang J, Zhu J (2020) Manipulating the phytoplankton community has the potential to create a stable bacterioplankton community in a shrimp rearing environment. Aquaculture 520:734789

Yao C, Ai J, Cao X, Xue S, Zhang W (2012) Enhancing starch production of a marine green microalga Tetraselmis subcordiformis through nutrient limitation. Bioresour Technol 118:438–444

Yousef EA, Hegab MH (2017) Evaluation of two novel feeding protocols utilizing alive and dried *Chlorella vulgaris* to grow *Heterocypris salina* (Ostracoda: Crustacea). J Zoo Res 1(1):10–15

Zhang L, Liao W, Huang Y, Wen Y, Chu Y, Zhao C (2022) Global seaweed farming and processing in the past 20 years. Food Prod Process Nutr 4(1):23

Zheng Y, Jin R, Zhang X, Wang Q, Wu J (2019) The considerable environmental benefits of seaweed aquaculture in China. Stoch Env Res Risk A 33(4):1203–1221

Zhou R, Zhou R, Zhang X, Fang Z, Wang X, Speight R, Wang H, Doherty W, Cullen PJ, Ostrikov K (2019) High-performance plasma-enabled biorefining of microalgae to value-added products. ChemSusChem 12(22):4976–4985

Algae as a Functional Food: A Case Study on Spirulina

Agnese Stunda-Zujeva and Megija Berele

Abstract Spirulina (*Arthrospira* spp.) is a blue-green microalga within the phylum Cyanophyta, also known as Cyanobacteria. It has superior nutritional value, clinically proven health benefits, and beneficial interaction in fermentation and preservation processes. Therefore, Spirulina is a functional ingredient and a key ingredient in functional foods for particular dietary use. Nutritional facts show that Spirulina has over 60% of protein, with relatively low fat and sugar contents. The clinical studies have highlighted that health benefits of Spirulina are caused by its high antioxidant activity, especially due to the presence of a unique blue pigment-protein complex—phycocyanin. Contrary to other edible algae, such as *Chlorella*, *Spirulina* cell membrane is easy to be digested, and nutrients have high bioavailability. Spirulina as a potential source for vitamin B_{12} is confirmed. The present chapter discusses some examples of cultivation parameters' affecting Spirulina's nutritional value. In addition, the most significant producers, mainly in Asia, with market share of powder, tablets, and capsules, are summarized. Fresh frozen Spirulina is a new, rapidly growing trend in Western countries due to its milder taste. Spirulina-containing products in the market are presented, and many commercial products of snacks, drinks, pasta, and dairy products are given. Studies on Spirulina as a functional ingredient that show its potential to boost fermentation process and prevent pathogen microbe development are summarized.

Keywords Spirulina · *Arthrospira* · Commercial products · Functional ingredient · SpirulinaNord

---

A. Stunda-Zujeva (✉)
Riga Technical University, Faculty of Materials Science and Applied Chemistry, Institute of General Chemical Engineering, Riga, Latvia

SIA SpirulinaNord, Riga, Latvia
e-mail: agnese.stunda-zujeva@rtu.lv

M. Berele
Riga Technical University, Faculty of Materials Science and Applied Chemistry, Institute of General Chemical Engineering, Riga, Latvia

A. Abomohra, S. Ende (eds.), *Value-added Products from Algae*,
https://doi.org/10.1007/978-3-031-42026-9_17

1 Introduction: Spirulina—New or Traditional Food?

The term Spirulina refers to the biomass of *Arthrospira platensis* and *Arthrospira maxima* (Habib et al. 2008). Spirulina (*Arthrospira* spp.) and *Chlorella* are the only two microalgal genera approved as a food in the EU (Probst et al. 2015). The modern story of Spirulina started in the mid-1960s by the report of J.Léonard, the Botanist of a French-Belgian expedition to Africa, described a blue-green cake sold in the food market in Chad. Sun-dried pure Spirulina cake is a traditional food of the Kanembu tribe living along the alkaline lakes of Chad and Niger (Habib et al. 2008).The other natural growth region was Mexico, where Spirulina was the traditional food "tecuitlatl" for Aztecs living in the Valley of Mexico (Karkos et al. 2011). The Sosa Texcoco developed Spirulina investigation epicenter around the middle of the sixteenth century (Vonshak 1997). The Institut Français du Pétrole studied a bloom of algae in the evaporation ponds of their sodium bicarbonate production facility in a lake near Mexico City, where Zarrouk did the first systematic study of Spirulina's growth requirements and physiology as part of his research study, and developed the typical growth medium recopies that is still the backbone of nowadays *Arthrospira* cultivation (Habib et al. 2008). This knowledge also has an input in the first large-scale Spirulina production plant development in the USA. In that context, the first Spirulina farm was developed in the hot desert in southeastern California in late 1970s by Earthrise Nutritionals Ltd. In 1981, Earthrise® formed a partnership with a Japanese company, Dainippon Ink and Chemicals (DIC), and began growing Spirulina in Thailand. Currently, they are still one of the largest producers in the World (https://www.earthrise.com/origins-history).

The United Nations World Food Conference in 1974 and the United States Department of Agriculture recognized Spirulina as the best food for the future (Soni et al. 2017). The UN-Food and Agriculture Organization (FAO) has also recommended the re-evaluation of Spirulina as a tool for both national governments and inter-governmental organizations to fulfill their own food security needs and as a mean of emergency response efforts for overseas development (Habib et al. 2008). World Health Organization (WHO) has also recognized the potential of Spirulina due to its high iron and protein content, in addition to its safety for human use (Madkour et al. 2012). Furthermore, only 10 grams of dried Spirulina can significantly reduce malnutrition symptoms, so it is also called the food of life.

Many projects worldwide introduce Spirulina cultivation in poor regions, so local people can grow this valuable food using minimal resources to obtain a fast-growing food supplementation system. Interestingly, Spirulina is also recommended to be used as food for astronauts. For example, European Space Agency (ESA) astronauts used algae-containing food on board the space station during the Futura Mission (2014/15) (https://outpost42.esa.int/futura/). NASA stated that astronauts could rely on microalgae to supply essentials including food, water, and oxygen on future long-duration spaceflight missions to the Moon and Mars. It could be essential to a hybrid life support system in such missions (https://www.nasa.gov/mission_pages/station/research/news/photobioreactor-better-life-support). Therefore, Spirulina is one of the oldest living organisms on the Earth, traditional food in some regions but with a high potential to become next worldwide famous product and maybe one day, it will be as usual daily routine as coffee is today.

2 The 10 Interesting Facts on Spirulina

1. Spirulina's name comes from its spiral shape; however, it is such tiny algal species that its spiral shape and can be seen only under a microscope. The cells are usually 3–5 μm in diameter and the filaments are up to 500 μm long (Sotiroudis and Sotiroudis 2013).
2. Spirulina is clinically well-studied, and no allergies are reported. Therefore, it is also recommended for children, pregnant ladies, and seniors' daily food (AlFadhly et al. 2022).
3. Raw Spirulina has no specific taste or smell, while dried powder usually has a specific smell and taste (www.spirulinanord.eu; Thomas et al. 2020).
4. Spirulina is a blue-green alga—the only known edible among other toxic blooming blue-green microalgae (Rzymski et al. 2015; Vonshak 1997).
5. Spirulina is also called cyanobacteria. It was classified as a bacteria, an ancient organism that is prokaryotic and is the only photosynthetic bacteria (Vonshak 1997).
6. From evolutionary aspect, cyanobacteria were the first photosynthetic organisms on the planet, and assumed that they caused one of the most significant climate crises in the history two billion years ago, known as the Oxygen Catastrophe. Before that, it is assumed that the atmosphere was based on relatively inert gases such as nitrogen and CO_2 (Schirrmeister et al. 2011). In comparison, oxygen is a very aggressive gas that is thought to oxidize many minerals and was toxic for many organisms living at that time (Kasting and Siefert 2002).
7. Spirulina is a powerful absorbent of heavy metals. Together with other species, e.g., *Chlorella*, it can be used for wastewater treatment (Al-Dhabi 2013). However, Spirulina is only suitable for food as a monoculture, cultivated under a highly controlled medium composition and clean environment.
8. Botanically, the genus *Spirulina* was changed to *Arthrospira* in 1974 when the term "Spirulina" as a food supplement had already become famous (Vonshak 1997). Therefore, the supplements under the name "Spirulina" describe the biomass of *Arthrospira* sp.
9. Spirulina is the best vegetable for several reasons:

   (a) No leaves, roots, peels, or other parts should be removed and discarded.
   (b) Due to low fiber content, it is easy to digest even for those who have digestive problems, and it will not create gases in the stomach (Mathur 2019).

(c) It is a vegetable of all colors as it contains green, blue, and orange pigments (Mathur 2019).
(d) It is low in sugars and fats while high in antioxidants and protein (Mathur 2019).

10. It has been estimated that more than 200,000 microalgal species exist. However, only Spirulina and *Chlorella* are authorized as safe for food consumption in European Union as whole cells (Caporgno and Mathys 2018). Only some other microalgae can be used to extract omega-3 oil or other compounds. However, most of the microalgae are poorly investigated (Ryckebosch et al. 2012).

3 Nutritional Value

Spirulina is a unique plant-based food due to its high content of protein (55–70 g protein per g dry matter) and one of the few vegan complete protein sources (AlFadhly et al. 2022). It also contains superoxide dismutase and gamma-linolenic acid (GLA), both are rare compounds (Ismail et al. 2015). Superoxide dismutase (SOD) is an antioxidant enzyme that protects the body from free radical damage (Sannasimuthu et al. 2018). By converting the harmful superoxide radicals into hydrogen peroxide, SOD helps to prevent oxidative stress, which is linked to many chronic diseases, including cancer, heart disease, and Alzheimer's disease (Forman and Zhang 2021). Spirulina's high SOD content makes it a valuable supplement for athletes and individuals who engage in high-intensity exercise, as it can help to reduce muscle damage and promote recovery (Kozakowska et al. 2015; Pappas et al. 2021). GLA is another essential compound found in Spirulina. This omega-6 fatty acid is important for maintaining healthy skin, hair, and nails. It also has anti-inflammatory effects and may help to reduce symptoms of eczema, rheumatoid arthritis, and other inflammatory conditions (Kawamura et al. 2011). GLA is relatively rare in plant-based foods, but Spirulina is one of the few sources of this important nutrient. This makes it a valuable addition to vegetarian or vegan diet, which may be lacking omega-6 fatty acids (Choopani et al. 2016). In clinical studies, C-phycocyanin is highlighted as the most promising Spirulina active ingredient showing antioxidant, anti-inflammatory, and immunomodulatory properties (Liu et al. 2016). Only cyanobacteria can produce C-phycocyanin, while Spirulina is the only edible cyanobacterium.

Spirulina contains 55–70% protein, 15–25% carbohydrates, 6–8% fats, 7–13% minerals, 3–7% moisture, and 8–10% dietary fibers (AlFadhly et al. 2022). The amount of protein is strongly influenced by the growth conditions, where stress restricts protein formation, and more carbs are generated in the cell (Udayan et al. 2022). Spirulina is one of the few plant-based sources of complete protein. It contains all essential amino acids—isoleucine 35 mg g^{-1}; leucine 54 mg g^{-1}, lysine 29 mg g^{-1}, methionine 14 mg g^{-1}, phenylalanine 28 mg g^{-1}, threonine 32 mg g^{-1}, valine 40 mg g^{-1}, and tryptophan 9 mg g^{-1}. Also, it contains non-essential amino acids, including cystine 6 mg g^{-1}, alanine 47 mg g^{-1}, aspartic acid 61 mg g^{-1}, serine

32 mg g^{-1}, glycine 32 mg g^{-1}, glutamic acid 91 mg g^{-1}, arginine 43 mg g^{-1} DM, tyrosine 30 mg g^{-1}, proline 27 mg g^{-1}, and histidine 10 mg g^{-1}(Aly and Esawy 2008). As discussed later in this chapter, the most valuable compounds in Spirulina are pigments (C-phycocyanin, chlorophyll-*a*, β-carotene, allophycocyanin, lutein, zeaxanthin, etc.). C-phycocyanin is well-studied, and various promising health benefits have been found (Anvar and Nowruzi 2021). Spirulina also contains various enzymes, as it is a rare vegan source of the SOD mentioned above as well as glutathione peroxidase (GPx), catalase (CAT), and ascorbate peroxidase (APX) (AlFadhly et al. 2022; Kumar et al. 2022).

Free radicals are the part of normal metabolites; however, physical and mental overload, stress viruses, pollution, and many other factors significantly increase free radical level (Ighodaro and Akinloye 2018). If free radicals in the body are too much or antioxidants for their neutralization are missing, they damage the outer membranes of healthy cells. That leads to cell degeneration and death, adversely affecting human health (Sharma 2014). SOD, CAT, GPx, and APX are primarily antioxidants that stop or inhibit the formation of free radicals and reactive species in the cells (Hasanuzzaman et al. 2020). Free radicals are various types, so various antioxidants neutralize each of them. SOD neutralizes superoxide anions to hydrogen peroxide (H_2O_2) and molecular oxygen (O_2). However, hydrogen peroxide is a precursor of hydroxyl radical ($^{\bullet}OH$), the most active and dangerous radical (Andrés et al. 2022; Ighodaro and Akinloye 2018). Therefore, CAT and APX enzymes the catalyze the degradation of hydrogen peroxide to water and molecular oxygen are also very important (Ighodaro and Akinloye 2018; Kaur et al. 2021). In addition, GPx is a group of selenium-dependent enzymes that breakdown hydrogen peroxide to water and lipid peroxide to corresponding alcohols (Ighodaro and Akinloye 2018).

Iron deficiency anemia is a common problem for almost half of the women on the planet (Miller 2013). Spirulina is suggested as an effective source of iron because it contains Fe^{3+} in the form of ferrihydrite which protects the cell from potentially harmful Fe^{2+}/Fe^{3+} redox cycling (Isani et al. 2022) and Spirulina can store iron in high concentrations (Isani et al. 2022). Spirulina has emerged as a promising candidate for preventing and treating iron-deficiency anemia, which is a global health concern (Selmi et al. 2011). Besides 3.4 mg g^{-1} Fe in Spirulina biomass, it also contains other minerals such as K 21.8 mg g^{-1}, Ca 10.4 mg g^{-1}, Cr 0.004 mg g^{-1}, Cu 0.01 mg g^{-1}, Mg 0.01 mg g^{-1}, Mn 0.07 mg g^{-1}, P 19.2 mg g^{-1}, Se 0,0005 mg g^{-1}, Na 15.1 mg g^{-1}, and Zn 0.04 mg g^{-1}(Mawla 2014). Some manufacturers have begun producing iron-fortified Spirulina supplements to increase their iron content. These fortified supplements typically contain higher iron levels, up to 20 mg g^{-1}.

Vitamins in Spirulina include β-carotene (pro-vitamin A) 5.8 mg g^{-1}, C 1.15 mg g^{-1}, E 0.41 mg g^{-1}, K 1.09 mg g^{-1}, B_1 0.048 mg g^{-1}, B_2 0.055 mg g^{-1}, B_3 0.15 mg g^{-1}, B_5 0.002 mg g^{-1}, B_6 0.008 mg g^{-1}, and B_9 0.0007 mg g^{-1} (Liestianty et al. 2019; Rahim et al. 2021). One of the most common misconceptions about spirulina is that it is a rich source of vitamin B_{12}. However, Spirulina itself does not produce vitamin B_{12}, but rather, some of the accompanying bacteria that grow alongside Spirulina culture are able to produce it (van den Berg et al. 1991).

Therefore, some Spirulina products may contain vitamin B_{12} while others do not. Various forms of vitamins also should be considered. Some studies have reported the presence of B_{12} analogs in Spirulina, which are molecules that structurally resemble B_{12} but do not have the same biological activity (Watanabe et al. 1999). Spirulina may contain up to 120 μg 100 g^{-1} of various vitamin B_{12} analogs. These analogs are primarily corrinoids, cobalt-containing compounds that share a corrin nucleus similar to biologically active cobalamin. However, they lack the 5,6-dimethylbenzimidazole group in the β-position to the Co^{3+} atom centered in the corrin nucleus (Kumudha et al. 2010; Sotiroudis and Sotiroudis 2013). While these analogs may be detected in Spirulina biomass, they are not considered a reliable source of vitamin B_{12} for human consumption (van den Oever and Mayer 2022; Watanabe et al. 1999). Some studies have suggested that certain strains of bacteria associated with Spirulina, such as *Pseudomonas* and *Lactobacillus*, may produce active vitamin B_{12}. However, the amount of vitamin these bacteria produce is unlikely to be consistent. Therefore, it is not recommended to rely on Spirulina as a sole source of vitamin B_{12} (Watanabe et al. 1999).

Spirulina biomass has relatively low-fat content, where the proportion of polyunsaturated fatty acids (PUFAs) is between 1.5% and 2% of the total fat content, and has high GLA content, an omega-6 fatty acid (Huarachi Olivera et al. 2015; Semih and Ruhsen 2001). GLA accounts for 36% of the total PUFA's (Hoseini et al. 2013). However, one nutrient Spirulina is often mistakenly thought to contain is omega-3 fatty acids. Probably, confusion is because Spirulina and *Chlorella* are often described together, and *Chlorella* can contain significant amounts of omega-3 fatty acids (Jahromi et al. 2022).

4 Effect of Cultivation Conditions on Spirulina Nutritional Value

As it is well known that weather and soil conditions affect the fruit and vegetable taste, cultivation conditions of Spirulina, such as nutrient availability, temperature, and light, can significantly affect the growth and biomass composition (AlFadhly et al. 2022).Typically, Spirulina grows at a relatively high pH level that restrains the growth of other algae in the system (Khan et al. 2018; Srivani et al. 2017). The optimal water temperature for Spirulina cultivation is 30–35 °C (Murugan and Rajesh 2014); the sunlight should be shaded in most cultivation systems at noon. Cold nights, cool spring, and autumn days can affect the growth and composition of Spirulina (Ciferri 1983). If Spirulina is cultivated under stress, its protein content may decrease significantly from 70% to 40%, while carbohydrate and lipid contents may increase (25% to 74% and 6% to 30%, respectively) as energy storage compounds. This response is believed to be an adaptation mechanism to improve the cell's energy metabolism and survival under stress conditions (Saxena et al. 2022).

Light intensity and temperature are essential factors affecting Spirulina's protein and carbohydrate content (AlFadhly et al. 2022). The increasing light intensity can decrease the protein content of Spirulina (by about 20%), while rising temperature can slightly increase Spirulina's protein content and decreases its carbohydrate content (Junique et al. 2021; Torzillo and Vonshak 1994). Also, elevated CO_2 concentration results in an increase in the carbohydrate content even as high as 50%, while reducing the protein (from 5 up to 30%) and pigment content (from 5 up to 20%) (Gordillo et al. 1999). The amount of pigments in Spirulina can vary depending on nutrient availability, including iron, phosphorus, and nitrogen (Saxena et al. 2022). Deficiencies in nitrogen can decrease phycocyanin by up to 50%, while iron limitation can affect chlorophyll content (from 50 to 70% of Chl amount), and phosphorus limitation can reduce protein content by up to 30% (Markou 2012; Nematollahi et al. 2020; Rout and Sahoo 2015). Increasing of media iron concentration during cultivation can reduce phycobiliproteins content, while increasing of light intensity enhances total carotenoid content of Spirulina for about 20–30% (Akbarnezhad et al. 2020; AlFadhly et al. 2022). Furthermore, it has been suggested that microalgae can undergo a carotenogenesis process in response to various environmental and cultural stresses such as light, temperature, salts, and nutrients, where the alga stops growing and changes its carotenoid metabolism dramatically, accumulating secondary carotenoids as an adaptation to extreme environment (Bhosale 2004). The influence of different growth conditions on microalgae cultivation as well as different cultivation systems are discussed in detail in Chap. 2. However, the aforementioned few examples are meant to draw readers' attention to the fact that it is essential to know where and how Spirulina is cultivated or request quality control information.

5 Effect of Drying on Spirulina Nutritional Value

About 99% of Spirulina biomass are dried to guarantee microbiological safety and long-term shelf-life of the dietary supplement, and they are light-weight and have no need for temperature control during transportation. In contrast, fresh-frozen biomass is heavy due to high water content (80–95%) and should be transported and stored under −18 °C. Therefore, the effect of drying on the nutritional value of Spirulina will be highlighted here in order to determine the optimal treatment method that could be chosen for various nutritional needs. From clinical studies, it can be concluded that phycocyanin is the most valuable component of Spirulina as a supplement.

The blue pigment—phycocyanin extract can be used as natural dyes. Blue dyes aren't often found in food. The main blue pigments are anthocyanins and phycocyanin. While anthocyanins can be extracted from various plants, phycocyanin can be recovered from cyanobacteria. Dyes must be approved by the EU and in the USA before they can be used in food (Olas et al. 2021). However, organic dyes are well-known to be sensitive to heat treatment. Some examples of the drying effect on

Table 1 Effect of drying method, temperature, and duration on phycocyanin loss

Drying regime	Phycocyanin loss, %	References
Oven drying (thin layer, oven at +80 °C, 7 h)	35.4	Güroy et al. (2017)
Freeze drying (−60 °C, 22 h)	0	Güroy et al. (2017)
Freeze drying (−60 °C, 3 h; 0.2 mbar 12 h)	0	Demarco et al. (2022)
Spray drying (150 °C, few seconds)	45	Sarada et al. (1999)
Oven drying (60 °C, 7 h)	46	Sarada et al. (1999)
Batch tray dryer: 50 °C, 3 mm, 192 min 60 °C, 3 mm, 150 min 70 °C, 3 mm, 133 min 50 °C, 5 mm, 369 min 60 °C, 5 mm, 210 min 70 °C, 5 mm, 145 min 50 °C, 7 mm, 608 min 60 °C, 7 mm, 300 min 70 °C, 7 mm, 214 min	44 44 75 56 40 76 82 85 93	Huarachi Olivera et al. (2015)
Vacuum drying (vacuum pressure of 13.3 kPa), 4 mm sample: 40 °C 600 min 50 °C 480 min 60 °C 420 min Oven drying 55 °C 540 min	26 66 72 81	Larrosa et al. (2018)

phycocyanin are summarized in Table 1. As it can be seen in the Table, the lower layer thickness decreases the drying time more effectively than increased temperature (Huarachi Olivera et al. 2015). Therefore, it reduces the exposure time to hot air and time of the product in increased temperature, which is essential to preserve maximum phycocyanin during drying (Huarachi Olivera et al. 2015). Comparing Larrosa et al. (2018) and Oliveira et al. (2010) results, temperature and time spent for drying are the most important parameters, not the vacuum or air used. The contact area of sample and air is the highest in spray drying, but as the time of drying is much shorter and sample do not reach the temperature of drying air, the degradation of phycocyanin is highly limited (Sarada et al. 1999). However, freeze-drying is the most appropriate method not to destroy the phycocyanin, while exposure to hot air leads to 26 up to 93% loss of phycocyanin.

Pigments are the most sensitive compounds to heat treatment, but also other compounds degrade during drying. For example, most enzymes are completely inactivated when exposed to temperatures close to 100 °C. Additionally, the loss of the SOD activity by over 50% at 80 °C has been reported in dried *Arthrospira* biomass (Ma et al. 2019). The total protein and sugar losses correlated with the used air temperature in the drying oven were described by Desmorieux and Hernandez (2004). More significant were total sugar losses (about 30%) than protein losses (10–20%). Protein losses are proportional to the drying air temperature in the range of 40 and 70 °C. However, the loss of total sugars was constant at the same time temperature range. In contrast, the freeze-drying resulted in a minimal loss (<10%) for proteins and total sugars. The highest loss of proteins and total sugars was obtained by convective and infrared drying in cylinders, while thin layer drying resulted in much lower losses. However, it is noted that air dried products did not have acceptable organoleptic properties (Desmorieux and Hernandez 2004).

The specific organoleptic properties of Spirulina powder are much discussed in various reports. Many people found specific flavors, characterized by a dirty pond smell and taste, as a significant problem using Spirulina daily (Thomas et al. 2020). There are several methods to overcome the barrier of this taste. The most popular solution is using Spirulina in capsules or tablets, while the other method is to mask Spirulina flavor with other ingredients, e.g., lemon, in smoothies. The third solution is a newly arising trend to offer fresh frozen Spirulina, which has no specific aroma or taste; however, after melting, it spoils within few hours at room temperature (https://www.spirulinanord.eu/for-food-producers). While no other conservation methods are developed, fresh Spirulina can be frozen for at least 24 months (https://www.freshspirulina.com.au/frequently-asked-questions-faq/).

It is essential to mention that the drying method impacts product quality, production time, and costs. Spray drying is the most widely used in large Spirulina farms as it is the most efficient and fast method (Neves et al. 2019). Although this method allows producing powders with a relatively high retention of functional components, it was demonstrated that due to cell structure degradation that occurs during drying, these components could be lost during storage under inadequate conditions (Neves et al. 2019). Although conventional drying of Spirulina biomass constitutes approximately 30% of the total production cost (Ma et al. 2019), freeze-drying is more expensive. So, alternative drying methods and ways to increase Spirulina's shelf life should be investigated.

Freeze-drying allows higher retention of functional components immediately after drying (Bhatta et al. 2020). Similarly, to spray drying the inappropriate storage of dried spirulina can significantly decrease, the content of functional components as the freeze-dried material has high porosity that facilitates contact with oxygen and air humidity, promoting oxidation during storage (Muhoza et al. 2023). In addition, freeze-drying is relatively expensive and energy-consuming compared to conventional drying; however, crunchy texture and fresh taste and smell of freeze dried products increase the popularity and offer such food rapidly (https://capitalmag.co.nz/2022/09/20/freeze-drying-is-the-new-hot-food-trend/). In spray drying, a particle dries within seconds, while freeze drying takes hours or even days depending on the moisture content (Schmitz-Schug et al. 2013; Vilas et al. 2020). As an intermediate method, air-drying has been discussed as one of the most studied cheap methods, spreading the wet biomass in a thin layer increases the evaporation rate allowing shorter drying time and using lower temperature, thus allowing higher retention of functional components in the final products (Desmorieux and Decaen 2005). Novel drying systems have been developed by combining thin layer drying under vacuum, reducing the drying temperature and time, as well as exposure to oxygen, therefore, obtaining higher-quality products (Neves et al. 2019). In general, thin-layer vacuum drying is cheaper than spray drying or freeze-drying in terms of equipment and operational costs (Neves et al. 2019). Mechanical dewatering of microalgae suspension is essential for saving costs and reducing the energy consumption spent for evaporation (Agbede et al. 2020).

6 Health Benefits of Spirulina

Spirulina is a well-known food supplement having antioxidant, anti-inflammatory, and immunomodulatory properties (Liu et al. 2016). In a fast screening, almost 3000 clinical studies on Spirulina health benefits have been published in the PubMed scientific database having Spirulina or *Arthrospira* in the title or abstract since 1967. The number rapidly increases with each year; as for example, 2000–2009 showed 530 articles published, which increased to 1234 articles in the period 2010–2019, and 880 articles in the last 2 years (2020–2022). However, only 10 meta-analyses of clinical studies of Spirulina were published in the PubMed scientific database during the last 10 years.

Reduction of body weight is one of the well-studied topics for Spirulina applications. Two systematic reviews and meta-analysis studies concluded that Spirulina supplementation can significantly reduce body weight, especially in obese individuals (Bohórquez-Medina et al. 2021; Moradi et al. 2019). Additionally, significant reductions in body fat and waist circumference were also observed, but not in the body mass index and waist-to-hip ratio (Moradi et al. 2019). Similar findings for obese patients were reported by Zarezadeh et al. 2021, where beneficial effects on weight and waist circumference as well as positive effect on the body mass index were revealed in the study that lasted for more than 12 weeks. In addition, significant lowering of systolic and diastolic blood pressure (by 4.6 and 7.0 mmHg on average) was reported by analyzing various studies that included 230 hypertension patients consuming Spirulina (Machowiec et al. 2021).

Glycemic control and glucose metabolism in patients with diabetes mellitus were analyzed in a systematic review and meta-analysis of seven clinical and 27 preclinical studies (Stettler et al. 2006). Pooled results of the clinical studies showed that Spirulina supplementation significantly reduced fasting blood sugar (FBS), total cholesterol (TC), and triglycerides (TG), with significant increase in high-density lipoprotein cholesterol (HDL-C). At the same time, it was not effective in reducing the glycated hemoglobin A1c (HbA1C) and low-density lipoprotein cholesterol (LDL-C). Pooled results of preclinical studies showed that Spirulina supplementation significantly reduced FBS and HbA1C in diabetic animals. However, the effective dose of reducing the FBS in humans was less than 2 grams consumed in less than two months. Therefore, Spirulina could be effective FBS, TG, TC, and HDL-C adjusting nutraceutical agent for diabetes mellitus (Serban et al. 2016). Similar results were found for patients with metabolic syndrome (MetS) and related disorders, where positive effect on insulin resistance was reported.

Other promising benefits include preventing skeletal muscle damage under high physical load conditions, immunomodulatory, and anti-inflammatory responses, and neuroprotective effects (Chen et al. 2022). It is still quite unclear how the complex composition of bioactive substances in Spirulina act in the molecular level, but many above-mentioned studies report that phycocyanin and β-carotene are important molecules. In addition, Spirulina activates cellular antioxidant enzymes, inhibits lipid peroxidation and DNA damage, scavenges free radicals, and increases the activity of superoxide dismutase and catalase (Wu et al. 2016). Oxidative stress and dysfunctional immunity cause many diseases, including atherosclerosis, cardiac hypertrophy, heart failure, and hypertension (Wu et al. 2016).

As discussed in detail in Chap. 16, microalgae showed high potential in aquaculture. Specifically, Spirulina has been studied extensively in fish feed, where its beneficial effects on growth were reported in omnivores and herbivores fishes, but not for carnivores fish (Tadesse 1998). The growth benefits were dose-dependent, and higher inclusion levels of *Spirulina* resulted in better growth; benefits were apparent from very modest inclusion levels, with 1% and less than 45% being the maximum *Spirulina* replacement considered (Trevi et al. 2023). Growth was improved in 71% of 17 species examined, but the best results were recorded in Cichlidae (Tilapia), Clariidae (airbreathing catfishes), and Mugilidae (mullets) species which are all herbivorous (Trevi et al. 2023). Studies that weighed fish individually were also more likely to reveal a positive effect of *Spirulina* on growth than those which used batch weighing. A negative impact (weight loss) was reported in 5% of studies and involved three species: Nile tilapia, mullet, and rainbow trout (Trevi et al. 2023). Almost half of the studies showed no effect on fish growth. Highly variable outcomes are common in microalgal studies in aquaculture as discussed in Chap. 16. As an example, the inclusion of *Chlorella vulgaris* in aquafeeds gave 36% significant improvements in 11 studies that examined changes in growth or fillet quality, 36% with no discernible benefit, and 27% negative effects (Ahmad et al. 2020). The high variability of results is explained by other factors in the aquaculture system, e.g., feed composition and fish stocking density, as well as the nutritional value of microalgae that differs between strains and producers (Trevi et al. 2023).

Spirulina in a feed of beef cattle, sheep, goats, llamas, and alpaca also, improved growth, health, and product quality (Holman and Malau-Aduli 2013). In addition, studies have shown that adding Spirulina to the diets of livestock can increase weight gain (Holman and Malau-Aduli 2013), milk production (Otto and Malau Aduli 2017), and reproductive performance (Iatrou et al. 2022). However, Spirulina is relatively expensive to produce, and purchase compared to other animal feeds, and therefore, Spirulina could not be considered as a competitive alternative cheap protein source to soy, but as a valuable additive containing antioxidants, vitamins, and minerals (Altmann and Rosenau 2022).

7 Supply and Demand of Spirulina

Spirulina is the world's most produced microalga, followed by Chlorella, *Dunaliella*, and *Haematococcus*. The Spirulina market is expected to reach $968.6 million by 2028, at a CAGR of 13.2% from 2021 to 2028 (https://www.meticulousresearch.com/product/Spirulina-market-5070). The increasing awareness of the health benefits of Spirulina, particularly its high protein and antioxidant content, has led to a surge in demand for this algal species as a dietary supplement. The market has also been buoyed by increasing demand for natural and organic products, as Spirulina is a natural alternative to synthetic supplements (Janda-Milczarek et al. 2023).

So far, the largest producers of Spirulina are in Asia, particularly in China and India (Table 2). In general, Asian countries have long been involved in Spirulina production and have a well-established infrastructure for cultivating and processing microalgae. There are more than 500 large Spirulina producers in China and it is the world's largest Spirulina producer, accounting for more than 50% of the global production. India is the second largest producer, with a significant portion of its Spirulina production used for domestic consumption. The increasing demand for Spirulina in Asia region can be attributed to the growing awareness of its health benefits and its use in traditional medicine. The United States is also an essential player in the Spirulina market, with several large producers. One of the biggest producers of Spirulina in the USA is Cyanotech Corporation, based in Hawaii. The company has been involved in Spirulina cultivation for more than 30 years, producing Spirulina as powder and tablets.

In recent years, following the demand, a significant increase in Spirulina production in Europe can be recorded. However, Asia remains the dominant player in the market. The JRC algae database (Araújo et al. 2021) is a valuable source for understanding the current state of Spirulina production in Europe. With 213 Spirulina-producing enterprises spread across 15 countries, cultivating this nutrient-rich alga is a growing industry in the region. France leads the way in Spirulina production, with 134 enterprises dedicated to its cultivation. The French Spirulina industry has a long history, with the first commercial operation established in the 1970s. Many of these producers focus on organic and sustainable cultivation practices, reflecting growing consumer demand for eco-friendly and healthy products. However, most of the producers are small businesses and sell locally. Italy, Spain, and Germany are important players in the European Spirulina market. Italy has 20 Spirulina-producing enterprises, and the industry has experienced significant growth in recent years. Spanish producers are also on the rise, with 18 enterprises cultivating Spirulina, while Germany has 14 enterprises focusing on high-quality and organic Spirulina for use in health supplements and functional foods. Table 2 summarizes some of the largest Spirulina manufacturers in the world. In North America and Europe, the demand for Spirulina is driven by the increasing popularity of plant-based diets and the growing demand for natural and organic food products. Spirulina market in these regions is expected to grow significantly in the coming years, driven

Table 2 Summary of some global leading Spirulina producers

Company/enterprise	Country	Foundation Year	Pond area, ha	Annual yield
Yunnan Green A Biological Project Co., Ltd	China, Yunnan province	1997	–	3000 ton powder, tablet, and extract-phycocyanin (spirulina, chlorella, and Haematococcus)
Inner Mongolia Rejuve Biotech Co. Ltd.	China, Wukan	2006	63	Spirulina powder is 1100 tons, spirulina tablet is 100 tons
Earthrise Nutritionals LLC. Operates as a subsidiary of DIC Corporation.	Headq. at California, U.S.;	1976	44	
Fuqing King Dnarmsa Spirulina Co. Ltd	FuQing, China	1995	–	1600 tons of spirulina and 400 tons of chlorella (spirulina powder and tablets)
Zhejiang Comp Spirulina Co	Jiangshan, China	1997	–	Spirulina powder—800 tons and chlorella –50 tons. Powder, tablets, and capsules, + feed grade
Dongtai City Spirulina Bio-engineering Co. Ltd.	Dongtai, China	1994	60	Tablet, powder, and phycocyanin powder
Qingdao Haizhijiao Biotechnology Co., Ltd	China	1998	100	1000 tons of spirulina and 150 tons of chlorella
Far East Microalgae Industries Co., Ltd (FEMICO)	Taipei, Taiwan	1976	14	200 tons of spirulina, 1000 tons of chlorella
Cyanotech Corporation	Kailua Kona, U.S.	1984	36	Microalgae (Bioastin) and spirulina
E.I.D.—Parry (India) Limited	Chennai, India	1788	52	Spirulina, chlorella, and microalgal products
DIC Corporation (Japan)	Japan	1908	30	1000 ton spirulina powder
Algene Biotech (India)	India	2011	20	Spirulina capsules
Tianjin Norland Biotech Co., Ltd.	China	2008	35	1100 tons microalgae (spirulina, chlorella, Haematococcus)
Cyanotech corporation (U.S.)	Hawaii	1983	36	Spirulina powder, tablets
Australian spirulina	Australia	1996	–	1100 ton spirulina powder
Spiruline du Soleil	France	2007	1000 m^2	Spirulina powder, tablets, flakes
Spiruline La Capitelle	France	2001	300 m^2	300 kg spirulina

by the increasing demand for natural food and dietary supplements. In terms of application, Spirulina market is segmented into dietary supplements, food and beverages, animal and aquaculture feed, and others.

8 Spirulina Laws and Regulations

Spirulina is gaining popularity, and many questions regarding the quality and safety of the product arise. Standardization and regulation serve both producers and consumers by ensuring consistent quality and safety standards. Algae are currently not a typical food in many countries, so laws and regulations that govern its production and marketing are under investigation. In the United States, Food and Drug Administration (FDA) has stated that Spirulina is "generally regarded as safe (GRAS)" and classifies Spirulina as a dietary supplement (Ersyah et al. 2022). According to the Novel Food Catalogue of the European Commission, Spirulina is among the foods that have been consumed to a significant degree in at least one Member State of the European Union before 15 May 1997. Therefore, it is considered as non-novel (or traditional) food, and can be placed in the EU market without a novel food pre-market authorization.

According to the European Algae Biomass association analysis (Vieira et al. 2022), cyanobacteria is a group of microalgae that carry out photosynthesis and, therefore, it can be classified as a plant-based food. If algae are cultivated (inoculated, managed cultivation process, protected from predators, etc.), then it is belonging to the *aquaculture* sector, while algae harvesting is not. Spirulina cultivation is also considered as *agriculture* sector—referring to the art and science of growing plants and raising animals for food, feed, and many other economic activities. Therefore, algae cultivation-based products are considered as *crops*.

Algae production in reactors requires specific production processes and, therefore, *industrial products*. Microalgae production is analogous to bacteria and yeast invasion at an industrial scale, with industrial processes ranging from closed dark fermentations to indoor autotrophic production and large-scale outdoor ponds.

Although all Spirulina should be cultivated without the use of pesticides and herbicides, *organic* refers to products that are cultivated without synthetic fertilizers also. The main difference is that organic fertilizers are applied in the soil that is not in direct contact with the crops (e.g., corn and fruits) in organic agriculture. Therefore, organic seaweed regulations cannot be used for microalgae (Vieira et al. 2022). Finding suitable organic fertilizers for algaculture is a challenge. It could be a very unsustainable choice for some ingredients available only in far countries while synthetic analog is produced in the closest neighborhood. Furthermore, organic regulations designed around soil-based systems do not transfer well into algaculture or aquaculture. That is why the production of organic algae requires specific aquaculture-based regulations. However, organic Spirulina from China and India is already widely available.

The quality and safety of various foodstuffs differ—unified standard needs for Spirulina. Spirulina is a microalga that does not have peels and is an excellent absorbent and perishable, then safety during production and storage differs from general vegetable storage (Lucas et al. 2018; Tiburcio et al. 2007). General quality requirements show the parameters included are protein, total carotenoids, chlorophyll-*a*, phycocyanin, moisture, standard plate count, mold, pathogens (e.g. with *Saccharomyces* sp., Coliform bacteria, *Staphylococcus aureus, Salmonella enteritidis)*, heavy metals (lead, mercury, cadmium, arsenic), insect fragment, and rodent hair (Ma et al. 2019).

9 Spirulina Products in the Market

There are various products containing Spirulina; 99% of them contain dried Spirulina and is sold as a powder or tablets, with a new direction to introduce fresh Spirulina (https://www.meticulousresearch.com/product/natural-food-colors-market-5088). Some of different selling forms of Spirulina are discussed in this section.

9.1 Powder

The most common form of dry Spirulina is sold as a powder form and is the most prevalent (53–55% of the global produced Spirulina). This is because dried products are easy to store and transport. Spirulina powder is found in health food stores, drugstore as well as in most of the largest supermarket chain supplement shelves. The powder can be mixed into water, juice, or other beverages and can also be added to smoothies or used as a natural food coloring. Tablets and capsules are another common form of Spirulina sold around the world. About ~25% are sold in the form of tablets and ~ 15% in the form of capsules (https://www.meticulousresearch.com/product/natural-food-colors-market-5088). Such products are convenient for people who want to take Spirulina as a supplement but do not want to deal with the taste or mess of the powder form. Flakes are a less common form of Spirulina products representing less than 9% of sold Spirulina (https://www.meticulousresearch.com/product/natural-food-colors-market-5088). Spirulina flakes are made by drying Spirulina in a thin layer and then breaking it into small pieces. They can be added to salads or any ready meal and are a good option for people who want to consume Spirulina in a more natural form.

When it comes to the global market for Spirulina, Asia-Pacific region is the largest consumer of Spirulina products by volume. In China, 40% of Spirulina (by volume) is sold. In contrast, the United States is the largest Spirulina market in the world by value –50% of Spirulina by value and 25% by volume is sold. This shows that the average price of Spirulina is fourfold higher in the USA comparing to China. Despite Europe consuming relatively low amount of only 10% of the global

Fig. 1 Examples of Spirulina powders

Spirulina produced, the superfood market is rapidly growing due to the increase in healthy food and self-care trends. Some examples of Spirulina powders are given in Fig. 1.

9.2 *Fresh Frozen*

Fresh frozen Spirulina is a relatively new product in the market, which is still in its early stages. It is expected to grow significantly in the coming years due to the much better taste and nutritional value in comparison with Spirulina powder (Agustini et al. 2015). Frozen Spirulina is often a trend of small Spirulina farms, and selling locally as the transportation of frozen goods are a more exclusive option than the delivery of Spirulina powder. Spirulina is offered frozen in one-portion ice cubes or balls. Due to the small production volumes of small farms, the price of fresh frozen Spirulina usually is several times higher. Fresh frozen Spirulina are for example sold by companies such as New Farmers, OASIA farm, Spira, and Greenspring Farms in USA, SpiraVeg in Canada, SIMPLIIGOOD in Israel, Umamiz, Spirul'in Vosges, and SpirulinaNord in Europe. Some examples of fresh frozen Spirulina are given in Fig. 2.

9.3 *Spirulina in Snacks*

Spirulina-based snacks have become increasingly popular due to their high nutritional value and convenience. These snacks are typically made by mixing Spirulina powder with other healthy ingredients such as nuts, seeds, and dried fruits. One example of such a product is Spirulina energy balls, which are made with dates fruit, almonds, cocoa powder, and Spirulina powder (https://naturya.com/blogs/recipes/spirulina-cacao-energy-balls). They are a healthy snack option that is easy to make at

Fig. 2 Examples of fresh frozen Spirulina

home and can be taken on the go. Other popular Spirulina-based snack options include energy bars, which are often made with nuts, dried fruits, and popcorn, which is often coated with Spirulina powder for added flavor and nutrition. Such a way of serving Spirulina was chosen by the European Space Agency offering musli bar with Spirulina and goji berry bars for astronauts (https://www.esa.int/ESA_Multimedia/Images/2016/09/Researching_spirulina).

One of the main advantages of Spirulina-based snacks is their high protein content. Spirulina is a complete protein source, meaning that it contains all the essential amino acids that human body needs. This makes Spirulina an ideal ingredient for snacks that can provide a quick protein boost on the go. Consuming protein-rich snacks helps to increase feelings of fullness, which can prevent overeating and contribute to weight management (Moon and Koh 2021). Another benefit of Spirulina-based snacks is their convenience. Many of these snacks are easy to carry around and can provide a quick and healthy snack option. This is especially important for people with busy lifestyles who may not have time to prepare healthy meals or snacks throughout the day. Spirulina-based snacks can be an excellent alternative to unhealthy snacks that are high in sugar and preservatives, providing a healthier option for snacking (Lucas et al. 2019).

Spirulina snack market is growing rapidly, with many companies offering a variety of products to choose from. Some of the most popular Spirulina snacks on the market include energy bars, popcorn, chips, and crackers. These products are

Fig. 3 Examples of commercial snacks containing Spirulina

often marketed as healthy snack options that provide a quick energy boost without the added sugar and preservatives found in many traditional snacks. Some examples of Spirulina-containing bars are given in Fig. 3, and a list of ingredients of some products is given Table 3. These products typically contain a combination of healthy ingredients, such as nuts, seeds, and dried fruits, in addition to Spirulina powder.

9.4 Spirulina in Drinks

Spirulina drinks can come in a variety of forms, such as smoothies, juices, and energy drinks. They are typically made by mixing Spirulina powder with other healthy ingredients such as fruits, vegetables, and herbs. Spirulina drinks are often marketed as a convenient and nutritious way to add more superfoods to your diet. One of the most common types of Spirulina drinks is Spirulina smoothies, made by blending Spirulina powder with fruits and/or other healthy ingredients such as yogurt, almond milk, and chia seeds (https://www.simplysqueezed.co.nz/our-range/smoothies/). This combination creates a delicious and creamy drink that is packed with protein, fiber, and antioxidants. Another type of Spirulina drinks is Spirulina juice, made by mixing Spirulina powder with water and lemon juice or

Table 3 Ingredients of some commercial products containing Spirulina

Products/Ingredients	Size	Producer	Country	References
Bars				
Almonds 10%, sunflower seeds 14%, raisins 15%, chicory syrup 19.8%, dates 15%, fava bean protein 8%, roasted pumpkin seed protein 12%, lemon oil 1.2%, **spirulina 5%**	50 g	Green bite OÜ	Estonia	https://greenbite.ee/shop/superfood-bars/girl-power-superfood-bar
Banana*, sprouted sesame seeds*, dates*, coconut (Unsulphured)*, **spirulina***, sea salt	48 and 14 g	Go raw	Northbrook, IL	https://goraw.com/products/spirulina-sesame-sprouted-bars
*dates 43%, *ALMONDS, *sultana raisins (*sultana raisins, *sunflower seed oil), *spirulina 5%, *lemon powder 1.3% (concentrated lemon juice 50%, *corn flour), *lemon essential oil 0.3%.	40 g	Lubs GmbH	Germany	https://www.lubs.de/en/produkte/fruit-plus-bars/
Drinks				
Banana puree 42.7%, kiwifruit puree 5.1%, apricot puree 1.1%, orange juice 24.1%, apple juice 15%, water, **spirulina powder 0.5%**	350 and 800 mL	Simply squeezed	New Zealand	https://www.simplysqueezed.co.nz/our-range/smoothies/
Water, **aqueous extract of spirulina 7%,** lemon juice 3%, beet sugar 3%	240 mL	Spirulina Becagli	Italy	https://spirulinabecagli.it/en/collections/bevande
Raw kombucha (water, sugar, green tea, tarter cultures), lime juice concentrates 0.5%, **spirulina powder 0.5%,** probiotic cultures, and stevia leaf extract.	340 mL	Dr. Tom's Kombucha	Republic of South Africa	https://drtoms.co.za/
Pasta				
Durum wheat 96%, **dry spirulina 4%**	500 g	Spring spirulina	Singapore	https://springspirulina.com/th/product/premium-pasta-penne-spirulina/
Whole rice flour 97%, **spirulina 3%**	250 g	Sarafino	USA	https://store.sarafino.com/collections/grains/products/spirulina-penne-pasta
Cheese (dairy)				
Organic pasteurized sheep's milk, Volterra salt, **spirulina 0.1%**, rennet	250 and 500 g	Spirulina Becagli	Italy	https://spirulinabecagli.it/en/products/pecorino-bio-con-spirulina

(continued)

Table 3 (continued)

Products/Ingredients	Size	Producer	Country	References
Raw goat Milk, **spirulina**	100 g	Happy days dairy	Canada	https://www.kornapet.com/Raw-Goat-Milk-Cheese-with-Spirulina-100g-784725651913-49760/
Pasteurized Milk, garlic, **spirulina**, microbial rennet, salt	200 g	Old Hill	India	https://www.anata.in/collections/keto-diet/products/gouda-with-spirulina-and-garlic-cheese
Cheese (vegan)				
Cashews 63%, culture, coconut oil, miso (soybeans, brown rice, salt, water), citric acid, salt, **spirulina powder**.	120 g	Kinda co.	England	https://thekindaco.com/products/spirulina-blue
Cashew* (68%), water, **spirulina* (2.6%),** salt, chilli* (0.05%), vegan fermentation cultures	120 g	Happy Cheeze	Germany	https://www.pour-nourrir-demain.fr/happy-cheeze-un-fromage-vegan-bleu
Ice-cream				
Coconut milk* (coconut, water, stabilizer: Guar gum), agave syrup*, **spirulina*(4%),** salty licorice syrup (water, cane sugar, glucose, licorice powder (11.7%), salmiak, salt) 4%, orange juice*, ginger*, peppermint oil.	125 mL	Landgren lab	Sweden	https://landgrenlab.se/

*organic ingredient

other citrus fruits (https://nutejuice.com/products/blue-spirulina). This type of drink is ideal for those who prefer a lighter and more refreshing taste. Spirulina juice can be a quick and easy way to incorporate Spirulina into diet, where lemon juice provides additional health benefits due to its high vitamin C content.

Spirulina energy drinks are also becoming more popular, which are typically made by mixing Spirulina powder with other natural energy-boosting ingredients such as green tea extract, ginseng, and guarana (https://fulcompany.com/). Spirulina energy drinks are often marketed as healthy alternative to traditional energy drinks that are high in sugar and caffeine. These drinks can provide an instant energy boost without the crash that is often associated with traditional energy drinks. In addition to these types of Spirulina drinks, there are also several more unique and specialized options available. For example, some companies offer Spirulina-infused coconut water or sparkling water. These products offer a convenient and refreshing way to consume Spirulina, especially for those who may not enjoy the taste of Spirulina on its own. Spirulina sparkling water is a great option for those who prefer a carbonated

Fig. 4 Examples of commercial drinks containing Spirulina

beverage but want to avoid the added sugar and artificial flavors found in many soda drinks. Spirulina drinks can help boost energy levels, support immune function, and promote overall health and well-being (Karkos et al. 2011). Some examples of commercial Spirulina-containing drinks are given in Fig. 4, and more details about the ingredients are shown in Table 3.

9.5 *Spirulina in Pasta*

The pastas are foods that can be used to incorporate additives due to their simple manufacturing process, low cost, nutritive value, and finally due to their high acceptance. Lemes et al. (2012) has reported that even 10% of Spirulina can be added to fresh pasta; however, elasticity was reduced comparing to pasta with 5% spirulina. For conventional pasta, up to 20% of Spirulina could be added technically, but various parameters, e.g., chewiness and firmness, were significantly affected (De Marco et al. 2014). In market, usually no product contains more than 5% of dried spirulina due to organoleptic changes (Koli et al. 2022).

Spirulina pasta provides a healthy twist to traditional pasta dishes while also providing a range of nutritional benefits. Pasta with spirulina can exhibit high phenolic compounds content and antioxidant activity compared to control pasta; however, it is dependent on a microstructure of pasta, as phenolic compounds can leach into the cooking medium and degrade due to thermal treatment (De Marco et al. 2014). In terms of taste, Spirulina pasta has a slightly earthy nutty taste

Fig. 5 Examples of commercial pasta containing Spirulina

(Manzocchi et al. 2020; Matos et al. 2022) that complements many different types of sauces and seasonings. It is important to note that Spirulina pasta has a deep green color, which people may find unusual—attractive or frightening. It can also be a great way to include some extra nutrients for picky eaters or those who may not be getting enough vegetables in their diet.

There are different types of Spirulina pasta available on the market, usually containing 2–4% of spirulina powder. Various shapes pastas are available, e.g., spaghetti, linguine, fusilli, spirals, or shells. Pasta containing spirulina is offered by companies in various countries, like felicia (Italy), Explore Cuisine (USA), and Spring spirulina (USA), interestingly that comparing to other products, pasta containing spirulina often is sold in e-shops without indication of manufacturer. Some examples of Spirulina containing pasta are given in Fig. 5, with more details about their ingredients in Table 3.

9.6 Spirulina in Dairy Products

One category of food products where Spirulina has been successfully incorporated is dairy products. Spirulina can be added to a variety of dairy products, including cheese, yogurt, and ice cream, providing a range of nutritional benefits. One of the most popular Spirulina-based dairy products is Spirulina yogurt. Spirulina yogurt is

Fig. 6 Examples of commercial dairy/fermented products containing Spirulina

typically made by adding Spirulina powder to yogurt (https://www.bareorganics.com/blogs/recipes/spirulina-yogurt-parfait). The resulting product is high in protein, antioxidants, and vitamins, providing a great nutritional boost. Spirulina yogurt combines the benefits of probiotics with the nutritional benefits of Spirulina, making it a great addition to a healthy diet. Spirulina yogurt can be consumed as a snack or used in various recipes, such as smoothie bowls or dips. Another Spirulina-based dairy product is Spirulina cheese. Spirulina cheese is typically made by adding Spirulina powder to cheese (https://spirulinabecagli.it/en/products/pecorino-bio-con-spirulina). The resulting product is high in protein, calcium, and antioxidants, providing a healthier alternative to traditional cheese. Cheese is a popular dairy product that is often high in saturated fatty acids and sodium. By adding Spirulina powder to cheese, the nutritional profile of the cheese is improved, making it a healthier option for consumers (Golmakani et al. 2019; Mazinani et al. 2016). Spirulina has also been incorporated into ice cream, providing a healthier alternative to traditional ice cream. Spirulina ice cream is typically made by adding Spirulina powder to the ice cream base (https://landgrenlab.se/). Ice cream is a popular dessert that is often high in calories and sugar. By adding Spirulina powder to ice cream, the nutritional profile of the dessert is improved, making it a healthier option for consumers (Agustini et al. 2016). Spirulina ice cream can be enjoyed as a standalone dessert or used in various recipes.

Apart from these traditional dairy products, Spirulina has also been used to create non-dairy alternatives such as Spirulina-based plant milk and yogurts. These non-dairy alternatives are a great option for people who are lactose intolerant or following a vegan diet. Spirulina-based plant milk can be made by blending Spirulina powder with various plant-based milk alternatives, such as almond milk or coconut milk. Spirulina-based plant yogurts can be made by adding Spirulina powder to a plant-based yogurt alternative, such as soy yogurt or coconut yogurt. Some examples of Spirulina-containing cheese are given in Fig. 6, with more details about their ingredients in Table 3.

10 Impact of Spirulina as a Functional Ingredient

Addition of Spirulina to food products can affect several factors, e.g., the nutritional profile, taste, texture, and color of the final product (Raczyk et al. 2022). In most cases, Spirulina is added as a powder and, therefore, adding too much Spirulina powder can negatively affect the taste of the product due to its specific taste (Grahl et al. 2020). The acceptable level of Spirulina in food products varies depending on the specific food application. However, for most of the commercially available Spirulina-containing products, Spirulina is not added by more than 5% (Koli et al. 2022).

In dairy products: Adding Spirulina to dairy products such as cheese, yogurt, and ice cream can enhance their nutritional value and provide a unique, natural color that can make these products more visually appealing (Golmakani et al. 2019; Martelli et al. 2020; Mazinani et al. 2016). Dairy products are a rich source of essential nutrients such as calcium, proteins, and vitamins, which are necessary for maintaining overall health (Górska-Warsewicz et al. 2019). Spirulina can help to enhance the nutritional value of dairy products without increasing their fat or calorie content. It can improve their nutritional profile by boosting their protein and micronutrient content, such as iron and calcium (Agustini et al. 2016, 2017). An increase in protein content and a decrease in fat content was observed by Agustini et al. by adding Spirulina at 2.29% to 3.55% for yoghurt and 3.79% to 22.62% for soft cheese (2016, 2017). In addition, Spirulina improved the water-holding capacity of cheese and yogurt, which led to reduced whey formation, while it did not significantly alter the overall sensory quality of the products. *Lactobacillus acidophilus* bacteria viability increase was observed by Patel et al. (2019). Its amount increased by more than 29%, while fermentation time decreased by 20% in carotenoid-rich probiotic yogurt (Patel et al. 2019).

The viable counts of *Lacticaseibacillus casei* were maintained more successfully in the probiotic BAF cheese, which contained 0.5 or 1.0% Spirulina; the values of titratable acidity, dry matter, and protein contents of Spirulina containing cheese were higher than the control. Additionally, these samples exhibited softer textures, which led to an easier disintegration and chewing of the BAF cheese (Golmakani et al. 2019). Similarly, the study of Mazinani et al. (2016) confirmed a positive effects of *A. platensis* on the survival of *Lactobacillus acidophilus* during the storage of cheeses. This study also described that adding *A. platensis* biomass significantly increased ($P < 0.05$) the amount of iron, protein, and hardness of probiotic feta cheese during refrigerated storage at 4 °C. Lightness and redness decreased by increasing the content of *Spirulina*, while *b* value (yellowness) increased. Overall, the use of Spirulina in dairy products shows promising results in improving the nutritional content and probiotic properties.

Antimicrobiological abilities: Spirulina has also been shown to possess antimicrobial properties, which can help delay spoilage in food products. The antimicrobial activity of Spirulina is mainly attributed to its phycocyanin, chlorophyll, carotenoids, and polysaccharides content. Antibacterial potential against drug-resistant food-borne bacterial pathogens such as *Escherichia coli*, *Klebsiella pneumoniae*, *Pseudomonas aeruginosa*, *Enterococcus faecalis*, *Staphylococcus aureus*, and *Streptococcus pyogenes* was evaluated by Alshuniaber et al. (2021). They reported that Spirulina metabolites are active against both Gram-positive and Gram-negative pathogens, and phenolic compounds from *Spirulina* provide a natural and sustainable source of food preservatives. Another study investigated the effect of Spirulina extract on the growth of *Pseudomonas fluorescens* and *Serratia liquefaciens* in a model dairy system. The results showed that the Spirulina extract effectively reduces the growth of both bacteria, with a higher concentration of extract resulting in more significant reduction in bacterial growth. In another study, 1.15% *w/v* of *A. platensis* biomass added to sterilized apple juice effectively inhibited the growth of *Aureobasidium pullulans* LW14 and the metabolism of all analyzed sugars (glucose, sucrose, and fructose) (Wajda et al. 2020). Further analysis revealed that the protein fraction present in *A. platensis* significantly contributed to its antifungal properties. Furthermore, when protein fraction of Spirulina was added to unpasteurized apple juice, similar results were obtained, but with the additional effect of inhibiting the growth of some bacteria (Wajda et al. 2020). Therefore, Spirulina is a promising natural preservative inhibiting microorganism growth and degradation of the juice.

11 Conclusions

Spirulina has high nutrition value; therefore, its popularity and demand increases. There are many clinical studies showing various promising health benefits for spirulina users; however, for most of benefits, further studies are needed with wider population. Currently, metabolic improvement, lowering of systolic and diastolic blood pressure for hypertension patients, and diabetes mellitus symptoms relief are proven by meta-analyses of enough clinical studies. The number of clinical studies exploring Spirulina clinically increases with each year almost exponentially; therefore, the demand for spirulina and spirulina containing products will increase. Currently, most of largest producers are in Asia; however, there are a lot of small producers in Europe and the USA that has a potential to grow. Strong limitation of spirulina consumption is its specific odor and flavor. It is solved by offering fresh-frozen spirulina and by incorporating Spirulina in various products, like snacks, drinks, pasta, and dairy, usually in low amounts (max 5%).

Acknowledgments This work has been supported by the European Regional Development Fund within Activity 1.1.1.2 "Post-doctoral Research Aid" of the Specific Aid Objective 1.1.1 "To increase the research and innovative capacity of scientific institutions of Latvia and the ability to attract external financing, investing in human resources and infrastructure" of the Operational Programme "Growth and Employment" (No.1.1.1.2/VIAA/3/19/427).

References

Agbede OO, Oke EO, Akinfenwa SI, Wahab KT, Ogundipe S, Aworanti OA, Arinkoola AO, Agarry SE, Ogunleye OO, Osuolale FN, Babatunde KA (2020) Thin layer drying of green microalgae (*Chlorella sp.*) paste biomass: drying characteristics, energy requirement and mathematical modeling. Bioresour Technol Rep 11:100467. https://doi.org/10.1016/j.biteb.2020.100467

Agustini TW, Suzery M, Sutrisnanto D, Ma'ruf WF, Hadiyanto (2015) Comparative study of bioactive substances extracted from fresh and dried *Spirulina sp*. Procedia Environ Sci 23:282–289. https://doi.org/10.1016/j.proenv.2015.01.042

Agustini TW, Ma'ruf WF, Widayat W, Suzery M, Hadiyanto H, Benjakul S (2016) Application of *Spirulina platensis* on ice cream and soft cheese with respect to their nutritional and sensory perspectives. J Teknol 78:245–251. https://doi.org/10.11113/jt.v78.8216

Agustini TW, Soetrisnanto D, Ma'ruf WF (2017) Study on chemical, physical, microbiological and sensory of yoghurt enriched by *Spirulina platensis*. Int Food Res J 24:367–371

Ahmad MT, Shariff M, Yusoff FM, Goh YM, Banerjee S (2020) Applications of microalga *Chlorella vulgaris* in aquaculture. Rev Aquac 12:328–346. https://doi.org/10.1111/raq.12320

Akbarnezhad M, Shamsaie Mehrgan M, Kamali A, Javaheri Baboli M (2020) Effects of microelements (Fe, Cu, Zn) on growth and pigment contents of *Arthrospira* (spirulina) *platensis*. Iran J Fish Sci 19:653–668. https://doi.org/10.22092/ijfs.2019.120614

Al-Dhabi NA (2013) Heavy metal analysis in commercial spirulina products for human consumption. Saudi J Biol Sci 20:383–388. https://doi.org/10.1016/j.sjbs.2013.04.006

AlFadhly NKZ, Alhelfi N, Altemimi AB, Verma DK, Cacciola F (2022) Tendencies affecting the growth and cultivation of genus spirulina: an investigative review on current trends. Plan Theory 11:1–21. https://doi.org/10.3390/plants11223063

Alshuniaber MA, Krishnamoorthy R, AlQhtani W (2021) Antimicrobial activity of polyphenolic compounds from Spirulina against food-borne bacterial pathogens. Saudi J Biol Sci, 28(1):459–464

Altmann BA, Rosenau S (2022) Spirulina as animal feed: opportunities and challenges. Foods 11(7):965. https://doi.org/10.3390/foods11070965

Aly MS, Esawy MA (2008) Evaluation of spirulina platensis as bio stimulator for organic farming systems. J Genet Eng Biotechnol 6:1–7

Andrés CMC, Pérez de la Lastra JM, Juan CA, Plou FJ, Pérez-Lebeña E (2022) Chemistry of hydrogen peroxide formation and elimination in mammalian cells, and its role in various pathologies. Stress 2:256–274. https://doi.org/10.3390/stresses2030019

Anvar AA, Nowruzi B (2021) Bioactive properties of spirulina: a review. Microb Bioact 4:134–142. https://doi.org/10.25163/microbbioacts.412117b0719110521

Araújo R, Vázquez Calderón F, Sánchez López J, Azevedo IC, Bruhn A, Fluch S, Garcia Tasende M, Ghaderiardakani F, Ilmjärv T, Laurans M, Mac Monagail M, Mangini S, Peteiro C, Rebours C, Stefansson T, Ullmann J (2021) Current status of the algae production industry in Europe: an emerging sector of the blue bioeconomy. Front Mar Sci 7:1–24. https://doi.org/10.3389/fmars.2020.626389

Bhatta S, Janezic TS, Ratti C (2020) Freeze-drying of plant-based foods. Foods 9:1–22

Bhosale P (2004) Environmental and cultural stimulants in the production of carotenoids from microorganisms. Appl Microbiol Biotechnol 63:351–361. https://doi.org/10.1007/s00253-003-1441-1

Bohórquez-Medina SL, Bohórquez-Medina AL, Benites Zapata VA, Ignacio-Cconchoy FL, Toro-Huamanchumo CJ, Bendezu-Quispe G, Pacheco-Mendoza J, Hernandez AV (2021) Impact of spirulina supplementation on obesity-related metabolic disorders: a systematic review and meta-analysis of randomized controlled trials. NFS J 25:21–30. https://doi.org/10.1016/j.nfs.2021.09.003

Caporgno MP, Mathys A (2018) Trends in microalgae incorporation into innovative food products with potential health benefits. Front Nutr 5:1–10. https://doi.org/10.3389/fnut.2018.00058

Chen J, Zhou R, Feng Y, Cheng L (2022) Molecular mechanisms of exercise contributing to tissue regeneration. Signal Transduct Target Ther 7:383. https://doi.org/10.1038/s41392-022-01233-2

Choopani A, Poorsoltan M, Fazilati M, Latifi AM, Salavati H (2016) Spirulina: a source of gamma-linoleic acid and its applications. J Appl Biotechnol. Reports 3:483–488

Ciferri O (1983) Spirulina, the edible microorganism. Microbiol Rev 47:551–578. https://doi.org/10.1128/mmbr.47.4.551-578.1983

De Marco ER, Steffolani ME, Martinez CS, Leon AE (2014) Effects of spirulina biomass on the technological and nutritional quality of bread wheat pasta. LWT - Food Sci Tehnol. https://doi.org/10.1016/j.lwt.2014.02.054

Demarco M, de Moraes JO, Ferrari MC, Neves F d F, Laurindo JB, Tribuzi G (2022) Production of spirulina (*Arthrospira platensis*) powder by innovative and traditional drying techniques. J Food Process Eng 45:1–10. https://doi.org/10.1111/jfpe.13919

Desmorieux H, Decaen N (2005) Convective drying of spirulina in thin layer. J Food Eng 66:497–503. https://doi.org/10.1016/j.jfoodeng.2004.04.021

Desmorieux H, Hernandez F (2004) Biochemical and physical criteria of spirulina after different drying processes. In: Proceedings of the 14th International Drying Symposium (IDS 2004), Sào Paulo, 22–25 August 2004, 900–907

Ersyah D, Jaziri AA, Setijawati D (2022) Effect of spirulina (*Arthrospira platensis*) powder on the Physico-chemical and sensory characterization of dry noodle. J Aquac Fish Heal 11:277–288

Forman HJ, Zhang H (2021) Targeting oxidative stress in disease: promise and limitations of antioxidant therapy. Nat Rev Drug Discov 20:689–709. https://doi.org/10.1038/s41573-021-00233-1

Golmakani MT, Soleimanian-Zad S, Alavi N, Nazari E, Eskandari MH (2019) Effect of spirulina (*Arthrospira platensis*) powder on probiotic bacteriologically acidified feta-type cheese. J Appl Phycol 31:1085–1094. https://doi.org/10.1007/s10811-018-1611-2

Gordillo FJL, Jimenez C, Figueroa FL, Niell XF (1999) Effects of increased atmospheric CO2 and N supply on photosynthesis, growth and cell composition of the cyanobacterium *Spirulina platensis* (*Arthrospira*). J Appl Phycol 10:461–469. https://doi.org/10.1023/a:1008090402847

Górska-Warsewicz H, Rejman K, Laskowski W, Czeczotko M (2019) Milk and dairy products and their nutritional contribution to the average polish diet. Nutrients 11(8):1771. https://doi.org/10.3390/nu11081771

Grahl S, Strack M, Mensching A, Mörlein D (2020) Alternative protein sources in Western diets: food product development and consumer acceptance of spirulina-filled pasta. Food Qual Prefer 84:103933. https://doi.org/10.1016/j.foodqual.2020.103933

Güroy B, Karadal O, Mantoğlu S, Irmak Cebeci O (2017) Effects of different drying methods on C-phycocyanin content of *Spirulina platensis* powder. Ege J Fish Aquat Sci 34:129–132. https://doi.org/10.12714/egejfas.2017.34.2.02

Habib AMB, Parvin M, Huntington TC, Hasan MR (2008) A review on culture, production and use of spirulina as food for humans and feeds for, aquaculture. FAO, Rome. isbn 978-92-5-106106-0

Hasanuzzaman M, Bhuyan MHMB, Zulfiqar F, Raza A, Mohsin SM, Al Mahmud J, Fujita M, Fotopoulos V (2020) Reactive oxygen species and antioxidant defense in plants under abiotic stress: revisiting the crucial role of a universal defense regulator. Antioxidants 9:1–52. https://doi.org/10.3390/antiox9080681

Holman BWB, Malau-Aduli AEO (2013) Spirulina as a livestock supplement and animal feed. J Anim Physiol Anim Nutr (Berl) 97:615–623. https://doi.org/10.1111/j.1439-0396.2012.01328.x

Hoseini SM, Khosravi-Darani K, Mozafari MR (2013) Nutritional and medical applications of spirulina microalgae. Mini-Reviews Med Chem 13:1231–1237. https://doi.org/10.2174/1389557511313080009

Huarachi Olivera RE, Yapo Pari UA, Dueñas Gonza AP, Condori Huamanga J, Pacheco Salazar D, Soto Flores J (2015) Cultivation of *Arthrospira platensis* (spirulina) in curved doubly tubular photobioreactor to environmental conditions in the south of the Peru. Colomb J Biotechnol 17: 143–150. https://doi.org/10.15446/rev.colomb.biote.v17n1.50679

Iatrou AM, Papadopoulos GA, Giannenas I, Lymberopoulos A, Fortomaris P (2022) Effects of dietary inclusion of *Spirulina platensis* on the reproductive performance of female mink. Vet Sci 9:1–11. https://doi.org/10.3390/vetsci9080428

Ighodaro OM, Akinloye OA (2018) First line defence antioxidants-superoxide dismutase (SOD), catalase (CAT) and glutathione peroxidase (GPX): their fundamental role in the entire antioxidant defence grid. Alexandria J Med 54:287–293. https://doi.org/10.1016/j.ajme.2017.09.001

Isani G, Ferlizza E, Bertocchi M, Dalmonte T, Menotta S, Fedrizzi G, Andreani G (2022) Iron content, iron speciation and Phycocyanin in commercial samples of *Arthrospira* spp. Int J Mol Sci 23(22):13949. https://doi.org/10.3390/ijms232213949

Ismail M, Hossain MF, Tanu AR, Shekhar HU (2015) Effect of Spirulina intervention on oxidative stress, antioxidant status, and lipid profile in chronic obstructive pulmonary disease patients. Biomed Res Int 2015:1–7. https://doi.org/10.1155/2015/486120

Jahromi KG, Koochi ZH, Kavoosi G, Shahsavar A (2022) Manipulation of fatty acid profile and nutritional quality of *Chlorella vulgaris* by supplementing with citrus peel fatty acid. Sci Rep 12:1–11. https://doi.org/10.1038/s41598-022-12309-y

Janda-Milczarek K, Szymczykowska K, Jakubczyk K, Kupnicka P, Skonieczna-Żydecka K, Pilarczyk B, Tomza-Marciniak A, Ligenza A, Stachowska E, Dalewski B (2023) Spirulina supplements as a source of mineral nutrients in the daily diet. Appl Sci 13:1–11. https://doi.org/10.3390/app13021011

Junique L, Watier L, Lejeune H, Viudes F, Deblieck M, Watier D (2021) Determination by response surface methodology of optimal protein and phycocyanin productivity conditions in *Arthrospira* (spirulina) *platensis* under different combinations of photoperiod variation and lighting intensity. Bioresour Technol Rep 15:100763. https://doi.org/10.1016/j.biteb.2021.100763

Karkos PD, Leong SC, Karkos CD, Sivaji N, Assimakopoulos DA (2011) Spirulina in clinical practice: evidence-based human applications. Evid Based Complement Altern Med 2011:4–7. https://doi.org/10.1093/ecam/nen058

Kasting JF, Siefert JL (2002) Life and the evolution of Earth's atmosphere. Science (80-) 296:1066–1068. https://doi.org/10.1126/science.1071184

Kaur S, Prakash P, Bak D-H, Hong SH, Cho C, Chung M-S, Kim J-H, Lee S, Bai H-W, Lee SY, Chung BY, Lee SS (2021) Regulation of dual activity of ascorbate peroxidase 1 from Arabidopsis thaliana by conformational changes and posttranslational modifications. Front Plant Sci 12:1–12. https://doi.org/10.3389/fpls.2021.678111

Kawamura A, Ooyama K, Kojima K, Kachi H, Abe T, Amano K, Aoyama T (2011) Dietary supplementation of gammalinolenic acid improves skin parameters in subjects with dry skin and mild atopic dermatitis. J Oleo Sci 60:597–607. https://doi.org/10.5650/jos.60.597

Khan MI, Shin JH, Kim JD (2018) The promising future of microalgae: current status, challenges, and optimization of a sustainable and renewable industry for biofuels, feed, and other products. Microb Cell Factories 17:36. https://doi.org/10.1186/s12934-018-0879-x

Koli DK, Rudra SG, Bhowmik A, Pabbi S (2022) Nutritional, functional, textural and sensory evaluation of spirulina enriched green pasta: a potential dietary and health supplement. Foods 11(7):979. https://doi.org/10.3390/foods11070979

Kozakowska M, Pietraszek-Gremplewicz K, Jozkowicz A, Dulak J (2015) The role of oxidative stress in skeletal muscle injury and regeneration: focus on antioxidant enzymes. J. Muscle Res Cell Motil 36:377–393. https://doi.org/10.1007/s10974-015-9438-9

Kumar A, Ramamoorthy D, Verma DK, Kumar A, Kumar N, Kanak KR, Marwein BM, Mohan K (2022) Antioxidant and phytonutrient activities of *Spirulina platensis*. Energy Nexus 6:100070. https://doi.org/10.1016/j.nexus.2022.100070

Kumudha A, Kumar SS, Thakur MS, Ravishankar GA, Sarada R (2010) Purification, identification, and characterization of methylcobalamin from *Spirulina platensis*. J Agric Food Chem 58: 9925–9930. https://doi.org/10.1021/jf102159j

Larrosa APQ, Camara ÁS, Moura JM, Pinto LAA (2018) *Spirulina sp.* biomass dried/disrupted by different methods and their application in biofilms production. Food Sci Biotechnol 27:1659–1665. https://doi.org/10.1007/s10068-018-0397-y

Lemes A, Takeuchi KP, Carvalho JCM, Danesi EDG (2012) Fresh pasta production enriched with *Spirulina platensis* biomass. Brazilian Archieves Biology Technol 55:741–750

Liestianty D, Rodianawati I, Arfah RA, Assa A, Patimah S, Muliadi (2019) Nutritional analysis of *Spirulina* sp to promote as superfood candidate. IOP Conf Ser Mater Sci Eng 509:012031. https://doi.org/10.1088/1757-899X/509/1/012031

Liu Q, Huang Y, Zhang R, Cai T, Cai Y (2016) Medical application of *Spirulina platensis* derived C-Phycocyanin. Evidence-based Complement Altern Med. https://doi.org/10.1155/2016/7803846

Lucas BF, de Morais MG, Santos TD, Costa JAV (2018) Spirulina for snack enrichment: nutritional, physical and sensory evaluations. Lwt - Food Sci Technol 90:270–276. https://doi.org/10.1016/j.lwt.2017.12.032

Lucas BF, da Rosa APC, de Carvalho LF, de Morais MG, Santos TD, Costa JAV (2019) Snack bars enriched with spirulina for schoolchildren nutrition. Food Sci Technol 2061:6946–6946

Ma Z, Ahmed F, Yuan B, Zhang W (2019) Fresh living *Arthrospira* as dietary supplements: current status and challenges. Trends Food Sci Technol 88:439–444. https://doi.org/10.1016/j.tifs.2019.04.010

Machowiec P, Ręka G, Maksymowicz M, Piecewicz-Szczęsna H, Smoleń A (2021) Effect of spirulina supplementation on systolic and diastolic blood pressure: systematic review and meta-analysis of randomized controlled trials. Nutrients 13:3054. https://doi.org/10.3390/nu13093054

Madkour FF, Kamil AEW, Nasr HS (2012) Production and nutritive value of *Spirulina platensis* in reduced cost media. Egypt J Aquat Res 38:51–57. https://doi.org/10.1016/j.ejar.2012.09.003

Manzocchi E, Guggenbuhl B, Kreuzer M, Giller K (2020) Effects of the substitution of soybean meal by spirulina in a hay-based diet for dairy cows on milk composition and sensory perception. J Dairy Sci 103:11349–11362

Markou G (2012) Alteration of the biomass composition of *Arthrospira* (spirulina) *platensis* under various amounts of limited phosphorus. Bioresour Technol 116:533–535

Martelli F, Alinovi M, Bernini V, Gatti M, Bancalari E (2020) *Arthrospira platensis* as natural fermentation booster for milk and soy fermented beverages. Foods 9:350. https://doi.org/10.3390/foods9030350

Mathur M (2019) Bioactive molecules of spirulina: a food supplement. Ref Ser Phytochem 1621–1642. https://doi.org/10.1007/978-3-319-78030-6_97

Matos AP, Novelli E, Tribuzi G (2022) Use of algae food ingredient: sensory acceptance and commercial products. Front Food Sci Technol 2:989801. https://doi.org/10.3389/frfst.2022.989801

Mawla EA (2014) Production and evaluation of some extruded food products using spirulina algae. Ann Agric Sci Moshtohor 52:495–510. https://doi.org/10.21608/assjm.2014.111899

Mazinani S, Fadaei V, Khosravi-Darani K (2016) Impact of *Spirulina platensis* on physicochemical properties and viability of lactobacillus acidophilus of probiotic UF feta cheese. J Food Process Preserv 40:1318–1324. https://doi.org/10.1111/jfpp.12717

Miller JL (2013) Iron deficiency anemia: a common and curable disease. Cold Spring Harb Perspect Med 3:1–13. https://doi.org/10.1101/cshperspect.a011866

Moon J, Koh G (2021) Clinical evidence and mechanisms of high-protein diet-induced weight loss. J Obes Metab Syndr 29:166–173. https://doi.org/10.7570/jomes20028

Moradi S, Ziaei R, Foshati S, Mohammadi H, Nachvak SM, Rouhani MH (2019) Effects of spirulina supplementation on obesity: a systematic review and meta-analysis of randomized clinical trials. Complement Ther Med 47:102211. https://doi.org/10.1016/j.ctim.2019.102211

Muhoza B, Yuyang H, Uriho A, Harindintwali JD, Liu Q, Li Y (2023) Spray-and freeze-drying of microcapsules prepared by complex coacervation method: a review. Food Hydrocoll 140:108650. https://doi.org/10.1016/j.foodhyd.2023.108650

Murugan T, Rajesh R (2014) Cultivation of two species of spirulina (*Spirulina platensis* and *Spirulina platensis var lonar*) on sea water medium and extraction of C-phycocyanin. Eur J Exp Biol 4:93–97

Nematollahi MA, Laird DW, Hughes LJ, Raeisossadati M, Moheimani NR (2020) Effect of organic carbon source and nutrient depletion on the simultaneous production of a high value bioplastic and a specialty pigment by *Arthrospira platensis*. Algal Res 47:101844. https://doi.org/10.1016/j.algal.2020.101844

Neves FDF, Demarco M, Tribuzi G (2019) Drying and quality of microalgal powders for human alimentation. In: Microalgae - from Physilogy to application. IntechOpen, London, pp 1–20

Olas B, Białecki J, Urbańska K, Bryś M (2021) The effects of natural and synthetic blue dyes on human health: a review of current knowledge and therapeutic perspectives. Adv Nutr 12:2301–2311. https://doi.org/10.1093/advances/nmab081

Oliveira EG, Duarte JH, Moraes K, Crexi VT, Pinto LAA (2010) Optimization of Spirulina platensis convective drying: evaluation of phycocyanin loss and lipid oxidation. Int J Food Sci Technol, 45(8):1572–1578

Otto J, Malau Aduli A (2017) Spirulina platensis (*Arthrospira* spp.): a potential novel feed source for pasture-based dairy cows. J Fish Livest Prod 5(3):252–257. https://doi.org/10.4172/2332-2608.1000253

Pappas A, Tsiokanos A, Fatouros IG, Poulios A, Kouretas D, Goutzourelas N, Giakas G, Jamurtas AZ (2021) The effects of Spirulina supplementation on redox status and performance following a muscle damaging protocol. Int J Mol Sci 22. https://doi.org/10.3390/ijms22073559

Patel P, Jethani H, Radha C, Vijayendra SVN, Mudliar SN, Sarada R, Chauhan VS (2019) Development of a carotenoid enriched probiotic yogurt from fresh biomass of spirulina and its characterization. J Food Sci Technol 56:3721–3731. https://doi.org/10.1007/s13197-019-03844-0

Probst L, Frideres L, Pedersen B, Amato F (2015) Sustainable, safe and nutritious food: New nutrient sources (Business Innovation Observatory)

Raczyk M, Polanowska K, Kruszewski B, Grygier A, Michalowska D (2022) Effect of spirulina (*Arthrospira platensis*) supplementation on physical and chemical properties of semolina (Triticum durum) based fresh pasta. Molecules 27(2):355. https://doi.org/10.3390/molecules27020355

Rahim A, Çakir C, Ozturk M, Şahin B, Soulaimani A, Sibaoueih M, Nasser B, Eddoha R, Essamadi A, El Amiri B (2021) Chemical characterization and nutritional value of *Spirulina platensis* cultivated in natural conditions of Chichaoua region (Morocco). South African J Bot 141:235–242. https://doi.org/10.1016/j.sajb.2021.05.006

Rout GR, Sahoo S (2015) Role of iron in plant growth and metabolism. Rev Agric Sci 3:1–24. https://doi.org/10.7831/ras.3.1

Ryckebosch E, Bruneel C, Muylaert K, Foubert I (2012) Microalgae as an alternative source of omega-3 long chain polyunsaturated fatty acids. Lipid Technol 24:128–130. https://doi.org/10.1002/lite.201200197

Rzymski P, Niedzielski P, Kaczmarek N, Jurczak T, Klimaszyk P (2015) The multidisciplinary approach to safety and toxicity assessment of microalgae-based food supplements following clinical cases of poisoning. Harmful Algae 46:34–42. https://doi.org/10.1016/j.hal.2015.05.003

Sannasimuthu A, Kumaresan V, Pasupuleti M, Paray BA, Al-Sadoon MK, Arockiaraj J (2018) Radical scavenging property of a novel peptide derived from C-terminal SOD domain of superoxide dismutase enzyme in Arthrospira platensis. Algal Res 35:519–529. https://doi.org/10.1016/j.algal.2018.09.028

Sarada R, Pillai MG, Ravishankar GA (1999) Phycocyanin from *spirulina sp*: influence of processing of biomass on phycocyanin yield, analysis of efficacy of extraction methods and stability studies on phycocyanin. Process Biochem 34:795–801. https://doi.org/10.1016/S0032-9592(98)00153-8

Saxena R, Rodríguez-Jasso RM, Chávez-Gonzalez ML, Aguilar CN, Quijano G, Ruiz HA (2022) Strategy development for microalgae *Spirulina platensis* biomass cultivation in a bubble

Photobioreactor to promote high carbohydrate content. Fermentation 8(8):374. https://doi.org/10.3390/fermentation8080374

Schirrmeister BE, Antonelli A, Bagheri HC (2011) The origin of multicellularity in cyanobacteria. BMC Evol Biol 11:45. https://doi.org/10.1186/1471-2148-11-45

Schmitz-Schug I, Foerst P, Kulozik U (2013) Impact of the spray drying conditions and residence time distribution on lysine loss in spray dried infant formula. Dairy Sci Technol 93:443–462. https://doi.org/10.1007/s13594-013-0115-8

Selmi C, Leung PSC, Fischer L, German B, Yang CY, Kenny TP, Cysewski GR, Gershwin ME (2011) The effects of spirulina on anemia and immune function in senior citizens. Cell Mol Immunol 8:248–254. https://doi.org/10.1038/cmi.2010.76

Semih O, Ruhsen P (2001) Fatty acid composition of chlorella and spirulina microalgae species. J AOAC Int 84(6):1708–1714

Serban MC, Sahebkar A, Dragan S, Stoichescu-Hogea G, Ursoniu S, Andrica F, Banach M (2016) A systematic review and meta-analysis of the impact of spirulina supplementation on plasma lipid concentrations. Clin Nutr 35:842–851. https://doi.org/10.1016/j.clnu.2015.09.007

Sharma N (2014) Free radicals, antioxidants and disease. Biol Med 6:1–6. https://doi.org/10.4172/0974-8369.1000214

Soni RA, Sudhakar K, Rana RS (2017) Spirulina – from growth to nutritional product: a review. Trends Food Sci Technol 69:157–171. https://doi.org/10.1016/j.tifs.2017.09.010

Sotiroudis TG, Sotiroudis GT (2013) Health aspects of spirulina (*Arthrospira*) microalga food supplement. J Serbian Chem Soc 78:395–405. https://doi.org/10.2298/JSC121020152S

Srivani P, Tony K, Akhila S, Rao MN, Amaze M et al (2017) Simplified method for the cultivation of spirulina for domestic use. Int J Biotechnol Bioeng 3:107–108. https://doi.org/10.25141/2475-3432-2017-4.0107

Stettler C, Allemann S, Jüni P, Cull CA, Holman RR, Egger M, Krähenbühl S, Diem P (2006) Glycemic control and macrovascular disease in types 1 and 2 diabetes mellitus: meta-analysis of randomized trials. Am Heart J 152:27–38. https://doi.org/10.1016/j.ahj.2005.09.015

Tadesse Z (1998) Food and feeding ecology of tilapia, Oreochromis niloticus L. and effects of diet on the lipid quality of fish in some lakes in Ethiopia. Addis Ababa University, Addis Ababa

Thomas C, Symoneaux R, Pantin-Sohier G, Picouet P, Pantin-Sohier G, Maître I (2020) Perceptions of spirulina from French consumers of organic products

Tiburcio PC, Galvez FCF, Cruz LJ, Gavino VC (2007) Determination of shelf life of Spirulina platensis (MI2) grown in The Philippines. J Appl Phycol 19:727–731. https://doi.org/10.1007/s10811-007-9227-y

Torzillo G, Vonshak A (1994) Effect of light and temperature on the photosynthetic activity of the cyanobacterium *Spirulina platensis*. Biomass Bioenergy 6:399–403. https://doi.org/10.1016/0961-9534(94)00076-6

Trevi S, Uren Webster T, Consuegra S, Garcia de Leaniz C (2023) Benefits of the microalgae spirulina and Schizochytrium in fish nutrition: a meta-analysis. Sci Rep 13:1–23. https://doi.org/10.1038/s41598-023-29183-x

Udayan A, Pandey AK, Sirohi R, Sreekumar N, Sang BI, Sim SJ, Kim SH, Pandey A (2022) Production of microalgae with high lipid content and their potential as sources of nutraceuticals. Phytochem Rev 0123456789:1. https://doi.org/10.1007/s11101-021-09784-y

van den Berg H, Brandsen L, Sinkeldam BJ (1991) Vitamin B-12 content and bioavailability of spirulina and nori in rats. J Nutr Biochem 2:314–318. https://doi.org/10.1016/0955-2863(91)90073-E

van den Oever SP, Mayer HK (2022) Biologically active or just "pseudo"-vitamin B12 as predominant form in algae-based nutritional supplements? J Food Compos Anal 109:104464. https://doi.org/10.1016/j.jfca.2022.104464

Vieira VV, Cadoret JP, Acien FG, Benemann J (2022) Clarification of Most relevant concepts related to the microalgae production sector. PRO 10(1):175. https://doi.org/10.3390/pr10010175

Vilas C, Alonso AA, Balsa-Canto E, López-Quiroga E, Trelea IC (2020) Model-based real time operation of the freeze-drying process. PRO 8:1–21. https://doi.org/10.3390/pr8030325

Vonshak A (1997) Spirulina platensis (Arthrospira): physiology, cell-biology and biotechnology. Taylor & Francis Ltd, London

Wajda Ł, Rękas Z, Tarko T, Duda-Chodak A, Liebersbach A, Makarewicz M (2020) Dried Biomass of *Arthrospira platensis* inhibits growth of *Aureobasidium pullulans* LW14 and some bacteria when added to unpasteurised apple Juice. Indian J Microbiol 60:346–352. https://doi.org/10.1007/s12088-020-00871-w

Watanabe F, Katsura H, Takenaka S, Fujita T, Abe K, Tamura Y, Nakatsuka T, Nakano Y (1999) Pseudovitamin B12 is the predominant cobamide of an algal health food, Spirulina tablets. J Agric Food Chem 47:4736–4741. https://doi.org/10.1021/jf990541b

Wu Q, Liu L, Miron A, Klímová B, Wan D, Kuča K (2016) The antioxidant, immunomodulatory, and anti-inflammatory activities of spirulina: an overview. Arch Toxicol 90:1817–1840. https://doi.org/10.1007/s00204-016-1744-5

Zarezadeh M, Faghfouri AH, Radkhah N, Foroumandi E, Khorshidi M, Rasouli A, Zarei M, Mohammadzadeh Honarvar N, Hazhir Karzar N, Ebrahimi Mamaghani M (2021) Spirulina supplementation and anthropometric indices: a systematic review and meta-analysis of controlled clinical trials. Phyther Res 35:577–586. https://doi.org/10.1002/ptr.6834

Legislation and Biosecurity

Fernanda de Noronha Sertori, Gilvana Scoculi de Lira, Beatriz Jacob Furlan, Paulo Alexandre Silveira da Silva, Allana Rachel Dias, and Michelle Aparecida Coelho Moreira

Abstract As potential sources of biologically active compounds, the diversity and variability of the biochemical composition of the algal biomass make them a target of interest for a wide range of industrial applications, including agriculture and food industry. As discussed in the previous chapters, algae are known for their antioxidant and anti-cancer properties, besides being considered abundant in vitamins and minerals essential for human health with high protein content. Many foods obtained from algal biomass can provide therapeutic substances and compounds that are necessary for a nutrient-rich diet. With the growing demand for a healthy lifestyle, industries are seeking to innovate their products to meet human needs and interests, given the ease of producing these foods on a large-scale in a reduced physical and temporal space. However, in parallel, legislative bodies are increasingly concerned about the use of algal products in industry, as well as ensuring food safety during their marketing. This is due to the danger of contamination of algal species from growth to storage, and possible accumulation of heavy metals by many algal species due to their high absorption capacity. Thus, the overall objective of this chapter is to outline and summarise the legal issues related to the use of algae as well as food safety.

Keywords Algae · Bioactives · Biosafety · Biotechnology · Food industry

F. de Noronha Sertori (✉)
Postgraduate Program in Industrial Biotechnology, Positivo University (UP), Curitiba, PR, Brazil

G. S. de Lira · B. J. Furlan · P. A. S. da Silva · M. A. C. Moreira
Technology Sector, Postgraduate Program in Materials Science and Engineering (PIPE), and Sustainable Energy Research & Development Center (NPDEAS), Federal University of Paraná (UFPR), Curitiba, PR, Brazil

A. R. Dias
Bodebrown Brewery, Curitiba, PR, Brazil

A. Abomohra, S. Ende (eds.), *Value-added Products from Algae*,
https://doi.org/10.1007/978-3-031-42026-9_18

1 Introduction

Human population growth coupled with climate change have a major impact on the current agricultural practices and food production. As the population increases, the world is also looking to expand food production. In order to meet global demands, this expansion tends to continue. As far as the environmental impact is concerned, meat consumption has been causing very significant consequences. Meat production requires large amounts of land and water, emiting considerable amounts of greenhouse gases (GHGs). Thus, extensive research is focusing on emission reduction from agriculture sector, where algae can play an important role (Chapter " Potential of Seaweeds to Mitigate Methane Emissions"). The GHG emissions are not considered the only problem caused by current agricultural practices. Conventional farming systems also contribute to soil erosion, overexploitation of water resources, and water contamination (both surface and groundwater). Thus, there is a need to implement more efficient and sustainable food production systems in order to guarantee the environmental constancy of the planet (Diaz et al. 2023). In that context, algae are considered an alternative source to supply the human diet and, in parallel, have the potential to minimise the negative environmental impacts. Given the high algal biomass productivity in a reduced physical and temporal space, it can be estimated that algae will become not only an alternative source of protein food, but also one of the main sources of protein in the human diet in the future. Seaweeds are used as a food rich in vitamins, minerals, trace elements, proteins (with essential amino acids), carbohydrates, and low in fat (Campbell et al. 2020; Diaz et al. 2023). Therefore, seaweeds are already part of the diet of people in the far East, where low obesity rates can be attributed to this diet.

Algae are found worldwide and form the base of the food chain, serving as a source of nutrients, not only for a wide variety of aquatic organisms, but also for human consumption and animal feed (Marinho-Soriano and Carneiro 2021; Raven et al. 1996). Large-scale cultivation of algae has also made them a target of interest for food, cosmetic, and pharmaceutical industries (Souza and Ostrosky 2022), as they are considered important photosynthetic organisms and represent a significant source of biological compounds commonly of industrial interest. Many species of algae possess natural characteristics that attract attention as a food source for human consumption (Diaz et al. 2023). Different species of algae produce a wide variety of natural substances, referred to in this book as "phycochemicals", which may have antioxidant, antimicrobial, antiviral, antitumour, and anti-inflammatory potential when incorporated into daily diets (Gupta and Abu-Ghannam 2011; Kumar et al. 2021; Gutiérrez et al. 2017). Due to their nutritional potential, companies and food industries are increasingly considering the potential use of algae as functional foods and in the development of value-added products, such as, among others, canned tuna with algae, or salad pastes and sauces (Fleurence and Levine 2016). However, the use of algae in any industry requires the commitment and implementation of laws and regulatory conducts that promote the sustainable purpose of these resources, aligned with national and international socioeconomic and environmental agendas

(Andrade et al. 2020). Thus, the aim of this chapter is to outline and synthesize the legal issues related to the industrial use of algae as well as their food safety legislations.

2 GRAS (Generally Recognized as Safe) Status

GRAS (Generally Recognized as Safe) status refers to a designation given to certain substances used in food products. The GRAS status indicates that these are substances that are considered safe for consumption under the intended conditions of use. However, the GRAS status has been subject to criticism over the years, with attention focused on the rigidity and transparency of the assessment process. The GRAS status is a regulatory concept in the United States that determines the safety of substances used in food. It was established by the Federal Food, Drug, and Cosmetic Act (FD&C Act) and is overseen by the U.S. Food and Drug Administration (FDA). The main purpose of GRAS is to ensure that food ingredients do not pose risks to human health when used as intended (Faustman et al. 2021; Hallagan et al. 2020).

To achieve the GRAS status, substances can be evaluated through two routes: the FDA's voluntary GRAS notification program or by food manufacturers' own determination. Under the voluntary notification program, manufacturers can submit a GRAS notification to the FDA, providing scientific evidence and data to support the safety of the substance. The FDA reviews the notification and has the option to issue a letter indicating that it has no further concerns about the safety of the substance. Alternatively, food manufacturers can establish their own assessment for the purpose of establishing the GRAS status without FDA involvement. They can establish the safety of a substance through a self-affirmation or through consultation with experts. In the case of such self-affirmation, previously mentioned, the manufacturer determines a substance as GRAS based on the scientific literature and available database information. Despite the intent of companies to ensure the safety of food ingredients, the GRAS status has faced criticism, especially regarding the lack of transparency and potential conflicts of interest in this evaluation process (Faustman et al. 2021; Hallagan et al. 2020; Morissette et al. 2023).

The FDA has been criticized for its relatively uncooperative approach to the GRAS notification program, suggesting that the FDA does not review or approve GRAS substances prior to actual marketing. The GRAS evaluation process has a heavy reliance on the involvement of subject matter experts. These experts are often hired by the manufacturing companies. However, creating a conflict of interest, impartiality, and independence of the evaluation. Also, there are arguments related to the lack of transparency in the process, as the FDA does not have access to important data on GRAS determinations, which also hinders public confidence (Burdock and Carabin 2004). Overall, the GRAS status plays a very important role in determining the safety of substances used in food products. However, it has come under criticism with regard to the enforcement process. However, FDA recognizes the need for improvement and increased transparency in order to strengthen GRAS oversight. By implementing the above initiatives, we hope to promote greater public confidence in the safety assessments of GRAS substances.

2.1 Biosecurity of Edible Algae

The term 'biosecurity', as interpreted by FAO and also as described in the 'Biosecurity Toolkit' (2007), covers issues such as food security, zoonoses, introduction of diseases and infections, introduction and release of living modified organisms, and introduction/management of invasive alien species (FAO 2020). Algae are characterized as an excellent source of several bioactive compounds of great dietary interest, such as polysaccharides, dietary fibres, carotenoids, vitamins, minerals, carbohydrates, proteins, lipids, essential fatty acids, polyphenols, and amino acids. Algae polysaccharides play a fundamental role in various emerging areas of biotechnology, such as food, pharmaceuticals, and different industries. Many algal species are consumed as food by humans and commercially exploited worldwide. Moreover, algae also play a key role as producers in the marine environment (Pradhan et al. 2022). Specifically, the great importance of seaweeds for health is noted; however, the expansion is still limited due to the high prevalence of recalcitrant diseases and epiphytic pests, presenting potential for emergence and dissemination, especially due to the absence of effective national and international biosafety policies. Seaweeds have been increasingly arousing interest as a human food source due to the valuable nutritional properties reported in their composition, such as abundance of protein, fibre, vitamins, and minerals, with relatively low lipid content (Kazir et al. 2019; Fleurence and Levine 2016). Application of algal bioactive compounds is not only limited to food safety and human health, but also involves factors such as the costs, product efficacy, and compliance with current legislation, which are of vital importance in deciding on the viability and commercial production of a product. However, there are still some challenges to overcome for the commercialisation specifically of bioactive compounds extracted from algae. This includes the optimisation of extraction and purification processes, the development of storage alternatives and the implementation of technologies that improve the bioaccessibility and bioavailability of these compounds. Recently, different encapsulation or emulsification processes have been developed to ensure that bioactive compounds achieve their biological function in humans.

The lack of global uniformity in legislation between countries makes the process even more complex. Therefore, to enable the marketing of algae-derived products, modifications are required to ensure compliance with current legislation. In the European Union, for example, as seaweeds are not traditionally consumed as food before 1997, they are considered as a 'novel food' according to EU Regulation 2015/2283. This is also adopted for products obtained from algae, such as food supplements containing bioactive components or food additives, such as florotannins from *Ecklonia cava*. Therefore, for these products to be placed on the market, authorisation is required from the European Food Safety Authority (EFSA). This process also involves carrying out health risk assessment studies. In addition, food safety analyses must comply with the legislation in force in relation to food safety and hygiene, such as Regulation (EU) 178/2002 and Regulation (EU) 852/2004, which has the task of ensuring consumer food safety. Moreover, these products can be marketed as nutraceuticals without the need for scientific proof conducted by EFSA, according to Regulation EC No. 1924/2006. However, this regulation states that health claims made for these products must be supported by adequate and significant scientific evidence, which should be submitted to EFSA evaluation (Pereira et al. 2021).

The main species of seaweeds used for human consumption belong to the Ochrophyta group (brown algae), with the main genera being *Laminaria*, *Saccharina*, *Undaria,* and *Sargassum*, and the red algae group (Rhodophyta), with the main commercially available genera being *Porphyra*, *Palmaria,* and *Gracilaria* (Moreira et al. 2022; Blikra et al. 2021).

Saccharina and *Laminaria* genera are used for the production of dried seaweeds, commercially known in Japan as Kombu, and sold in pieces or flakes. In China, Kombu is used as a seasoning for soups and broths, while in Japan, it is consumed as a condiment for salmon or other fish. Species belonging to the genus *Laminaria* occur naturally in Korea, but are not widely used in the country because Koreans prefer Wakame (*Undaria pinnatifida*), another species of brown seaweeds also used as a seasoning for oysters or soups. As a curiosity, Wakame is considered a luxury delicacy in both Korea and Japan (McHugh 2003). When processed, *Porphyra umbilicalis* is known as Nori, a Japanese name and well known for its use in sushi, with cultivation originating mainly in Korea, Japan, and China. Dulse, made from *Palmaria palmata*, is commonly used as food in Ireland, Iceland, Canada, and the United States, and can be eaten raw, cooked with potatoes, in soups, or fish dishes. Due to the characteristic seafood flavour of *Palmaria palmata*, Dulse is highly appreciated and consumed as snacks, savoury cocktails, and as a condiment in salads (McHugh 2003; Vasconcelos and Gonçalves 2013).

Regarding the genus *Gracilaria*, the edible species are known as Ogonori or Ogo in Japanese cuisine. They are also consumed in other countries like Hawaii and Portugal in their raw form in salads or accompanied by sauces, being sold fresh, salted, or dehydrated (McHugh 2003). The green seaweeds are considered the least exploited, with the genera *Ulva* (sea lettuce) and *Monostroma* being the most prominent for human consumption. Known as Aonori in Japan, they have a soft texture and refreshing taste. These seaweeds can be used as a condiment in powder form for soups, cheeses, omelettes, and wraps, or raw as a salad (Moreira et al. 2020).

From the species *Phyllophora brodiaei*, carrageenan has been extracted and used in food applications for about 600 years. In addition to its hypoglycaemic effect, λ-carrageenan has potential for the prevention and treatment of metabolic diseases, such as diabetes. Carrageenan has a long history of safe use, which led to its recognition as "Generally Recognised As Safe" (GRAS) by the US Food and Drug Administration (21 CFR 182.7255). λ-carrageenan derived from the species *Phyllophora brodiaei* is also approved as a food additive (21 CFR 172.620). According to the World Health Organization (WHO), there is no need to specify an acceptable daily intake limit for carrageenans. Although degraded carrageenan has been classified as a possible human carcinogen by the International Agency for Research on Cancer, native carrageenan retains its GRAS status and remains unclassified as a human colon cancer-causing agent (Holdt and Kraan 2011). As explained above, seaweeds can be used as a source of food, medicines, and minerals. Seaweeds represent an excellent health food for humans due to their low calories and high amounts of dietary fibre, providing significant amounts of calcium (Ca) and iron (Fe) to the human body. The protein content of seaweeds can be higher than that of cereals and legumes. Most seaweeds contain essential amino acids, aspartic acid, and glutamic acid, and are widely used as food, particularly in Asian countries and less widely in the West (Pradhan et al. 2022).

Regarding microalgae, during the Second World War, the commercial cultivation of microalgae began in Germany, where they were used as a source of protein due to the shortage of animal meat (Olaizola 2003; Soeder 1986). However, the consumption of these organisms stands out in Asian countries as of the 1960s, where the green microalga *Chlorella* is marketed and consumed as a nutritional and medicinal product (Richmond 2004). Currently, Spirulina is receiving attention worldwide as a sustainable superfood (as discussed in Chapter " Algae as a Functional Food: A Case Study on Spirulina"). In the post-war period in Germany, microalgae were also used as a source of biomass to produce methane (CH_4). During the 1970s and 1980s, several groups of researchers in the USA, Japan, Germany, Czechoslovakia, Israel, Thailand, and France considered microalgae as promising substitutes for animal protein in human food (Chaumont 1993). About seaweeds, polysaccharides were the first commercial products extracted from algae and used as a source of ficocolloids (agar, carrageenan, and alginates). However, microalgae were only introduced into the market for this type of application after the development of the technology for cultivating these unicellular organisms (Ben-Amotz et al. 2009). In short, due to the high content of essential nutrients present in both seaweeds and microalgae, these organisms have high potential for use in food and manufacturing in a wide variety of industrial sectors, after considering the required legislations.

3 Legislation in Food Industry

Regulatory legislation in all industrial sectors, especially in the food industry, is fundamental to ensure food safety and protect consumers from health risks. Regulatory standards for the production and marketing of food are established by government regulatory bodies. In Brazil, for example, it is regulated by the National Health Surveillance Agency (NHSA), while in the USA, it is regulated by the Food and Drug Administration (FDA). These regulatory agencies set strict guidelines for the production, labelling, transportation, and storage of food to ensure that food is safe and of high quality (NHSA 2019; FDA 2022). Regarding the regulations governing the production and marketing of food from microalgae in Brazil and the USA, there is still no known legislation instituted specifically for this type of food, being necessary to abide by the general food regulations established by NHSA and FDA. NHSA's Resolution RDC No. 12/2001 establishes technical regulations for special purpose foods, which includes foods for controlled nutrient intake, weight control, athletes, among others. The manufacture, transport, and marketing of food must follow good hygienic-sanitary practices (NHSA 2001).

In addition, NHSA published Resolution RDC No. 331/2019, which approves the Technical Regulation to establish the minimum requirements for safety assessment of foods and food ingredients produced from genetically modified microorganisms (GMOs). This resolution establishes requirements for safety assessment studies, which must be performed before the marketing of foods and food ingredients produced from GMOs, including genetically modified microalgae (NHSA 2019).

Recently, FDA has released some food safety letter regarding the consumption of dietary supplements containing astaxanthin, a compound derived from the microalga *Haematococcus pluvialis*. The letter highlights that these supplements are safe for human consumption, provided they are manufactured in accordance with Good Manufacturing Practices (GMP) and meet the appropriate labelling requirements set out. FDA regulates most packaged foods sold in the USA, and has specific requirements regarding the information that must be included on the labelling. For a particular industry to conduct the sale of its food products, the company must comply with FDA packaging regulations (FDA 2019). Table 1 presents the elements required for compliance with the packaging legislation.

Regulations regarding the trade of seaweeds intended for human consumption may vary according to the specific country and region. However, in general, there are some common rules and regulations that address the safety and quality of foods, including algae, as shown in Table 2.

Broadly speaking, legislation in food industry is essential to ensure the safety and quality of food products reaching consumers and is constantly evolving to adapt to new trends and consumer concerns. It is important that the food industry continue to work collaboratively with regulatory bodies to ensure that standards are effective and applicable to all food producers, regardless of size or location (NHSA 2019).

3.1 Nagoya Protocol

Nagoya Protocol, adopted in 2010 as part of the Convention on Biological Diversity (CDB), plays a key role in the conservation of biodiversity and the sustainable use of genetic resources. In the context of legislation relating to algae in food for humans, the application of this protocol is essential to ensure the conservation of algal genetic resources and the fair/equitable distribution of benefits arising from their use, as well as to minimise the negative environmental impacts. In this section, the relationship between Nagoya Protocol and its application in algal legislation is examined, based on relevant discussions and references (CDB 2014).

Algae play a crucial role as key components of marine ecosystems, maintaining the resilience of these environments (Hoeneisen 2019). However, the rampant and

Table 1 The required packaging elements according to Food and Drug Administration (FDA)

Elements	Description
Statement of identity	This item should not be confused with the brand name of the food product. The identity declaration refers to the legal name, the common name or, when necessary, to a description of the foodstuff itself.
Net quantity of contents	Refers to the specific quantity of food present in the package. It can be indicated as weight, fluid measure or the count of individual items present in the package.
Nutrition facts label	This element is intended to provide information about the food that is being consumed by consumers. The FDA also regulates the format of the label. The label must be used on your product, having factors such as the size and content of the package.
Ingredient statement	An ingredient declaration is a list of ingredients, arranged in descending order by weight, presented in a font that is at least 1/16″ high and easily legible for the consumer.
Allergen declaration	Packaged food products are required by the Food Allergen and Consumer Protection Labelling Act 2004 (FALCPA) to clearly indicate the presence of major food allergens (such as milk, egg, fish, shellfish, nuts, wheat, peanuts, soy, and sesame) on product packaging. Easily understandable language should be used.
Name and address of the manufacturer, packer, or distributor	The packaging of foodstuffs must bear the name of the manufacturer/packer/distributor, together with a qualifying phrase clarifying the company's link to the product.
Nutrient content claims	Item which should contain statements regarding nutrient levels in foods, such as low fat, high fibre, sugar free and others.
Barcode	Government regulatory agencies do not mandate the need for a barcode on the food packaging, although it is commonly expected by most retail outlets. However, to ensure compliance, the barcode should be positioned so as to avoid any interference with the essential elements.
Expired date	Some states have regulations mandating the inclusion of dates on certain food products. It is important to investigate the specific regulations pertaining to your state and the food in question. Regarding the placement of the date on packaging, it should not obstruct any mandatory labelling elements. In addition, the date should clearly display the month, day, and year, next to an explanatory phrase such as 'best before' or 'sell by'.

Source: Adapted from FDA (2019)

unregulated exploitation of algae can lead to a decrease in genetic diversity and the depletion of natural resources. In turn, the Nagoya Protocol aims to promote the conservation of marine biodiversity by ensuring that the use of these algae is carried out in a sustainable manner by adopting protective measures (CDB 2014) and clear principles for access to genetic resources. The signatory countries of the protocol are

Table 2 Examples for the existing regulations that address the safety and quality of algae as human food

Topic/organization	Regulations
Health and food authorities	- Regulation and supervision of foodstuffs - Establishment of food safety standards - Inspection and licensing of companies
Hygiene and good manufacturing practices	- Compliance with hygiene guidelines and good manufacturing practices - Adequate sanitary conditions in the facilities - Quality control
Labelling	- Clear and precise labelling - Information on ingredients, origin, shelf life, storage instructions, and allergens present
Quality control and testing	- Performing laboratory tests to ensure quality and safety - Microbiological analyses and contamination tests
International norms and standards	- Standards and guidelines set by international organisations such as FAO and WHO - Adoption or adaptation by national governments

Source: Adapted from NHSA (2019)

responsible for establishing the legal control to regulate access to the genetic resources of algae (Manzur and Díaz 2003).

The Nagoya Protocol, which addresses access to genetic resources and the fair and equitable sharing of benefits arising from their utilization, was adopted during the tenth meeting of the Conference of the Parties to the CDB, held on 29 October 2010 in Nagoya City, Japan. In accordance with Article 32 of the protocol, it was opened for signature by parties to the convention from 2 February 2011 to 1 February 2012 at the United Nations Headquarters in New York. The Protocol entered into force on 12 October 2014, and currently has 139 countries considered as parties, 140 countries that have ratified, and 92 signatures (CDB 2023). One of the main objectives of the protocol is to ensure the fair and equitable distribution of benefits derived from the utilization of genetic resources (CDB 2014). In the context of algae in human food, this implies that countries and communities that hold the biodiversity of these source organisms should obtain fair benefit-sharing in relation to their use (Manzur and Díaz 2003). Benefit sharing may include technology transfer, sharing of scientific knowledge, and sharing of profits from revenues from algae-derived products. The effective implementation of the Nagoya Protocol on legislation concerning algae in food depends on the adoption of measures at the national level. Each participating country is responsible for establishing legislation appropriate to the principles and guidelines. The implementation of this protocol is extremely important, considering that algae are valuable genetic resources used in various areas, including the food industry. By applying the Nagoya Protocol on algae legislation, countries can exercise control over access to these genetic resources and ensure that it is done in a legal and transparent manner. This means that those who wish to use microalgae or seaweeds for food production must obtain permission (Nagoya Protocol 2011).

The application of Nagoya Protocol in the legislation on algae as a food can bring several benefits, such as contributing to the conservation of marine biodiversity, as access and use of algae will be regulated, avoiding overexploitation and negative impacts on the coastal ecosystems. The application of the protocol also promotes the valorisation of traditional knowledge associated with the use of algae, as many coastal communities have extensive knowledge of the properties of algae and their applications in food and medicine. By stimulating the scientific research and development of new products derived from algae and by ensuring legal certainty and fair benefit sharing, companies and entities would attract incentives to invest in studies and innovations related to these natural resources, which can protect the food production industry and promote sustainable development (Manzur and Díaz 2003).

3.2 Genetically Modified Microorganisms

Many species of algae are suitable to produce biomass with diverse applications, predominantly related to aquaculture, food production for human and animal consumption, treatment of effluents, and production of biofertilizers, biofuels, fine chemicals, and pharmaceuticals (Chaumont 1993; Becker 2017). Genetic engineering techniques can be considered as alternatives for the purpose of improving algal characteristics of interest. These techniques of genetic modification allow combining and selecting fundamental regulatory characteristics of algae, while it is possible to increase the production of compounds, characteristics or functionalities. These genetically modified (GM) algae can be used in several fields, including food as a whole biomass or food supplement. However, large-scale production is not so simple and depends on many factors and how each country considers a GMO (Beacham et al. 2017).

3.2.1 Legislations of GM Algae

Applications of techniques that reproduce natural phenomena, such as non-GMO approved mutagenesis under European guidelines, are not subject to GMO control measures or legislation (Francescon 2001). There are several techniques to enhance algal phenotypes, such as random mutagenesis, traditional recombinant nucleic acid delivery technologies and genome editing tools. In addition, it is possible to mention transcription activator-like effector nucleases (TALEN), zinc finger nucleases (ZFN), and RNA-guided engineered nucleases (RGEN) - derived from the bacterial regularly spaced clustered short palindromic repeat (CRISPR)-Cas9 system (Chavez-Granados et al. 2022).

Whether any of these new technologies will produce a "GMO" will depend on the regulations of the country involved. In European countries, the definition of GMO is associated with the synthetic introduction of genetic material into an organism to create a new organism using recombinant nucleic acid technologies. It is unclear how existing legislation around the world will adopt new developments and features around genome editing techniques such as CRISPR/Cas9. Directly guided RNA delivery with purified Cas 9 protein in microalgae cells, compared to plasmid-mediated delivery, for example, is likely to go against USA GMO legislation as the genome-editing complex is degraded in the recipient cell without leaving traces of foreign DNA (Kanchiswamy et al. 2015). Whether this technique will be integrated into GMO legislation in the European Union will depend on the interpretation of the 2001 Directive on the release of genetically modified organisms into the environment (Francescon 2001), which states that a genetic modification occurs by means of "*recombinant nucleic acid techniques involving the formation of new combinations of genetic material by the insertion of nucleic acid molecules produced by whatever means outside an organism, into any virus, bacterial plasmid or other vector system and their incorporation into a host organism in which they do not naturally occur but in which they are capable of continued propagation*".

This inability of regulations to keep up with technological advances in GMO development has created impediments to the implementation of new technologies on an industrial-scale and will impact any transmission of GM microalgae. Currently, in Europe, there is legislation covering GMO aspects from deliberate release (Francescon 2001), environmental protection and/or repair of environmental damage (Bajpai 2022), GMO in food (animal feed) (Arpaia et al. 2013), and in labelling. However, within the framework of these directives, each member state may take other measures to regulate, manage, and control the GMOs, where other countries around the world follow their own sets of legislative rules. Most of the legislation is based on the requirements of the Cartagena Protocol on Biosafety to the CDB (CDB 2000), which provides international guidance on the regulation and management of living modified organisms (LMOs). However, as with all genetically modified foods, GM algae are the subject of controversy over its safety and long-term viability. It is important that rigorous studies are carried out and that GM algae are properly regulated before being used in food products and other applications.

3.2.2 The Bottleneck in Large-Scale Production of GM Algae

The bottleneck for the industrial advancement of genetically modified algae is the lack of information and evaluation tools available to researchers, industrial promoters, or regulators about the risks associated with their large-scale control, as well as the lack of adequate facilities to carry out trials. There is a need to develop tools and negotiation to assist the technical aspects of cultivation, containment, and risk assessment of genetically modified algae and to consider the legislative and policy aspects of such activities (Xu et al. 2019). Their commercialization for industrial purposes will require large-scale cultivation with well-defined criteria regarding risk assessment and more stringent environmental management strategies than those used for conventionally grown unmodified wild types. The main applications of genetically modified algae are given in Table 3. Only a few small collaborative ventures have begun to produce GM algae, such as Plymouth Marine Laboratory and

Table 3 Representative applications of genetically modified (GM) algae

GM algae	Application	References
Chlamydomonas reindhartii	Production of mCherry fluorescent protein from the study of macronutrients contained in the culture medium.	Arias et al. (2021)
	Biofuel production.	Cortez (2022)
Phaeodactylum tricornutum	Increase the production of essential fats (EPA and DHA).	Guedes and Malcata (2012)
Nannochloropsis oceanica	Human nutrition and aquaculture.	Roy and Pal (2015)
Chlamydomonas reinhardtii	Reduction of cellular pigmentation and improvement of photosynthetic efficiency for greater biomass production.	Borowitzka (2013)

Source: Adapted from Beacham et al. (2017)

Rothamsted Research who used the genetically modified strain of *Phaeodactylum tricornutum* expressing heterologous Δ 5-elongase for the accumulation of long-chain omega-3 fatty acids high-value long line (Hamilton et al. 2015) and a commercial venture between Sapphire Energy and UC San Diego reach pilot-scale.

3.3 Biosafety in Food Industry

Until May 15, 1997, the Novel Foods Regulation (NFR) was a legislation that provided that microalgae approved for food purposes were assessed. This classification defined as "not used for human consumption", among the species mentioned in the old NFR, were *Chlorella* sp., *C. pyrenoidosa*, *C. vulgaris*, *C. luteoviridis*, and *Arthrospira* sp. However, with the new Regulation (EU) No 2015/2283, the European Union, through EFSA, can ensure that novel foods and food ingredients were subjected to a safety analysis through a unified procedure in order to protect public health. Thus, ensuring that the previously mentioned species were approved for human consumption. In some countries, there is a possibility that specific regulations may exist for the collection and cultivation of seaweeds, which may affect the production and distribution of seaweed-based foods (EFSA 2020). According to Gomes (2023), regulatory issues related to seaweeds may include some aspects presented in Fig. 1.

Another application for algae can be wastewater treatment coupled with possible food supplementation. However, the type of application of biomass determines the way it is cultivated and processed. When the objective is directed to be used as a raw material for food, algae cultivation should be done under proper sanitary conditions, such as care with the use of water and precautions in the preparation of the culture medium, as well as the analysis of the process to ensure that the growth is adequate (Zhou et al. 2022; Goswami et al. 2022). According to Yadav and Sharma (2023) regarding these precautions, other steps in the biomass obtaining process should be performed in order to avoid contamination. Table 4 presents some existing regulations regarding the storage and marketing of algae. Overall, the storage and

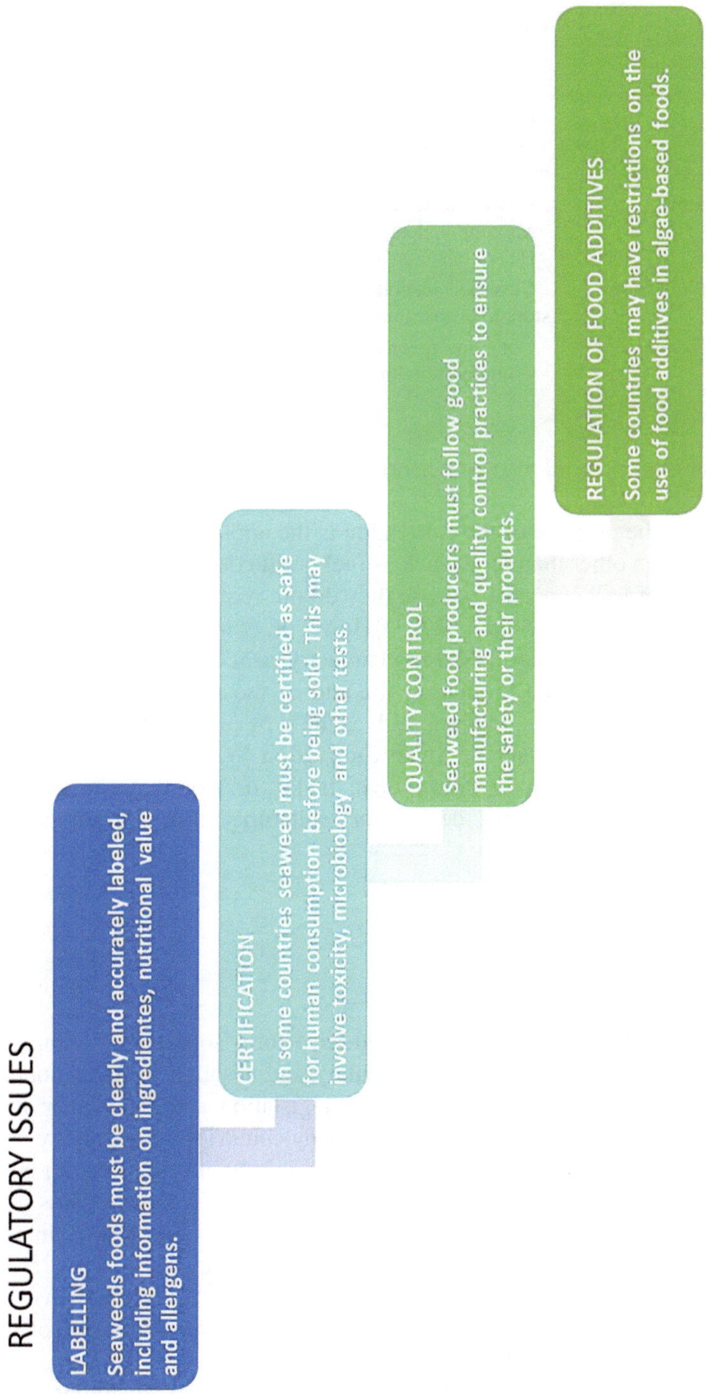

Fig. 1 Regulatory issues for the use of seaweeds in human food. Adapted from Gomes (2023)

Table 4 Current regulations about storage and commercialization of algal products

Topic	Regulations
Proper storage	- Controlled temperatures - Protection against light and humidity - Prevention of microorganism proliferation
Traceability	- Detailed records on the origin, cultivation, processing, and supply chain
Quality assurance	- Guidelines for nutritional composition - Microbiological contamination - Heavy metal content and contaminants
Labelling and consumer information	- Clear and accurate labels with information on ingredients, origin, storage instructions and allergens
Authorisations and permits	- Obtaining specific permits and licences - Compliance with regulations and sustainable practices

marketing of algae should also be performed according to the health regulations in force in each country.

However, if the purpose of the cultivation is the application of biomass as a raw material in sectors other than food, such as fuel production, the precautions and care to be taken are of less concern, because they do not require extreme care regarding contamination (Yadav and Sharma 2023). However, contamination is always harmful, even if the intention is not for the food industry, as it can cause losses in the yield of biomass and the desired product at the end of the process. Some aspects can influence the biomass contamination, such as the type of process and the bioreactor model. As discussed in Chapter "Algae Cultivation Systems", algal biomass production can be carried out in farms, open tanks, or closed reactors, which are classified according to their design or suitability for specific cultivation requirements.

4 Conclusions

Throughout history, algae have been used as a food source by various cultures around the world. Today, as interest in healthy and sustainable foods continues to grow, there has been increased attention to legislation and biosecurity measures regarding the use of algae in human food. In many countries, seaweeds are classified as a food product and is subject to specific regulations to ensure safety for consumption. These regulations and legal requirements can cover various aspects including algal biomass production, such as agricultural practices, water quality monitoring, responsible use of fertilisers, and minimising environmental impacts. In order to mitigate potential risks and ensure product quality control, comprehensive analyses are carried out to identify any potential hazards and ensure the safety of the final algae-based food products. The legislation requires that the packaging of algae-based foods provide accurate information on the ingredients, including nutritional values, allergens, and the origin of the product. In addition, maximum permitted limits are set for pesticide residues, heavy metals, among other contaminants possibly present in the edible product. Consequently, there are certifications and quality seals that guarantee adherence to specific regulations and standards, stating that the algal product complies with biosafety and quality guidelines. It is important to note that biosafety legislation and guidelines may vary between countries, so for current and detailed information on biosafety legislation and guidelines regarding algae in food, it is crucial to consult the specific legislation and competent authorities of each country.

References

Andrade HMMQ, Rosa LP, Souza FES, Silva NF, Cabral MC, Teixeira DI (2020) Seaweed production potential in the Brazilian northeast: a study on the eastern coast of the state of Rio Grande do Norte, RN, Brazil. *In Portuguese: Potencial Produção de Algas Marinhas no Nordeste Brasileiro: Um Estudo sobre o Litoral Leste do Estado do Rio Grande do Norte, RN, Brasil.* Sustain. 12, 780. https://doi.org/10.3390/su12030780

Arias CAD, Matsudo MC, Ferreira-Camargo LS, Molino JVD, Mayfield SP, de Carvalho JCM (2021) Semicontinuous system to produce recombinant mCherry protein in *Chlamydomonas reinhardtii. In Portuguese: Sistema semicontínuo para produção de proteína mCherry recombinante em Chlamydomonas reinhardtii.* Biotechnol Progress 37:e3101. https://doi.org/10.1002/btpr.3101

Arpaia S, Birch N, Chesson A, du Jardin P, Gathmann A, Gropp J, Tebbe C (2013) Scientific opinion on a request from the European Commission related to the emergency measure notified by Italy on genetically modified maize MON 810 according to article 34 of regulation (EC) no 1829/2003. EFSA J 11:3371. https://doi.org/10.2903/j.efsa.2013.3371

Bajpai P (2022) Regulations on cultivation and processing of genetically modified algae. In: Fourth Generation Biofuels. Springer, Singapore https://link.springer.com/book/10.1007/978-981-19-2001-1. Accessed 2 June 2023

Beacham TA, Sweet JB, Allen MJ (2017) Large scale cultivation of genetically modified microalgae: a new era for environmental risk assessment. Algal Res 25:90–100. https://doi.org/10.1016/j.algal.2017.04.028

Becker EW (2017) Microalgae as a source of protein. Biotechnol Adv 25:207–210. https://doi.org/10.1016/j.biotechadv.2006.11.002

Ben-Amotz A, Polle JEW, Rao DVS (2009) The alga Dunaliella: biodiversity, physiology, genomics and biotechnology.1st edn. Science Publishers, Londres

Blikra MJ, Altintzoglou T, Lovdal T, Rognsa G, Skipnes D, Skara T, Sivertsvik M, Fernandez NE (2021) Seaweed products for the future: using current tools to develop a sustainable food industry. Trends Food Sci Technol 118:765–776. https://doi.org/10.1016/j.tifs.2021.11.002

Borowitzka MA (2013) High-value products from microalgae-their development and commercialization. J Appl Phycol 25:743–756. https://doi.org/10.1007/s10811-013-9983-9

Burdock GA, Carabin IG (2004) Generally recognized as safe (GRAS): history and description. Toxicol Lett 150:3–18. https://doi.org/10.1016/j.toxlet.2003.07.004

Campbell I, Kambey CSB, Mateo JP, Rusekwa SB, Hurtado AQ, Msuya FE, Stentiford GD, Cottier-Cook EJ (2020) Biosecurity policy and legislation for the seaweed aquaculture industry. J Appl Phycol 32:2133–2146. https://doi.org/10.1007/s10811-019-02010-5

CDB (2000) Cartagena protocol on biosafety to the convention on biological diversity: text and annexes. https://bch.cbd.int/protocol/background/. Accessed 2 June 2023

CDB (2014) Secretary of the convention on biological diversity. https://www.gov.br/mma/pt-br/assuntos/biodiversidade/convencao-sobre-diversidade-biologica. Accessed 2 June 2023

CDB (2023) Convention biological diversity. Nagoya protocol on access and benefit-sharing – signatories – parties to the nagoya protocol. https://www.cbd.int/abs/nagoya-protocol/signatories/. Accessed 2 June 2023

Chaumont D (1993) Biotechnology of algal biomass production: a review of systems for outdoor mass culture. J Appl Phycol 5:593–604. https://doi.org/10.1007/BF02184638

Chavez-Granados PA, Manisekaran R, Acosta-Torres LS, Garcia-Contreras R (2022) CRISPR/Cas gene-editing technology and its advances in dentistry. Biochimie 194:96–107. https://doi.org/10.1016/j.biochi.2021.12.012

Cortez IG (2022) Microalgas, fábricas biológicas: Más allá de los biocombustibles.4th edn. Cidade do México, México

Diaz CJ, Douglas KJ, Kang K, Kolarik AL, Malinovski R, Torres-Tiji Y, Molino JV, Badary A, Mayfield SP (2023) Developing algae as sustainable food source. Front Nutr 9:1–36. https://doi.org/10.3389/fnut.2022.1029841

EFSA (2020) Panel on nutrition, novel foods and food allergens (NDA), Turck D, Castenmiller J, De Henauw S, Hirsch-Ernst K I, Kearney J, Knutsen H K (2020). Safety of *Schizochytrium* sp. oil as a novel food pursuant to regulation (EU) 2015/2283 (a). EFSA J 18, 6242. https://doi.org/10.2903/j.efsa.2020.6242

FAO (2020) Progress towards development of the progressive management pathway for improving aquaculture biosecurity (PMP/AB): highlights of 2019 activities. FAO Fisheries and Aquaculture Circular, 1211, Rome. https://doi.org/10.4060/cb0560en

Faustman C, Aaron D, Negowetti N, Leib EB (2021) Ten years post-GAO assessment, FDA remains uninformed of potentially harmful GRAS substances in foods. Crit Rev Food Sci Nutr 61:1260–1268. https://doi.org/10.1080/10408398.2020.1756217

FDA (2019) Food and drug administration. Responsible innovation in dietary supplements. https://www.fda.gov/food/workshops-meetings-webinars-food-and-dietary-supplements/public-meeting-discuss-responsible-innovation-dietary-supplements. Accessed 3 June 2023

FDA (2022) Food and drug administration. FOOD CODE 2022. https://www.fda.gov/food/fda-food-code/food-code-2022. Accessed 2 June 2023

Fleurence J, Levine I (2016) Seaweed in health and disease prevention, first ed. San Diego, California

Francescon S (2001) The new directive 2001/18/EC on the deliberate release of genetically modified organisms into the environment: changes and perspectives. Reciel 10:309. https://doi.org/10.1111/1467-9388.00290

Gomes MLDS (2023) Cassava (*Manihot esculenta* Crantz) biodegradable films: a review. In: Portuguese: Filmes biodegradáveis de mandioca (Manihot esculenta Crantz): uma revisão. UFAM. 20 Set 2022; 30f: 1–30

Goswami RK, Mehariya S, Karthikeyan OP, Grupta VK, Verma P (2022) Multifaceted application of microalgal biomass integrated with carbon dioxide reduction and wastewater remediation: a flexible concept for sustainable environment. J Clean Prod 339:130654. https://doi.org/10.1016/j.jclepro.2022.130654

Guedes AC, Malcata FX (2012) Nutritional value and uses of microalgae in aquaculture. Aquac 10:59–78. https://doi.org/10.5772/30576

Gupta S, Abu-Ghannam N (2011) Bioactive potential and possible health effects of edible brown seaweeds. Trends Food Sci Technol 22:315. https://doi.org/10.1016/j.tifs.2011.03.011

Gutiérrez R, Núñez R, Quintana L, Valdés O, González K, Rodríguez M, Hernández Y, Ortiz E (2017) Optimization of the extraction process of phenolic compounds from the brown algae *Sargassum fluitans* Børgesen (Børgesen). Biotecnol Apl 34:3301–3304

Hallagan JB, Hall RL, Drake J (2020) The GRAS provision - the FEMA GRAS program and the safety and regulation of flavors in the United States. FCT. 138:111236. https://doi.org/10.1016/j.fct.2020.111236

Hamilton ML, Warwick J, Terry A, Allen MJ, Napier JA, Sayanova O (2015) Towards the industrial production of omega-3 long chain polyunsaturated fatty acids from a genetically modified diatom Phaeodactylum tricornutum. PLoS One 10:12. https://doi.org/10.1371/journal.pone.0144054

Hoeneisen LP (2019) Lineamientos para una regulación sobre acceso a recursos genéticos y distribución de beneficios en Chile. RDA 12:29–55. https://doi.org/10.5354/0719-4633.2019.54165

Holdt SL, Kraan S (2011) Bioactive compounds in seaweed: functional food applications and legislation. J Appl Phycol 23:543–597. https://doi.org/10.1007/s10811-010-9632-5

Kanchiswamy CN, Malnoy M, Velasco R, Kim JS, Viola R (2015) Non-GMO genetically edited crop plants. Trends Biotechnol 33:489–491. https://doi.org/10.1016/j.tibtech.2015.04.002

Kazir M, Abuhassira Y, Robin A, Nahor O, Luo J, Israel A, Golberg A, Livney YD (2019) Extraction of proteins from two marine macroalgae, *Ulva* sp. and *Gracilaria* sp., for food application, and evaluating digestibility, amino acid composition and antioxidant properties of the protein concentrates. Food Hydrocoll 87:194–203. https://doi.org/10.1016/j.foodhyd.2018.07.047

Kumar Y, Tarafdar A, Badgujar PC (2021) Seaweed as a source of natural antioxidants: therapeutic activity and food applications. J Food Qual 2021:1. https://doi.org/10.1155/2021/5753391

Manzur MI, Díaz CL (2003) Acceso a recursos genéticos, Chile en el contexto mundial.1st edn. Santiago, Chile

Marinho-Soriano E, Carneiro MAA (2021) Marine macroalgae: biology, ecology and economic importance. In: Viana DL, Marine Sciences: from the oceans of the world to Northeast Brazil. In Portuguese: Macroalgas marinhas: biologia, ecologia e importância econômica, in: Viana, D.L., (2021), Ciências do Mar: dos oceanos do mundo ao Nordeste do Brasil. E-Publishing Via Design Publicações, Pernambuco, pp. 91–119

McHugh DJ (2003) A guide to the seaweed industry. FAO Fish Tech Pap 441:105. https://doi.org/10.4236/oalib.1100752

Moreira SSP, Costa E, Melo T, Sulpice R, Cardoso SM, Pitarma B, Pereira R, Abreu MH, Domingues P, Calado R, Domingues MR (2020) Seasonal plasticity of the polar lipidome of *Ulva rigida* cultivated in a sustainable integrated multi-trophic aquaculture. In Portuguese: Plasticidade sazonal do lipidoma polar de Ulva rigida cultivada em uma aquicultura multitrófica integrada sustentável. Algal Res. 49, 101958. https://doi.org/10.1016/j.algal.2020.101958

Moreira B, Vaz BDS, Cardias BB, Cruz CG, Almeida ACAD, Costa JAV, Morais MGD (2022) Microalgae polysaccharides: an alternative source for food production and sustainable agriculture. *In Portuguese: Polissacarídeos de microalgas: uma fonte alternativa para produção de alimentos e agricultura sustentável.* Polysaccharides 3:441–457. https://doi.org/10.3390/polysaccharides3020027

Morissette R, Mihalov J, Carlson SJ, Kaneko KJ (2023) Trends in ingredients added to infant formula: FDA's experiences in the GRAS. Food Chem Toxicol 178:113876. https://doi.org/10.1016/j.fct.2023.113876

Nagoya Protocol on Access and Benefit-sharing (2011). https://www.cbd.int/abs/doc/protocol/nagoya-protocol-en.pdf. Accessed 2 June 2023

NHSA (2001) National Health Surveillance Agency. RDC resolution no. 12, of January 2, 2001. Provides for the technical regulations of food for special purposes. *In Portuguese: Resolução RDC n° 12, de 2 de janeiro de 2001. Dispõe sobre os regulamentos técnicos de alimentos Para fins especiais*. Official Diary of the Union. 1, 42–46

NHSA (2019) National Health Surveillance Agency. RDC resolution no. 331, of April 23, 2019. Approves the technical regulation to establish the minimum requirements for assessing the safety of food and food ingredients produced from genetically modified microorganisms (GMOs). In Portuguese: Resolução RDC n° 331, de 23 de abril de 2019. Aprova o Regulamento Técnico Para fixar os requisitos mínimos Para a avaliação da segurança dos alimentos e ingredientes alimentares produzidos a partir de micro-organismos geneticamente modificados (OGMs). Official Diary of the Union 1, 59–61

Olaizola M (2003) Commercial development of microalgal biotechnology: from the test tube to the marketplace. Biomol Eng 20:459–466. https://doi.org/10.1016/S1389-0344(03)00076-5

Pereira AG, Otero P, Echave J, Carreira-Casais A, Chamorro F, Collazo N, Jaboui A, Lourenço-Lopes C, Simal-Gandara J, Prieto MA (2021) Xanthophylls from the sea: algae as source of bioactive carotenoids. In Portuguese: Xantofilas do mar: algas Como fonte de carotenóides bioativos. Mar Drugs 19:188. https://doi.org/10.3390/md19040188

Pradhan B, Bhuyan PP, Patra S, Nayak R, Behera PK, Behera C, Behera AK, Ki JS, Jena M (2022) Beneficial effects of seaweeds and seaweed-derived bioactive compounds: current evidence and future prospective. ISBAB 39:1–10. https://doi.org/10.1016/j.bcab.2021.102242

Raven PH, Evert RF, Eichhorn SE (1996) Algas verdes, vermelhas e pardas, fifth ed. Rio de Janeiro

Richmond A (2004) Handbook of microalgal culture: biotechnology and applied phycology.1st edn. Blackwell Science, London

Roy SS, Pal R (2015) Microalgae in aquaculture: a review with special references to nutritional value and fish dietetics. Proc Zool 68:1–8. https://doi.org/10.1007/s12595-013-0089-9

Soeder CA (1986) Handbook of microalgal culture: historical outline of applied phycology.1st edn. CRC Press, Boca Raton

Souza MRV, Ostrosky EA (2022) A review on the application of algae in pharmaceutical, cosmetic and food fields. In Portuguese: Uma revisão sobre a aplicação de algas nas áreas farmacêutica, cosmética e alimentícia. UFRN. 24 Dez 2022; 20f: 1-20

Vasconcelos BMF, Gonçalves AA (2013) Macroalgae and its uses – alternatives for Brazilian industries. In Portuguese: Macroalgas e seus usos – alternativas Para as indústrias brasileiras. Revista Verde 8:125–140

Xu X, Gu X, Wang Z, Shatner W, Wang Z (2019) Progress, challenges and solutions of research on photosynthetic carbon sequestration efficiency of microalgae. J Renew Sustain Energy 110:65–82. https://doi.org/10.1016/j.rser.2019.04.050

Yadav M, Sharma YC (2023) Approach to microalgal biodiesel production: insight review on recent advancements and future outlook. Biofpr 17:242–260. https://doi.org/10.1002/bbb.2434

Zhou JL, Yang L, Huang KX, Chen DZ, Gao F (2022) Mechanisms and application of microalgae on removing emerging contaminants from wastewater: a review. Bioresour Technol 364: 128049. https://doi.org/10.1016/j.biortech.2022.128049